Practical Business Math Procedures

The McGraw-Hill/Irwin Series in Operations and Decision Sciences

Supply Chain Management

Benton
Purchasing and Supply Chain Management
Third Edition

Burt, Petcavage, and Pinkerton
Supply Management
Eighth Edition

Bowersox, Closs, and Cooper
Supply Chain Logistics Management
Fifth Edition

Johnson, Leenders, and Flynn
Purchasing and Supply Management
Sixteenth Edition

Simchi-Levi, Kaminsky, and Simchi-Levi
Designing and Managing the Supply Chain: Concepts, Strategies, Case Studies
Third Edition

Stock and Manrodt
Fundamentals of Supply Chain Management

Project Management

Brown and Hyer
Managing Projects: A Team-Based Approach

Larson and Gray
Project Management: The Managerial Process
Seventh Edition

Service Operations Management

Fitzsimmons and Fitzsimmons
Service Management: Operations, Strategy, Information Technology
Ninth Edition

Management Science

Hillier and Hillier
Introduction to Management Science: A Modeling and Case Studies Approach with Spreadsheets
Sixth Edition

Business Research Methods

Schindler
Business Research Methods
Thirteenth Edition

Business Forecasting

Wilson and Keating
Forecasting and Predictive Analytics
Seventh Edition

Linear Statistics and Regression

Kutner, Nachtsheim, and Neter
Applied Linear Regression Models
Fourth Edition

Business Systems Dynamics

Sterman
Business Dynamics: Systems Thinking and Modeling for a Complex World
First Edition

Operations Management

Cachon and Terwiesch
Operations Management
Second Edition

Cachon and Terwiesch
Matching Supply with Demand: An Introduction to Operations Management
Fourth Edition

Jacobs and Chase
Operations and Supply Chain Management: The Core
Fifth Edition

Jacobs and Chase
Operations and Supply Chain Management
Fifteenth Edition

Schroeder, Goldstein, and Rungtusanatham
Operations Management in the Supply Chain: Decisions and Cases
Seventh Edition

Stevenson
Operations Management
Thirteenth Edition

Swink, Melnyk, Cooper, and Hartley
Managing Operations across the Supply Chain
Fourth Edition

Business Math

Slater/Wittry
Practical Business Math Procedures
Thirteenth Edition

Slater/Wittry
Math for Business and Finance: An Algebraic Approach
Second Edition

Business Statistics

Bowerman, O'Connell, and Murphree
Business Statistics in Practice
Ninth Edition

Doane and Seward
Applied Statistics in Business and Economics
Sixth Edition

Doane and Seward
Essential Statistics in Business and Economics
Third Edition

Jaggia and Kelly
Business Statistics: Communicating with Numbers
Third Edition

Jaggia and Kelly
Essentials of Business Statistics: Communicating with Numbers
Second Edition

Lind, Marchal, and Wathen
Basic Statistics for Business and Economics
Ninth Edition

Lind, Marchal, and Wathen
Statistical Techniques in Business and Economics
Seventeenth Edition

McGuckian
Connect Master: Business Statistics

Practical Business Math Procedures

Thirteenth Edition

JEFFREY SLATER
North Shore Community College
Danvers, Massachusetts

SHARON M. WITTRY
Pikes Peak Community College
Colorado Springs, Colorado

Mc Graw Hill Education

mheonline.com/advancedplacement

Send all inquiries to:
McGraw Hill
8787 Orion Place
Columbus, OH 43240

ISBN: 978-1-26-619947-9
MHID: 1-26-619947-0

Printed in the United States of America.

2 3 4 5 6 7 8 9 LKV 30 29 28 27 26 25 24

Dedication

To Shelley . . . My best pal.
Love, Jeff

To my mom, who did everything for me as a child.

To my dad, who paved the way.

Love, Sharon

Note to Students

ROADMAP TO SUCCESS

How to use this book and the Total Slater/Wittry Learning System.

Step 1: **Read "Your Guide to Successfully Completing This Chapter" at the beginning of each chapter. Each chapter is broken down into Learning Units. Read and master one Learning Unit at a time.**

How do I know whether I understand it?

- Try the Practice Quiz. All the worked-out solutions are provided. If you still have questions, watch the author videos in Connect, or get the information from your instructor and work each problem out.
- Repeat the above until you understand.

Once you feel confident with the subject matter, go on to the next Learning Unit in the chapter.

Step 2: **Review the Interactive Chapter Organizer at the end of the chapter.**

How do I know if I understand it?

- The third column, "You try it," gives you the chance to do additional practice.

Step 3: **Do assigned problems at the end of the chapter (or Appendix A). These may include discussion questions, drill, word problems, challenge problems, as well as projects from My Money and Kiplinger's magazine.**

Can I check my homework?

- Appendix B has check figures for all the odd-numbered problems.

Step 4: **Complete the "Interactive Video Worksheet" near the end of the chapter while completing the Summary Practice Test.**

Can I check my progress?

- Complete the Summary Practice Test. Check solutions from videos in Connect.

What do I do if I do not match check figures?

- Review the video tutorial in Connect, or through information from your instructor—the authors work out each problem.

To aid you in studying the book, we have developed the following color code:

Blue: Movement, cancellations, steps to solve, arrows, blueprints

Purple and yellow: Formulas and steps

Green: Tables and forms

Red: Key items we are solving for

If you have difficulty with any text examples, pay special attention to the red and the blue. These will help remind you of what you are looking for as well as what the procedures are.

FEATURES

The following are the features students have told us have helped them the most.

Blueprint Aid Boxes

For the first eight chapters (not in Chapter 4), blueprint aid boxes are available to help you map out a plan to solve a word problem. We know the hardest part of solving word problems is often figuring out where to start. Use the blueprint as a model to get started.

Business Math Handbook

This reference guide contains all the tables found in the text. It makes homework, exams, etc., easier to deal with than flipping back and forth through the text.

Interactive Chapter Organizer

At the end of each chapter is a quick reference guide called the Interactive Chapter Organizer, in which key points, formulas, and examples are provided. A list of vocabulary terms is also included. A column called "You try it" gives you a chance to do additional practice. And solutions are provided in Appendix B. (A complete glossary is found at the end of the text.) Think of the Interactive Chapter Organizer as your set of notes and use it as a reference when doing homework problems and reviewing before exams.

*For **extra help** from your authors–Sharon and Jeff–see the videos in Connect.*

Additionally, a series of author-created tutorial videos are available in Connect, or you can check with your instructor for more information. The videos cover all of the Learning Unit Practice Quizzes and Summary Practice Tests.

Your Guide to Successfully Completing This Chapter

Each chapter begins with a plan for you to follow to help you master the content.

Group Activity: Personal Finance, a Kiplinger Approach

In each chapter you can debate a business math issue based on a *Kiplinger's Personal Finance* magazine article. This is great for critical thinking, as well as improving your writing skills.

Spreadsheet Templates

Excel® templates are available for selected end-of-chapter problems. You can run these templates as-is or enter your own data. The templates also include an interest table feature that enables you to input any percentage rate and any terms. The program then generates table values for you.

Cumulative Reviews

At the end of Chapters 3, 8, and 13 are word problems that test your retention of business math concepts and procedures. Check figures for *all* cumulative review problems are in Appendix B.

Vocabulary

Each chapter includes highlighted words covering the key terms in the chapter. The Interactive Chapter Organizer includes a list of the terms. There's also a glossary at the end of the text.

Interactive Video Worksheet

At the end of each chapter is an interactive worksheet allowing you to work through the Summary Practice Test to success.

My Money

Each chapter has a personal finance page applying the concepts from the chapter toward personal finance success.

Photo: ©McGraw-Hill Education/Ryan McAndrews, photographer.

Acknowledgments

Academic Experts, Contributors

Dawn P. Addington
Sarah Alamilla
Tom Bilyeu
Katherine Broneck
Thomas Burke
Karmeleta Burnett
Patrick Cunningham
Linda Currie
James P. DeMeuse
Jennifer Euteneur
Peggy Fralick
Mary Frey
Joe Hanson
Johnny Howard
Edward Kavanaugh
Cynthia L. King
Jan La Bard
Lana Labruyere
Deborah Layton
Lynda L. Mattes
Jeannette Milius
Angela Deaton Mott
Joseph M. Nicassio
Mark Quinlan
Jo Ann Rawley
Karen Ruedinger
Kelly Russell
Marge Sunderland
Jason Tanner
Paul Tomko
Peter VanderWeyst

Company/*Applications*

Chapter 1

Visa; McDonald's—*Problem solving*
Google—*Reading and writing numbers*
Volkswagen—*Rounding numbers* and *Adding and subtracting numbers*
Star Wars—*Multiplying and dividing numbers*

Chapter 2

Health industry—*Introduction*
M&M'S/Mars—*Fractions and multiplication*

Chapter 3

McDonald's; Brexit—*Introduction*
Apple—*Decimal applications*
Toyota—*Multiplication and division shortcuts for decimals*

Chapter 4

ATMs— *Introduction*
Smartphones—*Checking account*
Apps—*Bank reconciliation*

Chapter 5

Big Food—*Unknowns*
Dunkin' Donuts—*Equations*

Chapter 6

Tesla, Hershey—*Introduction*
Procter & Gamble; M&M'S/Mars—*Percent increase and decrease*

Chapter 7

FedEx; Walmart, Amazon—*Introduction*
Michael's—*Discounts*
FedEx United Parcel Service—*Shipping*
New Hampshire Propane Co.—*Cash discounts*

Chapter 8

Gap; Amazon; Walmart—*Introduction*
Gap—*Markup on cost and selling price*

Chapter 9

Walmart—*Introduction*
Internal Revenue Service—*Circular E*

Chapter 10

Auto Lenders—*Introduction*
Penn—*Discounting*

Chapter 11

Treasury Department—*Treasury bills*

Chapter 12

Investing—*Introduction*

Chapter 13

Dunkin' Donuts— *Introduction; Compounding*

Chapter 14

Federal Reserve; Wells Fargo—*Introduction*
Federal Trade Commission—*Installments*
Citibank; MasterCard—*Finance charge*

Chapter 15

Bank of America—*Mortgages*

Chapter 16

Boeing—*Introduction*
Apple—*Financial statements*
Toys "R" Us; McDonald's—*Ratio analysis*

Chapter 17

Toyota; Mazda—*Introduction*
Big Lots—*Depreciation*

Chapter 18

Home Depot—*Introduction*
Fruit of the Loom, Inc.—*LIFO*

Chapter 19

Tax Foundation—*Sales tax*
Amazon—*Money tip*

Chapter 20

Lyft; Uber—*Auto insurance*

Chapter 21

Disney—*Introduction*
Amazon, Texaco, GM—*Stocks*
J. Crew—*Bonds*
American Funds—*Mutual funds*

Chapter 22

Apple—*Introduction*
U.S. Census Bureau—*Median*

Contents

Note to Students vi

CHAPTER 1 **Whole Numbers: How to Dissect and Solve Word Problems 2**

- **LU 1–1** Reading, Writing, and Rounding Whole Numbers 4
- **LU 1–2** Adding and Subtracting Whole Numbers 9
- **LU 1–3** Multiplying and Dividing Whole Numbers 12

CHAPTER 2 **Fractions 34**

- **LU 2–1** Types of Fractions and Conversion Procedures 36
- **LU 2–2** Adding and Subtracting Fractions 41
- **LU 2–3** Multiplying and Dividing Fractions 47

CHAPTER 3 **Decimals 68**

- **LU 3–1** Rounding Decimals; Fraction and Decimal Conversions 70
- **LU 3–2** Adding, Subtracting, Multiplying, and Dividing Decimals 75

Cumulative Review: A Word Problem Approach—Chapters 1–3 93

CHAPTER 4 **Banking 94**

- **LU 4–1** The Checking Account 96
- **LU 4–2** Bank Statement and Reconciliation Process; Latest Trends in Mobile Banking 99

CHAPTER 5 **Solving for the Unknown: A How-to Approach for Solving Equations 120**

- **LU 5–1** Solving Equations for the Unknown 121
- **LU 5–2** Solving Word Problems for the Unknown 127

CHAPTER 6 **Percents and Their Applications 148**

- **LU 6–1** Conversions 150
- **LU 6–2** Application of Percents—Portion Formula 154

CHAPTER 7 **Discounts: Trade and Cash 182**

- **LU 7–1** Trade Discounts—Single and Chain (Includes Discussion of Freight) 183
- **LU 7–2** Cash Discounts, Credit Terms, and Partial Payments 190

CHAPTER 8 **Markups and Markdowns: Perishables and Breakeven Analysis 214**

- **LU 8–1** Markups Based on Cost (100%) 216
- **LU 8–2** Markups Based on Selling Price (100%) 221
- **LU 8–3** Markdowns and Perishables 227
- **LU 8–4** Breakeven Analysis 230

Cumulative Review: A Word Problem Approach—Chapters 6–8 247

CHAPTER 9 **Payroll 248**

- **LU 9–1** Calculating Various Types of Employees' Gross Pay 249
- **LU 9–2** Computing Payroll Deductions for Employees' Pay; Employers' Responsibilities 253

CHAPTER 10 Simple Interest 272
LU 10–1 Calculation of Simple Interest and Maturity Value 273
LU 10–2 Finding Unknown in Simple Interest Formula 276
LU 10–3 U.S. Rule—Making Partial Note Payments before Due Date 278

CHAPTER 11 Promissory Notes, Simple Discount Notes, and the Discount Process 296
LU 11–1 Structure of Promissory Notes; the Simple Discount Note 297
LU 11–2 Discounting an Interest-Bearing Note before Maturity 301

CHAPTER 12 Compound Interest and Present Value 316
LU 12–1 Compound Interest (Future Value)—The Big Picture 317
LU 12–2 Present Value—The Big Picture 324

CHAPTER 13 Annuities and Sinking Funds 342
LU 13–1 Annuities: Ordinary Annuity and Annuity Due (Find Future Value) 344
LU 13–2 Present Value of an Ordinary Annuity (Find Present Value) 350
LU 13–3 Sinking Funds (Find Periodic Payments) 353
Cumulative Review: A Word Problem Approach—Chapters 10–13 367
Time-Value Relationship Appendix 369

CHAPTER 14 Installment Buying 374
LU 14–1 Cost of Installment Buying 375
LU 14–2 Revolving Charge Credit Cards 381

CHAPTER 15 The Cost of Home Ownership 398
LU 15–1 Types of Mortgages and the Monthly Mortgage Payment 399
LU 15–2 Amortization Schedule—Breaking Down the Monthly Payment 404

CHAPTER 16 How to Read, Analyze, and Interpret Financial Reports 418
LU 16–1 Balance Sheet—Report as of a Particular Date 420
LU 16–2 Income Statement—Report for a Specific Period of Time 425
LU 16–3 Trend and Ratio Analysis 429

CHAPTER 17 Depreciation 448
LU 17–1 Concept of Depreciation and the Straight-Line Method 450
LU 17–2 Units-of-Production Method 452
LU 17–3 Declining-Balance Method 453
LU 17–4 Modified Accelerated Cost Recovery System (MACRS) with Introduction to ACRS (1986, 1989, 2010, 2017) 454

CHAPTER 18 Inventory and Overhead 468
LU 18–1 Assigning Costs to Ending Inventory—Specific Identification; Weighted Average; FIFO; LIFO 470
LU 18–2 Retail Method; Gross Profit Method; Inventory Turnover; Distribution of Overhead 475

CHAPTER 19 Sales, Excise, and Property Taxes 494
LU 19–1 Sales and Excise Taxes 495
LU 19–2 Property Tax 498

CHAPTER 20 **Life, Fire, and Auto Insurance 510**

LU 20–1 Life Insurance 511

LU 20–2 Fire Insurance 516

LU 20–3 Auto Insurance 519

CHAPTER 21 **Stocks, Bonds, and Mutual Funds 538**

LU 21–1 Stocks 539

LU 21–2 Bonds 543

LU 21–3 Mutual Funds 545

CHAPTER 22 **Business Statistics 562**

LU 22–1 Mean, Median, and Mode 563

LU 22–2 Frequency Distributions and Graphs 567

LU 22–3 Measures of Dispersion (Optional) 573

APPENDIX A Additional Homework by Learning Unit A

APPENDIX B Check Figures B-1

APPENDIX C Metric System C-1

Glossary/Index G-1

Practical Business Math Procedures

CHAPTER 1

Whole Numbers: How to Dissect and Solve Word Problems

Visa Offers Merchants Cash to Go Cashless

By AnnaMaria Andriotis

Visa Inc. has a new offer for small merchants: Take thousands of dollars from the card giant to upgrade your payment technology and in return stop accepting cash from customers.

The company unveiled the initiative on Wednesday as part of a broader effort to steer Americans away from using old-fashioned money. Visa says it plans to give $10,000 apiece to as many as 50 restaurants and food vendors to cover technology and marketing costs, as long as the businesses pledge to start what Visa executive Jack Forestell calls a "journey to cashless."

LEARNING UNIT OBJECTIVES

LU 1–1: Reading, Writing, and Rounding Whole Numbers

1. Use place values to read and write numeric and verbal whole numbers.
2. Round whole numbers to the indicated position.
3. Use blueprint aid for dissecting and solving a word problem.

LU 1–2: Adding and Subtracting Whole Numbers

1. Add whole numbers; check and estimate addition computations.
2. Subtract whole numbers; check and estimate subtraction computations.

LU 1–3: Multiplying and Dividing Whole Numbers

1. Multiply whole numbers; check and estimate multiplication computations.
2. Divide whole numbers; check and estimate division computations.

Your Guide to Successfully Completing This Chapter

Traditional book or ebook

Check box as you complete each step.

Steps

☐ Read learning unit.

  ☐ Complete practice quiz at the end of the learning unit.

☐ Grade practice quiz using provided solutions. (For more help, watch the learning unit video in Connect and have a Study Session with the authors. Then complete the additional practice quiz in Connect.)

☐ Repeat above for each of the three learning units in Chapter 1.

  ☐ Review chapter organizer.

  ☐ Complete assigned homework.

    ☐ Finish summary practice test. (Go to Connect via the ebook link and do the interactive video worksheet to grade.)

☐ Complete instructor's exam.

GLOBAL

McDonald's Goes Digital

Company looks to lure millenials; responds to breakfast-all-day demand on Twitter

By Julie Jargon

OAK BROOK, ILL.—How many people does it take for a 61-year-old burger maker to tweet? At least a dozen.

Inside a high-tech room at **McDonald's** Corp.'s suburban Chicago headquarters, employees tap away at computers responding to tweets and crunching data on what's trending on social media, long a standard practice at most consumer companies.

Companies such as online retailer **Zappos.com** Inc., coffee giant **Starbucks** Corp. and discount airline **JetBlue Airways** Corp. have been using social media for years to manage customer complaints and generate ideas.

The *Wall Street Journal* clip "McDonald's Goes Digital" shows how important Tweets can be.

People of all ages make personal business decisions based on the answers to number questions. Numbers also determine most of the business decisions of companies. For example, go to the website of a company such as McDonald's and note the importance of numbers in the company's business decision-making process.

McDonald's has to use numbers to see

1. The effect of introducing healthy food choices.
2. The promptness of service orders.
3. The expenditures necessary for new-product development.
4. Ways to improve store layouts to achieve lower unit costs and better quality control.

Your study of numbers begins with a review of basic computation skills that focuses on speed and accuracy. You may think, "But I can use my calculator." Even if your instructor allows you to use a calculator, you still must know the basic computation skills. You need these skills to know what to calculate, how to interpret your calculations, how to make estimates to recognize errors you made in using your calculator, and how to make calculations when you do not have a calculator.

GLOBAL

The United States' numbering system is the **decimal system** or *base 10 system.* Your calculator gives the 10 single-digit numbers of the decimal system—0, 1, 2, 3, 4, 5, 6, 7, 8, and 9. The center of the decimal system is the **decimal point.** When you have a number with a decimal point, the numbers to the left of the decimal point are **whole numbers** and the numbers to the right of the decimal point are decimal numbers (discussed in Chapter 3). When you have a number *without* a decimal, the number is a whole number and the decimal is assumed to be after the number.

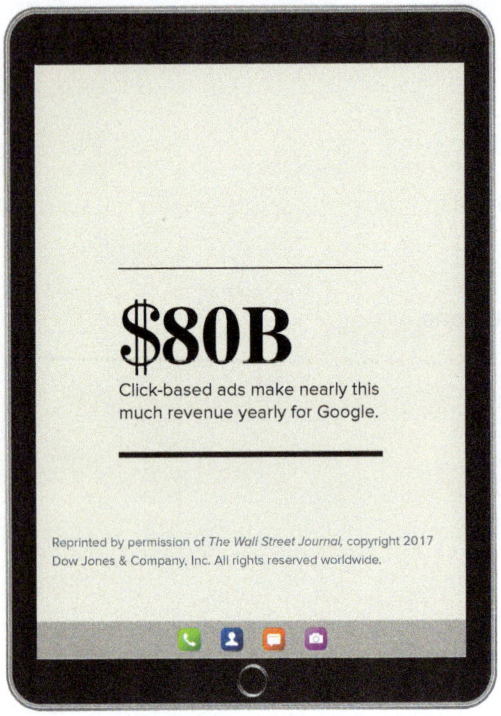

This chapter discusses reading, writing, and rounding whole numbers; adding and subtracting whole numbers; and multiplying and dividing whole numbers.

Learning Unit 1–1: Reading, Writing, and Rounding Whole Numbers

Wow! Did you know that back in 2017, $80 billion in sales resulted from click-on ads on Google? We will see how to read, write, and round whole numbers.

Now let's begin our study of whole numbers.

Reading and Writing Numeric and Verbal Whole Numbers

The decimal system is a *place-value system* based on the powers of 10. Any whole number can be written with the 10 digits of the decimal system because the position, or placement, of the digits in a number gives the value of the digits.

LO 1

To determine the value of each digit in a number, we use a place-value chart (Figure 1.1) that divides numbers into named groups of three digits, with each group separated by a comma. To separate a number into groups, you begin with the last digit in the number and insert commas every three digits, moving from right to left. This divides the number into the named groups (units, thousands, millions, billions, trillions) shown in the place-value chart. Within each group, you have a ones, tens, and hundreds place. Keep in mind that the leftmost group may have fewer than three digits.

In Figure 1.1, the numeric number 1,605,743,891,412 illustrates place values. When you study the place-value chart, you can see that the value of each place in the chart is 10 times the value of the place to the right. We can illustrate this by analyzing the last four digits in the number 1,605,743,891,412:

$$1{,}412 = (1 \times 1{,}000) + (4 \times 100) + (1 \times 10) + (2 \times 1)$$

So we can also say, for example, that in the number 745, the "7" means seven hundred (700); in the number 75, the "7" means 7 tens (70).

To read and write a numeric number in verbal form, you begin at the left and read each group of three digits as if it were alone, adding the group name at the end (except the last units group and groups of all zeros). Using the place-value chart in Figure 1.1, the number 1,605,743,891,412 is read as one trillion, six hundred five billion, seven hundred forty-three million, eight hundred ninety-one thousand, four hundred twelve. You do not read zeros. They fill vacant spaces as placeholders so that you can correctly state the number values. Also, the numbers twenty-one to ninety-nine must have a hyphen. And most important, when you read or write whole numbers in verbal form, do not use the word *and.* In the decimal system, *and* indicates the decimal, which we discuss in Chapter 3.

By reversing this process of changing a numeric number to a verbal number, you can use the place-value chart to change a verbal number to a numeric number. Remember that you must keep track of the place value of each digit. The place values of the digits in a number determine its total value.

Used by permission of Cartoon Features Syndicate

FIGURE 1.1

Whole number place-value chart

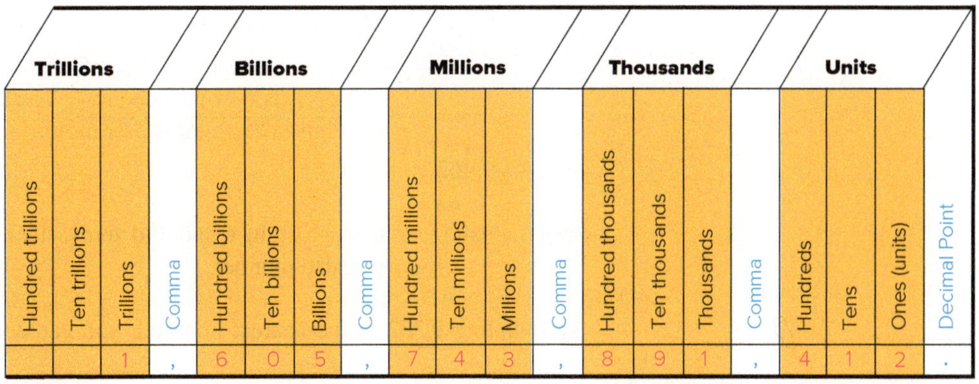

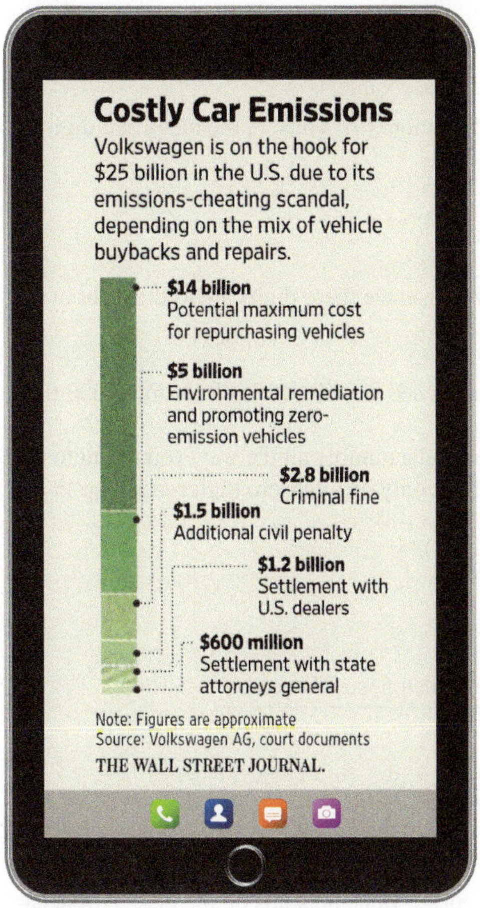

Before we look at how to round whole numbers, we should look at how to convert a number indicating parts of a whole number to a whole number. We will use the *Wall Street Journal* clip "Costly Car Emissions" as an example. Volkswagen paid a $1.2 billion settlement with U.S. dealers. This amount is 1 billion plus 200 million of an additional billion. The following steps explain how to convert decimal numbers into whole numbers.

CONVERTING PARTS OF A MILLION, BILLION, TRILLION, ETC., TO A REGULAR WHOLE NUMBER

Step 1. Drop the decimal point and insert a comma.

Step 2. Add zeros so the leftmost digit ends in the word name of the amount you want to convert. Be sure to add commas as needed.

EXAMPLE Convert 2.1 million to a regular whole number.

Step 1. 2.1 million

↓

2,1 Change the decimal point to a comma.

↓↓↓↓↓

Step 2. 2,100,000 Add zeros and commas so the whole number indicates million.

LO 2

Rounding Whole Numbers

Many of the whole numbers you read and hear are rounded numbers. Government statistics are usually rounded numbers. The financial reports of companies also use rounded numbers. All rounded numbers are *approximate* numbers. The more rounding you do, the more you approximate the number.

Rounded whole numbers are used for many reasons. With rounded whole numbers you can quickly estimate arithmetic results, check actual computations, report numbers that change quickly such as population numbers, and make numbers easier to read and remember.

Numbers can be rounded to any identified digit place value, including the first digit of a number (rounding all the way). To round whole numbers, use the following three steps:

ROUNDING WHOLE NUMBERS

Step 1. Identify the place value of the digit you want to round.

Step 2. If the digit to the right of the identified digit in Step 1 is 5 or more, increase the identified digit by 1 (round up). If the digit to the right is less than 5, do not change the identified digit.

Step 3. Change all digits to the right of the rounded identified digit to zeros.

EXAMPLE 1 Round 9,362 to the nearest hundred.

Step 1. 9,362 — The digit 3 is in the hundreds place value.

Step 2. → The digit to the right of 3 is 5 or more (6). Thus, 3, the identified digit in Step 1, is now rounded to 4. You change the identified digit only if the digit to the right is 5 or more.

9,462

Step 3. 9,400 — Change digits 6 and 2 to zeros, since these digits are to the right of 4, the rounded number.

By rounding 9,362 to the nearest hundred, you can see that 9,362 is closer to 9,400 than to 9,300.

Next, we show you how to round to the nearest thousand.

EXAMPLE 2 Round 67,951 to the nearest thousand.

Step 1. 67,951 — The digit 7 is in the thousands place value.

Step 2. → The digit to the right of 7 is 5 or more (9). Thus, 7, the identified digit in Step 1, is now rounded to 8.

68,951

Step 3. 68,000 — Change digits 9, 5, and 1 to zeros, since these digits are to the right of 8, the rounded number.

By rounding 67,951 to the nearest thousand, you can see that 67,951 is closer to 68,000 than to 67,000.

Now let's look at **rounding all the way.** To round a number all the way, you round to the first digit of the number (the leftmost digit) and have only one nonzero digit remaining in the number.

EXAMPLE 3 Round 7,843 all the way.

Step 1. 7,843 — Identified leftmost digit is 7.

Step 2. → Digit to the right of 7 is greater than 5, so 7 becomes 8.

8,843

Step 3. 8,000 — Change all other digits to zeros.

Rounding 7,843 all the way gives 8,000.

Remember that rounding a digit to a specific place value depends on the degree of accuracy you want in your estimate. For example, in the *Wall Street Journal* article "Costly Car Emissions," 1.2 billion rounded all the way would be 1 billion. Note the digit to the right of the identified digit is less than 5 so the identified digit (1) is kept at 1.

1,200,000,000 → 1,000,000,000

less than 5 (the 2)

Identified digit (the 1)

Before concluding this unit, let's look at how to dissect and solve a word problem.

How to Dissect and Solve a Word Problem

As a student, your author found solving word problems difficult. Not knowing where to begin after reading the word problem caused the difficulty. Today, students still struggle with word problems as they try to decide where to begin.

Solving word problems involves *organization* and *persistence.* Recall how persistent you were when you learned to ride a two-wheel bike. Do you remember the feeling of success you experienced when you rode the bike without help? Apply this persistence to word problems. Do not be discouraged. Each person learns at a different speed. Your goal must be to FINISH THE RACE and experience the success of solving word problems with ease.

To be organized in solving word problems, you need a plan of action that tells you where to begin—a blueprint aid. Like a builder, you will refer to this blueprint aid constantly until you know the procedure. The blueprint aid for dissecting and solving a word problem appears below. Note that the blueprint aid serves an important function—**it decreases your math anxiety.**

Aha!

Remember to RTDQ2: Read the darn question and then read it again before trying to solve it.

Blueprint Aid for Dissecting and Solving a Word Problem

BLUEPRINT	The facts	Solving for?	Steps to take	Key points

LO 3

©Ira Berger/Alamy Stock Photo

Now let's study this blueprint aid. The first two columns require that you *read* the word problem slowly. Think of the third column as the basic information you must know or calculate before solving the word problem. Often this column contains formulas that provide the foundation for the step-by-step problem solution. The last column reinforces the key points you should remember.

It's time now to try your skill at using the blueprint aid for dissecting and solving a word problem.

The Word Problem On the 100th anniversary of Tootsie Roll Industries, the company reported sharply increased sales and profits. Sales reached one hundred ninety-four million dollars and a record profit of twenty-two million, five hundred fifty-six thousand dollars. The company president requested that you round the sales and profit figures all the way.

Study the following blueprint aid and note how we filled in the columns with the information in the word problem. You will find the organization of the blueprint aid most helpful. Be persistent! You *can* dissect and solve word problems! When you are finished with the word problem, make sure the answer seems reasonable.

BLUEPRINT	The facts	Solving for?	Steps to take	Key points
	Sales: One hundred ninety-four million dollars. *Profit:* Twenty-two million, five hundred fifty-six thousand dollars.	Sales and profit rounded all the way.	Express each verbal form in numeric form. Identify leftmost digit in each number.	Rounding all the way means only the left-most digit will remain. All other digits become zeros.

MONEY tips

Do not carry your Social Security card in your wallet. Keep it and other important documents in a safe deposit box or fireproof container. Shred any document that contains personal information, such as anything with your Social Security number on it, old bank statements, applications for loans, and so on.

Steps to solving problem

1. Convert verbal to numeric.
   One hundred ninety-four million dollars ⟶ $194,000,000
   Twenty-two million, five hundred fifty-six thousand dollars ⟶ $ 22,556,000
2. Identify leftmost digit of each number.
   $194,000,000 $22,556,000
3. Round. ↓ $200,000,000 ↓ $20,000,000

Note that in the final answer, $200,000,000 and $20,000,000 have only one nonzero digit.

Remember that you cannot round numbers expressed in verbal form. You must convert these numbers to numeric form.

Now you should see the importance of the information in the third column of the blueprint aid. When you complete your blueprint aids for word problems, do not be concerned if the order of the information in your boxes does not follow the order given in the text boxes. Often you can dissect a word problem in more than one way.

Your first Practice Quiz follows. Be sure to study the paragraph that introduces the Practice Quiz.

LU 1–1 PRACTICE QUIZ

Complete this **Practice Quiz** to see how you are doing.

At the end of each learning unit, you can check your progress with a Practice Quiz. If you had difficulty understanding the unit, the Practice Quiz will help identify your area of weakness. Work the problems on scrap paper. Check your answers with the worked-out solutions that follow the quiz. Ask your instructor about specific assignments and the videos available in Connect for each unit Practice Quiz.

1. Write in verbal form:

a. 7,948 **b.** 48,775 **c.** 814,410,335,414

2. Round the following numbers as indicated:

Nearest ten	Nearest hundred	Nearest thousand	Rounded all the way
a. 92	**b.** 745	**c.** 8,341	**d.** 4,752

3. Kellogg's reported its sales as five million, one hundred eighty-one thousand dollars. The company earned a profit of five hundred two thousand dollars. What would the sales and profit be if each number were rounded all the way? (*Hint:* You might want to draw the blueprint aid since we show it in the solution.)

✓ Solutions

*For **extra help** from your authors–Sharon and Jeff–see the videos in Connect.*

1. a. Seven thousand, nine hundred forty-eight

b. Forty-eight thousand, seven hundred seventy-five

c. Eight hundred fourteen billion, four hundred ten million, three hundred thirty-five thousand, four hundred fourteen

2. a. 90 **b.** 700 **c.** 8,000 **d.** 5,000

3. Kellogg's sales and profit:

BLUEPRINT	The facts	Solving for?	Steps to take	Key points
	Sales: Five million, one hundred eighty-one thousand dollars. *Profit:* Five hundred two thousand dollars.	Sales and profit rounded all the way.	Express each verbal form in numeric form. Identify leftmost digit in each number.	Rounding all the way means only the leftmost digit will remain. All other digits become zeros.

Steps to solving problem

1. Convert verbal to numeric.
Five million, one hundred eighty-one thousand ⟶ $5,181,000
Five hundred two thousand ⟶ $ 502,000

2. Identify leftmost digit of each number.
$5,181,000 $502,000

3. Round.
$5,000,000 $500,000

Learning Unit 1–2: Adding and Subtracting Whole Numbers

LO 1

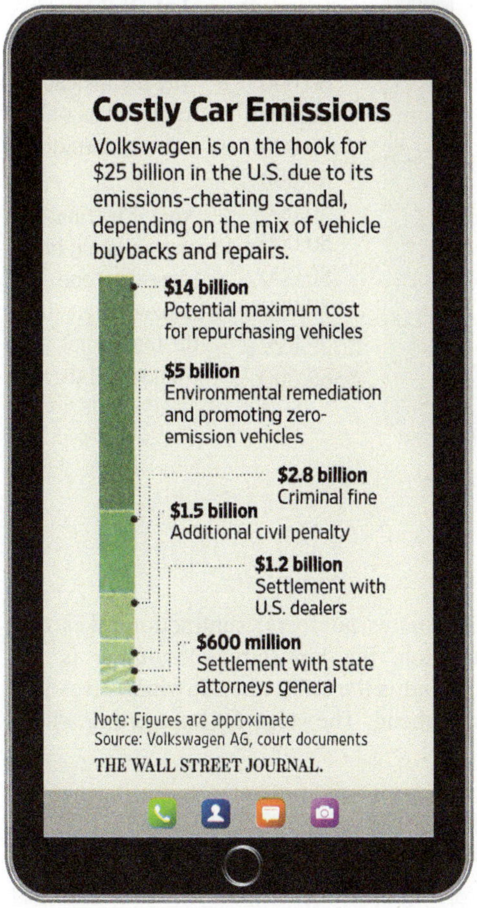

In the *Wall Street Journal* clip "Costly Car Emissions" reprinted from Learning Unit 1–1, note the difference between criminal fine and settlement with U.S. dealers.

Criminal fine	$2,800,000,000
U.S. dealer settlement	−1,200,000,000
	$1,600,000,000

This unit teaches you how to manually add and subtract whole numbers. When you least expect it, you will catch yourself automatically using this skill.

Addition of Whole Numbers

To add whole numbers, you unite two or more numbers called **addends** to make one number called a **sum,** *total,* or *amount.* The numbers are arranged in a column according to their place values—units above units, tens above tens, and so on. Then, you add the columns of numbers from top to bottom. To check the result, you re-add the columns from bottom to top. This procedure is illustrated in the steps that follow.

ADDING WHOLE NUMBERS

Step 1. Align the numbers to be added in columns according to their place values, beginning with the units place at the right and moving to the left.

Step 2. Add the units column. Write the sum below the column. If the sum is more than 9, write the units digit and carry the tens digit.

Step 3. Moving to the left, repeat Step 2 until all place values are added.

EXAMPLE

Adding top to bottom ↓

```
  2 11
  1,362
  5,913
  8,924
+ 6,594
 22,793
```

↑ Checking bottom to top

Alternate check
Add each column as a separate total and then combine. The end result is the same.

```
  1,362
  5,913
  8,924
+ 6,594
     13
    18
   26
  20
 22,793
```

How to Quickly Estimate Addition by Rounding All the Way In Learning Unit 1–1, you learned that rounding whole numbers all the way gives quick arithmetic estimates. Using the *Wall Street Journal* clip "Major Money" on page 10, note how you can round each number all the way and the total will not be rounded all the way. Remember that rounding all the way does not replace actual computations, but it is helpful in making quick commonsense decisions.

My Money

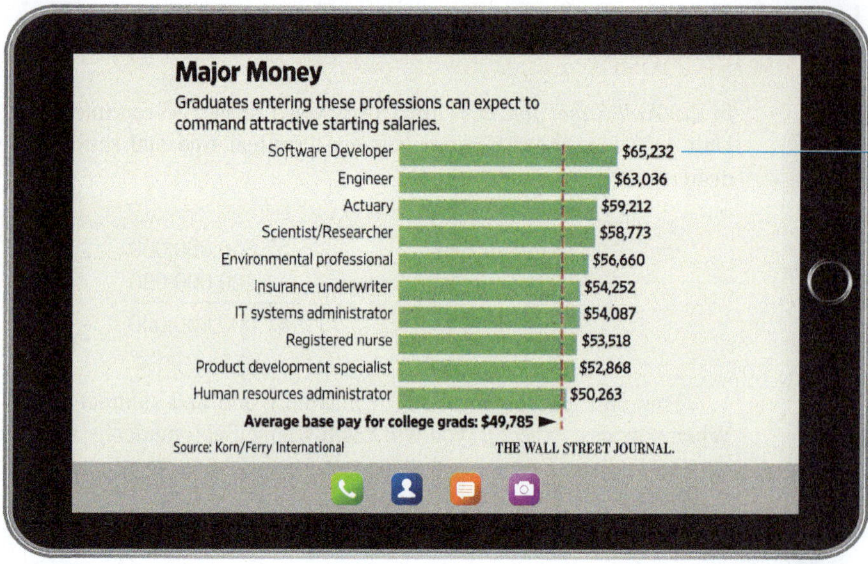

Rounded all the way

$70,000	← Rounding all the way means each number has only one nonzero digit.
60,000	
60,000	
60,000	
60,000	
50,000	
50,000	
50,000	
50,000	
+ 50,000	
$560,000	

Note: The final answer could have more than one nonzero digit since the total is not rounded all the way.

LO 2

Subtraction of Whole Numbers

Subtraction is the opposite of addition. Addition unites numbers; subtraction takes one number away from another number. In subtraction, the top (largest) number is the **minuend.** The number you subtract from the minuend is the **subtrahend,** which gives you the **difference** between the minuend and the subtrahend. The steps for subtracting whole numbers follow.

SUBTRACTING WHOLE NUMBERS

Step 1. Align the minuend and subtrahend according to their place values.

Step 2. Begin the subtraction with the units digits. Write the difference below the column. If the units digit in the minuend is smaller than the units digit in the subtrahend, borrow 1 from the tens digit in the minuend. One tens digit is 10 units.

Step 3. Moving to the left, repeat Step 2 until all place values in the subtrahend are subtracted.

©Rachel Frank/Corbis/Glow Images

EXAMPLE The previous *Wall Street Journal* clip about graduates illustrates the subtraction of whole numbers:

What is the difference in the starting salaries between a registered nurse and an engineer? As shown below you can use subtraction to arrive at the $9,518 difference.

$63,036	← Minuend (larger number)
−53,518	← Subtrahend
$ 9,518	← Difference

Check

$ 9,518
+53,518
$63,036

In subtraction, borrowing from the column at the left is often necessary. Remember that 1 ten = 10 units, 1 hundred = 10 tens, and 1 thousand = 10 hundreds.

In the ones column in the example above, 8 cannot be subtracted from 6 so we borrow from the tens column, resulting in 16 less 8 equals 8. In the hundreds column, we cannot subtract 5 from 0 so we borrow 10 tens from the hundreds column leaving 5 hundreds. 12 less 3 equals 9.

Checking subtraction requires adding the difference ($9,518) to the subtrahend ($53,518) to arrive at the minuend ($63,036).

How to Dissect and Solve a Word Problem

Accurate subtraction is important in many business operations. In Chapter 4 we discuss the importance of keeping accurate subtraction in your checkbook balance. Now let's check your progress by dissecting and solving a word problem.

The Word Problem Hershey's produced 25 million Kisses in one day. The same day, the company shipped 4 million to Japan, 3 million to France, and 6 million throughout the United States. At the end of that day, what is the company's total inventory of Kisses? What is the inventory balance if you round the number all the way?

MONEY tips

College *is* worth it! College graduates earn substantially more money each year than high school graduates *and* that wage premium is increasing steadily—almost twice as much. Stay in school.

BLUEPRINT	The facts	Solving for?	Steps to take	Key points
	Produced: 25 million. *Shipped:* Japan, 4 million; France, 3 million; United States, 6 million.	Total Kisses left in inventory. Inventory balance rounded all the way.	Total Kisses produced − Total Kisses shipped = Total Kisses left in inventory.	Minuend − Subtrahend = Difference. Rounding all the way means rounding to last digit on the left.

Steps to solving problem

1. Calculate the total Kisses shipped.

```
  4,000,000
  3,000,000
+ 6,000,000
 13,000,000
```

2. Calculate the total Kisses left in inventory.

```
 25,000,000
−13,000,000
 12,000,000
```

3. Rounding all the way.

Identified digit is 1. Digit to right of 1 is 2, which is less than 5. *Answer:* 10,000,000.

The Practice Quiz that follows will tell you how you are progressing in your study of Chapter 1.

LU 1–2 PRACTICE QUIZ

Complete this **Practice Quiz** to see how you are doing.

1. Add by totaling each separate column:

```
  8,974
  6,439
+ 6,941
```

2. Estimate by rounding all the way (do not round the total of estimate) and then do the actual computation:

```
  4,241
  8,794
+ 3,872
```

3. Subtract and check your answer:

```
  9,876
− 4,967
```

4. Jackson Manufacturing Company projected its year 2020 furniture sales at $900,000. During 2020, Jackson earned $510,000 in sales from major clients and $369,100 in sales from the remainder of its clients. What is the amount by which Jackson over- or underestimated its sales? Use the blueprint aid, since the answer will show the completed blueprint aid.

For **extra help** from your authors–Sharon and Jeff–see the videos in Connect.

✓ Solutions

1.

14
14
22
20
22,354

2.

Estimate	Actual
4,000	4,241
9,000	8,794
+ 4,000	+ 3,872
17,000	16,907

3.

	Check
8 18 6 16	
9,876	4,909
−4,967	+ 4,967
4,909	9,876

4. Jackson Manufacturing Company over- or underestimated sales:

BLUEPRINT	The facts	Solving for?	Steps to take	Key points
	Projected 2020 sales: $900,000. *Major clients:* $510,000. *Other clients:* $369,100.	How much were sales over- or underestimated?	Total projected sales − Total actual sales = Over- or underestimated sales.	Projected sales (minuend) − Actual sales (subtrahend) = Difference.

Steps to solving problem

1. Calculate total actual sales.

$ 510,000
+ 369,100
$ 879,100

2. Calculate overestimated or underestimated sales.

$ 900,000
− 879,100
$ 20,900 (overestimated)

LO 1

Learning Unit 1–3: Multiplying and Dividing Whole Numbers

The *Wall Street Journal* clip in the margin reveals that *Star Wars* movies grossed $6 billion worldwide in 2016. If the $6 billion figure were for 2 months the sales would be $3 billion per month. If you divide $6 billion by 2 months, you would get $3,000,000,000.

This unit will sharpen your skills in two important arithmetic operations—multiplication and division. These two operations frequently result in knowledgeable business decisions.

GLOBAL

Multiplication of Whole Numbers—Shortcut to Addition

From calculating the sales for 2 months you know that multiplication is a *shortcut to addition:*

$3,000,000,000 × 2 = $6,000,000,000

or

$3,000,000,000 + $3,000,000,000 = $6,000,000,000

Before learning the steps used to multiply whole numbers with two or more digits, you must learn some multiplication terminology.

Note in the following example that the top number (number we want to multiply) is the **multiplicand.** The bottom number (number doing the multiplying) is the **multiplier.** The final number (answer) is the **product.** The numbers between the multiplier and the product are **partial products.** Also note how we positioned the partial product 2090. This number is the result of multiplying 418 by 50 (the 5 is in the tens position). On each line in the partial products, we placed the first digit directly below the digit we used in the multiplication process.

EXAMPLE

	418	← Top number (multiplicand)	
	× 52	← Bottom number (multiplier)	
Partial products	836		2 × 418 = 836
	2090		50 × 418 = + 20,900
	21,736	← Product answer →	21,736

We can now give the following steps for multiplying whole numbers with two or more digits:

MULTIPLYING WHOLE NUMBERS WITH TWO OR MORE DIGITS

Step 1. Align the multiplicand (top number) and multiplier (bottom number) at the right. Usually, you should make the smaller number the multiplier.

Step 2. Begin by multiplying the right digit of the multiplier with the right digit of the multiplicand. Keep multiplying as you move left through the multiplicand. Your first partial product aligns at the right with the multiplicand and multiplier.

Step 3. Move left through the multiplier and continue multiplying the multiplicand. Your partial product right digit or first digit is placed directly below the digit in the multiplier that you used to multiply.

Step 4. Continue Steps 2 and 3 until you have completed your multiplication process. Then add the partial products to get the final product.

Checking and Estimating Multiplication We can check the multiplication process by reversing the multiplicand and multiplier and then multiplying. Let's first estimate 52 × 418 by rounding all the way.

EXAMPLE

$$\begin{array}{r} 50 \leftarrow \\ \times\ 400 \leftarrow \\ \hline 20{,}000 \end{array} \qquad \begin{array}{r} 52 \\ \times\ 418 \\ \hline 416 \\ 52 \\ 20\,8 \\ \hline 21{,}736 \end{array}$$

By estimating before actually working the problem, we know our answer should be about 20,000. When we multiply 52 by 418, we get the same answer as when we multiply 418 × 52—and the answer is about 20,000. Remember, if we had not rounded all the way, our estimate would have been closer. If we had used a calculator, the rounded estimate would have helped us check the calculator's answer. Our commonsense estimate tells us our answer is near 20,000—not 200,000.

Before you study the division of whole numbers, you should know (1) the multiplication shortcut with numbers ending in zeros and (2) how to multiply a whole number by a power of 10.

MULTIPLICATION SHORTCUT WITH NUMBERS ENDING IN ZEROS

Step 1. When zeros are at the end of the multiplicand or the multiplier, or both, disregard the zeros and multiply.

Step 2. Count the number of zeros in the multiplicand and multiplier.

Step 3. Attach the number of zeros counted in Step 2 to your answer.

EXAMPLE

$$\begin{array}{r} 65{,}000 \\ \times\ 420 \end{array} \qquad \begin{array}{r} 65 \\ \times\ \ 42 \\ \hline 1\,30 \\ 26\,0 \\ \hline 27{,}300{,}000 \end{array} \qquad \begin{array}{r} 3 \text{ zeros} \\ +\ 1 \text{ zero} \\ \hline 4 \text{ zeros} \end{array}$$

No need to multiply rows of zeros

$$\begin{array}{r} 65{,}000 \\ \times\ \ 420 \\ \hline 00\,000 \\ 1\,300\,00 \\ 26\,000\,0 \\ \hline 27{,}300{,}000 \end{array}$$

MONEY tips

Be vigilant about sharing personal information. Change passwords often and do not share them.

BLUEPRINT	The facts	Solving for?	Steps to take	Key points
	Sales per week: \$3,500. *Companies:* 4.	Total annual sales to all four companies. Yearly sales per company.	Sales per week × Weeks in year (52) = Total annual sales. Total annual sales ÷ Total companies = Yearly sales per company.	Division is the reverse of multiplication.

Steps to solving problem

1. Calculate total annual sales. — \$3,500 × 52 weeks = **\$182,000**
2. Calculate yearly sales per company. — \$182,000 ÷ 4 = **\$45,500**

Check

\$45,500 × 4 = \$182,000

It's time again to check your progress with a Practice Quiz.

LU 1–3 PRACTICE QUIZ

Complete this **Practice Quiz** to see how you are doing.

1. Estimate the actual problem by rounding all the way, work the actual problem, and check:

   Actual **Estimate** **Check**

   $$\begin{array}{r} 3{,}894 \\ \times\ 18 \end{array}$$

2. Multiply by shortcut method:

   $$\begin{array}{r} 77{,}000 \\ \times\ 1{,}800 \end{array}$$

3. Multiply by shortcut method:

   95 × 10,000

4. Divide by rounding all the way, complete the actual calculation, and check, showing remainder as a whole number.

   $26\overline{)5{,}325}$

5. Divide by shortcut method:

   $4{,}000\overline{)96{,}000}$

6. Assume General Motors produces 960 Chevrolets each workday (Monday through Friday). If the cost to produce each car is \$6,500, what is General Motors' total cost for the year? Check your answer.

*For **extra help** from your authors–Sharon and Jeff–see the videos in Connect.*

✓ Solutions

1. **Estimate**

   $$\begin{array}{r} 4{,}000 \\ \times\ \ 20 \\ \hline 80{,}000 \end{array}$$

   Actual

   $$\begin{array}{r} 3{,}894 \\ \times\ \ 18 \\ \hline 31\ 152 \\ 38\ 94 \\ \hline 70{,}092 \end{array}$$

   Check

   $$\begin{array}{rr} 8 \times 3{,}894 = & 31{,}152 \\ 10 \times 3{,}894 = & +\ 38{,}940 \\ \hline & 70{,}092 \end{array}$$

2. 77 × 18 = 1,386 + 5 zeros = **138,600,000**

3. 95 + 4 zeros = **950,000**

4. **Rounding**

   $$\begin{array}{r} 166\text{ R}20 \\ 30\overline{)5{,}000} \\ 3\ 0 \\ \hline 2\ 00 \\ 1\ 80 \\ \hline 200 \\ 180 \\ \hline 20 \end{array}$$

   Actual

   $$\begin{array}{r} 204\text{ R}21 \\ 26\overline{)5{,}325} \\ 5\ 2 \\ \hline 125 \\ 104 \\ \hline 21 \end{array}$$

   Check

   $$\begin{array}{r} 26 \times 204 = 5{,}304 \\ +\ \ \ 21 \\ \hline 5{,}325 \end{array}$$

5. Drop 3 zeros = $\begin{array}{r} 24 \\ 4\overline{)96} \end{array}$

6. General Motors' total cost per year:

	The facts	Solving for?	Steps to take	Key points
BLUEPRINT	*Cars produced each workday:* 960. *Workweek:* 5 days. *Cost per car:* $6,500.	Total cost per year.	Cars produced per week × 52 = Total cars produced per year. Total cars produced per year × Total cost per car = Total cost per year.	Whenever possible, use multiplication and division shortcuts with zeros. Multiplication can be checked by division.

Steps to solving problem

1. Calculate total cars produced per week. — 5 × 960 = 4,800 cars produced per week
2. Calculate total cars produced per year. — 4,800 cars × 52 weeks = 249,600 total cars produced per year
3. Calculate total cost per year. — 249,600 cars × $6,500 = **$1,622,400,000** (multiply 2,496 × 65 and add zeros)

Check

$1,622,400,000 ÷ 249,600 = $6,500
(drop 2 zeros before dividing)

INTERACTIVE CHAPTER ORGANIZER

Topic/Procedure/Formula	Example	You try it*
Reading and writing numeric and verbal whole numbers Placement of digits in a number gives the value of the digits (Figure 1.1). Commas separate every three digits, moving from right to left. Begin at left to read and write number in verbal form. Do not read zeros or use *and.* Hyphenate numbers twenty-one to ninety-nine. Reverse procedure to change verbal number to numeric.	462 → Four hundred sixty-two 6,741 → Six thousand, seven hundred forty-one	**Write in verbal form** 571 → 7,943 →
Rounding whole numbers **1.** Identify place value of the digit to be rounded. **2.** If digit to the right is 5 or more, round up; if less than 5, do not change. **3.** Change all digits to the right of rounded identified digit to zeros.	643 to nearest ten 4 in tens place value; 3 is not 5 or more Thus, 643 rounds to **640**.	**Round to nearest ten** 691
Rounding all the way Round to first digit of number. One nonzero digit remains. In estimating, you round each number of the problem to one nonzero digit. The final answer is not rounded.	468,451 → **500,000** The 5 is the only nonzero digit remaining.	**Round all the way** 429,685 →
Adding whole numbers **1.** Align numbers at the right. **2.** Add units column. If sum is more than 9, carry tens digit. **3.** Moving left, repeat Step 2 until all place values are added. Add from top to bottom. Check by adding bottom to top or adding each column separately and combining.	$\begin{array}{r} \overset{1}{6}5 \\ +\ 47 \\ \hline \mathbf{112} \end{array}$ $\begin{array}{r} 12 \\ +10 \\ \hline 112 \end{array}$ Checking sum of each digit	**Add** $\begin{array}{r} 76 \\ +38 \\ \hline \end{array}$

(continues)

INTERACTIVE CHAPTER ORGANIZER

Topic/Procedure/Formula	Example	You try it*
Subtracting whole numbers **1.** Align minuend and subtrahend at the right. **2.** Subtract units digits. If necessary, borrow 1 from tens digit in minuend. **3.** Moving left, repeat Step 2 until all place values are subtracted. Minuend less subtrahend equals difference.	5 18 ~~8~~~~8~~5 −492 **193** **Check** 193 +492 685	**Subtract** 629 −134
Multiplying whole numbers **1.** Align multiplicand and multiplier at the right. **2.** Begin at the right and keep multiplying as you move to the left. First partial product aligns at the right with multiplicand and multiplier. **3.** Move left through multiplier and continue multiplying multiplicand. Partial product right digit or first digit is placed directly below digit in multiplier. **4.** Continue Steps 2 and 3 until multiplication is complete. Add partial products to get final product. **Shortcuts:** (a) When multiplicand or multiplier, or both, end in zeros, disregard zeros and multiply; attach same number of zeros to answer. If zero is in center of multiplier, no need to show row of zeros. (b) If multiplying by power of 10, attach same number of zeros to whole number multiplied.	223 × 32 446 6 69 **7,136** a. 48,000 × 40 → 48 × 4 = **1,920,000** (3 zeros + 1 zero → 4 zeros) 524 × 206 3 144 104 8 **107,944** b. 14 × 10 = **140** (attach 1 zero) 14 × 1,000 = **14,000** (attach 3 zeros)	**Multiply** 491 × 28 **Multiply by shortcut** 13 × 10 = 13 × 1,000 =
Dividing whole numbers **1.** When divisor is divided into the dividend, the remainder is less than divisor. **2.** Drop zeros from dividend right to left by number of zeros found in the divisor. Even division has no remainder; uneven division has a remainder; divisor with one digit is short division; and divisor with more than one digit is long division.	1. 14)76 = **5 R6** 70 6 2. 5,000 ÷ 100 = 50 ÷ 1 = **50** 5,000 ÷ 1,000 = 5 ÷ 1 = **5**	**Divide** **1.** 16)95 **Divide by shortcut** **2.** 4,000 ÷ 100 4,000 ÷ 1,000

KEY TERMS			
	Addends	Minuend	Quotient
	Decimal point	Multiplicand	Remainder
	Decimal system	Multiplier	Rounding all the way
	Difference	Partial products	Subtrahend
	Dividend	Partial quotient	Sum
	Divisor	Product	Whole number

*Worked-out solutions are in Appendix B.

Critical Thinking Discussion Questions with Chapter Concept Check

1. List the four steps of the decision-making process. Do you think all companies should be required to follow these steps? Give an example.
2. Explain the three steps used to round whole numbers. Pick a whole number and explain why it should not be rounded.
3. How do you check subtraction? If you were to attend a movie, explain how you might use the subtraction check method.
4. Explain how you can check multiplication. If you visit a local supermarket, how could you show multiplication as a shortcut to addition?
5. Explain how division is the reverse of multiplication. Using the supermarket example in question 4, explain how division is a timesaving shortcut related to subtraction.
6. **Chapter Concept Check.** Using all the math you learned in Chapter 1, calculate the difference in cost and calories from dining at Subway versus McDonald's. Go online or visit these stores in your area to find current food prices.

END-OF-CHAPTER PROBLEMS

connect

Check figures for odd-numbered problems in Appendix B. Name ______________________ Date ________

DRILL PROBLEMS

Add the following: *LU 1-2(1)*

1–1. $\begin{array}{r} 90 \\ +15 \\ \hline \end{array}$ **1–2.** $\begin{array}{r} 900 \\ +\ 250 \\ \hline \end{array}$ **1–3.** $\begin{array}{r} 77 \\ +\ 77 \\ \hline \end{array}$ **1–4.** $\begin{array}{r} 88 \\ +\ 75 \\ \hline \end{array}$

1–5. $\begin{array}{r} 6{,}251 \\ +\ 7{,}329 \\ \hline \end{array}$ **1–6.** $\begin{array}{r} 59{,}481 \\ 51{,}411 \\ +\ 70{,}821 \\ \hline \end{array}$ **1–7.** $\begin{array}{r} 78{,}159 \\ 15{,}850 \\ +\ 19{,}681 \\ \hline \end{array}$

Subtract the following: *LU 1-2(2)*

1–8. $\begin{array}{r} 68 \\ -19 \\ \hline \end{array}$ **1–9.** $\begin{array}{r} 80 \\ -42 \\ \hline \end{array}$ **1–10.** $\begin{array}{r} 287 \\ -199 \\ \hline \end{array}$

1–11. $\begin{array}{r} 9{,}000 \\ -5{,}400 \\ \hline \end{array}$ **1–12.** $\begin{array}{r} 9{,}800 \\ -8{,}900 \\ \hline \end{array}$ **1–13.** $\begin{array}{r} 1{,}622 \\ -\ 548 \\ \hline \end{array}$

Multiply the following: *LU 1-3(1)*

1–14. $\begin{array}{r} 50 \\ \times\ 6 \\ \hline \end{array}$ **1–15.** $\begin{array}{r} 510 \\ \times\ \ 61 \\ \hline \end{array}$ **1–16.** $\begin{array}{r} 800 \\ \times\ \ 200 \\ \hline \end{array}$

1–17. $\begin{array}{r} 677 \\ \times\ \ 503 \\ \hline \end{array}$ **1–18.** $\begin{array}{r} 309 \\ \times\ 850 \\ \hline \end{array}$ **1–19.** $\begin{array}{r} 450 \\ \times\ 280 \\ \hline \end{array}$

Divide the following by short division: *LU 1-3(2)*

1–20. $4\overline{)1{,}600}$ **1–21.** $9\overline{)810}$ **1–22.** $4\overline{)164}$

Divide the following by long division. Show work and remainder. *LU 1-3(2)*

1–23. $6\overline{)520}$ **1–24.** $62\overline{)8{,}915}$

Add the following without rearranging: *LU 1-2(1)*

1–25. 95 + 310 **1–26.** 1,055 + 88

1–27. 666 + 950 **1–28.** 1,011 + 17

1–29. Add the following and check by totaling each column individually without carrying numbers: *LU 1-2(1)*

Check

$$\begin{array}{r} 8{,}539 \\ 6{,}842 \\ +\ 9{,}495 \\ \hline \end{array}$$

Estimate the following by rounding all the way and then do actual addition: *LU 1-1(2), LU 1-2(1)*

	Actual	Estimate		Actual	Estimate
1–30.	7,700 9,286 + 3,900		**1–31.**	6,980 3,190 + 7,819	

Subtract the following without rearranging: *LU 1-2(2)*

1–32. 190 – 66

1–33. 950 – 870

1–34. Subtract the following and check answer: *LU 1-2(2)*

$$\begin{array}{r} 591{,}001 \\ -375{,}956 \\ \hline \end{array}$$

Multiply the following horizontally: *LU 1-3(1)*

1–35. 19×7

1–36. 84×8

1–37. 27×8

1–38. $19 \times 5 =$

Divide the following and check by multiplication: *LU 1-2(2)*

1–39. $45\overline{)876}$ **Check**

1–40. $46\overline{)1{,}950}$ **Check**

Complete the following: *LU 1-2(2)*

1–41.

$$\begin{array}{r} 9{,}200 \\ -\ 1{,}510 \\ \hline \\ -\ \ \ 700 \\ \hline \end{array}$$

1–42.

$$\begin{array}{r} 3{,}000{,}000 \\ -\ \ 769{,}459 \\ \hline \\ -\ \ \ 68{,}541 \\ \hline \end{array}$$

1–43. Estimate the following problem by rounding all the way and then do the actual multiplication: *LU 1-1(2), LU 1-3(1)*

Actual **Estimate**

$$\begin{array}{r} 870 \\ \times\ \ 81 \\ \hline \end{array}$$

Divide the following by the shortcut method: *LU 1-3(2)*

1–44. $1{,}000\overline{)950{,}000}$

1–45. $100\overline{)70{,}000}$

1–46. Estimate actual problem by rounding all the way and do actual division: *LU 1-1(2), LU 1-3(2)*

Actual **Estimate**

$695\overline{)8{,}950}$

WORD PROBLEMS

1–47. *The Wall Street Journal* reported that the cost for lightbulbs over a 10-year period at a local Walmart parking lot in Kansas would be $248,134 if standard lightbulbs were used. If LED lightbulbs were used over the same period, the total cost would be $220,396. What would Walmart save by using LED bulbs? *LU 1-2(2)*

My Money

1–48. An education can be the key to higher earnings. In a U.S. Census Bureau study, high school graduates earned $30,400 per year. Associate's degree graduates averaged $38,200 per year. Bachelor's degree graduates averaged $52,200 per year. Assuming a 50-year work-life, calculate the lifetime earnings for a high school graduate, associate's degree graduate, and bachelor's degree graduate. What's the lifetime income difference between a high school and associate's degree? What about the lifetime difference between a high school and bachelor's degree? *LU 1-3(1), LU 1-2(2)*

1–49. Assume season-ticket prices in the lower bowl for the Buffalo Bills will rise from $480 for a 10-game package to $600. Fans sitting in the best seats in the upper deck will pay an increase from $440 to $540. Don Manning plans to purchase two season tickets for either lower bowl or upper deck. **(a)** How much more will two tickets cost for lower bowl? **(b)** How much more will two tickets cost for upper deck? **(c)** What will be his total cost for a 10-game package for lower bowl? **(d)** What will be his total cost for a 10-game package for upper deck? *LU 1-2(2), LU 1-3(1)*

1–50. Some ticket prices for *Lion King* on Broadway were $70, $95, $200, and $250. For a family of four, estimate the cost of the $95 tickets by rounding all the way and then do the actual multiplication: *LU 1-1(2), LU 1-3(1)*

1–51. Walt Disney World Resort and United Vacations got together to create a special deal. The air-inclusive package features accommodations for three nights at Disney's All-Star Resort, hotel taxes, and a four-day unlimited Magic Pass. Prices are $609 per person traveling from Washington, DC, and $764 per person traveling from Los Angeles. **(a)** What would be the cost for a family of four leaving from Washington, DC? **(b)** What would be the cost for a family of four leaving from Los Angeles? **(c)** How much more will it cost the family from Los Angeles? *LU 1-3(1)*

1–52. NTB Tires bought 910 tires from its manufacturer for $36 per tire. What is the total cost of NTB's purchase? If the store can sell all the tires at $65 each, what will be the store's gross profit, or the difference between its sales and costs (Sales – Costs = Gross profit)? *LU 1-3(1), LU 1-2(2)*

1–53. What was the total average number of visits for these websites? *LU 1-2(1), LU 1-3(2)*

Website	Average daily unique visitors
1. Orbitz.com	1,527,000
2. Mypoints.com	1,356,000
3. Americangreetings.com	745,000
4. Bizrate.com	503,000
5. Half.com	397,000

My Money

1–54. CNN.com reported in October 2017 that nearly 40 out of 100 adults in the United States are obese or overweight. Research has shown coffee has several health-related benefits. One such benefit is an antioxidant, chlorogenic acid (CGA), that may help protect against several obesity-related diseases. During a 15-week study, if 67 mice did not gain weight during the test period and an additional 48 demonstrated insulin resistance, how many mice were positively affected by the injection of the CGA solution? *LU 1-2(1)*

1–55. A report from the Center for Science in the Public Interest—a consumer group based in Washington, DC—released a study listing calories of various ice cream treats sold by six of the largest ice cream companies. The worst treat tested by the group was 1,270 total calories. People need roughly 2,200 to 2,500 calories per day. Using a daily average, how many additional calories should a person consume after eating ice cream? *LU 1-2(1), LU 1-3(2)*

1–56. At Rose State College, Alison Wells received the following grades in her online accounting class: 90, 65, 85, 80, 75, and 90. Alison's instructor, Professor Clark, said he would drop the lowest grade. What is Alison's average? *LU 1-2(1)*

1–57. The Bureau of Transportation's list of the 10 most expensive U.S. airports and their average fares is given below. Please use this list to answer the questions that follow. *LU 1-2(1, 2)*

Airport	Average fare
1. Houston, TX	$477
2. Huntsville, AL	473
3. Newark, NJ	470
4. Cincinnati, OH	466
5. Washington, DC	465
6. Charleston, SC	460
7. Memphis, TN	449
8. Knoxville, TN	449
9. Dallas–Fort Worth, TX	431
10. Madison, WI	429

a. What is the total of all the fares?

b. What would the total be if all the fares were rounded all the way?

c. How much does the actual number differ from the rounded estimate?

1–58. Ron Alf, owner of Alf's Moving Company, bought a new truck. On Ron's first trip, he drove 1,200 miles and used 80 gallons of gas. How many miles per gallon did Ron get from his new truck? On Ron's second trip, he drove 840 miles and used 60 gallons. What is the difference in miles per gallon between Ron's first trip and his second trip? *LU 1-3(2)*

My Money

1–59. Magnifymoney.com reported in September 2017 that 201 million Americans use credit cards and 125 million carry an average household balance per month of $8,158. If the average household balance per month last year was $7,946, what is the increase in the average balance? *LU 1-2(2)*

1–60. Assume BarnesandNoble.com has 289 business math texts in inventory. During one month, the online bookstore ordered and received 1,855 texts; it also sold 1,222 on the web. What is the bookstore's inventory at the end of the month? If each text costs $59, what is the end-of-month inventory cost? *LU 1-2(1), LU 1-2(2)*

1–61. Assume Cabot Company produced 2,115,000 cans of paint in August. Cabot sold 2,011,000 of these cans. If each can cost $18, what were Cabot's ending inventory of paint cans and its total ending inventory cost? *LU 1-2(2), LU 1-3(1)*

1–62. A local community college has 20 faculty members in the business department, 40 in psychology, 26 in English, and 140 in all other departments. What is the total number of faculty at this college? If each faculty member advises 25 students, how many students attend the local college? *LU 1-2(1), LU 1-3(1)*

1–63. Hometown Buffet had 90 customers on Sunday, 70 on Monday, 65 on Tuesday, and a total of 310 on Wednesday to Saturday. How many customers did Hometown Buffet serve during the week? If each customer spends $9, what were the total sales for the week? *LU 1-2(1), LU 1-3(1)*

If Hometown Buffet had the same sales each week, what were the sales for the year?

1–64. A local travel agency projected its year 2019 sales at $880,000. During 2019, the agency earned $482,900 sales from its major clients and $116,500 sales from the remainder of its clients. How much did the agency overestimate its sales? *LU 1-2(2)*

1–65. Ryan Seary works at US Airways and earned $71,000 last year before tax deductions. From Ryan's total earnings, his company subtracted $1,388 for federal income taxes, $4,402 for Social Security, and $1,030 for Medicare taxes. What was Ryan's actual, or net, pay for the year? *LU 1-2(1, 2)*

My Money

1–66. CompareCards.com lists credit card offers by such categories as low interest, no annual fee, cash back, and so on. A top card offers no interest payments for 18 months through 2020. If 11 credit card companies make this offer and 25,652 people are approved, on average how many new customers does each credit card company gain? *LU 1-3(2)*

1–67. Roger Company produces beach balls and operates three shifts. Roger produces 5,000 balls per shift on shifts 1 and 2. On shift 3, the company can produce 6 times as many balls as on shift 1. Assume a 5-day workweek. How many beach balls does Roger produce per week and per year? *LU 1-2(1), LU 1-3(1)*

1–68. Assume 6,000 children go to Disneyland today. How much additional revenue will Disneyland receive if it raises the cost of admission from $31 to $41? *LU 1-2(1), LU 1-3(1)*

1–69. Moe Brink has a $900 balance in his checkbook. During the week, Moe wrote the following checks: rent, $350; telephone, $44; food, $160; and entertaining, $60. Moe also made a $1,200 deposit. What is Moe's new checkbook balance? *LU 1-2(1, 2)*

1–70. A local Dick's Sporting Store, an athletic sports shop, bought and sold the following merchandise: *LU 1-2(1, 2)*

	Cost	Selling price
Tennis rackets	$2,900	$ 3,999
Tennis balls	70	210
Bowling balls	1,050	2,950
Sneakers	+8,105	+14,888

What was the total cost of the merchandise bought by Dick's Sporting Store? If the shop sold all its merchandise, what were the sales and the resulting gross profit (Sales – Costs = Gross profit)?

eXcel **1–71.** Rich Engel, the bookkeeper for Engel's Real Estate, and his manager are concerned about the company's telephone bills. Last year the company's average monthly phone bill was $32. Rich's manager asked him for an average of this year's phone bills. Rich's records show the following: *LU 1-2(1), LU 1-3(2)*

January	$ 34	July	$ 28
February	60	August	23
March	20	September	29
April	25	October	25
May	30	November	22
June	59	December	41

What is the average of this year's phone bills? Did Rich and his manager have a justifiable concern?

eXcel **1–72.** On Monday, a local True Value Hardware sold 15 paint brushes at $3 each, six wrenches at $5 each, seven bags of grass seed at $3 each, four lawn mowers at $119 each, and 28 cans of paint at $8 each. What were True Value's total dollar sales on Monday? *LU 1-2(1), LU 1-3(1)*

1–73. While redecorating, Lee Owens went to Carpet World and bought 150 square yards of commercial carpet. The total cost of the carpet was $6,000. How much did Lee pay per square yard? *LU 1-3(2)*

eXcel **1–74.** Washington Construction built 12 ranch houses for $115,000 each. From the sale of these houses, Washington received $1,980,000. How much gross profit (Sales − Costs = Gross profit) did Washington make on the houses? *LU 1-2(2), LU 1-3(1, 2)*

The four partners of Washington Construction split all profits equally. How much will each partner receive?

CHALLENGE PROBLEMS

1–75. A mall in Lexington has 18 stores. The following is a breakdown of what each store pays for rent per month. The rent is based on square footage.

5 department/computer stores	$1,250	2 bakeries	$ 500
5 restaurants	860	2 drugstores	820
3 bookstores	750	1 supermarket	1,450

Calculate the total rent that these stores pay annually. What would the answer be if it were rounded all the way? How much more each year do the drugstores pay in rent compared to the bakeries? *LU 1-2(2), LU 1-3(1)*

1–76. Paula Sanchez is trying to determine her 2019 finances. Paula's actual 2018 finances were as follows: *LU 1-1, LU 1-2, LU 1-3*

My Money

2018				
Income:		Assets:		
Gross income	$69,000	Checking account	$ 1,950	
Interest income	450	Savings account	8,950	
Total	$69,450	Automobile	1,800	
Expenses:		Personal property	14,000	
Living	$24,500	Total	$26,700	
Insurance premium	350	Liabilities:		
Taxes	14,800	Note to bank	4,500	
Medical	585	Net worth	$22,200	($26,700 – $4,500)
Investment	4,000			
Total	$44,235			

Net worth = Assets – Liabilities
(own) (owe)

Paula believes her gross income will double in 2019 but her interest income will decrease $150. She plans to reduce her 2019 living expenses by one-half. Paula's insurance company wrote a letter announcing that her insurance premiums would triple in 2019. Her accountant estimates her taxes will decrease $250 and her medical costs will increase $410. Paula also hopes to cut her investments expenses by one-fourth. Paula's accountant projects that her savings and checking accounts will each double in value. On January 2, 2019, Paula sold her automobile and began to use public transportation. Paula forecasts that her personal property will decrease by one-seventh. She has sent her bank a $375 check to reduce her bank note. Could you give Paula an updated list of her 2019 finances? If you round all the way each 2018 and 2019 asset and liability, what will be the difference in Paula's net worth?

Classroom Notes

SUMMARY PRACTICE TEST

Do you need help? Connect videos have step-by-step worked-out solutions.

1. Translate the following verbal forms to numbers and add. *LU 1-1(1), LU 1-2(1)*
   - **a.** Four thousand, eight hundred thirty-nine
   - **b.** Seven million, twelve
   - **c.** Twelve thousand, three hundred ninety-two
2. Express the following number in verbal form. *LU 1-1(1)*

   9,622,364
3. Round the following numbers. *LU 1-1(2)*

Nearest ten	**Nearest hundred**	**Nearest thousand**	**Round all the way**
a. 68	**b.** 888	**c.** 8,325	**d.** 14,821

4. Estimate the following actual problem by rounding all the way, work the actual problem, and check by adding each column of digits separately. *LU 1-1(2), LU 1-2(1)*

Actual	**Estimate**	**Check**
1,886 9,411 + 6,395		

5. Estimate the following actual problem by rounding all the way and then do the actual multiplication. *LU 1-1(2), LU 1-3(1)*

Actual	**Estimate**
8,843 × 906	

6. Multiply the following by the shortcut method. *LU 1-3(1)*

   829,412 × 1,000
7. Divide the following and check the answer by multiplication. *LU 1-3(1, 2)*

	Check
$39\overline{)14{,}800}$	

8. Divide the following by the shortcut method. *LU 1-3(2)*

   6,000 ÷ 60
9. Ling Wong bought a $299 iPod that was reduced to $205. Ling gave the clerk three $100 bills. What change will Ling receive? *LU 1-2(2)*
10. Sam Song plans to buy a $16,000 Ford Focus with an interest charge of $4,000. Sam figures he can afford a monthly payment of $400. If Sam must pay 40 equal monthly payments, can he afford the Ford Focus? *LU 1-2(1), LU 1-3(2)*
11. Lester Hal has the oil tank at his business filled 20 times per year. The tank has a capacity of 200 gallons. Assume **(a)** the price of oil fuel is $3 per gallon and **(b)** the tank is completely empty each time Lester has it filled. What is Lester's average monthly oil bill? Complete the following blueprint aid for dissecting and solving the word problem. *LU 1-3(1, 2)*

BLUEPRINT	The facts	Solving for?	Steps to take	Key points

Steps to solving problem

INTERACTIVE VIDEO WORKSHEET

GRADING THE SUMMARY PRACTICE TEST

Go to the summary practice test video in Connect (or click on it here in the ebook). Grade your summary practice test while viewing the video.

C for Correct/I for Incorrect

1. ________	5. ________	9. ________
2. ________	6. ________	10. ________
3. ________	7. ________	11. ________
4. ________	8. ________	

If you achieved 100%, you are ready for your instructor's exam.

If any of the problems were incorrect, list the questions you missed and show steps to solve the problem correctly.

Replay the video to see if you have made the correct fixes to your mistakes. If you have any questions, contact your instructor asap.

Notes on Watching Videos

MY MONEY

My Money = My Plan

What I need to know

There is an old adage that says: "If you fail to plan, then plan to fail." This is especially true as it relates to your finances. You need to develop a plan for your current financial situation as well as where your finances may take you in the future. Establishing this plan is crucial in order to realize your financial goals. Just as with anything new, creating good financial habits at an early age will make the process easier as your income grows in the future. Adhering to good financial habits now will pay huge dividends in the future and prepare you for the financial ups and downs you will most likely experience.

What I need to do

As a college student you are learning a great amount of knowledge that will serve you well in the future. Knowledge truly is power and knowing how your current financial situation is impacted by your earnings and expenses will go a long way in planning effectively. Establishing a solid financial foundation at an early age will help you in reaching your financial goals. Here is some practical advice to plan for the financial future you desire:

- Budgeting is important no matter your income level. By establishing a budget based on your current earnings you will begin to develop the skills you need to continue budgeting in the future. As part of this budgeting process, identify and document your current income and expenses. Insurance will most likely be a part of your expenses in your budget. These expenses are a way to manage your risk and avoid costly surprises should there be an incident with your car or home, for example. An established budget will allow you to address outstanding debts and reduce them over time, providing more opportunity to use your money for other activities such as investments.
- Retirement may seem like a lifetime away but now is the time to prepare for your retirement future! Starting your retirement planning early gives your money more time to work in your favor. Plan to set aside money now to help you meet the goals you have for your retirement. As a part of your retirement planning, it is helpful to consider how you would like to leave your estate for your family members so that they know your wishes. Effective estate planning helps to reduce much of the stress and anxiety loved ones encounter when no plan is in place.

Resources I can use

- Mint: Personal Finance & Money (mobile app)
- Creating a financial plan: https://www.wellsfargo.com/financial-education/basic-finances/build-the-future/short-long-term-planning/financial-plan/
- Learning from others – ask parents, friends, teachers about their financial planning
- Creating a budget: https://bettermoneyhabits.bankofamerica.com/en/saving-budgeting/creating-a-budget

MY MONEY ACTIVITY

- Create a budget: Document current expenses and earnings by tracking all activity for 1 month.
- Create a monthly budget from this documentation. Then create an annual budget from the monthly budget.
- Track actual versus estimated amounts and adjust the budget as necessary.

PERSONAL FINANCE

A KIPLINGER APPROACH

CRYPTOMANIA

What You Need to Know About Bitcoin

Investing in cryptocurrencies is dicey. Instead, consider the underlying technology. BY NELLIE S. HUANG

©REDPIXEL.PL/Shutterstock

WHEN "BITCOIN" APPEARED as a clue in the *New York Times* Sunday crossword puzzle earlier this year (the five-letter answer was "e-coin"), it was confirmation that the cryptocurrency had officially entered the zeitgeist. Such virtual currencies use computer-generated encryption to secure and track transactions, independent of a central bank. Bitcoin made headlines last year because of its meteoric rise from $963 per bitcoin at the start of 2017 to a high of nearly $20,000 in December. This year's news has been about bitcoin's descent, to a low of $7,000 in February, from which it was recently on the rebound.

The craze over bitcoin and its ilk—there are more than 1,000 digital currencies, although bitcoin is the best-known and the oldest—is as mythic as its beginning. (The true identity of bitcoin's creator or creators remains a hotly debated mystery.) Bitcoin was born in the aftermath of the financial crisis, when distrust of financial institutions was at a high, and a currency divorced from a central bank found wide appeal. The currency of choice for black-market buying crossed over to the mainstream with the purchase of a pizza in 2010. Techies glommed on, and by 2016, bitcoin was an investing phenomenon. Along the way, bitcoin has spawned as many skeptics as believers, while regulators grapple with how to oversee this newfangled asset and the potential for fraud that comes with it. One thing is sure: Whether bitcoin or any other cryptocurrency survives, the technology behind it is here to stay. Read on to learn more.

What can I buy with bitcoin? Just about anything. You can buy towels, bedding and other goods at Overstock.com. Or purchase movies and games at Microsoft's online Windows and Xbox stores. At Gyft.com you can apply your bitcoin toward gift cards to use at Starbucks, Target, Uber and Whole Foods. Expedia, the travel booking website, lets you pay for hotel reservations with bitcoin.

Are digital currencies such as bitcoin a good investment? Only for pure speculators who understand the risks. Bitcoin has no revenue, earnings or underlying asset value. Its price rise is driven by demand alone, and that shows signs of a frenzy. By contrast, when a stock price rises, even that of a speculative penny stock, the move is typically based on an expectation of earnings or revenue growth. With bitcoin, "you buy it in the hope that someone will pay you more for it later," says Mark Dodson, a money manager with Hays Advisory. At worst, that's a classic definition of Wall Street's "greater fool theory." At best, it describes speculation, not investment.

Of course, bitcoin bulls say that there's still room to run on the price of bitcoin and that the technology behind the digital currency—called blockchain—makes the coin a worthwhile investment. But there are better ways to bet on the technology (more on that later).

What if I still want to buy? If you must buy a cryptocurrency, invest only what you are willing to lose, and prepare for a wild ride. "Knock yourself out" if you want to buy bitcoin, Dodson tells friends and clients who ask about digital currencies, "but assume the price is going to zero." This is something you put your mad money into, not your savings. The IRS classifies digital currencies as property, so you'll owe capital gains tax on any profits that you earn when you sell or spend.

BUSINESS MATH ISSUE

Bitcoin is a worthwhile investment.

1. List the key points of the article and information to support your position.
2. Write a group defense of your position using math calculations to support your view. If you are in an online course, post to a discussion board.

Classroom Notes

CHAPTER 2

Fractions

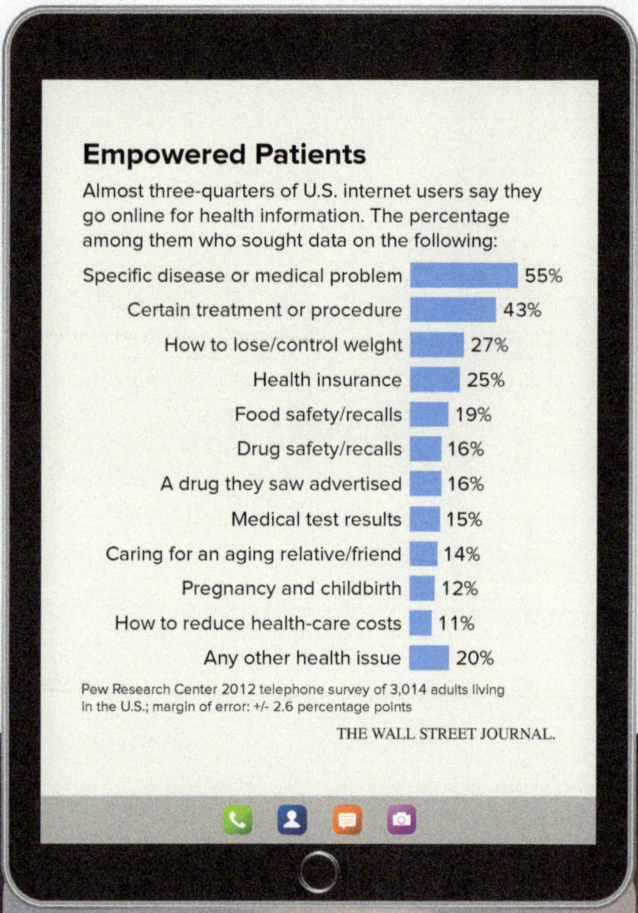

©demaerre/Getty Images

LEARNING UNIT OBJECTIVES

LU 2–1: Types of Fractions and Conversion Procedures

1. Recognize the three types of fractions.
2. Convert improper fractions to whole or mixed numbers and mixed numbers to improper fractions.
3. Convert fractions to lowest and highest terms.

LU 2–2: Adding and Subtracting Fractions

1. Add like and unlike fractions.
2. Find the least common denominator by inspection and prime numbers.
3. Subtract like and unlike fractions.
4. Add and subtract mixed numbers with the same or different denominators.

LU 2–3: Multiplying and Dividing Fractions

1. Multiply and divide proper fractions and mixed numbers.
2. Use the cancellation method in the multiplication and division of fractions.

Your Guide to Successfully Completing This Chapter

Traditional book or ebook

Check box as you complete each step.

Steps

- ☐ Read learning unit.
  - ☐ Complete practice quiz at the end of the learning unit.
- ☐ Grade practice quiz using provided solutions. (For more help, watch the learning unit video in Connect and have a Study Session with the authors. Then complete the additional practice quiz in Connect.)
- ☐ Repeat above for each of the three learning units in Chapter 2.
  - ☐ Review chapter organizer.
  - ☐ Complete assigned homework.
    - ☐ Finish summary practice test. (Go to Connect via the ebook link and do the interactive video worksheet to grade.)
- ☐ Complete instructor's exam.

My Money

The *Wall Street Journal* chapter opener clip "Empowered Patients" illustrates the use of a fraction. From the clipping you learn that almost $\frac{3}{4}$ of U.S Internet users say they go online for health information.

Now let's look at Milk Chocolate M&M'S® candies as another example of using fractions.

As you know, M&M'S® candies come in different colors. Do you know how many of each color are in a bag of M&M'S®? If you go to the M&M'S® website, you learn that a typical bag of M&M'S® contains approximately 17 brown, 11 yellow, 11 red, and 5 each of orange, blue, and green M&M'S®.[1]

The 1.69-ounce bag of M&M'S® shown on the next page contains 55 M&M'S®. In this bag, you will find the following colors:

18 yellow	9 blue	6 brown
10 red	7 orange	5 green

[1]Off 1 due to rounding.

55 pieces in the bag

©Food Tree Images/Alamy

The number of yellow candies in a bag might suggest that yellow is the favorite color of many people. Since this is a business math text, however, let's look at the 55 M&M'S® in terms of fractional arithmetic.

Of the 55 M&M'S® in the 1.69-ounce bag, 5 of these M&M'S® are green, so we can say that 5 parts of 55 represent green candies. We could also say that 1 out of 11 M&M'S® is green. Are you confused?

For many people, fractions are difficult. If you are one of these people, this chapter is for you. First you will review the types of fractions and the fraction conversion procedures. Then you will gain a clear understanding of the addition, subtraction, multiplication, and division of fractions.

LO 1

Learning Unit 2–1: Types of Fractions and Conversion Procedures

This chapter explains the parts of whole numbers called **fractions.** With fractions you can divide any object or unit—a whole—into a definite number of equal parts. For example, the bag of 55 M&M'S® described above contains 6 brown candies. If you eat only the brown M&M'S®, you have eaten 6 parts of 55, or 6 parts of the whole bag of M&M'S®. We can express this in the following fraction:

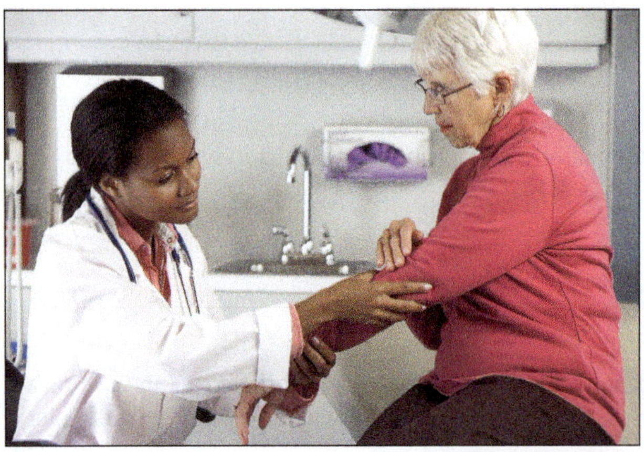

©Monkey Business Images/Shutterstock

$$\frac{6}{55}$$

6 is the **numerator,** or top of the fraction. The numerator describes the number of equal parts of the whole bag that you ate.

55 is the **denominator,** or bottom of the fraction. The denominator gives the total number of equal parts in the bag of M&M'S®.

Before reviewing the arithmetic operations of fractions, you must recognize the three types of fractions described in this unit. You must also know how to convert fractions to a workable form.

Types of Fractions

In the following *Wall Street Journal* clip, "Two-thirds of adults get a yearly checkup," the $\frac{2}{3}$ is a proper fraction.

PROPER FRACTIONS

A **proper fraction** has a value less than 1; its numerator is smaller than its denominator.

EXAMPLES $\frac{2}{3}, \frac{1}{4}, \frac{1}{2}, \frac{1}{10}, \frac{1}{12}, \frac{1}{3}, \frac{4}{7}, \frac{9}{10}, \frac{12}{13}, \frac{18}{55}, \frac{499}{1{,}000}, \frac{501}{1{,}000}$

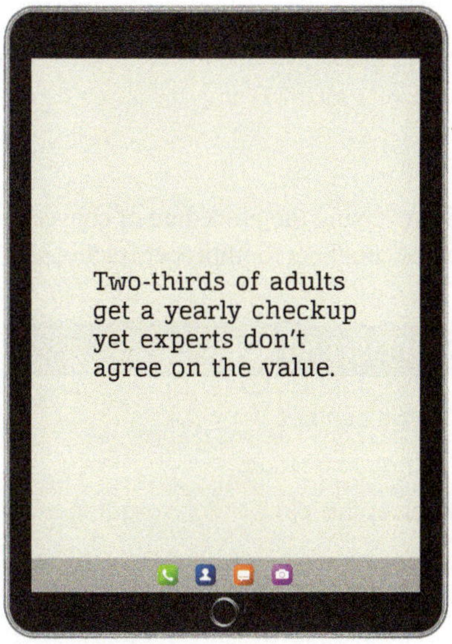

IMPROPER FRACTIONS

An **improper fraction** has a value equal to or greater than 1; its numerator is equal to or greater than its denominator.

EXAMPLES $\frac{15}{15}, \frac{9}{8}, \frac{15}{14}, \frac{22}{19}$

MIXED NUMBERS

A **mixed number** is the sum of a whole number greater than zero and a proper fraction.

EXAMPLES $7\frac{1}{8}, 5\frac{9}{10}, 8\frac{7}{8}, 33\frac{5}{6}, 139\frac{9}{11}$

Conversion Procedures

In Chapter 1 we worked with two of the division symbols (÷ and $\overline{)\ \ }$). The horizontal line (or the diagonal) that separates the numerator and the denominator of a fraction also indicates division. The numerator, like the dividend, is the number we are dividing into. The denominator, like the divisor, is the number we use to divide. Then, referring to the 6 brown M&M'S® in the bag of 55 M&M'S® ($\frac{6}{55}$) shown at the beginning of this unit, we can say that we are dividing 55 into 6, or 6 is divided by 55. Also, in the fraction $\frac{3}{4}$, we can say that we are dividing 4 into 3, or 3 is divided by 4. *Remember "The top dog gets the hat" when converting proper fractions to decimals. For example, in the fraction $\frac{3}{4}$, the 3 is the top dog. The division sign is the hat. Put the hat over the 3 and divide: $4\overline{)3}$ = .75.*

Working with the smaller numbers of simple fractions such as $\frac{3}{4}$ is easier, so we often convert fractions to their simplest terms. In this unit we show how to convert improper fractions to whole or mixed numbers, mixed numbers to improper fractions, and fractions to lowest and highest terms.

Converting Improper Fractions to Whole or Mixed Numbers

Business situations often make it necessary to change an improper fraction to a whole number or mixed number. You can use the following steps to make this conversion:

LO 2

CONVERTING IMPROPER FRACTIONS TO WHOLE OR MIXED NUMBERS

Step 1. Divide the numerator of the improper fraction by the denominator.

Step 2. **a.** If you have no remainder, the quotient is a whole number.

b. If you have a remainder, the whole number part of the mixed number is the quotient. The remainder is placed over the old denominator as the proper fraction of the mixed number.

EXAMPLES

$$\frac{15}{15} = 1 \qquad \frac{16}{5} = 3\frac{1}{5} \qquad \begin{array}{r} 3\text{ R}1 \\ 5\overline{)16} \\ \underline{15} \\ 1 \end{array}$$

Converting Mixed Numbers to Improper Fractions By reversing the procedure of converting improper fractions to mixed numbers, we can change mixed numbers to improper fractions.

CONVERTING MIXED NUMBERS TO IMPROPER FRACTIONS

Step 1. Multiply the denominator of the fraction by the whole number.

Step 2. Add the product from Step 1 to the numerator of the old fraction.

Step 3. Place the total from Step 2 over the denominator of the old fraction to get the improper fraction.

EXAMPLE

$$6\frac{1}{8} = \frac{(8 \times 6) + 1}{8} = \frac{49}{8}$$

Note that the denominator stays the same.

Converting (Reducing) Fractions to Lowest Terms

When solving fraction problems, you always reduce the fractions to their lowest terms. This reduction does not change the value of the fraction. For example, in the bag of M&M'S®, 5 out of 55 were green. The fraction for this is $\frac{5}{55}$. If you divide the top and bottom of the fraction by 5, you have reduced the fraction to $\frac{1}{11}$ without changing its value. Remember, we said in the chapter introduction that 1 out of 11 M&M'S® in the bag of 55 M&M'S® represents green candies. Now you know why this is true.

To reduce a fraction to its lowest terms, begin by inspecting the fraction, looking for the largest whole number that will divide into both the numerator and the denominator without leaving a remainder. This whole number is the **greatest common divisor,** which cannot be zero. When you find this largest whole number, you have reached the point where the fraction is reduced to its **lowest terms.** At this point, no number (except 1) can divide evenly into both parts of the fraction.

REDUCING FRACTIONS TO LOWEST TERMS BY INSPECTION

LO 3

Step 1. By inspection, find the largest whole number (greatest common divisor) that will divide evenly into the numerator and denominator (does not change the fraction value).

Step 2. Divide the numerator and denominator by the greatest common divisor. Now you have reduced the fraction to its lowest terms, since no number (except 1) can divide evenly into the numerator and denominator.

EXAMPLE

$$\frac{24}{30} = \frac{24 \div 6}{30 \div 6} = \frac{4}{5}$$

Using inspection, you can see that the number 6 in the above example is the greatest common divisor. When you have large numbers, the greatest common divisor is not so obvious. For large numbers, you can use the following step approach to find the greatest common divisor:

STEP APPROACH FOR FINDING GREATEST COMMON DIVISOR

Step 1. Divide the smaller number (numerator) of the fraction into the larger number (denominator).

Step 2. Divide the remainder of Step 1 into the divisor of Step 1.

Step 3. Divide the remainder of Step 2 into the divisor of Step 2. Continue this division process until the remainder is a 0, which means the last divisor is the greatest common divisor.

EXAMPLE

	Step 1	Step 2	
$\frac{24}{30}$	$24\overline{)30}$ = 1, 24, remainder 6	$6\overline{)24}$ = 4, 24, remainder 0	$\frac{24 \div 6}{30 \div 6} = \frac{4}{5}$

Reducing a fraction by inspection is to some extent a trial-and-error method. Sometimes you are not sure what number you should divide into the top (numerator) and bottom (denominator) of the fraction. The following reference table on divisibility tests will be helpful. Note that to reduce a fraction to lowest terms might result in more than one division.

	2	3	4	5	6	10
Will divide evenly into a number if the	Last digit is 0, 2, 4, 6, 8.	Sum of the digits is divisible by 3.	Last two digits can be divided by 4.	Last digit is 0 or 5.	The number is even and 3 will divide into the sum of digits.	The last digit is 0.
Examples	$\frac{12}{14} = \frac{6}{7}$	$\frac{36}{69} = \frac{12}{23}$ $3 + 6 = 9 \div 3 = 3$ $6 + 9 = 15 \div 3 = 5$	$\frac{140}{160} = \frac{1(40)}{1(60)}$ $= \frac{35}{40} = \frac{7}{8}$	$\frac{15}{20} = \frac{3}{4}$	$\frac{12}{18} = \frac{2}{3}$	$\frac{90}{100} = \frac{9}{10}$

Converting (Raising) Fractions to Higher Terms Later, when you add and subtract fractions, you will see that sometimes fractions must be raised to **higher terms.** Recall that when you reduced fractions to their lowest terms, you looked for the largest whole number (greatest common divisor) that would divide evenly into both the numerator and the denominator. When you raise fractions to higher terms, you do the opposite and multiply the numerator and the denominator by the same whole number. For example, if you want to raise the fraction $\frac{1}{4}$, you can multiply the numerator and denominator by 2.

EXAMPLE $\frac{1}{4} \times \frac{2}{2} = \frac{2}{8}$

The fractions $\frac{1}{4}$ and $\frac{2}{8}$ are **equivalent** in value. By converting $\frac{1}{4}$ to $\frac{2}{8}$, you only divided it into more parts.

Let's suppose that you have eaten $\frac{4}{7}$ of a pizza. You decide that instead of expressing the amount you have eaten in 7ths, you want to express it in 28ths. How would you do this?

To find the new numerator when you know the new denominator (28), use the steps that follow.

MONEY tips

As a good rule of thumb, students should not borrow more for their education than their expected starting salary after they graduate. Students who borrow more than \$25,000 for an associate's degree, \$45,000 for a bachelor's degree, \$75,000 for a master's degree, \$100,000 for a PhD, \$160,000 for a law degree, and \$215,000 for an MD are probably over-borrowing.

RAISING FRACTIONS TO HIGHER TERMS WHEN DENOMINATOR IS KNOWN

Step 1. Divide the *new* denominator by the *old* denominator to get the common number that raises the fraction to higher terms.

Step 2. Multiply the common number from Step 1 by the old numerator and place it as the new numerator over the new denominator.

EXAMPLE $\frac{4}{7} = \frac{?}{28}$

Step 1. Divide 28 by 7 = 4.

Step 2. Multiply 4 by the numerator 4 = 16.

Result:

$$\frac{4}{7} = \frac{16}{28} \quad \left(\textit{Note}\text{: This is the same as multiplying } \frac{4}{7} \times \frac{4}{4}.\right)$$

Note that $\frac{4}{7}$ and $\frac{16}{28}$ are equivalent in value, yet they are different fractions.

Now try the following Practice Quiz to check your understanding of this unit.

LU 2–1 PRACTICE QUIZ

Complete this **Practice Quiz** to see how you are doing.

1. Identify the type of fraction—proper, improper, or mixed:

   a. $\frac{4}{5}$ **b.** $\frac{6}{5}$ **c.** $19\frac{1}{5}$ **d.** $\frac{20}{20}$

2. Convert to a mixed number:

   $\frac{160}{9}$

3. Convert the mixed number to an improper fraction:

   $9\frac{5}{8}$

4. Find the greatest common divisor by the step approach and reduce to lowest terms:

   a. $\frac{24}{40}$ **b.** $\frac{91}{156}$

5. Convert to higher terms:

   a. $\frac{14}{20} = \frac{\quad}{200}$ **b.** $\frac{8}{10} = \frac{\quad}{60}$

*For **extra help** from your authors–Sharon and Jeff–see the videos in Connect.*

✓ Solutions

1. **a.** Proper
b. Improper
c. Mixed
d. Improper

2. $17\frac{7}{9}$

9)160
9
70
63
7

3. $\frac{(9 \times 8) + 5}{8} = \frac{77}{8}$

4. a.

24)40 = 1, 24, remainder 16 → 16)24 = 1, 16, remainder 8 → 8)16 = 2, 16, remainder 0

8 is greatest common divisor.

$\frac{24 \div 8}{40 \div 8} = \frac{3}{5}$

b.

91)156 = 1, 91, remainder 65 → 65)91 = 1, 65, remainder 26 → 26)65 = 2, 52, remainder 13 → 13)26 = 2, 26, remainder 0

13 is greatest common divisor.

$\frac{91 \div 13}{156 \div 13} = \frac{7}{12}$

5. a. 20)200 = 10 $\quad 10 \times 14 = 140 \quad \frac{14}{20} = \frac{140}{200}$

b. 10)60 = 6 $\quad 6 \times 8 = 48 \quad \frac{8}{10} = \frac{48}{60}$

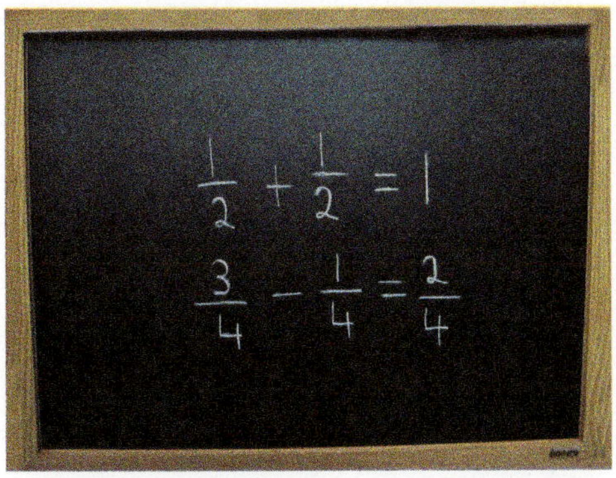

©Roberts Publishing Services

Learning Unit 2–2: Adding and Subtracting Fractions

More teachers are using online video-sharing sites that are modeled after Google Inc.'s YouTube. As you can see in the illustration, these fractions can be added because the fractions have the same denominator. These are called *like fractions.*

In this unit you learn how to add and subtract fractions with the same denominators **(like fractions)** and fractions with different denominators **(unlike fractions).** We have also included how to add and subtract mixed numbers.

Addition of Fractions

LO 1

When you add two or more quantities, they must have the same name or be of the same denomination. You cannot add 6 quarts and 3 pints unless you change the denomination of one or both quantities. You must either make the quarts into pints or the pints into quarts. The same principle also applies to fractions. That is, to add two or more fractions, they must have a **common denominator.**

Adding Like Fractions

Earlier we stated that because the fractions had the same denominator, or a common denominator, they were *like fractions.* Adding like fractions is similar to adding whole numbers.

ADDING LIKE FRACTIONS

Step 1. Add the numerators and place the total over the original denominator.

Step 2. If the total of your numerators is the same as your original denominator, convert your answer to a whole number; if the total is larger than your original denominator, convert your answer to a mixed number.

EXAMPLE $\frac{1}{7} + \frac{4}{7} = \frac{5}{7}$

The denominator, 7, shows the number of pieces into which some whole was divided. The two numerators, 1 and 4, tell how many of the pieces you have. So if you add 1 and 4, you get 5, or $\frac{5}{7}$.

Adding Unlike Fractions Since you cannot add *unlike fractions* because their denominators are not the same, you must change the unlike fractions to *like fractions*—fractions with the same denominators. To do this, find a denominator that is common to all the fractions you want to add. Then look for the **least common denominator (LCD).**[2] The LCD is the smallest nonzero whole number into which all denominators will divide evenly. You can find the LCD by inspection or with prime numbers.

Finding the Least Common Denominator (LCD) by Inspection The example that follows shows you how to use inspection to find an LCD (this will make all the denominators the same).

EXAMPLE $\frac{3}{7} + \frac{5}{21}$

Inspection of these two fractions shows that the smallest number into which denominators 7 and 21 divide evenly is 21. Thus, 21 is the LCD.

You may know that 21 is the LCD of $\frac{3}{7} + \frac{5}{21}$, but you cannot add these two fractions until you change the denominator of $\frac{3}{7}$ to 21. You do this by building (raising) the equivalent of $\frac{3}{7}$, as explained in Learning Unit 2–1. You can use the following steps to find the LCD by inspection:

Step 1. Divide the new denominator (21) by the old denominator (7): $21 \div 7 = 3$.

Step 2. Multiply the 3 in Step 1 by the old numerator (3): $3 \times 3 = 9$. The new numerator is 9.

[2] Often referred to as the *lowest common denominator.*

Result:

$$\frac{3}{7} = \frac{9}{21}$$

Now that the denominators are the same, you add the numerators.

$$\frac{9}{21} + \frac{5}{21} = \frac{14}{21} = \frac{2}{3}$$

Note that $\frac{14}{21}$ is reduced to its lowest terms $\frac{2}{3}$. Always reduce your answer to its lowest terms.

You are now ready for the following general steps for adding proper fractions with different denominators. These steps also apply to the following discussion on finding LCD by prime numbers.

ADDING UNLIKE FRACTIONS

LO 2

Step 1. Find the LCD.
Step 2. Change each fraction to a like fraction with the LCD.
Step 3. Add the numerators and place the total over the LCD.
Step 4. If necessary, reduce the answer to lowest terms.

Finding the Least Common Denominator (LCD) by Prime Numbers When you cannot determine the LCD by inspection, you can use the prime number method. First you must understand prime numbers.

PRIME NUMBERS

A **prime number** is a whole number greater than 1 that is only divisible by itself and 1. The number 1 is not a prime number.

©2010 Jef Mallett/Dist. by Universal Uclick.

EXAMPLES 2, 3, 5, 7, 11, 13, 17, 19, 23, 29, 31, 37, 41, 43

Note that the number 4 is not a prime number. Not only can you divide 4 by 1 and by 4, but you can also divide 4 by 2. A whole number that is greater than 1 and is only divisible by itself and 1 has become a source of interest to some people.

EXAMPLE $\frac{1}{3}+\frac{1}{8}+\frac{1}{9}+\frac{1}{12}$

Step 1. Copy the denominators and arrange them in a separate row.

3 8 9 12

Step 2. Divide the denominators in Step 1 by prime numbers. Start with the smallest number that will divide into at least two of the denominators. Bring down any number that is not divisible. Keep in mind that the lowest prime number is 2.

```
2 /3  8  9  12
   3  4  9   6
```

Note: The 3 and 9 were brought down, since they were not divisible by 2.

Step 3. Continue Step 2 until no prime number will divide evenly into at least two numbers.

Note: The 3 is used, since 2 can no longer divide evenly into at least two numbers.

```
2 /3  8  9  12
2 /3  4  9   6
3 /3  2  9   3
   1  2  3   1
```

Step 4. To find the LCD, multiply all the numbers in the divisors (2, 2, 3) and in the last row (1, 2, 3, 1).

$\boxed{2 \times 2 \times 3} \times \boxed{1 \times 2 \times 3 \times 1} = 72$ (LCD)

Divisors × Last row

Step 5. Raise each fraction so that each denominator will be 72 and then add fractions.

$\frac{1}{3} = \frac{?}{72}$ $72 \div 3 = 24$ $24 \times 1 = 24$

$\frac{1}{8} = \frac{?}{72}$ $72 \div 8 = 9$ $9 \times 1 = 9$

$$\frac{24}{72}+\frac{9}{72}+\frac{8}{72}+\frac{6}{72}=\frac{47}{72}$$

The above five steps used for finding LCD with prime numbers are summarized as follows:

FINDING LCD FOR TWO OR MORE FRACTIONS

Step 1. Copy the denominators and arrange them in a separate row.

Step 2. Divide the denominators by the smallest prime number that will divide evenly into at least two numbers.

Step 3. Continue until no prime number divides evenly into at least two numbers.

Step 4. Multiply all the numbers in divisors and last row to find the LCD.

Step 5. Raise all fractions so each has a common denominator and then complete the computation.

Adding Mixed Numbers

The following steps will show you how to add mixed numbers:

ADDING MIXED NUMBERS

Step 1. Add the fractions (remember that fractions need common denominators, as in the previous section).

Step 2. Add the whole numbers.

Step 3. Combine the totals of Steps 1 and 2. Be sure you do not have an improper fraction in your final answer. Convert the improper fraction to a whole or mixed number. Add the whole numbers resulting from the improper fraction conversion to the total whole numbers of Step 2. If necessary, reduce the answer to lowest terms.

Using prime numbers to find LCD of example

$$\begin{array}{r|rrr} 2 & 20 & 5 & 4 \\ 2 & 10 & 5 & 2 \\ 5 & 5 & 5 & 1 \\ \hline & 1 & 1 & 1 \end{array}$$

$2 \times 2 \times 5 = 20$ LCD

EXAMPLE

$$\begin{array}{rr} 4\frac{7}{20} & 4\frac{7}{20} \\ 6\frac{3}{5} & 6\frac{12}{20} \\ +\,7\frac{1}{4} & +\,7\frac{5}{20} \\ \hline \end{array}$$

$$\frac{3}{5} = \frac{?}{20}$$

$$20 \div 5 = 4$$

$$\begin{array}{r} 4 \\ \times 3 \\ \hline 12 \end{array}$$

Step 1 → $\frac{24}{20} = 1\frac{4}{20}$

Step 2 → $+\,17$

Step 3 → $= 18\frac{4}{20} = 18\frac{1}{5}$

Subtraction of Fractions

The subtraction of fractions is similar to the addition of fractions. This section explains how to subtract like and unlike fractions and how to subtract mixed numbers.

Subtracting Like Fractions To subtract like fractions, use the steps that follow.

LO 3

SUBTRACTING LIKE FRACTIONS

Step 1. Subtract the numerators and place the answer over the common denominator.

Step 2. If necessary, reduce the answer to lowest terms.

EXAMPLE

$$\frac{9}{10} - \frac{1}{10} = \frac{8 \div 2}{10 \div 2} = \frac{4}{5}$$

↑ Step 1 ↑ Step 2

Subtracting Unlike Fractions Now let's learn the steps for subtracting unlike fractions.

SUBTRACTING UNLIKE FRACTIONS

Step 1. Find the LCD.

Step 2. Raise the fraction to its equivalent value.

Step 3. Subtract the numerators and place the answer over the LCD.

Step 4. If necessary, reduce the answer to lowest terms.

EXAMPLE

$$\begin{array}{rr} \frac{5}{8} & \frac{40}{64} \\ -\,\frac{2}{64} & -\,\frac{2}{64} \\ \hline & \frac{38}{64} = \frac{19}{32} \end{array}$$

By inspection, we see that LCD is 64.
Thus $64 \div 8 = 8 \times 5 = 40$.

Subtracting Mixed Numbers When you subtract whole numbers, sometimes borrowing is not necessary. At other times, you must borrow. The same is true of subtracting mixed numbers.

LO 4

SUBTRACTING MIXED NUMBERS	
When Borrowing Is Not Necessary	*When Borrowing Is Necessary*
Step 1. Subtract fractions, making sure to find the LCD.	**Step 1.** Make sure the fractions have the LCD.
Step 2. Subtract whole numbers.	**Step 2.** Borrow from the whole number of the minuend (top number).
Step 3. Reduce the fraction(s) to lowest terms.	**Step 3.** Subtract the whole numbers and fractions.
	Step 4. Reduce the fraction(s) to lowest terms.

EXAMPLE Where borrowing is not necessary:

$$6\frac{1}{2} - \frac{3}{8}$$

Find LCD of 2 and 8. LCD is 8.

$$6\frac{4}{8} - \frac{3}{8} = 6\frac{1}{8}$$

MONEY tips

Create an emergency fund for the unexpected. Having four to six months of monthly expenses in a liquid account will provide you with a great cushion in the event of an unforeseen expense.

EXAMPLE Where borrowing is necessary:

$$3\frac{1}{2} = 3\frac{2}{4} = 2\frac{6}{4}\left(\frac{4}{4}+\frac{2}{4}\right)$$

$$-1\frac{3}{4} = -1\frac{3}{4} = -1\frac{3}{4}$$

LCD is 4.

$$1\frac{3}{4}$$

©Fancy Collection/SuperStock

Since $\frac{3}{4}$ is larger than $\frac{2}{4}$, we must borrow 1 from the 3. This is the same as borrowing $\frac{4}{4}$. A fraction with the same numerator and denominator represents a whole. When we add $\frac{4}{4}+\frac{2}{4}$, we get $\frac{6}{4}$. Note how we subtracted the whole number and fractions, being sure to reduce the final answer if necessary.

How to Dissect and Solve a Word Problem

Let's now look at how to dissect and solve a word problem involving fractions.

The Word Problem Albertsons grocery store has $550\frac{1}{4}$ total square feet of floor space. Albertsons' meat department occupies $115\frac{1}{2}$ square feet, and its deli department occupies $145\frac{7}{8}$ square feet. If the remainder of the floor space is for groceries, what square footage remains for groceries?

BLUEPRINT	The facts	Solving for?	Steps to take	Key points
	Total square footage: $550\frac{1}{4}$ sq. ft. *Meat department:* $115\frac{1}{2}$ sq. ft. *Deli department:* $145\frac{7}{8}$ sq. ft.	Total square footage for groceries.	Total floor space − Total meat and deli floor space = Total grocery floor space.	Denominators must be the same before adding or subtracting fractions. $\frac{8}{8}=1$ Never leave improper fraction as final answer.

Steps to solving problem

1. Calculate total square footage of the meat and deli departments.

$$\begin{aligned} \text{Meat: } & 115\tfrac{1}{2} = 115\tfrac{4}{8} \\ \text{Deli: } & +145\tfrac{7}{8} = +145\tfrac{7}{8} \\ \hline & \qquad 260\tfrac{11}{8} = 261\tfrac{3}{8} \text{ sq. ft.} \end{aligned}$$

2. Calculate total grocery square footage.

$$\begin{aligned} 550\tfrac{1}{4} &= 550\tfrac{2}{8} = 549\tfrac{10}{8} \quad \left(\tfrac{2}{8} + \tfrac{8}{8}\right) \\ -261\tfrac{3}{8} &= -261\tfrac{3}{8} = -261\tfrac{3}{8} \\ \hline & \qquad\qquad 288\tfrac{7}{8} \text{ sq. ft.} \end{aligned}$$

Check

$$\begin{aligned} & 261\tfrac{3}{8} \\ & +288\tfrac{7}{8} \\ \hline & 549\tfrac{10}{8} = 550\tfrac{2}{8} = 550\tfrac{1}{4} \text{ sq. ft.} \end{aligned}$$

Note how the above blueprint aid helped to gather the facts and identify what we were looking for. To find the total square footage for groceries, we first had to sum the areas for meat and deli. Then we could subtract these areas from the total square footage. Also note that in Step 1 above, we didn't leave the answer as an improper fraction. In Step 2, we borrowed from the 550 so that we could complete the subtraction.

It's your turn to check your progress with a Practice Quiz.

LU 2–2 PRACTICE QUIZ

Complete this Practice Quiz to see how you are doing.

1. Find LCD by the division of prime numbers:

   12, 9, 6, 4

2. Add and reduce to lowest terms if needed:

   a. $\frac{3}{40} + \frac{2}{5}$ **b.** $2\frac{3}{4} + 6\frac{1}{20}$

3. Subtract and reduce to lowest terms if needed:

   a. $\frac{6}{7} - \frac{1}{4}$ **b.** $8\frac{1}{4} - 3\frac{9}{28}$ **c.** $4 - 1\frac{3}{4}$

4. Computerland has $660\frac{1}{4}$ total square feet of floor space. Three departments occupy this floor space: hardware, $201\frac{1}{8}$ square feet; software, $242\frac{1}{4}$ square feet; and customer service, _________ square feet. What is the total square footage of the customer service area? You might want to try a blueprint aid, since the solution will show a completed blueprint aid.

✓ Solutions

*For **extra help** from your authors–Sharon and Jeff–see the videos in Connect.*

1.

2	/12	9	6	4
2	/6	9	3	2
3	/3	9	3	1
	1	3	1	1

$\text{LCD} = 2 \times 2 \times 3 \times 1 \times 3 \times 1 \times 1 = 36$

2. **a.** $\frac{3}{40} + \frac{2}{5} = \frac{3}{40} + \frac{16}{40} = \frac{19}{40}$ $\left(\frac{2}{5} = \frac{?}{40} \quad 40 \div 5 = 8 \times 2 = 16\right)$

   b.

$$\begin{aligned} 2\tfrac{3}{4} \quad & 2\tfrac{15}{20} \\ +6\tfrac{1}{20} \quad & +6\tfrac{1}{20} \\ \hline & 8\tfrac{16}{20} = 8\tfrac{4}{5} \end{aligned}$$

$\frac{3}{4} = \frac{?}{20}$ $20 \div 4 = 5 \times 3 = 15$

3. a.

$$\frac{6}{7} = \frac{24}{28}$$
$$-\frac{1}{4} = -\frac{7}{28}$$
$$\boxed{\frac{17}{28}}$$

b.

$$8\frac{1}{4} = 8\frac{7}{28} = 7\frac{35}{28} \leftarrow \left(\frac{28}{28} + \frac{7}{28}\right)$$
$$-3\frac{9}{28} = -3\frac{9}{28} = -3\frac{9}{28}$$
$$4\frac{26}{28} = \boxed{4\frac{13}{14}}$$

c.

$$3\frac{4}{4}$$
$$-1\frac{3}{4}$$
$$\boxed{2\frac{1}{4}}$$

Note how we showed the 4 as $3\frac{4}{4}$.

4. Computerland's total square footage for customer service:

BLUEPRINT	The facts	Solving for?	Steps to take	Key points
	Total square footage: $660\frac{1}{4}$ sq. ft. *Hardware:* $201\frac{1}{8}$ sq. ft. *Software:* $242\frac{1}{4}$ sq. ft.	Total square footage for for customer service.	Total floor space − Total hardware and software floor space = Total customer service floor space.	Denominators must be the same before adding or subtracting fractions.

Steps to solving problem

1. Calculate the total square footage of hardware and software.

$$201\frac{1}{8} = 201\frac{1}{8} \text{ (hardware)}$$
$$+242\frac{1}{4} = +242\frac{2}{8} \text{ (software)}$$
$$443\frac{3}{8}$$

2. Calculate the total square footage for customer service.

$$660\frac{1}{4} = 660\frac{2}{8} = 659\frac{10}{8} \text{ (total square footage)}$$
$$-443\frac{3}{8} = -443\frac{3}{8} = -443\frac{3}{8} \text{ (hardware plus software)}$$
$$\boxed{216\frac{7}{8} \text{ sq. ft. (customer service)}}$$

Learning Unit 2–3: Multiplying and Dividing Fractions

LO 1

The following recipe for Coconutty "M&M'S"® Brand Brownies makes 16 brownies. What would you need if you wanted to triple the recipe and make 48 brownies?

©Roberts Publishing Services

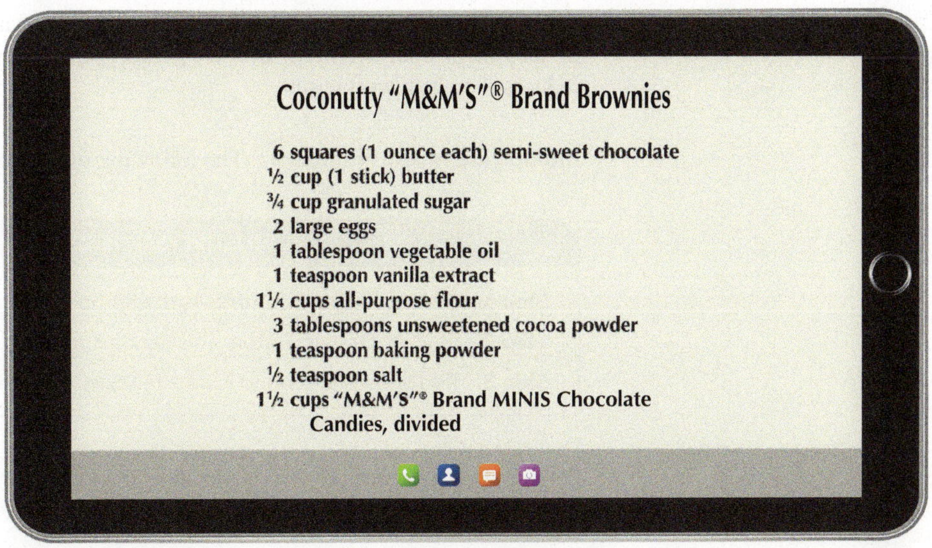

Source: Adapted from Mars, Inc.

In this unit you learn how to multiply and divide fractions.

LO 2

Multiplication of Fractions

Multiplying fractions is easier than adding and subtracting fractions because you do not have to find a common denominator. This section explains the multiplication of proper fractions and the multiplication of mixed numbers.

MULTIPLYING PROPER FRACTIONS[3]

Step 1. Multiply the numerators and the denominators.

Step 2. Reduce the answer to lowest terms or use the cancellation method.

First let's look at an example that results in an answer that we do not have to reduce.

EXAMPLE $\frac{1}{7} \times \frac{5}{8} = \boxed{\frac{5}{56}}$

In the next example, note how we reduce the answer to lowest terms.

EXAMPLE $\frac{5}{1} \times \frac{1}{6} \times \frac{4}{7} = \frac{20}{42} = \boxed{\frac{10}{21}}$ Keep in mind $\frac{5}{1}$ is equal to 5.

We can reduce $\frac{20}{42}$ by the step approach as follows:

$$20\overline{)42} = 2 \text{ remainder } 2 \quad (42 - 40 = 2)$$

$$2\overline{)20} = 10 \text{ remainder } 0 \quad (20 - 20 = 0)$$

$$\frac{20 \div 2}{42 \div 2} = \boxed{\frac{10}{21}}$$

We could also have found the greatest common divisor by inspection.

As an alternative to reducing fractions to lowest terms, we can use the **cancellation** technique. Let's work the previous example using this technique.

EXAMPLE $\frac{5}{1} \times \frac{1}{\overset{}{\cancel{6}}_{3}} \times \frac{\overset{2}{\cancel{4}}}{7} = \boxed{\frac{10}{21}}$ 2 divides evenly into 4 twice and into 6 three times.

Note that when we cancel numbers, we are reducing the answer before multiplying. We know that multiplying or dividing both numerator and denominator by the same number gives an equivalent fraction. So we can divide both numerator and denominator by any number that divides them both evenly. It doesn't matter which we divide first. Note that this division reduces $\frac{10}{21}$ to its lowest terms.

Multiplying Mixed Numbers The following steps explain how to multiply mixed numbers:

MULTIPLYING MIXED NUMBERS

Step 1. Convert the mixed numbers to improper fractions.

Step 2. Multiply the numerators and denominators.

Step 3. Reduce the answer to lowest terms or use the cancellation method.

[3]You would follow the same procedure to multiply improper fractions.

EXAMPLE $2\frac{1}{3}\times 1\frac{1}{2}=\frac{7}{\cancel{3}_1}\times\frac{\cancel{3}^1}{2}=\frac{7}{2}=3\frac{1}{2}$

Step 1 **Step 2** **Step 3**

Before we look at dividing fractions, reference the article below from the *Wall Street Journal*, "Seeing is Believing," showing research of the brain and its relationship to your fingers and math skills.

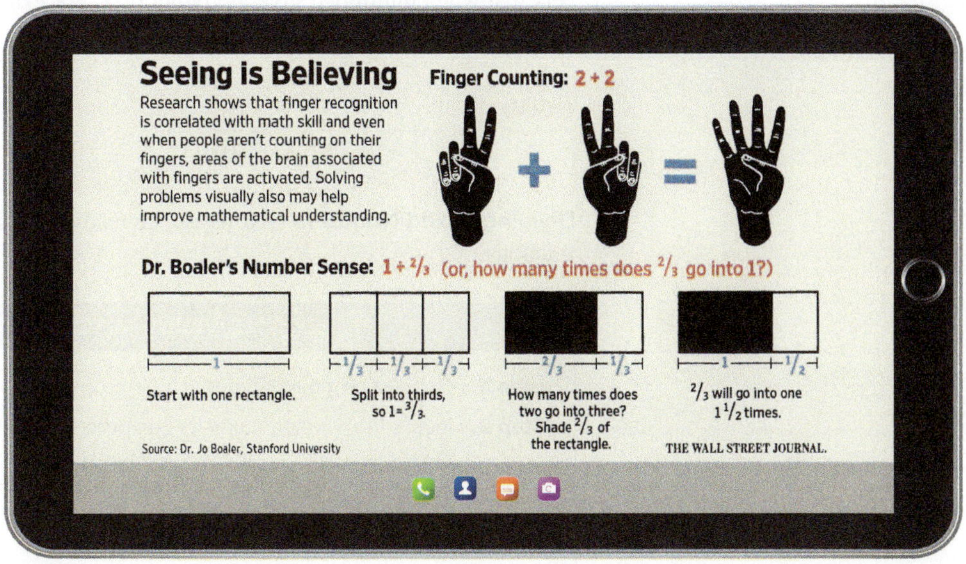

Division of Fractions

When you studied whole numbers in Chapter 1, you saw how multiplication can be checked by division. The multiplication of fractions can also be checked by division, as you will see in this section on dividing proper fractions and mixed numbers.

Dividing Proper Fractions The division of proper fractions introduces a new term—the **reciprocal.** To use reciprocals, we must first recognize which fraction in the problem is the divisor—the fraction that we divide by. Let's assume the problem we are to solve is $\frac{1}{8}\div\frac{2}{3}$. We read this problem as "$\frac{1}{8}$ divided by $\frac{2}{3}$." The divisor is the fraction after the division sign (or the second fraction). The steps that follow show how the divisor becomes a reciprocal.

DIVIDING PROPER FRACTIONS

Step 1. Invert (turn upside down) the divisor (the second fraction). The inverted number is the *reciprocal.*

Step 2. Multiply the fractions.

Step 3. Reduce the answer to lowest terms or use the cancellation method.

Do you know why the inverted fraction number is a reciprocal? Reciprocals are two numbers that when multiplied give a product of 1. For example, 2 (which is the same as $\frac{2}{1}$) and $\frac{1}{2}$ are reciprocals because multiplying them gives 1.

EXAMPLE $\frac{1}{8} \div \frac{2}{3}$ $\qquad \frac{1}{8} \times \frac{3}{2} = \boxed{\frac{3}{16}}$

Dividing Mixed Numbers Now you are ready to divide mixed numbers by using improper fractions.

DIVIDING MIXED NUMBERS

Step 1. Convert all mixed numbers to improper fractions.

Step 2. Invert the divisor (take its reciprocal) and multiply. If your final answer is an improper fraction, reduce it to lowest terms. You can do this by finding the greatest common divisor or by using the cancellation technique.

EXAMPLE $8\frac{3}{4} \div 2\frac{5}{6}$

Step 1. $\frac{35}{4} \div \frac{17}{6}$

Step 2. $\frac{35}{\cancel{4}_{2}} \times \frac{\cancel{6}^{3}}{17} = \frac{105}{34} = \boxed{3\frac{3}{34}}$ Here we used the cancellation technique.

How to Dissect and Solve a Word Problem

The Word Problem Jamie ordered $5\frac{1}{2}$ cords of oak. The cost of each cord is \$150. He also ordered $2\frac{1}{4}$ cords of maple at \$120 per cord. Jamie's neighbor, Al, said that he would share the wood and pay him $\frac{1}{5}$ of the total cost. How much did Jamie receive from Al?

Note how we filled in the blueprint aid columns. We first had to find the total cost of all the wood before we could find Al's share—$\frac{1}{5}$ of the total cost.

MONEY tips

Make good buying decisions. Do not spend more money than you make. In fact, remember to pay yourself first by putting away money each paycheck for your retirement—even \$10 each paycheck adds up.

BLUEPRINT	The facts	Solving for?	Steps to take	Key points
	Cords ordered: $5\frac{1}{2}$ at \$150 per cord; $2\frac{1}{4}$ at \$120 per cord. *Al's cost share:* $\frac{1}{5}$ the total cost.	What will Al pay Jamie?	Total cost of wood × $\frac{1}{5}$ = Al's cost	Convert mixed numbers to improper fractions when multiplying. Cancellation is an alternative to reducing fractions.

Steps to solving problem

1. Calculate the cost of oak. $\qquad 5\frac{1}{2} \times \$150 = \frac{11}{\cancel{2}_{1}} \times \cancel{\$150}^{\$75} = \$825$
2. Calculate the cost of maple. $\qquad 2\frac{1}{4} \times \$120 = \frac{9}{\cancel{4}_{1}} \times \cancel{\$120}^{\$30} = \$270$

   \$1,095 (total cost of wood)
3. What Al pays. $\qquad \frac{1}{\cancel{5}_{1}} \times \cancel{\$1,095}^{\$219} = \boxed{\$219}$

You should now be ready to test your knowledge of the final unit in the chapter.

LU 2–3 PRACTICE QUIZ

Complete this **Practice Quiz** to see how you are doing.

For ***extra help*** *from your authors–Sharon and Jeff–see the videos in Connect.*

1. Multiply (use cancellation technique):

a. $\frac{4}{8} \times \frac{4}{6}$ **b.** $35 \times \frac{4}{7}$

2. Multiply (do not use canceling; reduce by finding the greatest common divisor):

$\frac{14}{15} \times \frac{7}{10}$

3. Complete the following. Reduce to lowest terms as needed.

a. $\frac{1}{9} \div \frac{5}{6}$ **b.** $\frac{51}{5} \div \frac{5}{9}$

4. Jill Estes bought a mobile home that was $8\frac{1}{8}$ times as expensive as the home her brother bought. Jill's brother paid $16,000 for his mobile home. What is the cost of Jill's new home?

✓ Solutions

1. a. $\frac{\overset{1}{\cancel{\overset{2}{\cancel{4}}}}}{\underset{1}{\cancel{\underset{2}{\cancel{8}}}}} \times \frac{\overset{1}{\cancel{4}}}{\underset{3}{\cancel{6}}} = \boxed{\frac{1}{3}}$ **b.** $\overset{5}{\cancel{35}} \times \frac{4}{\underset{1}{\cancel{7}}} = \boxed{20}$

2. $\frac{14}{15} \times \frac{7}{10} = \frac{98 \div 2}{150 \div 2} = \boxed{\frac{49}{75}}$

$98\overline{)150}$ = 1, remainder 52 → $52\overline{)98}$ = 1, remainder 46 → $46\overline{)52}$ = 1, remainder 6 → $6\overline{)46}$ = 7, remainder 4 → $4\overline{)6}$ = 1, remainder 2 → $2\overline{)4}$ = 2, remainder 0

3. a. $\frac{1}{9} \times \frac{6}{5} = \frac{6 \div 3}{45 \div 3} = \boxed{\frac{2}{15}}$ **b.** $\frac{51}{5} \times \frac{9}{5} = \frac{459}{25} = \boxed{18\frac{9}{25}}$

4. Total cost of Jill's new home:

BLUEPRINT	The facts	Solving for?	Steps to take	Key points
	Jill's mobile home: $8\frac{1}{8}$ as expensive as her brother's. *Brother paid:* $16,000.	Total cost of Jill's new home.	$8\frac{1}{8}$ × Total cost of Jill's brother's mobile home = Total cost of Jill's new home.	Canceling is an alternative to reducing.

Steps to solving problem

1. Convert $8\frac{1}{8}$ to a mixed number. $\frac{65}{8}$

2. Calculate the total cost of Jill's home. $\frac{65}{\underset{1}{\cancel{8}}} \times \overset{\$2{,}000}{\cancel{\$16{,}000}} = \boxed{\$130{,}000}$

INTERACTIVE CHAPTER ORGANIZER

Topic/Procedure/Formula	Example	You try it*
Types of fractions *Proper:* Value less than 1; numerator smaller than denominator. *Improper:* Value equal to or greater than 1; numerator equal to or greater than denominator. *Mixed:* Sum of whole number greater than zero and a proper fraction.	$\frac{3}{5}, \frac{7}{9}, \frac{8}{15}$ $\frac{14}{14}, \frac{19}{18}$ $6\frac{3}{8}, 9\frac{8}{9}$	**Identify type of fraction** $\frac{3}{10}, \frac{9}{8}, 1\frac{4}{5}$
Fraction conversions *Improper to whole or mixed:* Divide numerator by denominator; place remainder over old denominator. *Mixed to improper:* $\frac{\text{Whole number} \times \text{Denominator} + \text{Numerator}}{\text{Old denominator}}$	$\frac{17}{4} = 4\frac{1}{4}$ $4\frac{1}{8} = \frac{32+1}{8} = \frac{33}{8}$	**Convert to mixed number** $\frac{18}{7}$ **Convert to improper fraction** $5\frac{1}{7}$
Reducing fractions to lowest terms **1.** Divide numerator and denominator by largest possible divisor (does not change fraction value). **2.** When reduced to lowest terms, no number (except 1) will divide evenly into both numerator and denominator.	$\frac{18 \div 2}{46 \div 2} = \frac{9}{23}$	**Reduce to lowest terms** $\frac{16}{24}$
Step approach for finding greatest common denominator **1.** Divide smaller number of fraction into larger number. **2.** Divide remainder into divisor of Step 1. Continue this process until no remainder results. **3.** The last divisor used is the greatest common divisor.	$\frac{15}{65}$ → 15)65 = 4; 60; remainder 5 5)15 = 3; 15; remainder 0 5 is greatest common divisor.	**Find greatest common denominator** $\frac{20}{50}$
Raising fractions to higher terms Multiply numerator and denominator by same number. Does not change fraction value.	$\frac{15}{41} = \frac{?}{410}$ $410 \div 40 = 10 \times 15 = 150$	**Raise to higher terms** $\frac{16}{31} = \frac{?}{310}$
Adding and subtracting like and unlike fractions When denominators are the same (like fractions), add (or subtract) numerators, place total over original denominator, and reduce to lowest terms. When denominators are different (unlike fractions), change them to like fractions by finding LCD using inspection or prime numbers. Then add (or subtract) the numerators, place total over LCD, and reduce to lowest terms.	$\frac{4}{9} + \frac{1}{9} = \frac{5}{9}$ $\frac{4}{9} - \frac{1}{9} = \frac{3}{9} = \frac{1}{3}$ $\frac{4}{5} + \frac{2}{7} = \frac{28}{35} + \frac{10}{35} = \frac{38}{35} = 1\frac{3}{35}$	**Add** $\frac{3}{7} + \frac{2}{7}$ **Subtract** $\frac{5}{7} - \frac{2}{7}$ **Add** $\frac{5}{8} + \frac{3}{40}$

(continues)

INTERACTIVE CHAPTER ORGANIZER

Topic/Procedure/Formula	Example	You try it*
Prime numbers Whole numbers larger than 1 that are only divisible by itself and 1.	2, 3, 5, 7, 11	**List the next two prime numbers after 11**
LCD by prime numbers **1.** Copy denominators and arrange them in a separate row. **2.** Divide denominators by smallest prime number that will divide evenly into at least two numbers. **3.** Continue until no prime number divides evenly into at least two numbers. **4.** Multiply all the numbers in the divisors and last row to find LCD. **5.** Raise fractions so each has a common denominator and complete computation.	$\frac{1}{3}+\frac{1}{6}+\frac{1}{8}+\frac{1}{12}+\frac{1}{9}$ 2 / 3 6 8 12 9 2 / 3 3 4 6 9 3 / 3 3 2 3 9 1 1 2 1 3 $2 \times 2 \times 3 \times 1 \times 1 \times 2 \times 1 \times 3 = 72$	**Find LCD** $\frac{1}{2}+\frac{1}{4}+\frac{1}{5}$
Adding mixed numbers **1.** Add fractions. **2.** Add whole numbers. **3.** Combine totals of Steps 1 and 2. If denominators are different, a common denominator must be found. Answer cannot be left as improper fraction.	$1\frac{4}{7}+1\frac{3}{7}$ Step 1: $\frac{4}{7}+\frac{3}{7}=\frac{7}{7}$ Step 2: $1+1=2$ Step 3: $2\frac{7}{7}=3$	**Add mixed numbers** $2\frac{1}{4}+3\frac{3}{4}$
Subtracting mixed numbers **1.** Subtract fractions. **2.** If necessary, borrow from whole numbers. **3.** Subtract whole numbers and fractions if borrowing was necessary. **4.** Reduce fractions to lowest terms. If denominators are different, a common denominator must be found.	$12\frac{2}{5}-7\frac{3}{5}$ $11\frac{7}{5}-7\frac{3}{5}$ $=4\frac{4}{5}$ Due to borrowing $\frac{5}{5}$ from number 12 $\frac{5}{5}+\frac{2}{5}=\frac{7}{5}$ The whole number is now 11.	**Subtract mixed numbers** $11\frac{1}{3}$ $-2\frac{2}{3}$
Multiplying proper fractions **1.** Multiply numerators and denominators. **2.** Reduce answer to lowest terms or use cancellation method.	$\frac{4}{\cancel{7}_1}\times\frac{\cancel{7}^1}{9}=\frac{4}{9}$	**Multiply and reduce** $\frac{4}{5}\times\frac{25}{26}$
Multiplying mixed numbers **1.** Convert mixed numbers to improper fractions. **2.** Multiply numerators and denominators. **3.** Reduce answer to lowest terms or use cancellation method.	$1\frac{1}{8}\times 2\frac{5}{8}$ $\frac{9}{8}\times\frac{21}{8}=\frac{189}{64}=2\frac{61}{64}$	**Multiply and reduce** $2\frac{1}{4}\times 3\frac{1}{4}$
Dividing proper fractions **1.** Invert divisor. **2.** Multiply. **3.** Reduce answer to lowest terms or use cancellation method.	$\frac{1}{4}\div\frac{1}{8}=\frac{1}{\cancel{4}_1}\times\frac{\cancel{8}^2}{1}=2$	**Divide** $\frac{1}{8}\div\frac{1}{4}$

(continues)

INTERACTIVE CHAPTER ORGANIZER

Topic/Procedure/Formula	Example	You try it*
Dividing mixed numbers **1.** Convert mixed numbers to improper fractions. **2.** Invert divisor and multiply. If final answer is an improper fraction, reduce to lowest terms by finding greatest common divisor or using the cancellation method.	$1\frac{1}{2} \div 1\frac{5}{8} = \frac{3}{2} \div \frac{13}{8}$ $= \frac{3}{\cancel{2}_1} \times \frac{\cancel{8}^4}{13}$ $= \frac{12}{13}$	**Dividing mixed numbers** $3\frac{1}{4} \div 1\frac{4}{5}$

KEY TERMS			
	Cancellation	Higher terms	Mixed numbers
	Common denominator	Improper fraction	Numerator
	Denominator	Least common denominator (LCD)	Prime numbers
	Equivalent	Like fractions	Proper fraction
	Fraction	Lowest terms	Reciprocal
	Greatest common divisor		Unlike fractions

*Worked-out solutions are in Appendix B.

Critical Thinking Discussion Questions with Chapter Concept Check

1. What are the steps to convert improper fractions to whole or mixed numbers? Give an example of how you could use this conversion procedure when you eat at Pizza Hut.

2. What are the steps to convert mixed numbers to improper fractions? Show how you could use this conversion procedure when you order doughnuts at Dunkin' Donuts.

3. What is the greatest common divisor? How could you use the greatest common divisor to write an advertisement showing that 35 out of 60 people prefer MCI to AT&T?

4. Explain the step approach for finding the greatest common divisor. How could you use the MCI–AT&T example in question 3 to illustrate the step approach?

5. Explain the steps of adding or subtracting unlike fractions. Using a ruler, measure the heights of two different-size cans of food and show how to calculate the difference in height.

6. What is a prime number? Using the two cans in question 5, show how you could use prime numbers to calculate the LCD.

7. Explain the steps for multiplying proper fractions and mixed numbers. Assume you went to Staples (a stationery superstore). Give an example showing the multiplying of proper fractions and mixed numbers.

8. **Chapter Concept Check.** Using all the information you have learned about fractions, search the web to find out how many cars are produced in the United States in a year and what fractional part represents cars produced by foreign-owned firms. Finally, present calculations using fractions.

2

END-OF-CHAPTER PROBLEMS

Check figures for odd-numbered problems in Appendix B. Name ______________ Date ________

DRILL PROBLEMS

Identify the following types of fractions: *LU 2-1(1)*

2–1. $\frac{9}{10}$ **2–2.** $\frac{12}{11}$ **2–3.** $\frac{25}{13}$

Convert the following to mixed numbers: *LU 2-1(2)*

2–4. $\frac{91}{10}$ **2–5.** $\frac{921}{15}$

Convert the following to improper fractions: *LU 2-1(2)*

2–6. $8\frac{7}{8}$ **2–7.** $19\frac{2}{3}$

Reduce the following to the lowest terms. Show how to calculate the greatest common divisor by the step approach. *LU 2-1(3)*

2–8. $\frac{16}{38}$ **2–9.** $\frac{44}{52}$

Convert the following to higher terms: *LU 2-1(3)*

2–10. $\frac{9}{10} = \frac{}{70}$

Determine the LCD of the following (a) by inspection and (b) by division of prime numbers: *LU 2-2(2)*

2–11. $\frac{3}{4}, \frac{7}{12}, \frac{5}{6}, \frac{1}{5}$ **Check**

Inspection

2–12. $\frac{5}{6}, \frac{7}{18}, \frac{5}{9}, \frac{2}{72}$ **Check**

Inspection

2–13. $\frac{1}{4}, \frac{3}{32}, \frac{5}{48}, \frac{1}{8}$ **Check**

Inspection

Add the following and reduce to lowest terms: *LU 2-2(1), LU 2-1(3)*

2–14. $\frac{3}{9} + \frac{3}{9}$ **2–15.** $\frac{3}{7} + \frac{4}{21}$

2–16. $6\frac{1}{8} + 4\frac{3}{8}$ **2–17.** $6\frac{3}{8} + 9\frac{1}{24}$

2–18. $9\frac{9}{10} + 6\frac{7}{10}$

Subtract the following and reduce to lowest terms: *LU 2-2(3), LU 2-1(3)*

2–19. $\frac{11}{12} - \frac{1}{12}$

2–20. $14\frac{3}{8} - 10\frac{5}{8}$

2–21. $12\frac{1}{9} - 4\frac{2}{3}$

Multiply the following and reduce to lowest terms. Do not use the cancellation technique for these problems. *LU 2-3(1), LU 2-1(3)*

2–22. $17 \times \frac{4}{2}$

2–23. $\frac{5}{6} \times \frac{3}{8}$

2–24. $8\frac{7}{8} \times 64$

Multiply the following. Use the cancellation technique. *LU 2-3(1), LU 2-1(2)*

2–25. $\frac{4}{10} \times \frac{30}{60} \times \frac{6}{10}$

2–26. $3\frac{3}{4} \times \frac{8}{9} \times 4\frac{9}{12}$

Divide the following and reduce to lowest terms. Use the cancellation technique as needed. *LU 2-3(2), LU 2-1(2)*

2–27. $\frac{12}{9} \div 4$

2–28. $18 \div \frac{1}{5}$

2–29. $4\frac{2}{3} \div 12$

2–30. $3\frac{5}{6} \div 3\frac{1}{2}$

WORD PROBLEMS

My Money

2–31. Michael Wittry has been investing in his Roth IRA retirement account for 20 years. Two years ago, his account was worth $215,658. After losing $\frac{1}{3}$ of its original value, it then gained $\frac{1}{2}$ of its new value back. What is the current value of his Roth IRA? *LU 2-3(1)*

2–32. Delta pays Pete Rose $180 per day to work in the maintenance department at the airport. Pete became ill on Monday and went home after $\frac{1}{6}$ of a day. What did he earn on Monday? Assume no work, no pay. *LU 2-3(1)*

2–33. Statista.com estimated the cumulative 2017 wind power capacity would reach 540,000 megawatts globally. If 2016 was $\frac{9}{10}$ of this, how much was the cumulative 2016 wind power capacity? *LU 2-3(1)*

2–34. Joy Wigens, who works at Putnam Investments, received a check for $1,600. She deposited $\frac{1}{4}$ of the check in her Citibank account. How much money does Joy have left after the deposit? *LU 2-3(1)*

2–35. Lee Jenkins worked the following hours as a manager for a local Pizza Hut: $14\frac{1}{4}$, $5\frac{1}{4}$, $8\frac{1}{2}$ and $7\frac{1}{4}$. How many total hours did Lee work? *LU 2-2(1)*

2–36. Lester bought a piece of property in Vail, Colorado. The sides of the land measure $115\frac{1}{2}$ feet, $66\frac{1}{4}$ feet, $106\frac{1}{8}$ feet, and $110\frac{1}{4}$ feet. Lester wants to know the perimeter (sum of all sides) of his property. Can you calculate the perimeter for Lester? *LU 2-2(1)*

2–37. Tiffani Lind got her new weekly course schedule from Roxbury Community College in Boston. Following are her classes and their length: Business Math, $2\frac{1}{2}$ hours; Introduction to Business, $1\frac{1}{2}$ hours; Microeconomics, $1\frac{1}{2}$ hours; Spanish, $2\frac{1}{4}$ hours; Marketing, $1\frac{1}{4}$ hours; and Business Statistics, $1\frac{3}{4}$ hours. How long will she be in class each week? *LU 2-2(1)*

2–38. Seventy-seven million people were born between 1946 and 1964. The U.S. Census classifies this group of individuals as baby boomers. It is said that today and every day for the next 18 years, 10,000 baby boomers will reach 65. If $\frac{1}{4}$ of the 65 and older age group uses e-mail, $\frac{1}{5}$ obtains the news from the Internet, and $\frac{1}{6}$ searches the Internet, find the LCD and determine total technology usage for this age group as a fraction. *LU 2-2(1, 2)*

2–39. At a local Walmart store, a Coke dispenser held $19\frac{1}{4}$ gallons of soda. During working hours, $12\frac{3}{4}$ gallons were dispensed. How many gallons of Coke remain? *LU 2-2(2, 3)*

My Money

2–40. CNBC.com reported in 2017 $\frac{35}{100}$ people have saved only a few hundred dollars for retirement and $\frac{34}{100}$ people have zero savings. If there are an estimated 42,729,344 persons more than 64 years old, how many people have not properly prepared for their retirement? Round to the nearest whole person. *LU 2-3(1)*

2–41. A local garden center charges \$250 per cord of wood. If Logan Grace orders $3\frac{1}{2}$ cords, what will the total cost be? *LU 2-3(1)*

2–42. A local Target store bought 90 pizzas at Pizza Hut for its holiday party. Each guest ate $\frac{1}{6}$ of a pizza and there was no pizza left over. How many guests did Target have for the party? *LU 2-3(1)*

2–43. Marc, Steven, and Daniel entered into a Subway sandwich shop partnership. Marc owns $\frac{1}{9}$ of the shop and Steven owns $\frac{1}{4}$. What part does Daniel own? *LU 2-2(1, 2)*

2–44. Lionel Sullivan works for Burger King. He is paid time and one-half for Sundays. If Lionel works on Sunday for 6 hours at a regular pay of \$8 per hour, what does he earn on Sunday? *LU 2-3(1)*

My Money

2–45. Financial analysts recommend people have an emergency fund covering up to six months of expenses. Money.cnn.com reported in 2017 $\frac{31}{100}$ people have such a fund. If you have monthly expenses of $2,100 and have saved only $\frac{2}{5}$ of your recommended six months of expenses, how much more do you have to save? *LU 2-3(1)*

eXcel

2–46. A trip to the White Mountains of New Hampshire from Boston will take you $2\frac{3}{4}$ hours. Assume you have traveled $\frac{1}{11}$ of the way. How much longer will the trip take? *LU 2-3(1, 2)*

eXcel

2–47. Andy, who loves to cook, makes apple cobbler for his family. The recipe (serves 6) calls for $1\frac{1}{2}$ pounds of apples, $3\frac{1}{4}$ cups of flour, $\frac{1}{4}$ cup of margarine, $2\frac{3}{8}$ cups of sugar, and 2 teaspoons of cinnamon. Since guests are coming, Andy wants to make a cobbler that will serve 15 (or increase the recipe $2\frac{1}{2}$ times). How much of each ingredient should Andy use? *LU 2-3(1, 2)*

2–48. Mobil allocates $1{,}692\frac{3}{4}$ gallons of gas per month to Jerry's Service Station. The first week, Jerry sold $275\frac{1}{2}$ gallons; second week, $280\frac{1}{4}$ gallons; and third week, $189\frac{1}{8}$ gallons. If Jerry sells $582\frac{1}{2}$ gallons in the fourth week, how close is Jerry to selling his allocation? *LU 2-2(4)*

2–49. A marketing class at North Shore Community College conducted a viewer preference survey. The survey showed that $\frac{5}{6}$ of the people surveyed preferred Apple's iPhone over the Blackberry. Assume 2,400 responded to the survey. How many favored using a Blackberry? *LU 2-3(1, 2)*

2–50. The price of a used Toyota LandCruiser has increased to $1\frac{1}{4}$ times its earlier price. If the original price of the LandCruiser was $30,000, what is the new price? *LU 2-3(1, 2)*

2–51. Tempco Corporation has a machine that produces $12\frac{1}{2}$ baseball gloves each hour. In the last 2 days, the machine has run for a total of 22 hours. How many baseball gloves has Tempco produced? *LU 2-3(2)*

2–52. Alicia, an employee of Dunkin' Donuts, receives $23\frac{1}{4}$ days per year of vacation time. So far this year she has taken $3\frac{1}{8}$ days in January, $5\frac{1}{2}$ days in May, $6\frac{1}{4}$ days in July, and $4\frac{1}{4}$ days in September. How many more days of vacation does Alicia have left? *LU 2-2(1, 2, 3)*

eXcel

2–53. A Hamilton multitouch watch was originally priced at $600. At a closing of the Alpha Omega Jewelry Shop, the watch is being reduced by $\frac{1}{4}$. What is the new selling price? *LU 2-3(1)*

2–54. Shelly Van Doren hired a contractor to refinish her kitchen. The contractor said the job would take $49\frac{1}{2}$ hours. To date, the contractor has worked the following hours:

Monday	$4\frac{1}{4}$
Tuesday	$9\frac{1}{8}$
Wednesday	$4\frac{1}{4}$
Thursday	$3\frac{1}{2}$
Friday	$10\frac{5}{8}$

How much longer should the job take to be completed? *LU 2-2(4)*

2–55. An issue of *Taunton's Fine Woodworking* included plans for a hall stand. The total height of the stand is $81\frac{1}{2}$ inches. If the base is $36\frac{5}{16}$ inches, how tall is the upper portion of the stand? *LU 2-2(4)*

2–56. Albertsons grocery planned a big sale on apples and received 750 crates from the wholesale market. Albertsons will bag these apples in plastic. Each plastic bag holds $\frac{1}{9}$ of a crate. If Albertsons has no loss to perishables, how many bags of apples can be prepared? *LU 2-3(1)*

2–57. Frank Puleo bought 6,625 acres of land in ski country. He plans to subdivide the land into parcels of $13\frac{1}{4}$ acres each. Each parcel will sell for \$125,000. How many parcels of land will Frank develop? If Frank sells all the parcels, what will be his total sales? *LU 2-3(1)*

If Frank sells $\frac{3}{5}$ of the parcels in the first year, what will be his total sales for the year?

2–58. A local Papa Gino's conducted a food survey. The survey showed that $\frac{1}{9}$ of the people surveyed preferred eating pasta to hamburger. If 5,400 responded to the survey, how many actually favored hamburger? *LU 2-3(1)*

2–59. Tamara, Jose, and Milton entered into a partnership that sells men's clothing on the web. Tamara owns $\frac{3}{8}$ of the company and Jose owns $\frac{1}{4}$. What part does Milton own? *LU 2-2(1, 3)*

2–60. *Quilters Newsletter Magazine* gave instructions on making a quilt. The quilt required $4\frac{1}{2}$ yards of white-on-white print, 2 yards blue check, $\frac{1}{2}$ yard blue-and-white stripe, $2\frac{3}{4}$ yards blue scraps, $\frac{3}{4}$ yard yellow scraps, and $4\frac{7}{8}$ yards lining. How many total yards are needed? *LU 2-2(1, 2)*

2–61. A trailer carrying supplies for a Krispy Kreme from Virginia to New York will take $3\frac{1}{4}$ hours. If the truck traveled $\frac{1}{5}$ of the way, how much longer will the trip take? *LU 2-3(1, 2)*

2–62. Land Rover has increased the price of a FreeLander by $\frac{1}{5}$ from the original price. The original price of the FreeLander was \$30,000. What is the new price? *LU 2-3(1, 2)*

CHALLENGE PROBLEMS

2–63. *Woodsmith* magazine gave instructions on how to build a pine cupboard. Lumber will be needed for two shelves $10\frac{1}{4}$ inches long, two base sides $12\frac{1}{2}$ inches long, and two door stiles $29\frac{1}{8}$ inches long. Your lumber comes in 6 foot lengths. **(a)** How many feet of lumber will you need? **(b)** If you want $\frac{1}{2}$ a board left over, is this possible with two boards? *LU 2-2(1, 2, 3, 4)*

2–64. Jack MacLean has entered into a real estate development partnership with Bill Lyons and June Reese. Bill owns $\frac{1}{4}$ of the partnership, while June has a $\frac{1}{5}$ interest. The partners will divide all profits on the basis of their fractional ownership. The partnership bought 900 acres of land and plans to subdivide each lot into $2\frac{1}{4}$ acres. Homes in the area have been selling for $240,000. By time of completion, Jack estimates the price of each home will increase by $\frac{1}{3}$ of the current value. The partners sent a survey to 12,000 potential customers to see whether they should heat the homes with oil or gas. One-fourth of the customers responded by indicating a 5-to-1 preference for oil. From the results of the survey, Jack now plans to install a 270-gallon oil tank at each home. He estimates that each home will need five fills per year. The current price of home heating fuel is $1 per gallon. The partnership estimates its profit per home will be $\frac{1}{8}$ the selling price of each home. From the above, please calculate the following: *LU 2-1(1, 2, 3), LU 2-2(1, 2, 3, 4), LU 2-3(1, 2)*

a. Number of homes to be built.

b. Selling price of each home.

c. Number of people responding to survey.

d. Number of people desiring oil.

e. Average monthly cost per house to heat using oil.

f. Amount of profit Jack will receive from the sale of homes.

SUMMARY PRACTICE TEST

Do you need help? Connect videos have step-by-step worked-out solutions.

Identify the following types of fractions. *LU 2-1(1)*

1. $5\frac{1}{8}$ **2.** $\frac{2}{7}$ **3.** $\frac{20}{19}$

4. Convert the following to a mixed number. *LU 2-1(2)*

$\frac{163}{9}$

5. Convert the following to an improper fraction. *LU 2-1(2)*

$8\frac{1}{8}$

6. Calculate the greatest common divisor of the following by the step approach and reduce to lowest terms. *LU 2-2(1, 2)*

$\frac{63}{90}$

7. Convert the following to higher terms. *LU 2-1(3)*

$\frac{16}{94} = \frac{?}{376}$

8. Find the LCD of the following by using prime numbers. Show your work. *LU 2-2(2)*

$\frac{1}{8} + \frac{1}{3} + \frac{1}{2} + \frac{1}{12}$

9. Subtract the following. *LU 2-2(4)*

$$\begin{array}{r} 15\frac{4}{5} \\ -8\frac{19}{20} \\ \hline \end{array}$$

Complete the following using the cancellation technique. *LU 2-3(1, 2)*

10. $\frac{3}{4} \times \frac{2}{4} \times \frac{6}{9}$ **11.** $7\frac{1}{9} \times \frac{6}{7}$ **12.** $\frac{3}{7} \div 6$

13. A trip to Washington from Boston will take you $5\frac{3}{4}$ hours. If you have traveled $\frac{1}{3}$ of the way, how much longer will the trip take? *LU 2-3(1)*

14. Quiznos produces 640 rolls per hour. If the oven runs $12\frac{1}{4}$ hours, how many rolls will the machine produce? *LU 2-3(1, 2)*

15. A taste-testing survey of Zing Farms showed that $\frac{2}{3}$ of the people surveyed preferred the taste of veggie burgers to regular burgers. If 90,000 people were in the survey, how many favored veggie burgers? How many chose regular burgers? *LU 2-3(1)*

16. Jim Janes, an employee of Enterprise Co., worked $9\frac{1}{4}$ hours on Monday, $4\frac{1}{2}$ hours on Tuesday, $9\frac{1}{4}$ hours on Wednesday, $7\frac{1}{2}$ hours on Thursday, and 9 hours on Friday. How many total hours did Jim work during the week? *LU 2-2(1, 2)*

17. JCPenney offered a $\frac{1}{3}$ rebate on its $39 hair dryer. Joan bought a JCPenney hair dryer. What did Joan pay after the rebate? *LU 2-3(1)*

INTERACTIVE VIDEO WORKSHEET

GRADING THE SUMMARY PRACTICE TEST

Go to the summary practice test video in Connect (or click on it here in the ebook). Grade your summary practice test while viewing the video.

C for Correct/I for Incorrect

1. ________	6. ________	11. ________	16. ________
2. ________	7. ________	12. ________	17. ________
3. ________	8. ________	13. ________	
4. ________	9. ________	14. ________	
5. ________	10. ________	15. ________	

If you achieved 100%, you are ready for your instructor's exam.

If any of the problems were incorrect, list the questions you missed and show steps to solve the problem correctly.

Replay the video to see if you have made the correct fixes to your mistakes. If you have any questions, contact your instructor asap.

Notes on Watching Videos

MY MONEY

Give Yourself a Run for Your Money!

What I need to know

As you work through your college courses you no doubt have your sights set on attaining a degree or credential that will get you the job, promotion, or career move you are seeking. As you prepare for that eventual career, it is important to understand and anticipate the level of salary you will earn upon your graduation. You should ask yourself three questions to determine if the salary expectations of your chosen career are a good fit with your financial goals:

1. What is the starting pay for the position for someone with my degree?
2. What other forms of earning could be a part of this position? (commission, bonus, etc.)
3. How does this salary compare to my financial goals?

What I need to do

Research, research, research! Be sure you know the salary expectations before committing to a desired course of study. Many times this information is available through your collegiate institution and is provided based upon the program of study you pursue. Although only a range may be given from the college, it will give you a rough estimate from which to determine your educational path. Compare the cost it will take to attain your degree to the expected salary to determine the cost effectiveness of each degree option you are considering.

Additionally, you could seek out professionals in the field you are considering to get some first-hand feedback on the position and the financial expectations of the position. Ask these professionals about their personal experiences within this career field. What do they like best about their chosen profession? What do they see as the future opportunities within this career? Are there other factors to consider outside of just salary such as, for example, benefits, personal growth, and contribution to a greater cause? If these professionals had it to do all over again, what might they do differently as it relates to career preparedness? Obtaining such valuable first-hand knowledge will go a long way in helping you make a decision about which discipline to pursue while in college—a decision that will help lead you to your desired career.

Ultimately you will want to determine whether or not a career field will fit into your financial plans. How does this salary range compare to your financial goals and will you be able to meet these goals with such earnings? Place the salary expectation against your budget to see how it will meet your expenses. Furthermore, determine what salary range will allow for spending opportunities outside of your expenses such as investments, savings, and entertainment.

Resources I can use

- Indeed Job Search (mobile app)
- https://www.payscale.com/ — get a report about your expected pay
- https://money.usnews.com/money/careers/articles/2015/01/13/do-you-know-how-much-youre-worth — helpful hints for determining your worth as you enter the job market

MY MONEY ACTIVITY

- Search for job openings in your local area related to your degree.
- Compare the expected salaries to obtain a range for this position in your area.

PERSONAL FINANCE

A KIPLINGER APPROACH

By the Numbers

How Much to Save for a Four-Year Degree

The table below shows the estimated total cost for four years of college for children at four different ages—for in-state and out-of-state public colleges as well as private colleges. The estimates are based on the current average annual sticker prices for tuition, fees, and room and board and assume that costs will continue to rise at rates similar to those seen over the past decade. The good news: You don't have to save the whole tab. Each sticker price is accompanied by a savings goal. Most experts recommend that you aim to save between one-fourth and one-third of the projected sticker price (we've based the savings goals on the larger of the two). The rest can come from financial aid, scholarships and student loans.

©jaroon/Getty Images

©Photodisc/Getty Images

©George Doyle/Stockbyte

©YAY Media AS/Alamy Stock Photo

Newborn	6-year-old	12-year-old	17-year-old
In-State Public College **TOTAL COST** **$137,757**	In-State Public College **TOTAL COST** **$116,722**	In-State Public College **TOTAL COST** **$98,900**	In-State Public College **TOTAL COST** **$86,145**
Savings Goal $45,919	Savings Goal $38,907	Savings Goal $32,967	Savings Goal $28,715
Out-of-State Public College **TOTAL COST** **$220,501**	Out-of-State Public College **TOTAL COST** **$192,378**	Out-of-State Public College **TOTAL COST** **$167,843**	Out-of-State Public College **TOTAL COST** **$149,804**
Savings Goal $73,500	Savings Goal $64,126	Savings Goal $55,948	Savings Goal $49,935
Private College **TOTAL COST** **$282,842**	Private College **TOTAL COST** **$246,769**	Private College **TOTAL COST** **$215,296**	Private College **TOTAL COST** **$192,158**
Savings Goal $94,281	Savings Goal $82,256	Savings Goal $71,765	Savings Goal $64,053

SOURCES: College Board Trends in College Pricing 2016, Saving for College's World's Simplest College Cost Calculator

BUSINESS MATH ISSUE

Saving $\frac{1}{4}$ to $\frac{1}{3}$ of sticker price shown means your college goals will be met.

1. List the key points of the article and information to support your position.
2. Write a group defense of your position using math calculations to support your view. If you are in an online course, post to a discussion board.

Classroom Notes

CHAPTER 3

Decimals

LEARNING UNIT OBJECTIVES

LU 3–1: Rounding Decimals; Fraction and Decimal Conversions

1. Explain the place values of whole numbers and decimals; round decimals.
2. Convert decimal fractions to decimals, proper fractions to decimals, mixed numbers to decimals, and pure and mixed decimals to decimal fractions.

LU 3–2: Adding, Subtracting, Multiplying, and Dividing Decimals

1. Add, subtract, multiply, and divide decimals.
2. Complete decimal applications in foreign currency.
3. Multiply and divide decimals by shortcut methods.

Your Guide to Successfully Completing This Chapter

Traditional book or ebook

Check box as you complete each step.

Steps

☐ Read learning unit.

  ☐ Complete practice quiz at the end of the learning unit.

☐ Grade practice quiz using provided solutions. (For more help, watch the learning unit video in Connect and have a Study Session with the authors. Then complete the additional practice quiz in Connect.)

☐ Repeat above for each of the two learning units in Chapter 3.

  ☐ Review chapter organizer.

  ☐ Complete assigned homework.

    ☐ Finish summary practice test. (Go to Connect via the ebook link and do the interactive video worksheet to grade.)

☐ Complete instructor's exam.

In the chapter opener a burger sells for \$9.00 and costs the restaurant \$1.86 to prepare it. The new tax law passed in 2017 charges significantly lower taxes for many businesses. Hopefully this new tax law will mean savings to customers. Check with the IRS for latest details.

Assume a Louis Vuitton bag costs \$801.99 in the U.K. The same bag in the U.S. costs \$969.50. The bag is \$167.51 higher in the U.S. We will look at currency exchange rates in this chapter.

$$\begin{array}{r} \$969.50 \\ -801.99 \\ \hline \$167.51 \end{array}$$

Chapter 2 introduced the 1.69-ounce bag of M&M'S®. In Table 3.1, the six colors in that 1.69-ounce bag of M&M'S® are given in fractions and their values expressed in decimal equivalents that are rounded to the nearest hundredths.

This chapter is divided into two learning units. The first unit discusses rounding decimals, converting fractions to decimals, and converting decimals to fractions. The second unit shows you how to add, subtract, multiply, and divide decimals, along with some shortcuts for multiplying and dividing decimals. Added to this unit is a global application of decimals dealing with foreign exchange rates. One of the most common uses of decimals occurs when we spend dollars and cents, which is a *decimal number*.

Decimals are decimal numbers with digits to the right of a *decimal point,* indicating that decimals, like fractions, are parts of a whole that are less than one. Thus, we can interchange the terms *decimals* and *decimal numbers*. Remembering this will avoid confusion between the terms *decimal, decimal number,* and *decimal point.*

TABLE 3-1

Analyzing a bag of M&M'S®

©usersam2007/123RF

Color*	Fraction	Decimal
Yellow	$\frac{18}{55}$	.33
Red	$\frac{10}{55}$	.18
Blue	$\frac{9}{55}$	.16
Orange	$\frac{7}{55}$	.13
Brown	$\frac{6}{55}$	.11
Green	$\frac{5}{55}$	.09
Total	$\frac{55}{55} = 1$	1.00

*The color ratios currently given are a sample used for educational purposes. They do not represent the manufacturer's color ratios.

LO 1

Learning Unit 3–1: Rounding Decimals; Fraction and Decimal Conversions

In Chapter 1 we stated that the **decimal point** is the center of the decimal numbering system. So far we have studied the whole numbers to the left of the decimal point and the parts of whole numbers called fractions. We also learned that the position of the digits in a whole number gives the place values of the digits (Figure 1.1). Now we will study the position (place values) of the digits to the right of the decimal point (Figure 3.1). Note that the words to the right of the decimal point end in *ths.*

You should understand why the decimal point is the center of the decimal system. If you move a digit to the left of the decimal point by place (ones, tens, and so on), *you increase its value 10 times for each place (power of 10).* If you move a digit to the right of the decimal point by place (tenths, hundredths, and so on), *you decrease its value 10 times for each place.*

EXAMPLES

$.06 → The 6 is in the hundred*ths* place value.

1.527 → The 5 is in the ten*ths* place value.

2.8394 → The 4 is in the ten thousand*ths* place value.

.33 → The thirty-three hundred*ths* represents the yellow M&M'S® in our M&M'S® bag of 55 M&M'S®.

1.69 oz. → The one ounce and sixty-nine hundred*ths* of another ounce is the weight of our bag of M&M'S®.

FIGURE 3.1

Decimal place-value chart

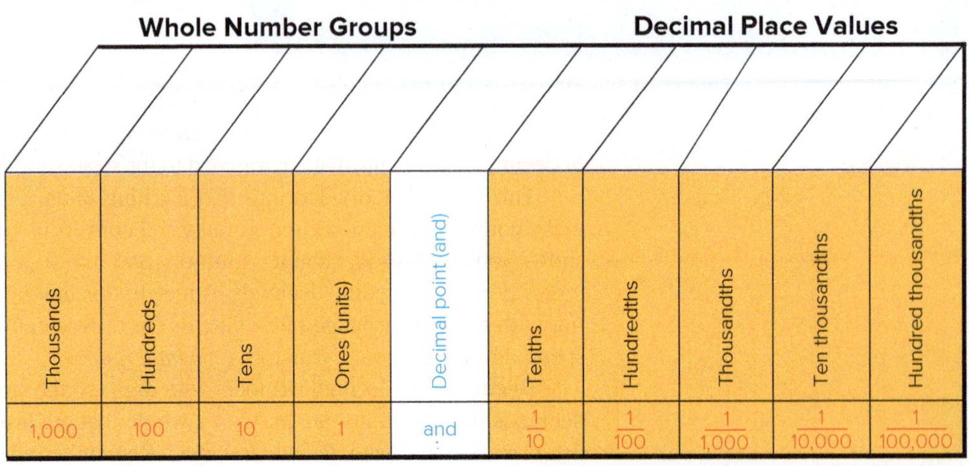

Do you recall from Chapter 1 how you used a place-value chart to read or write whole numbers in verbal form? To read or write decimal numbers, you read or write the decimal number as if it were a whole number. Then you use the name of the decimal place of the last digit as given in Figure 3.1. For example, you would read or write the decimal .0796 as seven hundred ninety-six ten thousandths (the last digit, 6, is in the ten thousandths place).

To read a decimal with four or fewer whole numbers, you can also refer to Figure 3.1. For larger whole numbers, refer to the whole number place-value chart in Chapter 1 (Figure 1.1). For example, from Figure 3.1 you would read the number 126.2864 as one hundred twenty-six and two thousand eight hundred sixty-four ten thousandths. *Remember to read the decimal point as* "and."

Now let's round decimals. Rounding decimals is similar to the rounding of whole numbers that you learned in Chapter 1.

Used by permission of Cartoon Features Syndicate

Rounding Decimals

From Table 3.1, you know that the 1.69-ounce bag of M&M'S® introduced in Chapter 2 contained $\frac{18}{55}$, or .33, yellow M&M'S®. The .33 was rounded to the nearest hundredth. **Rounding decimals** involves the following steps:

ROUNDING DECIMALS TO A SPECIFIED PLACE VALUE

Step 1. Identify the place value of the digit you want to round.

Step 2. If the digit to the right of the identified digit in Step 1 is 5 or more, increase the identified digit by 1. If the digit to the right is less than 5, do not change the identified digit.

Step 3. Drop all digits to the right of the identified digit.

Let's practice rounding by using the $\frac{18}{55}$ yellow M&M'S® that we rounded to .33 in Table 3.1. Before we rounded $\frac{18}{55}$ to .33, the number we rounded was .32727. This is an example of a **repeating decimal** since the 27 repeats itself.

EXAMPLE Round .3272727 to the nearest hundredth.

Step 1. .3272727 — The identified digit is 2, which is in the hundredths place (two places to the right of the decimal point).

Step 2. The digit to the right of 2 is more than 5 (7). Thus, 2, the identified digit in Step 1, is changed to 3.

.3372727

Step 3. .33 — Drop all other digits to the right of the identified digit 3.

We could also round the .3272727 M&M'S® to the nearest tenth or thousandth as follows:

	Tenth	or	Thousandth
.3272727 →	.3	.3272727 →	.327

OTHER EXAMPLES

Round to nearest dollar:	$166.39 →	$166
Round to nearest cent:	$1,196.885 →	$1,196.89
Round to nearest hundredth:	$38.563 →	$38.56
Round to nearest thousandth:	$1,432.9981 →	$1,432.998

The rules for rounding can differ with the situation in which rounding is used. For example, have you ever bought one item from a supermarket produce department that was marked "3 for $1" and noticed what the cashier charged you? One item marked "3 for $1" would not cost you $33\frac{1}{3}$ cents rounded to 33 cents. You will pay 34 cents. Many retail stores round to the next cent even if the digit following the identified digit is less than $\frac{1}{2}$ of a penny. In this text we round on the concept of 5 or more.

LO 2

Fraction and Decimal Conversions

In business operations we must frequently convert fractions to decimal numbers and decimal numbers to fractions. This section begins by discussing three types of fraction-to-decimal conversions. Then we discuss converting pure and mixed decimals to decimal fractions.

©2014 Jef Mallett/Dist. by Universal Uclick.

Converting Decimal Fractions to Decimals From Figure 3.1 you can see that a **decimal fraction** (expressed in the digits to the right of the decimal point) is a fraction with a denominator that has a power of 10, such as $\frac{1}{10}$, $\frac{17}{100}$, and $\frac{23}{1,000}$. To convert a decimal fraction to a decimal, follow these steps:

CONVERTING DECIMAL FRACTIONS TO DECIMALS

Step 1. Count the number of zeros in the denominator.

Step 2. Place the numerator of the decimal fraction to the right of the decimal point the same number of places as you have zeros in the denominator. (The number of zeros in the denominator gives the number of digits your decimal has to the right of the decimal point.) Do not go over the total number of denominator zeros.

Now let's change $\frac{3}{10}$ and its higher multiples of 10 to decimals.

EXAMPLES

Verbal form	Decimal fraction	Decimal[1]	Number of decimal places to right of decimal point
a. Three tenths	$\frac{3}{10}$	.3	1
b. Three hundredths	$\frac{3}{100}$	.03	2
c. Three thousandths	$\frac{3}{1,000}$	.003	3
d. Three ten thousandths	$\frac{3}{10,000}$	.0003	4

©McGraw-Hill Education/Mark Dierker, photographer

Note how we show the different values of the decimal fractions above in decimals. The zeros after the decimal point and before the number 3 indicate these values. If you add zeros after the number 3, you do not change the value. Thus, the numbers .3, .30, and .300 have the

[1]From .3 to .0003, the values get smaller and smaller, but if you go from .3 to .3000, the values remain the same.

same value. So 3 tenths of a pizza, 30 hundredths of a pizza, and 300 thousandths of a pizza are the same total amount of pizza. The first pizza is sliced into 10 pieces. The second pizza is sliced into 100 pieces. The third pizza is sliced into 1,000 pieces. Also, we don't need to place a zero to the left of the decimal point.

Converting Proper Fractions to Decimals Recall from Chapter 2 that proper fractions are fractions with a value less than 1. That is, the numerator of the fraction is smaller than its denominator. How can we convert these proper fractions to decimals? Since proper fractions are a form of division, it is possible to convert proper fractions to decimals by carrying out the division.

CONVERTING PROPER FRACTIONS TO DECIMALS

Step 1. Divide the numerator of the fraction by its denominator. (If necessary, add a decimal point and zeros to the number in the numerator.)

Step 2. Round as necessary.

EXAMPLES

$$\frac{3}{4} = 4\overline{)3.00} = .75 \qquad \begin{array}{r} 3.00 \\ \underline{2\,8} \\ 20 \\ \underline{20} \end{array}$$

$$\frac{3}{8} = 8\overline{)3.000} = .375 \qquad \begin{array}{r} 3.000 \\ \underline{2\,4} \\ 60 \\ \underline{56} \\ 40 \\ \underline{40} \end{array}$$

$$\frac{1}{3} = 3\overline{)1.000} = .33\overline{3} \qquad \begin{array}{r} 1.000 \\ \underline{9} \\ 10 \\ \underline{9} \\ 10 \\ \underline{9} \\ 1 \end{array}$$

Note that in the last example $\frac{1}{3}$, the 3 in the quotient keeps repeating itself (never ends). The short bar over the last 3 means that the number endlessly repeats.

Converting Mixed Numbers to Decimals

A mixed number, you will recall from Chapter 2, is the sum of a whole number greater than zero and a proper fraction. To convert mixed numbers to decimals, use the following steps:

CONVERTING MIXED NUMBERS TO DECIMALS

Step 1. Convert the fractional part of the mixed number to a decimal (as illustrated in the previous section).

Step 2. Add the converted fractional part to the whole number.

EXAMPLE

$$8\frac{2}{5} = \textbf{(Step 1)} \quad 5\overline{)2.0} = .4 \qquad \begin{array}{r} 2.0 \\ \underline{2.0} \end{array} \qquad \textbf{(Step 2)} = \begin{array}{r} 8.00 \\ +\ .40 \\ \hline 8.40 \end{array}$$

Now that we have converted fractions to decimals, let's convert decimals to fractions.

Converting Pure and Mixed Decimals to Decimal Fractions A **pure decimal** has no whole number(s) to the left of the decimal point (.43, .458, and so on). A **mixed decimal** is a combination of a whole number and a decimal. An example of a mixed decimal follows:

EXAMPLE 737.592 = Seven hundred thirty-seven and five hundred ninety-two thousandths

Note the following conversion steps for converting pure and mixed decimals to decimal fractions:

CONVERTING PURE AND MIXED DECIMALS TO DECIMAL FRACTIONS

Step 1. Place the digits to the right of the decimal point in the numerator of the fraction. Omit the decimal point. (For a decimal fraction with a fractional part, see examples **c** and **d** below.)

Step 2. Put a 1 in the denominator of the fraction.

Step 3. Count the number of digits to the right of the decimal point. Add the same number of zeros to the denominator of the fraction. For mixed decimals, add the fraction to the whole number.

If desired, you can reduce the fractions in Step 3.

EXAMPLES	Step 1	Step 2	Places	Step 3
a. .3	$\frac{3}{}$	$\frac{3}{1}$	1	$\frac{3}{10}$
b. .24	$\frac{24}{}$	$\frac{24}{1}$	2	$\frac{24}{100}$
c. $.24\frac{1}{2}$	$\frac{245}{}$	$\frac{245}{1}$	3	$\frac{245}{1{,}000}$

Before completing Step 1 in example **c,** we must remove the fractional part, convert it to a decimal ($\frac{1}{2} = .5$), and multiply it by .01 ($.5 \times .01 = .005$). We use .01 because the 4 of .24 is in the hundredths place. Then we add $.005 + .24 = .245$ (three places to right of the decimal) and complete Steps 1, 2, and 3.

d. $.07\frac{1}{4}$	$\frac{725}{}$	$\frac{725}{1}$	4	$\frac{725}{10{,}000}$

In example **d,** be sure to convert $\frac{1}{4}$ to .25 and multiply by .01. This gives .0025. Then add .0025 to .07, which is .0725 (four places), and complete Steps 1, 2, and 3.

e. 17.45	$\frac{45}{}$	$\frac{45}{1}$	2	$\frac{45}{100} = 17\frac{45}{100}$

Example **e** is a mixed decimal. Since we substitute *and* for the decimal point, we read this mixed decimal as seventeen and forty-five hundredths. Note that after we converted the .45 of the mixed decimal to a fraction, we added it to the whole number 17.

The Practice Quiz that follows will help you check your understanding of this unit.

MONEY tips

Formula for Financial Success: Reduce Spending + Decrease Debt + Increase Savings (Investing) = Healthy Net Worth

LU 3–1 PRACTICE QUIZ

Complete this Practice Quiz to see how you are doing.

Write the following as a decimal number.

1. Four hundred eight thousandths

Name the place position of the identified digit:

2. 6.8241 (arrow under 2) **3.** 9.3942 (arrow under 4)

Round each decimal to place indicated:

		Tenth	Thousandth
4.	.62768	**a.**	**b.**
5.	.68341	**a.**	**b.**

Convert the following to decimals:

6. $\frac{9}{10{,}000}$ **7.** $\frac{14}{100{,}000}$

Convert the following to decimal fractions (do not reduce):

8. .819 **9.** 16.93 **10.** $.05\frac{1}{4}$

Convert the following fractions to decimals and round answer to nearest hundredth:

11. $\frac{1}{6}$ **12.** $\frac{3}{8}$ **13.** $12\frac{1}{8}$

✓ Solutions

*For **extra help** from your authors–Sharon and Jeff–see the videos in Connect.*

1. .408 (3 places to right of decimal)
2. Hundredths
3. Thousandths
4. a. .6 (identified digit 6—digit to right less than 5)
   b. .628 (identified digit 7—digit to right greater than 5)
5. a. .7 (identified digit 6—digit to right greater than 5)
   b. .683 (identified digit 3—digit to right less than 5)
6. .0009 (4 places)
7. .00014 (5 places)
8. $\frac{819}{1{,}000}\left(\frac{819}{1 + 3 \text{ zeros}}\right)$
9. $16\frac{93}{100}$
10. $\frac{525}{10{,}000}\left(\frac{525}{1 + 4 \text{ zeros}} \quad \frac{1}{4} \times .01 = .0025 + .05 = .0525\right)$
11. $.16666 = .17$
12. $.375 = .38$
13. $12.125 = 12.13$

Learning Unit 3–2: Adding, Subtracting, Multiplying, and Dividing Decimals

The *Wall Street Journal* clip in the chapter opener uses decimals showing the average price of a burger is $9.00 while costing $1.86, thus earning a profit of $7.14.

$$\begin{array}{r} \$9.00 \\ -1.86 \\ \hline \$7.14 \end{array}$$

This learning unit shows you how to add, subtract, multiply, and divide decimals. You also make calculations involving decimals, including decimals used in foreign currency.

LO 1

Addition and Subtraction of Decimals

Since you know how to add and subtract whole numbers, to add and subtract decimal numbers you have only to learn about the placement of the decimals. The following steps will help you:

ADDING AND SUBTRACTING DECIMALS

Step 1. Vertically write the numbers so that the decimal points align. You can place additional zeros to the right of the decimal point if needed without changing the value of the number.

Step 2. Add or subtract the digits starting with the right column and moving to the left.

Step 3. Align the decimal point in the answer with the above decimal points.

EXAMPLES Add 4 + 7.3 + 36.139 + .0007 + 8.22.

Whole number to the right of the last digit is assumed to have a decimal. → 4.0000

Extra zeros have been added to make calculation easier.

$$\begin{array}{r} 4.0000 \\ 7.3000 \\ 36.1390 \\ .0007 \\ 8.2200 \\ \hline 55.6597 \end{array}$$

Subtract 45.3 − 15.273.

```
  2 9 10
 45.300
−15.273
 30.027
```

Subtract 7 − 6.9.

```
 6 10
 7.0
−6.9
  .1
```

Multiplication of Decimals

The multiplication of decimal numbers is similar to the multiplication of whole numbers except for the additional step of placing the decimal in the answer (product). The steps that follow simplify this procedure.

MULTIPLYING DECIMALS

Step 1. Multiply the numbers as whole numbers, ignoring the decimal points.

Step 2. Count and total the number of decimal places in the multiplier and multiplicand.

Step 3. Starting at the right in the product, count to the left the number of decimal places totaled in Step 2. Place the decimal point so that the product has the same number of decimal places as totaled in Step 2. If the total number of places is greater than the places in the product, insert zeros in front of the product.

EXAMPLES

```
         8.52   (2 decimal places)
Step 1 → × 6.7  (1 decimal place)  ← Step 2
         5964
        5112
Step 3 → 57.084
```

```
  2.36   (2 places)
 ×.016   (3 places)
  1416
  236
.03776   Need to add zero
```

Division of Decimals

If the divisor in your decimal division problem is a whole number, first place the decimal point in the quotient directly above the decimal point in the dividend. Then divide as usual. If the divisor has a decimal point, complete the steps that follow.

DIVIDING DECIMALS

Step 1. Make the divisor a whole number by moving the decimal point to the right.

Step 2. Move the decimal point in the dividend to the right the same number of places that you moved the decimal point in the divisor (Step 1). If there are not enough places, add zeros to the right of the dividend.

Step 3. Place the decimal point in the quotient above the new decimal point in the dividend. Divide as usual.

EXAMPLE

```
                 Step 3 ↓
           13.12
Step 1 → 2.5)32.8.00 ← Step 2
           25
            78
            75
             30
             25
              50
              50
```

Stop a moment and study the above example. Note that the quotient does not change when we multiply the divisor and the dividend by the same number. This is why we can move the decimal point in division problems and always divide by a whole number.

LO 2

GLOBAL

Decimal Applications in Foreign Currency

EXAMPLE

Hanna Lind, who lives in Canada, wanted to buy a new Apple watch. She went on eBay and found that the cost would be $700 in U.S. dollars. Wanting to know how much this would cost in Canadian dollars, Hanna consulted the following *Wall Street Journal* currency table on page 78 and found that a Canadian dollar was worth $.7945 in U.S. dollars. Therefore, for each Canadian dollar it would cost $1.2587 to buy a U.S. good.

Using this information, Hanna completed the following calculation to determine what an Apple watch would cost her:

$700 (cost of the watch in U.S. dollars)	× $1.2587	= $881.09 (cost of the watch in Canadian dollars)

To check her findings, Hanna did the following calculation:

$881.09 (cost of the Apple watch in Canadian dollars)	× $.7945 (what the Canadian dollar is worth against the U.S. dollar)	= $700.03 (off due to rounding) (U.S. selling price)

©Lukas Gojda/Shutterstock

GLOBAL

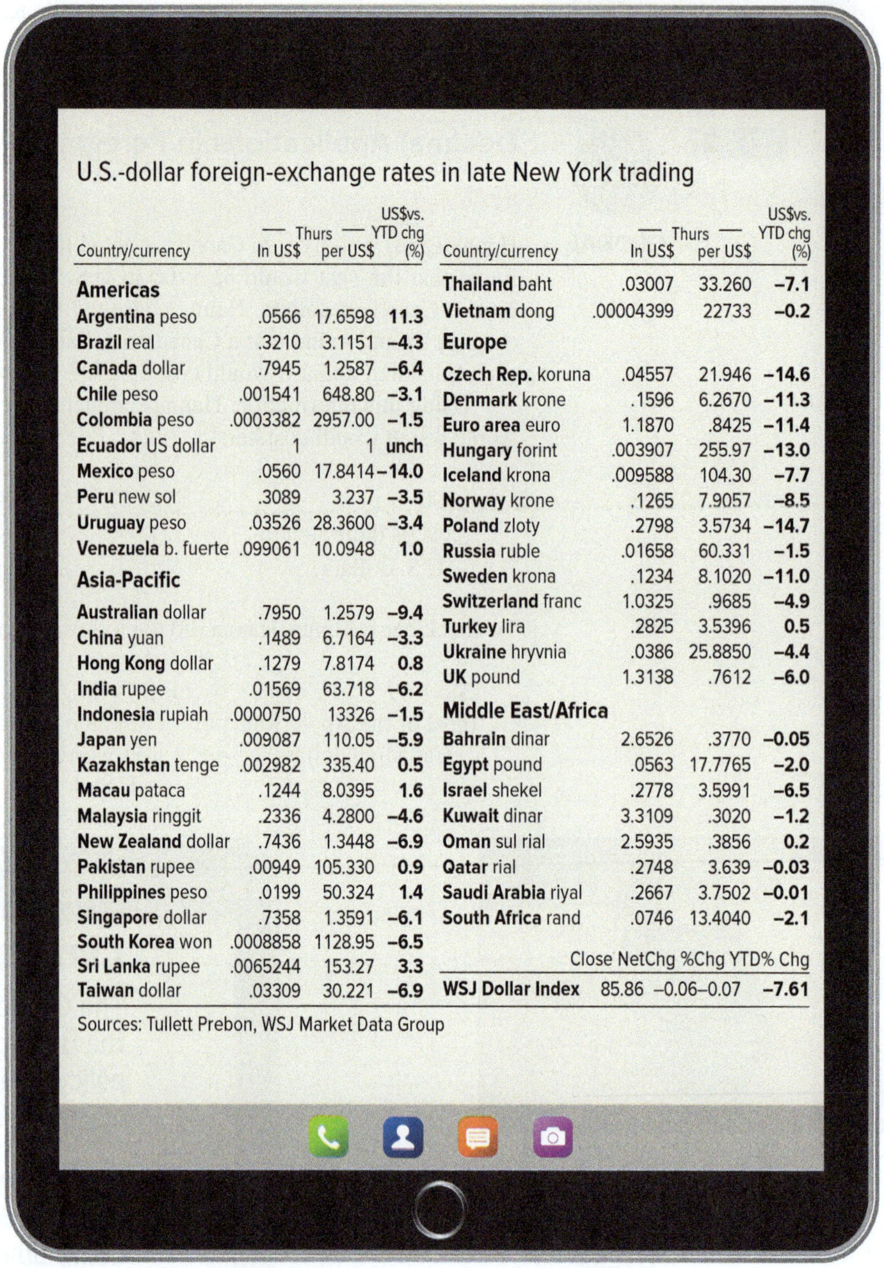

U.S.-dollar foreign-exchange rates in late New York trading

Country/currency	Thurs In US$	Thurs per US$	US$vs. YTD chg (%)
Americas			
Argentina peso	.0566	17.6598	**11.3**
Brazil real	.3210	3.1151	**−4.3**
Canada dollar	.7945	1.2587	**−6.4**
Chile peso	.001541	648.80	**−3.1**
Colombia peso	.0003382	2957.00	**−1.5**
Ecuador US dollar	1	1	**unch**
Mexico peso	.0560	17.8414	**−14.0**
Peru new sol	.3089	3.237	**−3.5**
Uruguay peso	.03526	28.3600	**−3.4**
Venezuela b. fuerte	.099061	10.0948	**1.0**
Asia-Pacific			
Australian dollar	.7950	1.2579	**−9.4**
China yuan	.1489	6.7164	**−3.3**
Hong Kong dollar	.1279	7.8174	**0.8**
India rupee	.01569	63.718	**−6.2**
Indonesia rupiah	.0000750	13326	**−1.5**
Japan yen	.009087	110.05	**−5.9**
Kazakhstan tenge	.002982	335.40	**0.5**
Macau pataca	.1244	8.0395	**1.6**
Malaysia ringgit	.2336	4.2800	**−4.6**
New Zealand dollar	.7436	1.3448	**−6.9**
Pakistan rupee	.00949	105.330	**0.9**
Philippines peso	.0199	50.324	**1.4**
Singapore dollar	.7358	1.3591	**−6.1**
South Korea won	.0008858	1128.95	**−6.5**
Sri Lanka rupee	.0065244	153.27	**3.3**
Taiwan dollar	.03309	30.221	**−6.9**
Thailand baht	.03007	33.260	**−7.1**
Vietnam dong	.00004399	22733	**−0.2**
Europe			
Czech Rep. koruna	.04557	21.946	**−14.6**
Denmark krone	.1596	6.2670	**−11.3**
Euro area euro	1.1870	.8425	**−11.4**
Hungary forint	.003907	255.97	**−13.0**
Iceland krona	.009588	104.30	**−7.7**
Norway krone	.1265	7.9057	**−8.5**
Poland zloty	.2798	3.5734	**−14.7**
Russia ruble	.01658	60.331	**−1.5**
Sweden krona	.1234	8.1020	**−11.0**
Switzerland franc	1.0325	.9685	**−4.9**
Turkey lira	.2825	3.5396	**0.5**
Ukraine hryvnia	.0386	25.8850	**−4.4**
UK pound	1.3138	.7612	**−6.0**
Middle East/Africa			
Bahrain dinar	2.6526	.3770	**−0.05**
Egypt pound	.0563	17.7765	**−2.0**
Israel shekel	.2778	3.5991	**−6.5**
Kuwait dinar	3.3109	.3020	**−1.2**
Oman sul rial	2.5935	.3856	**0.2**
Qatar rial	.2748	3.639	**−0.03**
Saudi Arabia riyal	.2667	3.7502	**−0.01**
South Africa rand	.0746	13.4040	**−2.1**

	Close	NetChg	%Chg	YTD% Chg
WSJ Dollar Index	85.86	−0.06	−0.07	**−7.61**

Sources: Tullett Prebon, WSJ Market Data Group

GLOBAL

LO 3

Multiplication and Division Shortcuts for Decimals

The shortcut steps that follow show how to solve multiplication and division problems quickly involving multiples of 10 (10, 100, 1,000, 10,000, etc.).

SHORTCUTS FOR MULTIPLES OF 10

Multiplication

Step 1. Count the zeros in the multiplier.

Step 2. Move the decimal point in the multiplicand the same number of places to the right as you have zeros in the multiplier.

Division

Step 1. Count the zeros in the divisor.

Step 2. Move the decimal point in the dividend the same number of places to the left as you have zeros in the divisor.

©Zoran Karapancev/Shutterstock

In multiplication, the answers are *larger* than the original number.

EXAMPLE If Toyota spends $60,000 for magazine advertising, what is the total value if it spends this same amount for 10 years? What would be the total cost?

$60,000 × 10 = **$600,000** (1 place to the right)

OTHER EXAMPLES

6.89 × 10 = **68.9** (1 place to the right)

6.89 × 100 = **689.** (2 places to the right)

6.89 × 1,000 = **6,890.** (3 places to the right)

In division, the answers are *smaller* than the original number.

EXAMPLES

6.89 ÷ 10 = **.689** (1 place to the left)

6.89 ÷ 100 = **.0689** (2 places to the left)

6.89 ÷ 1,000 = **.00689** (3 places to the left)

6.89 ÷ 10,000 = **.000689** (4 places to the left)

Next, let's dissect and solve a word problem.

How to Dissect and Solve a Word Problem

The Word Problem May O'Mally went to Sears to buy wall-to-wall carpet. She needs 101.3 square yards for downstairs, 16.3 square yards for the upstairs bedrooms, and 6.2 square yards for the halls. The carpet cost $14.55 per square yard. The padding cost $3.25 per square yard. Sears quoted an installation charge of $6.25 per square yard. What was May O'Mally's total cost?

By completing the following blueprint aid, we will slowly dissect this word problem. Note that before solving the problem, we gather the facts, identify what we are solving for, and list the steps that must be completed before finding the final answer, along with any key points we should remember. Let's go to it!

MONEY tips

Set up automatic payments for the minimum payment due on your debt and eliminate late fees. Create an alert with your smartphone or computer for each bill. Pay more than the minimum whenever possible.

BLUEPRINT	The facts	Solving for?	Steps to take	Key points
	Carpet needed: 101.3 sq. yd.; 16.3 sq. yd.; 6.2 sq. yd. *Costs:* Carpet, $14.55 per sq. yd.; padding, $3.25 per sq. yd.; installation, $6.25 per sq. yd.	Total cost of carpet, padding, and installation.	Total square yards × Cost per square yard = Total cost.	Align decimals. Round answer to nearest cent.

Steps to solving problem

1. Calculate the total number of square yards.

```
101.3
 16.3
  6.2
-----
123.8 square yards
```

2. Calculate the total cost per square yard.

```
$14.55
  3.25
  6.25
------
$24.05
```

3. Calculate the total cost of carpet, padding, and installation. 123.8 × $24.05 = **$2,977.39**

It's time to check your progress.

LU 3–2 PRACTICE QUIZ

Complete this **Practice Quiz** to see how you are doing.

1. Rearrange vertically and add:
14, .642, 9.34, 15.87321

2. Rearrange and subtract:
28.1549 − .885

3. Multiply and round the answer to the nearest tenth:
28.53 × 17.4

4. Divide and round to the nearest hundredth:
2,182 ÷ 2.83

Complete by the shortcut method:

5. 14.28 × 100 **6.** 9,680 ÷ 1,000 **7.** 9,812 ÷ 10,000

8. Could you help Mel decide which product is the "better buy"?
Dog food A: $9.01 for 64 ounces **Dog food B:** $7.95 for 50 ounces

Round to the nearest cent as needed:

9. At Avis Rent-A-Car, the cost per day to rent a medium-size car is $39.99 plus 29 cents per mile. What will it cost to rent this car for 2 days if you drive 602.3 miles? Since the solution shows a completed blueprint, you might use a blueprint also.

10. A trip to Mexico cost 6,000 pesos. What would this be in U.S. dollars? Check your answer.

*For **extra help** from your authors–Sharon and Jeff–see the videos in Connect.*

✓ Solutions

1.
```
 14.00000
   .64200
  9.34000
 15.87321
 39.85521
```

2.
```
   7 101414
 28.1549
 −.8850
 27.2699
```

3.
```
    28.53
  × 17.4
   11 412
  199 71
  285 3
  496.422 = 496.4
```

4.
```
           771.024 = 771.02
 2.83)218200.000
      1981
       2010
       1981
         290
         283
          7 00
          5 66
          1 340
          1 132
```

5. 14.28 = 1,428 **6.** 9,680 = 9.680 **7.** 9,812 = .9812

8. **A:** $9.01 ÷ 64 = $.14 **B:** $7.95 ÷ 50 = $.16 Buy A.

9. Avis Rent-A-Car total rental charge:

BLUEPRINT

The facts	Solving for?	Steps to take	Key points
Cost per day, $39.99. 29 cents per mile. Drove 602.3 miles. 2-day rental.	Total rental charge.	Total cost for 2 days' rental + Total cost of driving = Total rental charge.	In multiplication, count the number of decimal places. Starting from right to left in the product, insert decimal in appropriate place. Round to nearest cent.

Steps to solving problem

1. Calculate total cost for 2 days' rental. $39.99 × 2 = $79.98
2. Calculate the total cost of driving. $.29 × 602.3 = $174.667 = $174.67
3. Calculate the total rental charge.
```
  $ 79.98
 +174.67
 $254.65
```

10. 6,000 × $.0560 = $336

Check $336 × 17.8414 = 5,994.71 pesos due to rounding

INTERACTIVE CHAPTER ORGANIZER

Topic/Procedure/Formula	Example	You try it*
Identifying place value $10, 1, \frac{1}{10}, \frac{1}{100}, \frac{1}{1,000}$, etc.	.439 in thousandths place value	**Identify place value** .8256
Rounding decimals **1.** Identify place value of digit you want to round. **2.** If digit to right of identified digit in Step 1 is 5 or more, increase identified digit by 1; if less than 5, do not change identified digit. **3.** Drop all digits to right of identified digit.	.875 rounded to nearest tenth = .9 ↑ Identified digit	**Round to nearest tenth** .841
Converting decimal fractions to decimals **1.** Decimal fraction has a denominator with multiples of 10. Count number of zeros in denominator. **2.** Zeros show how many places are in the decimal.	$\frac{8}{1,000} = .008$ $\frac{6}{10,000} = .0006$	**Convert to decimal** $\frac{9}{1,000}$ $\frac{3}{10,000}$
Converting proper fractions to decimals **1.** Divide numerator of fraction by its denominator. **2.** Round as necessary.	$\frac{1}{3}$ (to nearest tenth) = .3	**Convert to decimal (to nearest tenth)** $\frac{1}{7}$
Converting mixed numbers to decimals **1.** Convert fractional part of the mixed number to a decimal. **2.** Add converted fractional part to whole number.	$6\frac{1}{4}$ $\frac{1}{4} = .25 + 6 = 6.25$	**Convert to decimal** $5\frac{4}{5}$
Converting pure and mixed decimals to decimal fractions **1.** Place digits to right of decimal point in numerator of fraction. **2.** Put 1 in denominator. **3.** Add zeros to denominator, depending on decimal places of original number. For mixed decimals, add fraction to whole number.	.984 (3 places) **1.** $\frac{984}{}$ **2.** $\frac{984}{1}$ **3.** $\frac{984}{1,000}$	**Convert to fraction** .865
Adding and subtracting decimals **1.** Vertically write and align numbers on decimal points. **2.** Add or subtract digits, starting with right column and moving to the left. **3.** Align decimal point in answer with above decimal points.	Add 1.3 + 2 + .4 $\begin{array}{r} 1.3 \\ 2.0 \\ .4 \\ \hline 3.7 \end{array}$ Subtract 5 − 3.9 $\begin{array}{r} \overset{4}{\cancel{5}}.\overset{10}{\cancel{0}} \\ -3.9 \\ \hline 1.1 \end{array}$	**Add** 1.7 + 3 + .8 **Subtract** 6 − 4.1
Multiplying decimals **1.** Multiply numbers, ignoring decimal points. **2.** Count and total number of decimal places in multiplier and multiplicand. **3.** Starting at right in the product, count to the left the number of decimal places totaled in Step 2. Insert decimal point. If number of places greater than space in answer, add zeros.	$\begin{array}{r} 2.48 \\ \times\ .018 \\ \hline 1984 \\ 248 \\ \hline .04464 \end{array}$ (2 places) (3 places)	**Multiply** $\begin{array}{r} 3.49 \\ \times .015 \\ \hline \end{array}$

(continues)

INTERACTIVE CHAPTER ORGANIZER

Topic/Procedure/Formula	Example	You try it*
Dividing a decimal by a whole number 1. Place decimal point in quotient directly above the decimal point in dividend. 2. Divide as usual.	1.1 42)46.2 42 42 42	**Divide (to nearest tenth)** 33)49.5
Dividing if the divisor is a decimal 1. Make divisor a whole number by moving decimal point to the right. 2. Move decimal point in dividend to the right the same number of places as in Step 1. 3. Place decimal point in quotient above decimal point in dividend. Divide as usual.	14.3 2.9)41.39 29 123 116 79 58 21	**Divide (to nearest tenth)** 3.2)1.48
Shortcuts on multiplication and division of decimals When multiplying by 10, 100, 1,000, and so on, move decimal point in multiplicand the same number of places to the right as you have zeros in multiplier. For division, move decimal point to the left.	$4.85 \times 100 = 485$ $4.85 \div 100 = .0485$	**Multiply by shortcut** 6.92×100 **Divide by shortcut** $6.92 \div 100$

KEY TERMS			
	Decimals Decimal fraction Decimal point	Mixed decimal Pure decimal Repeating decimal	Rounding decimals

Note: For how to dissect and solve a word problem, see Learning Unit 3-2.

*Worked-out solutions are in Appendix B.

Critical Thinking Discussion Questions with Chapter Concept Check

1. What are the steps for rounding decimals? Federal income tax forms allow the taxpayer to round each amount to the nearest dollar. Do you agree with this?
2. Explain how to convert fractions to decimals. If 1 out of 20 people buys a Land Rover, how could you write an advertisement in decimals?
3. Explain why .07, .70, and .700 are not equal. Assume you take a family trip to Disney World that covers 500 miles. Show that $\frac{8}{10}$ of the trip, or .8 of the trip, represents 400 miles.
4. Explain the steps in the addition or subtraction of decimals. Visit a car dealership and find the difference between two sticker prices. Be sure to check each sticker price for accuracy. Should you always pay the sticker price?
5. **Chapter Concept Check.** Visit a publisher's website and calculate the difference between the prices for a printed text and an ebook. Estimate what you think the profit is to the publisher based on your research.

3

END-OF-CHAPTER PROBLEMS

connect

Check figures for odd-numbered problems in Appendix B. Name ____________ Date ________

DRILL PROBLEMS

Identify the place value for the following: *LU 3-1(1)*

3–1. 7.5328 (arrow under 5) **3–2.** 229.448 (arrow under 4)

Round the following as indicated: *LU 3-1(1)*

		Tenth	Hundredth	Thousandth
3–3.	.7391			
3–4.	6.8629			
3–5.	5.8312			
3–6.	6.8415			
3–7.	6.5555			
3–8.	75.9913			

Round the following to the nearest cent: *LU 3-1(1)*

3–9. $4,822.775 **3–10.** $4,892.046

Convert the following types of decimal fractions to decimals (round to nearest hundredth as needed): *LU 3-1(2)*

3–11. $\frac{8}{100}$ **3–12.** $\frac{3}{10}$ **3–13.** $\frac{61}{1,000}$ **3–14.** $\frac{610}{1,000}$

3–15. $\frac{91}{100}$ **3–16.** $\frac{979}{1,000}$ **3–17.** $16\frac{61}{100}$

Convert the following decimals to fractions. Do not reduce to lowest terms. *LU 3-1(2)*

3–18. .9 **3–19.** .71 **3–20.** .009 **3–21.** .0125

3–22. .609 **3–23.** .825 **3–24.** .9999 **3–25.** .7065

Convert the following to mixed numbers. Do not reduce to the lowest terms. *LU 3-1(2)*

3–26. 7.1 **3–27.** 28.48 **3–28.** 6.025

Write the decimal equivalent of the following: *LU 3-1(2)*

3–29. Five thousandths **3–30.** Three hundred three and two hundredths

3–31. Eighty-five ten thousandths **3–32.** Seven hundred seventy-five thousandths

Rearrange the following and add: *LU 3-2(1)*

3–33. .115, 10.8318, 4.7, 802.4811 **3–34.** .005, 2,002.181, 795.41, 14.0, .184

Rearrange the following and subtract: *LU 3-2(1)*

3–35. 9.2 – 5.8 **3–36.** 7 – 2.0815 **3–37.** 3.4 – 1.08

Estimate by rounding all the way and multiply the following (do not round final answer): *LU 3-2(1)*

3–38. 6.24 × 3.9 =
Estimate

3–39. .413 × 3.07 =
Estimate

3–40. 675 × 1.92 =
Estimate

3–41. 4.9 × .825 =
Estimate

Divide the following and round to the nearest hundredth: *LU 3-2(1)*

3–42. $.8931 \div 3$

3–43. $29.432 \div .0012$

3–44. $.0065 \div .07$

3–45. $7{,}742.1 \div 48$

3–46. $8.95 \div 1.81$

3–47. $2{,}600 \div 381$

Convert the following to decimals and round to the nearest hundredth: *LU 3-1(2)*

3–48. $\frac{1}{8}$

3–49. $\frac{1}{25}$

3–50. $\frac{5}{6}$

3–51. $\frac{5}{8}$

Complete these multiplications and divisions by the shortcut method (do not do any written calculations): *LU 3-2(3)*

3–52. $96.7 \div 10$

3–53. $258 \div 100$

3–54. $8.51 \times 1{,}000$

3–55. $.86 \div 100$

3–56. 9.015×100

3–57. 48.6×10

3–58. 750×10

3–59. $3{,}950 \div 1{,}000$

3–60. $8.45 \div 10$

3–61. $7.9132 \times 1{,}000$

WORD PROBLEMS

As needed, round answers to the nearest cent.

3–62. A Chevy Volt costs $30,000 in the United States. Using the exchange rate given in the WSJ currency table on page 78, what would it cost in Canada? Check your answer. *LU 3-2(2)*

3–63. Dustin Pedroia got 7 hits out of 12 at bats. What was his batting average to the nearest thousandths place? *LU 3-1(2)*

3–64. Pete Ross read in a *Wall Street Journal* article that the cost of parts and labor to make an Apple iPhone 4S were as follows: *LU 3-2(1)*

Display	$37.00	Wireless	$23.54
Memory	$28.30	Camera	$17.60
Labor	$ 8.00	Additional items	$81.56

Assuming Pete pays $649 for an iPhone 4S, how much profit does the iPhone generate?

3–65. At the Party Store, JoAnn Greenwood purchased 21.50 yards of ribbon. Each yard cost 91 cents. What was the total cost of the ribbon? Round to the nearest cent. *LU 3-2(1)*

3–66. Douglas Noel went to Home Depot and bought four doors at $42.99 each and six bags of fertilizer at $8.99 per bag. What was the total cost to Douglas? If Douglas had $300 in his pocket, what does he have left to spend? *LU 3-2(1)*

3–67. The stock of Intel has a high of $48.50 today. It closed at $47.75. How much did the stock drop from its high? *LU 3-2(1)*

My Money

3–68. If you net $14.25 per hour and work 40 hours a week, 4 weeks per month, and have monthly expenses of: rent $825.50, car payment $458.79, utilities $110, food $150, gas $105, phone $125.25, savings $225, and insurance $118.36, what do you have left to invest for your retirement? *LU 3-2(1)*

3–69. Mark Ogara rented a truck from Avis Rent-A-Car for the weekend (2 days). The base rental price was $29.95 per day plus $14\frac{1}{2}$ cents per mile. Mark drove 410.85 miles. How much does Mark owe? *LU 3-2(1)*

My Money

3–70. Nursing home costs are on the rise as consumeraffairs.com reports in its quarterly newsletter. The average cost is around $192 a day with an average length of stay of 2.5 years. Calculate the cost of the average nursing home stay. *LU 3-2(1)*

3–71. Bob Ross bought a smartphone on the web for $89.99. He saw the same smartphone in the mall for $118.99. How much did Bob save by buying on the web? *LU 3-2(1)*

3–72. Russell is preparing the daily bank deposit for his coffee shop. Before the deposit, the coffee shop had a checking account balance of $3,185.66. The deposit contains the following checks:

No. 1	$ 99.50	No. 3	$8.75
No. 2	110.35	No. 4	6.83

Russell included $820.55 in currency with the deposit. What is the coffee shop's new balance, assuming Russell writes no new checks? *LU 3-2(1)*

eXcel

3–73. The United Nations claims India will overtake China as the world's most populous country within seven years. If China has 1.436 billion people and India has 1.345 billion people, what is the difference in population? *LU 3-2(1)*

3–74. Randi went to Lowe's to buy wall-to-wall carpeting. She needs 110.8 square yards for downstairs, 31.8 square yards for the halls, and 161.9 square yards for the bedrooms upstairs. Randi chose a shag carpet that costs $14.99 per square yard. She ordered foam padding at $3.10 per square yard. The carpet installers quoted Randi a labor charge of $3.75 per square yard. What will the total job cost Randi? *LU 3-2(1)*

3–75. Paul Rey bought four new Dunlop tires at Goodyear for $95.99 per tire. Goodyear charged $3.05 per tire for mounting, $2.95 per tire for valve stems, and $3.80 per tire for balancing. If Paul paid no sales tax, what was his total cost for the four tires? *LU 3-2(1)*

eXcel

3–76. Shelly is shopping for laundry detergent, mustard, and canned tuna. She is trying to decide which of two products is the better buy. Using the following information, can you help Shelly? *LU 3-2(1)*

Laundry detergent A	**Mustard A**	**Canned tuna A**
$2.00 for 37 ounces	$.88 for 6 ounces	$1.09 for 6 ounces
Laundry detergent B	**Mustard B**	**Canned tuna B**
$2.37 for 38 ounces	$1.61 for $12\frac{1}{2}$ ounces	$1.29 for $8\frac{3}{4}$ ounces

eXcel

3–77. Roger bought season tickets for weekend professional basketball games. The cost was $945.60. The season package included 36 home games. What is the average price of the tickets per game? Round to the nearest cent. Marcelo, Roger's friend, offered to buy four of the tickets from Roger. What is the total amount Roger should receive? *LU 3-2(1)*

3–78. A nurse was to give each of her patients a 1.32-unit dosage of a prescribed drug. The total remaining units of the drug at the hospital pharmacy were 53.12. The nurse has 38 patients. Will there be enough dosages for all her patients? *LU 3-2(1)*

3–79. Jill Horn went to Japan and bought an animation cel of Spongebob. The price was 25,000 yen. Using the WSJ currency table on page 78, what is the price in U.S. dollars? Check your answer. *LU 3-2(2)*

3–80. Bitcoin, the first and by far the largest cryptocurrency, is a digital currency created by an unknown person in 2009. Nasdaq will begin trading bitcoin futures on its commodities trading platform in 2018. If there were 16.7 million coins in circulation valued at $189 billion as of the first week of December 2017, what is the value of each coin? Round to the nearest hundredth. *LU 3-2(1)*

My Money

3–81. BloombergMarkets.com reported U.S. consumer confidence surged to its highest level since December 2000. If 36.3 people surveyed out of 100 felt "jobs are plentiful" in 2017 compared to 34.5 people out of 100 in 2001, what was the change in confidence levels? *LU 3-2(1)*

3–82. Morris Katz bought four new tires at Goodyear for $95.49 per tire. Goodyear also charged Morris $2.50 per tire for mounting, $2.40 per tire for valve stems, and $3.95 per tire for balancing. Assume no tax. What was Morris's total cost for the four tires? *LU 3-2(1)*

3–83. The *Denver Post* reported that Xcel Energy is revising customer charges for monthly residential electric bills and gas bills. Electric bills will increase $3.32. Gas bills will decrease $1.74 a month. **(a)** What is the resulting new monthly increase for the entire bill? **(b)** If Xcel serves 2,350 homes, how much additional revenue will Xcel receive each month? *LU 3-2(1)*

My Money

3–84. If your car gets 28 miles per gallon and you travel 30 miles round-trip to work five days a week, how much do you pay each 4-week month if gas is $3.05 a gallon? Round each calculation to the hundredth. *LU 3-2(1)*

3–85. Gracie went to Home Depot to buy wall-to-wall carpeting for her house. She needs 104.8 square yards for downstairs, 17.4 square yards for halls, and 165.8 square yards for the upstairs bedrooms. Gracie chose a shag carpet that costs $13.95 per square yard. She ordered foam padding at $2.75 per square yard. The installers quoted Gracie a labor cost of $5.75 per square yard in installation. What will the total job cost Gracie? *LU 3-2(1)*

CHALLENGE PROBLEMES

eXcel

3–86. Fred and Winnie O'Callahan have put themselves on a very strict budget. Their goal at the end of the year is to buy a car for $14,000 in cash. Their budget includes the following per dollar:

$.40 food and lodging
.20 entertainment
.10 educational

Fred earns $2,000 per month and Winnie earns $2,500 per month. After 1 year will Fred and Winnie have enough cash to buy the car? *LU 3-2(1)*

3–87. Jill and Frank decided to take a long weekend in New York. City Hotel has a special getaway weekend for $79.95. The price is per person per night, based on double occupancy. The hotel has a minimum two-night stay. For this price, Jill and Frank will receive $50 credit toward their dinners at City's Skylight Restaurant. Also included in the package is a $3.99 credit per person toward breakfast for two each morning.

Since Jill and Frank do not own a car, they plan to rent a car. The car rental agency charges $19.95 a day with an additional charge of $.22 a mile and $1.19 per gallon of gas used. The gas tank holds 24 gallons.

From the following facts, calculate the total expenses of Jill and Frank (round all answers to nearest hundredth or cent as appropriate). Assume no taxes. *LU 3-2(1)*

Car rental (2 days):		Dinner cost at Skylight	$182.12
Beginning odometer reading	4,820	Breakfast for two:	
Ending odometer reading	4,940	Morning No. 1	24.17
Beginning gas tank: $\frac{3}{4}$ full		Morning No. 2	26.88
Gas tank on return: $\frac{1}{2}$ full		Hotel room	79.95
Tank holds 24 gallons			

SUMMARY PRACTICE TEST

Do you need help? Connect videos have step-by-step worked-out solutions.

1. Add the following by translating the verbal form to the decimal equivalent. *LU 3-1(1), LU 3-2(1)*

   Three hundred thirty-eight and seven hundred five thousandths
   Nineteen and fifty-nine hundredths
   Five and four thousandths
   Seventy-five hundredths
   Four hundred three and eight tenths

Convert the following decimal fractions to decimals. *LU 3-1(2)*

2. $\frac{7}{10}$ 3. $\frac{7}{100}$ 4. $\frac{7}{1,000}$

Convert the following to proper fractions or mixed numbers. Do not reduce to the lowest terms. *LU 3-1(2)*

5. .9 6. 6.97 7. .685

Convert the following fractions to decimals (or mixed decimals) and round to the nearest hundredth as needed. *LU 3-1(2)*

8. $\frac{2}{7}$ 9. $\frac{1}{8}$ 10. $4\frac{4}{7}$ 11. $\frac{1}{13}$

12. Rearrange the following decimals and add. *LU 3-2(1)*

    5.93, 11.862, 284.0382, 88.44

13. Subtract the following and round to the nearest tenth. *LU 3-2(1)*

    13.111 − 3.872

14. Multiply the following and round to the nearest hundredth. *LU 3-2(1)*

    7.4821 × 15.861

15. Divide the following and round to the nearest hundredth. *LU 3-2(1)*

    203,942 ÷ 5.88

Complete the following by the shortcut method. *LU 3-2(3)*

16. 62.94 × 1,000
17. 8,322,249.821 × 100
18. The average pay of employees is $795.88 per week. Lee earns $820.44 per week. How much is Lee's pay over the average? *LU 3-2(1)*
19. Lowes reimburses Ron $.49 per mile. Ron submitted a travel log for a total of 1,910.81 miles. How much will Lowes reimburse Ron? Round to the nearest cent. *LU 3-2(1)*
20. Lee Chin bought two new car tires from Michelin for $182.11 per tire. Michelin also charged Lee $3.99 per tire for mounting, $2.50 per tire for valve stems, and $4.10 per tire for balancing. What is Lee's final bill? *LU 3-2(1)*
21. Could you help Judy decide which of the following products is cheaper per ounce? *LU 3-2(1)*

Canned fruit A	**Canned fruit B**
$.37 for 3 ounces	$.58 for $3\frac{3}{4}$ ounces

22. Paula Smith bought a computer tablet for 400 shekels in Israel. Using the WSJ currency table on page 78, what is this price in U.S. dollars? *LU 3-2(2)*
23. Google stock traded at a high of $522.00 and closed at $518.55. How much did the stock fall from its high? *LU 3-2(1)*

INTERACTIVE VIDEO WORKSHEET

GRADING THE SUMMARY PRACTICE TEST

Go to the summary practice test video in Connect (or click on it here in the ebook). Grade your summary practice test while viewing the video.

C for Correct/I for Incorrect

1. ________	7. ________	13. ________	19. ________
2. ________	8. ________	14. ________	20. ________
3. ________	9. ________	15. ________	21. ________
4. ________	10. ________	16. ________	22. ________
5. ________	11. ________	17. ________	23. ________
6. ________	12. ________	18. ________	

If you achieved 100%, you are ready for your instructor's exam.

If any of the problems were incorrect, list the questions you missed and show steps to solve the problem correctly.

Replay the video to see if you have made the correct fixes to your mistakes. If you have any questions, contact your instructor asap.

Notes on Watching Videos

MY MONEY

Interviewing for Success!

What I need to know

Interviews can be intimidating and cause anxiety on the part of the interviewee. The best course of action to reduce this anxiety is to prepare yourself for the interview long before you step foot into the interviewer's room. The interview is an opportunity for both the interviewer and interviewee to determine whether there is a "good fit." The actual interview itself is conducted within a relatively short amount of time so making sure you are well prepared will allow you to make a good first impression. You want to show the potential employer that you are the best candidate for the job and match nicely to the type of candidate they are seeking. Keep in mind that you are most likely interviewing within a pool of other candidates and you want to be remembered as the employer decides whom to hire.

What I need to do

The old adage "practice makes perfect" really does hold some credence as it relates to interviewing. The more exposure you have to the interview environment, the more comfortable and effective you will become. Take part in mock interviews that are offered by your institution. Many of these mock interview events utilize local employers who can provide vital feedback on your interview skills. You should also enlist the help of your friends and family to either conduct a mock interview for you or put you into contact with their employers who may be able to offer a practice interview at their firm. The feedback you receive from this practice will be invaluable for you to improve your skills and perform at your best when you begin to interview for real.

Another key interviewing issue for college students is the lack of experience students have with the new career they are about to enter. As a student studying for a professional career, you most likely have worked in jobs that are not directly related to this new career path. Therefore, you need to find a way to transfer the experiences you have gained to the job you desire. For example, if you worked in a retail environment during college you had many interactions with customers and these skills can transfer to a new job in which customer service is a key requirement. By identifying your transferable skills, you can show an employer that you understand the job requirements and can meet these expectations to benefit the organization.

An interview typically comprises many questions to assess the ability of the candidate to perform the required tasks. It is vital, as the interviewee, that you have questions to ask of the employer. These questions allow you the opportunity to determine whether this employer is a good fit for you. Asking questions also lets your interviewer know that you are serious about the position and have a genuine interest in the job and the organization.

Resources I can use

- Glassdoor Job Search (mobile app)
- https://www.thebalance.com/top-job-interview-questions-2061228 — top interview questions and answers

MY MONEY ACTIVITY

- Select 10 questions from the top interview questions link above.
- Find a partner and take turns interviewing each other and providing feedback.

PERSONAL FINANCE

A KIPLINGER APPROACH

STICKER SHOCK

CHECK ALL LIVING COSTS BEFORE MOVING

Home prices are just one factor you should consider when choosing a city.

IF YOU'RE A RETIREE LOOKING for a change of scenery, or you've been offered a job in another city, you may want to consider more than just the cost of housing. Prices for groceries, utilities, health care and other necessities can also vary significantly.

For example, the average cost of a doctor's appointment in Rockford, Ill., is $139, compared with $75 in Bloomington, Ind., according to the Council for Community and Economic Research's cost of living index, or COLI. The general rule of thumb is that you should spend no more than 50% of your budget on necessities such as housing, food and transportation. But relocating to a high-cost area may make this guideline unworkable.

The table below shows what you would pay for commonly used products and services in four cities that made the list of 20 finalists for Amazon's second headquarters—which is expected to add 50,000 jobs. We added Orlando because that area is attractive to retirees.

You can find a COLI calculator at www.payscale.com/cost-of-living-calculator. **RIVAN STINSON**

A SAMPLING OF COSTS IN FIVE CITIES

	LOS ANGELES	ORLANDO	INDIANAPOLIS	PHILADELPHIA	DALLAS
STEAK	$12.10	$9.24	$12.44	$11.82	$10.44
DENTIST	$106.80	$79.99	$91.22	$96.14	$96.42
PRESCRIPTION DRUGS	$410.11	$449.88	$443.70	$387.28	$428.97
BEER	$9.88	$10.34	$8.62	$8.63	$8.73
DRY CLEANING	$13.15	$10.60	$13.20	$12.97	$10.03
COMPOSITE INDEX	147.8	95.3	92.4	117.0	101.9

FOR 2017. COMPOSITE INDEX NATIONAL AVERAGE: 100. SOURCE: THE COUNCIL FOR COMMUNITY AND ECONOMIC RESEARCH

©Daniel Dempster Photography/Alamy Stock Photo

PAY TO PLAY

POPULAR PARKS GET PRICIER

The National Park Service plans to more than double the peak-season entrance fees at 17 of its most popular parks, including Grand Canyon, Sequoia and Yosemite, as early as May 1. The proposal, which is expected to be approved by the NPS, will hike fees from about $25 or $30 per carload to $70 during the park's busiest five months.

Only 118 of the country's 417 national park sites charge an entrance fee, so most will still be free to enter. If you frequent national parks that charge admission, consider buying an annual pass ($80 at federal recreation sites, $85 at Store.usgs.gov/pass). Park visitors who are 62 or older can purchase a lifetime pass for $80 in person or $90 online or by mail. Current members of the military and their dependents can receive a free annual pass by showing a military ID at a federal recreation site that issues passes.

If you visit parks with admission fees only occasionally, look for fee-free days. The National Park Service waives entrance fees several times a year at parks that charge for admission, but be prepared for crowds. To learn more, visit NPS.gov/planyourvisit. **KAITLIN PITSKER**

BUSINESS MATH ISSUE

Amazon should pick the city with the lowest costs for its new second headquarters.

1. List the key points of the article and information to support your position.
2. Write a group defense of your position using math calculations to support your view. If you are in an online course, post to a discussion board.

CUMULATIVE REVIEW

A Word Problem Approach—Chapters 1, 2, 3

1. The top rate at the Waldorf Towers Hotel in New York is $754. The top rate at the Ritz Carlton in Boston is $730. If John spends 9 days at one of these hotels, how much can he save if he stays at the Ritz? *LU 1-2(2), LU 1-3(1)*

2. Robert Half Placement Agency was rated best by 4 to 1 in an independent national survey. If 250,000 responded to the survey, how many rated Robert Half the best? *LU 2-3(1)*

3. Of the 63.2 million people who watch professional football, only $\frac{1}{5}$ watch the commercials. How many viewers do not watch the commercials? *LU 2-3(1)*

4. AT&T advertised a 500-minute prepaid domestic calling card for $25. Diamante sells a 500-minute prepaid domestic calling card for $12.25. Assuming Bill Splat needs two 500-minute cards, how much could he save by buying Diamante's? *LU 3-2(1)*

5. A square foot of rental space in New York City, Boston, and Providence costs as follows: New York City, $6.25; Boston, $5.75; and Providence, $3.75. If Hewlett Packard wants to rent 112,500 square feet of space, what will Hewlett Packard save by renting in Providence rather than Boston? *LU 3-2(1)*

6. American Airlines has a frequent-flier program. Coupon brokers who buy and sell these awards pay between 1 and $1\frac{1}{2}$ cents for each mile earned. Fred Dietrich earned a 50,000-mile award (worth two free tickets to any city). If Fred decided to sell his award to a coupon broker, approximately how much would he receive? *LU 3-2(1)*

7. Lillie Wong bought four new Firestone tires at $82.99 each. Firestone also charged $2.80 per tire for mounting, $1.95 per tire for valves, and $3.15 per tire for balancing. Lillie turned her four old tires in to Firestone, which charged $1.50 per tire to dispose of them. What was Lillie's final bill? *LU 3-2(1)*

8. Tootsie Roll Industries bought Charms Company for $65 million. Some analysts believe that in 4 years the purchase price could rise to three times as much. If the analysts are right, how much did Tootsie Roll save by purchasing Charms immediately? *LU 1-3(1)*

9. Today the average business traveler will spend $47.73 a day on food. The breakdown is dinner, $22.26; lunch, $10.73; breakfast, $6.53; tips, $6.23; and tax, $1.98. If Clarence Donato, an executive for Kroger, spends only .33 of the average, what is Clarence's total cost for food for the day? If Clarence wanted to spend $\frac{1}{3}$ more than the average on the next day, what would be his total cost on the second day? Round to the nearest cent. *LU 2-3(1), LU 3-2(1)*

Be sure you use the fractional equivalent in calculating $.3\overline{3}$.

CHAPTER 4

Banking

ATMs Try to Stop Wave of Cyberthieft

By Robin Sidel

The ATM is the newest front in the war against cyberthieves.

A year after millions of U.S. merchants began installing equipment at the checkout line to accept credit and debit cards with security chips, the automated teller machine is getting similar technology.

The move comes as thieves increasingly target ATMs. While chip-enabled credit cards are expected to slow growth in fraud at the checkout counter, the number of ATMs compromised by criminals jumped more than sixfold from 2014, according to a recent report from FICO, a credit-score provider and analytics firm. FICO says the number of 2015 compromises was the highest it ever recorded, though it declined to disclose specific numbers.

The burst of ATM-related fraud also was the largest one-year increase since FICO started keeping track of such data about a dozen years ago. Meanwhile, rates of credit-card fraud, a more popular scheme for criminals, have largely leveled off.

©Michael Spring/Alamy Stock Photo

LEARNING UNIT OBJECTIVES

LU 4–1: The Checking Account

1. Define and state the purpose of signature cards, checks, deposit slips, check stubs, check registers, and endorsements.
2. Correctly prepare deposit slips and write checks.

LU 4–2: Bank Statement and Reconciliation Process; Latest Trends in Mobile Banking

1. Explain trends in the banking industry.
2. Define and state the purpose of the bank statement.
3. Complete a check register and a bank reconciliation.
4. Explain the trends in mobile banking.

Your Guide to Successfully Completing This Chapter

Traditional book or ebook

Check box as you complete each step.

Steps

- ☐ Read learning unit.
    - ☐ Complete practice quiz at the end of the learning unit.
- ☐ Grade practice quiz using provided solutions. (For more help, watch the learning unit video in Connect and have a Study Session with the authors. Then complete the additional practice quiz in Connect.)
- ☐ Repeat above for each of the two learning units in Chapter 4.
    - ☐ Review chapter organizer.
    - ☐ Complete assigned homework.
        - ☐ Finish summary practice test. (Go to Connect via the ebook link and do the interactive video worksheet to grade.)
- ☐ Complete instructor's exam.

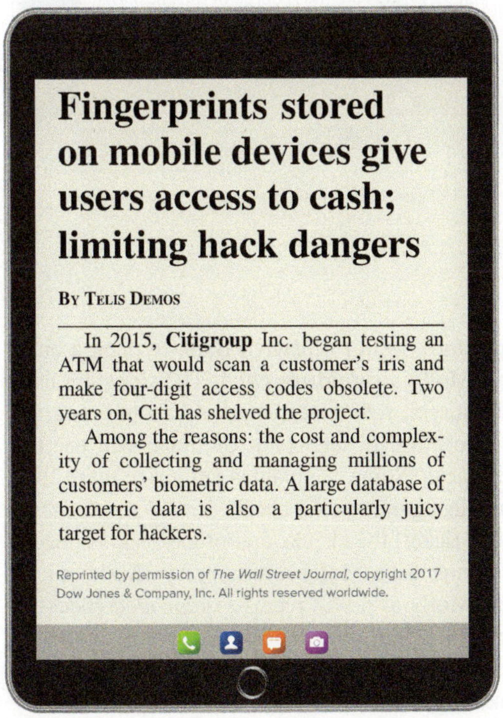

Fingerprints stored on mobile devices give users access to cash; limiting hack dangers

By Telis Demos

In 2015, **Citigroup** Inc. began testing an ATM that would scan a customer's iris and make four-digit access codes obsolete. Two years on, Citi has shelved the project.

Among the reasons: the cost and complexity of collecting and managing millions of customers' biometric data. A large database of biometric data is also a particularly juicy target for hackers.

The *Wall Street Journal* clip "ATMs Try to Stop Wave of Cybertheft" in the chapter opener shows how banks are trying to combat an increase in fraud. To the left, the *Wall Street Journal* clip "Fingerprints stored on mobile devices give users access to cash; limiting hack dangers" shows how banks are testing the latest technology to combat hackers.

The new tax act of 2017 will reap many benefits for banks because they will be paying lower tax rates. If you are a bank customer you will not be able to deduct on your tax return any fees you pay for a safe deposit box. Please check with the IRS for the latest details. Now let's look at a big concern for banks—cybertheft.

In this chapter we will look at how to do banking transactions manually, followed by a look at the latest trends in banking.

An important fixture in today's banking is the **automatic teller machine (ATM).** The ability to get instant cash is a convenience many bank customers enjoy.

The effect of using an ATM card is the same as using a **debit card**—both transactions result in money being immediately deducted from your checking account balance. As a result, debit cards have been called enhanced ATM cards or *check cards.* Often banks charge fees for these card transactions. The frequent complaints of bank customers have made many banks offer their ATMs as a free service, especially if customers use an ATM in the same network as their bank. Some banks charge fees for using another bank's ATM.

Remember that the use of debit cards involves planning. As *check cards,* you must be aware of your bank balance every time you use a debit card. Also, if you use a credit card instead of a debit card, you can only be held responsible for $50 of illegal charges; and during the time the credit card company investigates the illegal charges, they are removed from your account. However, with a debit card, this legal limit only applies if you report your card lost or stolen within two business days.

This chapter begins with a discussion of the checking account. You will follow Molly Kate as she opens a checking account for Gracie's Natural Superstore and performs her banking transactions. Pay special attention to the procedure used by Gracie's to reconcile its checking account and bank statement. This information will help you reconcile your checkbook records with the bank's record of your account. The chapter concludes by discussing the latest technology trends in banking.

Learning Unit 4–1: The Checking Account

LO 1

With an increase in mobile banking some have predicted the end of paper checks. Note in the *Wall Street Journal* clip "Go Ahead, Write a Check for Coffee, I've Got All Day" more than 17 billion checks are written each year. In LU 4–2, we will look at how to pay with smartphones. A **check** or **draft** is a written order instructing a bank, credit union, or savings and loan institution to pay a designated amount of your money on deposit to a person or an organization. Checking accounts are offered to individuals and businesses. Note that the business checking account usually receives more services than the personal checking account but may come with additional fees.

Most small businesses depend on a checking account for efficient record keeping. In this learning unit you will follow the checking account procedures of a newly organized small business. You can use many of these procedures in your personal check writing. You will also learn about e-checks.

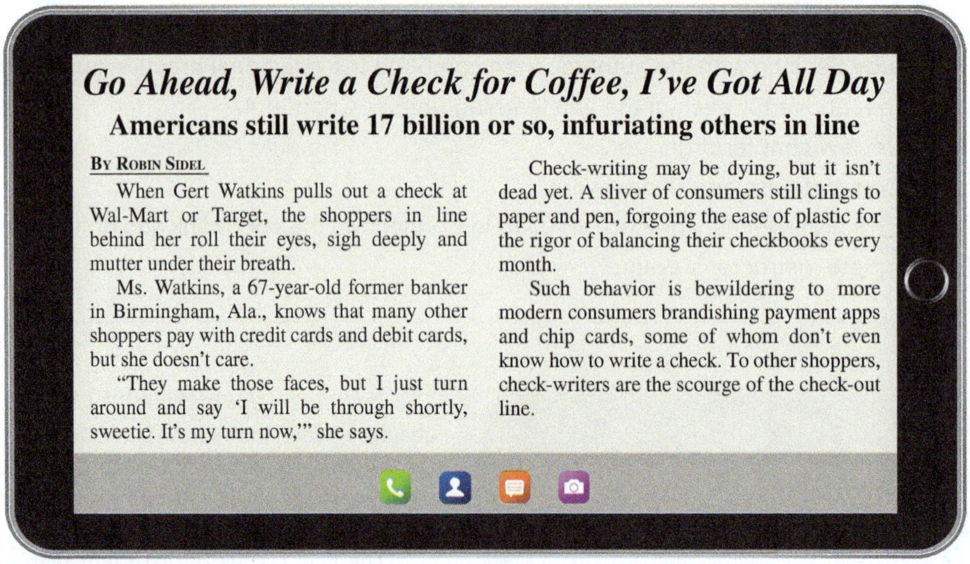

Go Ahead, Write a Check for Coffee, I've Got All Day

Americans still write 17 billion or so, infuriating others in line

By Robin Sidel

When Gert Watkins pulls out a check at Wal-Mart or Target, the shoppers in line behind her roll their eyes, sigh deeply and mutter under their breath.

Ms. Watkins, a 67-year-old former banker in Birmingham, Ala., knows that many other shoppers pay with credit cards and debit cards, but she doesn't care.

"They make those faces, but I just turn around and say 'I will be through shortly, sweetie. It's my turn now,'" she says.

Check-writing may be dying, but it isn't dead yet. A sliver of consumers still clings to paper and pen, forgoing the ease of plastic for the rigor of balancing their checkbooks every month.

Such behavior is bewildering to more modern consumers brandishing payment apps and chip cards, some of whom don't even know how to write a check. To other shoppers, check-writers are the scourge of the check-out line.

Opening the Checking Account

Molly Kate, treasurer of Gracie's Natural Superstore, went to Ipswich Bank to open a business checking account. The bank manager gave Molly a **signature card.** The signature card contained space for the company's name and address, references, type of account, and the signature(s) of the person(s) authorized to sign checks. If necessary, the bank will use the signature card to verify that Molly signed the checks. Some companies authorize more than one person to sign checks or require more than one signature on a check.

Molly then lists on a **deposit slip** (or deposit ticket) the checks and/or cash she is depositing in her company's business account. The bank gave Molly a temporary checkbook to use until the company's printed checks arrived. Molly also will receive *preprinted* checking account deposit slips like the one shown in Figure 4.1. Since the deposit slips are in duplicate, Molly can keep a record of her deposit. Note that the increased use of ATM machines has made it more convenient for people to make their deposits.

FIGURE 4.1 Deposit slip

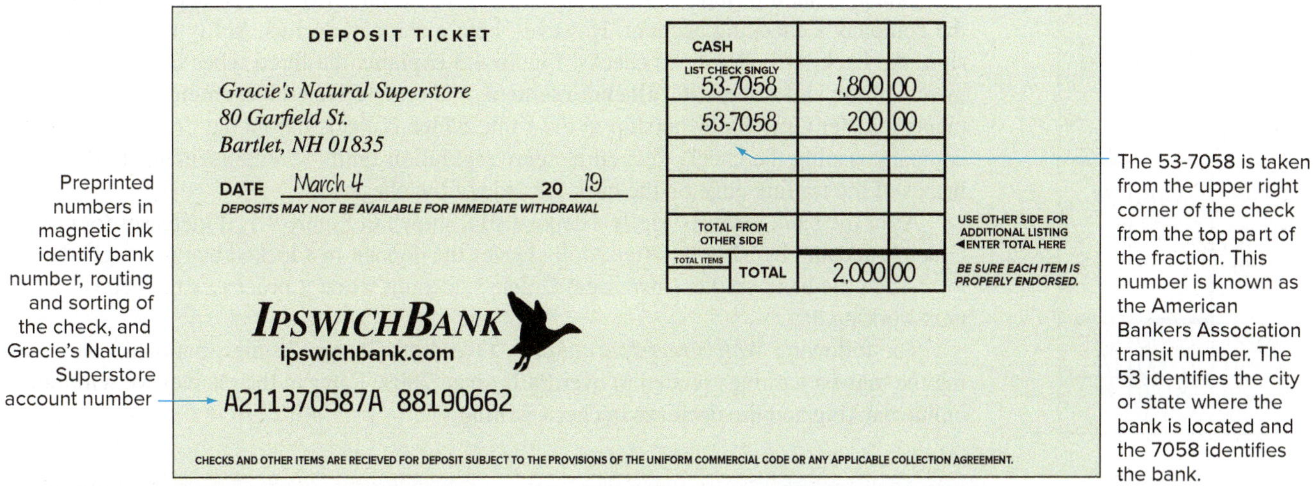

FIGURE 4.2 The structure of a check

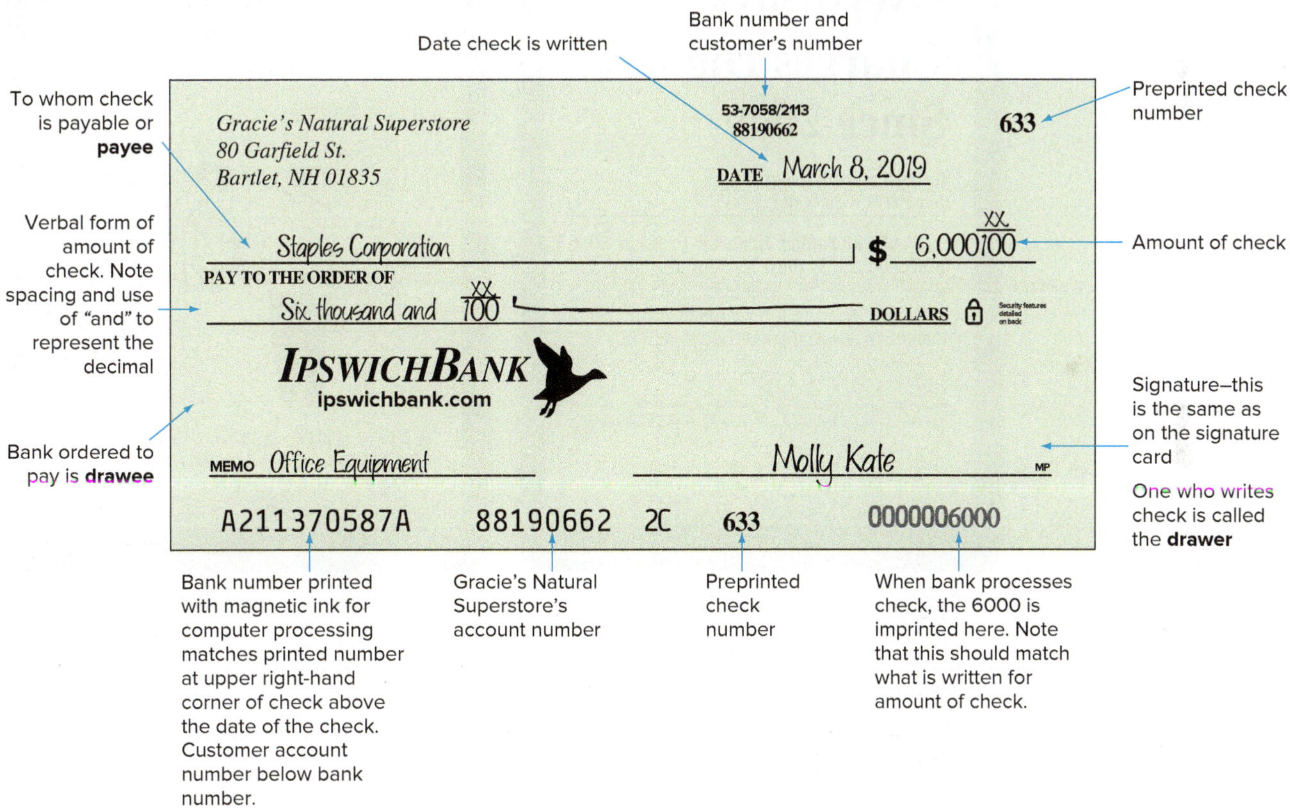

Writing business checks is similar to writing personal checks. Before writing any checks, however, you must understand the structure of a check and know how to write a check. Carefully study Figure 4.2. Note that the verbal amount written in the check should match the figure amount. If these two amounts are different, by law the bank uses the verbal amount. Also, note the bank imprint on the bottom right section of the check. When processing the check, the bank imprints the check's amount. This makes it easy to detect bank errors.

LO 2

Using the Checking Account

Once the check is written, the writer must keep a record of the check. Knowing the amount of your written checks and the amount in the bank should help you avoid writing a bad check. Business checkbooks usually include attached **check stubs** to keep track of written checks. The sample check stub in the margin on page 98 shows the information that the check writer will want to record. Some companies use a **check register** to keep their check records instead of check

Check Stub

It should be completed before the check is written.

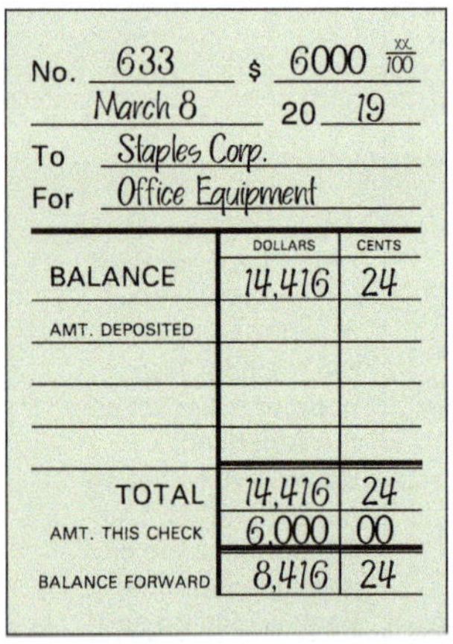

No. 633 $ 6000 xx/100

March 8 20 19

To Staples Corp.

For Office Equipment

	DOLLARS	CENTS
BALANCE	14,416	24
AMT. DEPOSITED		
TOTAL	14,416	24
AMT. THIS CHECK	6,000	00
BALANCE FORWARD	8,416	24

stubs. Figure 4.6 later in the chapter shows a check register with a ✓ column that is often used in balancing the checkbook with the bank statement (Learning Unit 4–2).

Gracie's Natural Superstore has had a busy week, and Molly must deposit its checks in the company's checking account. However, before she can do this, Molly must **endorse,** or sign, the back left side of the checks. Figure 4.3 explains the three types of check endorsements: **blank endorsement, full endorsement,** and **restrictive endorsement.** These endorsements transfer Gracie's ownership to the bank, which collects the money from the person or company issuing the check. Federal Reserve regulation limits all endorsements to the top $1\frac{1}{2}$ inches of the trailing edge on the back left side of the check.

After the bank receives Molly's deposit slip, shown in Figure 4.1, it increases (or credits) Gracie's account by $2,000. Often Molly leaves the deposit in a locked bag in a night depository. Then the bank credits (increases) Gracie's account when it processes the deposit on the next working day.

The following *Wall Street Journal* clip "Overdraft Charges" states that banks were having the highest earnings related to overdrafts since 2009. Later in the chapter we will look at online banking and the decrease in check writing.

MONEY tips

Conduct an annual check of your bank's interest rates and fees. You may find better rates and lower fees at a credit union.

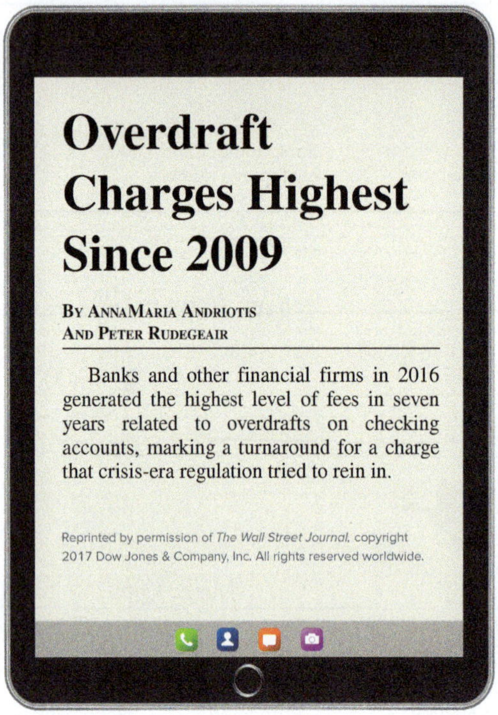

Overdraft Charges Highest Since 2009

BY ANNAMARIA ANDRIOTIS AND PETER RUDEGEAIR

Banks and other financial firms in 2016 generated the highest level of fees in seven years related to overdrafts on checking accounts, marking a turnaround for a charge that crisis-era regulation tried to rein in.

Used by permission of Cartoon Features Syndicate

FIGURE 4.3

Types of common endorsements

A. Blank Endorsement

Gracie's Natural Superstore
88190662

→ The company stamp or a signature alone on the back left side of a check legally makes the check payable to anyone holding the check. It can be *further* endorsed. This is not a safe type of endorsement.

B. Full Endorsement

Pay to the order of
Ipswich Bank
Gracie's Natural Superstore
88190662

→ Safer type of endorsement since Gracie's Natural Superstore indicates the name of the company or person to whom the check is to be payable. Only the person or company named in the endorsement can transfer the check to someone else.

C. Restrictive Endorsement

Pay to the order of
Ipswich Bank
For deposit only
Gracie's Natural Superstore
88190662

→ Safest endorsement for businesses. Gracie's stamps the back of the check so that this check must be deposited in the firm's bank account. This limits any further negotiation of the check.

Let's check your understanding of this unit.

LU 4–1 PRACTICE QUIZ

Complete this Practice Quiz to see how you are doing.

Complete the following check and check stub for Long Company. Note the $9,500.60 balance brought forward on check stub No. 113. You must make a $690.60 deposit on May 3. Sign the check for Roland Small.

Date	Check no.	Amount	Payable to	For
June 5, 2019	113	$83.76	Angel Corporation	Rent

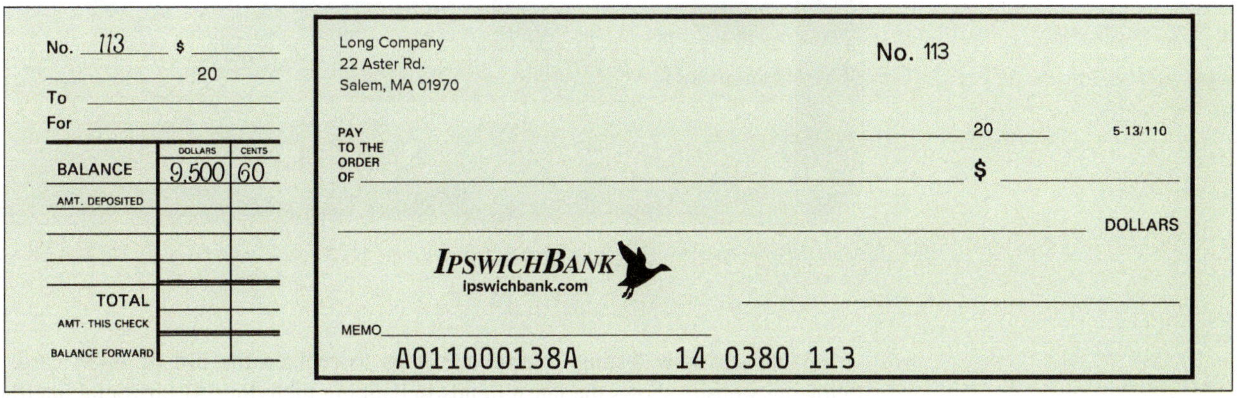

*For **extra help** from your authors–Sharon and Jeff–see the videos in Connect.*

✓ Solution

LO 1

Learning Unit 4–2: Bank Statement and Reconciliation Process; Latest Trends in Mobile Banking

Trends in Banking Industry

Today more and more people are using smartphone apps from lenders to do their banking transactions. In the *Wall Street Journal* clip on page 100, "More Ways to Use Smartphones," we see the latest trend in **mobile banking.**

Aha!

Be vigilant in safeguarding business and personal information to bank safely electronically. The rest of this learning unit is divided into two sections: (1) bank statement and reconciliation process, and (2) latest trends in mobile banking. The bank statement discussion will teach you why it was important for Gracie's Natural Superstore to reconcile its checkbook balance with the balance reported on its bank statement. Note that you can also use this reconciliation process in reconciling your personal checking account to avoid the expensive error of an overdrawn account.

Banks are now testing smartphone apps to replace the use of ATM cards. Note the following website shows the latest trends in banking including AI (artificial intelligence).

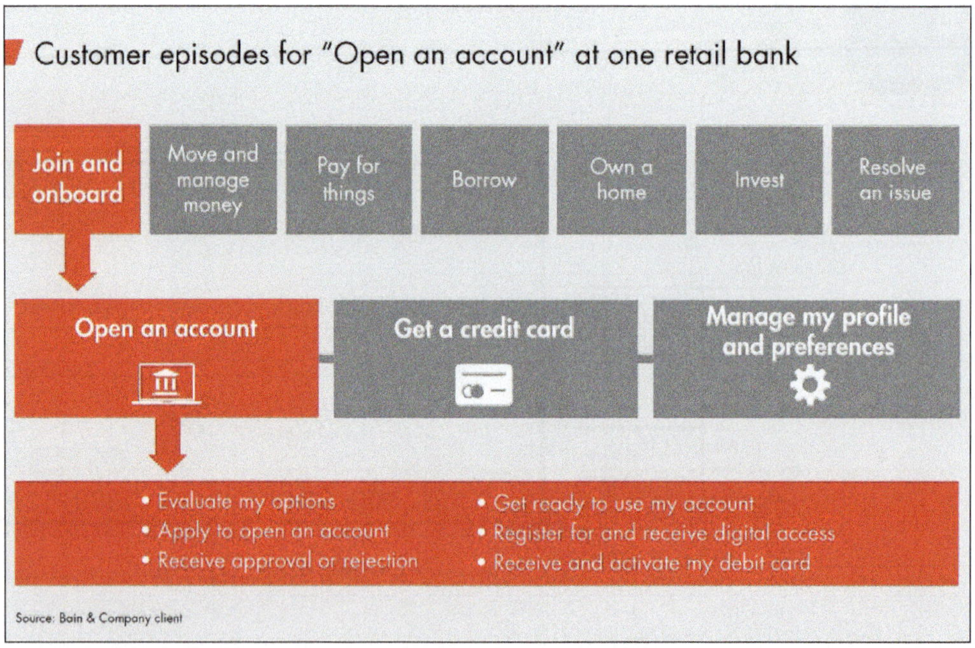

Used with permission from Bain & Company. http://www.bain.com/publications/articles/evolving-the-customer-experience-in-banking.aspx

LO 2

Bank Statement and Reconciliation Process

Each month, Ipswich Bank sends Gracie's Natural Superstore a **bank statement** (Figure 4.4). We are interested in the following:

1. Beginning bank balance.
2. Total of all the account increases. Each time the bank increases the account amount, it *credits* the account.
3. Total of all account decreases. Each time the bank decreases the account amount, it *debits* the account.
4. Final ending balance.

FIGURE 4.4

Bank statement

Online Banking-Windows Internet Explorer

https//www.onlinebanking.com

File Edit View Favorites Tools Help

Ipswich Online Banking

Privacy | Security | Sign off

Last Sign On: Thursday, March 12, 2019 at 11: 32 a.m

Ipswich Online Banking

My Accounts | Transfer Funds

Account Statement

Ipswich Bank
1 Pleasant St.
Bartlett, NH 01835

Gracie's Natural Superstore
80 Garfield St.
Bartlett, NH 01835

Checking Account: 881900662

Checking Account Summary as of 3/31/19

Beginning Balance	Total Deposits	Total Withdrawals	Service Charge	Ending Balance
$13,112.24	**$8,705.28**	**$9,926.00**	**$28.50**	**$11,863.02**

Checking Account Transactions

Deposits	Date	Amount
Deposit	3/05	2,000.00
Deposit	3/05	224.00
Deposit	3/09	389.20
EFT leasing: Bakery dept.	3/18	1,808.06
EFT leasing: Meat dept.	3/27	4,228.00
Interest	3/31	56.02

Charges	Date	Amount
Service charge: Check printing	3/31	28.50
EFT: Health insurance	3/21	722.00
NSF	3/21	104.00

Checks

Number	Date	Amount
301	3/07	200.00
633	3/13	6,000.00
634	3/13	300.00
635	3/11	200.00
636	3/18	200.00
637	3/31	2,200.00

Daily Balance

Date	Balance	Date	Balance
2/28	13,112.24	3/18	10,529.50
3/05	15,232.24	3/21	9,807.50
3/07	14,832.24	3/28	14,035.50
3/09	15,221.44	3/31	11,863.02
3/11	15,021.44		
3/13	8,721.44		

FIGURE 4.5

Reconciling checkbook with bank statement

Checkbook balance		Bank balance
+ EFT (electronic funds transfer)	− NSF check	+ Deposits in transit
+ Interest earned	− Online fees	− Outstanding checks
+ Notes collected^	− Automatic payments*	± Bank errors
+ Credit memo		
+ Direct deposits	− Overdrafts†	
− ATM withdrawals	− Service charges	
− Automatic withdrawals	− Stop payments‡	
− Debit memo	± Book errors§	

^Notes (money) collected by a financial institution on behalf of a customer.

*Preauthorized payments for utility bills, mortgage payments, insurance, etc.

†**Overdrafts** occur when the customer has no overdraft protection and a check bounces back to the company or person who received the check because the customer has written a check without enough money in the bank to pay for it.

‡A stop payment is issued when the writer of the check does not want the receiver to cash the check.

§If a $60 check is recorded at $50, the checkbook balance must be decreased by $10.

Due to differences in timing, the bank balance on the bank statement frequently does not match the customer's checkbook balance. Also, the bank statement can show transactions that have not been entered in the customer's checkbook. Figure 4.5 tells you what to look for when comparing a checkbook balance with a bank balance.

Gracie's Natural Superstore is planning to offer to its employees the option of depositing their checks directly into each employee's checking account. This is accomplished through the **electronic funds transfer (EFT)**—a computerized operation that electronically transfers funds among parties without the use of paper checks. Gracie's, which sublets space in the store, receives rental payments by EFT. Gracie's also has the bank pay the store's health insurance premiums by EFT.

LO 3

My Money

To reconcile the difference between the amount on the bank statement and in the checkbook, the customer should complete a **bank reconciliation.** Today, many companies and home computer owners are using software such as Quicken and QuickBooks to complete their bank reconciliation. Also, we have mentioned the increased use of **banking apps** available to customers. However, you should understand the following steps for manually reconciling a bank statement.

RECONCILING A BANK STATEMENT

Step 1. Identify the outstanding checks (checks written but not yet processed by the bank). You can use the ✓ column in the check register (Figure 4.6) to check the canceled checks listed in the bank statement against the checks you wrote in the check register. The unchecked checks are the outstanding checks.

Step 2. Identify the deposits in transit (deposits made but not yet processed by the bank), using the same method in Step 1.

Step 3. Analyze the bank statement for transactions not recorded in the check stubs or check registers (like EFT).

Step 4. Check for recording errors in checks written, in deposits made, or in subtraction and addition.

Step 5. Compare the adjusted balances of the checkbook and the bank statement. If the balances are not the same, repeat Steps 1–4.

Molly uses a check register (Figure 4.6) to keep a record of Gracie's checks and deposits. By looking at Gracie's check register, you can see how to complete Steps 1 and 2 above. The explanation that follows for the first four bank statement reconciliation steps will help you understand the procedure.

Step 1. Identify Outstanding Checks **Outstanding checks** are checks that Gracie's Natural Superstore has written but Ipswich Bank has not yet recorded for payment when it sends out the bank statement. Gracie's treasurer identifies the following checks written on 3/31 as outstanding:

No. 638	$572.00
No. 639	638.94
No. 640	166.00
No. 641	406.28
No. 642	917.06

Step 2. Identify Deposits in Transit **Deposits in transit** are deposits that did not reach Ipswich Bank by the time the bank prepared the bank statement. The March 30 deposit of $3,383.26 did not reach Ipswich Bank by the bank statement date. You can see this by comparing the company's bank statement with its check register.

Step 3. Analyze Bank Statement for Transactions Not Recorded in Check Stubs or Check Register The bank statement of Gracie's Natural Superstore (Figure 4.4) begins with the deposits, or increases, made to Gracie's bank account. Increases to accounts are

FIGURE 4.6

Gracie's Natural Superstore check register

NUMBER	DATE 2019	DESCRIPTION OF TRANSACTION	PAYMENT/DEBIT (−)	√	FEE (IF ANY) (−)	DEPOSIT/CREDIT (+)	BALANCE
		RECORD ALL CHARGES OR CREDITS THAT AFFECT YOUR ACCOUNT					$ 12,912 24
	3/04	Deposit	$		$	$ 2,000 00	+ 2,000 00
							14,912 24
	3/04	Deposit				224 00	+ 224 00
							15,136 24
633	3/08	Staples Company	6,000 00	✓			− 6,000 00
							9,136 24
634	3/09	Health Foods Inc.	1,020 00	✓			− 1,020 00
							8,116 24
	3/09	Deposit				389 20	+ 389 20
							8,505 44
635	3/10	Liberty Insurance	200 00	✓			− 200 00
							8,305 44
636	3/18	Ryan Press	200 00	✓			− 200 00
							8,105 44
637	3/29	Logan Advertising	2,200 00	✓			− 2,200 00
							5,905 44
	3/30	Deposit				3,383 26	+ 3,383 26
							9,288 70
638	3/31	Sears Roebuck	572 00				− 572 00
							8,716 70
639	3/31	Flynn Company	638 94				− 638 94
							8,077 76
640	3/31	Lynn's Farm	166 00				− 166 00
							7,911 76
641	3/31	Ron's Wholesale	406 28				− 406 28
							7,505 48
642	3/31	Grocery Natural, Inc.	917 06				− 917 06
							$6,588 42

REMEMBER TO RECORD AUTOMATIC PAYMENTS/DEPOSITS ON DATE AUTHORIZED.

MONEY tips

Always review your monthly bank statement to ensure there are no errors. The earlier you catch an error, the easier it is to remedy.

known as credits. These are the result of a **credit memo (CM).** Gracie's received the following increases or credits in March:

1. *EFT leasing:* $1,808.06 and $4,228.00. — Each month the bakery and meat departments pay for space they lease in the store.
2. *Interest credited:* $56.02. — Gracie's has a checking account that pays interest; the account has earned $56.02.

When Gracie's has charges against its bank account, the bank decreases, or debits, Gracie's account for these charges. Banks usually inform customers of a debit transaction by a **debit memo (DM).** The following items will result in debits to Gracie's account:

1. *Service charge:* $28.50. — The bank charged $28.50 for printing Gracie's checks.
2. *EFT payment:* $722. — The bank made a health insurance payment for Gracie's.
3. *NSF check:* $104. — One of Gracie's customers wrote Gracie's a check for $104. Gracie's deposited the check, but the check bounced for **nonsufficient funds (NSF).** Thus, Gracie's has $104 less than it figured.

Step 4. Check for Recording Errors The treasurer of Gracie's Natural Superstore, Molly Kate, recorded check No. 634 for the wrong amount—$1,020 (see the check register in Figure 4.6). The bank statement showed that check No. 634 cleared for $300. To reconcile Gracie's checkbook balance with the bank balance, Gracie's must add $720 to its checkbook balance. Neglecting to record a deposit also results in an error in the company's checkbook balance. As you can see, reconciling the bank's balance with a checkbook balance is a necessary part of business and personal finance.

Step 5. Completing the Bank Reconciliation Now we can complete the bank reconciliation on the back side of the bank statement as shown in Figure 4.7. This form is usually on the back of a bank statement. If necessary, however, the person reconciling the bank statement can construct a bank reconciliation form similar to Figure 4.8.

FIGURE 4.7

Reconciliation process

Used by permission of Cartoon Features Syndicate

CHECKS OUTSTANDING (NOT YET CHARGED TO ACCOUNT)		
NUMBER OR DATE	DOLLARS	CENTS
638	$572	00
639	638	94
640	166	00
641	406	28
642	917	06
TOTAL	$2,700	28

CHECKING ACCOUNT RECONCILEMENT	
Enter the new balance shown on the other side of this statement.	$ 11,863.02
ADD Deposits not shown	3,383.26
ADD Advances and Transfers to Checking not shown.	
SUBTOTAL	15,246.28
DEDUCT Checks outstanding	2,700.28
SUBTOTAL	12,546.00
DEDUCT Transfers from Checking not shown	
This amount should agree with the balance in your checkbook register.	$ 12,546.00

Checkbook balance		$6,588.42
+ EFT leasing	$6,036.06	
+ Interest	56.02	
+ Checkbook error	720.00	6,812.08
− Service charge	$ 28.50	
− EFT: health insurance	722.00	
− NSF	104.00	854.50
Ending checkbook balance		$12,546.00

FIGURE 4.8

Bank reconciliation

GRACIE'S NATURAL SUPERSTORE
Bank Reconciliation as of March 31, 2019

Checkbook balance			Bank balance		
Gracie's checkbook balance		$6,588.42	Bank balance		$11,863.02
Add:			Add:		
EFT leasing: Bakery dept.	$1,808.06		Deposit in transit, 3/30		3,383.26
					$15,246.28
EFT leasing: Meat dept.	4,228.00				
Interest	56.02				
Error: Overstated check No. 634	720.00	$ 6,812.08			
		$13,400.50			
Deduct:			Deduct:		
Service charge	$ 28.50		Outstanding checks:		
NSF check	104.00		No. 638	$572.00	
EFT health insurance payment	722.00	854.50	No. 639	638.94	
			No. 640	166.00	
			No. 641	406.28	
			No. 642	917.06	2,700.28
Reconciled balance		$12,546.00	Reconciled balance		$12,546.00

Trends in Mobile Banking

LO 4

The *Wall Street Journal* clip below shows four apps to help you with your saving and banking needs.

My Money

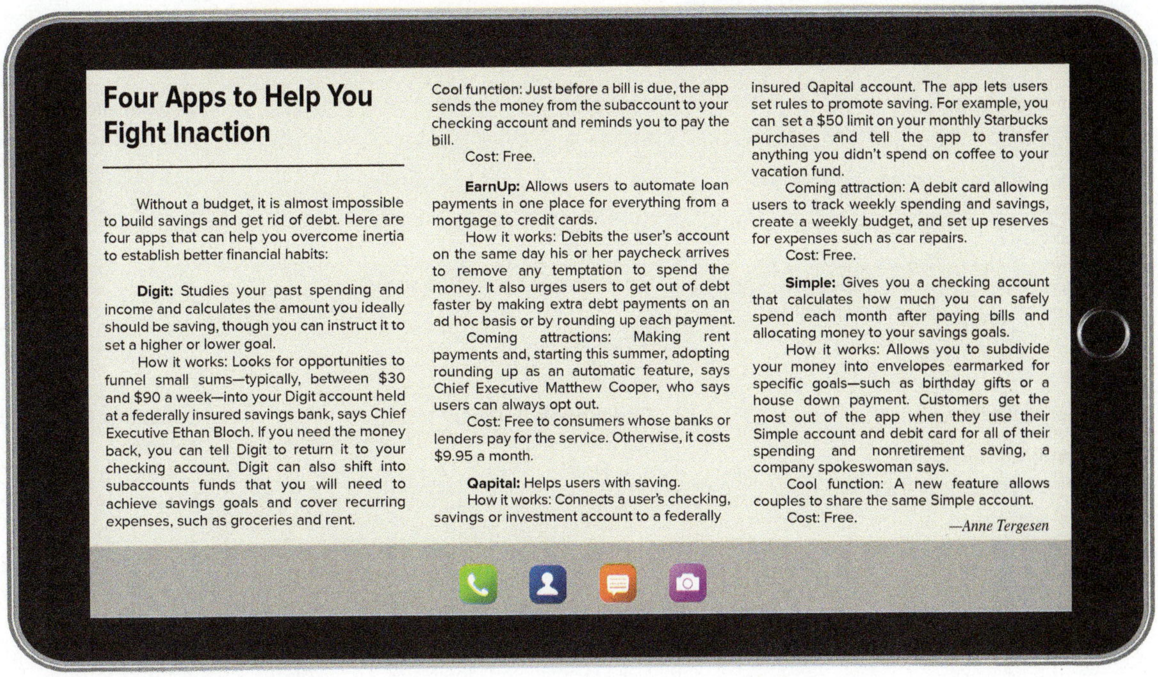

Four Apps to Help You Fight Inaction

Without a budget, it is almost impossible to build savings and get rid of debt. Here are four apps that can help you overcome inertia to establish better financial habits:

Digit: Studies your past spending and income and calculates the amount you ideally should be saving, though you can instruct it to set a higher or lower goal.

How it works: Looks for opportunities to funnel small sums—typically, between $30 and $90 a week—into your Digit account held at a federally insured savings bank, says Chief Executive Ethan Bloch. If you need the money back, you can tell Digit to return it to your checking account. Digit can also shift into subaccounts funds that you will need to achieve savings goals and cover recurring expenses, such as groceries and rent.

Cool function: Just before a bill is due, the app sends the money from the subaccount to your checking account and reminds you to pay the bill.

Cost: Free.

EarnUp: Allows users to automate loan payments in one place for everything from a mortgage to credit cards.

How it works: Debits the user's account on the same day his or her paycheck arrives to remove any temptation to spend the money. It also urges users to get out of debt faster by making extra debt payments on an ad hoc basis or by rounding up each payment.

Coming attractions: Making rent payments and, starting this summer, adopting rounding up as an automatic feature, says Chief Executive Matthew Cooper, who says users can always opt out.

Cost: Free to consumers whose banks or lenders pay for the service. Otherwise, it costs $9.95 a month.

Qapital: Helps users with saving.

How it works: Connects a user's checking, savings or investment account to a federally insured Qapital account. The app lets users set rules to promote saving. For example, you can set a $50 limit on your monthly Starbucks purchases and tell the app to transfer anything you didn't spend on coffee to your vacation fund.

Coming attraction: A debit card allowing users to track weekly spending and savings, create a weekly budget, and set up reserves for expenses such as car repairs.

Cost: Free.

Simple: Gives you a checking account that calculates how much you can safely spend each month after paying bills and allocating money to your savings goals.

How it works: Allows you to subdivide your money into envelopes earmarked for specific goals—such as birthday gifts or a house down payment. Customers get the most out of the app when they use their Simple account and debit card for all of their spending and nonretirement saving, a company spokeswoman says.

Cool function: A new feature allows couples to share the same Simple account.

Cost: Free.

—Anne Tergesen

LU 4–2 PRACTICE QUIZ

Complete this Practice Quiz to see how you are doing.

Rosa Garcia received her February 3, 2019, bank statement showing a balance of $212.80. Rosa's checkbook has a balance of $929.15. The bank statement showed that Rosa had an ATM fee of $12.00 and a deposited check returned fee of $20.00. Rosa earned interest of $1.05. She had three outstanding checks: No. 300, $18.20; No. 302, $38.40; and No. 303, $68.12. A deposit for $810.12 was not on her bank statement. Prepare Rosa Garcia's bank reconciliation.

*For **extra help** from your authors–Sharon and Jeff–see the videos in Connect.*

✓ Solution

ROSA GARCIA
Bank Reconciliation as of February 3, 2019

Checkbook balance			Bank balance		
Rosa's checkbook balance		$929.15	Bank balance		$ 212.80
Add:			Add:		
Interest		1.05	Deposit in transit		810.12
		$930.20			$1,022.92
Deduct:			Deduct:		
			Outstanding checks:		
Deposited check returned fee	$20.00		No. 300	$18.20	
ATM	12.00	32.00	No. 302	38.40	
			No. 303	68.12	124.72
Reconciled balance		$898.20	Reconciled balance		$ 898.20

INTERACTIVE CHAPTER ORGANIZER

Topic/Procedure/Formula	Examples	You try it*
Types of endorsements *Blank:* Not safe; can be further endorsed. *Full:* Only person or company named in endorsement can transfer check to someone else. *Restrictive:* Check must be deposited. Limits any further negotiation of the check.	Jones Co. 21-333-9 Pay to the order of Regan Bank Jones Co. 21-333-9 Pay to the order of Regan Bank. For deposit only. Jones Co. 21-333-9	**Write a sample of a blank, full, and restrictive endorsement.** Use Pete Co. Acct. # 24-111-9
Bank reconciliation **Checkbook balance** + EFT (electronic funds transfer) + Interest earned + Notes collected + Direct deposits − ATM withdrawals − NSF check − Online fees − Automatic withdrawals − Overdrafts − Service charges − Stop payments ± Book errors (see note, below) CM—adds to balance DM—deducts from balance **Bank balance** + Deposits in transit − Outstanding checks ± Bank errors	**Checkbook balance** Balance $800 − NSF 40 $760 − Service charge 4 **$756** **Bank balance** Balance $ 632 + Deposits in transit 416 $1,048 − Outstanding checks 292 **$ 756**	**Calculate ending checkbook balance** 1. Beg. checkbook bal.: $300 2. NSF: $50 3. Deposit in transit: $100 4. Outstanding check: $60 5. ATM service charge: $20

KEY TERMS

Automatic teller machine (ATM)	Debit card	Full endorsement
Bank reconciliation	Debit memo (DM)	Mobile banking
Bank statement	Deposit slip	Nonsufficient funds (NSF)
Banking apps	Deposits in transit	Outstanding checks
Blank endorsement	Draft	Overdrafts
Check	Drawee	Payee
Check register	Drawer	Restrictive endorsement
Check stub	Electronic funds transfer (EFT)	Signature card
Credit memo (CM)	Endorse	

Note: If a $60 check is recorded as $50, we must decrease checkbook balance by $10.

*Worked-out solutions are in Appendix B.

Critical Thinking Discussion Questions with Chapter Concept Check

1. Explain the structure of a check. The trend in bank statements is not to return the canceled checks. Do you think this is fair?
2. List the three types of endorsements. Endorsements are limited to the top $1\frac{1}{2}$ inches of the trailing edge on the back left side of your check. Why do you think the Federal Reserve made this regulation?
3. List the steps in reconciling a bank statement. Today, many banks charge a monthly fee for certain types of checking accounts. Do you think all checking accounts should be free? Please explain.
4. What are some of the trends in mobile banking? Will we become a cashless society in which all transactions are made with some type of credit card?
5. What do you think of the government's intervention in trying to bail out banks? Should banks be allowed to fail?
6. **Chapter Concept Check.** Create your own company and provide needed data to prepare a bank reconciliation. Then go to a bank website and explain how you would use the bank's app versus the manual system of banking.

END-OF-CHAPTER PROBLEMS

connect

Check figures for odd-numbered problems in Appendix B. Name ______________________ Date ________

DRILL PROBLEMS

4–1. Fill out the check register that follows with this information for August 2019: *LU 4-1(1)*

Aug 7	Check No. 959	AT&T	$143.50
15	Check No. 960	Staples	66.10
19	Deposit		800.00
20	Check No. 961	West Electric	451.88
24	Check No. 962	Bank of America	319.24
29	Deposit		400.30

RECORD ALL CHARGES OR CREDITS THAT AFFECT YOUR ACCOUNT

NUMBER	DATE 2019	DESCRIPTION OF TRANSACTION	PAYMENT/DEBIT (−)	√	FEE (IF ANY) (−)	DEPOSIT/CREDIT (+)	BALANCE
							$ 4,500 75
			$		$	$	

4–2. November 1, 2019, Payroll.com, an Internet company, has a $10,481.88 checkbook balance. Record the following transactions for Payroll.com by completing the two checks and check stubs provided. Sign the checks Garth Scholten, controller. *LU 4-1(2)*

a. November 8, 2019, deposited $688.10

b. November 8, check No. 190 payable to Staples for office supplies—$766.88

c. November 15, check No. 191 payable to Best Buy for computer equipment—$3,815.99.

No. ________ $ ________
________ 20 ____
To ________
For ________

	DOLLARS	CENTS
BALANCE		
AMT. DEPOSITED		
TOTAL		
AMT. THIS CHECK		
BALANCE FORWARD		

PAYROLL.COM
1 LEDGER RD.
ST. PAUL, MN 55113

No. 190

________ 20 ____ 5-13/110

PAY TO THE ORDER OF ________________ $ ________

________________ DOLLARS

IPSWICHBANK
ipswichbank.com

MEMO________________

A011000138A 25 11103 190

No. ________ $ ________
________ 20 ____
To ________
For ________

	DOLLARS	CENTS
BALANCE		
AMT. DEPOSITED		
TOTAL		
AMT. THIS CHECK		
BALANCE FORWARD		

PAYROLL.COM
1 LEDGER RD.
ST. PAUL, MN 55113

No. 191

________ 20 ____ 5-13/110

PAY TO THE ORDER OF ________________ $ ________

________________ DOLLARS

IPSWICHBANK
ipswichbank.com

MEMO________________

A011000138A 25 11103 191

4–3. Using the check register in Problem 4–1 and the following bank statement, prepare a bank reconciliation for Lee.com. *LU 4-2(3)*

BANK STATEMENT			
Date	**Checks**	**Deposits**	**Balance**
8/1 balance			$4,500.75
8/18	$143.50		4,357.25
8/19		$800.00	5,157.25
8/26	319.24		4,838.01
8/30	15.00 SC		4,823.01

WORD PROBLEMS

eXcel

4–4. The World Bank forecasts growth of world trade to be 2.9%, up from 2.7% in 2017. This change has caused Galapagos Islands Resort to analyze its current financial situation, beginning with reconciling its accounts. Galapagos Islands Resort received its bank statement showing a balance of $8,788. Its checkbook balance is $15,252. Deposits in transit are $3,450 and $6,521. There is a service charge of $45 and interest earned of $3. Notes collected total $1,575. Outstanding checks are No. 1021 for $1,260 and No. 1022 for $714. All numbers are in U.S. dollars. Help Galapagos Islands Resort reconcile its balances. *LU 4-2(3)*

My Money

4–5. The U.S. Chamber of Commerce provides a free monthly bank reconciliation template at business.uschamber.com/tools/bankre_m.asp. Riley Whitelaw just received her bank statement notice online. She wants to reconcile her checking account with her bank statement and has chosen to reconcile her accounts manually. Her checkbook shows a balance of $698. Her bank statement reflects a balance of $1,348. Checks outstanding are No. 2146, $25; No. 2148, $58; No. 2152, $198; and No. 2153, $464. Deposits in transit are $100 and $50. There is a $15 service charge and $5 ATM charge in addition to notes collected of $50 and $25. Reconcile Riley's balances. *LU 4-2(3)*

4–6. A local bank began charging $2.50 each month for returning canceled checks. The bank also has an $8.00 "maintenance" fee if a checking account slips below $750. Donna Sands likes to have copies of her canceled checks for preparing her income tax returns. She has received her bank statement with a balance of $535.85. Donna received $2.68 in interest and has been charged for the canceled checks and the maintenance fee. The following checks were outstanding: No. 94, $121.16; No. 96, $106.30; No. 98, $210.12; and No. 99, $64.84. A deposit of $765.69 was not recorded on Donna's bank statement. Her checkbook shows a balance of $806.94. Prepare Donna's bank reconciliation. *LU 4-2(3)*

4–7. Ben Luna received his bank statement with a $27.04 fee for a bounced check (NSF). He has an $815.75 monthly mortgage payment paid through his bank. There was also a $3.00 teller fee and a check printing fee of $3.50. His ATM card fee was $6.40. There was also a $530.50 deposit in transit. The bank shows a balance of $119.17. The bank paid Ben $1.23 in interest. Ben's checkbook shows a balance of $1,395.28. Check No. 234 for $80.30 and check No. 235 for $28.55 were outstanding. Prepare Ben's bank reconciliation. *LU 4-2(3)*

4–8. Kameron Gibson's bank statement showed a balance of $717.72. Kameron's checkbook had a balance of $209.50. Check No. 104 for $110.07 and check No. 105 for $15.55 were outstanding. A $620.50 deposit was not on the statement. He has his payroll check electronically deposited to his checking account—the payroll check was for $1,025.10. There was also a $4 teller fee and an $18 service charge. Prepare Kameron Gibson's bank reconciliation. *LU 4-2(3)*

My Money

4–9. Banks are finding more ways to charge fees, such as a $25 overdraft fee. Sue McVickers has an account in Fayetteville; she has received her bank statement with this $25 charge. Also, she was charged a $6.50 service fee; however, the good news is she earned $5.15 interest. Her bank statement's balance was $315.65, but it did not show the $1,215.15 deposit she had made. Sue's checkbook balance shows $604.30. The following checks have not cleared: No. 250, $603.15; No. 253, $218.90; and No. 254, $130.80. Prepare Sue's bank reconciliation. *LU 4-2(3)*

4–10. Carol Stokke receives her April 6 bank statement showing a balance of $859.75; her checkbook balance is $954.25. The bank statement shows an ATM charge of $25.00, NSF fee of $27.00, earned interest of $2.75, and Carol's $630.15 refund check, which was processed by the IRS and deposited to her account. Carol has two checks that have not cleared—No. 115 for $521.15 and No. 116 for $205.50. There is also a deposit in transit for $1,402.05. Prepare Carol's bank reconciliation. *LU 4-2(3)*

4–11. Lowell Bank reported the following checking account fees: $2 to see a real-live teller, $20 to process a bounced check, and $1 to $3 if you need an original check to prove you paid a bill or made a charitable contribution. This past month you had to transact business through a teller six times—a total $12 cost to you. Your bank statement shows a $305.33 balance; your checkbook shows a $1,009.76 balance. You received $1.10 in interest. An $801.15 deposit was not recorded on your statement. The following checks were outstanding: No. 413, $28.30; No. 414, $18.60; and No. 418, $60.72. Prepare your bank reconciliation. *LU 4-2(3)*

eXcel

4–12. According to the *Portland Business Journal,* Jim Houser, a Portland auto specialist, landed a key Small Business Administration appointment. Help Jim reconcile Remington's Auto Clinic's checkbook and bank balance according to the following: bank statement balance, $18,769; checkbook balance, $22,385,015; interest earned, $3,948; deposits in transit, $100,656 and $22,375,000; ATM card fees, $150; outstanding checks—No. 10189, $55,678; No. 10192, $15,287; No. 10193, $22,350; and No. 10194, $12,297. *LU 4-2(3)*

My Money

4–13. Identity Theft Resource Center (ITRC) provides consumer and victim support, public education, and advice. Marlena's grandmother is concerned her identity has been stolen. Help Marlena reconcile her grandmother's checkbook and bank statement. The checkbook reflects a balance of $1,245. The bank statement shows a balance of $207. Notes collected were $100 and $210. The bank charged a $25 service fee. Outstanding checks were No. 255, $985; No. 261, $233; and No. 262, $105. There is a deposit in transit of $2,646. *LU 4-2(2)*

CHALLENGE PROBLEMS

4–14. Carolyn Crosswell, who banks in New Jersey, wants to balance her checkbook, which shows a balance of $985.20. The bank shows a balance of $1,430.33. The following transactions occurred: $135.20 automatic withdrawal to the gas company, $6.50 ATM fee, $8.00 service fee, and $1,030.05 direct deposit from the IRS. Carolyn used her debit card five times and was charged 45 cents for each transaction; she was also charged $3.50 for check printing. A $931.08 deposit was not shown on her bank statement. The following checks were outstanding: No. 235, $158.20; No. 237, $184.13; No. 238, $118.12; and No. 239, $38.83. Carolyn received $2.33 interest. Prepare Carolyn's bank reconciliation. *LU 4-2(3)*

eXcel

4–15. Melissa Jackson, bookkeeper for Kinko Company, cannot prepare a bank reconciliation. From the following facts, can you help her complete the June 30, 2020, reconciliation? The bank statement showed a $2,955.82 balance. Melissa's checkbook showed a $3,301.82 balance. Melissa placed a $510.19 deposit in the bank's night depository on June 30. The deposit did not appear on the bank statement. The bank included two DMs and one CM with the returned checks: $690.65 DM for NSF check, $8.50 DM for service charges, and $400.00 CM (less $10 collection fee) for collecting a $400.00 non-interest-bearing note. Check No. 811 for $110.94 and check No. 912 for $82.50, both written and recorded on June 28, were not with the returned checks. The bookkeeper had correctly written check No. 884, $1,000, for a new cash register, but she recorded the check as $1,069. The May bank reconciliation showed check No. 748 for $210.90 and check No. 710 for $195.80 outstanding on April 30. The June bank statement included check No. 710 but not check No. 748. *LU 4-2(3)*

SUMMARY PRACTICE TEST

Do you need help? Connect videos have step-by-step worked-out solutions.

1. Walgreens has a $12,925.55 beginning checkbook balance. Record the following transactions in the check stubs provided. *LU 4-1(2)*
   a. November 4, 2019, check No. 180 payable to Ace Medical Corporation, $1,700.88 for drugs.
   b. $5,250 deposit—November 24.
   c. November 24, 2019, check No. 181 payable to John's Wholesale, $825.55 merchandise.

No. ______ $ ______		
______ 20 ______		
To ______		
For ______		
	DOLLARS	CENTS
BALANCE		
AMT. DEPOSITED		
TOTAL		
AMT. THIS CHECK		
BALANCE FORWARD		

No. ______ $ ______		
______ 20 ______		
To ______		
For ______		
	DOLLARS	CENTS
BALANCE		
AMT. DEPOSITED		
TOTAL		
AMT. THIS CHECK		
BALANCE FORWARD		

2. On April 1, 2019, Lester Company received a bank statement that showed a balance of $8,950. Lester showed an $8,000 checking account balance. The bank did not return check No. 115 for $750 or check No. 118 for $370. A $900 deposit made on March 31 was in transit. The bank charged Lester $20 for check printing and $250 for NSF checks. The bank also collected a $1,400 note for Lester. Lester forgot to record a $400 withdrawal at the ATM. Prepare a bank reconciliation. *LU 4-2(3)*

3. Felix Babic banks at Role Federal Bank. Today he received his March 31, 2019, bank statement showing a $762.80 balance. Felix's checkbook shows a balance of $799.80. The following checks have not cleared the bank: No. 140, $130.55; No. 149, $66.80; and No. 161, $102.90. Felix made an $820.15 deposit that is not shown on the bank statement. He has his $617.30 monthly mortgage payment paid through the bank. His $1,100.20 IRS refund check was mailed to his bank. Prepare Felix Babic's bank reconciliation. *LU 4-2(3)*

4. On June 30, 2019, Wally Company's bank statement showed a $7,500.10 bank balance. Wally has a beginning checkbook balance of $9,800.00. The bank statement also showed that it collected a $1,200.50 note for the company. A $4,500.10 June 30 deposit was in transit. Check No. 119 for $650.20 and check No. 130 for $381.50 are outstanding. Wally's bank charges $.40 cents per check. This month, 80 checks were processed. Prepare a reconciled statement. *LU 4-2(3)*

INTERACTIVE VIDEO WORKSHEET

GRADING THE SUMMARY PRACTICE TEST

Go to the summary practice test video in Connect (or click on it here in the ebook). Grade your summary practice test while viewing the video.

C for Correct/I for Incorrect

1. a. ______ 2. ______
   b. ______ 3. ______
   c. ______ 4. ______

If you achieved 100%, you are ready for your instructor's exam.

If any of the problems were incorrect, list the questions you missed and show steps to solve the problem correctly.

Replay the video to see if you have made the correct fixes to your mistakes.
If you have any questions, contact your instructor asap.

Notes on Watching Videos

MY MONEY

Breaking (down) the Bank!

What I need to know

Today there are so many choices when it comes to banking it is important to do your homework and find the one that best fits your needs. After you graduate college and move into your career, your earnings will undoubtedly take a significant positive turn. The bank that got you through your high school and college years may no longer be the best option for you going forward. It is in your best interest to take some time and consider how a bank can satisfy your needs and assist you in meeting your financial goals. Not all banks are created equally. Some will offer more incentives and financial services than others and some will even pay you interest on your money! Understanding the banking options available will help you find the best bank for you.

What I need to do

What should you consider when selecting your bank? It is important to consider the types of services you may be interested in when choosing where to bank. It is helpful to take a look at your financial goals and find a bank that will align with these goals. For instance, if you have plans to someday purchase a house, it may be important to select a bank that will be able to provide financing for home purchases. Additionally, you may want to consider whether mobile banking is of importance to you and exactly to what level you want to be able to access and transact business with your bank via a mobile device. Electronic bill pay is another feature that provides considerable convenience and ease of use in managing your expenses.

Banks come in all shapes and sizes. Which will provide you with the best fit? Maybe a small local bank is right for you if you desire to have a personal connection to your finances and help in meeting your financial goals. A credit union is established to benefit its members through low fees and low rates while emphasizing a personal touch. Smaller local banks will also strive to take a personal interest in you and your financial plans. On the other hand, large national banks have the benefit of accessibility with access to their physical locations and with ATMs that are available in many locations.

What if your banking needs change in the future? As your banking needs change over time, it never hurts to take a look into what other banks are providing compared to your current bank. The banking industry is a competitive one and you may be surprised at what you can find through a quick comparison. As these banks vie for your business, you may discover it beneficial to make a switch.

Resources I can use

- https://www.nerdwallet.com/blog/banking/how-to-choose-a-bank/ — tips for choosing a bank
- https://www.findabetterbank.com/ — interactive guide to finding your bank options

MY MONEY ACTIVITY

- Use https://www.findabetterbank.com/ and search for banks in your local area.
- Choose the features that are important to you in your bank choice.
- Compare results to the offerings of your current bank and determine if it is time to make a change.

A KIPLINGER APPROACH

BANKING

Mobile Deposit: Give It a Shot

MOBILE CHECK DEPOSIT IS catching on. In the past year, one in three Americans submitted a check to their bank account using a smartphone or other device.

If you have never tried mobile deposit, note that the process is usually fast and easy: Photograph the front and back of the check using the bank's mobile app, and choose the account where you want the money to go. For security, photographs of checks that you take through your bank's app are never stored on your device, says Michael Diamond, general manager of payments for Mitek, a maker of mobile-deposit software. To prevent a check from being deposited twice, write "deposited" on the back after you submit it.

Many banks cap the amount of money you may deposit. At Citibank, the limit is $500 daily for new customers and $1,000 daily after you've had the account for several months; the monthly limit is $3,000. Among 15 large banks, seven don't allow a mobile deposit of $5,000 or more—typically because of monthly or per-check limits—according to a report from Mitek and consultant Futurion.

Most large banks don't charge fees for mobile deposit, but U.S. Bank charges 50 cents per check and Regions Bank charges from 50 cents to $5 or more. If you need to, say, cover a payment that's due, find out when you must submit a check for it to be considered received the same day and how long you must wait until you can withdraw the deposited funds, recommends the Federal Deposit Insurance Corp.

Many banks suggest that you keep the check in a safe place for two weeks or so, in case it doesn't clear. (After that, shred it.) In the days after you deposit a check, monitor your account online or through the bank's app to ensure that the funds have been credited. LISA GERSTNER

Kiplinger.com

RATE UPDATES

For the latest savings yields and loan rates, visit kiplinger.com/finances/yields.

YIELD BENCHMARKS	Yield	Month-ago	Year-ago
U.S. Series EE savings bonds	0.10%	0.10%	0.10%
U.S. Series I savings bonds	2.76	0.26	1.64
Six-month Treasury bills	0.52	0.46	0.32
Five-year Treasury notes	1.29	1.26	1.73
Ten-year Treasury notes	1.83	1.73	2.34

SOURCES FOR TREASURIES: Bloomberg, U.S. Treasury.

As of November 7, 2016.
- EE savings bonds purchased after May 1, 2005, have a fixed rate of interest.
- Bonds bought between May 1, 1995, and May 1, 2005, earn a market-based rate from date of purchase.
- Bonds purchased before May 1, 1995, earn a minimum of 4% or a market-based rate from date of purchase.

TOP-YIELDING DEPOSIT ACCOUNTS

No-Fee Interest Checking (Minimum balance may be required)	Annual yield as of Nov. 7	For balances above	Website (www.)
GTE Financial (Fla.)#	2.49%	$0	gtefinancial.org
Langley Federal Credit Union (Va.)#	1.61	0	langleyfcu.org
EverBank (Fla.)*	1.11	0	everbank.com
Aspiration (Calif.)*	1.00	2,500	aspiration.com
NATIONAL AVERAGE	0.13%		

High-Yield Checking (Must meet activity requirements‡)	Annual yield as of Nov. 7	For balances up to†	Website (www.)
America's Credit Union (Wash.)#	5.00%	$1,000	youracu.org
Northpointe Bank (Mich.)	5.00	5,000	northpointe.com
Consumers Credit Union (Ill.)#	4.59	20,000	myconsumers.org
La Capitol FCU (La.)#	4.25	5,000	lacapfcu.org
NATIONAL AVERAGE	1.77%		

Savings	Annual yield as of Nov. 7	Min. deposit	Website (www.)
Popular Direct (Fla.)*	1.26%	$5,000	populardirect.com
Northpointe Bank (Mich.)	1.12	10,000	northpointe.com
Incredible Bank (Wis.)*	1.11	none	incrediblebank.com
Salem Five Direct (Mass.)*§	1.10	none	salemfivedirect.com
NATIONAL AVERAGE	0.18%		

*Internet only. ‡To earn the maximum rate, you must meet requirements such as using your debit card several times monthly and receiving electronic statements. †Portion of the balance higher than the maximum earns a lower rate or no interest. #Must be a member; to become a member, see website. §Palladian PrivateBank offers a similar yield. SOURCES: Bankrate, DepositAccounts.

TOP-YIELDING CERTIFICATES OF DEPOSIT

1-Year	Annual yield as of Nov. 7	Min. amount	Website (www.)
VirtualBank (Fla.)*	1.31%	$10,000	virtualbank.com
Connexus Credit Union (Wis.)†	1.30	10,000	connexuscu.org
Live Oak Bank (N.C.)*	1.30	2,500	liveoakbank.com
Pacific National Bank (Fla.)*	1.27	1,000	pnb.com
NATIONAL AVERAGE	0.49%		

5-Year	Annual yield as of Nov. 7	Min. amount	Website (www.)
Signal Financial FCU (Md.)†	2.16%	$0	signalfinancialfcu.org
Melrose Credit Union (N.Y.)†	2.12	5,000	melrosecu.org
Blue Federal Credit Union (Wyo.)†‡	2.10	2,000	bluefcu.com
RiverLand Fed Credit Union (La.)†‡	2.10	1,000	riverlandcu.org
NATIONAL AVERAGE	1.40%		

*Internet only. †Must be a member; to become a member, see website. ‡NuVision FCU offers a similar yield. SOURCES: Bankrate, DepositAccounts.

LOW-RATE CREDIT CARDS

Issuer	Rate as of Nov. 7*	Annual fee	Late fee	Website (www.)
Lake Mich Credit Union Prime (P)§	6.50%	none	$27†	lmcu.org
Simmons Bank Visa (P)	7.50	none	27†	simmonsfirst.com
First Command Bank Visa (P)	8.50	none	27†	firstcommandbank.com

RETAIL REBATE CREDIT CARDS

Issuer	Rate as of Nov. 7*	Annual fee	Rebate earned Store/Other	Web site (www.)
Amazon Rewards Visa	14.49%	none	3%/1%#	amazon.com/rewards
Sam's Club MasterCard	15.15	none§	1/1^	samsclub.com
Costco Anywhere Visa	15.49	none§	2/1∂	costco.com

Rates are adjustable. *If you do not qualify for this interest rate, the issuer will offer a higher-rate card. (P) Platinum. §Must be a member. †$37 if late more than once in 6 months. #2% at gas stations, restaurants, office supply stores and drugstores; 1% on all other purchases. ^5% on gas (up to $6,000 spent annually; 1% thereafter); 3% on travel and dining; 1% on all other purchases. ∂4% on gas (up to $7,000 spent annually; 1% thereafter); 3% on travel and dining; 2% on Costco purchases; 1% on all other purchases. SOURCE: Bankrate

BUSINESS MATH ISSUE

Mobile deposits are really a risky type of deposit.

1. List the key points of the article and information to support your position.
2. Write a group defense of your position using math calculations to support your view. If you are in an online course, post to a discussion board.

Classroom Notes

Solving for the Unknown: A How-to Approach for Solving Equations

Formula for Success

A logarithmic equation can determine exactly how long it takes for an investment to double in value, but the much simpler Rule of 72 can be used to make a good estimate.

	Years to double the investment			
Interest rate	Using rule of 72	Using logarithmic formula	Difference	
2%	**36.00**	35.00	1.00	
3	**24.00**	23.45	0.55	
4	**18.00**	17.67	0.33	
5	**14.40**	14.21	0.19	
6	**12.00**	11.90	0.10	
10	**7.20**	7.27	0.07	
13	**5.54**	5.67	0.13	
15	**4.80**	4.96	0.16	
20	**3.60**	3.80	0.20	

Source: WSJ calculations based on formulas

THE WALL STREET JOURNAL.

LEARNING UNIT OBJECTIVES

LU 5–1: Solving Equations for the Unknown

1. Explain the basic procedures used to solve equations for the unknown.
2. List the five rules and the mechanical steps used to solve for the unknown in seven situations; know how to check the answers.

LU 5–2: Solving Word Problems for the Unknown

1. List the steps for solving word problems.
2. Complete blueprint aids to solve word problems; check the solutions.

Your Guide to Successfully Completing This Chapter

Traditional book or ebook

Check box as you complete each step.

Steps

☐ Read learning unit.

  ☐ Complete practice quiz at the end of the learning unit.

☐ Grade practice quiz using provided solutions. (For more help, watch the learning unit video in Connect and have a Study Session with the authors. Then complete the additional practice quiz in Connect.)

☐ Repeat above for each of the two learning units in Chapter 5.

  ☐ Review chapter organizer.

  ☐ Complete assigned homework.

    ☐ Finish summary practice test. (Go to Connect via the ebook link and do the interactive video worksheet to grade.)

☐ Complete instructor's exam.

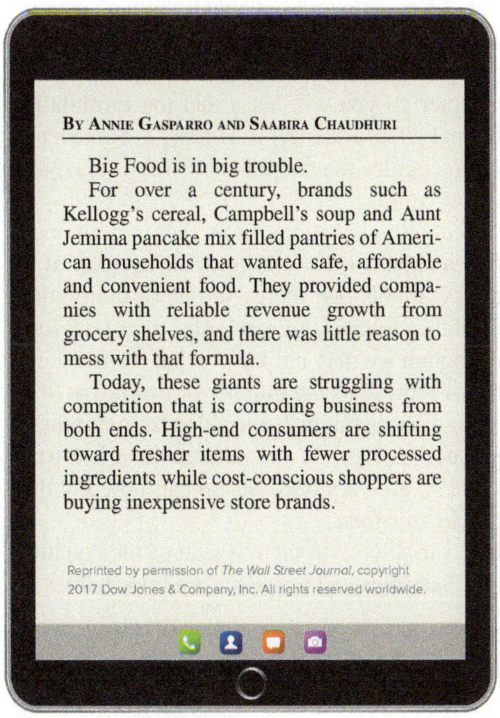

By Annie Gasparro and Saabira Chaudhuri

Big Food is in big trouble.

For over a century, brands such as Kellogg's cereal, Campbell's soup and Aunt Jemima pancake mix filled pantries of American households that wanted safe, affordable and convenient food. They provided companies with reliable revenue growth from grocery shelves, and there was little reason to mess with that formula.

Today, these giants are struggling with competition that is corroding business from both ends. High-end consumers are shifting toward fresher items with fewer processed ingredients while cost-conscious shoppers are buying inexpensive store brands.

The equation used in the chapter opener *Wall Street Journal* clip, "Formula for Success," on how long it could take an investment to double in value might change based on interest rates that are affected by the new tax law act of 2017. If the economy expands the Federal Reserve may raise rates. Check with the IRS for the latest tax updates.

When you use your GPS, there are usually several routes to reach a destination. In this chapter, we will be looking at how to solve math problems in different ways.

Learning Unit 5–1 explains how you can solve for unknowns in equations. In Learning Unit 5–2 you learn how to solve for unknowns in word problems. When you complete these learning units, you will not have to memorize as many formulas to solve business and personal math applications. The *Wall Street Journal* clip to the left shows some companies have used the same formula for many years to promote their products, causing competitive challenges.

LO 1

Learning Unit 5–1: Solving Equations for the Unknown

The following Rose Smith letter is based on a true story. Note how Rose states that the blueprint aids, the lesson on repetition, and the chapter organizers were important factors in the successful completion of her business math course.

Many of you are familiar with the terms *variables* and *constants.* If you are planning to prepare for your retirement by saving only what you can afford each year, your saving is a *variable;* if you plan to save the same amount each year, your saving is a *constant.* Now you can also say that you cannot buy clothes by

Rose Smith
15 Locust Street
Lynn, MA 01915

Flowers.net
Decorating Service

Dear Professor Slater,

Thank you for helping me get through your Business Math class. When I first started, my math anxiety level was real high. I felt I had no head for numbers. When you told us we would be covering the chapter on solving equations, I'll never forget how I started to shake. I started to panic. I felt I could never solve a word problem. I thought I was having an algebra attack.

Now that it's over (90 on the chapter on unknowns), I'd like to tell you what worked for me so you might pass this on to other students. It was your blueprint aids. Drawing boxes helped me to think things out. They were a tool that helped me more clearly understand how to dissect each word problem. They didn't solve the problem for me, but gave me the direction I needed. Repetition was the key to my success. At first I got them all wrong but after the third time, things started to click. I felt more confident. Your chapter organizers at the end of the chapter were great. Thanks for your patience – your repetition breeds success – now students are asking me to help them solve a word problem. Can you believe it!

Best,

Rose

Rose Smith

Used by permission of Cartoon Features Syndicate

size because of the many variables involved. This unit explains the importance of mathematical variables and constants when solving equations.

Basic Equation-Solving Procedures

Do you know the difference between a mathematical expression, equation, and formula? A mathematical **expression** is a meaningful combination of numbers and letters called *terms.* Operational signs (such as + or −) within the expression connect the terms to show a relationship between them. For example, $6 + 2$ and $6A - 4A$ are mathematical expressions. An **equation** is a mathematical statement with an equals sign showing that a mathematical expression on the left equals the mathematical expression on the right. An equation has an equals sign; an expression does not have an equals sign. A **formula** is an equation that expresses in symbols a general fact, rule, or principle. Formulas are shortcuts for expressing a word concept. For example, in Chapter 10 you will learn that the formula for simple interest is Interest (I) = Principal (P) × Rate (R) × Time (T). This means that when you see $I = P \times R \times T$, you recognize the simple interest formula. Now let's study basic equations.

As a mathematical statement of equality, equations show that two numbers or groups of numbers are equal. For example, $6 + 4 = 10$ shows the equality of an equation. Equations also use letters as symbols that represent one or more numbers. These symbols, usually a letter of the alphabet, are **variables** that stand for a number. We can use a variable even though we may not know what it represents. For example, $A + 2 = 6$. The variable A represents the number or **unknown** (4 in this example) for which we are solving. We distinguish variables from numbers, which have a fixed value. Numbers such as 3 or -7 are **constants** or **knowns,** whereas A and $3A$ (this means 3 times the variable A) are variables. So we can now say that variables and constants are *terms of mathematical expressions.*

Usually in solving for the unknown, we place variable(s) on the left side of the equation and constants on the right. The following rules for variables and constants are important.

VARIABLES AND CONSTANTS RULES

1. If no number is in front of a letter, it is a 1: $B = 1B$; $C = 1C$.
2. If no sign is in front of a letter or number, it is a +: $C = +C$; $4 = +4$.

You should be aware that in solving equations, the meaning of the symbols +, −, ×, and ÷ has not changed. However, some variations occur. For example, you can also write $A \times B$ (A times B) as $A \bullet B$, $A(B)$, or AB. Also, A divided by B is the same as A/B. Remember that to solve an equation, you must find a number that can replace the unknown in the equation and make it a true statement. Now let's take a moment to look at how we can change verbal statements into variables.

Assume Dick Hersh, an employee of Nike, is 50 years old. Let's assign Dick Hersh's changing age to the symbol A. The symbol A is a variable.

Verbal statement	Variable A (age)
Dick's age 8 years ago	$A - 8$
Dick's age 8 years from today	$A + 8$
Four times Dick's age	$4A$
One-fifth Dick's age	$A/5$

FIGURE 5.1

Equality in equations

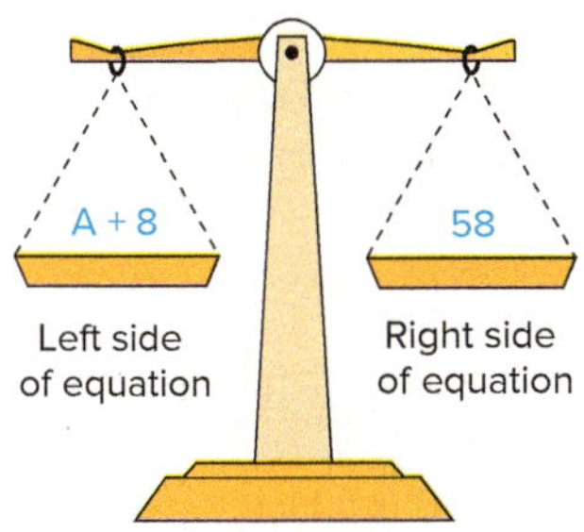

Dick's age in 8 years will equal 58.

To visualize how equations work, think of the old-fashioned balancing scale shown in Figure 5.1. The pole of the scale is the equals sign. The two sides of the equation are the two pans of the scale. In the left pan or left side of the equation, we have $A + 8$; in the right pan or right side of the equation, we have 58. To solve for the unknown (Dick's present age), we isolate or place the unknown (variable) on the left side and the numbers on the right. We will do this soon. For now, remember that to keep an equation (or scale) in balance, we must perform mathematical operations (addition, subtraction, multiplication, and division) to *both* sides of the equation.

SOLVING FOR THE UNKNOWN RULE

Whatever you do to one side of an equation, you must do to the other side.

How to Solve for Unknowns in Equations

LO 2

This section presents seven drill situations and the rules that will guide you in solving for unknowns in these situations. We begin with two basic rules—the opposite process rule and the equation equality rule.

OPPOSITE PROCESS RULE

If an equation indicates a process such as addition, subtraction, multiplication, or division, solve for the unknown or variable by using the opposite process. For example, if the equation process is addition, solve for the unknown by using subtraction.

EQUATION EQUALITY RULE

You can add the same quantity or number to both sides of the equation and subtract the same quantity or number from both sides of the equation without affecting the equality of the equation. You can also divide or multiply both sides of the equation by the same quantity or number *(except zero)* without affecting the equality of the equation.

To check your answer(s), substitute your answer(s) for the letter(s) in the equation. The sum of the left side should equal the sum of the right side.

Drill Situation 1: Subtracting Same Number from Both Sides of Equation

Example	Mechanical steps	Explanation
$A + 8 = 58$ Dick's age A plus 8 equals 58.	$A + 8 = 58$ $\underline{-8 \quad -8}$ $A = 50$	8 is subtracted from *both* sides of the equation to isolate variable A on the left. **Check** $50 + 8 = 58$ $58 = 58$

Note: Since the equation process used *addition,* we use the opposite process rule and solve for variable A with *subtraction.* We also use the equation equality rule when we subtract the same quantity from both sides of the equation.

Drill Situation 2: Adding Same Number to Both Sides of Equation

Example	Mechanical steps	Explanation
$B - 50 = 80$ Some number B less 50 equals 80.	$B - 50 = 80$ $\underline{+50 \quad +50}$ $B = 130$	50 is added to *both* sides to isolate variable B on the left. **Check** $130 - 50 = 80$ $80 = 80$

Note: Since the equation process used *subtraction,* we use the opposite process rule and solve for variable B with *addition.* We also use the equation equality rule when we add the same quantity to both sides of the equation.

Drill Situation 3: Dividing Both Sides of Equation by Same Number

Example	Mechanical steps	Explanation
$7G = 35$ Some number G times 7 equals 35.	$7G = 35$ $\frac{\cancel{7}G}{\cancel{7}} = \frac{35}{7}$ $G = 5$	By dividing both sides by 7, G equals 5. **Check** $7(5) = 35$ $35 = 35$

Note: Since the equation process used *multiplication,* we use the opposite process rule and solve for variable G with *division.* We also use the equation equality rule when we divide both sides of the equation by the same quantity.

Drill Situation 4: Multiplying Both Sides of Equation by Same Number

Example	Mechanical steps	Explanation
$\frac{V}{5} = 70$ Some number V divided by 5 equals 70.	$\frac{V}{5} = 70$ $\cancel{5}\left(\frac{V}{\cancel{5}}\right) = 70(5)$ $V = 350$	By multiplying both sides by 5, V is equal to 350. **Check** $\frac{350}{5} = 70$ $70 = 70$

Note: Since the equation process used *division,* we use the opposite process rule and solve for variable V with *multiplication.* We also use the equation equality rule when we multiply both sides of the equation by the same quantity.

Drill Situation 5: Equation That Uses Subtraction and Multiplication to Solve for Unknown

MULTIPLE PROCESSES RULE

When solving for an unknown that involves more than one process, do the addition and subtraction before the multiplication and division.

Example

$$\frac{H}{4} + 2 = 5$$

When we divide unknown H by 4 and add the result to 2, the answer is 5.

Mechanical steps

$$\frac{H}{4} + 2 = 5$$

$$\frac{H}{4} + 2 = 5$$
$$\underline{-2 \quad -2}$$
$$\frac{H}{4} = 3$$
$$\not{4}\left(\frac{H}{\not{4}}\right) = 4(3)$$
$$H = 12$$

Explanation

1. Move constant to right side by subtracting 2 from both sides.
2. To isolate H, which is divided by 4, we do the opposite process and multiply 4 times *both* sides of the equation.

Check

$$\frac{12}{4} + 2 = 5$$
$$3 + 2 = 5$$
$$5 = 5$$

MONEY tips

Negotiate. Over 9 out of 10 customers who ask for a discount on items such as electronics, appliances, furniture, and medical bills receive one. Just ask. You could be saving money.

Drill Situation 6: Using Parentheses in Solving for Unknown

PARENTHESES RULE

When equations contain parentheses (which indicate grouping together), you solve for the unknown by first multiplying each item inside the parentheses by the number or letter just outside the parentheses. Then you continue to solve for the unknown with the opposite process used in the equation. Do the additions and subtractions first; then the multiplications and divisions.

Example

$$5(P - 4) = 20$$

The unknown P less 4, multiplied by 5 equals 20.

Mechanical steps

$$5(P - 4) = 20$$
$$5P - 20 = 20$$
$$\underline{+20 \quad +20}$$
$$\frac{\not{5}P}{\not{5}} = \frac{40}{5}$$
$$P = 8$$

Explanation

1. Parentheses tell us that everything inside parentheses is multiplied by 5. Multiply 5 by P and 5 by -4.
2. Add 20 to both sides to isolate $5P$ on left.
3. To remove 5 in front of P, divide both sides by 5 to result in P equals 8.

Check

$$5(8 - 4) = 20$$
$$5(4) = 20$$
$$20 = 20$$

Drill Situation 7: Combining Like Unknowns

LIKE UNKNOWNS RULE

To solve equations with like unknowns, you first combine the unknowns and then solve with the opposite process used in the equation.

Example	Mechanical steps	Explanation
$4A + A = 20$	$4A + A = 20$ $\frac{\cancel{5}A}{\cancel{5}} = \frac{20}{5}$ $A = 4$	To solve this equation: $4A + 1A = 5A$. Thus, $5A = 20$. To solve for A, divide both sides by 5, leaving A equals 4. **Check** $4(4) + 4 = 20$ $20 = 20$

Aha!

Always, always, always do a logic check on your answer. Reread the word problem to see if your answer makes sense. If it does, move on. If it does not, review your strategy for solving the problem and make any needed adjustments.

Before you go to Learning Unit 5–2, let's check your understanding of this unit.

LU 5–1 PRACTICE QUIZ

Complete this Practice Quiz to see how you are doing.

1. Write equations for the following (use the letter Q as the variable). Do not solve for the unknown.
   a. Nine less than one-half a number is fourteen.
   b. Eight times the sum of a number and thirty-one is fifty.
   c. Ten decreased by twice a number is two.
   d. Eight times a number less two equals twenty-one.
   e. The sum of four times a number and two is fifteen.
   f. If twice a number is decreased by eight, the difference is four.
2. Solve the following:
   a. $B + 24 = 60$ **b.** $D + 3D = 240$ **c.** $12B = 144$
   d. $\frac{B}{6} = 50$ **e.** $\frac{B}{4} + 4 = 16$ **f.** $3(B - 8) = 18$

✓ Solutions

*For **extra help** from your authors–Sharon and Jeff–see the videos in Connect.*

1. a. $\frac{1}{2}Q - 9 = 14$ **b.** $8(Q + 31) = 50$ **c.** $10 - 2Q = 2$
d. $8Q - 2 = 21$ **e.** $4Q + 2 = 15$ **f.** $2Q - 8 = 4$

2. a. $B + 24 = 60$; $-24 \quad -24$; $B = 36$

b. $\frac{\cancel{4}D}{\cancel{4}} = \frac{240}{4}$; $D = 60$

c. $\frac{\cancel{12}B}{\cancel{12}} = \frac{144}{12}$; $B = 12$

d. $$\cancel{6}\left(\frac{B}{\cancel{6}}\right) = 50(6)$$
$$B = \boxed{300}$$

e. $$\begin{aligned} \frac{B}{4} + 4 &= 16 \\ -4 &\quad -4 \\ \frac{B}{4} &= 12 \\ \cancel{4}\left(\frac{B}{\cancel{4}}\right) &= 12(4) \\ B &= \boxed{48} \end{aligned}$$

f. $$\begin{aligned} 3(B-8) &= 18 \\ 3B - 24 &= 18 \\ +24 &\quad +24 \\ \frac{\cancel{3}B}{\cancel{3}} &= \frac{42}{3} \\ B &= \boxed{14} \end{aligned}$$

Learning Unit 5–2: Solving Word Problems for the Unknown

LO 1

When you buy a candy bar such as a Snickers, you should turn the candy bar over and carefully read the ingredients and calories contained on the back of the candy bar wrapper. For example, on the back of the Snickers wrapper you will read that there are "170 calories per piece." You could misread this to mean that the entire Snickers bar has 170 calories. However, look closer and you will see that the Snickers bar is divided into three pieces, so if you eat the entire bar, instead of consuming 170 calories, you will consume 510 calories. Making errors like this could result in a weight gain that you cannot explain.

©McGraw-Hill Education/Mark Dierker, photographer

$$\frac{1}{3}S = 170 \text{ calories}$$
$$\cancel{3}\left(\frac{1}{\cancel{3}}S\right) = 170 \times 3$$
$$S = \boxed{510} \text{ calories per bar}$$

In this unit, we use blueprint aids in six different situations to help you solve for unknowns. Be patient and *persistent.* Remember that the more problems you work, the easier the process becomes. Do not panic! Repetition is the key. Study the five steps that follow. They will help you solve for unknowns in word problems.

SOLVING WORD PROBLEMS FOR UNKNOWNS

Step 1. Carefully read the entire problem. You may have to read it several times.

Step 2. Ask yourself: What is the problem looking for?

Step 3. When you are sure what the problem is asking, let a variable represent the unknown. If the problem has more than one unknown, represent the second unknown in terms of the same variable. For example, if the problem has two unknowns, Y is one unknown. The second unknown is $4Y$—4 times the first unknown.

Step 4. Visualize the relationship between unknowns and variables. Then set up an equation to solve for the unknown(s).

Step 5. Check your result to see if it is accurate.

The clip on page 128 from *The Wall Street Journal,* "How to Ace That Test," may also help you in the process of solving word problems.

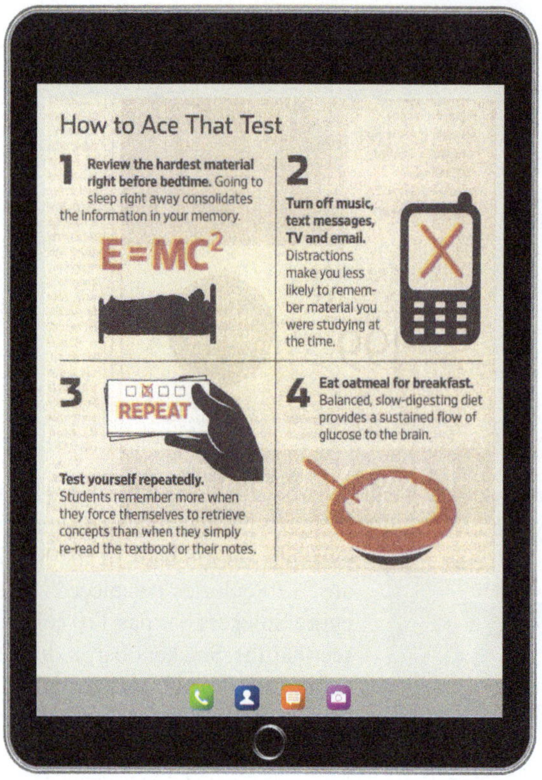

Word Problem Situation 1: Number Problems Today on sale at a local Stop & Shop supermarket, the price of a bag of Dunkin' Donuts coffee is \$9.99. This is a \$2 savings. What was the original price of the bag of coffee?

©Sara Stathas/Alamy Stock Photo

LO 2

BLUEPRINT	Unknown(s)	Variable(s)	Relationship*
	Original price of Dunkin' Donuts	P	$P - \$2 =$ New price

*This column will help you visualize the equation before setting up the actual equation.

Mechanical steps

$$\begin{aligned} P - 2 &= \$9.99 \\ +2 & \quad +2 \\ \hline P &= \$11.99 \end{aligned}$$

Explanation

The original price less $2 = $9.99. Note that we added $2 to both sides to isolate P on the left. Remember, $1P = P$.

Check

$\$11.99 - 2 = \$9.99$
$\$9.99 = \$9.99$

Word Problem Situation 2: Finding the Whole When Part Is Known A local Burger King budgets $\frac{1}{8}$ of its monthly profits on salaries. Salaries for the month were $12,000. What were Burger King's monthly profits?

BLUEPRINT	Unknown(s)	Variable(s)	Relationship
	Monthly profits	P	$\frac{1}{8}P$ Salaries = $12,000

Mechanical steps

$$\frac{1}{8}P = \$12,000$$
$$\cancel{8}\left(\frac{P}{\cancel{8}}\right) = \$12,000(8)$$
$$P = \$96,000$$

Explanation

$\frac{1}{8}P$ represents Burger King's monthly salaries. Since the equation used division, we solve for P by multiplying both sides by 8.

Check

$\frac{1}{8}(\$96,000) = \$12,000$
$\$12,000 = \$12,000$

Word Problem Situation 3: Difference Problems ICM Company sold 4 times as many computers as Ring Company. The difference in their sales is 27. How many computers of each company were sold?

BLUEPRINT	Unknown(s)	Variable(s)	Relationship
	ICM Ring	$4C$ C	$4C$ $\underline{-C}$ 27

Note: If problem has two unknowns, assign the variable to smaller item or one who sells less. Then assign the other unknown using the same variable. *Use the same letter.*

Mechanical steps

$$4C - C = 27$$
$$\frac{\cancel{3}C}{\cancel{3}} = \frac{27}{3}$$
$$C = 9$$

Ring = 9 computers
ICM = 4(9)
= 36 computers

Explanation

The variables replace the names ICM and Ring. We assigned Ring the variable C, since it sold fewer computers. We assigned ICM $4C$, since it sold 4 times as many computers.

Check

36 computers
$\underline{-9}$
27 computers

Word Problem Situation 4: Calculating Unit Sales Together Barry Sullivan and Mitch Ryan sold a total of 300 homes for Regis Realty. Barry sold 9 times as many homes as Mitch. How many did each sell?

BLUEPRINT	Unknown(s)	Variable(s)	Relationship
	Homes sold: B. Sullivan M. Ryan	 $9H$ H*	 $9H$ $\underline{+H}$ 300 homes

*Assign H to Ryan since he sold less.

Mechanical steps

$$9H + H = 300$$
$$\frac{\cancel{10}H}{\cancel{10}} = \frac{300}{10}$$
$$H = 30$$

Ryan: 30 homes
Sullivan: 9(30) = 270 homes

Explanation

We assigned Mitch H, since he sold fewer homes. We assigned Barry $9H$, since he sold 9 times as many homes. Together Barry and Mitch sold 300 homes.

Check

$30 + 270 = 300$

Word Problem Situation 5: Calculating Unit and Dollar Sales (Cost per Unit) When Total Units Are Not Given Andy sold watches ($9) and alarm clocks ($5) at a flea market. Total sales were $287. People bought 4 times as many watches as alarm clocks. How many of each did Andy sell? What were the total dollar sales of each?

BLUEPRINT	Unknown(s)	Variable(s)	Price	Relationship
	Unit sales:			
	Watches	$4C$	$9	$36C$
	Clocks	C	5	$+5C$
				$287 total sales

Mechanical steps

$36C + 5C = 287$

$\frac{41C}{41} = \frac{287}{41}$

$C = 7$

7 clocks

$4(7) = 28$ watches

Explanation

Number of watches times $9 sales price plus number of alarm clocks times $5 sales price equals $287 total sales.

Check

$7(\$5) + 28(\$9) = \$287$

$\$35 + \$252 = \$287$

$\$287 = \$287$

Word Problem Situation 6: Calculating Unit and Dollar Sales (Cost per Unit) When Total Units Are Given Andy sold watches ($9) and alarm clocks ($5) at a flea market. Total sales for 35 watches and alarm clocks were $287. How many of each did Andy sell? What were the total dollar sales of each?

MONEY tips

Go green. Recycle plastics, paper, cardboard, glass, steel, aluminum cans and foil, plastic bags, motor oil, tires, batteries, computer printers, compost, household cleaners, and so on. Reduce the volume of packaging you require by using a canvas bag. Do your part to help solve the unknown environmental challenges of the future.

BLUEPRINT	Unknown(s)	Variable(s)	Price	Relationship
	Unit sales:			
	Watches	W*	$9	$9W$
	Clocks	$35 - W$	5	$+5(35 - W)$
				$287 total sales

*The more expensive item is assigned to the variable first only for this situation to make the mechanical steps easier to complete.

Mechanical steps

$9W + 5(35 - W) = 287$

$9W + 175 - 5W = 287$

$4W + 175 = 287$

$\quad -175 \qquad -175$

$\frac{4W}{4} = \frac{112}{4}$

$W = 28$

Watches = 28

Clocks = 35 − 28 = 7

Explanation

Number of watches (W) times price per watch plus number of alarm clocks times price per alarm clock equals $287. Total units given was 35.

Check

$28(\$9) + 7(\$5) = \$287$

$\$252 + \$35 = \$287$

$\$287 = \$287$

Why did we use $35 - W$? Assume we had 35 pizzas (some cheese, others meatball). If I said that I ate all the meatball pizzas (5), how many cheese pizzas are left? Thirty? Right, you subtract 5 from 35. Think of $35 - W$ as meaning one number.

Note in Word Problem Situations 5 and 6 that the situation is the same. In Word Problem Situation 5, we were not given total units sold (but we were told which sold better). In Word Problem Situation 6, we were given total units sold, but we did not know which sold better.

Now try these six types of word problems in the Practice Quiz. Be sure to complete blueprint aids and the mechanical steps for solving the unknown(s).

LU 5–2 PRACTICE QUIZ

Complete this Practice Quiz to see how you are doing.

Situations

1. An L. L. Bean sweater was reduced $30. The sale price was $90. What was the original price?
2. Kelly Doyle budgets $\frac{1}{8}$ of her yearly salary for entertainment. Kelly's total entertainment bill for the year is $6,500. What is Kelly's yearly salary?
3. Micro Knowledge sells 5 times as many computers as Morse Electronics. The difference in sales between the two stores is 20 computers. How many computers did each store sell?

4. Susie and Cara sell stoves at Elliott's Appliances. Together they sold 180 stoves in January. Susie sold 5 times as many stoves as Cara. How many stoves did each sell?
5. Pasquale's Pizza sells meatball pizzas ($6) and cheese pizzas ($5). In March, Pasquale's total sales were $1,600. People bought 2 times as many cheese pizzas as meatball pizzas. How many of each did Pasquale's sell? What were the total dollar sales of each?
6. Pasquale's Pizza sells meatball pizzas ($6) and cheese pizzas ($5). In March, Pasquale's sold 300 pizzas for $1,600. How many of each did Pasquale's sell? What was the dollar sales price of each?

*For **extra help** from your authors–Sharon and Jeff–see the videos in Connect.*

✓ Solutions

1.

BLUEPRINT	Unknown(s)	Variable(s)	Relationship
	Original price	P^*	$P - \$30 =$ Sale price Sale price = $90

*p = Original price.

Mechanical steps

$$P - \$30 = \$90$$
$$\underline{\quad + 30 \qquad + 30}$$
$$P = \$120$$

2.

BLUEPRINT	Unknown(s)	Variable(s)	Relationship
	Yearly salary	S^*	$\frac{1}{8}S$ Entertainment = $6,500

*S = Salary.

Mechanical steps

$$\frac{1}{8}S = \$6{,}500$$
$$\cancel{8}\left(\frac{S}{\cancel{8}}\right) = \$6{,}500(8)$$
$$S = \$52{,}000$$

3.

BLUEPRINT	Unknown(s)	Variable(s)	Relationship
	Micro Morse	$5C^*$ C	$5C$ $\underline{-C}$ 20 computers

*C = Computers.

Mechanical steps

$$5C - C = 20$$
$$\frac{\cancel{4}C}{\cancel{4}} = \frac{20}{4}$$
$$C = 5 \text{ (Morse)}$$
$$5C = 25 \text{ (Micro)}$$

4.

BLUEPRINT	Unknown(s)	Variable(s)	Relationship
	Stoves sold: Susie Cara	 $5S^*$ S	 $5S$ $\underline{+S}$ 180 stoves

*S = Stoves.

Mechanical steps

$$5S + S = 180$$
$$\frac{\cancel{6}S}{\cancel{6}} = \frac{180}{6}$$
$$S = 30 \text{ (Cara)}$$
$$5S = 150 \text{ (Susie)}$$

5.

BLUEPRINT	Unknown(s)	Variable(s)	Price	Relationship
	Meatball Cheese	M $2M$	$6 5	$6M$ $\underline{+\ 10M}$ $1,600 total sales

Mechanical steps

$$6M + 10M = 1{,}600$$
$$\frac{\cancel{16}M}{\cancel{16}} = \frac{1{,}600}{16}$$
$$M = 100 \text{ (meatball)}$$
$$2M = 200 \text{ (cheese)}$$

Check

$$(100 \times \$6) + (200 \times \$5) = \$1{,}600$$
$$\$600 + \$1{,}000 = \$1{,}600$$
$$\$1{,}600 = \$1{,}600$$

6.

BLUEPRINT	Unknown(s)	Variable(s)	Price	Relationship
	Unit sales:			
	Meatball	M*	\$6	$6M$
	Cheese	$300 - M$	5	$+ 5(300 - M)$
				\$1,600 total sales

*We assign the variable to the most expensive item to make the mechanical steps easier to complete.

Mechanical steps

$$\begin{aligned} 6M + 5(300 - M) &= 1{,}600 \\ 6M + 1{,}500 - 5M &= 1{,}600 \\ M + 1{,}500 &= 1{,}600 \\ -1{,}500 &\quad -1{,}500 \\ M &= \boxed{100} \end{aligned}$$

Meatball = 100

Cheese = 300 − 100 = 200

Check

100(\$6) + 200(\$5) = \$600 + \$1,000
= \$1,600

INTERACTIVE CHAPTER ORGANIZER

Solving for unknowns from basic equations	Mechanical steps to solve unknowns	Key point(s)	You try it*
Situation 1: Subtracting same number from both sides of equation	$D + 10 = 12$ $\quad -10 \quad -10$ $D = \boxed{2}$	Subtract 10 from both sides of equation to isolate variable D on the left. Since equation used addition, we solve by using opposite process—subtraction.	**Solve** $E + 15 = 14$
Situation 2: Adding same number to both sides of equation	$L - 24 = 40$ $\quad +24 \quad +24$ $L = \boxed{64}$	Add 24 to both sides to isolate unknown L on left. We solve by using opposite process of subtraction—addition.	**Solve** $B - 40 = 80$
Situation 3: Dividing both sides of equation by same number	$6B = 24$ $\frac{\not{6}B}{\not{6}} = \frac{24}{\not{6}}$ $B = \boxed{4}$	To isolate B on the left, divide both sides of the equation by 6. Thus, the 6 on the left cancels—leaving B equal to 4. Since equation used multiplication, we solve unknown by using opposite process—division.	**Solve** $5C = 75$
Situation 4: Multiplying both sides of equation by same number	$\frac{R}{3} = 15$ $\not{3}\left(\frac{R}{\not{3}}\right) = 15(3)$ $R = \boxed{45}$	To remove denominator, multiply both sides of the equation by 3—the 3 on the left side cancels, leaving R equal to 45. Since equation used division, we solve unknown by using opposite process—multiplication.	**Solve** $\frac{A}{6} = 60$
Situation 5: Equation that uses subtraction and multiplication to solve for unknown	$\frac{B}{3} + 6 = 13$ $\quad -6 \quad -6$ $\frac{B}{3} = 7$ $\not{3}\left(\frac{B}{\not{3}}\right) = 7(3)$ $B = \boxed{21}$	**1.** Move constant 6 to right side by subtracting 6 from both sides. **2.** Isolate B on left by multiplying both sides by 3.	**Solve** $\frac{C}{4} + 10 = 17$

(continues)

INTERACTIVE CHAPTER ORGANIZER

Solving for unknowns from basic equations	Mechanical steps to solve unknowns	Key point(s)	You try it*
Situation 6: Using parentheses in solving for unknown	$6(A-5)=12$ $6A-30=12$ $+30 \quad +30$ $\frac{\not 6A}{\not 6}=\frac{42}{6}$ $A=7$	Parentheses indicate multiplication. Multiply 6 times A and 6 times −5. Result is $6A - 30$ on left side of the equation. Now add 30 to both sides to isolate $6A$ on left. To remove 6 in front of A, divide both sides by 6, to result in A equal to 7. Note that when deleting parentheses, we did not have to multiply the right side.	**Solve** $7(B-10)=35$
Situation 7: Combining like unknowns	$6A+2A=64$ $\frac{\not 8A}{\not 8}=\frac{64}{8}$ $A=8$	$6A + 2A$ combine to $8A$. To solve for A, we divide both sides by 8.	**Solve** $5B+3B=16$

Solving for unknowns from word problems	Blueprint aid	Mechanical steps to solve unknown with check	You try it*
Situation 1: Number problems **U.S. Air reduced its airfare to California by \$60. The sale price was \$95. What was the original price?**	BLUEPRINT Unknown(s): Original price Variable(s): P Relationship: P − \$60 = Sale price; Sale price = \$95	$P-\$60=\$95$ $+60 \quad +60$ $P=\$155$ **Check** $\$155-\$60=\$95$ $\$95=\$95$	**Solve** U.S. Air reduced its airfare to California by \$53. The sale price was \$110. What was the original price?
Situation 2: Finding the whole when part is known **K. McCarthy spends 1/8 of her budget for school. What is the total budget if school costs \$5,000?**	BLUEPRINT Unknown(s): Total budget Variable(s): B Relationship: ⅛B; School = \$5,000	$\frac{1}{8}B=\$5{,}000$ $\not 8\left(\frac{B}{\not 8}\right)=\$5{,}000(8)$ $B=\$40{,}000$ **Check** $\frac{1}{8}(\$40{,}000)=\$5{,}000$ $\$5{,}000=\$5{,}000$	**Solve** K. McCarthy spends ⅐ of her budget for school. What is the total budget if school costs \$6,000?
Situation 3: Difference problems **Moe sold 8 times as many suitcases as Bill. The difference in their sales is 280 suitcases. How many suitcases did each sell?**	BLUEPRINT Unknown(s): *Suitcases sold:* Moe; Bill Variable(s): $8S$; S Relationship: $8S$ − S = 280 suitcases	$8S-S=280$ $\frac{\not 7S}{\not 7}=\frac{280}{7}$ $S=40$ (Bill) $8(40)=320$ (Moe) **Check** $320-40=280$ $280=280$	**Solve** Moe sold 9 times as many suitcases as Bill. The difference in their sales is 640 suitcases. How many suitcases did each sell?
Situation 4: Calculating unit sales **Moe sold 8 times as many suitcases as Bill. Together they sold a total of 360. How many did each sell?**	BLUEPRINT Unknown(s): *Suitcases sold:* Moe; Bill Variable(s): $8S$; S Relationship: $8S$ + S = 360 suitcases	$8S+S=360$ $\frac{\not 9S}{\not 9}=\frac{360}{9}$ $S=40$ (Bill) $8(40)=320$ (Moe) **Check** $320+40=360$ $360=360$	**Solve** Moe sold 9 times as many suitcases as Bill. Together they sold a total of 640. How many did each sell?

(continues)

INTERACTIVE CHAPTER ORGANIZER

<table>
<tr><th>Solving for unknowns from word problems</th><th>Blueprint aid</th><th>Mechanical steps to solve unknown with check</th><th>You try it*</th></tr>
<tr>
<td>Situation 5: Calculating unit and dollar sales (cost per unit) when total units not given

Blue Furniture Company ordered sleepers ($300) and nonsleepers ($200) that cost $8,000. Blue expects sleepers to outsell nonsleepers 2 to 1. How many units of each were ordered? What were the dollar costs of each?</td>
<td>BLUEPRINT<table><tr><th>Unknown(s)</th><th>Variable(s)</th><th>Price</th><th>Relationship</th></tr><tr><td>Sleepers
Nonsleepers</td><td>2N
N</td><td>$300
200</td><td>600N
+ 200N
$8,000 total cost</td></tr></table></td>
<td>$600N + 200N = 8,000$
$\frac{800N}{800} = \frac{8,000}{800}$
$N = 10$ (nonsleepers)
$2N = 20$ (sleepers)

Check
$10 \times \$200 = \$2,000$
$20 \times \$300 = \underline{\ \ 6,000}$
$= \$8,000$</td>
<td>Solve
Blue Furniture Company ordered sleepers ($400) and nonsleepers ($300) that cost $15,000. Blue expects sleepers to outsell nonsleepers 3 to 1. How many units of each were ordered? What were the dollar costs of each?</td>
</tr>
<tr>
<td>Situation 6: Calculating unit and dollar sales (cost per unit) when total units given

Blue Furniture Company ordered 30 sofas (sleepers and nonsleepers) that cost $8,000. The wholesale unit cost was $300 for the sleepers and $200 for the nonsleepers. How many units of each were ordered? What were the dollar costs of each?</td>
<td>BLUEPRINT<table><tr><th>Unknown(s)</th><th>Variable(s)</th><th>Price</th><th>Relationship</th></tr><tr><td>Unit costs
Sleepers
Nonsleepers</td><td>
S
30 − S</td><td>
$300
200</td><td>
300S
+200(30 − S)
$8,000 total cost</td></tr></table>
Note: When the total units are given, the higher-priced item (sleepers) is assigned to the variable first. This makes the mechanical steps easier to complete.</td>
<td>$300S + 200(30 - S) = 8,000$
$300S + 6,000 - 200S = 8,000$
$100S + 6,000 = 8,000$
$\quad - 6,000 \quad -6,000$
$\frac{100S}{100} = \frac{2,000}{100}$
$S = 20$
Nonsleepers $= 30 - 20$
$= 10$

Check
$20(\$300) + 10(\$200) = \$8,000$
$\$6,000 + \$2,000 = \$8,000$
$\$8,000 = \$8,000$</td>
<td>Solve
Blue Furniture Company ordered 40 sofas (sleepers and nonsleepers) that cost $15,000. The wholesale unit cost was $400 for the sleepers and $300 for the nonsleepers. How many units of each were ordered? What were the dollar costs of each?</td>
</tr>
<tr>
<td>KEY TERMS</td>
<td>Constants
Equation
Expression</td>
<td>Formula
Knowns
Unknown</td>
<td>Variables</td>
</tr>
</table>

*Worked-out solutions are in Appendix B.

Critical Thinking Discussion Questions with Chapter Concept Check

1. Explain the difference between a variable and a constant. What would you consider your monthly car payment—a variable or a constant?
2. How does the opposite process rule help solve for the variable in an equation? If a Mercedes costs 3 times as much as a Saab, how could the opposite process rule be used? The selling price of the Mercedes is $60,000.
3. What is the difference between Word Problem Situations 5 and 6 in Learning Unit 5–2? Show why the more expensive item in Word Problem Situation 6 is assigned to the variable first.
4. **Chapter Concept Check.** Go to a weight-loss website and create several equations on how to lose weight. Be sure to create a word problem and specify the steps you need to take to solve this weight-loss problem.

END-OF-CHAPTER PROBLEMS

Connect

5

Check figures for odd-numbered problems in Appendix B. Name ______________________ Date ________

DRILL PROBLEMS (First of Three Sets)

Solve the unknown from the following equations: *LU 5-1(2)*

5–1. $X - 40 = 400$ **5–2.** $A + 64 = 98$ **5–3.** $Q + 100 = 400$ **5–4.** $Q - 60 = 850$

5–5. $5Y = 75$ **5–6.** $\frac{P}{6} = 92$ **5–7.** $8Y = 96$ **5–8.** $\frac{N}{16} = 5$

5–9. $4(P - 9) = 64$ 5–10. $3(P - 3) = 27$

WORD PROBLEMS (First of Three Sets)

5–11. Lee and Fred are elementary school teachers. Fred works for a charter school in Pacific Palisades, California, where class size reduction is a goal for 2019. Lee works for a noncharter school where funds do not allow for class size reduction policies. Lee's fifth-grade class has 1.4 times as many students as Fred's. If there are a total of 60 students, how many students does Fred's class have? How many students does Lee's class have? *LU 5-2(2)*

excel **5–12.** A car that originally cost \$3,668 in 1955 is valued today at \$62,125 if in excellent condition, which is $1\frac{3}{4}$ times as much as a car in very nice condition—if you can find an owner willing to part with one for any price. What would be the value of the car in very nice condition? *LU 5-2(2)*

5–13. Jessica and Josh are selling Entertainment Books to raise money for the art room at their school. One book sells for \$15. Jessica received the prize for selling the most books in the school. Jessica sold 15 times more books than Josh. Together they sold 256 books. How many did each one of them sell? *LU 5-2(1)*

5–14. Nanda Yueh and Lane Zuriff sell homes for ERA Realty. Over the past 6 months they sold 120 homes. Nanda sold 3 times as many homes as Lane. How many homes did each sell? *LU 5-2(2)*

5–15. Dots sells T-shirts ($2) and shorts ($4). In April, total sales were $600. People bought 4 times as many T-shirts as shorts. How many T-shirts and shorts did Dots sell? Check your answer. *LU 5-2(2)*

5–16. Dots sells a total of 250 T-shirts ($2) and shorts ($4). In April, total sales were $600. How many T-shirts and shorts did Dots sell? Check your answer. *Hint:* Let S = Shorts. *LU 5-2(2)*

DRILL PROBLEMS (Second of Three Sets)

Solve the unknown from the following equations: *LU 5-1(2)*

5–17. $7B = 490$

5–18. $7(A - 5) = 63$

5–19. $\frac{N}{9} = 7$

5–20. $18(C - 3) = 162$

5–21. $9Y - 10 = 53$

5–22. $7B + 5 = 26$

WORD PROBLEMS (Second of Three Sets)

5–23. On a flight from Boston to San Diego, American reduced its Internet price by $190.00. The new sale price was $420.99. What was the original price? *LU 5-2(2)*

5–24. Jill, an employee at Old Navy, budgets $\frac{1}{5}$ of her yearly salary for clothing. Jill's total clothing bill for the year is $8,000. What is her yearly salary? *LU 5-2(2)* eXcel **5–25.** Bill's Roast Beef sells 5 times as many sandwiches as Pete's Deli. The difference between their sales is 360 sandwiches. How many sandwiches did each sell? *LU 5-2(2)*

My Money **5–26.** Kathy and Mark Smith believe investing in their retirement is critical. Kathy began investing 20% of each paycheck in a retirement account when she was 20 years old. She has saved four times more than Mark, who began saving when he was 35. If their total retirement savings equals $1,450,000, how much is Kathy's investment worth? *LU 5-2(2)*

5–27. A local Ace Hardware sells batteries ($3) and small boxes of pens ($5). In August, total sales were $960. Customers bought 5 times as many batteries as boxes of pens. How many of each did Ace Hardware sell? Check your answer. *LU 5-2(2)*

5–28. Staples sells boxes of pens ($10) and rubber bands ($4). Leona ordered a total of 24 cartons for $210. How many boxes of each did Leona order? Check your answer. *Hint:* Let P = Pens. *LU 5-2(2)*

DRILL PROBLEMS (Third of Three Sets)

Solve the unknown from the following equations: *LU 5-1(2)*

5–29. $A + 90 - 15 = 210$

5–30. $5Y + 15(Y + 1) = 35$

5–31. $3M + 20 = 2M + 80$

5–32. $20(C - 50) = 19{,}000$

WORD PROBLEMS (Third of Three Sets)

5–33. If Colorado Springs, Colorado, has 1.2 times as many days of sunshine as Boston, Massachusetts, how many days of sunshine does each city have if there are a total of 464 days of sunshine between the two in a year? (Round to the nearest day.) *LU 5-2(2)*

5–34. Ben and Jerry's sells 4 times more ice cream cones ($3) than shakes ($8). If last month's sales totaled $4,800, how many of each were sold? Check your answer. *LU 5-2(1)*

5–35. Ivy Corporation gave 84 people a bonus. If Ivy had given 2 more people bonuses, Ivy would have rewarded $\frac{2}{3}$ of the workforce. How large is Ivy's workforce? *LU 5-2(2)*

My Money

5–36. Jim Murray and Phyllis Lowe received a total of $50,000 from a deceased relative's estate. They decided to put $10,000 in a trust for their nephew and divide the remainder. Phyllis received $\frac{3}{4}$ of the remainder; Jim received $\frac{1}{4}$. How much did Jim and Phyllis receive? *LU 5-2(2)*

5–37. The first shift of GME Corporation produced $1\frac{1}{2}$ times as many lanterns as the second shift. GME produced 5,600 lanterns in November. How many lanterns did GME produce on each shift? *LU 5-2(2)*

5–38. Banana Republic at the Orlando, Florida, outlet store sells casual jeans for $40 and dress jeans for $60. If customers bought 5 times more casual than dress jeans and last month's sales totaled $6,500, how many of each type of jeans were sold? Check your answer. *LU 5-2(2)*

5–39. Lowe's sells boxes of wrenches ($100) and hammers ($300). Howard ordered 40 boxes of wrenches and hammers for $8,400. How many boxes of each are in the order? Check your answer. *LU 5-2(2)*

5–40. The Susan Hansen Group in St. George, Utah, sells $16,000,000 of single-family homes and townhomes a year. If single-family homes, with an average selling price of $250,000, sell 3.5 times more often than townhomes, with an average selling price of $190,000, how many of each are sold? (Round to nearest whole.) *LU 5-2(2)*

My Money

5–41. Want to donate to a better cause? Consider micro-lending. Micro-lending is a process where you lend directly to entrepreneurs in developing countries. You can lend starting at $25. Kiva.org boasts a 99% repayment rate. The average loan to an entrepreneur is $388.44 and the average loan amount is $261.14. With a total amount loaned of $283,697,150, how many people are lending money if the average number of loans per lender is 8? (Round final answer to nearest whole lender.) *LU 5-2(2)*

CHALLENGE PROBLEMS

5–42. Myron Corporation is sponsoring a walking race at its company outing. Leona Jackson and Sam Peterson love to walk. Leona walks at the rate of 5 miles per hour. Sam walks at the rate of 6 miles per hour. Assume they start walking from the same place and walk in a straight line. Sam starts $\frac{1}{2}$ hour after Leona. Answer the questions that follow. *Hint:* Distance = Rate × Time. *LU 5-2(2)*

a. How long will it take Sam to meet Leona?

b. How many miles would each have walked?

c. Assume Leona and Sam meet in Lonetown Station where two buses leave along parallel routes in opposite directions. The bus traveling east has a 60 mph speed. The bus traveling west has a 40 mph speed. In how many hours will the buses be 600 miles apart?

5–43. Bessy has 6 times as much money as Bob, but when each earns $6, Bessy will have 3 times as much money as Bob. How much does each have before and after earning the $6? *LU 5-2(2)*

Classroom Notes

SUMMARY PRACTICE TEST

Do you need help? Connect videos have step-by-step worked-out solutions.

1. Delta reduced its round-trip ticket price from Portland to Boston by $140. The sale price was $401.90. What was the original price? *LU 5-2(2)*

2. David Role is an employee of Google. He budgets $\frac{1}{7}$ of his salary for clothing. If David's total clothing for the year is $12,000, what is his yearly salary? *LU 5-2(2)*

3. A local Best Buy sells 8 times as many iPods as Sears. The difference between their sales is 490 iPods. How many iPods did each sell? *LU 5-2(2)*

4. Working at Staples, Jill Reese and Abby Lee sold a total of 1,200 calculators. Jill sold 5 times as many calculators as Abby. How many did each sell? *LU 5-2(2)*

5. Target sells sets of pots ($30) and dishes ($20) at the local store. On the July 4 weekend, Target's total sales were $2,600. People bought 6 times as many pots as dishes. How many of each did Target sell? Check your answer. *LU 5-2(2)*

6. A local Dominos sold a total of 1,600 small pizzas ($9) and pasta dinners ($13) during the Super Bowl. How many of each did Dominos sell if total sales were $15,600? Check your answer. *LU 5-2(2)*

INTERACTIVE VIDEO WORKSHEET

GRADING THE SUMMARY PRACTICE TEST

Go to the summary practice test video in Connect (or click on it here in the ebook). Grade your summary practice test while viewing the video.

C for Correct/I for Incorrect

1. ________	4. ________
2. ________	5. ________
3. ________	6. ________

If you achieved 100%, you are ready for your instructor's exam.

If any of the problems were incorrect, list the questions you missed and show steps to solve the problem correctly.

Replay the video to see if you have made the correct fixes to your mistakes. If you have any questions, contact your instructor asap.

Notes on Watching Videos

MY MONEY

Personal and Financial Goals

What I need to know

Goal setting can appear to be quite the daunting task for many. In fact, when considering the time it takes to effectively plan goals for yourself, you may feel it is not worth the effort. However, the time you invest in establishing and setting realistic, attainable goals will be time well spent as you are more able to achieve your desired results. Goals assist you in determining where you would like to be in the future. With clear goals in mind you are able to create a roadmap toward achievement!

Financial goals allow you the opportunity to identify where you would like to be in the future when it comes to your finances. Maybe there are certain purchases you would like to make in the future like a home or a new car. Or you may want to travel during your retirement. These are all goals that will take some serious planning now in order to reach them in the future. In addition it is important to set personal goals for yourself. You might want to set a goal of increasing your physical activity to improve your overall health. You may also want to pursue a degree that allows you to advance your career and expand your knowledge base. These are personal goals that will set a course for your life toward a sense of achievement and success.

What I need to do

Goal setting requires significant attention and commitment in order to provide the motivation you need to reach the goals you have set for yourself. A guideline to use in setting goals for yourself is to keep the goals realistic, measureable, and a bit of a stretch for you to achieve. A realistic goal will allow you the belief that this can be achieved and help to maintain your motivation in reaching the goal. You also need to be able to measure your goal so that you know whether or not progress is being made. Finally, set goals that require you to put forth an amount of effort that challenges you to achieve.

So, are you ready to set some goals? Here are a couple examples to get you started on this process: Goal #1: Completion of a master's degree in business administration within the next four years. By establishing a time frame for completion of this goal you are able to measure your progress through the courses you are completing. Also, having a stated completion date will place some pressure on you for completion. Goal #2: Save $2,000 toward the purchase of a new car in 18 months. This goal is measurable in both the amount of money to be saved and the time frame in which to complete. Some simple math shows that the monetary amount of this goal can be achieved by saving just over $100 per month ($2,000 divided by 18).

Resources I can use

- Strides Habit Tracker (mobile app)
- https://www.bbvacompass.com/moneyfit/savings-and-budgeting/5-simple-steps-to-setting-financial-goals.html — setting financial goals
- https://www.mindtools.com/page6.html — setting personal goals

MY MONEY ACTIVITY

- Set one financial and one personal goal for yourself for the next month.
- Track your progress and make notes about how you performed.
- Use what you learned to set some longer-range goals for yourself.

A KIPLINGER APPROACH

"How to Solve the Password Problem" by Kaitlin Pitsker from *Kiplinger's*, January 2018, p. 70. Used by permission of The Kiplinger Washington Editors, Inc.

©designer491/Shutterstock

TECH

How to Solve the Password Problem

Forget sign-ons that look like hieroglyphics. You can beat hackers with simple words and phrases.

SECURITY EXPERTS HAVE warned for years that to protect our online accounts we need to change passwords frequently and make sure that those passwords are "complex"—meaning filled with letters, numbers and random characters. But that advice may have done more harm than good. Such passwords are nearly impossible to remember (try recalling something like "Tri3cer&top$"). So many people continue to rely on weak passwords, such as "123456," "password" and "qwerty."

Now, new research shows that not only are complex passwords user-unfriendly, but they're also not hacker-proof. That's partly because once people finally commit passwords to memory, they often reuse them for multiple accounts. That makes "passphrases"—long, easy-to-remember strings of words—a better deterrent to the bad guys.

Creating a strong passphrase. Start by picking a series of unrelated common words—such as *cloud tomato history bridge*—or a phrase that may be obscure but that you can remember. Length is more important than randomness, although many websites currently limit you to, say, a dozen characters. Put capital letters, numbers or special characters within the passphrase, not just at the beginning or the end, says Lorrie Cranor, a computer science professor at Carnegie Mellon University. For example, you could use "Cloud!Tomato2History Bridge." Avoid repetitive or sequential characters, such as "777" or "XYZ," or even using letters that form a pattern on the keyboard.

Lock them up. Still, the average internet user has more than 100 accounts to keep track of. And even the best passwords are easily compromised if you write them down—which is what 73% of people do, according to a 2017 survey by the Pew Research Center.

One solution is to sign up with a password manager that will store all of them behind one master login—the only password you'll need to remember. A password manager can also help you create strong, unique passwords for each of your accounts. Passwords generated by the service will still be long, unpredictable and impossible to remember. But that's okay because you'll never need to type them in yourself.

For example, if you want password manager LastPass (free) to generate a password for you, log into LastPass and then visit the site that you want to add to your LastPass account. Ask to reset your password, then use the LastPass browser extension to generate a new password. Change your password on the site and log in to that account using the newly generated password. A pop-up will ask if you'd like to add the new password to LastPass. After that, LastPass will fill in the new password automatically. The service's premium option ($24 a year) adds a few features, including priority tech support, some multifactor authentication options and 1 gigabyte of encrypted storage. The family plan ($48 a year) allows up to six people to use the service and share log-in information with one another for shared accounts.

To add another layer to your security network, enable two-factor or multistep authentication on any account that allows you to. You'll enter your username and password as usual, but the account will then confirm your identity by asking you to enter a code that has been sent to your smartphone or e-mail address. The extra step deters hackers, and you'll know if an intruder attempts to log in with your password.

KAITLIN PITSKER
kpitsker@kiplinger.com

BUSINESS MATH ISSUE

Solving the password problem by the password manager results in a constant password.

1. List the key points of the article and information to support your position.
2. Write a group defense of your position using math calculations to support your view. If you are in an online course, post to a discussion board.

Classroom Notes

CHAPTER 6

Percents and Their Applications

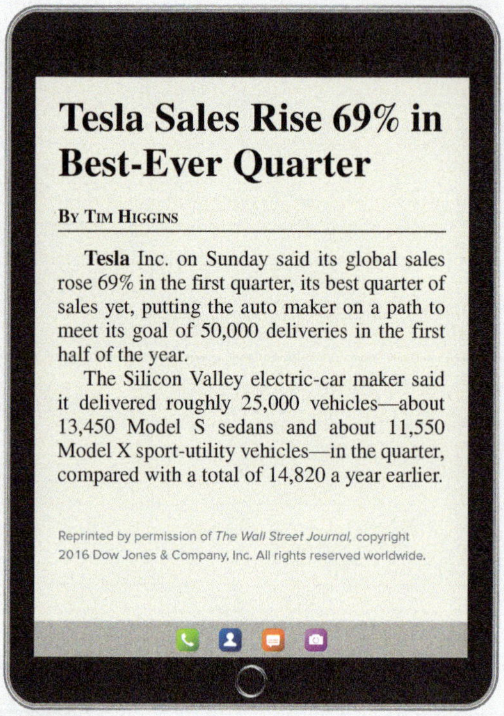

Tesla Sales Rise 69% in Best-Ever Quarter

By Tim Higgins

Tesla Inc. on Sunday said its global sales rose 69% in the first quarter, its best quarter of sales yet, putting the auto maker on a path to meet its goal of 50,000 deliveries in the first half of the year.

The Silicon Valley electric-car maker said it delivered roughly 25,000 vehicles—about 13,450 Model S sedans and about 11,550 Model X sport-utility vehicles—in the quarter, compared with a total of 14,820 a year earlier.

©Hadrian/Shutterstock

©Jef Mallett/Dist. by Universal Uclick.

LEARNING UNIT OBJECTIVES

LU 6–1: Conversions

1. Convert decimals to percents (including rounding percents), percents to decimals, and fractions to percents.
2. Convert percents to fractions.

LU 6–2: Application of Percents—Portion Formula

1. List and define the key elements of the portion formula.
2. Solve for one unknown of the portion formula when the other two key elements are given.
3. Calculate the rate of percent increases and decreases.

Your Guide to Successfully Completing This Chapter

Traditional book or ebook

Check box as you complete each step.

Steps

☐ Read learning unit.

  ☐ Complete practice quiz at the end of the learning unit.

☐ Grade practice quiz using provided solutions. (For more help, watch the learning unit video in Connect and have a Study Session with the authors. Then complete the additional practice quiz in Connect.)

☐ Repeat above for each of the two learning units in Chapter 6.

  ☐ Review chapter organizer.

  ☐ Complete assigned homework.

    ☐ Finish summary practice test. (Go to Connect via the ebook link and do the interactive video worksheet to grade.)

☐ Complete instructor's exam.

Effective in 2018 the new tax act will help many companies post a profit due to the favorable tax rule changes. The corporate tax rate is now 21% versus an old rate of 35%.

The following *Wall Street Journal* clips illustrate the use of percents to show relationships between numbers: Hershey cutting workers by 15% while Twitter trims its workforce by 9%.

GLOBAL

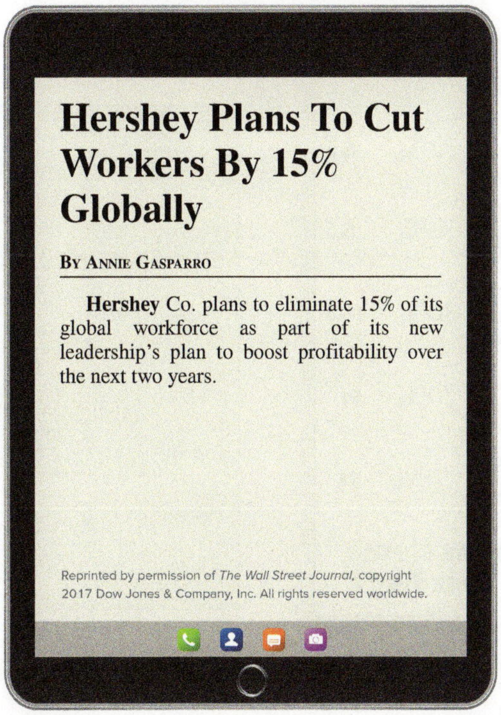

Hershey Plans To Cut Workers By 15% Globally

By Annie Gasparro

Hershey Co. plans to eliminate 15% of its global workforce as part of its new leadership's plan to boost profitability over the next two years.

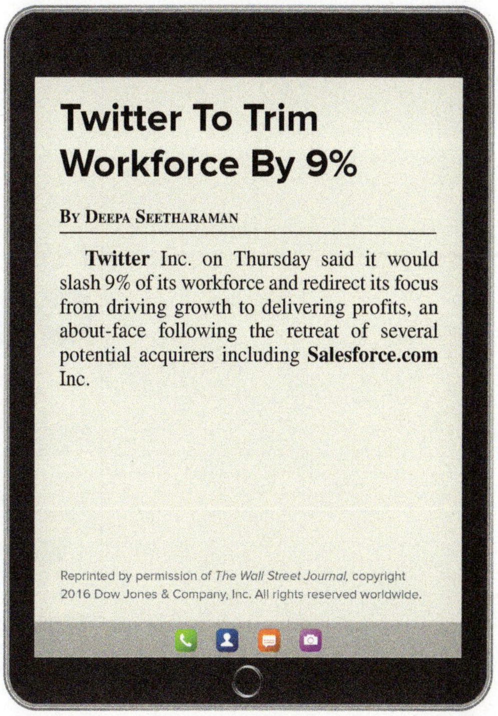

Twitter To Trim Workforce By 9%

By Deepa Seetharaman

Twitter Inc. on Thursday said it would slash 9% of its workforce and redirect its focus from driving growth to delivering profits, an about-face following the retreat of several potential acquirers including **Salesforce.com** Inc.

To understand percents, you should first understand the conversion relationship between decimals, percents, and fractions as explained in Learning Unit 6–1. Then, in Learning Unit 6–2, you will be ready to apply percents to personal and business events.

Learning Unit 6–1: Conversions

LO 1

When we described parts of a whole in previous chapters, we used fractions and decimals. Percents also describe parts of a whole. The word *percent* means per 100. The percent symbol (%) indicates hundredths (division by 100). **Percents** are the result of expressing numbers as part of 100.

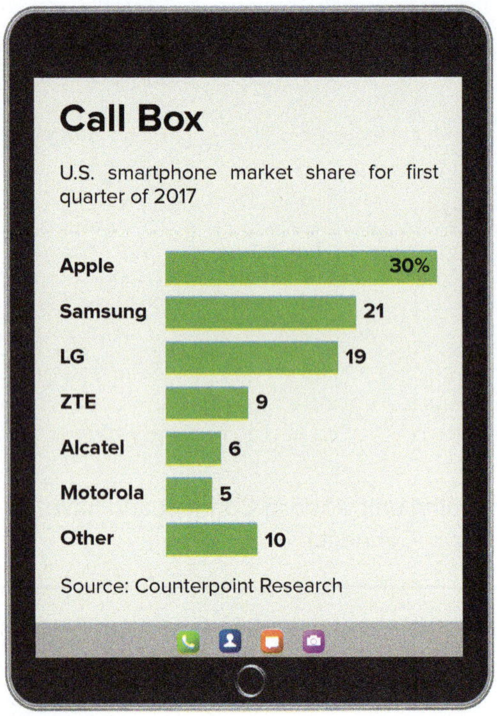

Reprinted by permission of *The Wall Street Journal*, copyright 2017 Dow Jones & Company, Inc. All rights reserved worldwide.

©Anthony Brown/Alamy Stock Photo

Percents can provide some revealing information. The *Wall Street Journal* clip "Call Box" shows Apple in 2017 has 30% of the smartphone market in the U.S.

Let's return to the M&M'S® example from earlier chapters. In Table 6.1, we use our bag of 55 M&M'S® to show how fractions, decimals, and percents can refer to the same parts of a whole. For example, the bag of 55 M&M'S® contains 18 yellow M&M'S®. As you can see in Table 6.1, the 18 candies in the bag of 55 can be expressed as a fraction $(\frac{18}{55})$, decimal (.33), and percent (32.73%). If you visit the M&M'S® website, you will see that the standard is 11 yellow M&M'S®. The clip (below) "What Colors Come in Your Bag?" shows an M&M'S® Milk Chocolate Candies Color Chart.

In this unit we discuss converting decimals to percents (including rounding percents), percents to decimals, fractions to percents, and percents to fractions. You will see when you study converting fractions to percents why you should first learn how to convert decimals to percents.

Converting Decimals to Percents

The *Wall Street Journal* clip "Call Box" (above) shows Apple has 30% of the smartphone market in the U.S. If the clipping had stated the 30% as a decimal (.30), could you give its equivalent in percent? The decimal .30 in decimal fraction is $\frac{30}{100}$. As you know, percents are the result of expressing numbers as part of 100, so $30\% = \frac{30}{100}$. You can now conclude that $.30 = \frac{30}{100} = 30\%$.

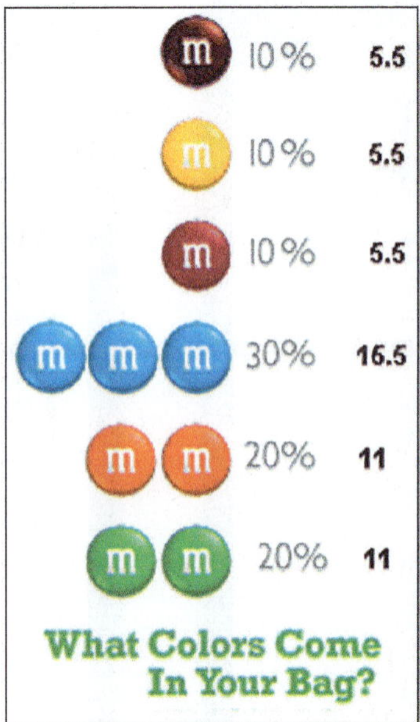

Information adapted from http://us.mms.com/us/about/products/milkchocolate/

TABLE 6.1 Analyzing a bag of M&M'S®

Color	Fraction	Decimal (hundredth)	Percent (hundredth)
Yellow	$\frac{18}{55}$	.33	32.73%
Red	$\frac{10}{55}$	.18	18.18
Blue	$\frac{9}{55}$	.16	16.36
Orange	$\frac{7}{55}$	.13	12.73
Brown	$\frac{6}{55}$	.11	10.91
Green	$\frac{5}{55}$	.09	9.09
Total	$\frac{55}{55} = 1$	1.00	100.00%

Used by permission of Cartoon Features Syndicate

The steps for converting decimals to percents are as follows:

CONVERTING DECIMALS TO PERCENTS

Step 1. Move the decimal point two places to the right. You are multiplying by 100. If necessary, add zeros. This rule is also used for whole numbers and mixed decimals.

Step 2. Add a percent symbol at the end of the number.

EXAMPLES

.30 = .30. = **30%**   .8 = .80. = **80%**   8 = 8.00. = **800%**

(.8: Add 1 zero to make two places.) (8: Add 2 zeros to make two places.)

.425 = .42.5 = **42.5%**   .007 = .00.7 = **.7%**   2.51 = 2.51. = **251%**

Caution: One percent means 1 out of every 100. Since .7% is less than 1%, it means $\frac{7}{10}$ of 1%—a very small amount. Less than 1% is less than .01. To show a number less than 1%, you must use more than two decimal places and add 2 zeros. Example: .7% = .007.

Aha!

Use "D2P" to help you remember how to change a decimal to a percent. "D" stands for "decimal," "2" tells you to move the decimal two places, and "P" stands for "percent." Since P is to the right of D in D2P, we move the decimal two places to the right and add a percent sign:

.159 = 15.9%

Rounding Percents

When necessary, percents should be rounded. Rounding percents is similar to rounding whole numbers. Use the following steps to round percents:

ROUNDING PERCENTS

Step 1. When you convert from a fraction or decimal, be sure your answer is in percent before rounding.

Step 2. Identify the specific digit. If the digit to the right of the identified digit is 5 or greater, round up the identified digit.

Step 3. Delete digits to the right of the identified digit.

For example, Table 6.1 shows that the 18 yellow M&M'S® rounded to the nearest hundredth percent is 32.73% of the bag of 55 M&M'S®. Let's look at how we arrived at this figure.

Step 1.	$\frac{18}{55}$ = .3272727 = 32.72727%	Note that the number is in percent! Identify the hundredth percent digit.
Step 2.	32.73727%	Digit to the right of the identified digit is greater than 5, so the identified digit is increased by 1.
Step 3.	**32.73%**	Delete digits to the right of the identified digit.

When using a calculator, you press [18] [÷] [55] [%]. This allows you to go right to percent, avoiding the decimal step.

Converting Percents to Decimals

Note in the following *Wall Street Journal* clip "What's the Motive" (page 152), 51% of people surveyed strike out on their own to set their own hours.

To convert percents to decimals, you reverse the process used to convert decimals to percents. In our earlier discussion on converting decimals to percents, we asked if the 30% in the "Call Box" clip had been in decimal and not percent, could you convert the decimal to the 30%? Once again, the definition of percent states that $30\% = \frac{30}{100}$. The fraction $\frac{30}{100}$ can be written in decimal form as .30. You can conclude that $30\% = \frac{30}{100} = .30$. Now you can see this procedure in the following conversion steps:

CONVERTING PERCENTS TO DECIMALS

Step 1. Drop the percent symbol.

Step 2. Move the decimal point two places to the left. You are dividing by 100. If necessary, add zeros.

Aha!

Remember our D2P trick for converting decimals to percents? Well good news! It works in reverse for converting percents to decimals. Because we are changing a percent to a decimal, read D2P from right to left, that is, P2D. Start by removing the percent sign and then move the decimal two places to the left because P is to the left of D in P2D.

15.9% = 15.9 = .159

EXAMPLES

Note that when a percent is less than 1%, the decimal conversion has at least two leading zeros before the number .004.

.4% = .00.4 = **.004** — Add 2 zeros to make two places.

2% = .02. = **.02** — Add 1 zero to make two places.

.83% = .00.83 = **.0083** — Add 2 zeros to make two places.

49% = .49. = **.49**

54.5% = .54.5 = **.545**

824.4% = 8.24.4 = **8.244**

Now we must explain how to change fractional percents such as $\frac{1}{5}\%$ to a decimal. Remember that fractional percents are values less than 1%. For example, $\frac{1}{5}\%$ is $\frac{1}{5}$ of 1%. Fractional percents can appear singly or in combination with whole numbers. To convert them to decimals, use the following steps:

CONVERTING FRACTIONAL PERCENTS TO DECIMALS

Step 1. Convert a single fractional percent to its decimal equivalent by dividing the numerator by the denominator. If necessary, round the answer.

Step 2. If a fractional percent is combined with a whole number (mixed fractional percent), convert the fractional percent first. Then combine the whole number and the fractional percent.

Step 3. Drop the percent symbol; move the decimal point two places to the left (this divides the number by 100).

EXAMPLES

$\frac{1}{5}\% = .20\% = .00.20 = \mathbf{.0020}$

$\frac{1}{4}\% = .25\% = .00.25 = \mathbf{.0025}$

$7\frac{3}{4}\% = 7.75\% = .07.75 = \mathbf{.0775}$

$6\frac{1}{2}\% = 6.5\% = .06.5 = \mathbf{.065}$

Think of $7\frac{3}{4}\%$ as

7% =	.07
$+\frac{3}{4}\%$ =	+.0075
$7\frac{3}{4}\%$ =	.0775

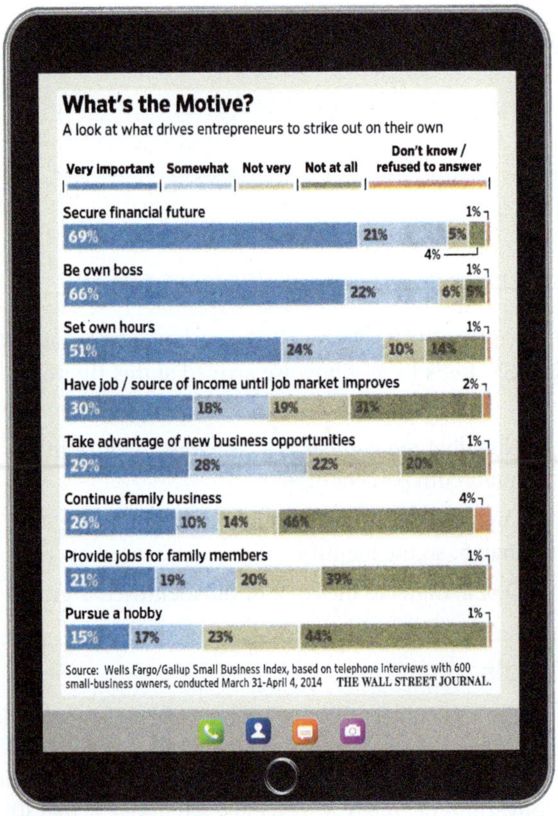

Converting Fractions to Percents

When fractions have denominators of 100, the numerator becomes the percent. Other fractions must be first converted to decimals; then the decimals are converted to percents.

CONVERTING FRACTIONS TO PERCENTS

Step 1. Divide the numerator by the denominator to convert the fraction to a decimal.

Step 2. Move the decimal point two places to the right; add the percent symbol.

EXAMPLES

$\frac{3}{4} = .75 = .75. = 75\%$ $\quad \frac{1}{5} = .20 = .20. = 20\%$ $\quad \frac{1}{20} = .05 = .05. = 5\%$

LO 2

Converting Percents to Fractions

Using the definition of percent, you can write any percent as a fraction whose denominator is 100. Thus, when we convert a percent to a fraction, we drop the percent symbol and write the number over 100, which is the same as multiplying the number by $\frac{1}{100}$. This method of multiplying by $\frac{1}{100}$ is also used for fractional percents.

MONEY tips

Nearly half, 47%, of adult Americans have no life insurance coverage. Consider the impact on survivors. At a minimum, carry burial insurance and a letter of last instruction stating your burial wishes.

CONVERTING A WHOLE PERCENT (OR A FRACTIONAL PERCENT) TO A FRACTION

Step 1. Drop the percent symbol.

Step 2. Multiply the number by $\frac{1}{100}$.

Step 3. Reduce to lowest terms.

EXAMPLES

$$76\% = 76 \times \frac{1}{100} = \frac{76}{100} = \frac{19}{25} \qquad \frac{1}{8}\% = \frac{1}{8} \times \frac{1}{100} = \frac{1}{800}$$

$$156\% = 156 \times \frac{1}{100} = \frac{156}{100} = 1\frac{56}{100} = 1\frac{14}{25}$$

Sometimes a percent contains a whole number and a fraction such as $12\frac{1}{2}\%$ or 22.5%. Extra steps are needed to write a mixed or decimal percent as a simplified fraction.

CONVERTING A MIXED OR DECIMAL PERCENT TO A FRACTION

Step 1. Drop the percent symbol.

Step 2. Change the mixed percent to an improper fraction.

Step 3. Multiply the number by $\frac{1}{100}$.

Step 4. Reduce to lowest terms.

Note: If you have a mixed or decimal percent, change the decimal portion to its fractional equivalent and continue with Steps 1 to 4.

EXAMPLES

$$12\frac{1}{2}\% = \frac{25}{2} \times \frac{1}{100} = \frac{25}{200} = \frac{1}{8}$$

$$12.5\% = 12\frac{1}{2}\% = \frac{25}{2} \times \frac{1}{100} = \frac{25}{200} = \frac{1}{8}$$

$$22.5\% = 22\frac{1}{2}\% = \frac{45}{2} \times \frac{1}{100} = \frac{45}{200} = \frac{9}{40}$$

It's time to check your understanding of Learning Unit 6–1.

LU 6–1 PRACTICE QUIZ

Complete this Practice Quiz to see how you are doing.

Convert to percents (round to the nearest tenth percent as needed):

1. .6666 ______ **2.** .832 ______

3. .004 ______ **4.** 8.94444 ______

Convert to decimals (remember, decimals representing less than 1% will have at least 2 leading zeros before the number):

5. $\frac{1}{4}\%$ ______ **6.** $6\frac{3}{4}\%$ ______

7. 87% ______ **8.** 810.9% ______

Convert to percents (round to the nearest hundredth percent):

9. $\frac{1}{7}$ ______ **10.** $\frac{2}{9}$ ______

Convert to fractions (remember, if it is a mixed number, first convert to an improper fraction):

11. 19% ______ **12.** $71\frac{1}{2}\%$ ______ **13.** 130% ______

14. $\frac{1}{2}\%$ ______ **15.** 19.9% ______

*For **extra help** from your authors–Sharon and Jeff–see the videos in Connect.*

✓ Solutions

1. .66.66 = **66.7%** **2.** .83.2 = **83.2%**

3. .00.4 = **.4%** **4.** 8.94.444 = **894.4%**

5. $\frac{1}{4}\% = .25\% =$ **.0025** **6.** $6\frac{3}{4}\% = 6.75\% =$ **.0675**

7. 87% = .87. = **.87** **8.** 810.9% = 8.10.9 = **8.109**

9. $\frac{1}{7} = .14.285 =$ **14.29%** **10.** $\frac{2}{9} = .22.2\overline{2} =$ **22.22%**

11. $19\% = 19 \times \frac{1}{100} = \frac{19}{100}$ **12.** $71\frac{1}{2}\% = \frac{143}{2} \times \frac{1}{100} = \frac{143}{200}$

13. $130\% = 130 \times \frac{1}{100} = \frac{130}{100} = 1\frac{30}{100} = 1\frac{3}{10}$ **14.** $\frac{1}{2}\% = \frac{1}{2} \times \frac{1}{100} = \frac{1}{200}$

15. $19\frac{9}{10}\% = \frac{199}{10} \times \frac{1}{100} = \frac{199}{1,000}$

Learning Unit 6–2: Application of Percents—Portion Formula

LO 1

The bag of M&M'S® we have been studying contains Milk Chocolate M&M'S®. M&M/Mars also makes Peanut M&M'S® and some other types of M&M'S®. To study the application of percents to problems involving M&M'S®, we make two key assumptions:

1. Total sales of Milk Chocolate M&M'S®, Peanut M&M'S®, and other M&M'S® chocolate candies are $400,000.
2. Eighty percent of M&M'S® sales are Milk Chocolate M&M'S®. This leaves the Peanut and other M&M'S® chocolate candies with 20% of sales (100% – 80%).

80% M&M'S® Milk Chocolate M&M'S®	+	20% M&M'S® Peanut and other chocolate candies	=	100% Total sales ($400,000)

Before we begin, you must understand the meaning of three terms—*base, rate,* and *portion.* These terms are the key elements in solving percent problems.

- Base (***B***). The **base** is the beginning whole quantity or value (100%) with which you will compare some other quantity or value. Often the problems give the base after the word *of.* For example, the whole (total) sales of M&M'S®—Milk Chocolate M&M'S, Peanut, and other M&M'S® chocolate candies—are $400,000.

- Rate (***R***). The **rate** is a percent, decimal, or fraction that indicates the part of the base that you must calculate. The percent symbol often helps you identify the rate. For example, Milk Chocolate M&M'S® currently account for 80% of sales. So the rate is 80%. Remember that 80% is also $\frac{4}{5}$, or .80.
- **Portion (*P*).** The **portion** is the amount or part that results from the base multiplied by the rate. For example, total sales of M&M'S® are $400,000 (base); $400,000 times .80 (rate) equals $320,000 (portion), or the sales of Milk Chocolate M&M'S®. *A key point to remember is that portion is a number and not a percent. In fact, the portion can be larger than the base if the rate is greater than 100%.*

Solving Percents with the Portion Formula

LO 2

In problems involving portion, base, and rate, we give two of these elements. You must find the third element. Remember the following key formula:

Portion (*P*) = Base (*B*) × Rate (*R*)

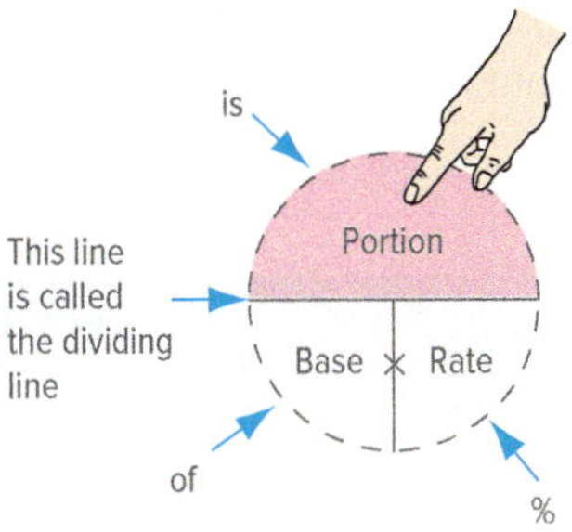

To help you solve for the portion, base, and rate, this unit shows pie charts. The shaded area in each pie chart indicates the element that you must solve for. For example, since we shaded *portion* in the pie chart at the left, you must solve for portion. To use the pie charts, put your finger on the shaded area (in this case portion). The formula that remains tells you what to do. So in the pie chart at the left, you solve the problem by multiplying base by the rate. Note the circle around the pie chart is broken since we want to emphasize that portion can be larger than base if rate is greater than 100%. The horizontal line in the pie chart is called the dividing line, and we will use it when we solve for base or rate.

The following example summarizes the concept of base, rate, and portion. Assume that you received a small bonus check of $100. This is a gross amount—your company did not withhold any taxes. You will have to pay 20% in taxes.

Base: 100%—whole. Usually given after the word *of*—but not always.	**Rate:** Usually expressed as a percent but could also be a decimal or fraction.	**Portion:** A number—not a percent and not the whole.
$100 bonus check	20% taxes	$20 taxes

First decide what you are looking for. You want to know how much you must pay in taxes—the portion. How do you get the portion? From the portion formula Portion (*P*) = Base (*B*) × Rate (*R*), you know that you must multiply the base ($100) by the rate (20%). When you do this, you get $100 × .20 = $20. So you must pay $20 in taxes.

Let's try our first word problem by taking a closer look at the M&M'S® example to see how we arrived at the $320,000 sales of Milk Chocolate M&M'S® given earlier. We will be using blueprint aids to help dissect and solve each word problem.

Solving for Portion

The Word Problem Sales of Milk Chocolate M&M'S® are 80% of the total M&M'S® sales. Total M&M'S® sales are $400,000. What are the sales of Milk Chocolate M&M'S®?

BLUEPRINT

The facts	Solving for?	Steps to take	Key points
Milk Chocolate M&M'S® sales: 80%. *Total M&M'S® sales:* $400,000.	Sales of Milk Chocolate M&M'S®.	Identify key elements. *Base:* $400,000. *Rate:* .80. *Portion:* ? Portion = Base × Rate.	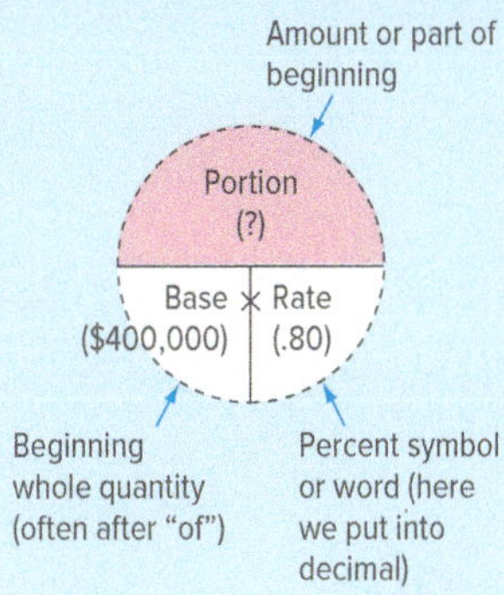 Portion and rate must relate to same piece of base.

Steps to solving problem

1. Set up the formula. Portion = Base × Rate
2. Calculate portion (sales of Milk Chocolate M&M'S®). $P = \$400{,}000 \times .80$
   $P = \$320{,}000$

In the first column of the blueprint aid, we gather the facts. In the second column, we state that we are looking for sales of Milk Chocolate M&M'S®. In the third column, we identify each key element and the formula needed to solve the problem. Review the pie chart in the fourth column. *The portion and rate must relate to the same piece of the base.* In this word problem, we can see from the solution below the blueprint aid that sales of Milk Chocolate M&M'S® are $320,000. The $320,000 does indeed represent 80% of the base. Note here that the portion ($320,000) is less than the base of $400,000 since the rate is less than 100%.

Now let's work another word problem that solves for the portion.

The Word Problem Sales of Milk Chocolate M&M'S® are 80% of the total M&M'S® sales. Total M&M'S® sales are $400,000. What are the sales of Peanut and other M&M'S® chocolate candies?

BLUEPRINT	The facts	Solving for?	Steps to take	Key points
	Milk Chocolate M&M'S® sales: 80%. *Total M&M'S® sales:* $400,000.	Sales of Peanut and other M&M'S® chocolate candies.	Identify key elements. *Base:* $400,000. *Rate:* .20 (100% − 80%). *Portion:* ? Portion = Base × Rate.	If 80% of sales are Milk Chocolate M&M'S, then 20% are Peanut and other M&M'S® chocolate candies. 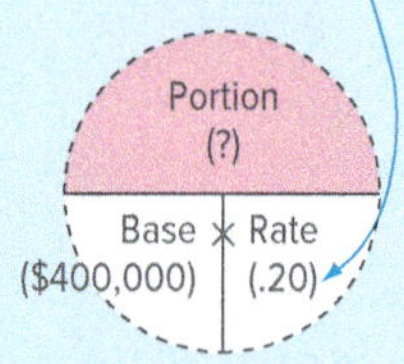 Portion and rate must relate to same piece of base.

Steps to solving problem

1. Set up the formula. Portion = Base × Rate
2. Calculate portion (sale of Peanut and other M&M'S® chocolate candies). $P = \$400{,}000 \times .20$
   $P = \$80{,}000$

In the previous blueprint aid, *note that we must use a rate that agrees with the portion so the portion and rate refer to the same piece of the base.* Thus, if 80% of sales are Milk Chocolate M&M'S®, 20% must be Peanut and other M&M'S® chocolate candies (100% − 80% = 20%). So we use a rate of .20.

In Step 2, we multiplied $400,000 × .20 to get a portion of $80,000. This portion represents the part of the sales that were *not* Milk Chocolate M&M'S®. Note that the rate of .20 and the portion of $80,000 relate to the same piece of the base—$80,000 is 20% of $400,000. Also note that the portion ($80,000) is less than the base ($400,000) since the rate is less than 100%.

Take a moment to review the two blueprint aids in this section. Be sure you understand why the rate in the first blueprint aid was 80% and the rate in the second blueprint aid was 20%.

Solving for Rate

The Word Problem Sales of Milk Chocolate M&M'S® are $320,000. Total M&M'S® sales are $400,000. What is the percent of Milk Chocolate M&M'S® sales compared to total M&M'S® sales?

	The facts	Solving for?	Steps to take	Key points
BLUEPRINT	*Milk Chocolate M&M'S® sales:* \$320,000. *Total M&M'S® sales:* \$400,000.	Percent of Milk Chocolate M&M'S® sales to total M&M'S® sales.	Identify key elements. *Base:* \$400,000. *Rate:* ? *Portion:* \$320,000 $\text{Rate} = \frac{\text{Portion}}{\text{Base}}$	Since portion is less than base, the rate must be less than 100% 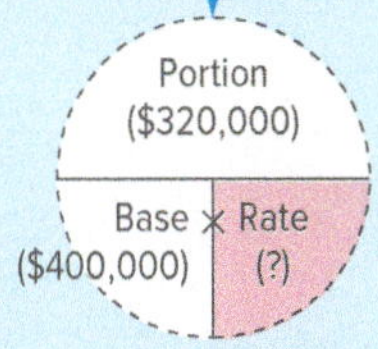 Portion and rate must relate to the same piece of base.

Steps to solving problem

1. Set up the formula. $\text{Rate} = \frac{\text{Portion}}{\text{Base}}$

2. Calculate rate (percent of Milk Chocolate M&M'S® sales). $R = \frac{\$320{,}000}{\$400{,}000}$

$R = 80\%$

Note that in this word problem, the rate of 80% and the portion of \$320,000 refer to the same piece of the base.

The Word Problem Sales of Milk Chocolate M&M'S® are \$320,000. Total sales of Milk Chocolate M&M'S, Peanut, and other M&M'S® chocolate candies are \$400,000. What percent of Peanut and other M&M'S® chocolate candies are sold compared to total M&M'S® sales?

	The facts	Solving for?	Steps to take	Key points
BLUEPRINT	*Milk Chocolate M&M'S® sales:* \$320,000. *Total M&M'S® sales:* \$400,000.	Percent of Peanut and other M&M'S® chocolate candies sales compared to total M&M'S® sales.	Identify key elements. *Base:* \$400,000. *Rate:* ? *Portion:* \$80,000 (\$400,000 − \$320,000). $\text{Rate} = \frac{\text{Portion}}{\text{Base}}$	Represents sales of Peanut and other M&M'S® chocolate candies 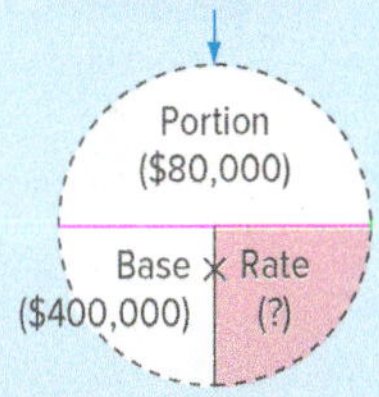 When portion becomes \$80,000, the portion and rate now relate to same piece of base.

Steps to solving problem

1. Set up the formula. $\text{Rate} = \frac{\text{Portion}}{\text{Base}}$

2. Calculate rate. $R = \frac{\$80{,}000}{\$400{,}000}$ (\$400,000 − \$320,000)

$R = 20\%$

The word problem asks for the rate of candy sales that are *not* Milk Chocolate M&M'S. Thus, \$400,000 of total candy sales less sales of Milk Chocolate M&M'S® (\$320,000) allows us to arrive at sales of Peanut and other M&M'S® chocolate candies (\$80,000). The \$80,000 portion represents 20% of total candy sales. The \$80,000 portion and 20% rate refer to the same piece of the \$400,000 base. Compare this blueprint aid with the blueprint aid for the previous word problem. Ask yourself why in the previous word problem the rate was 80% and in this word problem the rate is 20%. In both word problems, the portion was less than the base since the rate was less than 100%.

Now we go on to calculate the base. Remember to read the word problem carefully so that you match the rate and portion to the same piece of the base.

Solving for Base

The Word Problem Sales of Peanut and other M&M'S® chocolate candies are 20% of total M&M'S® sales. Sales of Milk Chocolate M&M'S® are $320,000. What are the total sales of all M&M'S®?

BLUEPRINT	The facts	Solving for?	Steps to take	Key points
	Peanut and other M&M'S® chocolate candies sales: 20%. *Milk Chocolate M&M'S® sales:* $320,000.	Total M&M'S® sales.	Identify key elements. *Base:* ? *Rate:* .80 (100% − 20%) *Portion:* $320,000 $\text{Base} = \frac{\text{Portion}}{\text{Rate}}$	Portion ($320,000) Base × Rate (?) (.80) (100% − 20%) Portion ($320,000) and rate (.80) do relate to the same piece of base.

Steps to solving problem

1. Set up the formula. $\text{Base} = \frac{\text{Portion}}{\text{Rate}}$
2. Calculate the base. $B = \frac{\$320{,}000}{.80}$ ← $320,000 is 80% of base

   $B = \$400{,}000$

Note that we could not use 20% for the rate. The $320,000 of Milk Chocolate M&M'S® represents 80% (100% − 20%) of the total sales of M&M'S®. We use 80% so that the portion and rate refer to same piece of the base. Remember that the portion ($320,000) is less than the base ($400,000) since the rate is less than 100%.

Calculating Percent Increases and Decreases

LO 3

The following *Wall Street Journal* clip shows Procter & Gamble is keeping the price of Tide at $11.99 but reducing the number of loads the new product washes from 60 to only 48 loads. Let's calculate the cost per load (rounded to nearest cent) before and after the load change:

Before

$$\frac{\$11.99}{60 \text{ loads}} = \$.20 \text{ per load}$$

After

$$\frac{\$11.99}{48 \text{ loads}} = \$.25 \text{ per load}$$

Using this clip, let's look at how to calculate percent increases and decreases.

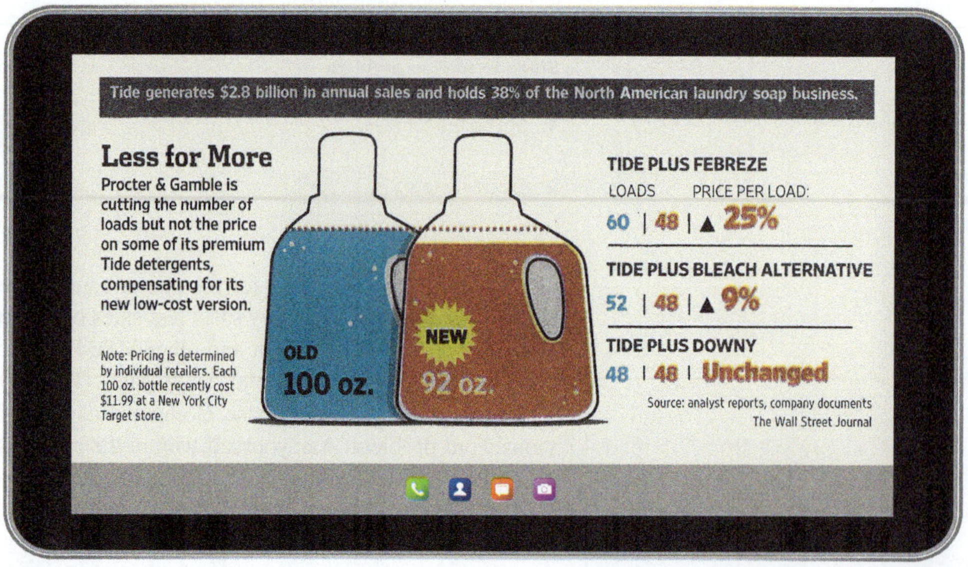

The Tide Example: Rate of Percent Increase in Price per Load Assume: per load cost increase from \$.20 to \$.25

$$\text{Rate} = \frac{\text{Portion}}{\text{Base}} \quad \begin{array}{l} \leftarrow \text{Difference between old and new price per load} \\ \leftarrow \text{Old price per load} \end{array}$$

$$R = \frac{\$.05}{\$.20} \qquad (\$.25 - \$.20)$$

$$R = 25\% \text{ increase}$$

Let's prove the 25% with a pie chart.

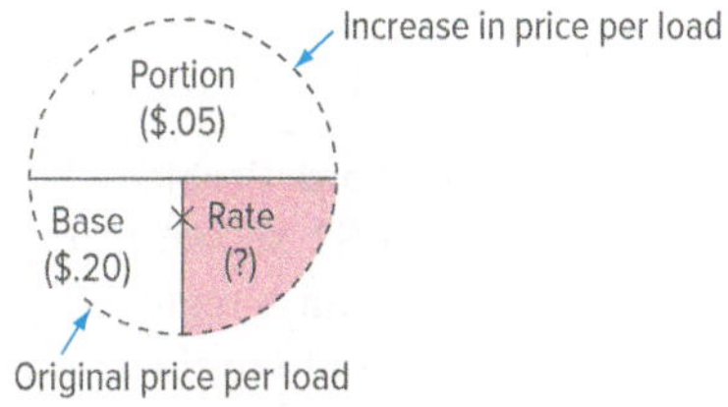

The formula for calculating **percent increase** is as follows:

Percent increase

$$\text{Percent of increase } (R) = \frac{\text{Amount of price per load increase } (P) \ (\$.05)}{\text{Original price per load } (B) \ (\$.20)} \quad (25\%)$$

Now let's look at how to calculate the math for a decrease in price per load for Tide.

The Tide Example: Rate of Percent Decrease Assume: Price of \$11.99 but loads increase from 60 to 70. The first step is to calculate the price per load (rounded to nearest cent) before and after:

Before	**After**
$\frac{\$11.99}{60} = \$.20$ per load	$\frac{\$11.99}{70} = \$.17$ per load

$$\text{Rate} = \frac{\text{Portion}}{\text{Base}} \quad \begin{array}{l} \leftarrow \text{Difference between old and new price per load} \\ \leftarrow \text{Old price per load} \end{array}$$

$$R = \frac{\$.03}{\$.20} \qquad (\$.20 - \$.17 = \$.03)$$

$$R = 15\% \text{ decrease}$$

Let's prove the 15% with a pie chart.

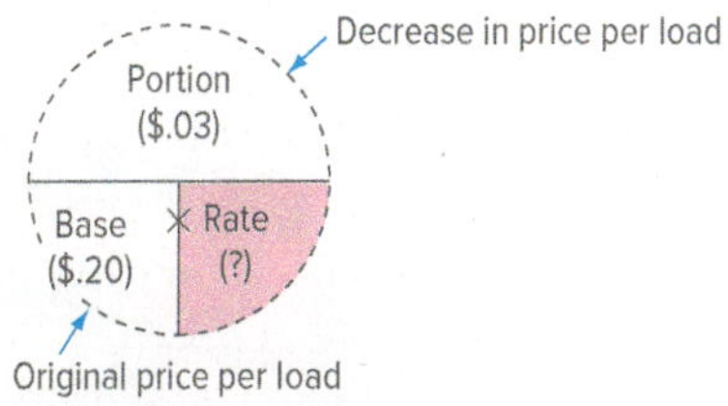

The formula for calculating **percent decrease** is as follows:

Percent decrease

$$\text{Percent of decrease } (R) = \frac{\text{Amount of price per load decrease } (P) \ (\$.03)}{\text{Original price per load } (B) \ (\$.20)} \quad (15\%)$$

In conclusion, the following steps can be used to calculate percent increases and decreases:

CALCULATING PERCENT INCREASES AND DECREASES

Step 1. Find the difference between amounts (such as sales).

Step 2. Divide Step 1 by the original amount (the base): $R = P \div B$. Be sure to express your answer in percent.

Before concluding this chapter, we will show how to calculate a percent increase and decrease using M&M'S® (Figure 6.1).

Additional Examples Using M&M'S

The Word Problem Sheila Leary went to her local supermarket and bought the bag of M&M'S® shown in Figure 6.1. The bag gave its weight as 18.40 ounces, which was 15% more than a regular 1-pound bag of M&M'S®. Sheila, who is a careful shopper, wanted to check and see if she was actually getting a 15% increase. Let's help Sheila dissect and solve this problem.

BLUEPRINT	The facts	Solving for?	Steps to take	Key points
	New bag of M&M'S®: 18.40 oz. 15% increase in weight. *Original bag of M&M'S®:* 16 oz. (1 lb.)	Checking percent increase of 15%.	Identify key elements. *Base:* 16 oz. *Rate:* ? *Portion:* 2.40 oz. (18.40 oz. − 16.00 = 2.40 oz.) $\text{Rate} = \frac{\text{Portion}}{\text{Base}}$	Difference between base and new weight Portion (2.40 oz.) Base (16 oz.) × Rate (?) Original amount sold

Steps to solving problem

1. Set up the formula. $\text{Rate} = \frac{\text{Portion}}{\text{Base}}$

2. Calculate the rate. $R = \frac{2.40 \text{ oz.}}{16.00 \text{ oz.}}$ ← Difference between base and new weight. ← Old weight equals 100%.

   $R = 15\%$ increase

The new weight of the bag of M&M'S® is really 115% of the old weight:

$$\begin{array}{rcl} 16.00 \text{ oz.} & = & 100\% \\ +\ 2.40 & = & +\ 15 \\ \hline 18.40 \text{ oz.} & = & 115\% = 1.15 \end{array}$$

FIGURE 6.1

Bag of 18.40-ounce M&M'S®

We can check this by looking at the following pie chart:

$$\text{Portion} = \text{Base} \times \text{Rate}$$

$$18.40 \text{ oz.} = 16 \text{ oz.} \times 1.15$$

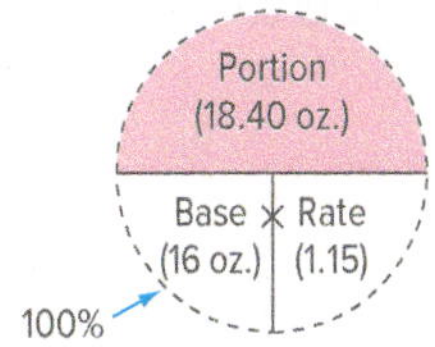

Why is the portion greater than the base? Remember that the portion can be larger than the base only if the rate is greater than 100%. Note how the portion and rate relate to the same piece of the base—18.40 oz. is 115% of the base (16 oz.).

Let's see what could happen if M&M/Mars has an increase in its price of sugar. This is an additional example to reinforce the concept of percent decrease.

MONEY tips

When planning for retirement, a rule of thumb is that you will need 70% of your preretirement pay to live comfortably. This number assumes your house is paid off and you are in good health. Automating your savings can be a huge factor in helping you reach your goals. So, begin planning early in life and start saving for a financially sound retirement.

The Word Problem The increase in the price of sugar caused the M&M/Mars company to decrease the weight of each 1-pound bag of M&M'S® to 12 ounces. What is the rate of percent decrease?

BLUEPRINT	The facts	Solving for?	Steps to take	Key points
	16-oz. bag of M&M'S®: reduced to 12 oz.	Rate of percent decrease.	Identify key elements. *Base:* 16 oz. *Rate:* ? *Portion:* 4 oz. (16 oz. − 12 oz.) $\text{Rate} = \frac{\text{Portion}}{\text{Base}}$	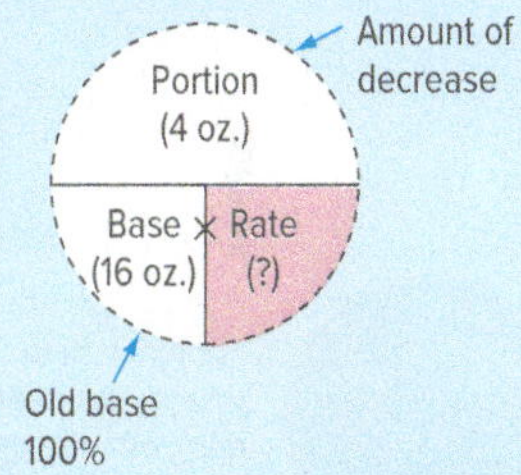

Steps to solving problem

1. Set up the formula. $\text{Rate} = \frac{\text{Portion}}{\text{Base}}$
2. Calculate the rate. $R = \frac{4 \text{ oz.}}{16.00 \text{ oz.}}$

   $R = 25\%$ decrease

The new weight of the bag of M&M'S® is 75% of the old weight:

$$\begin{array}{rr} 16 \text{ oz.} = & 100\% \\ -\ 4 & -\ 25 \\ \hline 12 \text{ oz.} = & 75\% \end{array}$$

We can check this by looking at the following pie chart:

$$\text{Portion} = \text{Base} \times \text{Rate}$$

$$12 \text{ oz.} = 16 \text{ oz.} \times .75$$

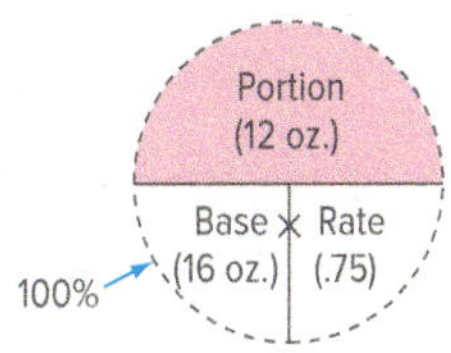

Note that the portion is smaller than the base because the rate is less than 100%. Also note how the portion and rate relate to the same piece of the base—12 ounces is 75% of the base (16 oz.).

After your study of Learning Unit 6–2, you should be ready for the Practice Quiz.

LU 6–2 PRACTICE QUIZ

Complete this **Practice Quiz** to see how you are doing.

Solve for portion:

1. 38% of 900. **2.** 60% of $9,000.

Solve for rate (round to the nearest tenth percent as needed):

3. 430 is ______% of 5,000. **4.** 200 is ______% of 700.

Solve for base (round to the nearest tenth as needed):

5. 55 is 40% of ______. **6.** 900 is $4\frac{1}{2}$% of ______.

Solve the following (blueprint aids are shown in the solution; you might want to try some on scrap paper):

7. Five out of 25 students in Professor Ford's class received an A grade. What percent of the class *did not* receive the A grade?

8. Abby Biernet has yet to receive 60% of her lobster order. Abby received 80 lobsters to date. What was her original order?

9. Assume in 2019 Dunkin' Donuts Company had $300,000 in doughnut sales. In 2020, sales were up 40%. What are Dunkin' Donuts sales for 2020?

10. The price of an Apple computer dropped from $1,600 to $1,200. What was the percent decrease?

11. In 1982, a ticket to the Boston Celtics cost $14. In 2020, a ticket cost $50. What is the percent increase to the nearest hundredth percent?

For ***extra help*** *from your authors–Sharon and Jeff–see the videos in Connect.*

✓ Solutions

1. $342 = 900 \times .38$
$(P) = (B) \times (R)$

2. $\$5,400 = \$9,000 \times .60$
$(P) = (B) \times (R)$

3. $\frac{(P)430}{(B)5,000} = .086 = 8.6\%\ (R)$

4. $\frac{(P)200}{(B)700} = .2857 = 28.6\%\ (R)$

5. $\frac{(P)55}{(R).40} = 137.5\ (B)$

6. $\frac{(P)900}{(R).045} = 20,000\ (B)$

7. Percent of Professor Ford's class that did not receive an A grade:

BLUEPRINT	The facts	Solving for?	Steps to take	Key points
	5 As. 25 in class.	Percent that did not receive A.	Identify key elements. *Base:* 25 *Rate:* ? *Portion:* 20 (25 – 5) $\text{Rate} = \frac{\text{Portion}}{\text{Base}}$	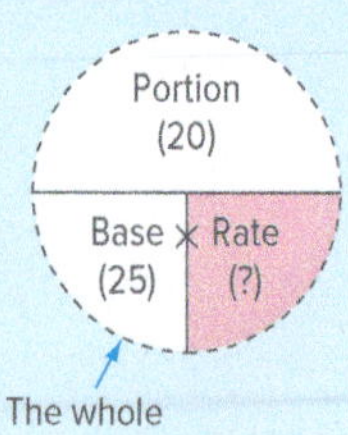 Portion and rate must relate to same piece of base.

Steps to solving problem

1. Set up the formula. $\text{Rate} = \frac{\text{Portion}}{\text{Base}}$

2. Calculate the base rate. $R = \frac{20}{25}$

$R = 80\%$

8. Abby Biernet's original order:

BLUEPRINT	The facts	Solving for?	Steps to take	Key points
	60% of the order not in. 80 lobsters received.	Total order of lobsters.	Identify key elements. *Base:* ? *Rate:* .40 (100% − 60%) *Portion:* 80 $\text{Base} = \frac{\text{Portion}}{\text{Rate}}$	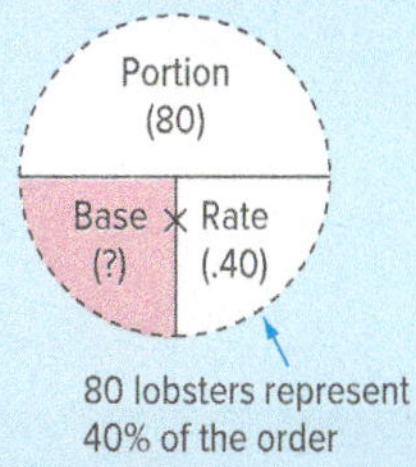 Portion and rate must relate to same piece of base.

Steps to solving problem

1. Set up the formula. $\text{Base} = \frac{\text{Portion}}{\text{Rate}}$
2. Calculate the base rate. $B = \frac{80}{.40}$ ← 80 lobsters is 40% of base.

$B = 200$ lobsters

9. Dunkin' Donuts Company sales for 2019:

BLUEPRINT	The facts	Solving for?	Steps to take	Key points
	2019: $300,000 sales. *2020:* Sales up 40% from 2019.	Sales for 2020.	Identify key elements. *Base:* $300,000. *Rate:* 1.40. Old year 100% New year + 40 140% *Portion:* ? Portion = Base × Rate.	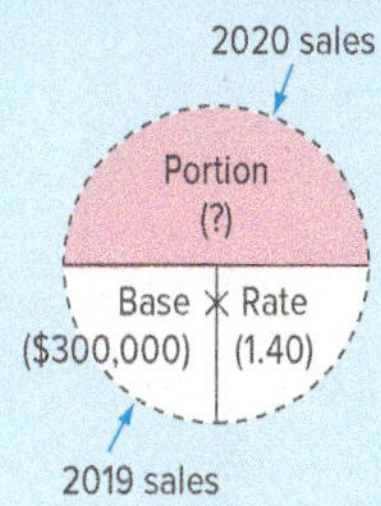 When rate is greater than 100%, portion will be larger than base.

Steps to solving problem

1. Set up the formula. Portion = Base × Rate
2. Calculate the portion. $P = \$300{,}000 \times 1.40$

$P = \$420{,}000$

10. Percent decrease in Apple computer price:

BLUEPRINT	The facts	Solving for?	Steps to take	Key points
	Apple computer was $1,600; now, $1,200.	Percent decrease in price.	Identify key elements. *Base:* $1,600. *Rate:* ? *Portion:* $400 ($1,600 − $1,200). $\text{Rate} = \frac{\text{Portion}}{\text{Base}}$	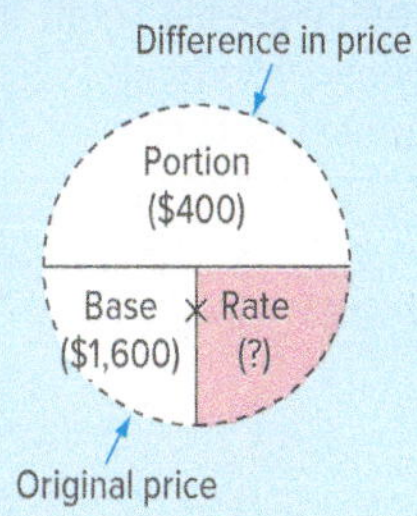

Steps to solving problem

1. Set up the formula. $\text{Rate} = \dfrac{\text{Portion}}{\text{Base}}$

2. Calculate the rate. $R = \dfrac{\$400}{\$1{,}600}$

$R = 25\%$

11. Percent increase in Boston Celtics ticket:

©Cal Sport Media/Alamy Stock

BLUEPRINT	The facts	Solving for?	Steps to take	Key points
	$14 ticket (old). $50 ticket (new).	Percent increase in price.	Identify key elements. *Base:* $14 *Rate:* ? *Portion:* $36 ($50 − $14) $\text{Rate} = \dfrac{\text{Portion}}{\text{Base}}$	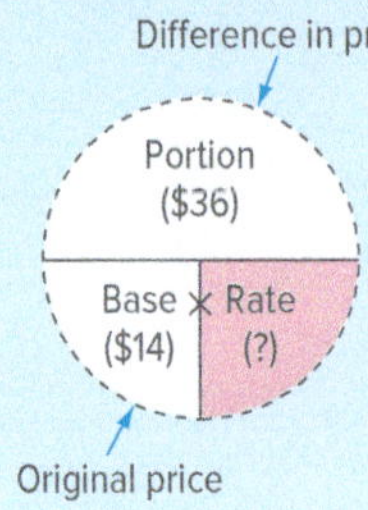 When portion is greater than base, rate will be greater than 100%.

Steps to solving problem

1. Set up the formula. $\text{Rate} = \dfrac{\text{Portion}}{\text{Base}}$

2. Calculate the rate. $R = \dfrac{\$36}{\$14}$

$R = 2.5714 = 257.14\%$

INTERACTIVE CHAPTER ORGANIZER

Topic/Procedure/Formula	Examples	You try it*
Converting decimals to percents **1.** Move decimal point two places to right. If necessary, add zeros. This rule is also used for whole numbers and mixed decimals. **2.** Add a percent symbol at end of number.	.81 = .81. = 81% .008 = .00.8 = .8% 4.15 = 4.15. = 415%	**Convert to percent** .92 .009 5.46
Rounding percents **1.** Answer must be in percent before rounding. **2.** Identify specific digit. If digit to right is 5 or greater, round up. **3.** Delete digits to right of identified digit.	Round to the nearest hundredth percent. $\frac{3}{7} = .4285714 = 42.85714 = 42.86\%$	**Round to the nearest hundredth percent** $\frac{2}{9}$

(continues)

INTERACTIVE CHAPTER ORGANIZER

Topic/Procedure/Formula	Examples	You try it*
Converting percents to decimals **1.** Drop percent symbol. **2.** Move decimal point two places to left. If necessary, add zeros. For fractional percents: **1.** Convert to decimal by dividing numerator by denominator. If necessary, round answer. **2.** If a mixed fractional percent, convert fractional percent first. Then combine whole number and fractional percent. **3.** Drop percent symbol; move decimal point two places to left.	.89% = .0089 95% = .95 195% = 1.95 $8\frac{3}{4}\% = 8.75\% = .0875$ $\frac{1}{4}\% = .25\% = .0025$ $\frac{1}{5}\% = .20\% = .0020$	**Convert to decimal** .78% 96% 246% $7\frac{3}{4}\%$ $\frac{3}{4}\%$ $\frac{1}{2}\%$
Converting fractions to percents **1.** Divide numerator by denominator. **2.** Move decimal point two places to right; add percent symbol.	$\frac{4}{5} = .80 = 80\%$	**Convert to percent** $\frac{3}{5}$
Converting percents to fractions Whole percent (or fractional percent) to a fraction: **1.** Drop percent symbol. **2.** Multiply number by $\frac{1}{100}$. **3.** Reduce to lowest terms. Mixed or decimal percent to a fraction: **1.** Drop percent symbol. **2.** Change mixed percent to an improper fraction. **3.** Multiply number by $\frac{1}{100}$. **4.** Reduce to lowest terms. If you have a mixed or decimal percent, change decimal portion to fractional equivalent and continue with Steps 1 to 4.	$64\% \rightarrow 64 \times \frac{1}{100} = \frac{64}{100} = \frac{16}{25}$ $\frac{1}{4}\% \rightarrow \frac{1}{4} \times \frac{1}{100} = \frac{1}{400}$ $119\% \rightarrow 119 \times \frac{1}{100} = \frac{119}{100} = 1\frac{19}{100}$ $16\frac{1}{4}\% \rightarrow \frac{65}{4} \times \frac{1}{100} = \frac{65}{400} = \frac{13}{80}$ $16.25\% \rightarrow 16\frac{1}{4}\% = \frac{65}{4} \times \frac{1}{100}$ $= \frac{65}{400} = \frac{13}{80}$	**Convert to fractions** 74% $\frac{1}{5}\%$ 121% $17\frac{1}{5}\%$ 17.75%
Solving for portion 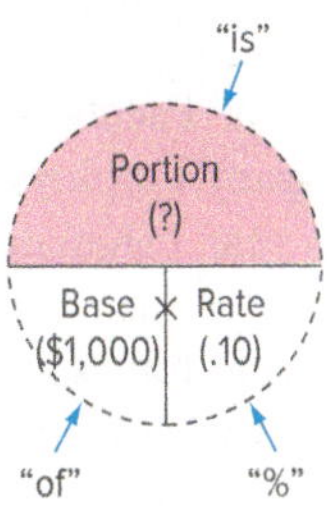	10% of Mel's paycheck of $1,000 goes for food. What portion is deducted for food? $100 = $1,000 × .10 *Note:* If question was what amount does not go for food, the portion would have been: $900 = $1,000 × .90 (100% − 10% = 90%)	**Find portion** Base $2,000 Rate 80%
Solving for rate 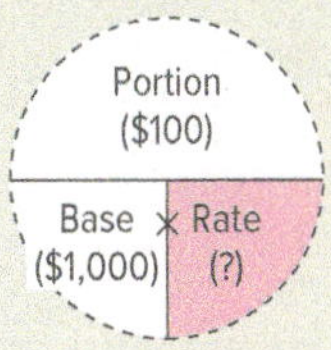	Assume Mel spends $100 for food from his $1,000 paycheck. What percent of his paycheck is spent on food? $\frac{\$100}{\$1,000} = .10 = 10\%$ *Note:* Portion is less than base since rate is less than 100%.	**Find rate** Base $2,000 Portion $500

(continues)

INTERACTIVE CHAPTER ORGANIZER

Topic/Procedure/Formula	Examples	You try it*
Solving for base 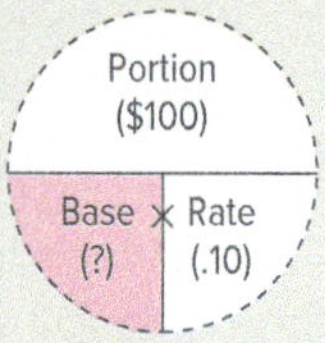	Assume Mel spends $100 for food, which is 10% of his paycheck. What is Mel's total paycheck? $\frac{\$100}{.10} = \$1,000$	**Find base** Rate 20% Portion $200
Calculating percent increases and decreases 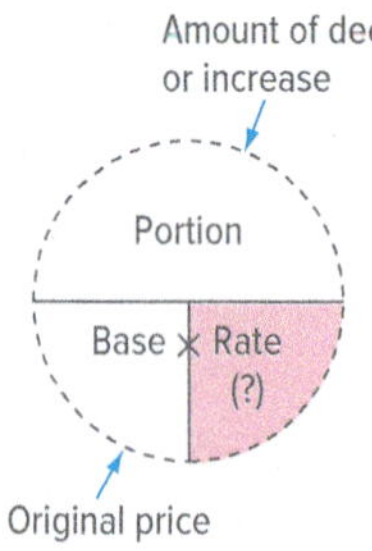	Stereo, $2,000 original price. Stereo, $2,500 new price. $\frac{\$500}{\$2,000} = .25 = 25\%$ increase **Check** $2,000 × 1.25 = $2,500 *Note:* Portion is greater than base since rate is greater than 100%. 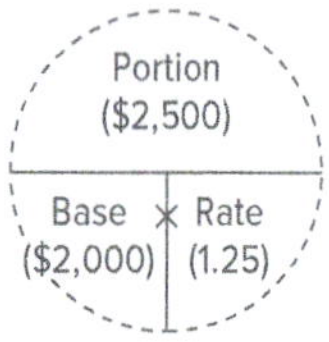	**Find percent increase** Old price $500 New price $600

KEY TERMS			
	Base Percent decrease	Percent increase Percents	Portion Rate

Note: For how to dissect and solve a word problem, see learning unit 6-2.

*Worked-out solutions are in Appendix B.

Critical Thinking Discussion Questions with Chapter Concept Check

1. In converting from a percent to a decimal, when will you have at least 2 leading zeros before the whole number? Explain this concept, assuming you have 100 bills of $1.

2. Explain the steps in rounding percents. Count the number of students who are sitting in the back half of the room as a percent of the total class. Round your answer to the nearest hundredth percent. Could you have rounded to the nearest whole percent without changing the accuracy of the answer?

3. Define portion, rate, and base. Create an example using Walt Disney World to show when the portion could be larger than the base. Why must the rate be greater than 100% for this to happen?

4. How do we solve for portion, rate, and base? Create an example using Apple computer sales to show that the portion and rate do relate to the same piece of the base.

5. Explain how to calculate percent decreases or increases. Many years ago, comic books cost 10 cents a copy. Visit a bookshop or newsstand. Select a new comic book and explain the price increase in percent compared to the 10-cent comic. How important is the rounding process in your final answer?

6. **Chapter Concept Check.** Go to the Google or Facebook site and find out how many people the company employs. Assuming a 10% increase in employment this year, calculate the total number of new employees by the end of the year, and identify the base, rate, and portion. If, in the following year, the 10% increase in employment fell by 5%, what would the total number of current employees be?

END-OF-CHAPTER PROBLEMS

connect

6

Check figures for odd-numbered problems in Appendix B. Name ______________________ Date ________

DRILL PROBLEMS

Convert the following decimals to percents: *LU 6-1(1)*

6–1. .88 **6–2.** .384 **6–3.** .4

6–4. 8.00 **6–5.** 3.561 **6–6.** 6.006

Convert the following percents to decimals: *LU 6-1(1)*

6–7. 4% **6–8.** 14% **6–9.** $64\frac{3}{10}\%$

6–10. 75.9% **6–11.** 119% **6–12.** 89%

Convert the following fractions to percents (round to the nearest tenth percent as needed): *LU 6-1(1)*

6–13. $\frac{1}{12}$ **6–14.** $\frac{1}{400}$

6–15. $\frac{7}{8}$ **6–16.** $\frac{11}{12}$

Convert the following percents to fractions and reduce to the lowest terms: *LU 6-1(2)*

6–17. 4% **6–18.** $18\frac{1}{2}\%$

6–19. $31\frac{2}{3}\%$ **6–20.** $61\frac{1}{2}\%$

6–21. 6.75% **6–22.** 182%

Solve for the portion (round to the nearest hundredth as needed): *LU 6-2(2)*

eXcel **6–23.** 7% of 150 **6–24.** 125% of 4,320 eXcel **6–25.** 25% of 410 eXcel

eXcel **6–26.** 119% of 128.9 **6–27.** 17.4% of 900 eXcel **6–28.** 11.2% of 85 eXcel

6–29. $12\frac{1}{2}\%$ of 919 **6–30.** 45% of 300

6–31. 18% of 90 **6–32.** 30% of 2,000

Solve for the base (round to the nearest hundredth as needed): *LU 6-2(2)*

6–33. 170 is 120% of ________ **6–34.** 36 is .75% of ________

6–35. 50 is .5% of ________ **6–36.** 10,800 is 90% of ________

6–37. 800 is $4\frac{1}{2}\%$ of ________

Solve for rate (round to the nearest tenth percent as needed): *LU 6-2(2)*

6–38. ________ of 80 is 50 **6–39.** ________ of 85 is 92

6–40. ________ of 250 is 65 **6–41.** 110 is ________ of 100

6–42. .09 is ________ of 2.25 **6–43.** 16 is ________ of 4

Solve the following problems. Be sure to show your work. Round to the nearest hundredth or hundredth percent as needed: *LU 6-2(2)*

6–44. What is 180% of 310?

6–45. 66% of 90 is what?

6–46. 40% of what number is 20?

6–47. 770 is 70% of what number?

6–48. 4 is what percent of 90?

6–49. What percent of 150 is 60?

Complete the following table: *LU 6-2(3)*

	Product	Selling price 2019	Selling price 2020	Amount of decrease or increase	Percent change (to nearest hundredth percent as needed)
6–50.	Apple iPad	$650	$500		
6–51.	Smartphone	$100	$120		

WORD PROBLEMS (First of Four Sets)

eXcel **6–52.** At a local Dunkin' Donuts, a survey showed that out of 1,200 customers eating lunch, 240 ordered coffee with their meal. What percent of customers ordered coffee? *LU 6-2(2)*

eXcel **6–53.** What percent of customers in Problem 6–52 did not order coffee? *LU 6-2(2)*

6–54. If the price of gas was on average $2.05 per gallon, and this was $1.20 cheaper than a year before, what is the price decrease? Round to the nearest hundredth percent. *LU 6-2(3)*

6–55. Wally Chin, the owner of an ExxonMobil station, bought a used Ford pickup truck, paying $2,000 as a down payment. He still owes 80% of the selling price. What was the selling price of the truck? *LU 6-2(2)*

6–56. Maria Fay bought four Dunlop tires at a local Goodyear store. The salesperson told her that her mileage would increase by 8%. Before this purchase, Maria was getting 24 mpg. What should her mileage be with the new tires? Round to the nearest hundredth. *LU 6-2(2)*

eXcel

My Money

6–57. The Social Security Administration announced the following rates to explain what percent of your Social Security benefits you will receive based on how old you are when you start receiving Social Security benefits.

Age	Percent of benefit
62	75
63	80
64	86.7
65	93.3
66	100

Assume Shelley Kate decides to take her Social Security at age 63. What amount of Social Security money will she receive each month, assuming she is entitled to $800 per month? *LU 6-2(2)*

eXcel

6–58. Assume that in the year 2019, 800,000 people attended the Christmas Eve celebration at Walt Disney World. In 2020, attendance for the Christmas Eve celebration is expected to increase by 35%. What is the total number of people expected at Walt Disney World for this event? *LU 6-2(2)*

6–59. Pete Smith found in his attic a Woody Woodpecker watch in its original box. It had a price tag on it for $4.50. The watch was made in 1949. Pete brought the watch to an antiques dealer and sold it for $35. What was the percent of increase in price? Round to the nearest hundredth percent. *LU 6-2(3)*

6–60. Christie's Auction sold a painting for $24,500. It charges all buyers a 15% premium of the final bid price. How much did the bidder pay Christie's? *LU 6-2(2)*

WORD PROBLEMS (Second of Four Sets)

6–61. Out of 9,000 college students surveyed, 540 responded that they do not eat breakfast. What percent of the students do not eat breakfast? *LU 6-2(2)*

6–62. What percent of college students in Problem 6–61 eat breakfast? *LU 6-2(2)*

My Money

6–63. You are saving for an emergency fund totaling six months' worth of expenses as well as investing 20% of each paycheck for retirement. If your monthly expenses amount to $1,465 and you want to put aside 30% of each month's expenses for your emergency fund and you earn $2,650 each month, how much do you need to set aside each month for your emergency fund and retirement savings? Can you afford to do this? *LU 6-2(2)*

6–64. Rainfall for January in Fiji averages 12″ according to *World Travel Guide*. This year it rained 5% less. How many inches (to the nearest tenth) did it rain this year? *LU 6-2(2)*

6–65. Jim and Alice Lange, employees at Walmart, have put themselves on a strict budget. Their goal at year's end is to buy a boat for $15,000 in cash. Their budget includes the following:

40% food and lodging   20% entertainment   10% educational

Jim earns $1,900 per month and Alice earns $2,400 per month. After 1 year, will Alice and Jim have enough cash to buy the boat? *LU 6-2(2)*

6–66. Assume there were 936,795 bankruptcy filings in 2018. If there were 310,061 Chapter 13 filings, what percent of filings were not Chapter 13? Round to the nearest whole percent. *LU 6-2(2)*

6–67. The Museum of Science in Boston estimated that 64% of all visitors came from within the state. On Saturday, 2,500 people attended the museum. How many attended the museum from out of state? *LU 6-2(2)*

6–68. Staples pays George Nagovsky an annual salary of $36,000. Today, George's boss informs him that he will receive a $4,600 raise. What percent of George's old salary is the $4,600 raise? Round to the nearest tenth percent. *LU 6-2(2)*

6–69. Assume in 2019, a local Dairy Queen had $550,000 in sales. In 2020, Dairy Queen's sales were up 35%. What were Dairy Queen's sales in 2020? *LU 6-2(2)*

6–70. Blue Valley College has 600 female students. This is 60% of the total student body. How many students attend Blue Valley College? *LU 6-2(2)*

6–71. Dr. Grossman was reviewing his total accounts receivable. This month, credit customers paid $44,000, which represented 20% of all receivables due (what customers owe). What was Dr. Grossman's total accounts receivable? *LU 6-2(2)*

My Money

6–72. Your city has a sales tax rate of 8.25%. If you just spent $20 on sales tax, how much were your purchases? *LU 6-2(2)*

6–73. The price of an antique doll increased from $600 to $800. What was the percent of increase? Round to the nearest tenth percent. *LU 6-2(3)*

6–74. A local Barnes and Noble bookstore ordered 80 marketing books but received 60 books. What percent of the order was missing? *LU 6-2(2)*

WORD PROBLEMS (Third of Four Sets)

My Money

6–75. According to the U.S. Census Bureau, in May 2017, the average American couple has only $5,000 in savings. In addition, only 33% of Americans are saving in a retirement account. Compare this with mainland Chinese who save over 50%. If the average American couple should save $1.3 million for retirement, what percent has actually been saved on average to date? Round to the nearest hundredth percent. *LU 6-2(2)*

6–76. Due to increased mailing costs, the new rate will cost publishers $50 million; this is 12.5% more than they paid the previous year. How much did it cost publishers last year? Round to the nearest hundreds. *LU 6-2(2)*

6–77. Jim Goodman, an employee at Walgreens, earned $45,900, an increase of 17.5% over the previous year. What were Jim's earnings the previous year? Round to the nearest cent. *LU 6-2(2)*

6–78. If the number of mortgage applications declined by 7% to 1,625,415, what had been the previous year's number of applications? *LU 6-2(2)*

6–79. If the price of a business math text rose to $150 and this was 8% more than the original price, what was the original selling price? Round to the nearest cent. *LU 6-2(2)*

6–80. Web Consultants, Inc., pays Alice Rose an annual salary of $48,000. Today, Alice's boss informs her that she will receive a $6,400 raise. What percent of Alice's old salary is the $6,400 raise? Round to the nearest tenth percent. *LU 6-2(2)*

6–81. Earl Miller, a lawyer, charges Lee's Plumbing, his client, 25% of what he can collect for Lee from customers whose accounts are past due. The attorney also charges, in addition to the 25%, a flat fee of $50 per customer. This month, Earl collected $7,000 from three of Lee's past-due customers. What is the total fee due to Earl? *LU 6-2(2)*

6–82. A local Petco ordered 100 dog calendars but received 60. What percent of the order was missing? *LU 6-2(2)*

6–83. Ray's Video uses MasterCard. MasterCard charges $2\frac{1}{2}$% on net deposits (credit slips less returns). Ray's made a net deposit of $4,100 for charge sales. How much did MasterCard charge Ray's? *LU 6-2(2)*

6–84. Internetlivestats.com reported in December 2017 around 40% of the world population has an Internet connection today. If there are 3,061,707,850 users, what is the world population? *LU 6-2(2)*

WORD PROBLEMS (Fourth of Four Sets)

6–85. Chevrolet raised the base price of its Volt by $1,200 to $33,500. What was the percent increase? Round to the nearest tenth percent. *LU 6-2(2)*

6–86. The sales tax rate is 8%. If Jim bought a new Buick and paid a sales tax of $1,920, what was the cost of the Buick before the tax? *LU 6-2(2)*

6–87. Puthina Unge bought a new Dell computer system on sale for $1,800. It was advertised as 30% off the regular price. What was the original price of the computer? Round to the nearest dollar. *LU 6-2(2)*

6–88. John O'Sullivan has just completed his first year in business. His records show that he spent the following in advertising:

Internet	$600	Radio	$650	Yellow Pages	$700	Local flyers	$400

What percent of John's advertising was spent on the Yellow Pages? Round to the nearest hundredth percent. *LU 6-2(2)*

6–89. Jay Miller sold his ski house at Attitash Mountain in New Hampshire for $35,000. This sale represented a loss of 15% off the original price. What was the original price Jay paid for the ski house? Round your answer to the nearest dollar. *LU 6-2(2)*

6–90. Out of 4,000 colleges surveyed, 60% reported that SAT scores were not used as a high consideration in viewing their applications. How many schools view the SAT as important in screening applicants? *LU 6-2(2)*

My Money

6–91. If refinishing your basement at a cost of $45,404 would add $18,270 to the resale value of your home, what percent of your cost is recouped? Round to the nearest percent. *LU 6-2(2)*

6–92. A major airline laid off 4,000 pilots and flight attendants. If this was a 12.5% reduction in the workforce, what was the size of the workforce after the layoffs? *LU 6-2(2)*

6–93. Assume 450,000 people line up on the streets to see the Macy's Thanksgiving Parade. If attendance is expected to increase 30% next year, what will be the number of people lined up on the street to see the parade? *LU 6-2(2)*

CHALLENGE PROBLEMS

6–94. Each Tuesday, Ryan Airlines reduces its one-way ticket from Fort Wayne to Chicago from $125 to $40. To receive this special $40 price, the customer must buy a round-trip ticket. Ryan has a nonrefundable 25% penalty fare for cancellation; it estimates that about nine-tenths of 1% will cancel their reservations. The airline also estimates this special price will cause a passenger traffic increase from 400 to 900. Ryan expects revenue for the year to be 55.4% higher than the previous year. Last year, Ryan's sales were $482,000. To receive the special rate, Janice Miller bought two round-trip tickets. On other airlines, Janice has paid $100 round trip (with no cancellation penalty). Calculate the following: *LU 6-2(2)*

a. Percent discount Ryan is offering.

b. Percent passenger travel will increase.

c. Sales for new year.

d. Janice's loss if she cancels one round-trip flight.

e. Approximately how many more cancellations can Ryan Airlines expect (after Janice's cancellation)?

6–95. A local Dunkin' Donuts shop reported that its sales have increased exactly 22% per year for the last 2 years. This year's sales were $82,500. What were Dunkin' Donuts' sales 2 years ago? Round each year's sales to the nearest dollar. *LU 6-2(2)*

Classroom Notes

SUMMARY PRACTICE TEST

Do you need help? Connect videos have step-by-step worked-out solutions.

Convert the following decimals to percents. *LU 6-1(1)*

1. .921 **2.** .4 **3.** 15.88 **4.** 8.00

Convert the following percents to decimals. *LU 6-1(1)*

5. 42% **6.** 7.98% **7.** 400% **8.** $\frac{1}{4}\%$

Convert the following fractions to percents. Round to the nearest tenth percent. *LU 6-1(1)*

9. $\frac{1}{6}$ **10.** $\frac{1}{3}$

Convert the following percents to fractions and reduce to the lowest terms as needed. *LU 6-1(2)*

11. $19\frac{3}{8}\%$ **12.** 6.2%

Solve the following problems for portion, base, or rate:

13. An Arby's franchise has a net income before taxes of $900,000. The company's treasurer estimates that 40% of the company's net income will go to federal and state taxes. How much will the Arby's franchise have left? *LU 6-2(2)*

14. Domino's projects a year-end net income of $699,000. The net income represents 30% of its annual sales. What are Domino's projected annual sales? *LU 6-2(2)*

15. Target ordered 400 iPhones. When Target received the order, 100 iPhones were missing. What percent of the order did Target receive? *LU 6-2(2)*

16. Matthew Song, an employee at Putnam Investments, receives an annual salary of $120,000. Today his boss informed him that he would receive a $3,200 raise. What percent of his old salary is the $3,200 raise? Round to the nearest hundredth percent. *LU 6-2(2)*

17. The price of a Delta airline ticket from Los Angeles to Boston increased to $440. This is a 15% increase. What was the old fare? Round to the nearest cent. *LU 6-2(2)*

18. Scupper Grace earns a gross pay of $900 per week at Office Depot. Scupper's payroll deductions are 29%. What is Scupper's take-home pay? *LU 6-2(2)*

19. Mia Wong is reviewing the total accounts receivable of Wong's department store. Credit customers paid $90,000 this month. This represents 60% of all receivables due. What is Mia's total accounts receivable? *LU 6-2(2)*

INTERACTIVE VIDEO WORKSHEET

GRADING THE SUMMARY PRACTICE TEST

Go to the summary practice test video in Connect (or click on it here in the ebook). Grade your summary practice test while viewing the video.

C for Correct/I for Incorrect

1. ________	6. ________	11. ________	16. ________
2. ________	7. ________	12. ________	17. ________
3. ________	8. ________	13. ________	18. ________
4. ________	9. ________	14. ________	19. ________
5. ________	10. ________	15. ________	

If you achieved 100%, you are ready for your instructor's exam.

If any of the problems were incorrect, list the questions you missed and show steps to solve the problem correctly.

Replay the video to see if you have made the correct fixes to your mistakes.
If you have any questions, contact your instructor asap.

Notes on Watching Videos

MY MONEY

Pay Yourself First!

What I need to know

Save for a rainy day, a penny saved is a penny earned, scrimp and save. There are so many idioms concerning saving that we have heard from countless people in our lives. What makes saving such an important financial goal? We have all most likely experienced a situation where we wished we had set aside some funds to cover a surprise expense. This is the quintessential case for establishing a savings plan and sticking to it. Savings is not something we do once we have enough money, it is something that should be implemented as soon as we enter the workforce and earn a paycheck. No matter the amount of savings we start with, the act of saving creates a healthy financial habit that will serve us well throughout our life.

What I need to do

Start saving right now as there is no time like the present. As a student you may feel there is no way you can save due to your limited earnings and significant expenses. However, the act of saving begins with a focus on creating a good financial habit versus a focus on the amount of money being saved. Start small by determining an amount you could set aside from each paycheck in order to establish your savings account. Even if your initial amount is only $10/month, you will find this grows over time. Over the course of just one year you will have saved $120.

If you have direct deposit for your paychecks, see if you are able to specify separate amounts to be deposited in multiple accounts for each payday. Identify the amount to be placed in savings (such as $10) and have the balance of your paycheck deposited in your checking account. This way the money is deposited into your savings account automatically, making it easier for you to stick to your savings goal. Also, since the money is taken directly from your direct deposit, you will find that once you move past the first couple deposits you don't even miss the money that used to go to your checking account. If your employer doesn't offer the ability to split your deposit into multiple accounts, you can accomplish the same result by creating an automatic transfer with your bank. You can still specify the amount to be transferred to your savings account and, in many cases, select the date on which the transfer should take place.

Once you have established a savings plan, you will want to modify the plan based on your changing earnings over time. Set a goal to increase your savings at a specified time in the future. It may be beneficial to set this plan to correspond with your annual performance review at work where your salary and potential raise is discussed. As your income increases each year, adjust your savings plan by increasing the amount based on the changes to your income. This will keep you focused on your long-term financial goals and also make the increase in savings less noticeable as it corresponds to a time in which your income is increasing anyway.

Resources I can use

- https://www.bankrate.com/banking/savings/how-to-build-a-successful-savings-plan-and-start-saving-money/ — set a savings plan

MY MONEY ACTIVITY

- Reference the budget you created in Chapter 1.
- Determine an amount you could allocate toward a savings account.
- Evaluate after three months to see if this savings amount needs to be adjusted.

PERSONAL FINANCE

A KIPLINGER APPROACH

"Retirement Reality Check: Are You Saving Enough?", *Kiplinger's*, March 2018, p. 29. Used by permission of The Kiplinger Washington Editors, Inc.

Retirement Reality Check: Are You Saving Enough?

To maintain a comfortable lifestyle in retirement, you should aim to replace about 75% to 80% of your current gross income. Social Security will supply about 30% of that amount for most middle-income workers. (You can get your personalized estimate of how much to expect from Social Security at www.ssa.gov/estimator.) The rest of your retirement income may come from pension benefits, a job or personal savings. To establish a nest egg target, you need a rough idea of how much money you'll want to draw from your savings each month.

How It Works	What to Do	Your Answers
STEP 1: **Meet Kate.** She's 55, plans to retire in 10 years and thinks she may live until 90. She estimates that she'll need $2,500 per month from her savings, and she assumes an average 6% return on her investments. Kate looks at table 1 and finds where 25 years and 6% intersect: $205,000. That's how much she'll need to produce $1,000 per month of retirement income. But because she needs $2,500 a month, she divides $2,500 by $1,000 and comes up with a factor of 2.5. Then she calculates her **target nest egg amount**: $205,000 x 2.5 = $512,500.	**Divide your monthly income needs by 1,000 and multiply it by the amount in table 1.**	TARGET NEEDS $______
STEP 2: Table 2 will help you calculate the **future value of your existing investments**. Kate has $250,000 in savings. With 10 years to go before retiring and an assumed rate of return of 6%, Kate's factor from table 2 is 1.79. When she multiplies her current balance of $250,000 by 1.79, the future value of her account at retirement is $447,500.	**Multiply your current account balance by the factor in table 2.**	$______
STEP 3: Table 3 shows the **future value of your ongoing monthly contributions**. Kate is also saving $500 per month and assumes her savings will earn 6% over the next ten years. The future value of her monthly contribution is $82,350 ($500 x 164.70 = $82,350).	**Multiply your monthly contributions by the factor in table 3.**	$______
STEP 4: Using Kate's example, her **projected savings** are step 2 ($447,500) + step 3 ($82,350) = $529, 850. That's more than her target nest egg amount, so Kate is on track.	**Add your answers in step 2 and step 3 together.**	PROJECTED SAVINGS $______

TABLE 1 (dollar values needed to produce $1,000 per month; assumes 3% annual inflation)

Rate of return	Years in retirement 5	10	15	20	25	30	35
2%	$60,000	$123,000	$189,000	$259,000	$331,000	$408,000	$488,000
4	57,000	111,000	162,000	211,000	258,000	302,000	344,000
6	54,000	100,000	140,000	175,000	**205,000**	231,000	254,000
8	51,000	91,000	122,000	147,000	167,000	182,000	194,000
10	48,000	83,000	108,000	125,000	138,000	148,000	154,000

TABLE 2 (current balance factor)

Time to retire.	Rate of return 4%	6%	8%	10%
5 years	1.22	1.34	1.47	1.61
10 years	1.48	**1.79**	2.16	2.60
15 years	1.80	2.34	3.17	4.18
20 years	2.19	3.21	4.66	6.73

TABLE 3 (monthly contribution factor)

Time to retire.	Rate of return 4%	6%	8%	10%
5 years	66.52	70.11	73.97	78.08
10 years	147.74	**164.70**	184.17	206.55
15 years	246.91	292.27	348.35	417.92
20 years	368.00	464.35	592.95	765.70

SOURCE: PHILIP LUBINSKI, PRESIDENT OF STRATEGIC DISTRIBUTION INSTITUTE

WHAT TO DO

If the step 4 figure is larger than the step 1 figure, congratulations! You're on track.

If step 4 is smaller than step 1, you need to make some changes to reach your nest egg goal. Consider working a few years longer, saving more each month, downsizing or moving to a cheaper locale. Or reconsider how much retirement income you will need.

In Kate's example, she is slightly ahead of her goal ($529,850 - $512,500 = $17,350). But because this is just a rough estimate, she should continue her current savings plan. Very few people have reached retirement regretting that they saved too much.

BUSINESS MATH ISSUE

Reaching retirement goals means you should start saving when you reach 50 years old.

1. List the key points of the article and information to support your position.
2. Write a group defense of your position using math calculations to support your view. If you are in an online course, post to a discussion board.

Classroom Notes

CHAPTER 7

Discounts: Trade and Cash

FedEx to Rein In Its Holiday Charges

By Paul Ziobro

FedEx Corp. won't charge additional fees for most orders during the holidays, in contrast with **United Parcel Service** Inc., undercutting its rival as they battle for e-commerce customers.

The decision by FedEx, announced Thursday, is a gamble that it can cover the extra costs during a period when daily volume can double to more than 26 million packages. It will charge extra fees for deliveries requiring additional handling, which include larger and irregularly shaped packages, as well as oversize packages, which now makeup 10% of its ground-shipping volume.

©Antonio Gravante/123RF

LEARNING UNIT OBJECTIVES

LU 7–1: Trade Discounts—Single and Chain (Includes Discussion of Freight)

1. Calculate single trade discounts with formulas and complements.
2. Explain the freight terms *FOB shipping point* and *FOB destination.*
3. Find list price when net price and trade discount rate are known.
4. Calculate chain discounts with the net price equivalent rate and single equivalent discount rate.

LU 7–2: Cash Discounts, Credit Terms, and Partial Payments

1. List and explain typical discount periods and credit periods that a business may offer.
2. Calculate outstanding balance for partial payments.

Your Guide to Successfully Completing This Chapter

Traditional book or ebook

Check box as you complete each step.

Steps

- ☐ Read learning unit.
  - ☐ Complete practice quiz at the end of the learning unit.
- ☐ Grade practice quiz using provided solutions. (For more help, watch the learning unit video in Connect and have a Study Session with the authors. Then complete the additional practice quiz in Connect.)
- ☐ Repeat above for each of the two learning units in Chapter 7.
  - ☐ Review chapter organizer.
  - ☐ Complete assigned homework.
    - ☐ Finish summary practice test. (Go to Connect via the ebook link and do the interactive video worksheet to grade.)
- ☐ Complete instructor's exam.

LO 1

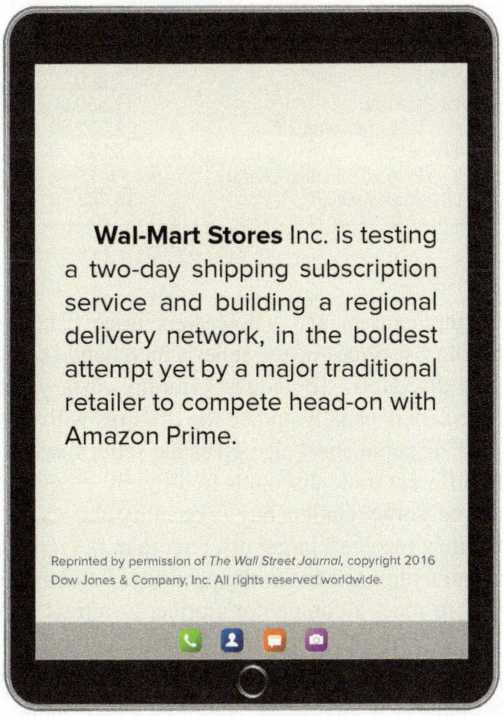

Wal-Mart Stores Inc. is testing a two-day shipping subscription service and building a regional delivery network, in the boldest attempt yet by a major traditional retailer to compete head-on with Amazon Prime.

The *Wall Street Journal* clip to the left talks about how Walmart plans to compete with Amazon Prime. We will look at various trade and shipping terms but first this chapter discusses two types of discounts taken by retailers—trade and cash. A **trade discount** is a reduction off the original selling price (list price) of an item and is not related to early payment. A **cash discount** is the result of an early payment based on the terms of the sale.

Learning Unit 7–1: Trade Discounts—Single and Chain (Includes Discussion of Freight)

The merchandise sold by retailers is bought from manufacturers and wholesalers who sell only to retailers and not to customers. These manufacturers and wholesalers offer retailer discounts so retailers can resell the merchandise at a profit. The discounts are off the manufacturers' and wholesalers' **list price** (suggested retail price), and the amount of discount that retailers receive off the list price is the **trade discount amount.** The following photo shows a Michael's discount online coupon. Keep in mind that retailers can track customer purchases and preferences. The smartphone is a great tool customers can use to find discounts and retailers can use to gather marketing data.

My Money

When you make a purchase, the retailer (seller) gives you a purchase **invoice.** Invoices are important business documents that help sellers keep track of sales transactions and buyers keep track of purchase transactions. North Shore Community College Bookstore is a retail seller of textbooks to students. The bookstore usually purchases its textbooks directly from publishers. Figure 7.1 shows a sample of what a textbook invoice from McGraw-Hill Higher Education to the North Shore Community College Bookstore might look like. Note that the trade discount amount is given in percent. This is the **trade discount rate,** which is a percent off the list price that retailers can deduct. The following formula for calculating a trade discount amount gives the numbers from the Figure 7.1 invoice:

©iPhone/Alamy

TRADE DISCOUNT AMOUNT FORMULA

Trade discount amount = List price × Trade discount rate

$2,887.50 $11,550 25%

The price that the retailer (bookstore) pays the manufacturer (publisher) or wholesaler is the **net price.** The following formula for calculating the net price gives the numbers from the Figure 7.1 invoice:

NET PRICE FORMULA

Net price = List price − Trade discount amount

$8,662.50 $11,550 $2,887.50

FIGURE 7.1

Bookstore invoice showing a trade discount

Invoice No.: 5582

McGraw-Hill Higher Education
1333 Burr Ridge Parkway
Burr Ridge, Illinois 60527

Date: July 8, 2020
Ship: Two-day UPS
Terms: 2/10, n/30

Sold to: North Shore Community College Bookstore
1 Ferncroft Road
Danvers, MA 01923

Description	Unit list price	Total amount
50 Financial Management—Block/Hirt	$195	$9,750.00
10 Introduction to Business—Nichols	180	1,800.00
	Total List Price	11,550.00
	Less: Trade Discount 25%	2,887.50
	Net Price	8,662.50
	Plus: Prepaid Shipping Charge	+125.00
	Total Invoice Amount	$8,787.50

©MikeDotta/Shutterstock

Frequently, manufacturers and wholesalers issue catalogs to retailers containing list prices of the seller's merchandise and the available trade discounts. To reduce printing costs when prices change, these sellers usually update the catalogs with new *discount sheets.* The discount sheet also gives the seller the flexibility of offering different trade discounts to different classes of retailers. For example, some retailers buy in quantity and service the products. They may receive a larger discount than the retailer who wants the manufacturer to service the products. Sellers may also give discounts to meet a competitor's price, to attract new retailers, and to reward the retailers who buy product-line products. Sometimes the ability of the retailer to negotiate with the seller determines the trade discount amount.

Retailers cannot take trade discounts on freight, returned goods, sales tax, and so on. Trade discounts may be single discounts or a chain of discounts. Before we discuss single trade discounts, let's study freight terms.

Freight Terms

The most common **freight terms** are *FOB shipping point* and *FOB destination.* These terms determine how the freight will be paid. The key words in the terms are *shipping point* and *destination.* Note in the chapter opener how FedEx did not charge additional fees during the holiday season in contrast to competitor United Parcel Service, which did.

LO 2

FOB shipping point means free on board at shipping point; that is, the buyer pays the freight cost of getting the goods to the place of business.

For example, assume that IBM in San Diego bought goods from Argo Suppliers in Boston. Argo ships the goods FOB Boston by plane. IBM takes title to the goods when the aircraft in Boston receives the goods, so IBM pays the freight from Boston to San Diego. Frequently, the seller (Argo) prepays the freight and adds the amount to the buyer's (IBM) invoice. When paying the invoice, the buyer takes the cash discount off the net price and adds the freight cost. FOB shipping point can be illustrated as follows:

FOB shipping point (Boston)

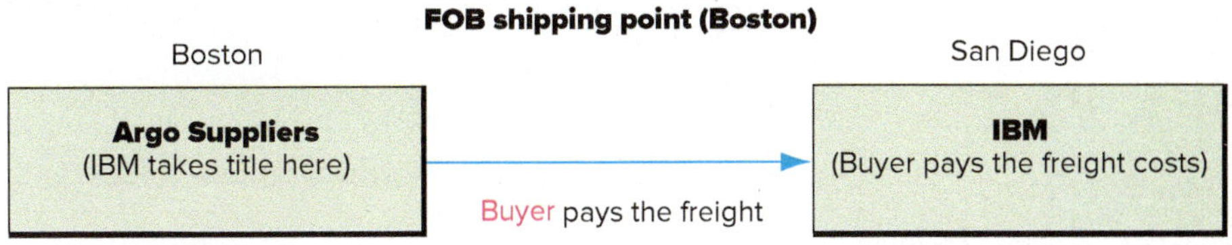

My Money

FOB destination means the seller pays the freight cost until it reaches the buyer's place of business. If Argo ships its goods to IBM FOB destination or FOB San Diego, the title to the goods remains with Argo. Then it is Argo's responsibility to pay the freight from Boston to IBM's place of business in San Diego. FOB destination can be illustrated as follows:

FOB destination (San Diego)

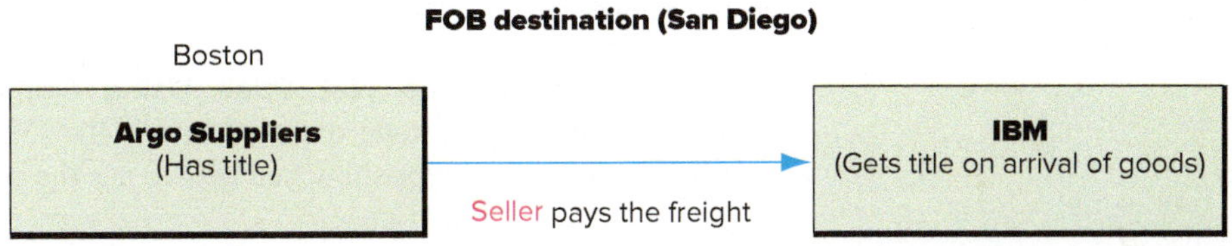

The following *Wall Street Journal* clip shows some new trends in shipping for the future.

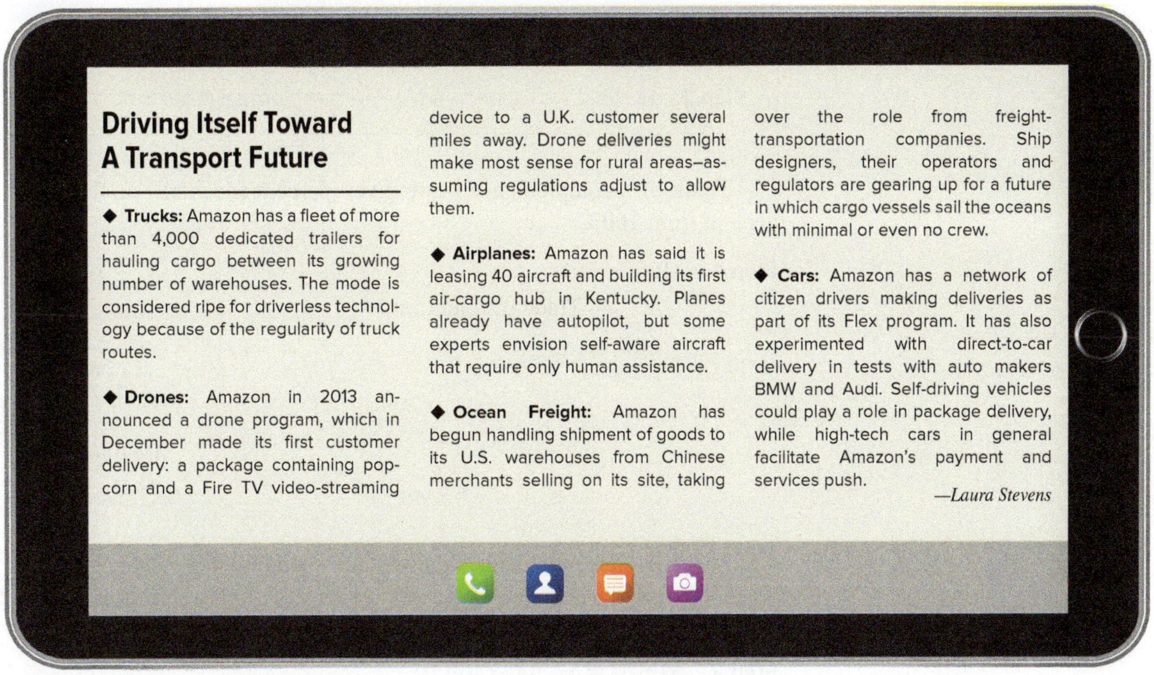

Driving Itself Toward A Transport Future

◆ **Trucks:** Amazon has a fleet of more than 4,000 dedicated trailers for hauling cargo between its growing number of warehouses. The mode is considered ripe for driverless technology because of the regularity of truck routes.

◆ **Drones:** Amazon in 2013 announced a drone program, which in December made its first customer delivery: a package containing popcorn and a Fire TV video-streaming device to a U.K. customer several miles away. Drone deliveries might make most sense for rural areas–assuming regulations adjust to allow them.

◆ **Airplanes:** Amazon has said it is leasing 40 aircraft and building its first air-cargo hub in Kentucky. Planes already have autopilot, but some experts envision self-aware aircraft that require only human assistance.

◆ **Ocean Freight:** Amazon has begun handling shipment of goods to its U.S. warehouses from Chinese merchants selling on its site, taking over the role from freight-transportation companies. Ship designers, their operators and regulators are gearing up for a future in which cargo vessels sail the oceans with minimal or even no crew.

◆ **Cars:** Amazon has a network of citizen drivers making deliveries as part of its Flex program. It has also experimented with direct-to-car delivery in tests with auto makers BMW and Audi. Self-driving vehicles could play a role in package delivery, while high-tech cars in general facilitate Amazon's payment and services push.

—Laura Stevens

Now you are ready for the discussion on single trade discounts.

Single Trade Discount

In the introduction to this unit, we showed how to use the trade discount amount formula and the net price formula to calculate the McGraw-Hill Higher Education textbook sale to the North Shore Community College Bookstore. Since McGraw-Hill Higher Education gave the bookstore only one trade discount, it is a **single trade discount.** In the following word problem, we use the formulas to solve another example of a single trade discount. Again, we will use a blueprint aid to help dissect and solve the word problem.

The Word Problem The list price of a Macintosh computer is $2,700. The manufacturer offers dealers a 40% trade discount. What are the trade discount amount and the net price?

BLUEPRINT	The facts	Solving for?	Steps to take	Key points
	List price: $2,700. *Trade discount rate:* 40%.	Trade discount amount. Net price.	Trade discount amount = List price × Trade discount rate. Net price = List price − Trade discount amount.	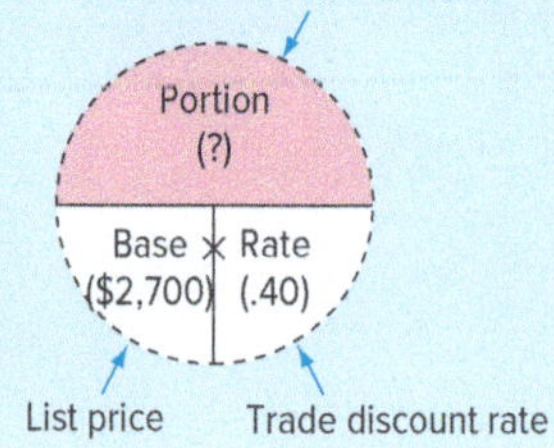

Steps to solving problem

1. Calculate the trade discount amount. $2,700 × .40 = $1,080
2. Calculate the net price. $2,700 − $1,080 = $1,620

Now let's learn how to check the dealers' net price of $1,620 with an alternate procedure using a complement.

How to Calculate the Net Price Using Complement of Trade Discount Rate

The **complement** of a trade discount rate is the difference between the discount rate and 100%. The following steps show you how to use the complement of a trade discount rate:

CALCULATING NET PRICE USING COMPLEMENT OF TRADE DISCOUNT RATE

Step 1. To find the complement, subtract the single discount rate from 100%.

Step 2. Multiply the list price times the complement (from Step 1).

Think of a complement of any given percent (decimal) as the result of subtracting the percent from 100%.

Step 1. 100%
− 40 ← Trade discount rate
60% or .60

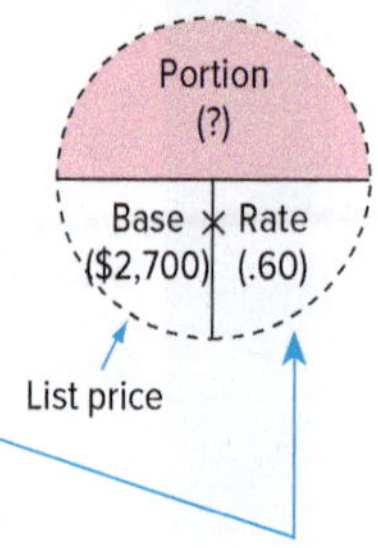

The complement means that we are spending 60 cents per dollar because we save 40 cents per dollar. Since we planned to spend $2,700, we multiply .60 by $2,700 to get a net price of $1,620.

Step 2. $1,620 = $2,700 × .60

Note how the portion ($1,620) and rate (.60) relate to the same piece of the base ($2,700). The portion ($1,620) is smaller than the base, since the rate is less than 100%.

Be aware that some people prefer to use the trade discount amount formula and the net price formula to find the net price. Other people prefer to use the complement of the trade discount rate to find the net price. The result is always the same.

LO 3

Finding List Price When You Know Net Price and Trade Discount Rate The following formula has many useful applications:

CALCULATING LIST PRICE WHEN NET PRICE AND TRADE DISCOUNT RATE ARE KNOWN

$$\text{List price} = \frac{\text{Net price}}{\text{Complement of trade discount rate}}$$

Next, let's see how to dissect and solve a word problem calculating list price.

The Word Problem A Macintosh computer has a $1,620 net price and a 40% trade discount. What is its list price?

BLUEPRINT	The facts	Solving for?	Steps to take	Key points
	Net price: $1,620. *Trade discount rate:* 40%.	List price.	$\text{List price} = \dfrac{\text{Net price}}{\text{Complement of trade discount rate}}$	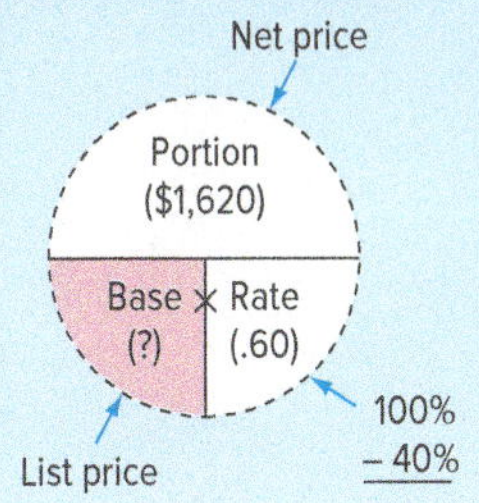

Steps to solving problem

1. Calculate the complement of the trade discount. $100\% - 40 = 60\% = .60$
2. Calculate the list price. $\dfrac{\$1,620}{.60} = \$2,700$

Note that the portion ($1,620) and rate (.60) relate to the same piece of the base.

Let's return to the McGraw-Hill Higher Education invoice in Figure 7.1 and calculate the list price using the formula for finding list price when the net price and trade discount rate are known. The net price of the textbooks is $8,662.50. The complement of the trade discount rate is 100% − 25% = 75% = .75. Dividing the net price $8,662.50 by the complement .75 equals $11,550, the list price shown in the McGraw-Hill Higher Education invoice. We can show this as follows:

$$\frac{\$8,662.50}{.75} = \$11,550, \text{ the list price}$$

LO 4

Chain Discounts

Frequently, manufacturers want greater flexibility in setting trade discounts for different classes of customers, seasonal trends, promotional activities, and so on. To gain this flexibility, some sellers give **chain** or **series discounts**—trade discounts in a series of two or more successive discounts.

Sellers list chain discounts as a group, for example, 20/15/10. Let's look at how Mick Company arrives at the net price of office equipment with a 20/15/10 chain discount.

EXAMPLE The list price of the office equipment is $15,000. The chain discount is 20/15/10. The long way to calculate the net price is as follows:

Step 1	Step 2	Step 3	Step 4
$15,000	$15,000	$12,000	$10,200
× .20	−3,000	− 1,800	− 1,020
$ 3,000	$12,000	$10,200	**$ 9,180** net price
	× .15	× .10	
	$ 1,800	$ 1,020	

Note how we multiply the percent (in decimal) times the new balance after we subtract the previous trade discount amount. *Never add the 20/15/10 together.* For example, in Step 3, we change the last discount, 10%, to decimal form and multiply times $10,200. Remember that each percent is multiplied by a successively *smaller* base. You could write the 20/15/10 discount rate in any order and still arrive at the same net price. Thus, you would get the $9,180 net price if the discount were 10/15/20 or 15/20/10. However, sellers usually give the larger discounts first. *Never try to shorten this step process by adding the discounts.* Your net price will be incorrect because, when done properly, each percent is calculated on a different base.

Net Price Equivalent Rate In the example above, you could also find the $9,180 net price with the **net price equivalent rate**—a shortcut method. Let's see how to use this rate to calculate net price.

CALCULATING NET PRICE USING NET PRICE EQUIVALENT RATE

Step 1. Subtract each chain discount rate from 100% (find the complement) and convert each percent to a decimal.

Step 2. Multiply the decimals. Do not round off decimals, since this number is the net price equivalent rate.

Step 3. Multiply the list price times the net price equivalent rate (Step 2).

The following word problem with its blueprint aid illustrates how to use the net price equivalent rate method.

The Word Problem The list price of office equipment is $15,000. The chain discount is 20/15/10. What is the net price?

BLUEPRINT	The facts	Solving for?	Steps to take	Key points
	List price: $15,000. *Chain discount:* 20/15/10.	Net price.	Net price equivalent rate. Net price = List price × Net price equivalent rate.	Do not round net price equivalent rate.

Steps to solving problem

1. Calculate the complement of each rate and convert each percent to a decimal.

100%	100%	100%
− 20	− 15	−10
80%	85%	90%
↓	↓	↓
.8	.85	.9

2. Calculate the net price equivalent rate. (Do not round.) .8 × .85 × .9 = .612 Net price equivalent rate. For each $1, you are spending about 61 cents.

3. Calculate the net price (actual cost to buyer). $15,000 × .612 = **$9,180**

Next we see how to calculate the trade discount amount with a simpler method.

In the previous word problem, we could calculate the trade discount amount as follows:

$15,000 ← List price
− 9,180 ← Net price
$ 5,820 ← Trade discount amount

Single Equivalent Discount Rate You can use another method to find the trade discount by using the **single equivalent discount rate.**

MONEY tips

Double-check invoices. On average 9 out of 10 invoices contain an error.

CALCULATING TRADE DISCOUNT AMOUNT USING SINGLE EQUIVALENT DISCOUNT RATE

Step 1. Subtract the net price equivalent rate from 1. This is the single equivalent discount rate.

Step 2. Multiply the list price times the single equivalent discount rate. This is the trade discount amount.

Let's now do the calculations.

Step 1. 1.000 ← If you are using a calculator, just press 1.
 − .612
 .388 ← This is the single equivalent discount rate.

Step 2. $15,000 × .388 = **$5,820** → This is the trade discount amount.

Remember that when we use the net price equivalent rate, the buyer of the office equipment pays $.612 on each $1 of list price. Now with the single equivalent discount rate, we can say that the buyer saves $.388 on each $1 of list price. The .388 is the single equivalent discount rate for the 20/15/10 chain discount. Note how we use the .388 single equivalent discount rate as if it were the only discount.

Knowing the terminology for what you pay and what you save is an important step in understanding how to calculate net price and trade discount amounts. The pie charts show the terminology relating to each.

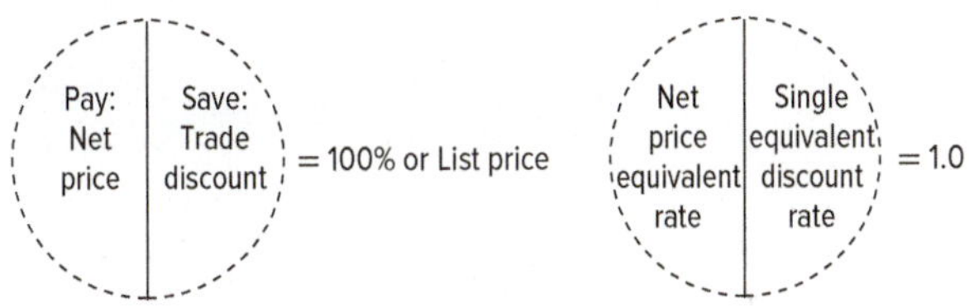

It's time to try the Practice Quiz.

LU 7–1 PRACTICE QUIZ

Complete this **Practice Quiz** to see how you are doing.[1]

1. The list price of a dining room set with a 40% trade discount is $12,000. What are the trade discount amount and net price? (Use the complement method for net price.)
2. The net price of a video system with a 30% trade discount is $1,400. What is the list price?
3. Lamps Outlet bought a shipment of lamps from a wholesaler. The total list price was $12,000 with a 5/10/25 chain discount. Calculate the net price and trade discount amount. (Use the net price equivalent rate and single equivalent discount rate in your calculation.)

*For **extra help** from your authors–Sharon and Jeff–see the videos in Connect.*

✓ Solutions

1. Dining room set trade discount amount and net price:

BLUEPRINT	The facts	Solving for?	Steps to take	Key points
	List price: $12,000. *Trade discount rate:* 40%.	Trade discount amount. Net price.	Trade discount amount = List price × Trade discount rate. Net price = List price × Complement of trade discount rate.	Trade discount amount Portion (?) Base ($12,000) × Rate (.40) List price Trade discount rate

[1]For all three problems we will show blueprint aids. You might want to draw them on scrap paper.

Steps to solving problem

1. Calculate the trade discount. $12,000 × .40 = **$4,800** Trade discount amount
2. Calculate the net price. $12,000 × .60 = **$7,200** (100% − 40% = 60%)

2. Video system list price:

BLUEPRINT	The facts	Solving for?	Steps to take	Key points
	Net price: $1,400. *Trade discount rate:* 30%.	List price.	List price = $\frac{\text{Net price}}{\text{Complement of trade discount}}$	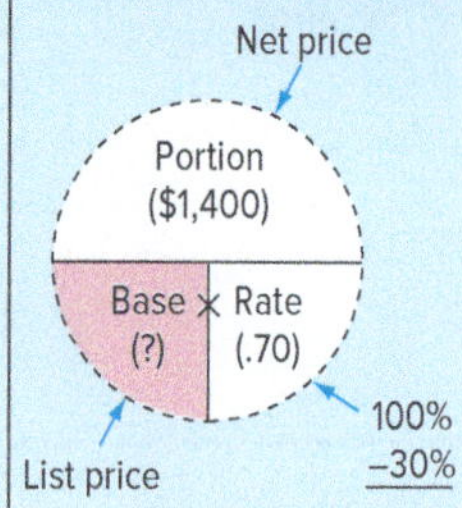

Steps to solving problem

1. Calculate the complement of trade discount.

$$\begin{array}{r} 100\% \\ -\ 30 \\ \hline 70\% = .70 \end{array}$$

2. Calculate the list price. $\frac{\$1,400}{.70} =$ **$2,000**

3. Lamps Outlet's net price and trade discount amount:

BLUEPRINT	The facts	Solving for?	Steps to take	Key points
	List price: $12,000. *Chain discount:* 5/10/25.	Net price. Trade discount amount.	Net price = List price × Net price equivalent rate. Trade discount amount = List price × Single equivalent discount rate.	Do not round off net price equivalent rate or single equivalent discount rate.

Steps to solving problem

1. Calculate the complement of each chain discount.

$$\begin{array}{r} 100\% \\ -\ 5 \\ \hline 95\% \end{array} \quad \begin{array}{r} 100\% \\ -\ 10 \\ \hline 90\% \end{array} \quad \begin{array}{r} 100\% \\ -\ 25 \\ \hline 75\% \end{array}$$

2. Calculate the net price equivalent rate. .95 × .90 × .75 = .64125
3. Calculate the net price. $12,000 × .64125 = **$7,695**
4. Calculate the single equivalent discount rate.

$$\begin{array}{r} 1.00000 \\ -\ .64125 \\ \hline .35875 \end{array}$$

5. Calculate the trade discount amount. $12,000 × .35875 = **$4,305**

LO 1

Learning Unit 7–2: Cash Discounts, Credit Terms, and Partial Payments

To introduce this learning unit, we will use the New Hampshire Propane Company invoice that follows. The invoice shows that if you pay your bill early, you will receive a 19-cent discount. Every penny counts.

©Vale Stock/Shutterstock

New Hampshire Propane Company

Date	Description	Qty.	Price	Total
	Previous Balance			$0.00
06/24/20	PROPANE	3.60	$3.40	$12.24

Invoice No. 004433L

Invoice Date 6/26/20

Totals this invoice: $12.24

AMOUNT DUE: $12.24

Prompt Pay Discount: $0.19

Net Amount Due if RECEIVED by 07/10/20: $12.05

Due Date 7/26/20

Now let's study cash discounts.

Cash Discounts

In the New Hampshire Propane Company invoice, we receive a cash discount of 19 cents. This amount is determined by the **terms of the sale,** which can include the credit period, cash discount, discount period, and freight terms.

My Money

Buyers can often benefit from buying on credit. The time period that sellers give buyers to pay their invoices is the **credit period.** Frequently, buyers can sell the goods bought during this credit period. Then, at the end of the credit period, buyers can pay sellers with the funds from the sales of the goods. When buyers can do this, they can use the consumer's money to pay the invoice instead of their money.

A cash discount is for prompt payment. A trade discount is not.

Sellers can also offer a cash discount, or reduction from the invoice price, if buyers pay the invoice within a specified time. This time period is the **discount period,** which is part of the total credit period. Sellers offer this cash discount because they can use the dollars to better advantage sooner than later. Buyers who are not short of cash like cash discounts because the goods will cost them less and, as a result, provide an opportunity for larger profits.

Trade discounts should be taken before cash discounts.

Remember that buyers do not take cash discounts on freight, returned goods, sales tax, and trade discounts. Buyers take cash discounts on the *net price* of the invoice. Before we discuss how to calculate cash discounts, let's look at some aids that will help you calculate credit **due dates** and **end of credit periods.**

Aids in Calculating Credit Due Dates

Sellers usually give credit for 30, 60, or 90 days. Not all months of the year have 30 days. So you must count the credit days from the date of the invoice. The trick is to remember the number of days in each month. You can choose one of the following three options to help you do this.

Every year that is exactly divisible by 4 is a leap year except for years that are exactly divisible by 100 years unless the centurial year is exactly divisible by 400.

Option 1: Days-in-a-Month Rule You may already know this rule. Remember that every 4 years is a leap year.

> Thirty days has September, April, June, and November; all the rest have 31 except February has 28, and 29 in leap years.

Option 2: Knuckle Months Some people like to use the knuckles on their hands to remember which months have 30 or 31 days. Note in the following diagram that each knuckle represents a month with 31 days. The short months are in between the knuckles.

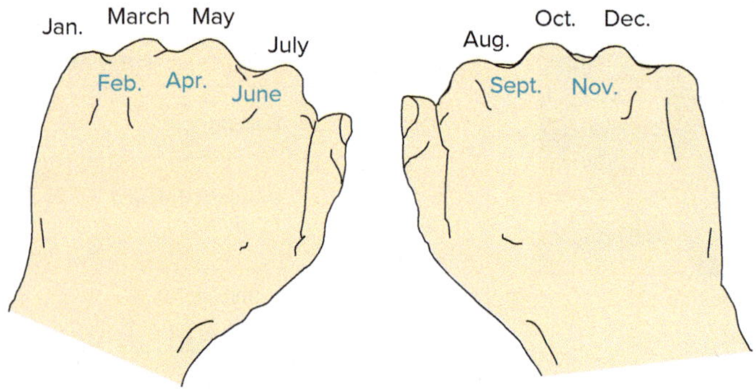

31 days: Jan., March, May, July, Aug., Oct., Dec.

A financial calculator can calculate maturity date, the number of days between dates, and loan date.

Option 3: Days-in-a-Year Calendar The days-in-a-year calendar (excluding leap year) is another tool to help you calculate dates for discount and credit periods (Table 7.1). For example, let's use Table 7.1 to calculate 90 days from August 12.

EXAMPLE By Table 7.1: August 12 = 224 days
+ 90
314 days

Search for day 314 in Table 7.1. You will find that day 314 is November 10. In this example, we stayed within the same year. Now let's try an example in which we overlap from year to year.

When using the days-in-a-year calendar, always put the number of days in the numerator and 365 (366 in leap years or 360 for ordinary interest) as the denominator.

EXAMPLE What date is 80 days after December 5?

Table 7.1 shows that December 5 is 339 days from the beginning of the year. Subtracting 339 from 365 (the end of the year) tells us that we have used up 26 days by the end of the year. This leaves 54 days in the new year. Go back in the table and start with the beginning of the year and search for 54 (80 − 26) days. The 54th day is February 23.

By table

365 days in year
− 339 days until December 5
26 days used in year

80 days from December 5
− 26 days used in year
54 days in new year or February 23

Without use of table

December 31
− December 5
26
+ 31 days in January
57
+ 23 due date (February 23)
80 total days

When you know how to calculate credit due dates, you can understand the common business terms sellers offer buyers involving discounts and credit periods. Remember that discount and credit terms vary from one seller to another.

Common Credit Terms Offered by Sellers

My Money

The common credit terms sellers offer buyers include *ordinary dating, receipt of goods (ROG),* and *end of month (EOM).* In this section we examine these credit terms. To determine the due dates, we use the exact days-in-a-year calendar (Table 7.1).

Ordinary Dating Today, businesses frequently use the **ordinary dating** method. It gives the buyer a cash discount period that begins with the invoice date. The credit terms of two common ordinary dating methods are 2/10, n/30 and 2/10, 1/15, n/30.

TABLE 7.1 Exact days-in-a-year calendar (excluding leap year)*

Day of month	31 Jan.	28 Feb.	31 Mar.	30 Apr.	31 May	30 June	31 July	31 Aug.	30 Sept.	31 Oct.	30 Nov.	31 Dec.
1	1	32	60	91	121	152	182	213	244	274	305	335
2	2	33	61	92	122	153	183	214	245	275	306	336
3	3	34	62	93	123	154	184	215	246	276	307	337
4	4	35	63	94	124	155	185	216	247	277	308	338
5	5	36	64	95	125	156	186	217	248	278	309	339
6	6	37	65	96	126	157	187	218	249	279	310	340
7	7	38	66	97	127	158	188	219	250	280	311	341
8	8	39	67	98	128	159	189	220	251	281	312	342
9	9	40	68	99	129	160	190	221	252	282	313	343
10	10	41	69	100	130	161	191	222	253	283	314	344
11	11	42	70	101	131	162	192	223	254	284	315	345
12	12	43	71	102	132	163	193	224	255	285	316	346
13	13	44	72	103	133	164	194	225	256	286	317	347
14	14	45	73	104	134	165	195	226	257	287	318	348
15	15	46	74	105	135	166	196	227	258	288	319	349
16	16	47	75	106	136	167	197	228	259	289	320	350
17	17	48	76	107	137	168	198	229	260	290	321	351
18	18	49	77	108	138	169	199	230	261	291	322	352
19	19	50	78	109	139	170	200	231	262	292	323	353
20	20	51	79	110	140	171	201	232	263	293	324	354
21	21	52	80	111	141	172	202	233	264	294	325	355
22	22	53	81	112	142	173	203	234	265	295	326	356
23	23	54	82	113	143	174	204	235	266	296	327	357
24	24	55	83	114	144	175	205	236	267	297	328	358
25	25	56	84	115	145	176	206	237	268	298	329	359
26	26	57	85	116	146	177	207	238	269	299	330	360
27	27	58	86	117	147	178	208	239	270	300	331	361
28	28	59	87	118	148	179	209	240	271	301	332	362
29	29	—	88	119	149	180	210	241	272	302	333	363
30	30	—	89	120	150	181	211	242	273	303	334	364
31	31	—	90	—	151	—	212	243	—	304	—	365

*Often referred to as a Julian calendar.

2/10, n/30 Ordinary Dating Method The 2/10, n/30 is read as "two ten, net thirty." Buyers can take a 2% cash discount off the gross amount of the invoice if they pay the bill within 10 days from the invoice date. If buyers miss the discount period, the net amount—without a discount—is due between day 11 and day 30. *Freight, returned goods, sales tax, and trade discounts must be subtracted from the gross before calculating a cash discount.*

EXAMPLE $400 invoice dated July 5: terms 2/10, n/30; no freight; paid on July 11.

Step 1. Calculate end of 2% discount period:

July 5 date of invoice
+ 10 days
July 15 end of 2% discount period

Step 2. Calculate end of credit period:

July 5 by Table 7.1 → 186 days
+ 30
216 days

Search in Table 7.1 for 216 → August 4 → end of credit period

Step 3. Calculate payment on July 11:

$.02 \times \$400 = \$8$ cash discount

$\$400 - \$8 =$ $392 paid

> *Note:* A 2% cash discount means that you save 2 cents on the dollar and pay 98 cents on the dollar. Thus, $\$.98 \times \$400 =$ $392.

The following time line illustrates the 2/10, n/30 ordinary dating method beginning and ending dates of the above example:

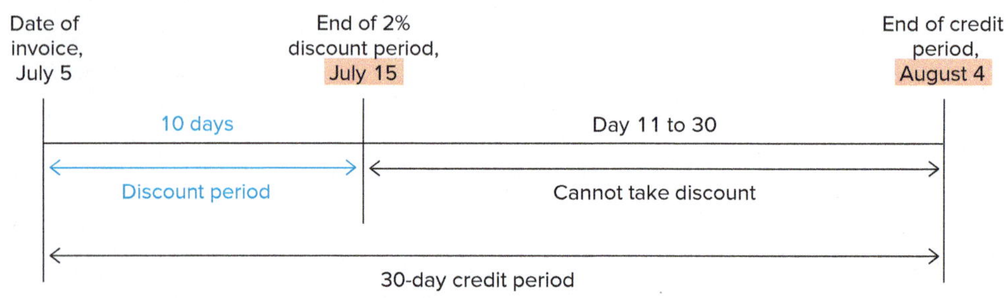

2/10, 1/15, n/30 Ordinary Dating Method The 2/10, 1/15, n/30 is read "two ten, one fifteen, net thirty." The seller will give buyers a 2% (2 cents on the dollar) cash discount if they pay within 10 days of the invoice date. If buyers pay between day 11 and day 15 from the date of the invoice, they can save 1 cent on the dollar. If buyers do not pay on day 15, the net or full amount is due 30 days from the invoice date.

EXAMPLE $600 invoice dated May 8; $100 of freight included in invoice price; paid on May 22. Terms 2/10, 1/15, n/30.

Step 1. Calculate the end of the 2% discount period:

May 8 date of invoice
+ 10 days
May 18 end of 2% discount period

Step 2. Calculate end of 1% discount period:

May 18 end of 2% discount period
+ 5 days
May 23 end of 1% discount period

Step 3. Calculate end of credit period:

May 8 by Table 7.1 → 128 days
+ 30
158 days

Search in Table 7.1 for 158 → June 7 → end of credit period

Step 4. Calculate payment on May 22 (14 days after date of invoice):

$$
\begin{array}{rl}
\$600 & \text{invoice} \\
-100 & \text{freight} \\
\hline
\$500 & \\
\times\quad .01 & \\
\hline
\$5.00 &
\end{array}
$$

$500 − $5.00 + $100 freight = **$595**

> A 1% discount means we pay $.99 on the dollar or
> $500 × $.99 = $495 + $100 freight = **$595.**
> *Note:* Freight is added back since no cash discount is taken on freight.

The following time line illustrates the 2/10, 1/15, n/30 ordinary dating method beginning and ending dates of the above example:

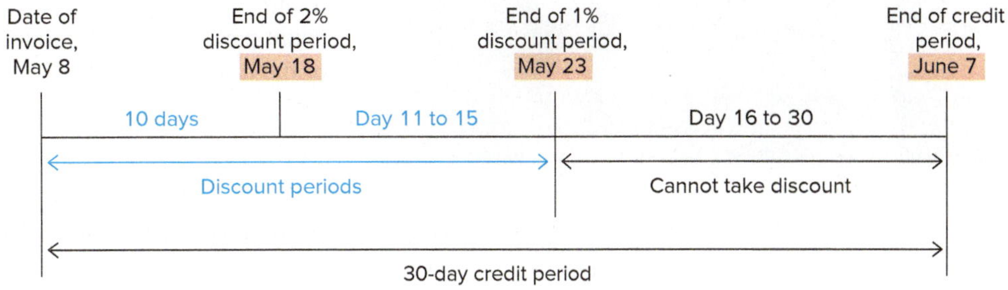

Receipt of Goods (ROG)

3/10, n/30 ROG With the **receipt of goods (ROG),** the cash discount period begins when the buyer receives the goods, *not* the invoice date. Industry often uses the ROG terms when buyers cannot expect delivery until a long time after they place the order. Buyers can take a 3% discount within 10 days *after* receipt of goods. The full amount is due between day 11 and day 30 if the cash discount period is missed.

EXAMPLE $900 invoice dated May 9; no freight or returned goods; the goods were received on July 8; terms 3/10, n/30 ROG; payment made on July 20.

Step 1. Calculate the end of the 3% discount period:

July 8 date goods arrive
\+ 10 days
July 18 end of 3% discount period

Step 2. Calculate the end of the credit period:

July 8 by Table 7.1 → 189 days
\+ 30
219 days

Search in Table 7.1 for 219 → **August** 7 → end of credit period

Step 3. Calculate payment on July 20:

Missed discount period and paid net or full amount of $900.

The following time line illustrates 3/10, n/30 ROG beginning and ending dates of the above example:

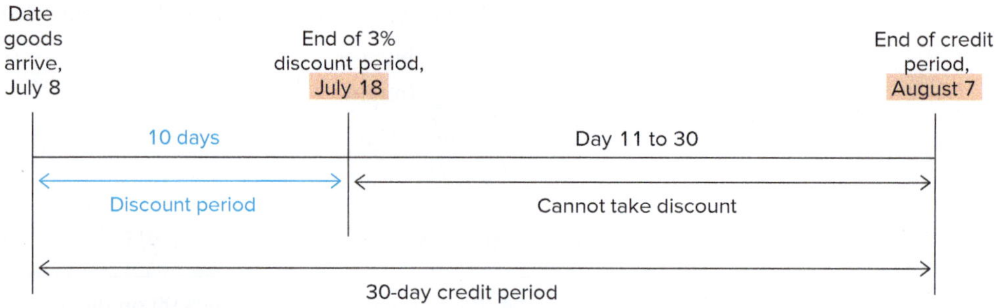

©DreamPictures/Getty Images

End of Month (EOM)[2] In this section we look at invoices involving **end of month (EOM)** terms. If an invoice is dated the *25th or earlier* of a month, we follow one set of rules. If an invoice is dated after the 25th of the month, a new set of rules is followed. Let's look at each situation.

Invoice Dated 25th or Earlier in Month, 1/10 EOM If sellers date an invoice on the 25th or earlier in the month, buyers can take the cash discount if they pay the invoice by the first 10 days of the month following the sale (next month). If buyers miss the discount period, the full amount is due within 20 days after the end of the discount period.

EXAMPLE $600 invoice dated July 6; no freight or returns; terms 1/10 EOM; paid on August 8.

Step 1. Calculate the end of the 1% discount period:
August 10 ← First 10 days of month following sale

Step 2. Calculate the end of the credit period:
August 10
+ 20 days
August 30 → Credit period is 20 days after discount period.

Step 3. Calculate payment on August 8:
$.99 \times \$600 = \$594$

The following time line illustrates the beginning and ending dates of the EOM invoice of the previous example:

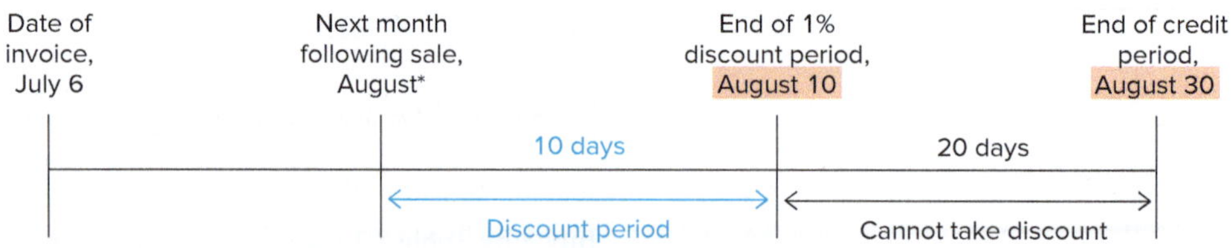

*Even though the discount period begins with the next month following the sale, if buyers wish, they can pay before the discount period (date of invoice until the discount period).

Invoice Dated after 25th of Month, 2/10 EOM When sellers sell goods *after* the 25th of the month, buyers gain an additional month. The cash discount period ends on the 10th day of the second month that follows the sale. Why? This occurs because the seller guarantees the 15 days' credit of the buyer. If a buyer bought goods on August 29, September 10 would be only 12 days. So the buyer gets the extra month.

[2]Sometimes the Latin term *proximo* is used. Other variations of EOM exist, but the key point is that the seller guarantees the buyer 15 days' credit. We assume a 30-day month.

EXAMPLE $800 invoice dated April 29; no freight or returned goods; terms 2/10 EOM; payment made on June 18.

Step 1. Calculate the end of the 2% discount period:
June 10 ← First 10 days of second month following sale

Step 2. Calculate the end of the credit period:

June 10
+ 20 days
June 30 ← Credit period is 20 days after discount period.

Step 3. Calculate the payment on June 18:
No discount; $800 paid.

The following time line illustrates the beginning and ending dates of the EOM invoice of the above example:

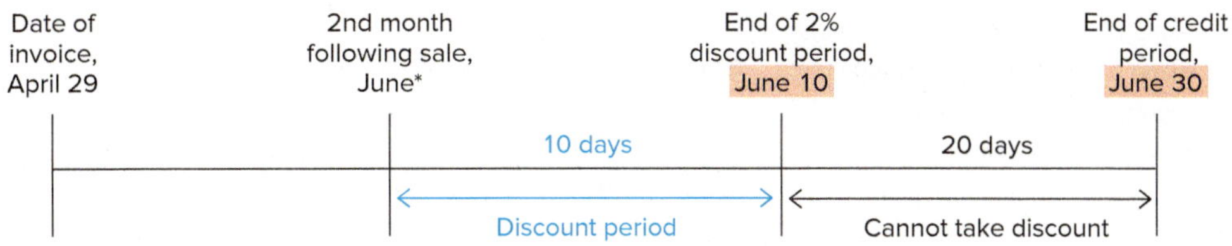

*Even though the discount period begins with the second month following the sale, if buyers wish, they can pay before the discount date (date of invoice until the discount period).

Solving a Word Problem with Trade and Cash Discount

Now that we have studied trade and cash discounts, let's look at a combination that involves both a trade and a cash discount.

The Word Problem Hardy Company sent Regan Corporation an invoice for office equipment with a $10,000 list price. Hardy dated the invoice July 29 with terms of 2/10 EOM (end of month). Regan receives a 30% trade discount and paid the invoice on September 6. Since terms were FOB destination, Regan paid no freight charge. What was the cost of office equipment for Regan?

BLUEPRINT	The facts	Solving for?	Steps to take	Key points
	List price: $10,000. *Trade discount rate:* 30%. *Terms:* 2/10 EOM. *Invoice date:* 7/29. *Date paid:* 9/6.	Cost of office equipment.	Net price = List price × Complement of trade discount rate. After 25th of month for EOM. Discount period is first 10 days of second month that follows sale.	Trade discounts are deducted before cash discounts are taken. Cash discounts are not taken on freight or returns.

Steps to solving problem

1. Calculate the net price. $10,000 × .70 = $7,000 (100% − 30% (trade discount))

2. Calculate the discount period. Sale: 7/29 Month 1: Aug. Month 2: Sept 10 → Paid on Sept. 6—is entitled to 2% off.

3. Calculate the cost of office equipment. $7,000 × .98 = $6,860 (100% − 2%) If you save 2 cents on a dollar, you are spending 98 cents.

LO 2

My Money

MONEY tips

The formula for determining your credit score is roughly: 35% from your payment history (may include library fines and parking tickets); 30% determined by your debt to available credit ratio; 15% on the length of your credit history—the fewer and older the accounts, the better; 10% based on how many recent accounts were opened along with the number of inquiries made by lenders on your credit report; and, finally, 10% for the types of credit used.

Partial Payments

Often buyers cannot pay the entire invoice before the end of the discount period. To calculate partial payments and outstanding balance, use the following steps:

CALCULATING PARTIAL PAYMENTS AND OUTSTANDING BALANCE

Step 1. Calculate the complement of a discount rate.

Step 2. Divide partial payments by the complement of a discount rate (Step 1). This gives the amount credited.

Step 3. Subtract Step 2 from the total owed. This is the outstanding balance.

EXAMPLE Molly McGrady owed $400. Molly's terms were 2/10, n/30. Within 10 days, Molly sent a check for $80. The actual credit the buyer gave Molly is as follows:

Step 1. 100% − 2% = 98% → .98

Step 2. $\frac{\$80}{.98} = \$81.63$ $\qquad \frac{\$80}{1-.02}$ ← Discount rate

Step 3.
$400.00
− 81.63 partial payment—although sent in $80
$318.37 outstanding balance

Note: We do not multiply .02 × $80 because the seller did not base the original discount on $80. When Molly makes a payment within the 10-day discount period, 98 cents pays each $1 she owes. Before buyers take discounts on partial payments, they must have permission from the seller. Not all states allow partial payments.

You have completed another unit. Let's check your progress.

LU 7–2 PRACTICE QUIZ

Complete this **Practice Quiz** to see how you are doing.

Complete the following table:

	Date of invoice	Date goods received	Terms	Last day* of discount period	End of credit period
1.	July 6		2/10, n/30		
2.	February 19	June 9	3/10, n/30 ROG		
3.	May 9		4/10, 1/30, n/60		
4.	May 12		2/10 EOM		
5.	May 29		2/10 EOM		

*If more than one discount, assume date of last discount.

6. Metro Corporation sent Vasko Corporation an invoice for equipment with an $8,000 list price. Metro dated the invoice May 26. Terms were 2/10 EOM. Vasko receives a 20% trade discount and paid the invoice on July 3. What was the cost of equipment for Vasko? (A blueprint aid will be in the solution to help dissect this problem.)

7. Complete amount to be credited and balance outstanding:

Amount of invoice: $600
Terms: 2/10, 1/15, n/30
Date of invoice: September 30
Paid October 3: $400

*For **extra help** from your authors–Sharon and Jeff–see the videos in Connect.*

✓ Solutions

1. End of discount period: July 6 + 10 days = **July 16**

End of credit period: By Table 7.1, July 6 = 187 days
+ 30 days
217 → search → **Aug. 5**

2. End of discount period: June 9 + 10 days = **July 19**

End of credit period: By Table 7.1, June 9 = 160 days
+ 30 days
190 → search → **July 9**

3. End of discount period: By Table 7.1, May 9 = 129 days
+ 30 days
159 → search → **June 8**

End of credit period: By Table 7.1, May 9 = 129 days
+ 60 days
189 → search → **July 8**

4. End of discount period: **June 10**

End of credit period: June 10 + 20 = **June 30**

5. End of discount period: **July 10**

End of credit period: July 10 + 20 = **July 30**

6. Vasko Corporation's cost of equipment:

BLUEPRINT	The facts	Solving for?	Steps to take	Key points
	List price: \$8,000. *Trade discount rate:* 20%. *Terms:* 2/10 EOM. *Invoice date:* 5/26. *Date paid:* 7/3.	Cost of equipment.	Net price = List price × Complement of trade discount rate. *EOM before 25th:* Discount period is first 10 days of month that follows sale.	Trade discounts are deducted before cash discounts are taken. Cash discounts are not taken on freight or returns.

Steps to solving problem

1. Calculate the net price. — \$8,000 × .80 = \$6,400 (100% − 20%)
2. Calculate the discount period. — Until July 10
3. Calculate the cost of office equipment. — \$6,400 × .98 = **\$6,272** (100% − 2%)

7. $\frac{\$400}{.98}$ = \$408.16, amount credited.

\$600 − \$408.16 = **\$191.84,** balance outstanding.

INTERACTIVE CHAPTER ORGANIZER

Topic/Procedure/Formula	Examples	You try it*
Trade discount amount Trade discount amount = List price × Trade discount rate	$600 list price 30% trade discount rate Trade discount amount = $600 × .30 = $180	**Calculate trade discount amount** $700 list price 20% trade discount
Calculating net price Net price = List price − Trade discount amount or List price × Complement of trade discount price	$600 list price 30% trade discount rate Net price = $600 × .70 = $420 1.00 − .30 .70	**Calculate net price** $700 list price 20% trade discount
Freight FOB shipping point—buyer pays freight. FOB destination—seller pays freight.	Moose Company of New York sells equipment to Agee Company of Oregon. Terms of shipping are FOB New York. Agee pays cost of freight since terms are FOB shipping point.	**Calculate freight** If a buyer in Boston buys equipment with shipping terms of FOB destination, who will pay cost of freight?
Calculating list price when net price and trade discount rate are known List price = $\frac{\text{Net price}}{\text{Complement of trade discount rate}}$	40% trade discount rate Net price, $120 $\frac{\$120}{.60}$ = $200 list price .60 ← (1.00 − .40)	**Calculate list price** 60% trade discount rate Net price, $240
Chain discounts Successively lower base.	5/10 on a $100 list item $ 100 × .05 = $ 5.00 $ 95 × .10 = $9.50 (running balance) $95.00 − 9.50 $85.50 net price	**Calculate net price** 6/8 on $200 list item
Net price equivalent rate Actual cost to buyer = List price × Net price equivalent rate Take complement of each chain discount and multiply—do not round. Trade discount amount = List price − Actual cost to buyer	Given: 5/10 on $1,000 list price Take complement: .95 × .90 = .855 (net price equivalent) $1,000 × .855 = $855 (actual cost or net price) $1,000 − 855 $ 145 trade discount amount	**Calculate net price equivalent rate, net price, and trade discount amount** 6/8 on $2,000 list
Single equivalent discount rate Trade discount amount = List price × 1 − Net price equivalent rate	See preceding example for facts: 1 − .855 = .145 .145 × $1,000 = $145	**From the above You Try It, calculate single equivalent discount**
Cash discounts Cash discounts, due to prompt payment, are not taken on freight, returns, etc.	Gross $1,000 (includes freight) Freight $25 Returns $25 Terms 2/10, n/30 Purchased: Sept. 9; paid Sept. 15 Cash discount = $950 × .02 = $19	**Calculate cash discount** Gross $2,000 (includes freight) Freight $40 Returns $40 Terms 2/10, n/30 Purchased: Sept. 2; paid Sept. 8

INTERACTIVE CHAPTER ORGANIZER

Topic/Procedure/Formula	Examples	You try it*
Calculating due dates *Option 1:* Thirty days has September, April, June, and November; all the rest have 31 except February has 28, and 29 in leap years. *Option 2:* Knuckles—31-day month; in between knuckles are short months. *Option 3:* Days-in-a-year table.	Invoice $500 on March 5; terms 2/10, n/30 March 5 *End of discount* + 10 *period:* → March 15 *End of credit* March 5 = 64 days *period by* + 30 *Table 7.1:* → 94 days Search in Table 7.1 April 4	**Calculate end of discount and end of credit periods** Invoice $600 on April 2; terms 2/10, n/30
Common terms of sale **a. Ordinary dating** Discount period begins from date of invoice. Credit period ends 20 days from the end of the discount period unless otherwise stipulated; example, 2/10, n/60—the credit period ends 50 days from end of discount period.	Invoice $600 (freight of $100 included in price) dated March 8; payment on March 16; 3/10, n/30. March 8 *End of discount* + 10 *period:* → March 18 *End of credit* March 8 = 67 days *period by* + 30 *Table 7.1:* → 97 days Search in Table 7.1 April 7 *If paid on March 16:* .97 × $500 = $485 + 100 freight $585	**Calculate amount paid** Invoice $700 (freight of $100 included in price) dated May 7; payment on May 15; 2/10, n/30
b. Receipt of goods (ROG) Discount period begins when goods are received. Credit period ends 20 days from end of discount period.	4/10, n/30, ROG. $600 invoice; no freight; dated August 5; goods received October 2, payment made October 20. October 2 *End of discount* + 10 *period:* → October 12 *End of* October 2 = 275 *credit period* + 30 *by Table 7.1:* → 305 Search in Table 7.1 November 1 *Payment on October 20:* No discount, pay $600	**Calculate amount paid** 3/10, n/30, ROG. $700 invoice; no freight; dated September 6; goods received September 20; payment made October 15.
c. End of month (EOM) On or before 25th of the month, discount period is 10 days after month following sale. After 25th of the month, an additional month is gained.	$1,000 invoice dated May 12; no freight or returns; terms 2/10 EOM. *End of discount period* → June 10 *End of credit period* → June 30	**Calculate end of discount and end of credit periods** $2,000 invoice dated October 11; terms 2/10 EOM
Partial payments $\text{Amount credited} = \dfrac{\text{Partial payment}}{1 - \text{Discount rate}}$	$200 invoice; terms 2/10, n/30; dated March 2; paid $100 on March 5. $\dfrac{\$100}{1 - .02} = \dfrac{\$100}{.98} = \$102.04$	**Calculate amount credited** $400 invoice; terms 2/10, n/30; dated May 4; paid $300 on May 7.

(continues)

INTERACTIVE CHAPTER ORGANIZER

Topic/Procedure/Formula	Examples		You try it*
KEY TERMS	Cash discount Chain discounts Complement Credit period Discount period Due dates End of credit period End of month (EOM) FOB destination	FOB shipping point Freight terms Invoice List price Net price Net price equivalent rate Ordinary dating Receipt of goods (ROG) Series discounts	Single equivalent discount rate Single trade discount Terms of the sale Trade discount Trade discount amount Trade discount rate

*Worked-out solutions are in Appendix B.

Critical Thinking Discussion Questions with Chapter Concept Check

1. What is the net price? June Long bought a jacket from a catalog company. She took her trade discount off the original price plus freight. What is wrong with June's approach? Who would benefit from June's approach—the buyer or the seller?
2. How do you calculate the list price when the net price and trade discount rate are known? A publisher tells the bookstore its net price of a book along with a suggested trade discount of 20%. The bookstore uses a 25% discount rate. Is this ethical when textbook prices are rising?
3. If Jordan Furniture ships furniture FOB shipping point, what does that mean? Does this mean you get a cash discount?
4. What are the steps to calculate the net price equivalent rate? Why is the net price equivalent rate not rounded?
5. What are the steps to calculate the single equivalent discount rate? Is this rate off the list or net price? Explain why this calculation of a single equivalent discount rate may not always be needed.
6. What is the difference between a discount and credit period? Are all cash discounts taken before trade discounts? Do you agree or disagree? Why?
7. Explain the following credit terms of sale:
   a. 2/10, n/30.
   b. 3/10, n/30 ROG.
   c. 1/10 EOM (on or before 25th of month).
   d. 1/10 EOM (after 25th of month).
8. Explain how to calculate a partial payment. Whom does a partial payment favor—the buyer or the seller?
9. **Chapter Concept Check.** Search Facebook to find out what customer discounts companies offer to Facebook users. Be sure to talk about shipping charges and trade and cash discounts. What kind of savings can you find?

7

END-OF-CHAPTER PROBLEMS

connect

Check figures for odd-numbered problems in Appendix B. Name ________________ Date ________

DRILL PROBLEMS

For all problems, round your final answer to the nearest cent. Do not round net price equivalent rates or single equivalent discount rates.

Complete the following: *LU 7-1(4)*

	Item	List price	Chain discount	Net price equivalent rate (in decimals)	Single equivalent discount rate (in decimals)	Trade discount	Net price
7–1.	Surface computer	$1,200	4/1				
7–2.	Samsung Blu-Ray player	$199	8/4/3				
7–3.	Canon document scanner	$269	7/3/1				

Complete the following: *LU 7-1(4)*

	Item	List price	Chain discount	Net price	Trade discount
7–4.	Trotter treadmill	$3,000	9/4		
7–5.	Maytag dishwasher	$450	8/5/6		
7–6.	Sony digital camera	$320	3/5/9		
7–7.	Land Rover roofrack	$1,850	12/9/6		

eXcel

7–8. Which of the following companies, A or B, gives a higher discount? Use the single equivalent discount rate to make your choice (convert your equivalent rate to the nearest hundredth percent).

Company A
8/10/15/3

Company B
10/6/16/5

Complete the following: *LU 7-2(1)*

	Invoice	Date goods are received	Terms	Last day* of discount period	Final day bill is due (end of credit period)
7–9.	June 18		1/10, n/30		
7–10.	Nov. 27		2/10 EOM		
7–11.	May 15	June 5	3/10, n/30, ROG		
7–12.	April 10		2/10, 1/30, n/60		
7–13.	June 12		3/10 EOM		
7–14.	Jan. 10	Feb. 3 (no leap year)	4/10, n/30, ROG		

*If more than one discount, assume date of last discount.

Complete the following by calculating the cash discount and net amount paid: *LU 7-2(1)*

	Gross amount of invoice (freight charge already included)	Freight charge	Date of invoice	Terms of invoice	Date of payment	Cash discount	Net amount paid
7–15.	$7,000	$100	4/8	2/10, n/60	4/15		
7–16.	$600	None	8/1	3/10, 2/15, n/30	8/13		
7–17.	$200	None	11/13	1/10 EOM	12/3		
7–18.	$500	$100	11/29	1/10 EOM	1/4		

Complete the following: *LU 7-2(2)*

	Amount of invoice	Terms	Invoice date	Actual partial payment made	Date of partial payment	Amount of payment to be credited	Balance outstanding
7–19.	$700	2/10, n/60	5/6	$400	5/15		
7–20.	$600	4/10, n/60	7/5	$400	7/14		

WORD PROBLEMS (Round to Nearest Cent as Needed)

7–21. The list price of a smartphone is $299. A local Samsung dealer receives a trade discount of 20%. Find the trade discount amount and the net price. *LU 7-1(1)*

eXcel

7–22. A model NASCAR race car lists for $79.99 with a trade discount of 40%. What is the net price of the car? *LU 7-1(1)*

My Money

7–23. Lucky you! You went to couponcabin.com and found a 20% off coupon to your significant other's favorite store. Armed with that coupon, you went to the store only to find a storewide sale offering 10% off everything in the store. In addition, your credit card has a special offer that allows you to save 10% if you use your credit card for all purchases that day. Using your credit card, what will you pay before tax for the $155 gift you found? Use the single equivalent discount to calculate how much you save and then calculate your final price. *LU 7-1(4)*

7–24. Levin Furniture buys a living room set with a $4,000 list price and a 55% trade discount. Freight (FOB shipping point) of $50 is not part of the list price. What is the delivered price (including freight) of the living room set, assuming a cash discount of 2/10, n/30, ROG? The invoice had an April 8 date. Levin received the goods on April 19 and paid the invoice on April 25. *LU 7-1(1, 2)*

7–25. A manufacturer of skateboards offered a 5/2/1 chain discount to many customers. Bob's Sporting Goods ordered 20 skateboards for a total $625 list price. What was the net price of the skateboards? What was the trade discount amount? *LU 7-1(4)*

eXcel

7–26. Home Depot wants to buy a new line of fertilizers. Manufacturer A offers a 21/13 chain discount. Manufacturer B offers a 26/8 chain discount. Both manufacturers have the same list price. What manufacturer should Home Depot buy from? *LU 7-1(4)*

7–27. Maplewood Supply received a $5,250 invoice dated 4/15/20. The $5,250 included $250 freight. Terms were 4/10, 3/30, n/60. **(a)** If Maplewood pays the invoice on April 27, what will it pay? **(b)** If Maplewood pays the invoice on May 21, what will it pay? *LU 7-2(1)*

eXcel

7–28. A local Dick's Sporting Goods ordered 50 pairs of tennis shoes from Nike Corporation. The shoes were priced at $85 for each pair with the following terms: 4/10, 2/30, n/60. The invoice was dated October 15. Dick's Sporting Goods sent in a payment on October 28. What should have been the amount of the check? *LU 7-2(1)*

eXcel

7–29. Macy of New York sold LeeCo. of Chicago office equipment with a $6,000 list price. Sale terms were 3/10, n/30 FOB New York. Macy agreed to prepay the $30 freight. LeeCo. pays the invoice within the discount period. What does LeeCo. pay Macy? *LU 7-2(2)*

7–30. Royal Furniture bought a sofa for $800. The sofa had a $1,400 list price. What was the trade discount rate Royal received? Round to the nearest hundredth percent. *LU 7-2(1)*

My Money

7–31. After researching the cost of a shredder for your online business, you settle on a more expensive shredder because the credit terms are better and cash flow is a challenge for you during the summer months. If you buy a shredder for $1,399 with 3/15, net 30 terms on August 22, how much do you need to pay on September 5? *LU 7-2(1)*

7–32. Bally Manufacturing sent Intel Corporation an invoice for machinery with a $14,000 list price. Bally dated the invoice July 23 with 2/10 EOM terms. Intel receives a 40% trade discount. Intel pays the invoice on August 5. What does Intel pay Bally? *LU 7-2(1)*

7–33. On August 1, Intel Corporation (Problem 7–32) returns $100 of the machinery due to defects. What does Intel pay Bally on August 5? Round to nearest cent. *LU 7-2(1)*

7–34. Stacy's Dress Shop received a $1,050 invoice dated July 8 with 2/10, 1/15, n/60 terms. On July 22, Stacy's sent a $242 partial payment. What credit should Stacy's receive? What is Stacy's outstanding balance? *LU 7-2(2)*

7–35. On March 11, Jangles Corporation received a $20,000 invoice dated March 8. Cash discount terms were 4/10, n/30.On March 15, Jangles sent an $8,000 partial payment. What credit should Jangles receive? What is Jangles' outstanding balance? *LU 7-2(2)*

7–36. A used Porsche Macan Turbo starts at a consumer price of $72,300. If a dealership can purchase 10 with a 15/10/5 chain discount, what is the net price for the dealership? *LU 7-1(4)*

7–37. A local Barnes and Noble paid a $79.99 net price for each hardbound atlas. The publisher offered a 20% trade discount. What was the publisher's list price? *LU 7-1(3)*

7–38. Rocky Mountain Chocolate Factory (RMCF) founder and president Frank Crail employs 220 people in 361 outlets in the United States, Canada, United Arab Emirates, Japan, and South Korea. If RMCF purchases 20 kilograms of premium dark chocolate at $16.25 per kilo, what is the net price with a 10/5 chain discount? Round to the nearest cent. *LU 7-1(1)*

7–39. Vail Ski Shop received a $1,201 invoice dated July 8 with 2/10, 1/15, n/60 terms. On July 22, Vail sent a $485 partial payment. What credit should Vail receive? What is Vail's outstanding balance? *LU 7-2(2)*

7–40. True Value received an invoice dated 4/15/20. The invoice had a $5,500 balance that included $300 freight. Terms were 4/10, 3/30, n/60. True Value pays the invoice on April 29. What amount does True Value pay? *LU 7-1(1, 2)*

7–41. Baker's Financial Planners purchased seven new computers for $850 each. It received a 15% discount because it purchased more than five and an additional 6% discount because it took immediate delivery. Terms of payment were 2/10, n/30. Baker's pays the bill within the cash discount period. How much should the check be? Round to the nearest cent. *LU 7-1(4)*

7–42. On May 14, Talbots of Boston sold Forrest of Los Angeles $7,000 of fine clothes. Terms were 2/10 EOM FOB Boston. Talbots agreed to prepay the $80 freight. If Forrest pays the invoice on June 8, what will Forrest pay? If Forrest pays on June 20, what will Forrest pay? *LU 7-1(2), LU 7-2(1)*

7–43. Sam's Snowboards.com offers 5/4/1 chain discounts to many of its customers. The Ski Hut ordered 20 snowboards with a total list price of $1,200. What is the net price of the snowboards? What was the trade discount amount? Round to the nearest cent. *LU 7-1(4)*

7–44. Majestic Manufacturing sold Jordans Furniture a living room set for an $8,500 list price with 35% trade discount. The $100 freight (FOB shipping point) was not part of the list price. Terms were 3/10, n/30 ROG. The invoice date was May 30. Jordans received the goods on July 18 and paid the invoice on July 20. What was the final price (include cost of freight) of the living room set? *LU 7-1(1, 2), LU 7-2(1)*

7–45. Boeing Truck Company received an invoice showing 8 tires at $110 each, 12 tires at $160 each, and 15 tires at $180 each. Shipping terms are FOB shipping point. Freight is $400; trade discount is 10/5; and a cash discount of 2/10, n/30 is offered. Assuming Boeing paid within the discount period, what did Boeing pay? *LU 7-1(4)*

7–46. Verizon offers to sell cellular phones listing for $99.99 with a chain discount of 15/10/5. Cellular Company offers to sell its cellular phones that list at $102.99 with a chain discount of 25/5. If Irene is to buy six phones, how much could she save if she buys from the lower-priced company? *LU 7-1(4)*

7–47. Living Ornaments is offering a special for wedding planners. Wedding flower orders totaling over $500 receive a 10% discount, over $750 a 15% discount, over $1,000 a 20% discount. All orders $1,500 and above receive a 25% discount. The delivery charge is $75 on weekdays and $125 on weekends. Terms are 2/10 EOM. WeddingsRUs placed an order for Thursday, June 1, delivery. The list price on the chosen flowers totals $848.50. Calculate the trade discount and the net price for the flowers to be delivered. How much does WeddingsRUs owe if it pays the invoice on July 10? (Round to the nearest cent.) *LU 7-2(1)*

CHALLENGE PROBLEMS

7–48. The original price of a 2018 Honda Shadow to the dealer is $17,995, but the dealer will pay only $16,495 after rebate. If the dealer pays Honda within 15 days, there is a 1% cash discount. **(a)** How much is the rebate? **(b)** What percent is the rebate? Round to nearest hundredth percent. **(c)** What is the amount of the cash discount if the dealer pays within 15 days? **(d)** What is the dealer's final price? **(e)** What is the dealer's total savings? Round answer to the nearest hundredth. *LU 7-1(1), LU 7-2(1)*

7–49. On March 30, Century Link received an invoice dated March 28 from ACME Manufacturing for 50 televisions at a cost of $125 each. Century received a 10/4/2 chain discount. Shipping terms were FOB shipping point. ACME prepaid the $70 freight. Terms were 2/10 EOM. When Century received the goods, 3 sets were defective. Century returned these sets to ACME. On April 8, Century sent a $150 partial payment. Century will pay the balance on May 6. What is Century's final payment on May 6? Assume no taxes. *LU 7-1(1, 2, 4), LU 7-2(1)*

SUMMARY PRACTICE TEST

Do you need help? Connect videos have step-by-step worked-out solutions.

(Round to the Nearest Cent as Needed)

Complete the following: *LU 7-1(1)*

	Item	List price	Single trade discount	Net price
1.	Apple iPod	$350	5%	
2.	Palm Pilot		10%	$190

Calculate the net price and trade discount (use net price equivalent rate and single equivalent discount rate) for the following: *LU 7-1(4)*

	Item	List price	Chain discount	Net price	Trade discount
3.	Sony HD flat-screen TV	$899	5/4		

4. From the following, what is the last date for each discount period and credit period? *LU 7-1(1)*

	Date of invoice	Terms	End of discount period	End of credit period
a.	Nov. 4	2/10, n/30		
b.	Oct. 3, 2015	3/10, n/30 ROG (Goods received March 10, 2016)		
c.	May 2	2/10 EOM		
d.	Nov. 28	2/10 EOM		

5. Best Buy buys an iPad from a wholesaler with a $300 list price and a 5% trade discount. What is the trade discount amount? What is the net price of the iPad? *LU 7-1(1)*

6. Jordan's of Boston sold Lee Company of New York computer equipment with a $7,000 list price. Sale terms were 4/10, n/30 FOB Boston. Jordan's agreed to prepay the $400 freight. Lee pays the invoice within the discount period. What does Lee pay Jordan's? *LU 7-1(2), LU 7-2(1)*

7. Julie Ring wants to buy a new line of Tonka trucks for her shop. Manufacturer A offers a 14/8 chain discount. Manufacturer B offers a 15/7 chain discount. Both manufacturers have the same list price. Which manufacturer should Julie buy from? *LU 7-1(4)*

8. Office.com received an $8,000 invoice dated April 10. Terms were 2/10, 1/15, n/60. On April 14, Office.com sent a $1,900 partial payment. What credit should Office.com receive? What is Office.com's outstanding balance? Round to the nearest cent. *LU 7-2(2)*

9. Logan Company received from Furniture.com an invoice dated September 29. Terms were 1/10 EOM. List price on the invoice was $8,000 (freight not included). Logan receives an 8/7 chain discount. Freight charges are Logan's responsibility, but Furniture.com agreed to prepay the $300 freight. Logan pays the invoice on November 7. What does Logan Company pay Furniture.com? *LU 7-1(4)*

INTERACTIVE VIDEO WORKSHEET

GRADING THE SUMMARY PRACTICE TEST

Go to the summary practice test video in Connect (or click on it here in the ebook). Grade your summary practice test while viewing the video.

C for Correct/I for Incorrect

1. ________	4. ________	7. ________
2. ________	5. ________	8. ________
3. ________	6. ________	9. ________

If you achieved 100%, you are ready for your instructor's exam.

If any of the problems were incorrect, list the questions you missed and show steps to solve the problem correctly.

Replay the video to see if you have made the correct fixes to your mistakes. If you have any questions, contact your instructor asap.

Notes on Watching Videos

MY MONEY

Buying Into It!

What I need to know

As consumers we make a multitude of different purchases throughout our lives. We can all agree that when making purchases we like to get the most benefit from making the purchase. This is even more crucial when making large purchases (i.e., new car, house, or even furniture). In order to ensure we are making wise decisions it is important to do our research concerning the large purchase. Are we getting the best price? Are there incentives available to lower the price in the form of coupons or rebates? Additionally there could be other indicators such as product warranties, company reputation, or quality level that should play a role in finding the best purchase for our money. If we take the time to learn about all of our purchase options we can create more value in the purchase and lessen the occurrence of experiencing regret from our purchase.

What I need to do

Much like doing research for one of your classes, you want to ensure you know where to look to find the information that will assist you in making these larger purchases. Obviously the Internet will play a vital role with an enormous amount of information available to you. Use a simple Internet search to compare prices for the item or service you are seeking. Be sure to compare like products to one another and investigate any price differences to ensure it is representative of the same exact product. Be sure to take note of all influencers on the final price including tax, service fees, delivery charges, handling fees, and so on. By understanding all of the fees involved, you can make a more informed decision as to which is actually the lowest price based on the overall price.

In your search for the best price, you also want to consider any promotional offers. Some sellers may be offering a percentage/dollar off the ticket price for a limited time sale offer. Additionally others may offer free shipping on the product which could equate to a substantial cost savings. Another offer may include rebates from the manufacturer or retailer of the item you intend to purchase. Although these rebates will normally come in the form of a payment received at a later date, it can still provide you with a significant reduction in the overall expenditure you incur for the item.

A final component to your purchase is the consideration of quality and the reputation of the company from which you purchase. Utilize online reviews to get an indication of the experiences others have had with the organization to determine whether it is a good fit for you. Also, you can learn more about the quality of the item you are purchasing from users who have bought and used the item on a personal level. This is a way to compare the item description from the seller with those experiences of actual users of the product to determine if the actual use of the product lives up to its description.

Resources I can use

- RetailMeNot: Coupons & Savings (mobile app)
- https://www.mindtools.com/page6.html — online coupons
- https://www.thebalance.com/before-you-make-large-purchases-2385817 — things to consider before making a large purchase

MY MONEY ACTIVITY

Conduct an Internet search for a new laptop computer. Note the price differences and incentives offered from different retailers.

PERSONAL FINANCE

A KIPLINGER APPROACH

"Shop Smarter", *Kiplinger's*, April 2018, p. 34. Used by permission of The Kiplinger Washington Editors, Inc.

©supparsorn/Shutterstock

Shop Smarter

Hunt for discounts. You can find store discount codes, often worth up to 20%, at sites such as **RETAILMENOT.COM** and **PROMOCODES.COM**. But if you're not in the mood to conduct the search yourself, **HONEY** (www.joinhoney.com), a browser extension available for Chrome, Safari and Firefox, searches for coupons at thousands of sites and applies the discounts at checkout. Honey also searches Amazon.com for the lowest price.

Don't assume Amazon has the best deal. LendEDU, a comparison marketplace for loans and other financial products, compared a shopping cart of 50 identical items from Amazon and Walmart in five categories: home goods, kitchen/appliances, tech, food and beverage, and miscellaneous items. Walmart bested Amazon in every category except food and beverage.

You can score deals at the supermarket without clipping coupons. If you sign up for a loyalty card, you can save up to 30% on sale items, and you may be able to add virtual coupons to your card if you register it on the grocer's website.

Avoid the shipping fee. If you're paying pesky shipping fees, maybe you're not shopping right. The coupon sites mentioned above may turn up free-shipping codes. Or search **BRADSDEAL.COM** for more than 50 stores—including Coach, Dell, DKNY, Microsoft, Neiman Marcus, Nordstrom, North Face, Ray Ban and Zappos—that always offer free shipping with no minimum purchase. But compare what you buy at sites with shipping fees to make sure you're getting the best overall deal. Another option: Best Buy, Nordstrom, Target and other large retailers offer free shipping to nearby stores, so you can take advantage of the online price. Or contact customer service through the chat box to ask about free-shipping codes.

Skip the extended warranty. Don't pay an extra 20% on an extra warranty—your credit card may have your back. American Express offers an extended warranty on all of its cards, and most cards from Visa and MasterCard also include this perk (Discover recently discontinued it). To see what your card issuer offers, go to WalletHub.com and search for "extended warranty." The extra warranty typically lasts up to a year beyond the manufacturer's warranty and has coverage up to $10,000.

Buy When Things Are On Sale

JANUARY: linens and bedding, exercise equipment, winter clothing

FEBRUARY: TVs, tax-filing software, cookware and small kitchen appliances

MARCH: skis and snowboards, high-end fashion, luggage

APRIL: tires, cruise vacations, hotel discounts on tax day

MAY: mattresses, smart-home hubs, office furniture

JUNE: gardening tools, gym memberships, camping gear

JULY: swimwear, air conditioners

AUGUST: laptops, school supplies, shoes

SEPTEMBER: older iPhone models, grills, summer clothing

OCTOBER: outgoing models of sedans and SUVs, Halloween costumes, patio furniture

NOVEMBER: gaming systems, tablets, large appliances

DECEMBER: jewelry, toys, wedding dresses

BUSINESS MATH ISSUE

A sharp shopper always buys the lowest priced item and does not take out warranties.

1. List the key points of the article and information to support your position.
2. Write a group defense of your position using math calculations to support your view. If you are in an online cource, post to a discussion board.

Classroom Notes

CHAPTER 8

Markups and Markdowns: Perishables and Breakeven Analysis

©Evlied/Shutterstock

LEARNING UNIT OBJECTIVES

LU 8–1: Markups[1] Based on Cost (100%)

1. Calculate dollar markup and percent markup on cost.
2. Calculate selling price when you know the cost and percent markup on cost.
3. Calculate cost when you know the selling price and percent markup on cost.

LU 8–2: Markups Based on Selling Price (100%)

1. Calculate dollar markup and percent markup on selling price.
2. Calculate selling price when cost and percent markup on selling price are known.
3. Calculate cost when selling price and percent markup on selling price are known.
4. Convert from percent markup on cost to percent markup on selling price and vice versa.

LU 8–3: Markdowns and Perishables

1. Calculate markdowns; compare markdowns and markups.
2. Price perishable items to cover spoilage loss.

LU 8–4: Breakeven Analysis

1. Calculate contribution margin.
2. Calculate breakeven point.

Your Guide to Successfully Completing This Chapter

Traditional book or ebook

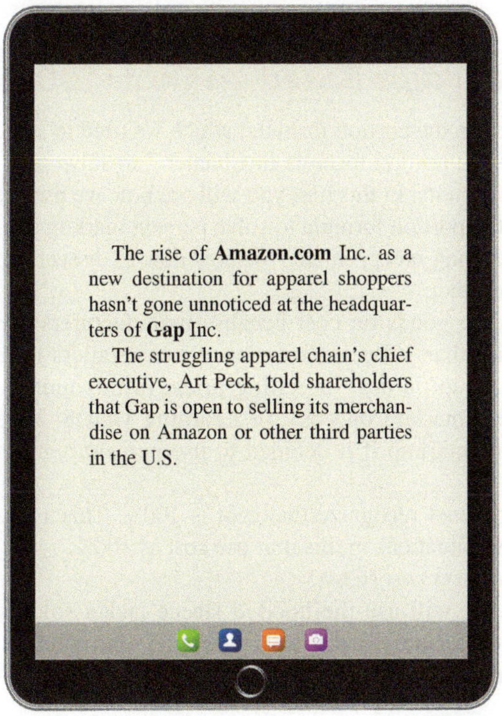

The rise of **Amazon.com** Inc. as a new destination for apparel shoppers hasn't gone unnoticed at the headquarters of **Gap** Inc.

The struggling apparel chain's chief executive, Art Peck, told shareholders that Gap is open to selling its merchandise on Amazon or other third parties in the U.S.

Check box as you complete each step.

Steps

- ☐ Read learning unit.
  - ☐ Complete practice quiz at the end of the learning unit.
- ☐ Grade practice quiz using provided solutions. (For more help, watch the learning unit video in Connect and have a Study Session with the authors. Then complete the additional practice quiz in Connect.)
- ☐ Repeat above for each of the four learning units in Chapter 8.
  - ☐ Review chapter organizer.
  - ☐ Complete assigned homework.
    - ☐ Finish summary practice test. (Go to Connect via the ebook link and do the interactive video worksheet to grade.)
- ☐ Complete instructor's exam.

Let's turn our attention to a store you may be familiar with—Gap. Note in the *Wall Street Journal* clip to the left, Gap is considering selling its merchandise on Amazon. Amazon has 6.79% of the apparel market while Gap has 4.6%.

We will look at some of Gap's pricing options for its fleece hoody jackets. Before we study the two pricing methods available to Gap (percent markup on cost and percent markup on selling price), we must know the following terms:

- **Selling price.** The price retailers charge consumers. The total selling price of all the goods sold by a retailer (like Gap) represents the retailer's total sales.

[1]Some texts use the term *markon* (selling price minus cost).

©TY Lim/Shutterstock

- **Cost.** The price retailers pay to a manufacturer or supplier to bring the goods into the store.
- **Markup, margin,** or **gross profit.** These three terms refer to the difference between the cost of bringing the goods into the store and the selling price of the goods.
- **Operating expenses** or **overhead.** The regular expenses of doing business such as wages, rent, utilities, insurance, and advertising.
- **Net profit** or **net income.** The profit remaining after subtracting the cost of bringing the goods into the store and the operating expenses from the sale of the goods (including any returns or adjustments). In Learning Unit 8–4 we will take a closer look at the point at which costs and expenses are covered. This is called the *breakeven* point.

From these definitions, we can conclude that *markup* represents the amount that retailers must add to the cost of the goods to cover their operating expenses and make a profit.[2]

Let's assume Gap plans to sell hooded fleece jackets for $23 that cost $18.[3]

Basic selling price formula

Selling price (*S*)	=	Cost (*C*)	+	Markup (*M*)
$23	=	$18 (price paid to bring fleece jackets into store)	+	$5 (amount in dollars to cover operating expenses and make a profit)

Reprinted by permission of *The Wall Street Journal*, copyright 2016 Dow Jones & Company, Inc. All rights reserved worldwide.

In the Gap example, the markup is a dollar amount, or a **dollar markup.** Markup is also expressed in percent. When expressing markup in percent, retailers can choose a percent based on *cost* (Learning Unit 8–1) or a percent based on *selling price* (Learning Unit 8–2).

Before looking at markups, look at the *Wall Street Journal* clip "Price Check" which shows that Walmart is the largest apparel retailer.

Learning Unit 8–1: Markups Based on Cost (100%)

In Chapter 6 you were introduced to the portion formula, which we used to solve percent problems. We also used the portion formula in Chapter 7 to solve problems involving trade and cash discounts. In this unit you will see how we use the basic selling price formula and the portion formula to solve percent markup situations based on cost. We will be using blueprint aids to show how to dissect and solve all word problems in this chapter.

Many manufacturers mark up goods on cost because manufacturers can get cost information more easily than sales information. Since retailers have the choice of using percent markup on cost or selling price, in this unit we assume Gap has chosen percent markup on cost. In Learning Unit 8–2 we show how Gap would determine markup if it decided to use percent markup on selling price.

Businesses that use **percent markup on cost** recognize that cost is 100%. This 100% represents the base of the portion formula. All situations in this unit use cost as 100%.

For markups based on cost, the base is always cost (C).

To calculate percent markup on cost, we will use the hooded fleece jacket sold by Gap and begin with the basic selling price formula given in the chapter introduction. When we know the dollar markup, we can use the portion formula to find the percent markup on cost.

[2]In this chapter, we concentrate on the markup of retailers. Manufacturers and suppliers also use markup to determine selling price.

[3]These may not be actual store prices but we assume these prices in our examples.

Markup expressed in dollars:

Selling price (\$23) = Cost (\$18) + Markup (\$5)

Markup expressed as a percent markup on cost:

Cost	100.00%
+ Markup	+ 27.78
= Selling price	127.78%

Cost is 100%—the base. Dollar markup is the portion, and percent markup on cost is the rate.

In Situation 1 (below) we show why Gap has a 27.78% markup (dollar markup [\$5] divided by cost [\$18]) based on cost by presenting the hooded fleece jacket as a word problem. We solve the problem with the blueprint aid used in earlier chapters. In the second column, however, you will see footnotes after two numbers. These refer to the steps we use below the blueprint aid to solve the problem. Throughout the chapter, the numbers that we are solving for are in red. Remember that cost is the base for this unit.

LO 1

Situation 1: Calculating Dollar Markup and Percent Markup on Cost

Dollar markup is calculated with the basic selling price formula $S = C + M$. When you know the cost and selling price of goods, reverse the formula to $M = S - C$. Subtract the cost from the selling price, and you have the dollar markup.

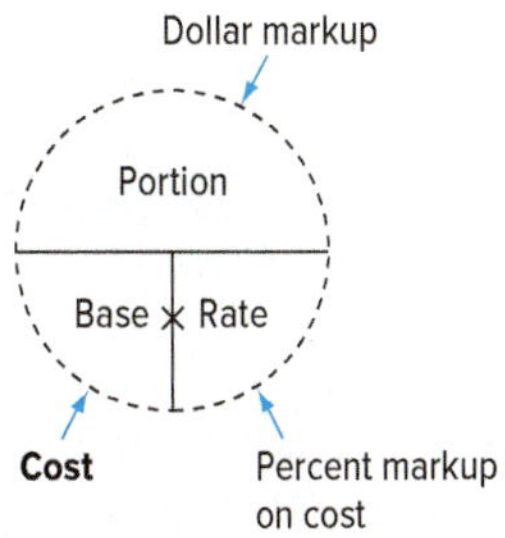

The percent markup on cost is calculated with the portion formula. For Situation 1 the *portion* (*P*) is the dollar markup, which you know from the selling price formula. In this unit the *rate* (*R*) is always the percent markup on cost and the *base* (*B*) is always the cost (100%). To find the percent markup on cost (*R*), use the portion formula $R = \frac{P}{B}$ and divide the dollar markup (*P*) by the cost (*B*). Convert your answer to a percent and round if necessary.

Now we will look at the Gap example to see how to calculate the 27.78% markup on cost.

The Word Problem The Gap pays \$18 for a hooded fleece jacket, which the store plans to sell for \$23. What is Gap's dollar markup? What is the percent markup on cost (rounded to the nearest hundredth percent)?

BLUEPRINT	The facts	Solving for?	Steps to take	Key points
	Hooded fleece jacket cost: \$18. *Hooded fleece jacket selling price:* \$23.	% \$ *C* 100.00% \$18 *+M* 27.78[2] 5[1] *= S* 127.78% \$23	Dollar markup = Selling price − Cost Percent markup on cost = Dollar markup / Cost	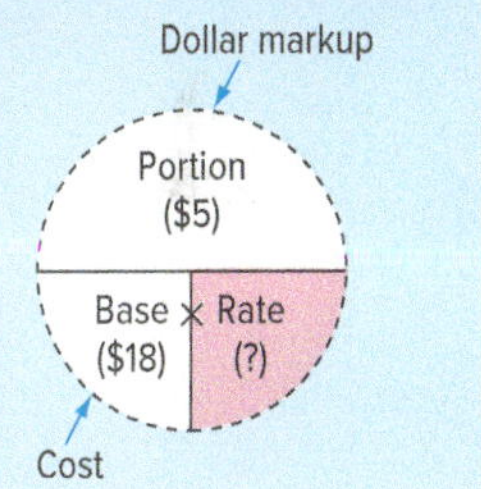

[1]Dollar markup. See Step 1, below.

[2]Percent markup on cost. See Step 2, below.

Steps to solving problem

1. Calculate the dollar markup. Dollar markup = Selling price − Cost

   \$5 = \$23 − \$18

2. Calculate the percent markup on cost. $\text{Percent markup on cost} = \frac{\text{Dollar markup}}{\text{Cost}}$

   $= \frac{\$5}{\$18} = 27.78\%$

To check the percent markup on cost, you can use the basic selling price formula $S = C + M$. Convert the percent markup on cost found with the portion formula to a decimal and multiply it by the cost. This gives the dollar markup. Then add the cost and the dollar markup to get the selling price of the goods.

You could also check the cost (*B*) by dividing the dollar markup (*P*) by the percent markup on cost (*R*).

Check

Selling price = Cost + Markup

or

Cost $(B) = \dfrac{\text{Dollar markup } (P)}{\text{Percent markup on cost } (R)}$

$\$23 = \$18 + .2778(\$18)$
$\$23 = \$18 + \$5$
$\$23 = \$23$

$= \dfrac{\$5}{.2778} = \$18$

Parentheses mean that you multiply the percent markup on cost in decimal by the cost.

LO 2

Situation 2: Calculating Selling Price When You Know Cost and Percent Markup on Cost

When you know the cost and the percent markup on cost, you calculate the selling price with the basic selling formula $S = C + M$. Remember that when goods are marked up on cost, the cost is the base (100%). So you can say that the selling price is the cost plus the markup in dollars (percent markup on cost times cost).

Now let's look at Mel's Furniture where we calculate Mel's dollar markup and selling price.

The Word Problem Mel's Furniture bought a lamp that cost $100. To make Mel's desired profit, he needs a 65% markup on cost. What is Mel's dollar markup? What is his selling price?

BLUEPRINT	The facts	Solving for?	Steps to take	Key points
	Lamp cost: $100. *Markup on cost:* 65%.	% / \$ C 100% \$100 $+M$ 65 65[1] $=S$ 165% \$165[2]	Dollar markup: $S = C + M$ or $S = \text{Cost} \times (1 + \text{Percent markup on cost})$	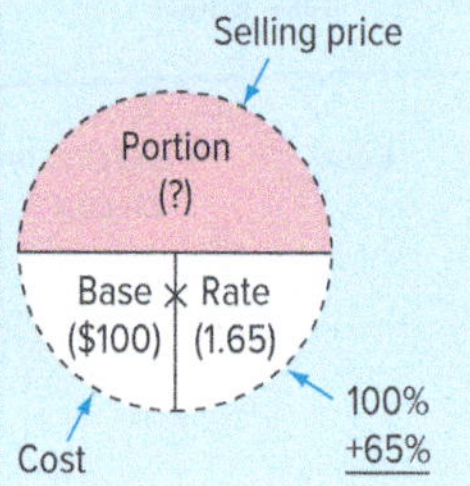

[1]Dollar markup. See Step 1, below.
[2]Selling price. See Step 2, below.

Steps to solving problem

1. Calculate the dollar markup.
   $S = C + M$
   $S = \$100 + .65(\$100)$ ← Parentheses mean you multiply the percent markup in decimal by the cost.
   $S = \$100 + \$65$ ← Dollar markup
2. Calculate the selling price.
   $S = \$165$

You can check the selling price with the formula $P = B \times R$. You are solving for the portion (P)—the selling price. Rate (R) represents the 100% cost plus the 65% markup on cost. Since in this unit the markup is on cost, the base is the cost. Convert 165% to a decimal and multiply the cost by 1.65 to get the selling price of $165.

Check

Selling price (P) = Cost (B) × (1 + Percent markup on cost) (R)

$= \$100 \times 1.65 = \$165$

LO 3

Situation 3: Calculating Cost When You Know Selling Price and Percent Markup on Cost

When you know the selling price and the percent markup on cost, you calculate the cost with the basic selling formula $S = C + M$. Since goods are marked up on cost, the percent markup on cost is added to the cost.

Let's see how this is done in the following Jill Sport example.

The Word Problem Jill Sport, owner of Sports, Inc., sells tennis rackets for $50. To make her desired profit, Jill needs a 40% markup on cost. What do the tennis rackets cost Jill? What is the dollar markup?

BLUEPRINT	The facts	Solving for?	Steps to take	Key points
	Selling price: $50. *Markup on cost:* 40%.	% | $ C 100% | $35.71[1] $+M$ 40 | 14.29[2] $=S$ 140% | $50.00	$S = C + M$ or $\text{Cost} = \dfrac{\text{Selling price}}{\text{Percent } 1 + \text{markup on cost}}$ $M = S - C$	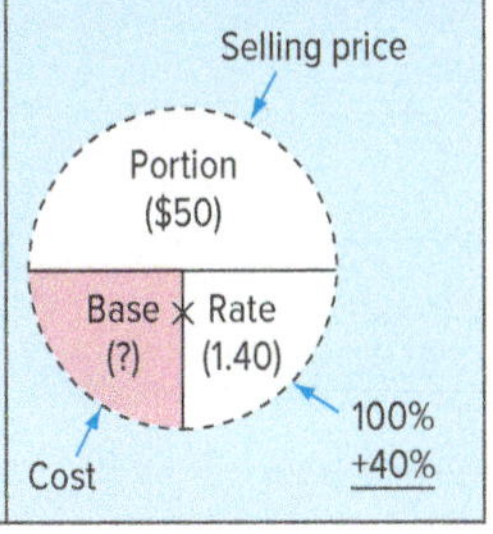

[1]Cost. See Step 1, below.

[2]Dollar markup. See Step 2, below.

Steps to solving problem

1. Calculate the cost.

$$S = C + M$$
$$\$50.00 = C + .40C$$ ← This means 40% times cost. C is the same as $1C$. Adding $.40C$ to $1C$ gives the percent markup on cost of $1.40C$ in decimal.
$$\frac{\$50.00}{1.40} = \frac{\cancel{1.40}C}{\cancel{1.40}}$$
$$\$35.71 = C$$

2. Calculate the dollar markup.

$$M = S - C$$
$$M = \$50.00 - \$35.71$$
$$M = \$14.29$$

MONEY tips

Automate your savings. Save more money from each paycheck starting NOW. If you have increments of 1% automatically taken out of your check, you will never miss it. Over time, it can add up to six figures and will help you start building a financially healthy retirement TODAY.

You can check your cost answer with the portion formula $B = \frac{P}{R}$. Portion (P) is the selling price. Rate (R) represents the 100% cost plus the 40% markup on cost. Convert the percents to decimals and divide the portion by the rate to find the base, or cost.

Check

$$\text{Cost } (B) = \frac{\text{Selling price } (P)}{1 + \text{Percent markup on cost } (R)} = \frac{\$50.00}{1.40} = \$35.71$$

Now try the following Practice Quiz to check your understanding of this unit.

LU 8–1 PRACTICE QUIZ

Complete this Practice Quiz to see how you are doing.

Solve the following situations (markups based on cost):

1. Irene Westing bought a desk for $400 from an office supply house. She plans to sell the desk for $600. What is Irene's dollar markup? What is her percent markup on cost? Check your answer.
2. Suki Komar bought dolls for her toy store that cost $12 each. To make her desired profit, Suki must mark up each doll 35% on cost. What is the dollar markup? What is the selling price of each doll? Check your answer.
3. Jay Lyman sells calculators. His competitor sells a new calculator line for $14 each. Jay needs a 40% markup on cost to make his desired profit, and he must meet price competition. At what cost can Jay afford to bring these calculators into the store? What is the dollar markup? Check your answer.

*For **extra help** from your authors–Sharon and Jeff–see the videos in Connect.*

✓ Solutions

1. Irene's dollar markup and percent markup on cost:

BLUEPRINT	The facts	Solving for?	Steps to take	Key points
	Desk cost: $400. *Desk selling price:* $600.	% \$ C 100% \$400 +M 50[2] 200[1] =S 150% \$600	Dollar markup = Selling price − Cost Percent markup on cost = Dollar markup / Cost	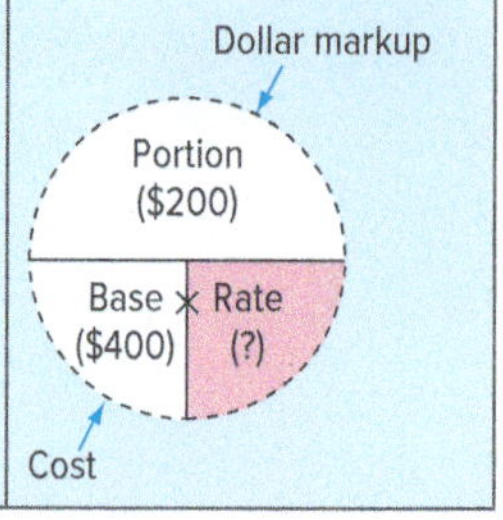

[1]Dollar markup. See Step 1, below.
[2]Percent markup on cost. See Step 2, below.

Steps to solving problem

1. Calculate the dollar markup.
   $$\text{Dollar markup} = \text{Selling price} - \text{Cost}$$
   $$\$200 = \$600 - \$400$$
2. Calculate the percent markup on cost.
   $$\text{Percent markup on cost} = \frac{\text{Dollar markup}}{\text{Cost}} = \frac{\$200}{\$400} = 50\%$$

Check

$$\text{Selling price} = \text{Cost} + \text{Markup}$$
$$\$600 = \$400 + .50(\$400)$$
$$\$600 = \$400 + \$200$$
$$\$600 = \$600$$

or

$$\text{Cost } (B) = \frac{\text{Dollar markup } (P)}{\text{Percent markup on cost } (R)} = \frac{\$200}{.50} = \$400$$

2. Dollar markup and selling price of doll:

BLUEPRINT	The facts	Solving for?	Steps to take	Key points
	Doll cost: $12 each. *Markup on cost:* 35%.	% \$ C 100% \$12.00 +M 35 4.20[1] =S 135% \$16.20[2]	Dollar markup: $S = C + M$ or $S = \text{Cost} \times \left(1 + \text{Percent markup on cost}\right)$	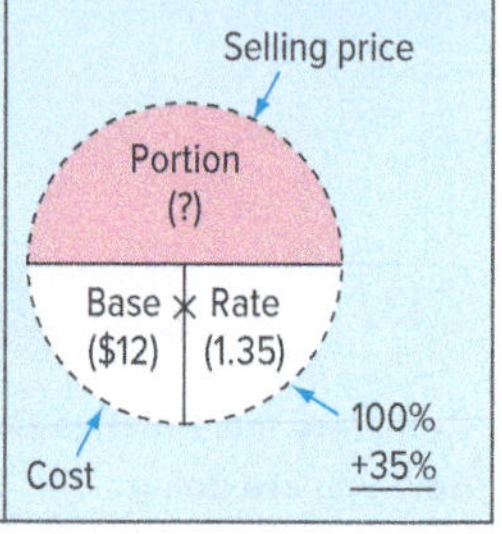

[1]Dollar markup. See Step 1, below.
[2]Selling price. See Step 2, below.

Steps to solving problem

1. Calculate the dollar markup.
   $$S = C + M$$
   $$S = \$12.00 + .35(\$12.00)$$
   $$S = \$12.00 + \$4.20 \leftarrow \text{Dollar markup}$$
2. Calculate the selling price.
   $$S = \$16.20$$

Check

$$\underset{(P)}{\text{Selling price}} = \underset{(B)}{\text{Cost}} \times \underset{(R)}{(1 + \text{Percent markup on cost})} = \$12.00 \times 1.35 = \$16.20$$

3. Cost and dollar markup:

BLUEPRINT	The facts	Solving for?	Steps to take	Key points
	Selling price: $14. *Markup on cost:* 40%.	% $ C 100% $10[1] $+M$ 40 4[2] $= S$ 140% $14	$S = C + M$ or $\text{Cost} = \dfrac{\text{Selling price}}{\text{Percent } 1 + \text{markup on cost}}$ $M = S - C$	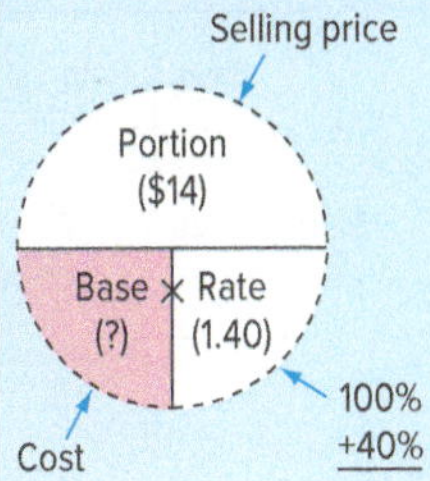

[1]Cost. See Step 1, below.

[2]Dollar markup. See Step 2, below.

Steps to solving problem

1. Calculate the cost.

$$S = C + M$$
$$\$14 = C + .40C$$
$$\frac{\$14}{1.40} = \frac{1.40C}{1.40}$$
$$\$10 = C$$

2. Calculate the dollar markup.

$$M = S - C$$
$$M = \$14 - \$10$$
$$M = \$4$$

Check

$$\text{Cost }(B) = \frac{\text{Selling price }(P)}{1 + \text{Percent markup on cost }(R)} = \frac{\$14}{1.40} = \$10$$

Learning Unit 8–2: Markups Based on Selling Price (100%)

Many retailers mark up their goods on the selling price since sales information is easier to get than cost information. These retailers use retail prices in their inventory and report their expenses as a percent of sales.

Aha!

For markups based on selling price, the base is always selling price (S).

Businesses that mark up their goods on selling price recognize that selling price is 100%. We begin this unit by assuming Gap has decided to use percent markup based on selling price. We repeat Gap's selling price formula expressed in dollars.

©Sorbis/Shutterstock

Markup expressed in dollars:

Selling price ($23) = Cost ($18) + Markup ($5)

Markup expressed as **percent markup on selling price:**

Cost	78.26%
+ Markup	+ 21.74
= Selling price	100.00%

Selling price is 100%—the base. Dollar markup is the portion, and percent markup on selling price is the rate.

In Situation 1 (below) we show why Gap has a 21.74% markup based on selling price. In the last unit, markups were based on *cost.* In this unit, markups are based on *selling price.*

LO 1

Situation 1: Calculating Dollar Markup and Percent Markup on Selling Price

The dollar markup is calculated with the selling price formula used in Situation 1, Learning Unit 8–1: $M = S - C$. To find the percent markup on selling price, use the portion formula $R = \frac{P}{B}$, where rate (the percent markup on selling price) is found by dividing the portion (dollar markup) by the base (selling price). Note that when solving for percent markup on cost in Situation 1, Learning Unit 8–1, you divided the dollar markup by the cost.

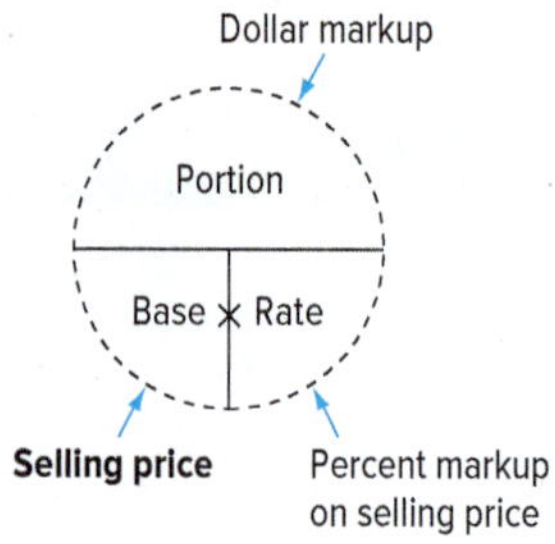

The Word Problem The cost to Gap for a hooded fleece jacket is \$18; the store then plans to sell the jacket for \$23. What is Gap's dollar markup? What is its percent markup on selling price? (Round to the nearest hundredth percent.)

BLUEPRINT	The facts	Solving for?	Steps to take	Key points
	Hooded fleece jacket cost: \$18. *Hooded fleece jacket price:* \$23.	% | \$ C 78.26% | \$18 $+M$ 21.74[2] | 5[1] $= S$ 100.00% | \$23	Dollar markup = Selling price − Cost Percent markup on selling price = Dollar markup / Selling price	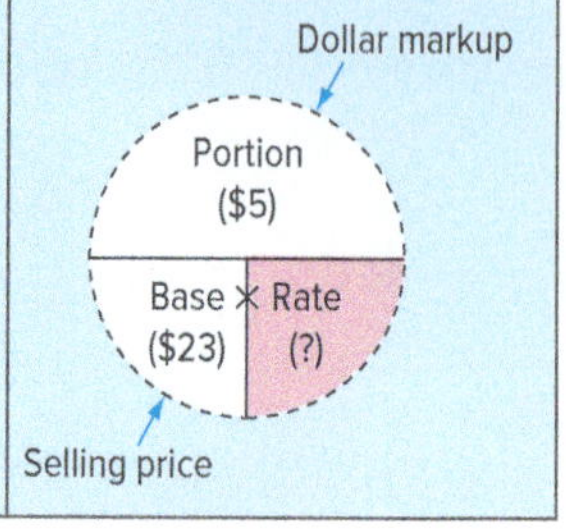

[1]Dollar markup. See Step 1, below.

[2]Percent markup on selling price. See Step 2, below.

Steps to solving problem

1. Calculate the dollar markup.

$$\text{Dollar markup} = \text{Selling price} - \text{Cost}$$
$$\$5 = \$23 - \$18$$

2. Calculate the percent markup on selling price.

$$\text{Percent markup on selling price} = \frac{\text{Dollar markup}}{\text{Selling price}}$$
$$= \frac{\$5}{\$23} = 21.74\%$$

You can check the percent markup on selling price with the basic selling price formula $S = C + M$. You can also use the portion formula by dividing the dollar markup (P) by the percent markup on selling price (R).

Check

Selling price = Cost + Markup

or

$$\text{Selling price } (B) = \frac{\text{Dollar markup } (P)}{\text{Percent markup on selling price } (R)}$$

$$\$23 = \$18 + .2174(\$23)$$
$$\$23 = \$18 + \$5$$
$$\$23 = \$23$$

$$= \frac{\$5}{.2174} = \$23$$

Parentheses mean you multiply the percent markup on selling price in decimal by the selling price.

LO 2

Situation 2: Calculating Selling Price When You Know Cost and Percent Markup on Selling Price

When you know the cost and percent markup on selling price, you calculate the selling price with the basic selling formula $S = C + M$. Remember that when goods are marked up on selling price, the selling price is the base (100%). Since you do not know the selling price, the percent markup is based on the unknown selling price. To find the dollar markup after you find the selling price, use the selling price formula $M = S - C$.

The Word Problem Mel's Furniture bought a lamp that cost \$100. To make Mel's desired profit, he needs a 65% markup on selling price. What are Mel's selling price and his dollar markup?

BLUEPRINT	The facts	Solving for?	Steps to take	Key points
	Lamp cost: $100. *Markup on selling price:* 65%.	% | \$ C | 35% | \$100.00 +M | 65 | 185.71[2] = S | 100% | \$285.71[1]	$S = C + M$ or $S = \dfrac{\text{Cost}}{1 - \text{Percent markup on selling price}}$	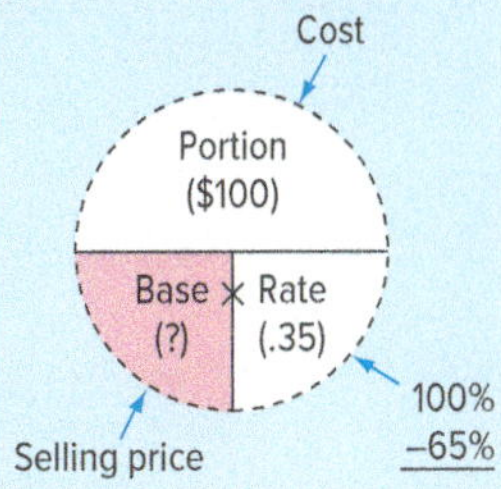

[1]Selling price. See Step 1, below.

[2]Dollar markup. See Step 2, below.

Steps to solving problem

1. Calculate the selling price.

$$S = C + M$$
$$S = \$100.00 + .65S$$
$$-.65S \qquad -.65S$$
$$\frac{.35S}{.35} = \frac{\$100.00}{.35}$$
$$S = \$285.71$$

$1.00S - .65S = .35S$

Do not multiply the .65 times $100.00. The 65% is based on selling price not cost.

2. Calculate the dollar markup.

$$M = S - C$$
$$\$185.71 = \$285.71 - \$100.00$$

You can check your selling price with the portion formula $B = \frac{P}{R}$. To find the selling price (B), divide the cost (P) by the rate (100% − Percent markup on selling price).

Check

$$\text{Selling price } (B) = \frac{\text{Cost } (P)}{1 - \text{Percent markup on selling price } (R)}$$

$$= \frac{\$100.00}{1 - .65} = \frac{\$100.00}{.35} = \$285.71$$

LO 3

Situation 3: Calculating Cost When You Know Selling Price and Percent Markup on Selling Price

When you know the selling price and the percent markup on selling price, you calculate the cost with the basic formula $S = C + M$. To find the dollar markup, multiply the markup percent by the selling price. When you have the dollar markup, subtract it from the selling price to get the cost.

The Word Problem Jill Sport, owner of Sports, Inc., sells tennis rackets for $50. To make her desired profit, Jill needs a 40% markup on the selling price. What is the dollar markup? What do the tennis rackets cost Jill?

BLUEPRINT	The facts	Solving for?	Steps to take	Key points
	Selling price: $50. *Markup on selling price:* 40%.	% | \$ C | 60% | \$30[2] +M | 40 | 20[1] = S | 100% | \$50	$S = C + M$ or $\text{Cost} = \text{Selling price} \times \left(1 - \dfrac{\text{Percent markup}}{\text{on selling price}}\right)$	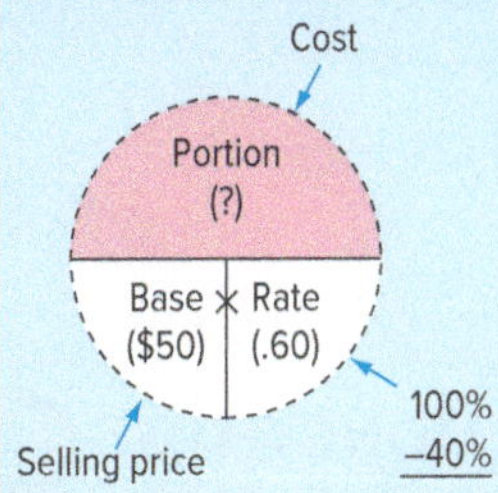

[1]Dollar markup. See Step 1.

[2]Cost. See Step 2.

Steps to solving problem

1. Calculate the dollar markup.
$$S = C + M$$
$$\$50 = C + .40(\$50)$$
$$\$50 = C + \$20 \leftarrow \text{Dollar markup}$$
$$-20 \qquad -20$$
2. Calculate the cost.
$$\$30 = C$$

To check your cost, use the portion formula Cost (P) = Selling price (B) × (100% selling price − Percent markup on selling price) (R).

Check

$$\underset{(P)}{\text{Cost}} = \underset{(B)}{\text{Selling price}} \times \left(1 - \underset{(R)}{\text{Percent markup on selling price}}\right) = \$50 \times .60 = \$30$$

(1.00 − .40)

In Table 8.1, we compare percent markup on cost with percent markup on retail (selling price). This table is a summary of the answers we calculated from the word problems in Learning Units 8–1 and 8–2. The word problems in the units were the same except in Learning Unit 8–1, we assumed markups were on cost, while in Learning Unit 8–2, markups were on selling price. Note that in Situation 1, the dollar markup is the same $5, but the percent markup is different.

Let's now look at how to convert from percent markup on cost to percent markup on selling price and vice versa. We will use Situation 1 from Table 8.1.

LO 4

Formula for Converting Percent Markup on Cost to Percent Markup on Selling Price

To convert percent markup on cost to percent markup on selling price:

$$\frac{\text{Percent markup on cost}}{1 + \text{Percent markup on cost}}$$

$$\frac{.2778}{1 + .2778} = 21.74\%$$

TABLE 8.1

Comparison of markup on cost versus markup on selling price

Markup based on cost—Learning Unit 8–1	Markup based on selling price—Learning Unit 8–2
Situation 1: Calculating dollar amount of markup and percent markup on cost. Hooded fleece jacket cost, $18. Hooded fleece jacket selling price, $23. $M = S - C$ $M = \$23 - \$18 = \$5$ markup $M \div C = \$5 \div \$18 = 27.78\%$	*Situation 1: Calculating dollar amount of markup and percent markup on selling price.* Hooded fleece jacket cost, $18. Hooded fleece jacket selling price, $23. $M = S - C$ $M = \$23 - \$18 = \$5$ markup $M \div S = \$5 \div \$23 = 21.74\%$
Situation 2: Calculating selling price on cost. Lamp cost, $100. 65% markup on cost $S = C \times (1 + \text{Percent markup on cost})$ $S = \$100 \times 1.65 = \$165$ (100% + 65% = 165% = 1.65)	*Situation 2: Calculating selling price on selling price.* Lamp cost, $100. 65% markup on selling price $S = C \div (1 - \text{Percent markup on selling price})$ $S = \$100.00 \div .35$ (100% − 65% = 35% = .35) $S = \$285.71$
Situation 3: Calculating cost on cost. Tennis racket selling price, $50. 40% markup on cost $C = S \div (1 + \text{Percent markup on cost})$ $C = \$50.00 \div 1.40$ (100% + 40% = 140% = 1.40) $C = \$35.71$	*Situation 3: Calculating cost on selling price.* Tennis racket selling price, $50. 40% markup on selling price $C = S \times (1 - \text{Percent markup on selling price})$ $C = \$50 \times .60 = \$30$ (100% − 40% = 60% = .60)

MONEY tips

When analyzing a job offer, make sure you include the value of the benefits. Salary alone will not let you know the true value of the offer.

Formula for Converting Percent Markup on Selling Price to Percent Markup on Cost

To convert percent markup on selling price to percent markup on cost:

$$\frac{\text{Percent markup on selling price}}{1 - \text{Percent markup on selling price}}$$

$$\frac{.2174}{1 - .2174} = 27.78\%$$

Key point: A 21.74% markup on selling price or a 27.78% markup on cost results in the same dollar markup of $5.

Now let's test your knowledge of Learning Unit 8–2.

LU 8–2 PRACTICE QUIZ

Complete this Practice Quiz to see how you are doing.

Solve the following situations (markups based on selling price). Note numbers 1, 2, and 3 are parallel problems to those in Practice Quiz LU 8–1.

1. Irene Westing bought a desk for $400 from an office supply house. She plans to sell the desk for $600. What is Irene's dollar markup? What is her percent markup on selling price (rounded to the nearest tenth percent)? Check your answer. Selling price will be slightly off due to rounding.
2. Suki Komar bought dolls for her toy store that cost $12 each. To make her desired profit, Suki must mark up each doll 35% on the selling price. What is the selling price of each doll? What is the dollar markup? Check your answer.
3. Jay Lyman sells calculators. His competitor sells a new calculator line for $14 each. Jay needs a 40% markup on the selling price to make his desired profit, and he must meet price competition. What is Jay's dollar markup? At what cost can Jay afford to bring these calculators into the store? Check your answer.
4. Dan Flow sells wrenches for $10 that cost $6. What is Dan's percent markup on cost? Round to the nearest tenth percent. What is Dan's percent markup on selling price? Check your answer.

*For **extra help** from your authors–Sharon and Jeff–see the videos in Connect.*

✓ Solutions

1. Irene's dollar markup and percent markup on selling price:

BLUEPRINT

The facts	Solving for?	Steps to take	Key points
Desk cost: $400. *Desk selling price:* $600.	% $ *C* 66.7% $400 +*M* 33.3[2] 200[1] = *S* 100% $600	$\text{Dollar markup} = \text{Selling price} - \text{Cost}$ $\text{Percent markup on selling price} = \frac{\text{Dollar markup}}{\text{Selling price}}$	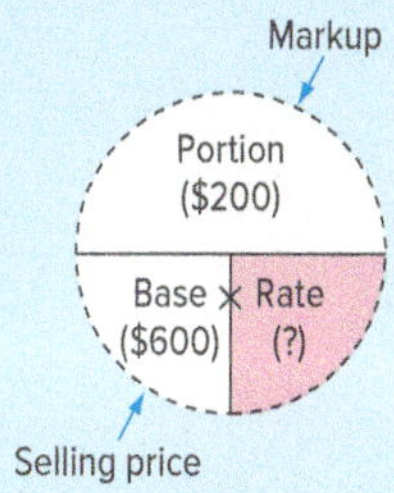

[1]Dollar markup. See Step 1, below.

[2]Percent markup on selling price. See Step 2, below.

Steps to solving problem

1. Calculate the dollar markup.

$$\text{Dollar markup} = \text{Selling price} - \text{Cost}$$
$$\$200 = \$600 - \$400$$

2. Calculate the percent markup on selling price.

$$\text{Percent markup on selling price} = \frac{\text{Dollar markup}}{\text{Selling price}}$$
$$= \frac{\$200}{\$600} = 33.3\%$$

Check

$$\text{Selling price} = \text{Cost} + \text{Markup} \quad \textbf{or} \quad \text{Selling price } (B) = \frac{\text{Dollar markup } (P)}{\text{Percent markup on selling price } (R)}$$

$$\$600 = \$400 + .333(\$600)$$

$$\$600 = \$400 + \$199.80 \qquad = \frac{\$200}{.333} = \$600.60$$

$$\$600 = \$599.80 \text{ (off due to rounding)} \qquad \text{(not exactly \$600 due to rounding)}$$

2. Selling price of doll and dollar markup:

BLUEPRINT	The facts	Solving for?	Steps to take	Key points
	Doll cost: $12 each. *Markup on selling price:* 35%.	% / \$ C: 65% / \$12.00 $+M$: 35 / 6.46[2] $=S$: 100% / \$18.46[1]	$S = C + M$ or $S = \dfrac{\text{Cost}}{1 - \text{Percent markup on selling price}}$	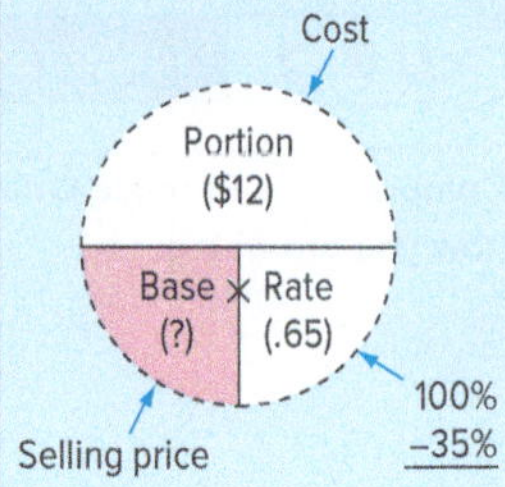

[1]Selling price. See Step 1, below.
[2]Dollar markup. See Step 2, below.

Steps to solving problem

1. Calculate the selling price.

$$S = C + M$$
$$S = \$12.00 + .35S$$
$$-.35S \qquad -.35S$$
$$\frac{.65S}{.65} = \frac{\$12.00}{.65}$$
$$S = \$18.46$$

2. Calculate the dollar markup.

$$M = S - C$$
$$\$6.46 = \$18.46 - \$12.00$$

Check

$$\text{Selling price } (B) = \frac{\text{Cost } (P)}{1 - \text{Percent markup on selling price } (R)} = \frac{\$12.00}{.65} = \$18.46$$

3. Dollar markup and cost:

BLUEPRINT	The facts	Solving for?	Steps to take	Key points
	Selling price: $14. *Markup on selling price:* 40%.	% / \$ C: 60% / \$ 8.40[2] $+M$: 40 / 5.60[1] $=S$: 100% / \$14.00	$S = C + M$ or $\text{Cost} = \text{Selling price} \times \left(1 - \begin{array}{c}\text{Percent markup}\\ \text{on selling price}\end{array}\right)$	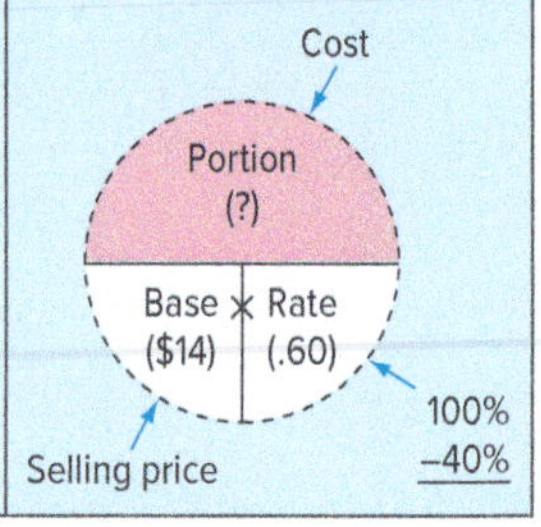

[1]Dollar markup. See Step 1, below.
[2]Cost. See Step 2, below.

Steps to solving problem

1. Calculate the dollar markup.

$$S = C + M$$
$$\$14.00 = C + .40(\$14.00)$$
$$\$14.00 = C + \$5.60 \leftarrow \text{Dollar markup}$$

2. Calculate the cost.

$$-5.60 \qquad -5.60$$
$$\$8.40 = C$$

Check

$$\underset{(P)}{\text{Cost}} = \underset{(B)}{\text{Selling price}} \times \underset{(R)}{(1 - \text{Percent markup on selling price})} = \$14.00 \times .60 = \$8.40$$

$(1.00 - .40)$

4. $\text{Cost} = \dfrac{\$4}{\$6} = 66.7\%$ $\qquad \dfrac{.40}{1 - .40} = \dfrac{.40}{.60} = \dfrac{2}{3} = 66.7\%$

$\text{Selling price} = \dfrac{\$4}{\$10} = 40\%$ $\qquad \dfrac{.667}{1 + .667} = \dfrac{.667}{1.667} = 40\%$ (due to rounding)

Learning Unit 8–3: Markdowns and Perishables

The following *Wall Street Journal* clips show how Gap and other stores are marking down products.

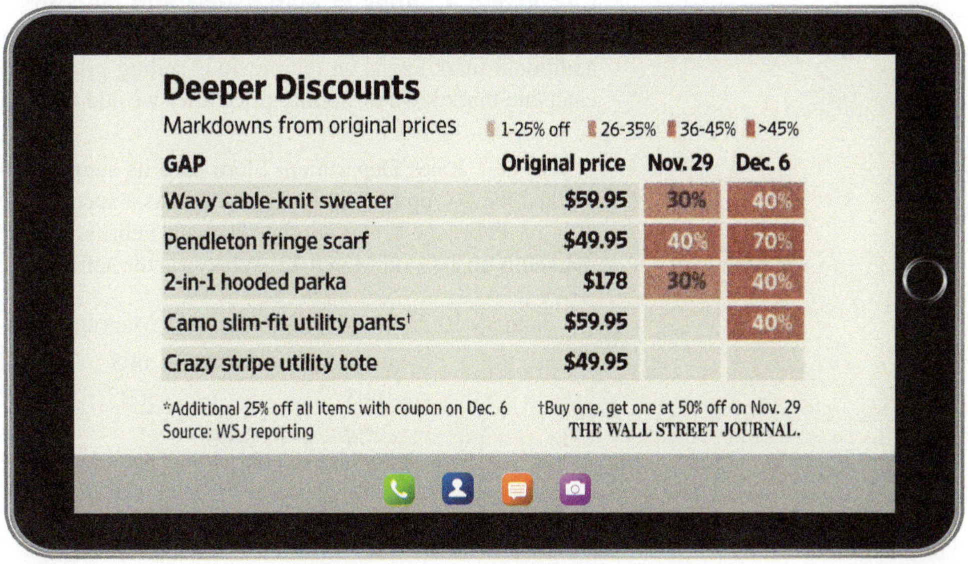

Deeper Discounts

Markdowns from original prices — 1-25% off | 26-35% | 36-45% | >45%

GAP	Original price	Nov. 29	Dec. 6
Wavy cable-knit sweater	$59.95	30%	40%
Pendleton fringe scarf	$49.95	40%	70%
2-in-1 hooded parka	$178	30%	40%
Camo slim-fit utility pants†	$59.95		40%
Crazy stripe utility tote	$49.95		

*Additional 25% off all items with coupon on Dec. 6 †Buy one, get one at 50% off on Nov. 29

Source: WSJ reporting

THE WALL STREET JOURNAL.

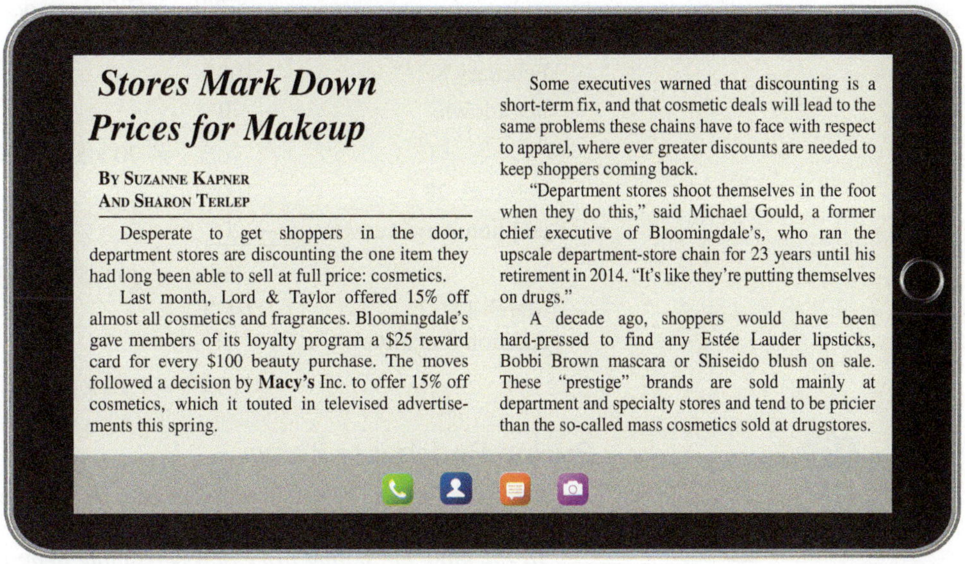

Stores Mark Down Prices for Makeup

By Suzanne Kapner And Sharon Terlep

Desperate to get shoppers in the door, department stores are discounting the one item they had long been able to sell at full price: cosmetics.

Last month, Lord & Taylor offered 15% off almost all cosmetics and fragrances. Bloomingdale's gave members of its loyalty program a $25 reward card for every $100 beauty purchase. The moves followed a decision by **Macy's** Inc. to offer 15% off cosmetics, which it touted in televised advertisements this spring.

Some executives warned that discounting is a short-term fix, and that cosmetic deals will lead to the same problems these chains have to face with respect to apparel, where ever greater discounts are needed to keep shoppers coming back.

"Department stores shoot themselves in the foot when they do this," said Michael Gould, a former chief executive of Bloomingdale's, who ran the upscale department-store chain for 23 years until his retirement in 2014. "It's like they're putting themselves on drugs."

A decade ago, shoppers would have been hard-pressed to find any Estée Lauder lipsticks, Bobbi Brown mascara or Shiseido blush on sale. These "prestige" brands are sold mainly at department and specialty stores and tend to be pricier than the so-called mass cosmetics sold at drugstores.

My Money

This learning unit focuses your attention on how to calculate markdowns. Then you will learn how a business prices perishable items that may spoil before customers buy them.

LO 1

Markdowns

Markdowns are reductions from the original selling price caused by seasonal changes, special promotions, style changes, and so on. We calculate the markdown percent as follows:

$$\text{Markdown percent} = \frac{\text{Dollar markdown}}{\text{Selling price (original)}}$$

Let's look at the following Kmart example:

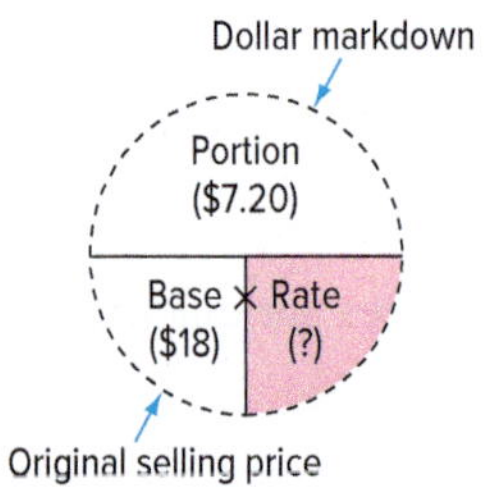

EXAMPLE Kmart marked down an $18 video to $10.80. Calculate the **dollar markdown** and the markdown percent.

$18.00	Original selling price
− 10.80	Sale price
$ 7.20	Markdown

$$\frac{\text{Dollar markdown, } \$7.20}{\text{Selling price (original), } \$18.00} = 40\%$$

Calculating a Series of Markdowns and Markups Often the final selling price is the result of a series of markdowns (and possibly a markup in between markdowns). We calculate additional markdowns on the previous selling price. Note in the following example how we calculate markdown on selling price after we add a markup.

EXAMPLE Jones Department Store paid its supplier $400 for a TV. On January 10, Jones marked the TV up 60% on selling price. As a special promotion, Jones marked the TV down 30% on February 8 and another 20% on February 28. No one purchased the TV, so Jones marked it up 10% on March 11. What was the selling price of the TV on March 11?

January 10: Selling price = Cost + Markup

$$S = \$400 + .60S$$
$$-.60S \qquad\qquad -.60S$$
$$\frac{.40S}{.40} = \frac{\$400}{.40}$$
$$S = \$1{,}000$$

Check

$$S = \frac{\text{Cost}}{1 - \text{Percent markup on selling price}}$$

$$S = \frac{\$400}{1 - .60} = \frac{\$400}{.40} = \$1{,}000$$

February 8 markdown:
100%
− 30
70% → $.70 \times \$1{,}000 = \$700$ selling price

February 28 additional markdown:
100%
− 20
80% → $.80 \times \$700 = \$560$

March 11 additional markup:
100%
+ 10
110% → $1.10 \times \$560 = \$616$

LO 2

Pricing Perishable Items

The following formula can be used to determine the price of goods that have a short shelf life such as fruit, flowers, and pastry. (We limit this discussion to obviously **perishable** items.)

To calculate the selling price of a perishable item: (1) Calculate selling price based on cost or based on selling price (total dollar sales). (2) Divide total dollar sales by the number of units available for sale (after accounting for spoilage).

$$\text{Selling price of perishables} = \frac{\text{Total dollar sales}}{\text{Number of units produced} - \text{Spoilage}}$$

MONEY tips

Shopping for perishable items? Seek a quantity discount with the retailer especially towards the end of the business day.

The Word Problem Audrey's Bake Shop baked 20 dozen bagels. Audrey expects 10% of the bagels to become stale and not salable. The bagels cost Audrey $1.20 per dozen. Audrey wants a 60% markup on cost. What should Audrey charge for each dozen bagels so she will make her profit? Round to the nearest cent.

BLUEPRINT	The facts	Solving for?	Steps to take	Key points
	Bagels cost: $1.20 per dozen. *Not salable:* 10%. *Baked:* 20 dozen. *Markup on cost:* 60%.	Price of a dozen bagels.	Total cost. Total dollar markup. Total selling price. Bagel loss. $TS = TC + TM$	Markup is based on cost.

Steps to solving problem

1. Calculate the total cost. $TC = 20 \text{ dozen} \times \$1.20 = \$24.00$
2. Calculate the total dollar markup. $TS = TC + TM$
   $TS = \$24.00 + .60(\$24.00)$
   $TS = \$24.00 + \$14.40$ ← Total dollar markup
3. Calculate the total selling price. $TS = \$38.40$ ← Total selling price
4. Calculate the bagel loss. $20 \text{ dozen} \times .10 = 2 \text{ dozen}$
5. Calculate the selling price for a dozen bagels. $\frac{\$38.40}{18} = \$2.13$ per dozen ($20 - 2 = 18$)

It's time to try the Practice Quiz.

LU 8–3 PRACTICE QUIZ

Complete this Practice Quiz to see how you are doing.

1. Sunshine Music Shop bought a stereo for $600 and marked it up 40% on selling price. To promote customer interest, Sunshine marked the stereo down 10% for 1 week. Since business was slow, Sunshine marked the stereo down an additional 5%. After a week, Sunshine marked the stereo up 2%. What is the new selling price of the stereo to the nearest cent? What is the markdown percent based on the original selling price to the nearest hundredth percent?
2. Alvin Rose owns a fruit and vegetable stand. He knows that he cannot sell all his produce at full price. Some of his produce will be markdowns, and he will throw out some produce. Alvin must put a high enough price on the produce to cover markdowns and rotted produce and still make his desired profit. Alvin bought 300 pounds of tomatoes at 14 cents per pound. He expects a 5% spoilage and marks up tomatoes 60% on cost. What price per pound should Alvin charge for the tomatoes?

*For **extra help** from your authors–Sharon and Jeff–see the videos in Connect.*

✓ Solutions

1.

$$S = C + M$$
$$S = \$600 + .40S$$
$$-.40S \qquad -.40S$$
$$\frac{.60S}{.60} = \frac{\$600}{.60}$$
$$S = \$1{,}000$$

Check

$$S = \frac{\text{Cost}}{1 - \text{Percent markup on selling price}}$$
$$S = \frac{\$600}{1 - .40} = \frac{\$600}{.60} = \$1{,}000$$

First markdown: $.90 \times \$1{,}000 = \$900$ selling price

Second markdown: $.95 \times \$900 = \$855$ selling price

Markup: $1.02 \times \$855 = \$872.10$ final selling price

$$\$1{,}000 - \$872.10 = \frac{\$127.90}{\$1{,}000} = 12.79\%$$

2. Price of tomatoes per pound:

BLUEPRINT	The facts	Solving for?	Steps to take	Key points
	300 lb. tomatoes at \$.14 per pound. *Spoilage:* 5%. *Markup on cost:* 60%.	Price of tomatoes per pound.	Total cost. Total dollar markup. Total selling price. Spoilage amount. $TS = TC + TM$	Markup is based on cost.

Steps to solving problem

1. Calculate the total cost. $TC = 300 \text{ lb.} \times \$.14 = \$42.00$
2. Calculate the total dollar markup. $TS = TC + TM$
   $TS = \$42.00 + .60(\$42.00)$
   $TS = \$42.00 + \$25.20$ ← Total dollar markup
3. Calculate the total selling price. $TS = \$67.20$ ← Total selling price
4. Calculate the tomato loss. 300 pounds × .05 = 15 pounds spoilage
5. Calculate the selling price per pound of tomatoes. $\frac{\$67.20}{285} = \$.24$ per pound (rounded to nearest hundredth)
   (300 − 15)

Learning Unit 8–4: Breakeven Analysis

So far in this chapter, cost is the price retailers pay to a manufacturer or supplier to bring the goods into the store. In this unit, we view costs from the perspective of manufacturers or suppliers who produce goods to sell in units, such as polo shirts, pens, calculators, lamps, and so on. These manufacturers or suppliers deal with two costs—fixed costs (*FC*) and variable costs (*VC*).

To understand how the owners of manufacturers or suppliers that produce goods per unit operate their businesses, we must understand fixed costs (*FC*), variable costs (*VC*), contribution margin (*CM*), and breakeven point (*BE*). Carefully study the following definitions of these terms:

- **Fixed costs (*FC*).** Costs that *do not change* with increases or decreases in sales; they include payments for insurance, a business license, rent, a lease, utilities, labor, and so on.
- **Variable costs (*VC*).** Costs that *do change* in response to changes in the volume of sales; they include payments for material, some labor, and so on.
- **Selling price (*S*).** In this unit we focus on manufacturers and suppliers who produce goods to sell in units.
- **Contribution margin (*CM*).** The difference between selling price (*S*) and variable costs (*VC*). This difference goes *first* to pay off total fixed costs (*FC*); when they are covered, *profits (or losses)* start to accumulate.
- **Breakeven point (*BE*).** The point at which the seller has covered all expenses and costs of a unit and has not made any profit or suffered any loss. Every unit sold after the breakeven point (*BE*) will bring some profit or cause a loss.

Learning Unit 8–4 is divided into two sections: calculating a contribution margin (*CM*) and calculating a breakeven point (*BE*). You will learn the importance of these two concepts and the formulas that you can use to calculate them. Study the example given for each concept to help you understand why the success of business owners depends on knowing how to use these two concepts.

LO 1

Calculating a Contribution Margin (*CM*)

Before we calculate the breakeven point, we must first calculate the contribution margin. The formula is as follows:

Contribution margin (*CM*) = Selling price (*S*) − Variable cost (*VC*)

EXAMPLE Assume Jones Company produces pens that have a selling price (*S*) of $2.00 and a variable cost (*VC*) of $.80. We calculate the contribution margin (*CM*) as follows:

$$\text{Contribution margin } (CM) = \$2.00\ (S) - \$.80\ (VC)$$
$$CM = \$1.20$$

This means that for each pen sold, $1.20 goes to cover fixed costs (*FC*) and results in a profit. It makes sense to cover fixed costs (*FC*) first because the nature of a fixed cost is that it does not change with increases or decreases in sales.

Now we are ready to see how Jones Company will reach a breakeven point (*BE*).

LO 2

Calculating a Breakeven Point (*BE*)

Sellers like Jones Company can calculate their profit or loss by using a concept called the **breakeven point (*BE*).** This important point results after sellers have paid all their expenses and costs. Study the following formula and the example:

$$\text{Breakeven point } (BE) = \frac{\text{Fixed costs } (FC)}{\text{Contribution margin } (CM)}$$

MONEY tips

What is your financial breakeven point? Create and stick to a budget. Knowing where your money goes can help you save on unnecessary expenditures and allows you to meet your financial goals. To help you track your spending and create a budget, check out mint.com.

EXAMPLE Jones Company produces pens. The company has a fixed cost (*FC*) of $60,000. Each pen sells for $2.00 with a variable cost (*VC*) of $.80 per pen.

Fixed cost (*FC*)	$60,000
Selling price (*S*) per pen	$2.00
Variable cost (*VC*) per pen	$.80

$$\text{Breakeven point } (BE) = \frac{\$60{,}000\ (FC)}{\$2.00\ (S) - \$.80\ (VC)} = \frac{\$60{,}000\ (FC)}{\$1.20\ (CM)} = 50{,}000 \text{ units (pens)}$$

At 50,000 units (pens), Jones Company is just covering its costs. Each unit after 50,000 brings in a profit of $1.20 (*CM*).

It is time to try the Practice Quiz.

LU 8–4 PRACTICE QUIZ

Complete this **Practice Quiz** to see how you are doing.

Blue Company produces holiday gift boxes. Given the following, calculate (1) the contribution margin (*CM*) and (2) the breakeven point (*BE*) for Blue Company.

Fixed cost (*FC*)	$45,000
Selling price (*S*) per gift box	$20
Variable cost (*VC*) per gift box	$8

*For **extra help** from your authors–Sharon and Jeff–see the videos in Connect.*

✓ Solutions

1. Contribution margin (*CM*) = $20 (*S*) – $8 (*VC*) = $12

2. $$\text{Breakeven point } (BE) = \frac{\$45{,}000\ (FC)}{\$20\ (S) - \$8\ (VC)} = \frac{\$45{,}000\ (FC)}{\$12\ (CM)} = 3{,}750 \text{ units (gift boxes)}$$

INTERACTIVE CHAPTER ORGANIZER

Topic/Procedure/Formula	Examples	You try it*
Markups based on cost: **Cost is 100% (base)** Selling price (S) = Cost (C) + Markup (M)	$\$400 = \$300 + \$100$ $S = C + M$	**Calculate selling price** Cost, $400; Markup, $200
Percent markup on cost $\frac{\text{Dollar markup (portion)}}{\text{Cost (base)}} = \frac{\text{Percent markup}}{\text{on cost (rate)}}$ **Cost** $C = \frac{\text{Dollar markup}}{\text{Percent markup on cost}}$	$\frac{\$100}{\$300} = \frac{1}{3} = 33\frac{1}{3}\%$ $\frac{\$100}{.33} = \$303$ Off slightly due to rounding	**Calculate percent markup on cost** Dollar markup, $50; Cost, $200 **Calculate cost** Dollar markup, $50; Percent markup on cost, 25%
Calculating selling price $S = C + M$ **Check** $S = \text{Cost} \times (1 + \text{Percent markup on cost})$	Cost, $6; percent markup on cost, 20% $S = \$6 + .20(\$6)$ $S = \$6 + \$1.20$ $S = \$7.20$ **Check** → $\$6 \times 1.20 = \$7.20$	**Calculate selling price** Cost, $8; Percent markup on cost, 10%
Calculating cost $S = C + M$ **Check** $\text{Cost} = \frac{\text{Selling price}}{1 + \text{Percent markup on cost}}$	$S = \$100$; $M = 70\%$ of cost $S = C + M$ (*Remember,* $C = 1.00C$) $\$100 = C + .70C$ $\$100 = 1.7C$ $\frac{\$100}{1.7} = C$ $\$58.82 = C$ **Check** → $\frac{\$100}{1 + .70} = \$58.82$	**Calculate cost** Selling price, $200; Markup on cost, 60%
Markups based on selling price: **selling price is 100% (Base)** Dollar markup = Selling price − Cost	$M = S - C$ $\$600 = \$1{,}000 - \$400$	**Calculate dollar markup** Cost, $2,000; Selling price, $4,500
Percent markup on selling price $\frac{\text{Dollar markup (portion)}}{\text{Selling price (base)}} = \frac{\text{Percent markup on}}{\text{selling price (rate)}}$ **Selling price** $S = \frac{\text{Dollar markup}}{\text{Percent markup on selling price}}$	$\frac{\$600}{\$1{,}000} = 60\%$ $\frac{\$600}{.60} = \$1{,}000$	**Calculate percent markup on selling price** Dollar markup, $700; Selling price, $2,800 **Calculate selling price** Dollar markup, $700; Percent markup on selling price, 50%
Calculating selling price $S = C + M$ **Check** $\text{Selling price} = \frac{\text{Cost}}{1 - \text{Percent markup on selling price}}$	Cost, $400; percent markup on S, 60% $S = C + M$ $S = \$400 + .60S$ $S - .60S = \$400 + .60S - .60S$ $\frac{.40S}{.40} = \frac{\$400}{.40}$ $S = \$1{,}000$ **Check** → $\frac{\$400}{1 - .60} = \frac{\$400}{.40} = \$1{,}000$	**Calculate selling price** Cost, $800; Markup on selling price, 40%
Calculating cost $S = C + M$ **Check** $\text{Cost} = \text{Selling price} \times \left(1 - \text{Percent markup on selling price}\right)$	$\$1{,}000 = C + 60\%(\$1{,}000)$ $\$1{,}000 = C + \$600$ $\$400 = C$ **Check** → $\$1{,}000 \times (1 - .60)$ $\$1{,}000 \times .40 = \$400$	**Calculate cost** Selling price, $2,000; 70% markup on selling price

(continues)

INTERACTIVE CHAPTER ORGANIZER

Topic/Procedure/Formula	Examples	You try it*
Conversion of markup percent Percent markup on cost to Percent markup on selling price $\frac{\text{Percent markup on cost}}{1 + \text{Percent markup on cost}}$ Percent markup on selling price to Percent markup on cost $\frac{\text{Percent markup on selling price}}{1 - \text{Percent markup on selling price}}$	*Round to nearest percent:* 54% markup on cost → 35% markup on selling price $\frac{.54}{1+.54} = \frac{.54}{1.54} = 35\%$ 35% markup on selling price → 54% markup on cost $\frac{.35}{1-.35} = \frac{.35}{.65} = 54\%$	**Calculate percent markup on selling price** Convert 47% markup on cost to markup on selling price. Round to nearest percent.
Markdowns $\text{Markdown percent} = \frac{\text{Dollar markdown}}{\text{Selling price (original)}}$	\$40 selling price 10% markdown \$40 × .10 = \$4 markdown $\frac{\$4}{\$40} = 10\%$	**Calculate markdown percent** Selling price, \$50; Markdown, 20%
Pricing perishables 1. Calculate total cost and total selling price. 2. Calculate selling price per unit by dividing total sales in Step 1 by units expected to be sold after taking perishables into account.	50 pastries cost 20 cents each; 10 will spoil before being sold. Markup is 60% on cost. 1. $TC = 50 \times \$.20 = \$10$ $TS = TC + TM$ $TS = \$10 + .60(\$10)$ $TS = \$10 + \$6$ $TS = \$16$ 2. $\frac{\$16}{40 \text{ pastries}} = \$.40$ per pastry	**Calculate cost of each pastry** 30 pastries cost 30 cents each; 15 will spoil; markup is 30% on cost.
Breakeven point (*BE*) $BE = \frac{\text{Fixed cost } (FC)}{\text{Contribution margin } (CM)}$ (Selling price, S – Variable cost, VC)	Fixed cost (*FC*) \$60,000 Selling price (*S*) \$90 Variable cost (*VC*) \$30 $BE = \frac{\$60{,}000}{\$90 - \$30} = \frac{\$60{,}000}{\$60} = 1{,}000 \text{ units}$	**Calculate *BE*** Fixed cost (*FC*) \$70,000 Selling price (*S*) \$80 Variable cost (*VC*) \$60

KEY TERMS

Breakeven point	Margin	Percent markup on selling price
Contribution margin	Markdowns	Perishables
Cost	Markup	Selling price
Dollar markdown	Net profit (net income)	Variable cost
Dollar markup	Operating expenses (overhead)	
Fixed cost	Percent markup on cost	
Gross profit		

*Worked-out solutions are in Appendix B.

Critical Thinking Discussion Questions with Chapter Concept Check

1. Assuming markups are based on cost, explain how the portion formula could be used to calculate cost, selling price, dollar markup, and percent markup on cost. Pick a company and explain why it would mark goods up on cost rather than on selling price.
2. Assuming markups are based on selling price, explain how the portion formula could be used to calculate cost, selling price, dollar markup, and percent markup on selling price. Pick a company and explain why it would mark up goods on selling price rather than on cost.
3. What is the formula to convert percent markup on selling price to percent markup on cost? How could you explain that a 40% markup on selling price, which is a 66.7% markup on cost, would result in the same dollar markup?

4. Explain how to calculate markdowns. Do you think stores should run 1-day-only markdown sales? Would it be better to offer the best price "all the time"?
5. Explain the five steps in calculating a selling price for perishable items. Recall a situation where you saw a store that did *not* follow the five steps. How did it sell its items?
6. Explain how Walmart uses breakeven analysis. Give an example.
7. **Chapter Concept Check.** Visit a retailer's website and find out how that retailer marks up goods and marks down specials. Present calculations based on this chapter to support your findings.

END-OF-CHAPTER PROBLEMS

Check figures for odd-numbered problems in Appendix B. Name ____________ Date ________

DRILL PROBLEMS

Assume markups in Problems 8–1 to 8–6 are based on cost. Find the dollar markup and selling price for the following problems. Round answers to the nearest cent. *LU 8-1(1, 2)*

	Item	Cost	Markup percent	Dollar markup	Selling price
8–1.	Bell and Ross watch	$2,000	30%		
8–2.	Burberry men's watch	$425	200%		

Solve for cost (round to the nearest cent): *LU 8-1(3)*

8–3. Selling price of office furniture at Staples, $6,000

Percent markup on cost, 40%

Actual cost?

8–4. Selling price of lumber at Home Depot, $4,000

Percent markup on cost, 30%

Actual cost?

Complete the following: *LU 8-1(1)*

	Cost	Selling price	Dollar markup	Percent markup on cost*
8–5.	$15.10	$22.00	?	?
8–6.	?	?	$4.70	102.17%

*Round to the nearest hundredth percent.

Assume markups in Problems 8–7 to 8–12 are based on selling price. Find the dollar markup and cost (round answers to the nearest cent): *LU 8-2(1, 2)*

	Item	Selling price	Markup percent	Dollar markup	Cost
8–7.	Sony LCD TV	$1,000	45%		
8–8.	Canon scanner	$80	30%		

Solve for the selling price (round to the nearest cent): *LU 8-2(3)*

8–9. Selling price of a complete set of pots and pans at Walmart
40% markup on selling price
Cost, actual, $66.50

8–10. Selling price of a dining room set at Macy's
55% markup on selling price
Cost, actual, $800

Complete the following: *LU 8-2(1)*

	Cost	Selling price	Dollar markup	Percent markup on selling price (round to nearest tenth percent)
8–11.	$14.80	$49.00	?	?
8–12.	?	?	$4	20%

By conversion of the markup formula, solve the following (round to the nearest whole percent as needed): *LU 8-2(4)*

	Percent markup on cost	Percent markup on selling price
8–13.	12.4%	?
8–14.	?	13%

Complete the following: *LU 8-3(1, 2)*

8–15. Calculate the final selling price to the nearest cent and markdown percent to the nearest hundredth percent:

Original selling price	First markdown	Second markdown	Markup	Final markdown
$5,000	20%	10%	12%	5%

	Item	Total quantity bought	Unit cost	Total cost	Percent markup on cost	Total selling price	Percent that will spoil	Selling price per brownie
8–16.	Brownies	20	$.79	?	60%	?	10%	?

Complete the following: *LU 8-4(1, 2)*

	Breakeven point	Fixed cost	Contribution margin	Selling price per unit	Variable cost per unit
8–17.		$65,000		$5.00	$1.00
8–18.		$90,000		$9.00	$4.00

WORD PROBLEMS

8–19. Bari Jay, a gown manufacturer, received an order for 600 prom dresses from China. Her cost is $35 a gown. If her markup based on selling price is 79%, what is the selling price of each gown? Round to the nearest cent. *LU 8-2(2)*

8–20. Brian May, guitarist for Queen, does not know how to price his signature Antique Cherry Special that cost him £280 to make. He knows he wants 85% markup on cost. What price should Brian May ask for the guitar? *LU 8-1(2)*

My Money

8–21. You are buying and reselling items found at your local thrift shop. You found an antique pitcher for sale. If you need a 40% markup on cost and know most people will not pay more than $20 for it, what is the most you can pay for the pitcher? Round to the nearest cent. *LU 8-1(4)*

8–22. Macy's was selling Calvin Klein jean shirts that were originally priced at $58.00 for $8.70. **(a)** What was the amount of the markdown? **(b)** Based on the selling price, what is the percent markdown? *LU 8-3(1)*

8–23. Brownsville, Texas, boasts being the southernmost international seaport and the largest city in the lower Rio Grande Valley. Ben Supple, an importer in Brownsville, has just received a shipment of Peruvian opals that he is pricing for sale. He paid $150 for the shipment. If he wants a 75% markup, calculate the selling price based on selling price. Then calculate the selling price based on cost. *LU 8-1(2), LU 8-2(2)*

8–24. Front Range Cabinet Distributors in Colorado Springs, Colorado, sells to its contractors with a 42% markup on cost. If the selling price for cabinets is $9,655, what is the cost to contractors based on cost? Round to the nearest tenth. Check your answer. *LU 8-1(3)*

8–25. Misu Sheet, owner of the Bedspread Shop, knows his customers will pay no more than $120 for a comforter. Misu wants a 30% markup on selling price. What is the most that Misu can pay for a comforter? *LU 8-2(4)*

8–26. Assume Misu Sheet (Problem 8–25) wants a 30% markup on cost instead of on selling price. What is Misu's cost? Round to the nearest cent. *LU 8-1(4)*

8–27. Misu Sheet (Problem 8–25) wants to advertise the comforter as "percent markup on cost." What is the equivalent rate of percent markup on cost compared to the 30% markup on selling price? Check your answer. Is this a wise marketing decision? Round to the nearest hundredth percent. *LU 8-2(4)*

8–28. DeWitt Company sells a kitchen set for $475. To promote July 4, DeWitt ran the following advertisement:

Beginning each hour up to 4 hours we will mark down the kitchen set 10%. At the end of each hour, we will mark up the set 1%.

Assume Ingrid Swenson buys the set 1 hour 50 minutes into the sale. What will Ingrid pay? Round each calculation to the nearest cent. What is the markdown percent? Round to the nearest hundredth percent. *LU 8-3(1)*

8–29. Angie's Bake Shop makes birthday chocolate chip cookies that cost $2 each. Angie expects that 10% of the cookies will crack and be discarded. Angie wants a 60% markup on cost and produces 100 cookies. What should Angie price each cookie? Round to the nearest cent. *LU 8-3(2)*

8–30. Assume that Angie (Problem 8–29) can sell the cracked cookies for $1.10 each. What should Angie price each cookie? *LU 8-3(2)*

8–31. Jane Corporation produces model toy cars. Each sells for $29.99. Its variable cost per unit is $14.25. What is the breakeven point for Jane Corporation assuming it has a fixed cost of $314,800? *LU 8-4(2)*

8–32. Aunt Sally's "New Orleans Most Famous Pralines" sells pralines costing $1.10 each to make. If Aunt Sally's wants a 35% markup based on selling price and produces 45 pralines with an anticipated 15% spoilage (rounded to the nearest whole number), what should each praline be sold for? *LU 8-3(2)*

My Money

8–33. On Black Friday, Amazon.com featured an Echo 2nd Generation for $35 normally selling for $79. Calculate the dollar markdown and the markdown percent based on the selling price to the nearest whole percent. *LU 8-3(1)*

8–34. The Food Co-op Club boasts that it has 5,000 members and a 200% increase in sales. Its markup is 36% based on cost. What would be its percent markup if selling price were the base? Round to the nearest hundredth percent. *LU 8-2(4)*

8–35. At Bed Bath and Beyond, the manager, Jill Roe, knows her customers will pay no more than $300 for a bedspread. Jill wants a 35% markup on selling price. What is the most that Jill can pay for a bedspread? *LU 8-2(4)*

8–36. Kathleen Osness purchased a $60,000 RV with a 40 percent markup on selling price. **(a)** What was the amount of the dealer's markup? **(b)** What was the dealer's original cost? *LU 8-2(4)*

8–37. John's Smoothie Stand at Utah's Wasatch County's Demolition Derby sells bananas. If John bought 50 lbs. of bananas at $.23 per pound expecting 10% to spoil, how should he price his bananas to achieve 57% on selling price? *LU 8-2(4)* eXcel **8–38.** Arley's Bakery makes fat-free cookies that cost $1.50 each. Arley expects 15% of the cookies to fall apart and be discarded. Arley wants a 45% markup on cost and produces 200 cookies. What should Arley price each cookie? Round to the nearest cent. *LU 8-3(2)*

8–39. Assume that Arley (Problem 8–38) can sell the broken cookies for $1.40 each. What should Arley price each cookie? *LU 8-3(2)*

8–40. An Apple Computer store sells computers for $1,258.60. Assuming the computers cost $10,788 per dozen, find for each computer the **(a)** dollar markup, **(b)** percent markup on cost, and **(c)** percent markup on selling price, to the nearest hundredth percent. *LU 8-1(1), LU 8-2(1)*

Prove (**b**) and (**c**) of the above problem using the equivalent formulas.

8–41. Pete Corporation produces bags of peanuts. Its fixed cost is $17,280. Each bag sells for $2.99 with a unit cost of $1.55. What is Pete's breakeven point? *LU 8-4(2)*

CHALLENGE PROBLEMS

8–42. Nissan Appliances bought two dozen camcorders at a cost of $4,788. The markup on the camcorders is 25% of the selling price. What was the original selling price of each camcorder? *LU 8-2(3)*

8–43. On July 8, Leon's Kitchen Hut bought a set of pots with a $120 list price from Lambert Manufacturing. Leon's receives a 25% trade discount. Terms of the sale were 2/10, n/30. On July 14, Leon's sent a check to Lambert for the pots. Leon's expenses are 20% of the selling price. Leon's must also make a profit of 15% of the selling price. A competitor marked down the same set of pots 30%. Assume Leon's reduces its selling price by 30%. *LU 8-2(3)*

a. What is the sale price at Kitchen Hut?

b. What was the operating profit or loss?

SUMMARY PRACTICE TEST

Do you need help? Connect videos have step-by-step worked-out solutions.

1. Sunset Co. marks up merchandise 40% on cost. A DVD player costs Sunset $90. What is Sunset's selling price? Round to the nearest cent. *LU 8-1(2)*

2. JCPenney sells jeans for $49.50 that cost $38.00. What is the percent markup on cost? Round to the nearest hundredth percent. Check the cost. *LU 8-1(1)*

3. Best Buy sells a flat-screen high-definition TV for $700. Best Buy marks up the TV 45% on cost. What is the cost and dollar markup of the TV? *LU 8-1(1, 3)*

4. Sports Authority marks up New Balance sneakers $30 and sells them for $109. Markup is on cost. What are the cost and percent markup to the nearest hundredth percent? *LU 8-1(1, 3)*

5. The Shoe Outlet bought boots for $60 and marks up the boots 55% on the selling price. What is the selling price of the boots? Round to the nearest cent. *LU 8-2(2)*

6. Office Max sells a desk for $450 and marks up the desk 35% on the selling price. What did the desk cost Office Max? Round to the nearest cent. *LU 8-2(4)*

7. Zales sells diamonds for $1,100 that cost $800. What is Zales's percent markup on selling price? Round to the nearest hundredth percent. Check the selling price. *LU 8-2(1)*

8. Earl Miller, a customer of J. Crew, will pay $400 for a new jacket. J. Crew has a 60% markup on selling price. What is the most that J. Crew can pay for this jacket? *LU 8-2(4)*

9. Home Liquidators marks up its merchandise 35% on cost. What is the company's equivalent markup on selling price? Round to the nearest tenth percent. *LU 8-2(4)*

10. The Muffin Shop makes no-fat blueberry muffins that cost $.70 each. The Muffin Shop knows that 15% of the muffins will spoil. If The Muffin Shop wants 40% markup on cost and produces 800 muffins, what should The Muffin Shop price each muffin? Round to the nearest cent. *LU 8-3(2)*

11. Angel Corporation produces calculators selling for $25.99. Its unit cost is $18.95. Assuming a fixed cost of $80,960, what is the breakeven point in units? *LU 8-4(2)*

INTERACTIVE VIDEO WORKSHEET

GRADING THE SUMMARY PRACTICE TEST

Go to the summary practice test video in Connect (or click on it here in the ebook). Grade your summary practice test while viewing the video.

C for Correct/I for Incorrect

1. ________	5. ________	9. ________
2. ________	6. ________	10. ________
3. ________	7. ________	11. ________
4. ________	8. ________	

If you achieved 100%, you are ready for your instructor's exam.

If any of the problems were incorrect, list the questions you missed and show steps to solve the problem correctly.

Replay the video to see if you have made the correct fixes to your mistakes.
If you have any questions, contact your instructor asap.

Notes on Watching Videos

MY MONEY

Getting Credit Where Credit Is Due

What I need to know

Your credit score is something that will be with you for your entire adult life. It is an important component in many decisions made about you and your creditworthiness. Because of the vital role it plays in your financial life, it is important you understand your credit score/report. Part of this understanding is to determine where your score is today and where you would like it to be in the future. What activities have an impact on your score? How and why does your credit score fluctuate over time? What are the benefits of having a good credit score and solid credit report? With this knowledge in hand you will find your financial situation becomes a bit easier to understand and control.

What I need to do

Know where you stand today. A credit score is similar to a photograph in that it gives you an image of something at a point in time. Taking a look at your credit score today provides you with a baseline assessment of your creditworthiness at this point in your life. Based on this current score you can determine a course of action to help improve your credit score. Since financial decisions, such as financing a new car, are impacted by your credit score, you need to know your current score and realize that score will ultimately determine the interest rate, terms, and amount of interest you will pay on such a purchase. Therefore, you can save money by understanding your credit score and improving it prior to financing a large purchase.

Know how your credit score changes over time. You need to understand the financial decisions you make today will ultimately determine your credit score tomorrow. Life happens and understanding how common situations will impact your credit is a valuable life lesson to learn. For instance, missing a scheduled payment of your rent, utilities, or even your cell phone will impact your credit negatively. Carrying high balances compared to your credit limit on your accounts can also be viewed as a negative against your credit score. Knowledge truly is power as it relates to your credit score and knowing what situations to avoid will help you to minimize the negative impact on your credit.

Know the benefits of a good credit score. If it seems maintaining a good credit score takes some work, you are right. So, what do you gain from all this work? The benefits of a good credit score are many and provide you with the financial incentives to manage your money effectively. You will receive lower interest rates on money you borrow, saving you money over the course of your loan. You will also gain greater access to credit accounts with larger purchase limits and easier approval on new credit accounts. Knowing your good score provides you with the ability to negotiate your purchases in your favor because you are now a more attractive customer to the seller.

Resources I can use

- Credit Karma (mobile app)
- https://www.credit.com/credit-repair/how-to-improve-credit-score/ — improving your credit score

MY MONEY ACTIVITY

Retrieve your current credit report/score. Identify some practical ways you could improve your score.

PERSONAL FINANCE

A KIPLINGER APPROACH

"Online Merchants That Don't Deliver", *Kiplinger's*, November 2017, p. 11. Used by permission of The Kiplinger Washington Editors, Inc.

©Glow Images/Getty Images

NAUGHTY OR NICE?

ONLINE MERCHANTS THAT DON'T DELIVER

How to avoid fraud and late shipments when you're doing your holiday shopping.

WITH THE WINTER HOLIDAYS approaching, there's a good chance you'll start chipping away at your gift list online. But the rise in e-commerce has also led to an increase in complaints from unhappy buyers. In 2016, the FBI reported that more than 80,000 people lost more than $138 million to online fraud. The Consumer Federation of America's list of top 10 consumer complaints has included "internet sales" every year for the past decade. The most common complaints involved misrepresentations or failure to deliver purchases, the CFA said.

Fraud isn't the only thing that can derail a good deal. As more companies offer their merchandise online, some aren't up to the task. If a seller hasn't put in the time and money to process online transactions, it may not be able to fulfill customers' orders.

Limiting your purchases to major sites doesn't mean you'll avoid problems. Thousands of third-party sellers on websites such as Amazon Marketplace pitch their own deals to customers. Although Amazon guarantees the delivery of all purchases made on its site, a late or missing delivery from an unreliable business can be a pain, even if you get your money back.

Emily Hough of Fox River Grove, Ill., says she spent months waiting for linens she bought from a Chinese company through Amazon. She says she probably won't purchase from a third-party seller again.

To protect yourself against fraud and delivery mishaps, use a credit card when you make online purchases. That way, you'll be able to take advantage of consumer protections, such as the right to dispute transactions, that other systems don't offer. Look for merchants with positive customer reviews. And it doesn't hurt to do a search for the company's name with words such as *complaint* or *scam*. **BRENDAN PEDERSEN**

SOURCES: KIPLINGER RESEARCH, NATIONAL RETAIL FEDERATION

NO REPRIEVE

DON'T STOP PAYING STUDENT LOANS

Judges recently dismissed dozens of lawsuits against private student loan borrowers because of shoddy record keeping by companies claiming to own the debt. That could be promising news for borrowers, particularly those with loans held by the National Collegiate Student Loan Trusts, one of the largest owners of private student loan debt and the organization at the center of recent court cases. But the chances that your loan will be wiped out are slim. So far, courts have looked only at cases in which National Collegiate sued borrowers. Until you learn otherwise, continue making payments.

If you're contacted by an unfamiliar lender, ask for a copy of the master promissory note. The company is not legally required to provide the information, but if it doesn't, you can file a complaint with the Consumer Financial Protection Bureau and your local attorney general's office.

To avoid questions about who owns your loans, keep a copy of the signed promissory note for each loan as well as any communication you receive about the sale of your loans to another company. Borrowers with federal student loans can track them through the National Student Loan System at www.nslds.ed.gov. **KAITLIN PITSKER**

BUSINESS MATH ISSUE

Online shopping means going to the mall is no longer necessary.

1. List the key points of the article and information to support your position.
2. Write a group defense of your position using math calculations to support your view. If you are in an online course, post to a discussion board.

CUMULATIVE REVIEW

A Word Problem Approach—Chapters 6, 7, 8

1. Assume Kellogg's produced 715,000 boxes of Corn Flakes this year. This was 110% of the annual production last year. What was last year's annual production? *LU 6-2(2)*

2. A new Sony camera has a list price of $420. The trade discount is 10/20 with terms of 2/10, n/30. If a retailer pays the invoice within the discount period, what is the amount the retailer must pay? *LU 7-1(4), LU 7-2(1)*

3. JCPenney sells loafers with a markup of $40. If the markup is 30% on cost, what did the loafers cost JCPenney? Round to the nearest dollar. *LU 8-1(3)*

4. Aster Computers received from Ring Manufacturers an invoice dated August 28 with terms 2/10 EOM. The list price of the invoice is $3,000 (freight not included). Ring offers Aster a 9/8/2 trade chain discount. Terms of freight are FOB shipping point, but Ring prepays the $150 freight. Assume Aster pays the invoice on October 9. How much will Ring receive? *LU 7-1(4), LU 7-2(1)*

5. Runners World marks up its Nike jogging shoes 25% on selling price. The Nike shoes sell for $65. How much did the store pay for them? *LU 8-2(4)*

6. Ivan Rone sells antique sleds. He knows that the most he can get for a sled is $350. Ivan needs a 35% markup on cost. Since Ivan is going to an antiques show, he wants to know the maximum he can offer a dealer for an antique sled. *LU 8-1(3)*

7. Bonnie's Bakery bakes 60 loaves of bread for $1.10 each. Bonnie's estimates that 10% of the bread will spoil. Assume a 60% markup on cost. What is the selling price of each loaf? If Bonnie's can sell the old bread for one-half the cost, what is the selling price of each loaf? *LU 8-3(2)*

CHAPTER 9

Payroll

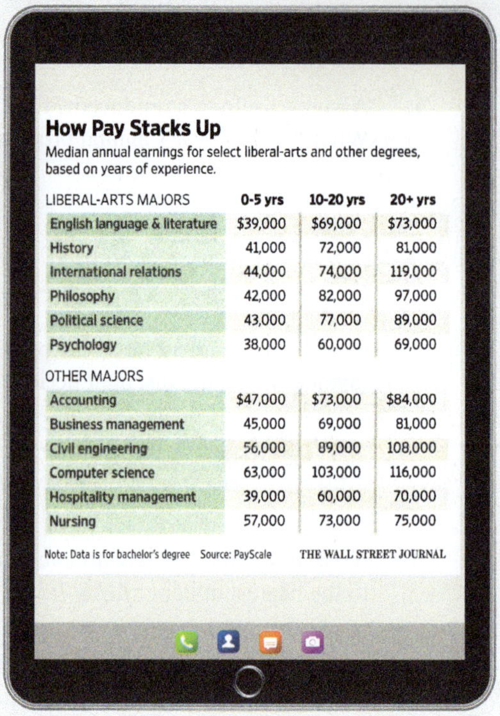

How Pay Stacks Up

Median annual earnings for select liberal-arts and other degrees, based on years of experience.

LIBERAL-ARTS MAJORS	0-5 yrs	10-20 yrs	20+ yrs
English language & literature	$39,000	$69,000	$73,000
History	41,000	72,000	81,000
International relations	44,000	74,000	119,000
Philosophy	42,000	82,000	97,000
Political science	43,000	77,000	89,000
Psychology	38,000	60,000	69,000
OTHER MAJORS			
Accounting	$47,000	$73,000	$84,000
Business management	45,000	69,000	81,000
Civil engineering	56,000	89,000	108,000
Computer science	63,000	103,000	116,000
Hospitality management	39,000	60,000	70,000
Nursing	57,000	73,000	75,000

Note: Data is for bachelor's degree Source: PayScale THE WALL STREET JOURNAL

LEARNING UNIT OBJECTIVES

LU 9–1: Calculating Various Types of Employees' Gross Pay

1. Define, compare, and contrast weekly, biweekly, semimonthly, and monthly pay periods.
2. Calculate gross pay with overtime on the basis of time.
3. Calculate gross pay for piecework, differential pay schedule, straight commission with draw, variable commission scale, and salary plus commission.

LU 9–2: Computing Payroll Deductions for Employees' Pay; Employers' Responsibilities

1. Prepare and explain the parts of a payroll register.
2. Explain and calculate federal and state unemployment taxes.

Your Guide to Successfully Completing This Chapter

Traditional book or ebook

Check box as you complete each step.

Steps

☐ Read learning unit.
  ☐ Complete practice quiz at the end of the learning unit.
☐ Grade practice quiz using provided solutions. (For more help, watch the learning unit video in Connect and have a Study Session with the authors. Then complete the additional practice quiz in Connect.)
☐ Repeat above for each of the two learning units in Chapter 9.
  ☐ Review chapter organizer.
  ☐ Complete assigned homework.
    ☐ Finish summary practice test. (Go to Connect via the ebook link and do the interactive video worksheet to grade.)
☐ Complete instructor's exam.

Wal-Mart Stores Inc. is trying out a scheduling system in hundreds of stores to give employees more control over their hours, as it seeks to address worker discontent around unpredictable scheduling without boosting labor costs.

Starting in February 2018 your take-home pay has increased due to the new tax act. The new withholding tables are now taking less out of your paycheck. In this chapter we will use the withholding tables based on the new tax act for all of our calculations.

The *Wall Street Journal* clip to the left discusses how Walmart is trying to maintain labor costs while giving employees more control over their work hours.

This chapter discusses (1) the type of pay people work for, (2) how employers calculate paychecks and deductions, and (3) what employers must report and pay in taxes.

Learning Unit 9–1: Calculating Various Types of Employees' Gross Pay

LO 1

My Money

Logan Company manufactures dolls of all shapes and sizes. These dolls are sold worldwide. We study Logan Company in this unit because of the variety of methods Logan uses to pay its employees.

Companies usually pay employees **weekly, biweekly, semimonthly,** or **monthly.** How often employers pay employees can affect how employees manage their money. Some employees prefer a weekly paycheck that spreads the inflow of money. Employees who have monthly bills may find the twice-a-month or monthly paycheck more convenient. All employees would like more money to manage.

Let's assume you earn $50,000 per year. The following table shows what you would earn each pay period. Remember that 13 weeks equals one quarter. Four quarters or 52 weeks equals a year.

Salary paid	Period (based on a year)	Earnings for period (dollars)
Weekly	52 times (once a week)	$961.54 ($50,000 ÷ 52)
Biweekly	26 times (every two weeks)	$1,923.08 ($50,000 ÷ 26)
Semimonthly	24 times (twice a month)	$2,083.33 ($50,000 ÷ 24)
Monthly	12 times (once a month)	$4,166.67 ($50,000 ÷ 12)

Aha!

You can estimate an annual salary by doubling the full-time hourly rate and then multiplying by 1,000. Example: $15 an hour, $15 × 2 × 1,000 = $30,000. You can estimate an hourly full-time rate by dividing an annual salary by 1,000 and then dividing by 2. Example: $30,000/1,000 = $30; 30/2 = $15.

Now let's look at some pay schedule situations and examples of how Logan Company calculates its payroll for employees of different pay status.

LO 2

Situation 1: Hourly Rate of Pay; Calculation of Overtime

The **Fair Labor Standards Act** sets minimum wage standards and overtime regulations for employees of companies covered by this federal law. The law provides that employees working for an hourly rate receive time-and-a-half pay for hours worked in excess of their regular 40-hour week. Many managerial people, however, are exempt from the time-and-a-half pay for all hours in excess of a 40-hour week. Other workers may also be exempt.

The current federal hourly minimum wage is $7.25. Various states have passed their own minimum wages. The *Wall Street Journal* clips below show how 20 states and D.C. plan to raise the minimum wage. South Korea will increase its minimum wage 16.4%.

GLOBAL

My Money

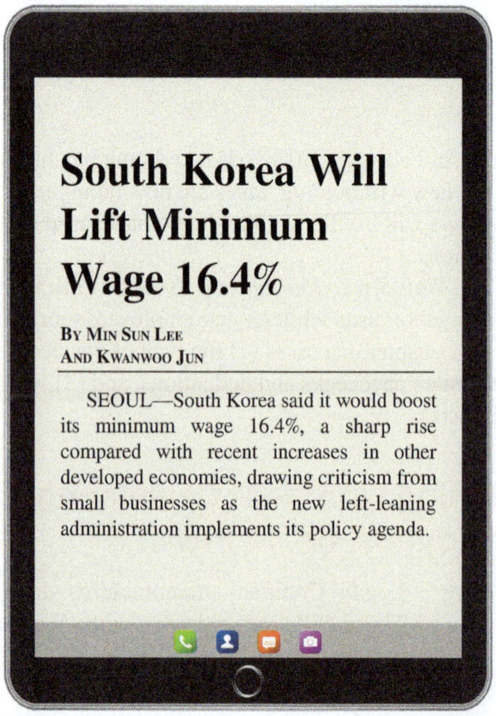

South Korea Will Lift Minimum Wage 16.4%

By Min Sun Lee and Kwanwoo Jun

SEOUL—South Korea said it would boost its minimum wage 16.4%, a sharp rise compared with recent increases in other developed economies, drawing criticism from small businesses as the new left-leaning administration implements its policy agenda.

Now we return to our Logan Company example. Logan Company is calculating the weekly pay of Ramon Valdez, who works in its manufacturing division. For the first 40 hours Ramon works, Logan calculates his **gross pay** (earnings before **deductions**) as follows:

Gross pay = Hours employee worked × Rate per hour

Ramon works more than 40 hours in a week. For every hour over his 40 hours, Ramon must be paid an **overtime** pay of at least 1.5 times his regular pay rate. The following formula is used to determine Ramon's overtime:

Hourly overtime pay rate = Regular hourly pay rate × 1.5

Logan Company must include Ramon's overtime pay with his regular pay. To determine Ramon's gross pay, Logan uses the following formula:

Gross pay = Earnings for 40 hours + Earnings at time-and-a-half rate (1.5)

We are now ready to calculate Ramon's gross pay from the following data:

EXAMPLE

Employee	M	T	W	Th	F	S	Total hours	Rate per hour
Ramon Valdez	13	$8\frac{1}{2}$	10	8	$11\frac{1}{4}$	$10\frac{3}{4}$	$61\frac{1}{2}$	$9

$61\frac{1}{2}$ total hours
−40 regular hours
$21\frac{1}{2}$ hours overtime[1]   Time-and-a-half pay: $9 × 1.5 = $13.50

Gross pay = (40 hours × $9) + ($21\frac{1}{2}$ hours × $13.50)
= $360 + $290.25
= **$650.25**

Note that the $13.50 overtime rate came out even. However, throughout the text, *if an overtime rate is greater than two decimal places, do not round it. Round only the final answer. This gives greater accuracy.*

LO 3

Situation 2: Straight Piece Rate Pay

Some companies, especially manufacturers, pay workers according to how much they produce. Logan Company pays Ryan Foss for the number of dolls he produces in a week. This gives Ryan an incentive to make more money by producing more dolls. Ryan receives $.96 per doll, less any defective units. The following formula determines Ryan's gross pay:

Gross pay = Number of units produced × Rate per unit

Companies may also pay a guaranteed hourly wage and use a piece rate as a bonus. However, Logan uses straight piece rate as wages for some of its employees.

EXAMPLE During the last week of April, Ryan Foss produced 900 dolls. Using the above formula, Logan Company paid Ryan $864.

Gross pay = 900 dolls × $.96
= **$864**

[1]Some companies pay overtime for time over 8 hours in one day; Logan Company pays overtime for time over 40 hours per week.

Situation 3: Differential Pay Schedule

Some of Logan's employees can earn more than the $.96 straight piece rate for every doll they produce. Logan Company has set up a **differential pay schedule** for these employees. The company determines the rate these employees make by the amount of units the employees produce at different levels of production.

EXAMPLE Logan Company pays Abby Rogers on the basis of the following schedule:

	Units produced	Amount per unit
First 50 →	1–50	$ .50
Next 100 →	51–150	.62
Next 50 →	151–200	.75
	Over 200	1.25

MONEY tips

Understand the costs of credit. Do not spend money you do not currently have—especially if it is for entertainment. Avoid reaching or coming close to the maximum on your credit cards. Be careful to fully understand the terms of any credit card you use. Read the fine print.

Last week Abby produced 300 dolls. What is Abby's gross pay?

Logan calculated Abby's gross pay as follows:

$$(50 \times \$.50) + (100 \times \$.62) + (50 \times \$.75) + (100 \times \$1.25)$$
$$\$25 + \$62 + \$37.50 + \$125 = \$249.50$$

Now we will study some of the other types of employee commission payment plans.

Situation 4: Straight Commission with Draw

Companies frequently use **straight commission** to determine the pay of salespersons. This commission is usually a certain percentage of the amount the salesperson sells. An example of one group of companies ceasing to pay commissions is the rental-car companies.

Companies such as Logan Company allow some of their salespersons to draw against their commission at the beginning of each month. A **draw** is an advance on the salesperson's commission. Logan subtracts this advance later from the employee's commission earned based on sales. When the commission does not equal the draw, the salesperson owes Logan the difference between the draw and the commission.

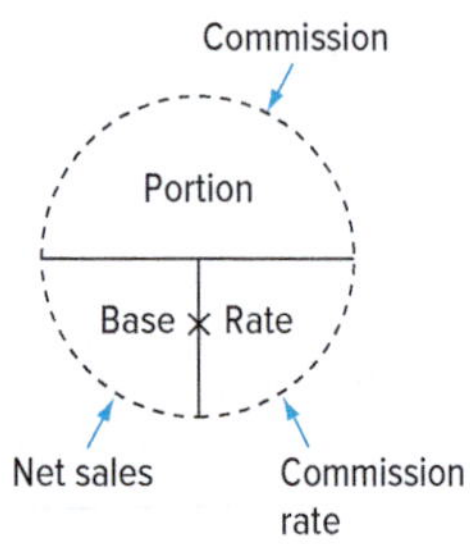

EXAMPLE Logan Company pays Jackie Okamoto a straight commission of 15% on her net sales (net sales are total sales less sales returns). In May, Jackie had net sales of $56,000. Logan gave Jackie a $600 draw in May. What is Jackie's gross pay?

Logan calculated Jackie's commission minus her draw as follows:

$$\$56{,}000 \times .15 = \$8{,}400$$
$$- \ 600$$
$$\$7{,}800$$

Logan Company pays some people in the sales department on a variable commission scale. Let's look at this, assuming the employee had no draw.

Situation 5: Variable Commission Scale

A company with a **variable commission scale** uses different commission rates for different levels of net sales.

EXAMPLE Last month, Jane Ring's net sales were $160,000. What is Jane's gross pay based on the following schedule?

Up to $35,000	4%
Excess of $35,000 to $45,000	6%
Over $45,000	8%

$$\text{Gross pay} = (\$35{,}000 \times .04) + (\$10{,}000 \times .06) + (\$115{,}000 \times .08)$$
$$= \$1{,}400 + \$600 + \$9{,}200$$
$$= \$11{,}200$$

Situation 6: Salary Plus Commission

Logan Company pays Joe Roy a \$3,000 monthly salary plus a 4% commission for sales over \$20,000. Last month Joe's net sales were \$50,000. Logan calculated Joe's gross monthly pay as follows:

$$\begin{aligned}\text{Gross pay} &= \text{Salary} + (\text{Commission} \times \text{Sales over } \$20{,}000)\\ &= \$3{,}000 + (.40 \times \$30{,}000)\\ &= \$3{,}000 + \$1{,}200\\ &= \$4{,}200\end{aligned}$$

Before you take the Practice Quiz, you should know that many managers today receive **overrides.** These managers receive a commission based on the net sales of the people they supervise.

LU 9–1 PRACTICE QUIZ

Complete this Practice Quiz to see how you are doing.

1. Jill Foster worked 52 hours in one week for Delta Airlines. Jill earns \$10 per hour. What is Jill's gross pay, assuming overtime is at time-and-a-half?
2. Matt Long had \$180,000 in sales for the month. Matt's commission rate is 9%, and he had a \$3,500 draw. What was Matt's end-of-month commission?
3. Bob Meyers receives a \$1,000 monthly salary. He also receives a variable commission on net sales based on the following schedule (commission doesn't begin until Bob earns \$8,000 in net sales):

\$8,000–\$12,000	1%	Excess of \$20,000 to \$40,000	5%
Excess of \$12,000 to \$20,000	3%	More than \$40,000	8%

Assume Bob earns \$40,000 net sales for the month. What is his gross pay?

*For **extra help** from your authors–Sharon and Jeff–see the videos in Connect.*

✓ Solutions

1. $40 \text{ hours} \times \$10.00 = \$400.00$
$12 \text{ hours} \times \$15.00 = \underline{180.00}$ $(\$10.00 \times 1.5 = \$15.00)$
$\$580.00$

2. $\$180{,}000 \times .09 = \$16{,}200$
$- \underline{3{,}500}$
$\$12{,}700$

3. $$\begin{aligned}\text{Gross pay} &= \$1{,}000 + (\$4{,}000 \times .01) + (\$8{,}000 \times .03) + (\$20{,}000 \times .05)\\ &= \$1{,}000 + \$40 + \$240 + \$1{,}000\\ &= \$2{,}280\end{aligned}$$

MONEY tips

Open a flexible-spending account. Eighty-five percent of employers offer them and they help you save by using pretax dollars for several expense categories.

Learning Unit 9–2: Computing Payroll Deductions for Employees' Pay; Employers' Responsibilities

The following *Wall Street Journal* clip, "Perception Gap," on page 254, discusses how employers and employees see things differently when considering topics such as being paid fairly, being valued at work, and pay transparency.

This unit begins by dissecting a paycheck. Then we give you an insight into the tax responsibilities of employers.

LO 1

My Money

Computing Payroll Deductions for Employees

Companies often record employee payroll information in a multicolumn form called a **payroll register.** The increased use of computers in business has made computerized registers a time-saver for many companies. In 2017, a new tax act was passed greatly updating withholding taxes. This became effective in February 2018. Each year the rates will change.

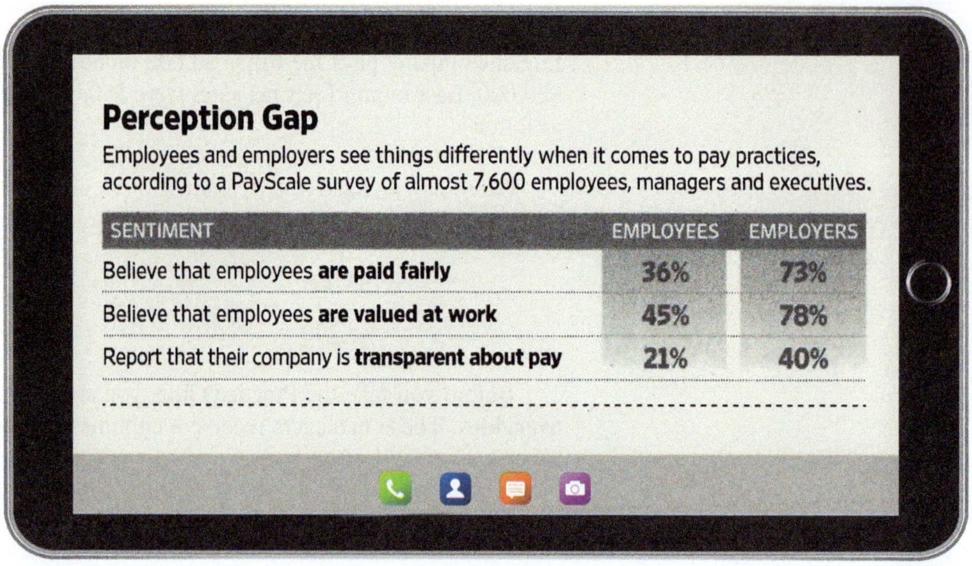

Glo Company uses a multicolumn payroll register. Below is Glo's partial payroll register showing the payroll information for Janet Wong during week 49. Let's check each column to see if Janet's take-home pay of $1,935.34 is correct. Note how the circled letters in the register correspond to the explanations that follow.

GLO COMPANY
Payroll Register
Week #49

Employee name	Allow. & marital status	Cum. earn.	Sal. per week	Earning Reg.	Earning Ovt.	Earning Gross	Cum. earn.	FICA Taxable Earning S.S.	FICA Taxable Earning Med.	Deductions FICA S.S.	Deductions FICA Med.	Deductions FIT	Deductions SIT	Deductions Health ins.	Net pay
Wong, Janet	M-2	$127,200	2,650	2,650	—	2,650	129,850	1,200	2,650	74.40	38.43	342.83	159.00	100	$1,935.34
	Ⓐ	Ⓑ	Ⓒ		Ⓓ		Ⓔ	Ⓕ	Ⓖ	Ⓗ	Ⓘ	Ⓙ	Ⓚ	Ⓛ	Ⓜ

Payroll Register Explanations
Ⓐ—Allowance and marital status
Ⓑ, Ⓒ, Ⓓ—Cumulative earnings before payroll, salaries, earnings
Ⓔ—Cumulative earnings after payroll

When Janet was hired, she completed the **W-4 (Employee's Withholding Allowance Certificate)** form shown in Figure 9.1 stating that she is married and claims an allowance (exemption) of 2. Glo Company will need this information to calculate the federal income tax Ⓙ.

Before this pay period, Janet had earned $127,200 (48 weeks × $2,650 salary per week). Since Janet receives no overtime, her $2,650 salary per week represents her gross pay (pay before any deductions).

After this pay period, Janet has earned $129,850 ($127,200 + $2,650).

FIGURE 9.1

Employee's W-4 form

Form **W-4** Department of the Treasury Internal Revenue Service

Employee's Withholding Allowance Certificate

▶ Whether you are entitled to claim a certain number of allowances or exemption from withholding is subject to review by the IRS. Your employer may be required to send a copy of this form to the IRS.

OMB No. 1545-0074 **20XX**

1 Your first name and middle initial: Janet | Last name: Wong | 2 Your social security number: 987-65-4321

Home address (number and street or rural route): 1234 Rolling Hills Lane | 3 ☐ Single ☒ Married ☐ Married, but withhold at higher Single rate. Note. If married, but legally separated, or spouse is a nonresident alien, check the "Single" box.

City or town, state, and ZIP code: Reading, PA 19606 | 4 If your last name differs from that shown on your social security card, check here. You must call 1-800-772-1213 for a replacement card. ▶ ☐

5 Total number of allowances you are claiming (from line **H** above **or** from the applicable worksheet on page 2) 5 2

6 Additional amount, if any, you want withheld from each paycheck 6 $

7 I claim exemption from withholding for 2012, and I certify that I meet **both** of the following conditions for exemption.
- Last year I had a right to a refund of **all** federal income tax withheld because I had **no** tax liability, **and**
- This year I expect a refund of **all** federal income tax withheld because I expect to have **no** tax liability.

If you meet both conditions, write "Exempt" here . . . ▶ 7

Under penalties of perjury, I declare that I have examined this certificate and, to the best of my knowledge and belief, it is true, correct, and complete.

Employee's signature (This form is not valid unless you sign it.) ▶ Janet Wong **Date** ▶ 1/1/20XX

8 Employer's name and address (Employer: Complete lines 8 and 10 only if sending to the IRS.) | 9 Office code (optional) | 10 Employer identification number (EIN)

Cat. No. 10220Q

MONEY tips

Many people expect the new tax laws to reduce their federal tax burden. However, there may be a risk of not having enough money withheld throughout the year as many will receive more after-tax income, potentially leaving more money owed at tax time. Note: You may need to revise your W-4 to avoid this underpayment.

Used by permission of Cartoon Features Syndicate

The **Federal Insurance Contribution Act (FICA)** funds the **Social Security** program. The program includes Old Age and Disability, Medicare, Survivor Benefits, and so on. The FICA tax requires separate reporting for Social Security and **Medicare.** We will use the following rates for Glo Company:

	Rate	Base
Social Security	6.20%	$128,400
Medicare	1.45	No base

These rates mean that Janet Wong will pay Social Security taxes on the first $128,400 she earns this year. After earning $128,400, Janet's wages will be exempt from Social Security. Note that Janet will be paying Medicare taxes on all wages since Medicare has no base cutoff.

To help keep Glo's record straight, the *taxable earnings column only shows what wages will be taxed. This amount is not the tax.* For example, in week 49, only $1,200 of Janet's salary will be taxable for Social Security.

$128,400 Social Security base
– $127,200 Ⓑ
$ 1,200

To calculate Janet's Social Security tax, we multiply $1,200 Ⓕ by 6.2%:

$1,200 × .062 = $74.40

Ⓕ, Ⓖ—Taxable earnings for Social Security and Medicare

Since Medicare has no base, Janet's entire weekly salary is taxed 1.45%, which is multiplied by $2,650.

Ⓗ—Social Security

$2,650 × .0145 = $38.43

Ⓘ—Medicare

Ⓙ—FIT

Using the W-4 form Janet completed, Glo deducts **federal income tax withholding (FIT).** The more allowances an employee claims, the less money Glo deducts from the employee's paycheck. Glo uses the percentage method to calculate FIT.[2]

The Percentage Method[3] Today, since many companies do not want to store the tax tables, they use computers for their payroll. These companies use the **percentage method.** For this method we use Table 9.1 and Table 9.2 (page 256) from Circular E to calculate Janet's FIT.

Step 1. In Table 9.1, locate the weekly withholding for one allowance. Multiply this number by 2.

$79.80 × 2 = $159.60

TABLE 9.1

Percentage method income tax withholding allowances

Payroll Period	One Withholding Allowance
Weekly	$ 79.80
Biweekly	159.60
Semimonthly	172.90
Monthly	345.80
Quarterly	1,037.50
Semiannually	2,075.00
Annually	4,150.00
Daily or miscellaneous (each day of the payroll period)	16.00

[2]The *Business Math Handbook* has a sample of the wage bracket method.

[3]An alternative method is the wage bracket method shown in the *Business Math Handbook*.

TABLE 9.2 Percentage method income tax withholding schedules

Percentage Method Tables for Income Tax Withholding
(For Wages Paid in 2018)

TABLE 1—WEEKLY Payroll Period

(a) SINGLE person (including head of household)—

If the amount of wages (after subtracting withholding allowances) is: Not over $71 .... The amount of income tax to withhold is: $0

Over—	But not over—	The amount of income tax to withhold is:	of excess over—
$71	—$254	$0.00 plus 10%	—$71
$254	—$815	$18.30 plus 12%	—$254
$815	—$1,658	$85.62 plus 22%	—$815
$1,658	—$3,100	$271.08 plus 24%	—$1,658
$3,100	—$3,917	$617.16 plus 32%	—$3,100
$3,917	—$9,687	$878.60 plus 35%	—$3,917
$9,687	....	$2,898.10 plus 37%	—$9,687

(b) MARRIED person—

If the amount of wages (after subtracting withholding allowances) is: Not over $222 .... The amount of income tax to withhold is: $0

Over—	But not over—	The amount of income tax to withhold is:	of excess over—
$222	—$588	$0.00 plus 10%	—$222
$588	—$1,711	$36.60 plus 12%	—$588
$1,711	—$3,395	$171.36 plus 22%	—$1,711
$3,395	—$6,280	$541.84 plus 24%	—$3,395
$6,280	—$7,914	$1,234.24 plus 32%	—$6,280
$7,914	—$11,761	$1,757.12 plus 35%	—$7,914
$11,761	....	$3,103.57 plus 37%	—$11,761

TABLE 2—BIWEEKLY Payroll Period

(a) SINGLE person (including head of household)—

If the amount of wages (after subtracting withholding allowances) is: Not over $142 .... The amount of income tax to withhold is: $0

Over—	But not over—	The amount of income tax to withhold is:	of excess over—
$142	—$509	$0.00 plus 10%	—$142
$509	—$1,631	$36.70 plus 12%	—$509
$1,631	—$3,315	$171.34 plus 22%	—$1,631
$3,315	—$6,200	$541.82 plus 24%	—$3,315
$6,200	—$7,835	$1,234.22 plus 32%	—$6,200
$7,835	—$19,373	$1,757.42 plus 35%	—$7,835
$19,373	....	$5,795.72 plus 37%	—$19,373

(b) MARRIED person—

If the amount of wages (after subtracting withholding allowances) is: Not over $444 .... The amount of income tax to withhold is: $0

Over—	But not over—	The amount of income tax to withhold is:	of excess over—
$444	—$1,177	$0.00 plus 10%	—$444
$1,177	—$3,421	$73.30 plus 12%	—$1,177
$3,421	—$6,790	$342.58 plus 22%	—$3,421
$6,790	—$12,560	$1,083.76 plus 24%	—$6,790
$12,560	—$15,829	$2,468.56 plus 32%	—$12,560
$15,829	—$23,521	$3,514.64 plus 35%	—$15,829
$23,521	....	$6,206.84 plus 37%	—$23,521

TABLE 3—SEMIMONTHLY Payroll Period

(a) SINGLE person (including head of household)—

If the amount of wages (after subtracting withholding allowances) is: Not over $154 .... The amount of income tax to withhold is: $0

Over—	But not over—	The amount of income tax to withhold is:	of excess over—
$154	—$551	$0.00 plus 10%	—$154
$551	—$1,767	$39.70 plus 12%	—$551
$1,767	—$3,592	$185.62 plus 22%	—$1,767
$3,592	—$6,717	$587.12 plus 24%	—$3,592
$6,717	—$8,488	$1,337.12 plus 32%	—$6,717
$8,488	—$20,988	$1,903.84 plus 35%	—$8,488
$20,988	....	$6,278.84 plus 37%	—$20,988

(b) MARRIED person—

If the amount of wages (after subtracting withholding allowances) is: Not over $481 .... The amount of income tax to withhold is: $0

Over—	But not over—	The amount of income tax to withhold is:	of excess over—
$481	—$1,275	$0.00 plus 10%	—$481
$1,275	—$3,706	$79.40 plus 12%	—$1,275
$3,706	—$7,356	$371.12 plus 22%	—$3,706
$7,356	—$13,606	$1,174.12 plus 24%	—$7,356
$13,606	—$17,148	$2,674.12 plus 32%	—$13,606
$17,148	—$25,481	$3,807.56 plus 35%	—$17,148
$25,481	....	$6,724.11 plus 37%	—$25,481

TABLE 4—MONTHLY Payroll Period

(a) SINGLE person (including head of household)—

If the amount of wages (after subtracting withholding allowances) is: Not over $308 .... The amount of income tax to withhold is: $0

Over—	But not over—	The amount of income tax to withhold is:	of excess over—
$308	—$1,102	$0.00 plus 10%	—$308
$1,102	—$3,533	$79.40 plus 12%	—$1,102
$3,533	—$7,183	$371.12 plus 22%	—$3,533
$7,183	—$13,433	$1,174.12 plus 24%	—$7,183
$13,433	—$16,975	$2,674.12 plus 32%	—$13,433
$16,975	—$41,975	$3,807.56 plus 35%	—$16,975
$41,975	....	$12,557.56 plus 37%	—$41,975

(b) MARRIED person—

If the amount of wages (after subtracting withholding allowances) is: Not over $963 .... The amount of income tax to withhold is: $0

Over—	But not over—	The amount of income tax to withhold is:	of excess over—
$963	—$2,550	$0.00 plus 10%	—$963
$2,550	—$7,413	$158.70 plus 12%	—$2,550
$7,413	—$14,713	$742.26 plus 22%	—$7,413
$14,713	—$27,213	$2,348.26 plus 24%	—$14,713
$27,213	—$34,296	$5,348.26 plus 32%	—$27,213
$34,296	—$50,963	$7,614.82 plus 35%	—$34,296
$50,963	....	$13,448.27 plus 37%	—$50,963

Source: Internal Revenue Service

Step 2. Subtract $159.60 in Step 1 from Janet's total pay.

$$
\begin{array}{r}
\$2{,}650.00 \\
-\ \ 159.60 \\
\hline
\$2{,}490.40
\end{array}
$$

Step 3. In Table 9.2, locate the married person's weekly pay table. The $2,490.40 falls between $1,711 and $3,395. The tax is $171.36 plus 22% of the excess over $1,711.00.

$$
\begin{array}{r}
\$2{,}490.40 \\
-\ 1{,}711.00 \\
\hline
\$779.40
\end{array}
$$

Tax $171.36 + .22($779.40)

$171.36 + $171.47 = $342.83

We assume a 6% **state income tax (SIT).**

(K)—SIT

$2,650 × .06 = $159

(L)—Health insurance

(M)—Net pay

Janet contributes $100 per week for health insurance.
Janet's **net pay** is her gross pay less all deductions.

	$2,650.00	gross
−	74.40	Social Security
−	38.43	Medicare
−	342.83	FIT
−	159.00	SIT
−	100.00	health insurance
=	$1,935.34	net pay

©DW labs Incorporated/Shutterstock

Employers' Responsibilities

In the first section of this unit, we saw that Janet contributed to Social Security and Medicare. Glo Company has the legal responsibility to match her contributions. Besides matching Social Security and Medicare, Glo must pay two important taxes that employees do not have to pay—federal and state unemployment taxes.

LO 2

MONEY tips

The IRS has twelve items it looks for on a return that raise the chances of an audit from the average rate of 1.1%: making too much money; failing to report all taxable income; taking large charitable deductions; claiming the home office deduction; claiming rental losses; deducting business meals, travel, and entertainment; claiming 100% business use of a vehicle; writing off a loss for a hobby; running a cash business; failing to report a foreign bank account; engaging in currency transactions; and taking higher-than-average deductions.

Federal Unemployment Tax Act (FUTA) The federal government participates in a joint federal-state unemployment program to help unemployed workers. At this writing, employers pay the government a 6% **FUTA** tax on the first $7,000 paid to employees as wages during the calendar year. Any wages in excess of $7,000 per worker are exempt wages and are not taxed for FUTA. If the total cumulative amount the employer owes the government is less than $100, the employer can pay the liability yearly (end of January in the following calendar year). If the tax is greater than $100, the employer must pay it within a month after the quarter ends.

Companies involved in a state unemployment tax fund can usually take a 5.4% credit against their FUTA tax. *In reality, then, companies are paying .6% (.006) to the federal unemployment program.* In all our calculations, FUTA is .006.

EXAMPLE Assume a company had total wages of $19,000 in a calendar year. No employee earned more than $7,000 during the calendar year. The FUTA tax is .6% (6% minus the company's 5.4% credit for state unemployment tax). How much does the company pay in FUTA tax?

The company calculates its FUTA tax as follows:

	6%	FUTA tax
−	5.4%	credit for SUTA tax
=	.6%	tax for FUTA

.006 × $19,000 = $114 FUTA tax due to federal government

State Unemployment Tax Act (SUTA) The current **SUTA** tax in many states is 5.4% on the first $7,000 the employer pays an employee. Some states offer a merit rating system that results in a lower SUTA rate for companies with a stable employment period. The federal government still allows 5.4% credit on FUTA tax to companies entitled to the lower SUTA rate. Usually states also charge companies with a poor employment record a higher SUTA rate. However, these companies cannot take any more than the 5.4% credit against the 6% federal unemployment rate.

EXAMPLE Assume a company has total wages of \$20,000 and \$4,000 of the wages are exempt from SUTA. What are the company's SUTA and FUTA taxes if the company's SUTA rate is 5.8% due to a poor employment record?

The exempt wages (over \$7,000 earnings per worker) are not taxed for SUTA or FUTA. So the company owes the following SUTA and FUTA taxes:

\$20,000
− 4,000 (exempt wages)
\$16,000 × .058 = **\$928** SUTA

Federal FUTA tax would then be:
\$16,000 × .006 = **\$96**

You can check your progress with the following Practice Quiz.

LU 9–2 PRACTICE QUIZ

Complete this **Practice Quiz** to see how you are doing.

1. Calculate Social Security taxes, Medicare taxes, and FIT for Joy Royce. Joy's company pays her a monthly salary of \$9,500. She is single and claims 1 deduction. Let's assume that due to some overtime, Joy's cumulative earnings were \$115,000. (Social Security maximum is 6.2% on \$128,400, and Medicare is 1.45%.) Calculate FIT by the percentage method.
2. Jim Brewer, owner of Arrow Company, has three employees who earn \$300, \$700, and \$900 a week. Assume a state SUTA rate of 5.1%. What will Jim pay for state and federal unemployment taxes for the first quarter?

*For **extra help** from your authors–Sharon and Jeff–see the videos in Connect.*

✓ Solutions

1. **Social Security**

\$128,400
− 115,000
\$ 13,400 × .062 = **\$830.80**

Medicare

\$9,500 × .0145 = **\$137.75**

FIT

Percentage method: \$9,500.00
\$345.80 × 1 = − 345.80 (Table 9.1)
\$9,154.20

\$7,183 to \$13,433 → \$1,174.12 plus 24% of excess over \$7,183 (Table 9.2)

\$9,154.20
− 7,183.00
\$1,971.20 × .24 = \$ 473.09
+ 1,174.12
\$1,647.21

2. 13 weeks × \$300 = \$ 3,900
13 weeks × \$700 = 9,100 (\$9,100 − \$7,000) → \$2,100
13 weeks × \$900 = 11,700 (\$11,700 − \$7,000) → 4,700
\$24,700 \$6,800
(\$2,100 + 4,700 = \$6,800: Exempt wages (not taxed for FUTA or SUTA))

\$24,700 − \$6,800 = \$17,900 taxable wages
SUTA = .051 × \$17,900 = **\$912.90**
FUTA = .006 × \$17,900 = **\$107.40**

Note: FUTA remains at .006 whether SUTA rate is higher or lower than standard.

INTERACTIVE CHAPTER ORGANIZER

<table>
<tr><th>Topic/Procedure/Formula</th><th>Examples</th><th>You try it*</th></tr>
<tr><td>Gross pay
Hours employee worked × Rate per hour</td><td>$6.50 per hour at 36 hours
Gross pay = 36 × $6.50 = $234</td><td>Calculate gross pay
$9.25 per hour; 38 hours</td></tr>
<tr><td>Overtime
Gross earnings (pay) = Regular pay + Earnings at overtime rate ($1\frac{1}{2}$)</td><td>$6 per hour; 42 hours
Gross pay = (40 × $6) + (2 × $9)
= $240 + $18 = $258</td><td>Calculate gross pay
$7 per hour; 43 hours</td></tr>
<tr><td>Straight piece rate
Gross pay = Number of units produced × Rate per unit</td><td>1,185 units; rate per unit, $.89
Gross pay = 1,185 × $.89
= $1,054.65</td><td>Calculate gross pay
2,250 units; $.79 per unit</td></tr>
<tr><td>Differential pay schedule
Rate on each item is related to the number of items produced.</td><td>1–500 at $.84; 501–1,000 at $.96; 900 units produced.
Gross pay = (500 × $.84) + (400 × $.96)
= $420 + $384 = $804</td><td>Calculate gross pay
1–600 at $.79; 601–1,000 at $.88; 900 produced</td></tr>
<tr><td>Straight commission
Total sales × Commission rate
Any draw would be subtracted from earnings.</td><td>$155,000 sales; 6% commission
$155,000 × .06 = $9,300</td><td>Calculate straight commission
$175,000 sales; 7% commission</td></tr>
<tr><td>Variable commission scale
Sales at different levels pay different rates of commission.</td><td>Up to $5,000, 5%; $5,001 to $10,000, 8%; over $10,000, 10%
Sold: $6,500
Solution:
($5,000 × .05) + ($1,500 × .08)
= $250 + $120 = $370</td><td>Calculate commission
Up to $6,000, 5%; $6,001 to $8,000, 9%; Over $8,000, 12%
Sold: $12,000</td></tr>
<tr><td>Salary plus commission
Regular wages (fixed) + Commissions earned</td><td>Base $400 per week + 2% on sales over $14,000
Actual sales: $16,000
$400 (base) + (.02 × $2,000) = $440</td><td>Calculate gross pay
Base $600 per week plus 4% on sales over $16,000. Actual sales $22,000.</td></tr>
<tr><td>Payroll register
Multicolumn form to record payroll.
Married and paid weekly. (Table 9.2)
Claims 1 allowance.
FICA rates from chapter.</td><td><table><tr><th rowspan="3">Earnings
Gross</th><th colspan="3">Deductions</th><th rowspan="3">Net pay</th></tr><tr><th colspan="2">FICA</th><th></th></tr><tr><th>S.S.</th><th>Med.</th><th>FIT</th></tr><tr><td>1,515</td><td>93.93</td><td>21.97</td><td>138.26</td><td>$1,260.84</td></tr></table></td><td>Calculate net pay
Gross pay, $490; Married, paid weekly. Claims, one allowance. Use rates in text for Social Security, Medicare, and FIT.</td></tr>
<tr><td>FICA
Social Security Medicare
6.2% on $128,400 (S.S.)
1.45% (Med.)</td><td>If John earns $120,000, what did he contribute for the year to Social Security and Medicare?
S.S.: $120,000 × .062 = $7,440
Med.: $120,000 × .0145 = $1,740.00</td><td>Calculate FICA
If John earns $150,000, what did he contribute to Social Security and Medicare?</td></tr>
<tr><td>FIT calculation (percentage method)
Facts:
Al Doe: Married
Claims: 2
Paid weekly: $1,600</td><td>$1,600.00
− 159.60 ($79.80 × 2) Table 9.1
$1,440.40
By Table 9.2
$1,440.40
− 588.00
$ 852.40
$36.60 + .12($852.40)
$36.60 + $102.29 = $138.89</td><td>Calculate FIT
Jim Smith, married, claims 3; Paid weekly, $1,400</td></tr>
</table>

(continues)

INTERACTIVE CHAPTER ORGANIZER

Topic/Procedure/Formula	Examples	You try it*
State and federal unemployment Employer pays these taxes. Rates are 6% on \$7,000 for federal and 5.4% for state on \$7,000 (6% − 5.4% = .6% federal rate after credit). If state unemployment rate is higher than 5.4%, no additional credit is taken. If state unemployment rate is less than 5.4%, the full 5.4% credit can be taken for federal unemployment.	Cumulative pay before payroll, \$6,400; this week's pay, \$800. What are state and federal unemployment taxes for employer, assuming a 5.2% state unemployment rate? State → .052 × \$600 = \$31.20 Federal → .006 × \$600 = \$3.60 (\$6,400 + \$600 = \$7,000 maximum)	**Calculate SUTA and FUTA** Cumulative pay before payroll, \$6,800. This week's payroll, \$9,000. State rate is 5.4%.

KEY TERMS			
	Biweekly Deductions Differential pay schedule Draw Employee's Withholding Allowance Certificate (W-4) Fair Labor Standards Act Federal income tax withholding (FIT) Federal Insurance Contribution Act (FICA)	Federal Unemployment Tax Act (FUTA) Gross pay Medicare Monthly Net pay Overrides Overtime Payroll register Percentage method Semimonthly	Social Security State income tax (SIT) State Unemployment Tax Act (SUTA) Straight commission Variable commission scale W-4 Weekly

*Worked-out solutions are in Appendix B.

Critical Thinking Discussion Questions with Chapter Concept Check

1. Explain the difference between biweekly and semimonthly. Explain what problems may develop if a retail store hires someone on straight commission to sell cosmetics.
2. Explain what each column of a payroll register records and how each number is calculated. Social Security tax is based on a specific rate and base; Medicare tax is based on a rate but has no base. Do you think this is fair to all taxpayers?
3. What taxes are the responsibility of the employer? How can an employer benefit from a merit-rating system for state unemployment?
4. **Chapter Concept Check.** Visit the Starbucks website to see what benefits the company provides for its employees. Discuss the responsibilities of the employee and the employer.

9

END-OF-CHAPTER PROBLEMS

connect

Check figures for odd-numbered problems in Appendix B. Name ______________________ Date ________

DRILL PROBLEMS

Complete the following table: *LU 9-1(2)*

	Employee	M	T	W	Th	F	Hours	Rate per hour	Gross pay
9–1.	Bernie Roy	9	6	9	7	6		$8.95	
9–2.	Kristina Shaw	5	9	10	8	8		$8.10	

Complete the following table (assume the overtime for each employee is a time-and-a-half rate after 40 hours): *LU 9-1(2)*

	Employee	M	T	W	Th	F	Sa	Total regular hours	Total overtime hours	Regular rate	Overtime rate	Gross earnings
9–3.	Blue	12	9	9	9	9	3			$8.00		
9–4.	Tagney	14	8	9	9	5	1			$7.60		

Calculate gross earnings: *LU 9-1(3)*

	Worker	Number of units produced	Rate per unit	Gross earnings
9–5.	Lang	480	$3.50	
9–6.	Swan	846	$ .58	

Calculate the gross earnings for each apple picker based on the following differential pay scale: *LU 9-1(3)*

1–1,000: $.03 each 1,001–1,600: $.05 each Over 1,600: $.07 each

	Apple picker	Number of apples picked	Gross earnings
9–7.	Ryan	1,600	
9–8.	Rice	1,925	

Calculate the end-of-month commission. *LU 9-1(3)*

	Employee	Total sales	Commission rate	Draw	End-of-month commission received
9–9.	Reese	$300,000	7%	$8,000	

Ron Company has the following commission schedule:

Commission rate	Sales
2%	Up to $80,000
3.5%	Excess of $80,000 to $100,000
4%	More than $100,000

Calculate the gross earnings of Ron Company's two employees: *LU 9-1(3)*

	Employee	Total sales	Gross earnings
9–10.	Bill Moore	$ 70,000	
9–11.	Ron Ear	$155,000	

Complete the following table, given that A Publishing Company pays its salespeople a weekly salary plus a 2% commission on all net sales over $5,000 (no commission on returned goods): *LU 9-1(3)*

		Employee	Gross sales	Return	Net sales	Given quota	Commission sales	Commission rates	Total commission	Regular wage	Total wage
eXcel	**9–12.**	Ring	$ 8,000	$ 25		$5,000		2%		$250	
eXcel	**9–13.**	Porter	$12,000	$100		$5,000		2%		$250	

Calculate the Social Security and Medicare deductions for the following employees (assume a tax rate of 6.2% on $128,400 for Social Security and 1.45% for Medicare): *LU 9-2(1)*

	Employee	Cumulative earnings before this pay period	Pay amount this period	Social Security	Medicare
9–14.	Logan	$128,300	$3,000		
9–15.	Rouche	$122,300	$7,000		
9–16.	Cleaves	$400,000	$6,000		

Complete the following payroll register. Calculate FIT by the percentage method for this weekly period; Social Security and Medicare are the same rates as in the previous problems. No one will reach the maximum for FICA. *LU 9-2(1)*

	Employee	Marital status	Allowances claimed	Gross pay	FIT	FICA S.S.	FICA Med.	Net pay
9–17.	Mike Rice	M	2	$2,000				
9–18.	Pat Brown	M	4	$2,500				

9–19. Given the following, calculate the state (assume 5.3%) and federal unemployment taxes that the employer must pay for each of the first two quarters. The federal unemployment tax is .6% on the first $7,000. *LU 9-2(2)*

	PAYROLL SUMMARY	
	Quarter 1	Quarter 2
Bill Adams	$4,000	$ 8,000
Rich Haines	8,000	14,000
Alice Smooth	3,200	3,800

WORD PROBLEMS

9–20. Lai Xiaodong, a 22-year-old college-educated man, accepted a job at Foxconn Technology (where the iPad was being produced for Apple) in Chengdu, China, for $22 a day at 12 hours a day, 6 days a week. A company perk included company housing in dorms for the 70,000 employees. It was common for 20 people to be assigned to the same three-bedroom apartment. What were Lai's hourly (rounded to the nearest cent), weekly, and annual gross pay? *LU 9-1(1)*

My Money

9–21. Rhonda Brennan found her first job after graduating from college through the classifieds of the *Miami Herald.* She was delighted when the offer came through at $18.50 per hour. She completed her W-4 stating that she is married with a child and claims an allowance of 3. Her company will pay her biweekly for 80 hours. Calculate her take-home pay for her first check. *LU 9-2(1)*

My Money

9–22. The Social Security Administration increased the taxable wage base from $127,400 to $128,400. The 6.2% tax rate is unchanged. Joe Burns earned over $130,000 each of the past two years. **(a)** What is the percent increase in the base? Round to the nearest hundredth percent. **(b)** What is Joe's increase in Social Security tax for the new year? *LU 9-2(1)*

9–23. Calculate Social Security taxes, Medicare taxes, and FIT for Jordon Barrett. He earns a monthly salary of $12,000. He is single and claims 1 deduction. Before this payroll, Barrett's cumulative earnings were $128,000. (Social Security maximum is 6.2% on $128,400 and Medicare is 1.45%.) Calculate FIT by the percentage method. *LU 9-2(1)*

9–24. Maggie Vitteta, single, works 40 hours per week at $12.00 an hour. How much is taken out for federal income tax with one withholding exemption? *LU 9-2(1)*

9–25. Robin Hartman earns $600 per week plus 3% of sales over $6,500. Robin's sales are $14,000. How much does Robin earn? *LU 9-1(3)*

9–26. Pat Maninen earns a gross salary of $3,000 each week. What are Pat's first week's deductions for Social Security and Medicare? Will any of Pat's wages be exempt from Social Security and Medicare for the calendar year? Assume a rate of 6.2% on $128,400 for Social Security and 1.45% for Medicare. *LU 9-2(1)*

9–27. Richard Gaziano is a manager for Health Care, Inc. Health Care deducts Social Security, Medicare, and FIT (by percentage method) from his earnings. Assume the same Social Security and Medicare rates as in Problem 9–26. Before this payroll, Richard is $1,000 below the maximum level for Social Security earnings. Richard is married, is paid weekly, and claims 2 exemptions. What is Richard's net pay for the week if he earns $1,300? *LU 9-2(1)*

My Money

9–28. Larren Buffett is concerned after receiving her weekly paycheck. She believes that her deductions for Social Security, Medicare, and federal income tax withholding (FIT) may be incorrect. Larren is paid a salary of $4,100 weekly. She is married, claims 3 deductions, and prior to this payroll check, has total earnings of $128,245. What are the correct deductions for Social Security, Medicare, and FIT? *LU 9-2(2)*

excel

9–29. Westway Company pays Suzie Chan $3,000 per week. By the end of week 52, how much did Westway deduct for Suzie's Social Security and Medicare for the year? Assume Social Security is 6.2% on $128,400 and 1.45% for Medicare. What state and federal unemployment taxes does Westway pay on Suzie's yearly salary? The state unemployment rate is 5.1%. FUTA is .6%. *LU 9-2(1, 2)*

9–30. Sarah Jones earns $525 per week selling life insurance for Farmer's Insurance plus 5% of sales over $5,750. Sarah's sales this month (four weeks) are $20,000. How much does Sarah earn this month? *LU 9-2(2)* eXcel **9–31.** Tiffani Lind earned \$1,200 during her biweekly pay period. She is married and claims 4 deductions. Her annual earnings to date are \$52,521. Calculate her net pay. *LU 9-2(1)*

CHALLENGE PROBLEMS

9–32. The San Bernardino County Fair hires about 150 people during fair time. Their hourly wages range from \$6.75 to \$8.00. California has a state income tax of 9%. Sandy Denny earns \$8.00 per hour; George Barney earns \$6.75 per hour (assume this is the current minimum wage). They both worked 35 hours this week. Both are married; however, Sandy claims 2 exemptions and George claims 1 exemption. Assume a rate of 6.2% on \$128,400 for Social Security and 1.45% for Medicare. **(a)** What is Sandy's net pay after FIT (use the tables in the text), Social Security tax, state income tax, and Medicare have been taken out? **(b)** What is George's net pay after the same deductions? **(c)** How much more is Sandy's net pay versus George's net pay? Round to the nearest cent. *LU 9-2(1)*

9–33. Bill Rose is a salesperson for Boxes, Inc. He believes his \$1,460.47 monthly paycheck is in error. Bill earns a \$1,400 salary per month plus a 9.5% commission on sales over \$1,500. Last month, Bill had \$8,250 in sales. Bill believes his traveling expenses are 16% of his weekly gross earnings before commissions. Monthly deductions include Social Security, \$126.56; Medicare, \$29.60; FIT, \$189.50; union dues, \$25.00; and health insurance, \$16.99. Calculate the following: **(a)** Bill's monthly take-home pay, and indicate the amount his check was under- or overstated, and **(b)** Bill's weekly traveling expenses. Round your final answer to the nearest dollar. *LU 9-2(1)*

SUMMARY PRACTICE TEST

Do you need help? Connect videos have step-by-step worked-out solutions.

1. Calculate Sam's gross pay (he is entitled to time-and-a-half). *LU 9-1(2)*

M	T	W	Th	F	Total hours	Rate per hour	Gross pay
$9\frac{1}{4}$	$9\frac{1}{4}$	$10\frac{1}{2}$	$8\frac{1}{2}$	$11\frac{1}{2}$		\$8.00	

2. Mia Kaminsky sells shoes for Macy's. Macy's pays Mia \$12 per hour plus a 5% commission on all sales. Assume Mia works 37 hours for the week and has \$7,000 in sales. What is Mia's gross pay? *LU 9-1(3)*

3. Lee Company pays its employees on a graduated commission scale: 6% on the first \$40,000 sales, 7% on sales from \$40,001 to \$80,000, and 13% on sales of more than \$80,000. May West, an employee of Lee, has \$230,000 in sales. What commission did May earn? *LU 9-1(3)*

4. Matty Kim, an accountant for Vernitron, earned \$120,900 from January to June. In July, Matty earned \$20,000. Assume a tax rate of 6.2% for Social Security on \$128,400 and 1.45% on Medicare. How much are the July taxes for Social Security and Medicare? *LU 9-2(1)*

5. Grace Kelley earns \$2,000 per week. She is married and claims 2 exemptions. What is Grace's income tax? Use the percentage method. *LU 9-2(1)*

6. Jean Michaud pays his two employees \$900 and \$1,200 per week. Assume a state unemployment tax rate of 5.7% and a federal unemployment tax rate of .6%. What state and federal unemployment taxes will Jean pay at the end of quarter 1 and quarter 2? *LU 9-2(2)*

INTERACTIVE VIDEO WORKSHEET

GRADING THE SUMMARY PRACTICE TEST

Go to the summary practice test video in Connect (or click on it here in the ebook). Grade your summary practice test while viewing the video.

C for Correct/I for Incorrect

1. ________ 4. ________
2. ________ 5. ________
3. ________ 6. ________

If you achieved 100%, you are ready for your instructor's exam.

If any of the problems were incorrect, list the questions you missed and show steps to solve the problem correctly.

Replay the video to see if you have made the correct fixes to your mistakes.
If you have any questions, contact your instructor asap.

Notes on Watching Videos

MY MONEY

My Paycheck—Where Is My Money Going?

What I need to know

Whether you are working in the payroll department or not, it is very important to understand how your pay is calculated and how deductions are made to arrive at your net pay. This will allow you to more effectively budget since you will be aware of your actual net pay, and you will know how much you have to spend and invest. As you enter the job market, it is also a good idea to understand how the tax brackets will impact your future earnings compared to the earnings you made during college. As your income changes, you may need to make adjustments to your withholdings in order to stay inline with your financial goals and tax responsibilities.

Insurance cost and other deductions are also important to take into consideration as they will have a direct impact on your net pay. Many employers will have a selection of insurance plans to choose from with differing costs and financial responsibilities. These are decisions not to be taken lightly and need your serious attention to ensure you are adequately covered and possess the understanding of the impact on your net pay. What is the cost to the employee for each insurance option offered? What will be the expense to the employee when health services are needed? How will you budget for out of pocket expenses such as medications, co-payments, and recurring treatments? These, and other considerations such as savings and investing, are critical parts of building financial health.

What I need to do

When considering your future career pay, you will need to consider the tax implications of your earnings. By taking the time to determine the financial impact of your tax responsibility, you will be able to better prepare yourself for future budgeting and stay on track with your financial goals. As you transition to a higher pay rate after college, the deductions from your paycheck become more pronounced than on your college job earnings. Understanding these calculations early on is crucial to evaluating job offers when it comes to salary since you will gain a better understanding of the net pay you will ultimately receive. This will allow you to determine if the salary amount under consideration will be enough to meet your expenses and provide opportunities for investments.

Ask for assistance from your employer in regards to the insurance options they offer to fully understand the impact this expense will have on your net pay. Additionally, there may be ancillary services offered by your employer to help with the insurance costs you will incur. Some employers will make flexible spending accounts available to you in order to offset the out of pocket costs you incur when utilizing health services. It is important for you to understand how these programs work and whether or not you feel you could benefit from utilizing these services. Be sure to note if the flexible spending account is a "use it or lose it" program to ensure you are using the funds you have set aside during the year and not missing out on the benefits of the program.

Resources I can use

- PaycheckFree (mobile app)
- https://www.thebalance.com/understanding-your-paycheck-withholdings-2386382 — understanding your paycheck

MY MONEY ACTIVITY

Use your expected career pay from the Chapter 2 My Money segment. Identify your FIT, Social Security, and Medicare deductions to arrive at your estimated net pay.

PERSONAL FINANCE

A KIPLINGER APPROACH

"The New Tax Law: What You Need to Know" by Sandra Block from *Kiplinger's*, March 2018, p. 35. Used by permission of The Kiplinger Washington Editors, Inc.

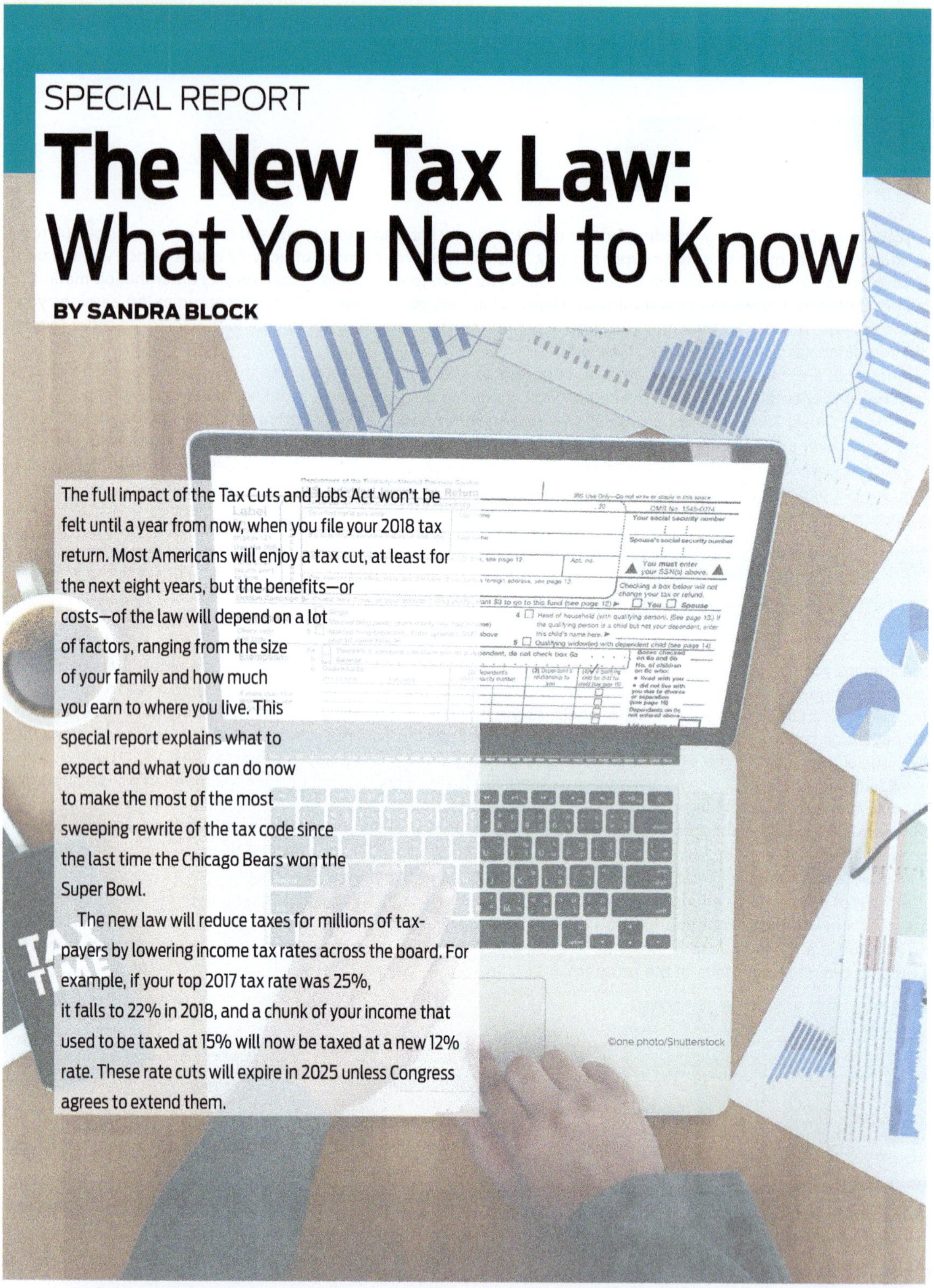

SPECIAL REPORT

The New Tax Law: What You Need to Know

BY SANDRA BLOCK

The full impact of the Tax Cuts and Jobs Act won't be felt until a year from now, when you file your 2018 tax return. Most Americans will enjoy a tax cut, at least for the next eight years, but the benefits—or costs—of the law will depend on a lot of factors, ranging from the size of your family and how much you earn to where you live. This special report explains what to expect and what you can do now to make the most of the most sweeping rewrite of the tax code since the last time the Chicago Bears won the Super Bowl.

The new law will reduce taxes for millions of taxpayers by lowering income tax rates across the board. For example, if your top 2017 tax rate was 25%, it falls to 22% in 2018, and a chunk of your income that used to be taxed at 15% will now be taxed at a new 12% rate. These rate cuts will expire in 2025 unless Congress agrees to extend them.

©one photo/Shutterstock

BUSINESS MATH ISSUE

The new tax law has little effect on employees and employers.

1. List the key points of the article and information to support your position.
2. Write a group defense of your position using math calculations to support your view. If you are in an online course, post to a discussion board.

Classroom Notes

CHAPTER 10

Simple Interest

High Debt Loads Shake Retailers

By Miriam Gottfried

Toys "R" Us Inc. will get a new website this summer as it struggles to compete online with cash-rich rivals. The toy retailer, laden with $5 billion in debt, has spent $100 million over the past several years to help boost its online sales.

That isn't likely to stop the toy chain from falling further behind its rivals. **Wal-Mart Stores** Inc. bought Jet.com last August for $3.3 billion and since then has purchased three more online retailers. **Target** Corp. is investing billions to lower prices and improve online sales.

Toys "R" Us is one of many retailers fighting to keep up. High debt loads, nervous lenders and falling sales make it impossible to invest enough to compete online. And when companies do generate online sales, the margins are so tight that future investments are harder to make.

©photocritical/Shutterstock

LEARNING UNIT OBJECTIVES

LU 10–1: Calculation of Simple Interest and Maturity Value

1. Calculate simple interest and maturity value for months and years.
2. Calculate simple interest and maturity value by **(a)** exact interest and **(b)** ordinary interest.

LU 10–2: Finding Unknown in Simple Interest Formula

1. Using the interest formula, calculate the unknown when the other two (principal, rate, or time) are given.

LU 10–3: U.S. Rule—Making Partial Note Payments before Due Date

1. List the steps to complete the U.S. Rule as well as calculate proper interest credits.

Your Guide to Successfully Completing This Chapter

Traditional book or ebook

Check box as you complete each step.

Steps

☐ Read learning unit.
  ☐ Complete practice quiz at the end of the learning unit.
☐ Grade practice quiz using provided solutions. (For more help, watch the learning unit video in Connect and have a Study Session with the authors. Then complete the additional practice quiz in Connect.)
☐ Repeat above for each of the three learning units in Chapter 10.
  ☐ Review chapter organizer.
  ☐ Complete assigned homework.
    ☐ Finish summary practice test. (Go to Connect via the ebook link and do the interactive video worksheet to grade.)
☐ Complete instructor's exam.

LO 1

Auto Lenders Get a New Reason to Worry

By AnnaMaria Andriotis

For auto lenders, there is trouble on the used-car lot.

Several large companies have warned that prices of used vehicles are likely to weaken, potentially leading to higher losses on loans on which cars are the collateral. That, combined with looser terms for loans and the growth of loans going to subprime borrowers, is sounding a warning for the long credit boom that has spurred auto sales.

The new tax act changes how much interest can be deducted by companies. The new law creates a cap on the interest deductions. Be sure to check with the IRS for specific details.

The chapter opening clip "High Debt Loads Shake Retailers" discusses how Toys "R" Us tried to stay in business. It did not work. In March 2018 Toys "R" Us closed all its stores.

In this chapter, you will study simple interest. The principles discussed apply whether you are paying interest or receiving interest. Note in the *Wall Street Journal* clip to the left how auto lenders can be affected by used car prices. Let's begin by learning how to calculate simple interest.

Learning Unit 10–1: Calculation of Simple Interest and Maturity Value

Hope Slater, a young attorney, rented an office in a professional building. Since Hope recently graduated from law school, she was short of cash. To purchase office furniture for her new office, Hope went to her bank and borrowed $40,000 for 6 months at a 4% annual interest rate. **Interest** expense is the cost of borrowing money.

The original amount Hope borrowed ($40,000) is the **principal** (face value) of the loan. Hope's price for using the $40,000 is the interest rate (4%) the bank charges on a yearly basis. Since Hope is borrowing the $40,000 for 6 months, Hope's loan will have a **maturity value** of $40,800—the principal

My Money

plus the interest on the loan. Thus, Hope's price for using the furniture before she can pay for it is \$800 interest, which is a percent of the principal for a specific time period. To make this calculation, we use the following formula:

$$\text{Maturity value } (MV) = \text{Principal } (P) + \text{Interest } (I)$$

$$\$40{,}800 = \$40{,}000 + \$800$$

Hope's furniture purchase introduces **simple interest**—the cost of a loan, usually for 1 year or less. Simple interest is only on the original principal or amount borrowed. Let's examine how the bank calculated Hope's \$800 interest.

Simple Interest Formula

To calculate simple interest, we use the following **simple interest formula:**

$$\text{Simple interest } (I) = \text{Principal } (P) \times \text{Rate } (R) \times \text{Time } (T)$$

In this formula, rate is expressed as a decimal, fraction, or percent; and time is expressed in years or a fraction of a year. In the *Wall Street Journal* clip to the lower left, note the low interest rates paid on interest-bearing deposits. In 2018 rates have increased almost 2 percent. Further rate increases are expected.

Do not round intermediate answers. Round only the final calculation.

In your calculator, multiply \$40,000 times .04 times 6. Divide your answer by 12. You could also use the % key—multiply \$40,000 times 4% times 6 and then divide your answer by 12.

EXAMPLE Hope Slater borrowed \$40,000 for office furniture. The loan was for 6 months at an annual interest rate of 4%. What are Hope's interest and maturity value?

Using the simple interest formula, the bank determined Hope's interest as follows:

Step 1. Calculate the interest.

$$I = \underset{(P)}{\$40{,}000} \times \underset{(R)}{.04} \times \underset{(T)}{\frac{6}{12}} = \$800$$

Step 2. Calculate the maturity value.

$$MV = \underset{(P)}{\$40{,}000} + \underset{(I)}{\$800} = \$40{,}800$$

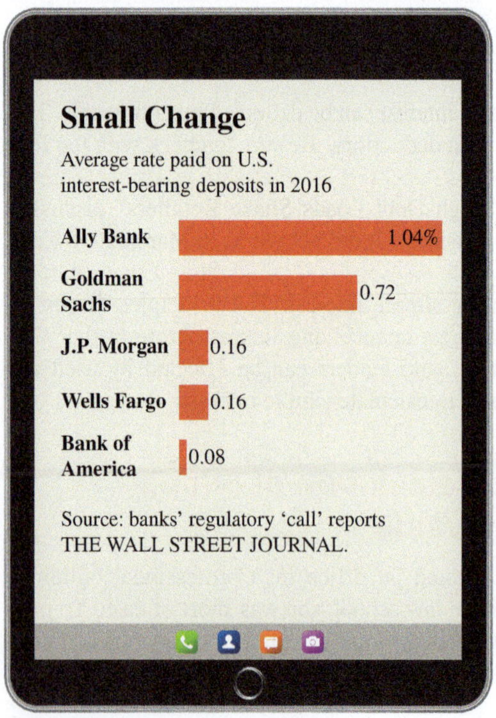

Now let's use the same example and assume Hope borrowed \$40,000 for 1 year. The bank would calculate Hope's interest and maturity value as follows:

Step 1. Calculate the interest.

$$I = \underset{(P)}{\$40{,}000} \times \underset{(R)}{.04} \times \underset{(T)}{1 \text{ year}} = \$1{,}600$$

Step 2. Calculate the maturity value.

$$MV = \underset{(P)}{\$40{,}000} + \underset{(I)}{\$1{,}600} = \$41{,}600$$

Let's use the same example again and assume Hope borrowed \$40,000 for 18 months[1]. Then Hope's interest and maturity value would be calculated as follows:

Step 1. Calculate the interest.

$$I = \underset{(P)}{\$40{,}000} \times \underset{(R)}{.04} \times \underset{(T)}{\frac{18}{12}} = \$2{,}400$$

Step 2. Calculate the maturity value.

$$MV = \underset{(P)}{\$40{,}000} + \underset{(I)}{\$2{,}400} = \$42{,}400$$

Next we'll turn our attention to two common methods we can use to calculate simple interest when a loan specifies its beginning and ending dates.

[1]This is the same as 1.5 years.

LO 2

Two Methods for Calculating Simple Interest and Maturity Value

Method 1: Exact Interest (365 Days) The Federal Reserve banks and the federal government use the **exact interest** method. The *exact interest* is calculated by using a 365-day year. For **time,** we count the exact number of days in the month that the borrower has the loan. The day the loan is made is not counted, but the day the money is returned is counted as a full day. This method calculates interest by using the following fraction to represent time in the formula:

$$\text{Time} = \frac{\text{Exact number of days}}{365} \leftarrow \text{Exact interest}$$

For this calculation, we use the exact days-in-a-year calendar from the *Business Math Handbook.* You learned how to use this calendar in Chapter 7.

From the *Business Math Handbook*

July 6	187th day
March 4	− 63rd day
	124 days (exact time of loan)
March	31
	− 4
	27
April	30
May	31
June	30
July	+ 6
	124 days

EXAMPLE On March 4, Joe Bench borrowed \$50,000 at 5% interest. Interest and principal are due on July 6. What are the interest cost and the maturity value?

Step 1. Calculate the interest.

$$I = P \times R \times T$$
$$= \$50{,}000 \times .05 \times \frac{124}{365}$$
$$= \$849.32 \text{ (rounded to nearest cent)}$$

Step 2. Calculate the maturity value.

$$MV = P + I$$
$$= \$50{,}000 + \$849.32$$
$$= \$50{,}849.32$$

Method 2: Ordinary Interest (360 Days) In the **ordinary interest** method, time in the formula $I = P \times R \times T$ is equal to the following:

$$\text{Time} = \frac{\text{Exact number of days}}{360} \leftarrow \text{Ordinary interest}$$

MONEY tips

Because debit cards draw money directly from your checking account, it is critical that you protect your account. Be cautious of the ATMs you use. Any outdoor transaction is at risk if the public has access to the machine—even at gas stations. Skimming devices can easily be used to pick up your information and leave your account at risk. Never use an ATM that has been tampered with.

Since banks commonly use the ordinary interest method, it is known as the **Banker's Rule.** Banks charge a slightly higher rate of interest because they use 360 days instead of 365 in the denominator. (Here's a hint: The word *ordinary* starts with an "O" and "360" ends with a "0.") By using 360 instead of 365, the calculation is supposedly simplified. Consumer groups, however, are questioning why banks can use 360 days, since this benefits the bank and not the customer. The use of computers and calculators no longer makes the simplified calculation necessary. For example, after a court case in Oregon, banks began calculating interest on 365 days except in mortgages.

Now let's replay the Joe Bench example we used to illustrate Method 1 to see the difference in bank interest when we use Method 2.

EXAMPLE On March 4, Joe Bench borrowed \$50,000 at 5% interest. Interest and principal are due on July 6. What are the interest cost and the maturity value?

Step 1. Calculate the interest.

$$I = \$50{,}000 \times .05 \times \frac{124}{360}$$
$$= \$861.11$$

Step 2. Calculate the maturity value.

$$MV = P + I$$
$$= \$50{,}000 + \$861.11$$
$$= \$50{,}861.11$$

Note: By using Method 2, the bank increases its interest by \$11.79.

$$\begin{array}{rl} \$861.11 & \leftarrow \text{Method 2} \\ -\ 849.32 & \leftarrow \text{Method 1} \\ \hline \$\ 11.79 & \end{array}$$

Aha!

Use ordinary interest any time a problem does not specify to use exact interest.

Now you should be ready for your first Practice Quiz in this chapter.

LU 10–1 PRACTICE QUIZ

Complete this Practice Quiz to see how you are doing.

Calculate simple interest (rounded to the nearest cent):

1. \$14,000 at 4% for 9 months
2. \$25,000 at 7% for 5 years
3. \$40,000 at $10\frac{1}{2}$% for 19 months
4. On May 4, Dawn Kristal borrowed \$15,000 at 8%. Dawn must pay the principal and interest on August 10. What are Dawn's simple interest and maturity value if you use the exact interest method?
5. What are Dawn Kristal's (Problem 4) simple interest and maturity value if you use the ordinary interest method?

*For **extra help** from your authors–Sharon and Jeff–see the videos in Connect.*

✓ Solutions

1. $\$14{,}000 \times .04 \times \frac{9}{12} = \$420$
2. $\$25{,}000 \times .07 \times 5 = \$8{,}750$
3. $\$40{,}000 \times .105 \times \frac{19}{12} = \$6{,}650$
4. August 10 → 222; May 4 → − 124; 98

   $\$15{,}000 \times .08 \times \frac{98}{365} = \$322.19$

   $MV = \$15{,}000 + \$322.19 = \$15{,}322.19$
5. $\$15{,}000 \times .08 \times \frac{98}{360} = \$326.67$

   $MV = \$15{,}000 + \$326.67 = \$15{,}326.67$

Learning Unit 10–2: Finding Unknown in Simple Interest Formula

LO 1

This unit begins with the formula used to calculate the principal of a loan. Then it explains how to find the *principal, rate,* and *time* of a simple interest loan. In all the calculations, we use 360 days and round only final answers.

Finding the Principal

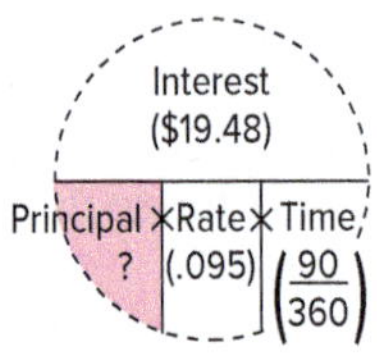

EXAMPLE Tim Jarvis paid the bank \$19.48 interest at 9.5% for 90 days. How much did Tim borrow using the ordinary interest method?

The following formula is used to calculate the principal of a loan:

$$\text{Principal} = \frac{\text{Interest}}{\text{Rate} \times \text{Time}}$$

Note how we illustrated this in the margin. The shaded area is what we are solving for. When solving for principal, rate, or time, you are dividing. Interest will be in the numerator, and the denominator will be the other two elements multiplied by each other.

Step 1. Set up the formula. $P = \dfrac{\$19.48}{.095 \times \frac{90}{360}}$

Step 1. When using a calculator, press .095 × 90 ÷ 360 M+.

Step 2. Multiply the denominator. .095 times 90 divided by 360 (do not round)

$P = \dfrac{\$19.48}{.02375}$

Step 2. When using a calculator, press 19.48 ÷ MR =.

Step 3. Divide the numerator by the result of Step 2. $P = \$820.21$

Step 4. Check your answer.

$$\underset{(I)}{\$19.48} = \underset{(P)}{\$820.21} \times \underset{(R)}{.095} \times \underset{(T)}{\frac{90}{360}}$$

Finding the Rate

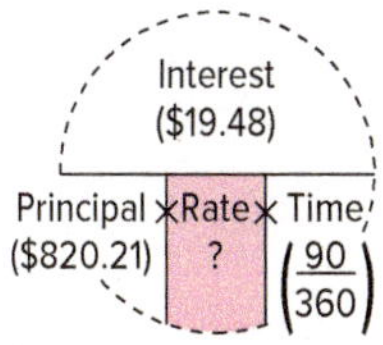

EXAMPLE Tim Jarvis borrowed $820.21 from a bank. Tim's interest is $19.48 for 90 days. What rate of interest did Tim pay using the ordinary interest method?

The following formula is used to calculate the rate of interest:

$$\text{Rate} = \frac{\text{Interest}}{\text{Principal} \times \text{Time}}$$

Step 1. Set up the formula. $R = \dfrac{\$19.48}{\$820.21 \times \dfrac{90}{360}}$

Step 2. Multiply the denominator. Do not round the answer. $R = \dfrac{\$19.48}{\$205.0525}$

Step 3. Divide the numerator by the result of Step 2. $R = 9.5\%$

Step 4. Check your answer. $\underset{(I)}{\$19.48} = \underset{(P)}{\$820.21} \times \underset{(R)}{.095} \times \underset{(T)}{\dfrac{90}{360}}$

MONEY tips

For each checking account you have, make certain to apply for overdraft protection to protect your account from human error and unintentional overdrafts.

Finding the Time

EXAMPLE Tim Jarvis borrowed $820.21 from a bank. Tim's interest is $19.48 at 9.5%. How much time does Tim have to repay the loan using the ordinary interest method?

The following formula is used to calculate time:

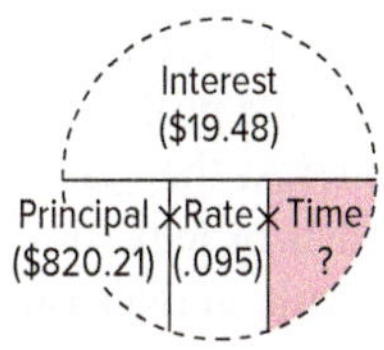

$$\text{Time (in years)} = \frac{\text{Interest}}{\text{Principal} \times \text{Rate}}$$

*Time is **always** over what makes up one year:* $\dfrac{\#\text{ of days}}{360 \text{ or } 365}, \dfrac{\#\text{ of weeks}}{52}, \dfrac{\#\text{ of months}}{12}, \dfrac{\#\text{ of quarters}}{4}.$

Step 1. Set up the formula. $T = \dfrac{\$19.48}{\$820.21 \times .095}$

Step 2. Multiply the denominator. Do not round the answer. $T = \dfrac{\$19.48}{\$77.91995}$

Step 3. Divide the numerator by the result of Step 2. $T = .25$ years

Step 4. Convert years to days (assume 360 days). $.25 \times 360 = 90$ days

Step 5. Check your answer. $\underset{(I)}{\$19.48} = \underset{(P)}{\$820.21} \times \underset{(R)}{.095} \times \underset{(T)}{\dfrac{90}{360}}$

Step 1. When using a calculator, press

[820.21] [×] [.095] [M+]

Step 2. When using a calculator, press

[19.48] [÷] [MR] [=]

Whole numbers in time represent full years. No adjustment to the time calculation is needed. However, when the calculation for number of days includes a decimal, multiply the decimal by 360 or 365 as indicated, to calculate the number of days the decimal represents in a year.

EXAMPLE $T = .37$ $.37 \times 365 = 135.05 = 136$ days

When dealing with a fraction (or part) of a day, always round up to a full day even if the number being rounded is less than 5.

Before we go on to Learning Unit 10–3, let's check your understanding of this unit.

LU 10–2 PRACTICE QUIZ

Complete this **Practice Quiz** to see how you are doing.

Complete the following (assume 360 days):

	Principal	Interest rate	Time (days)	Simple interest
1.	?	5%	90 days	$8,000
2.	$7,000	?	220 days	$350
3.	$1,000	8%	?	$300

*For **extra help** from your authors–Sharon and Jeff–see the videos in Connect.*

✓ Solutions

1. $\dfrac{\$8{,}000}{.05 \times \dfrac{90}{360}} = \dfrac{\$8{,}000}{.0125} = \$640{,}000$ $\qquad P = \dfrac{I}{R \times T}$

2. $\dfrac{\$350}{\$7{,}000 \times \dfrac{220}{360}} = \dfrac{\$350}{\$4{,}277.7777} = 8.18\%$ $\qquad R = \dfrac{I}{P \times T}$

(do not round) → $4,277.7777

3. $\dfrac{\$300}{\$1{,}000 \times .08} = \dfrac{\$300}{\$80} = 3.75 \times 360 = 1{,}350 \text{ days}$ $\qquad T = \dfrac{I}{P \times R}$

Learning Unit 10–3: U.S. Rule—Making Partial Note Payments before Due Date

LO 1

Often a person may want to pay off a debt in more than one payment before the maturity date. The **U.S. Rule** allows the borrower to receive proper interest credits. This rule states that any partial loan payment first covers any interest that has built up. The remainder of the partial payment reduces the loan principal. Courts or legal proceedings generally use the U.S. Rule. The Supreme Court originated the U.S. Rule in the case of *Story* v. *Livingston.*

My Money

EXAMPLE Jeff Edsell owes $5,000 on a 4%, 90-day note. On day 50, Jeff pays $600 on the note. On day 80, Jeff makes an $800 additional payment. Assume a 360-day year. What is Jeff's adjusted balance after day 50 and after day 80? What is the ending balance due?

To calculate $600 payment on day 50:

Step 1. Calculate interest on principal from date of loan to date of first principal payment. Round to nearest cent.

$I = P \times R \times T$

$I = \$5{,}000 \times .04 \times \dfrac{50}{360}$

$I = \$27.78$

Step 2. Apply partial payment to interest due. Subtract remainder of payment from principal. This is the **adjusted balance** (principal).

$600.00 payment
− 27.78 interest
$572.22

$5,000.00 principal
− 572.22
$4,427.78 adjusted balance—principal

To calculate $800 payment on day 80:

Step 3. Calculate interest on adjusted balance that starts from previous payment date and goes to new payment date. Then apply Step 2.

Compute interest on $4,427.78 for 30 days (80 − 50)

$I = \$4{,}427.78 \times .04 \times \dfrac{30}{360}$

$I = \$14.76$

$800.00 payment
− 14.76 interest
$785.24

$4,427.78
− 785.24
$3,642.54 adjusted balance

©Image Source/Getty Images

Step 4. At maturity, calculate interest from last partial payment. *Add* this interest to adjusted balance.

Ten days are left on note since last payment.

$$I = \$3{,}642.54 \times .04 \times \frac{10}{360}$$

$$I = \$4.05$$

$$\text{Balance owed} = \$3{,}646.59 \quad \begin{pmatrix} \$3{,}642.54 \\ + \quad 4.05 \end{pmatrix}$$

MONEY tips

Pay off debt instead of moving it around unless you have been offered 0% interest. Be wary of companies offering to consolidate your debt into a single loan. If you do, be certain to read and understand all the terms.

Note that when Jeff makes two partial payments, Jeff's total interest is \$46.59 (\$27.78 + \$14.76 + \$4.05). If Jeff had repaid the entire loan after 90 days, his interest payment would have been \$50—a total savings of \$3.41.

Let's check your understanding of the last unit in this chapter.

LU 10–3 PRACTICE QUIZ

Complete this Practice Quiz to see how you are doing.

Polly Flin borrowed \$5,000 for 60 days at 8%. On day 10, Polly made a \$600 partial payment. On day 40, Polly made a \$1,900 partial payment. What is Polly's ending balance due under the U.S. Rule (assuming a 360-day year)?

*For **extra help** from your authors–Sharon and Jeff–see the videos in Connect.*

✓ Solutions

$$\$5{,}000 \times .08 \times \frac{10}{360} = \$11.11$$

$$\begin{array}{r} \$600.00 \\ -\quad 11.11 \\ \hline \$588.89 \end{array} \qquad \begin{array}{r} \$5{,}000.00 \\ -\quad 588.89 \\ \hline \$4{,}411.11 \end{array}$$

$$\$4{,}411.11 \times .08 \times \frac{30}{360} = \$29.41$$

$$\begin{array}{r} \$1{,}900.00 \\ -\quad 29.41 \\ \hline \$1{,}870.59 \end{array} \qquad \begin{array}{r} \$4{,}411.11 \\ -\ 1{,}870.59 \\ \hline \$2{,}540.52 \end{array}$$

$$\$2{,}540.52 \times .08 \times \frac{20}{360} = \$11.29$$

$$\begin{array}{r} \$\quad 11.29 \\ +\ 2{,}540.52 \\ \hline \$2{,}551.81 \end{array}$$

INTERACTIVE CHAPTER ORGANIZER

Topic/Procedure/Formula	Examples	You try it*
Simple interest for months Interest = Principal × Rate × Time (I) (P) (R) (T)	\$2,000 at 9% for 17 months $I = \$2{,}000 \times .09 \times \frac{17}{12}$ $I = \$255$	**Calculate simple interest** \$4,000 at 3% for 18 months
Exact interest $T = \frac{\text{Exact number of days}}{365}$ $I = P \times R \times T$	\$1,000 at 10% from January 5 to February 20 $I = \$1{,}000 \times .10 \times \frac{46}{365}$ Feb. 20: 51 days Jan. 5: − 5 46 days $I = \$12.60$	**Calculate exact interest** \$3,000 at 4% from January 8 to February 22
Ordinary interest (Banker's Rule) $T = \frac{\text{Exact number of days}}{360}$ $I = P \times R \times T$ Higher interest costs	$I = \$1{,}000 \times .10 \times \frac{46}{360}$ (51 − 5) $I = \$12.78$	**Calculate ordinary interest** \$3,000 at 4% from January 8 to February 22

(continues)

INTERACTIVE CHAPTER ORGANIZER

Topic/Procedure/Formula	Examples	You try it*
Finding unknown in simple interest formula (use 360 days) $I = P \times R \times T$	Use this example for illustrations of simple interest formula parts: $1,000 loan at 9%, 60 days $I = \$1{,}000 \times .09 \times \frac{60}{360} = \$15$	**Calculate interest (use 360 days)** $2,000 loan at 4%, 90 days
Finding the principal $P = \frac{I}{R \times T}$ 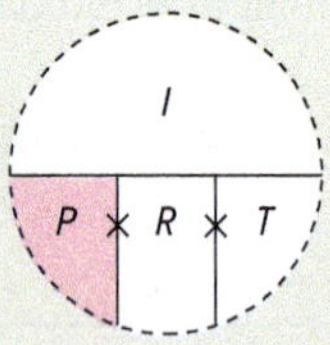	$P = \frac{\$15}{.09 \times \frac{60}{360}} = \frac{\$15}{.015} = \$1{,}000$	**Calculate principal** *Given:* interest, $20; rate, 4%; 90 days
Finding the rate $R = \frac{I}{P \times T}$ 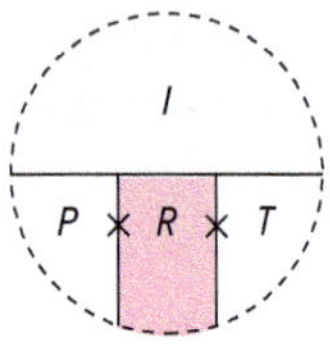	$R = \frac{\$15}{\$1{,}000 \times \frac{60}{360}} = \frac{\$15}{166.66666} = .09$ $= 9\%$ *Note*: We did not round the denominator.	**Calculate rate** *Given:* interest, $20; principal, $2,000; 90 days
Finding the time $T = \frac{I}{P \times R}$ (in years) 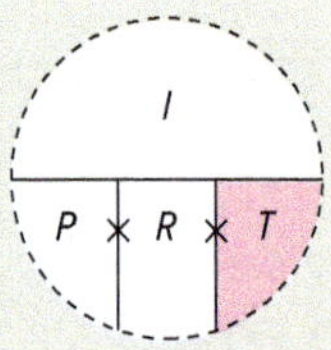 Multiply answer by 360 days to convert answer to days for ordinary interest.	$T = \frac{\$15}{\$1{,}000 \times .09} = \frac{\$15}{\$90} = .1666666$ $.1666666 \times 360 = 59.99 = 60$ days	**Calculate number of days** *Given:* principal, $2,000; rate, 4%; interest, $20
U.S. Rule (use 360 days) Calculate interest on principal from date of loan to date of first partial payment. Calculate adjusted balance by subtracting from principal the partial payment less interest cost. The process continues for future partial payments with the adjusted balance used to calculate cost of interest from last payment to present payment.	12%, 120 days, $2,000 *Partial payments:* On day 40: $250 On day 60: $200 *First payment:* $I = \$2{,}000 \times .12 \times \frac{40}{360}$ $I = \$26.67$ $250.00 payment − 26.67 interest $223.33 $2,000.00 principal − 223.33 $1,776.67 adjusted balance *Second payment:* $I = \$1{,}776.67 \times .12 \times \frac{20}{360}$ $I = \$11.84$ $200.00 payment − 11.84 interest $188.16 $1,776.67 − 188.16 $1,588.51 adjusted balance	**Calculate balance due and total interest** *Given:* $4,000; 4%; 90 days *Partial payments:* On day 30: $400 On day 70: $300

INTERACTIVE CHAPTER ORGANIZER

Topic/Procedure/Formula	Examples	You try it*
Balance owed equals last adjusted balance plus interest cost from last partial payment to final due date.	*60 days left:* $\$1{,}588.51 \times .12 \times \frac{60}{360} = \$31.77$ $\$1{,}588.51 + \$31.77 = $ **$1,620.28 balance due** Total interest = $26.67 11.84 + 31.77 $70.28	

KEY TERM			
	Adjusted balance Banker's Rule Exact interest Interest	Maturity value Ordinary interest Principal Simple interest	Simple interest formula Time U.S. Rule

*Worked-out solutions are in Appendix B.

Critical Thinking Discussion Questions with Chapter Concept Check

1. What is the difference between exact interest and ordinary interest? With the increase of computers in banking, do you think that the ordinary interest method is a dinosaur in business today?
2. Explain how to use the portion formula to solve the unknowns in the simple interest formula. Why would rounding the answer of the denominator result in an inaccurate final answer?
3. Explain the U.S. Rule. Why in the last step of the U.S. Rule is the interest added, not subtracted?
4. Do you believe the government bailout of banks is in the best interest of the country? Defend your position.
5. **Chapter Concept Check.** Prepare calculations based on the concepts in this chapter to prove credit unions would save you money in your personal life.

Classroom Notes

END-OF-CHAPTER PROBLEMS

connect

Check figures for odd-numbered problems in Appendix B. Name ____________________ Date ________

DRILL PROBLEMS

Calculate the simple interest and maturity value for the following problems. Round to the nearest cent as needed. *LU 10-1(1)*

	Principal	Interest rate	Time	Simple interest	Maturity value
10–1.	\$9,000	$2\frac{1}{4}\%$	18 mo.		
10–2.	\$4,500	3%	6 mo.		
10–3.	\$20,000	$6\frac{3}{4}\%$	9 mo.		

Complete the following, using ordinary interest: *LU 10-1(2)*

		Principal	Interest rate	Date borrowed	Date repaid	Exact time	Interest	Maturity value
eXcel	**10–4.**	\$1,000	8%	Mar. 8	June 9			
eXcel	**10–5.**	\$585	9%	June 5	Dec. 15			
eXcel	**10–6.**	\$1,200	12%	July 7	Jan. 10			

Complete the following, using exact interest: *LU 10-1(2)*

	Principal	Interest rate	Date borrowed	Date repaid	Exact time	Interest	Maturity value
10–7.	\$1,000	8%	Mar. 8	June 9			
10–8.	\$585	9%	June 5	Dec. 15			
10–9.	\$1,200	12%	July 7	Jan. 10			

Solve for the missing item in the following (round to the nearest hundredth as needed): *LU 10-2(1)*

	Principal	**Interest rate**	**Time (months or years)**	**Simple interest**
10–10.	\$400	5%	?	\$100
10–11.	?	7%	$1\frac{1}{2}$ years	\$200
10–12.	\$5,000	?	6 months	\$300

10–13. Use the U.S. Rule to solve for total interest costs, balances, and final payments (use ordinary interest). *LU 10-3(1)*

Given Principal: \$10,000, 8%, 240 days
Partial payments: On 100th day, \$4,000
On 180th day, \$2,000

WORD PROBLEMS

My Money

10–14. Nolan Walker decided to buy a used snowmobile since his credit union was offering such low interest rates. He borrowed \$2,700 at 3.5% on December 26, 2019, and paid it off February 21, 2021. How much did he pay in interest? (Assume ordinary interest and no leap year.) *LU 10-1(2)*

My Money

10–15. Harold Hill borrowed \$15,000 to pay for his child's education at Riverside Community College. Harold must repay the loan at the end of 9 months in one payment with $5\frac{1}{2}\%$ interest. How much interest must Harold pay? What is the maturity value? *LU 10-1(1)* eXcel **10–16.** On September 12, Jody Jansen went to Sunshine Bank to borrow $2,300 at 9% interest. Jody plans to repay the loan on January 27. Assume the loan is on ordinary interest. What interest will Jody owe on January 27? What is the total amount Jody must repay at maturity? *LU 10-1(2)*

eXcel **10–17.** Kelly O'Brien met Jody Jansen (Problem 10–16) at Sunshine Bank and suggested she consider the loan on exact interest. Recalculate the loan for Jody under this assumption. How much would she save in interest? *LU 10-1(2)*

10–18. On May 3, 2020, Leven Corp. negotiated a short-term loan of $685,000. The loan is due October 1, 2020, and carries a 6.86% interest rate. Use ordinary interest to calculate the interest. What is the total amount Leven would pay on the maturity date? *LU 10-1(2)*

eXcel **10–19.** Gordon Rosel went to his bank to find out how long it will take for $1,200 to amount to $1,650 at 8% simple interest. Please solve Gordon's problem. Round time in years to the nearest tenth. *LU 10-2(1)*

10–20. Lucky Champ owes $191.25 interest on a 6% loan he took out on his March 17 birthday to upgrade an oven in his Irish restaurant, Lucky's Pub and Grub. The loan is due on August 17. What is the principal (assume ordinary interest)? *LU 10-2(1)*

10–21. On April 5, 2019, Janeen Camoct took out an $8\frac{1}{2}$% loan for $20,000. The loan is due March 9, 2020. Use ordinary interest to calculate the interest. What total amount will Janeen pay on March 9, 2020? (Ignore leap year.) *LU 10-1(2)*

10–22. Sabrina Bowers took out the same loan as Janeen (Problem 10–21). Sabrina's terms, however, are exact interest. What is Sabrina's difference in interest? What will she pay on March 9, 2020? (Ignore leap year.) *LU 10-1(2)*

10–23. Max Wholesaler borrowed $2,000 on a 10%, 120-day note. After 45 days, Max paid $700 on the note. Thirty days later, Max paid an additional $630. What is the final balance due? Use the U.S. Rule to determine the total interest and ending balance due. Use ordinary interest. *LU 10-3(1)*

My Money

10–24. Johnny Rockefeller had a bad credit rating and went to a local cash center. He took out a $100 loan payable in two weeks for $115. What is the percent of interest paid on this loan? Do not round denominator before dividing. *LU 10-2(1)*

My Money

10–25. You decided it is important to pay off some of your debt to help build your credit score. If you paid $1,307 interest on $45,000 at 4.0%, what was the time, using exact interest (rounded up to the nearest day)? *LU 10-2(1)*

10–26. On September 14, Jennifer Rick went to Park Bank to borrow $2,500 at $11\frac{3}{4}\%$ interest. Jennifer plans to repay the loan on January 27. Assume the loan is on ordinary interest. What interest will Jennifer owe on January 27? What is the total amount Jennifer must repay at maturity? *LU 10-1(2)*

10–27. Steven Linden met Jennifer Rick (Problem 10–26) at Park Bank and suggested she consider the loan on exact interest. Recalculate the loan for Jennifer under this assumption. *LU 10-1(2)*

eXcel

10–28. Lance Lopes went to his bank to find out how long it will take for $1,000 to amount to $1,700 at 12% simple interest. Can you solve Lance's problem? Round time in years to the nearest tenth. *LU 10-2(1)*

10–29. Andres Michael bought a new boat. He took out a loan for $24,500 at 4.5% interest for 2 years. He made a $4,500 partial payment at 2 months and another partial payment of $3,000 at 6 months. How much is due at maturity? *LU 10-3(1)*

10–30. Shawn Bixby borrowed \$17,000 on a 120-day, 12% note. After 65 days, Shawn paid \$2,000 on the note. On day 89, Shawn paid an additional \$4,000. What is the final balance due? Determine total interest and ending balance due by the U.S. Rule. Use ordinary interest. *LU 10-3(1)*

My Money

10–31. Carol Miller went to Europe and forgot to pay her \$740 mortgage payment on her New Hampshire ski house. For her 59 days overdue on her payment, the bank charged her a penalty of \$15. What was the rate of interest charged by the bank? Round to the nearest hundredth percent. (Assume 360 days.) *LU 10-2(1)*

10–32. Evander Holyfield (the champion boxer who had part of his ear bitten off by Mike Tyson) made \$250 million during his boxing career but declared bankruptcy because of poor financial choices. His July interest at 15% was \$155. What was Evander's principal at the beginning of July (assume 360 days)? *LU 10-2(1)*

10–33. Kurt Busch won the 59th Daytona 500 in February 2017. If he paid back a \$6,800 loan with \$20 interest at 7.5%, what was the time of the loan (assume 360 days)? *LU 10-2(1)*

eXcel

10–34. Molly Ellen, bookkeeper for Keystone Company, forgot to send in the payroll taxes due on April 15. She sent the payment November 8. The IRS sent her a penalty charge of 8% simple interest on the unpaid taxes of \$4,100. Calculate the penalty. (Remember that the government uses exact interest.) *LU 10-1(2)*

10–35. Oakwood Plowing Company purchased two new plows for the upcoming winter. In 200 days, Oakwood must make a single payment of \$23,200 to pay for the plows. As of today, Oakwood has \$22,500. If Oakwood puts the money in a bank today, what rate of interest will it need to pay off the plows in 200 days? (Assume 360 days.) *LU 10-2(1)*

CHALLENGE PROBLEMS

10–36. Debbie McAdams paid 8% interest on a $12,500 loan balance. Jan Burke paid $5,000 interest on a $62,500 loan. Based on 1 year: **(a)** What was the amount of interest paid by Debbie? **(b)** What was the interest rate paid by Jan? **(c)** Debbie and Jan are both in the 28% tax bracket. Since the interest is deductible, how much would Debbie and Jan each save in taxes? *LU 10-2(1)*

10–37. Janet Foster bought a computer and printer at Computerland. The printer had a $600 list price with a $100 trade discount and 2/10, n/30 terms. The computer had a $1,600 list price with a 25% trade discount but no cash discount. On the computer, Computerland offered Janet the choice of (1) paying $50 per month for 17 months with the 18th payment paying the remainder of the balance or (2) paying 8% interest for 18 months in equal payments. *LU 10-1(2)*

a. Assume Janet could borrow the money for the printer at 8% to take advantage of the cash discount. How much would Janet save? (Assume 360 days.)

b. On the computer, what is the difference in the final payment between choices 1 and 2?

Classroom Notes

SUMMARY PRACTICE TEST

Do you need help? Connect videos have step-by-step worked-out solutions.

1. Lorna Hall's real estate tax of \$2,010.88 was due on December 14, 2019. Lorna lost her job and could not pay her tax bill until February 27, 2020. The penalty for late payment is $6\frac{1}{2}\%$ ordinary interest. *LU 10-1(1)*
   a. What is the penalty Lorna must pay?
   b. What is the total amount Lorna must pay on February 27?

2. Ann Hopkins borrowed \$60,000 for her child's education. She must repay the loan at the end of 8 years in one payment with $5\frac{1}{2}\%$ interest. What is the maturity value Ann must repay? *LU 10-1(1)*

3. On May 6, Jim Ryan borrowed \$14,000 from Lane Bank at $7\frac{1}{2}\%$ interest. Jim plans to repay the loan on March 11. Assume the loan is on ordinary interest. How much will Jim repay on March 11? *LU 10-1(2)*

4. Gail Ross met Jim Ryan (Problem 3) at Lane Bank. After talking with Jim, Gail decided she would like to consider the same loan on exact interest. Can you recalculate the loan for Gail under this assumption? *LU 10-1(2)*

5. Claire Russell is buying a car. Her November monthly interest was \$210 at $7\frac{3}{4}\%$ interest. What is Claire's principal balance (to the nearest dollar) at the beginning of November? Use 360 days. Do not round the denominator in your calculation. *LU 10-2(1)*

6. Comet Lee borrowed \$16,000 on a 6%, 90-day note. After 20 days, Comet paid \$2,000 on the note. On day 50, Comet paid \$4,000 on the note. What are the total interest and ending balance due by the U.S. Rule? Use ordinary interest. *LU 10-3(1)*

INTERACTIVE VIDEO WORKSHEET

GRADING THE SUMMARY PRACTICE TEST

Go to the summary practice test video in Connect (or click on it here in the ebook). Grade your summary practice test while viewing the video.

C for Correct/I for Incorrect

1. ________ 4. ________
2. ________ 5. ________
3. ________ 6. ________

If you achieved 100%, you are ready for your instructor's exam.

If any of the problems were incorrect, list the questions you missed and show steps to solve the problem correctly.

Replay the video to see if you have made the correct fixes to your mistakes. If you have any questions, contact your instructor asap.

Notes on Watching Videos

MY MONEY

Owe Yourself a Debt of Gratitude!

What I need to know

Debt is a part of life, and most people understand they will take on debt during their lifetime. However, understanding debt and managing debt are two vastly different subjects. When we take on debt we should already be considering the way in which we will be able to meet and satisfy the debt within the confines of our earnings. Taking on too much debt will have a significant impact on our ability to meet our recurring, and unexpected expenses, as well as meet our financial goals for the future.

Before you enter into any new debt, you should consider the plan you have to fully satisfy or pay off the debt. By doing the work ahead of time you will be better prepared to understand how much debt you can take on and whether or not this debt is a wise decision. Many times we find ourselves in a debt crisis situation because we acted on an impulse purchase decision as opposed to planning out debt appropriately.

What I need to do

Consider your budget and stay realistic. Taking the time to develop a financial budget that works with your current earnings and expenses is important. Don't take on too much risk with a large debt that will not coincide with the budget you have established. As you enter your career after college so many things are changing all at once. Avoid letting the excitement of these changes steer you in the wrong direction. For example, maybe you are considering a new vehicle, or other large purchase, to reward yourself for successfully completing your degree. Consider if purchasing a brand new car for $40,000 is as financially sound as purchasing a quality used car for $15,000. If you allow emotion to influence your decision you may find yourself in a fancy new ride that you really can't afford.

Know how you will pay off your debt prior to taking it on. Using some simple math can go a long way in assisting you with your debt decisions. Let's say you are considering a furniture purchase for your new residence. The total cost is $2,400 and the furniture company is offering no interest on purchases for 12 months. Sounds like an incredible deal, right? To avoid paying any interest on this purchase, you would need to pay off the entire balance within the 12-month window. This equals a monthly expense of $200. Prior to making such a purchase, however, you need to determine if your budget will allow for an additional $200 expense per month to pay off your debt before the time when you will incur interest charges on your purchase. By determining your payoff plan, you will be more prepared to make purchases that are in line with your budget.

Resources I can use

- Debt Payoff Planner Calculator (mobile app)
- https://www.thebalance.com/how-to-manage-your-debt-960856 — managing your debt

MY MONEY ACTIVITY

Consider a purchase you would make after college such as a car, furniture, or clothing.

- Determine the total purchase cost.
- Create your payoff plan for satisfying this debt in 6 months, 1 year, and 18 months.

A KIPLINGER APPROACH

LISA GERSTNER | Millennial Money

How to Tame Your Student Loans

Most retirees aren't concerned with building a credit history. By the same token, twenty- and thirtysomethings don't think much about annuities. That's why we're carving out a space to address the concerns of younger readers. It will be written by staffers who are members of the millennial generation—considered to be roughly ages 20 to 36—and who are themselves dealing with issues ranging from budgeting to buying a home to saving for long-term goals.

To kick off our column, we're tackling a topic that has become a defining characteristic of our generation: student loan debt. Bachelor's degree recipients are estimated to carry an average debt burden approaching $40,000 per borrower at graduation. Parents are shouldering some of the burden, too, with an average $21,000 of college debt.

Thanks to a scholarship, I was fortunate to graduate from college debt-free. But many of my peers express regret about the amount of debt they took on, saying that if they could do it all over again, they'd skip the advanced degree or consider alternative paths. "The loans feel like a constant burden and drain, with no end in sight," says Amber Richter, 32, whose undergraduate and law-school loans add up to six figures. "I worry all the time about saving—or lack thereof—for retirement and my children's education."

©Kiplinger Washington Editors, Inc.

To kick off our column, we're tackling a topic that has become a defining characteristic of our generation: student loan debt.

Nate Hamm, 34, racked up more than $300,000 in debt following podiatry school. Now a self-employed podiatric surgeon, his income and family size (he's married with two kids) qualify him to make no payments under an income-based plan. He'd pay nearly $3,000 a month if he were on the standard plan. But he is making payments on a $127,000 loan he took out to start his business. "I know I'm a lot more stressed than my parents ever were about their college education," he says. "It keeps me up at night sometimes."

Matthew Lesser, 32, carries $135,000 in student loan debt and has chosen to work for a nonprofit organization so he can qualify for the Public Service Loan Forgiveness program, which forgives the remaining balance on direct federal loans for borrowers who are employed at eligible organizations after they make 120 monthly payments. He has eight years to go. In the meantime, he's juggling his $385 monthly loan payment along with a mortgage, credit card debt and all his other expenses. (We explain federal student loan repayment options at kiplinger.com/links/rightway.)

Budgeting and student loans. Aggressively paying down student debt works for some but isn't the right move for everyone. Andrew Damcevski, a certified financial planner at RhineVest, in Cincinnati, says that some young people are inclined to "put every last penny toward student loans, and they completely neglect their future in terms of saving." At a minimum, aim to meet any match your employer provides on contributions to a 401(k) or other work-based retirement plan. You should also build an emergency fund that covers at least six months' worth of living expenses. If you have high-rate credit card debt, wipe it out before directing more cash to student loans.

Of course, your essential expenses are also part of the picture. A simple and popular budget guideline is the 50-20-30 rule: up to 50% of your take-home pay goes to essentials, 20% to whittling debt and adding to savings, and 30% to discretionary items, such as dining out, shopping, charitable giving, and streaming or cable TV. A budgeting app such as Mint can help you see where your money goes.

It may be worth getting help to hash out a plan. Certified financial planners who are members of the XY Planning Network (www.xyplanningnetwork.com) focus on younger clients, have no minimum asset requirements and offer virtual services (say, meeting via video chat). ■

TO SHARE THIS COLUMN, PLEASE GO TO KIPLINGER.COM/LINKS/MILLENNIALS. YOU CAN CONTACT THE AUTHOR AT LGERSTNER@KIPLINGER.COM.

BUSINESS MATH ISSUE

Student loan debt should always be paid off before planning for retirement.

1. List the key points of the article and information to support your position.
2. Write a group defense of your position using math calculations to support your view. If you are in an online course, post to a discussion board.

Classroom Notes

Design elements: Digital frame: (tablet) ©Radu Bercan/Shutterstock, (icons) ©12bit/Shutterstock; Globe: ©edge69/Getty Images; Money Tips: ©Pro Symbols/Shutterstock; Aha!: ©frender/Getty Images; Excel: ©McGraw-Hill Education; My Money: (tree) ©Cherkas/Shutterstock, (gold nuggets) ©JonahWong/Shutterstock; Video Worksheet: ©PureSolution/Shutterstock; author photo in circle: ©McGraw-Hill Education/Ryan McAndrews, photographer.

CHAPTER 11

Promissory Notes, Simple Discount Notes, and the Discount Process

©Courtney Keating/Getty Images

Firms' Tweak to Lift Credit Scores

By AnnaMaria Andriotis

Many tax liens and civil judgments soon will be taken off people's credit reports, the latest move to omit negative information from the powerful financial scorecards.

The decision by the three major credit-reporting firms—Equifax Inc., Experian PLC and TransUnion—could help boost credit scores for millions of U.S. consumers, but could pose risks for lenders. The reports and scores often help decide how much consumers can borrow for a new house or car as well as determine their credit-card spending limit.

The unusual move by the influential firms comes partially in response to regulatory concerns. The three reporting bureaus rarely tinker with the information that goes on credit reports and that lenders consult to gauge consumers' ability and willingness to pay back debts.

LEARNING UNIT OBJECTIVES

LU 11–1: Structure of Promissory Notes; the Simple Discount Note

1. Differentiate between interest-bearing and non-interest-bearing notes.
2. Calculate bank discount and proceeds for simple discount notes.
3. Calculate and compare the interest, maturity value, proceeds, and effective rate of a simple interest note with a simple discount note.
4. Explain and calculate the effective rate for a Treasury bill.

LU 11–2: Discounting an Interest-Bearing Note before Maturity

1. Calculate the maturity value, bank discount, and proceeds of discounting an interest-bearing note before maturity.
2. Identify and complete the four steps of the discounting process.

Your Guide to Successfully Completing This Chapter

Traditional book or ebook

Check box as you complete each step.

Steps

☐ Read learning unit.
  ☐ Complete practice quiz at the end of the learning unit.
☐ Grade practice quiz using provided solutions. (For more help, watch the learning unit video in Connect and have a Study Session with the authors. Then complete the additional practice quiz in Connect.)
☐ Repeat above for each of the two learning units in Chapter 11.
  ☐ Review chapter organizer.
  ☐ Complete assigned homework.
    ☐ Finish summary practice test. (Go to Connect via the ebook link and do the interactive video worksheet to grade.)
☐ Complete instructor's exam.

The following *Wall Street Journal* clip "More Older People Carry Student Debt" states more parents and grandparents are financing a great deal of student debt.

This chapter begins with a discussion of the structure of promissory notes and simple discount notes. We also look at the application of discounting with Treasury bills. The chapter concludes with an explanation of how to calculate the discounting of promissory notes.

Learning Unit 11–1: Structure of Promissory Notes; the Simple Discount Note

Although businesses frequently sign promissory notes, customers also sign promissory notes. For example, some student loans may require the signing of promissory notes. Appliance stores often ask customers to sign a promissory note when they buy large appliances on credit. In this unit, promissory notes usually involve interest payments.

LO 1

Structure of Promissory Notes

To borrow money, you must find a lender (a bank or a company selling goods on credit). You must also be willing to pay for the use of the money. In Chapter 10 you learned that interest is the cost of borrowing money for periods of time. Lenders charge interest as a rental fee on borrowing money.

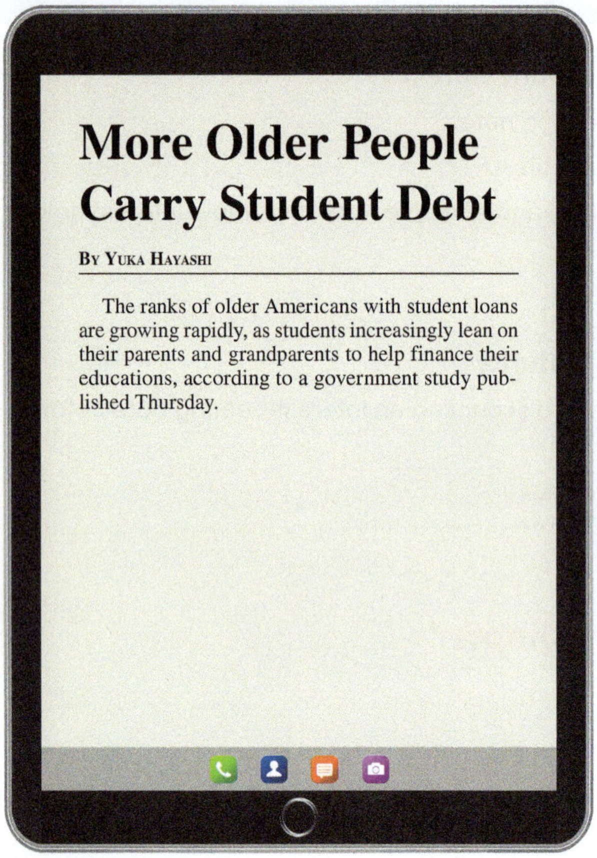

More Older People Carry Student Debt

By Yuka Hayashi

The ranks of older Americans with student loans are growing rapidly, as students increasingly lean on their parents and grandparents to help finance their educations, according to a government study published Thursday.

Money lenders usually require that borrowers sign a **promissory note.** This note states that the borrower will repay a certain sum at a fixed time in the future. The note often includes the charge for the use of the money, or the rate of interest. Figure 11.1 shows a sample promissory note with its terms identified and defined. Take a moment to look at each term.

In this section you will learn the difference between interest-bearing notes and non-interest-bearing notes.

Interest-Bearing versus Non-Interest-Bearing Notes A promissory note can be interest bearing or non–interest bearing. To be interest bearing, the note must state the rate of interest. Since the promissory note in Figure 11.1 states that its interest is 9%, it is an **interest-bearing note.** When the note matures, Regal Corporation will pay back the original amount **(face value)** borrowed plus interest. The simple interest formula (also known as the interest formula) and the maturity value formula from Chapter 10 are used for this transaction.

$$\text{Interest} = \text{Face value (principal)} \times \text{Rate} \times \text{Time}$$
$$\text{Maturity value} = \text{Face value (principal)} + \text{Interest}$$

If you sign a **non-interest-bearing** promissory note for \$10,000, you pay back \$10,000 at maturity. The maturity value of a non-interest-bearing note is the same as its face value. Usually, non-interest-bearing notes occur for short time periods under special conditions. For example, money borrowed from a relative could be secured by a non-interest-bearing promissory note.

LO 2

Simple Discount Note

The total amount due at the end of the loan, or the **maturity value (MV),** is the sum of the face value (principal) and interest. Some banks deduct the loan interest in advance. When banks do this, the note is a **simple discount note.**

In the simple discount note, the **bank discount** is the interest that banks deduct in advance and the **bank discount rate** is the percent of interest. The amount that the borrower receives after the bank deducts its discount from the loan's maturity value is the note's **proceeds.** Sometimes we refer to simple discount notes as non-interest-bearing notes. Remember, however, that borrowers *do* pay interest on these notes.

In the example that follows, Pete Runnels has the choice of a note with a simple interest rate (Chapter 10) or a note with a simple discount rate (Chapter 11). Table 11.1 provides a summary of the calculations made in the example and gives the key points that you should remember. Now let's study the example, and then you can review Table 11.1.

We will use 360 (not 365) days for all calculations in this chapter.

FIGURE 11.1

Interest-bearing promissory note

$10,000 a. LAWTON, OKLAHOMA October 2, 2020 c.
Sixty days b. AFTER DATE we PROMISE TO PAY TO
THE ORDER OF G.J. Equipment Company d.
Ten thousand and 00/100------------------DOLLARS.
PAYABLE AT Able National Bank
VALUE RECEIVED WITH INTEREST AT 9% e. REGAL CORPORATION f.
NO. 114 DUE December 1, 2020 g. J.M. Moore TREASURER

a. **Face value:** Amount of money borrowed—\$10,000. The face value is also the principal of the note.
b. **Term:** Length of time that the money is borrowed—60 days.
c. **Date:** The date that the note is issued—October 2, 2020.
d. **Payee:** The company extending the credit—G.J. Equipment Company.
e. **Rate:** The annual rate for the cost of borrowing the money—9%.
f. **Maker:** The company issuing the note and borrowing the money—Regal Corporation.
g. **Maturity date:** The date the principal and interest are due—December 1, 2020.

EXAMPLE Pete Runnels has a choice of two different notes that both have a face value (principal) of $14,000 for 60 days. One note has a simple interest rate of 8%, while the other note has a simple discount rate of 8%. For each type of note, calculate **(a)** interest owed, **(b)** maturity value, **(c)** proceeds, and **(d)** effective rate.

LO 3

Simple interest note—Chapter 10	Simple discount note—Chapter 11
Interest **a.** I = Face value (principal) × R × T $I = \$14{,}000 \times .08 \times \frac{60}{360}$ $I = \$186.67$	**Interest** **a.** I = Face value (principal) × R × T $I = \$14{,}000 \times .08 \times \frac{60}{360}$ $I = \$186.67$
Maturity value **b.** MV = Face value + Interest $MV = \$14{,}000 + \$186.67$ $MV = \$14{,}186.67$	**Maturity value** **b.** MV = Face value $MV = \$14{,}000$
Proceeds **c.** Proceeds = Face value = $14,000	**Proceeds** **c.** Proceeds = MV − Bank discount = $14,000 − $186.67 = $13,813.33
Effective rate **d.** Rate = $\frac{\text{Interest}}{\text{Proceeds} \times \text{Time}}$ $= \frac{\$186.67}{\$14{,}000 \times \frac{60}{360}}$ = 8%	**Effective rate** **d.** Rate = $\frac{\text{Interest}}{\text{Proceeds} \times \text{Time}}$ $= \frac{\$186.67}{\$13{,}813.33 \times \frac{60}{360}}$ = 8.11%

Aha!

Do not round intermediate answers. Round only the final calculation.

TABLE 11.1

Comparison of simple interest note and simple discount note (Calculations from the Pete Runnels example)

My Money

Simple interest note (Chapter 10)	Simple discount note (Chapter 11)
1. A promissory note for a loan with a term of usually less than 1 year. *Example:* 60 days.	**1.** A promissory note for a loan with a term of usually less than 1 year. *Example:* 60 days.
2. Paid back by one payment at maturity. Face value equals actual amount (or principal) of loan (this is not maturity value).	**2.** Paid back by one payment at maturity. Face value equals maturity value (what will be repaid).
3. Interest computed on face value or what is actually borrowed. *Example:* $186.67.	**3.** Interest computed on maturity value or what will be repaid and not on actual amount borrowed. *Example:* $186.67.
4. Maturity value = Face value + Interest. *Example:* $14,186.67.	**4.** Maturity value = Face value. *Example:* $14,000.
5. Borrower receives the face value. *Example:* $14,000.	**5.** Borrower receives proceeds = Face value − Bank discount. *Example:* $13,813.33.
6. Effective rate (true rate is same as rate stated on note). *Example:* 8%.	**6.** Effective rate is higher since interest was deducted in advance. *Example:* 8.11%.
7. Used frequently instead of the simple discount note.	**7.** Not used as much now because in 1969 congressional legislation required that the true rate of interest be revealed. Still used where legislation does not apply, such as personal loans.

Note that the interest of $186.67 is the same for the simple interest note and the simple discount note. The maturity value of the simple discount note is the same as the face value. In the simple discount note, interest is deducted in advance, so the proceeds are less than the face value. Note that the **effective rate** for a simple discount note is higher than the stated rate, since the bank calculated the rate on the face value of the note and not on what Pete received.

LO 4

Application of Discounting—Treasury Bills When the government needs money, it sells Treasury bills. A **Treasury bill** is a loan to the federal government for 28 days (4 weeks), 91 days (13 weeks), or 1 year. Note that the *Wall Street Journal* clipping "Treasury Auctions" announces a new sale.

Treasury Auctions

The Treasury Department will auction $175 billion in securities next week. Details (all with minimum denominations of $100):

◆ **Monday:** $39 billion in 13-week bills, a reopening of an issue first sold on April 27, 2017, maturing Oct. 26, 2017. Cusip number: 912796LZO.

Also Monday, $33 billion in 26-week bills, dated July 27, 2017, maturing Jan. 25, 2018. Cusip: 912796NPO.

Noncompetitive tenders for both issues must be received by 11 a.m. EDT Monday and competitive tenders, by 11:30 a.m.

◆ **Tuesday:** $26 billion in two-year notes, dated July 31, 2017, maturing July 31, 2019. Cusip: 9128282K5. Noncompetitive tenders must be received by noon EDT Tuesday; competitive tenders, by 1 p.m.

MONEY tips

Paying your rent on time can help improve your credit rating. Leaving a lease before it is up and bouncing checks to a landlord will reduce it.

Treasury bills can be bought over the phone or on the government website. The purchase price (or proceeds) of a Treasury bill is the value of the Treasury bill less the discount. For example, if you buy a $10,000, 13-week Treasury bill at 4%, you pay $9,900 since you have not yet earned your interest ($\$10{,}000 \times .04 \times \frac{13}{52} = \$100$). At maturity—13 weeks— the government pays you $10,000. You calculate your effective yield (4.04% rounded to the nearest hundredth percent) as follows:

$$(\$10{,}000 - \$100) \longrightarrow \frac{\$100}{\$9{,}900 \times \frac{13}{52}} = 4.04\% \text{ effective rate}$$

Now it's time to try the Practice Quiz and check your progress.

LU 11–1 PRACTICE QUIZ

Complete this Practice Quiz to see how you are doing.

1. Warren Ford borrowed $12,000 on a non-interest-bearing, simple discount, $9\frac{1}{2}\%$, 60-day note. Assume ordinary interest. What are **(a)** the maturity value, **(b)** the bank's discount, **(c)** Warren's proceeds, and **(d)** the effective rate to the nearest hundredth percent?
2. Jane Long buys a $10,000, 13-week Treasury bill at 6%. What is her effective rate? Round to the nearest hundredth percent.

*For **extra help** from your authors–Sharon and Jeff–see the videos in Connect.*

✓ Solutions

1. **a.** Maturity value = Face value = $12,000

b. Bank discount = $MV \times$ Bank discount rate $\times$ Time

$$= \$12{,}000 \times .095 \times \frac{60}{360}$$

$$= \$190$$

c. Proceeds = MV − Bank discount

$$= \$12{,}000 - \$190$$

$$= \$11{,}810$$

d. Effective rate $= \dfrac{\text{Interest}}{\text{Proceeds} \times \text{Time}}$

$$= \frac{\$190}{\$11{,}810 \times \frac{60}{360}}$$

$$= \$9.65\%$$

2. $\$10{,}000 \times .06 \times \frac{13}{52} = \$150$ interest $\qquad \dfrac{\$150}{\$9{,}850 \times \frac{13}{52}} = 6.09\%$

LO 1

Learning Unit 11–2: Discounting an Interest-Bearing Note before Maturity

Manufacturers frequently deliver merchandise to retail companies and do not request payment for several months. For example, Roger Company manufactures outdoor furniture that it delivers to Sears in March. Payment for the furniture is not due until September. Roger will have its money tied up in this furniture until September. So Roger requests that Sears sign promissory notes.

If Roger Company needs cash sooner than September, what can it do? Roger Company can take one of its promissory notes to the bank, assuming the company that signed the note is reliable. The bank will buy the note from Roger. Now Roger has discounted the note and has cash instead of waiting until September when Sears would have paid Roger.

Remember that when Roger Company discounts the promissory note to the bank, the company agrees to pay the note at maturity if the maker of the promissory note fails to pay the bank. The potential liability that may or may not result from discounting a note is called a **contingent liability.**

Think of **discounting a note** as a three-party arrangement. Roger Company realizes that the bank will charge for this service. The bank's charge is a **bank discount.** The actual amount Roger receives is the **proceeds** of the note. The four steps below and the formulas in the example that follows will help you understand this discounting process.

LO 2

DISCOUNTING A NOTE

Step 1. Calculate the interest and maturity value of the original simple interest note.
Step 2. Calculate the discount period (time the bank holds note).
Step 3. Calculate the bank discount.
Step 4. Calculate the proceeds.

EXAMPLE Roger Company sold the following promissory note to the bank:

Date of note	Face value of note	Length of note	Interest rate	Bank discount rate	Date of discount
March 8	$2,000	185 days	6%	5%	August 9

What are Roger's (1) interest and maturity value (MV)? What are the (2) discount period and (3) bank discount? (4) What are the proceeds?

1. *Calculate Roger's interest and maturity value (MV):*

MV = Face value (principal) + Interest

$$\text{Interest} = \$2{,}000 \times .06 \times \frac{185}{360} \leftarrow \text{Actual number of days over 360}$$

$$= \$61.67$$

$$MV = \$2{,}000 + \$61.67$$

$$= \$2{,}061.67$$

Calculating days without table:

March	31
	− 8
	23
April	30
May	31
June	30
July	31
August	9
	154

185 days—length of note
− 154 days Roger held note
31 days bank waits

2. *Calculate **discount period**:* Determine the number of days that the bank will have to wait for the note to come due (discount period).

August 9	221 days
March 8	− 67
	154 days passed before note is discounted
	185 days
	− 154
	31 days bank waits for note to come due

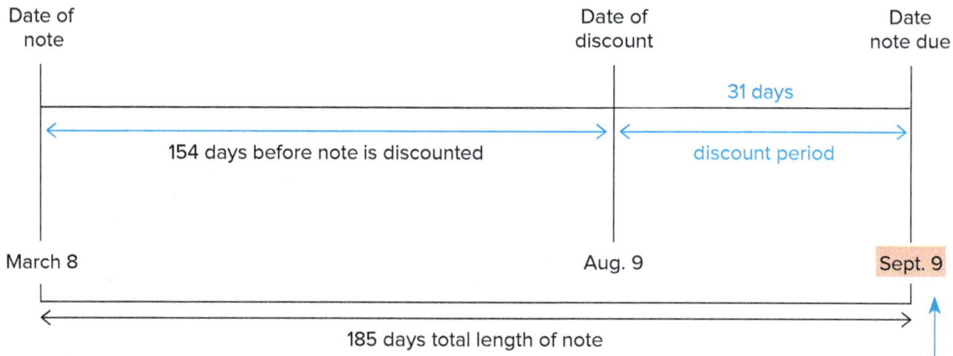

By table: March 8 = 67 days
+ 185
252 search in table

3. *Calculate bank discount (bank charge):*

$$\$2{,}061.67 \times .05 \times \frac{31}{360} = \$8.88$$

$$\text{Bank discount} = MV \times \text{Bank discount rate} \times \frac{\text{Number of days bank waits for note to come due}}{360}$$

4. *Calculate proceeds:*

$2,061.67
− 8.88
$2,052.79

Step 1 → Proceeds = *MV* − Bank discount (charge) ← **Step 3**

If Roger had waited until September 9, it would have received $2,061.67. Now, on August 9, Roger received $2,000 plus $52.79 interest.

MONEY tips

Protect your credit. At a minimum, review your credit report annually. Ideally, you can review your credit report for free three times per year if you stagger your free annual requests required by law from each of the three credit agencies: TransUnion, Equifax, and Experian. AnnualCreditReport.com has links to each.

Now let's assume Roger Company received a non-interest-bearing note. Then we follow the four steps for discounting a note except the maturity value is the amount of the loan. No interest accumulates on a non-interest-bearing note. Today, many banks use simple interest instead of discounting. Also, instead of discounting notes, many companies set up *lines of credit* so that additional financing is immediately available. The following *Wall Street Journal* clip shows that Pennsylvania got a credit line to have ready access to needed cash.

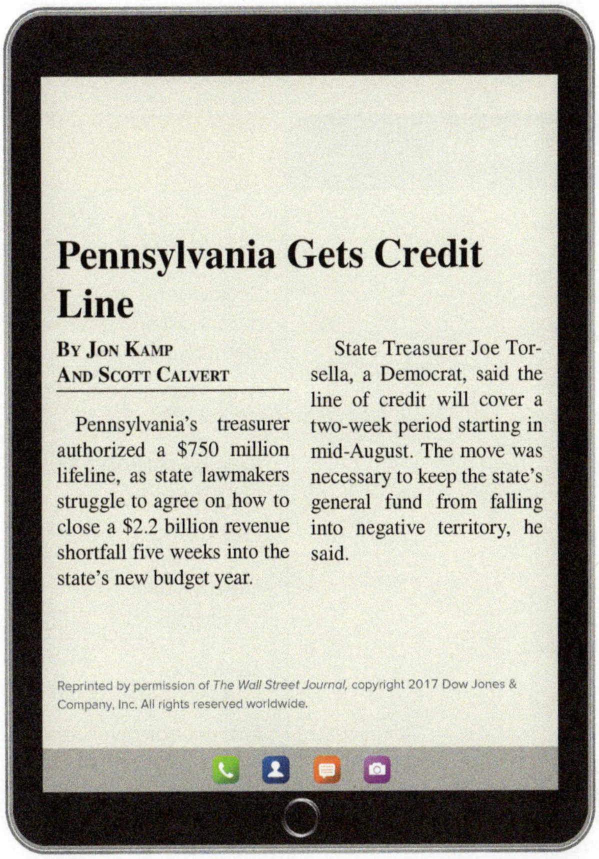

Pennsylvania Gets Credit Line

By Jon Kamp and Scott Calvert

Pennsylvania's treasurer authorized a $750 million lifeline, as state lawmakers struggle to agree on how to close a $2.2 billion revenue shortfall five weeks into the state's new budget year.

State Treasurer Joe Torsella, a Democrat, said the line of credit will cover a two-week period starting in mid-August. The move was necessary to keep the state's general fund from falling into negative territory, he said.

The Practice Quiz that follows will test your understanding of this unit.

LU 11–2 PRACTICE QUIZ

Complete this **Practice Quiz** to see how you are doing.

Date of note	Face value (principal) of note	Length of note	Interest rate	Bank discount rate	Date of discount
April 8	$35,000	160 days	11%	9%	June 8

From the above, calculate **(a)** interest and maturity value, **(b)** discount period, **(c)** bank discount, and **(d)** proceeds. Assume ordinary interest.

*For **extra help** from your authors–Sharon and Jeff–see the videos in Connect.*

✓ Solutions

a. $I = \$35{,}000 \times .11 \times \frac{160}{360} = \$1{,}711.11$

$MV = \$35{,}000 + \$1{,}711.11 = \$36{,}711.11$

b. Discount period = 160 − 61 = 99 days

April	30
	−8
	22
May	+31
	53
June	+8
	61

Or by table:

June 8	159
April 8	−98
	61

c. Bank discount $= \$36{,}711.11 \times .09 \times \frac{99}{360} = \$908.60$

d. Proceeds = $36,711.11 − $908.60 = $35,802.51

INTERACTIVE CHAPTER ORGANIZER

Topic/Procedure/Formula	Examples	You try it*
Simple discount note Bank discount (interest) $= MV \times$ Bank discount rate $\times$ Time Interest based on amount paid back and not on actual amount received.	$\$6{,}000 \times .09 \times \frac{60}{360} = \$90$ Borrower receives $5,910 (the proceeds) and pays back $6,000 at maturity after 60 days. A Treasury bill is a good example of a simple discount note.	**Calculate proceeds** $4,000 note at 2% for 30 days
Effective rate $\frac{\text{Interest}}{\text{Proceeds} \times \text{Time}}$ ↑ What borrower receives (Face value − Discount)	*Example:* $10,000 note, discount rate 12% for 60 days. $I = \$10{,}000 \times .12 \times \frac{60}{360} = \$200$ Effective rate: $\frac{\$200}{\$9{,}800 \times \frac{60}{360}} = \frac{\$200}{\$1{,}633.3333} = 12.24\%$ ↑ Amount borrower received	**Calculate effective rate** $15,000 note at 4% for 40 days
Discounting an interest-bearing note 1. Calculate interest and maturity value. I = Face value × Rate × Time MV = Face value + Interest 2. Calculate number of days bank will wait for note to come due (discount period). 3. Calculate bank discount (bank charge). $MV \times$ Bank discount rate $\times \frac{\text{Number of days bank waits}}{360}$ 4. Calculate proceeds. MV − Bank discount (charge)	*Example:* $1,000 note, 6%, 60 days, dated November 1 and discounted on December 1 at 8%. 1. $I = \$1{,}000 \times .06 \times \frac{60}{360} = \$10$ $MV = \$1{,}000 + \$10 = \$1{,}010$ 2. 30 days 3. $\$1{,}010 \times .08 \times \frac{30}{360} = \$6.73$ 4. $1{,}010 - \$6.73 = \$1{,}003.27$	**Calculate proceeds** $2,000 note, 3%, 60 days, dated November 5 and discounted on December 15 at 5%

KEY TERMS			
	Bank discount	Face value	Payee
	Bank discount rate	Interest-bearing note	Proceeds
	Contingent liability	Maker	Promissory note
	Discount period	Maturity date	Simple discount note
	Discounting a note	Maturity value (MV)	Treasury bill
	Effective rate	Non-interest-bearing note	

*Worked-out solutions are in Appendix B.

Critical Thinking Discussion Questions with Chapter Concept Check

1. What are the differences between a simple interest note and a simple discount note? Which type of note would have a higher effective rate of interest? Why?
2. What are the four steps of the discounting process? Could the proceeds of a discounted note be less than the face value of the note?
3. What is a line of credit? What could be a disadvantage of having a large credit line?
4. Discuss the impact of a slow economy on small business borrowing.
5. **Chapter Concept Check.** Go to the Internet and determine the current status of business loans. Include concepts you learned in this chapter in your review.

END-OF-CHAPTER PROBLEMS

connect

Check figures for odd-numbered problems in Appendix B. Name ______________________ Date ________

DRILL PROBLEMS

Complete the following table for these simple discount notes. Use the ordinary interest method. *LU 11-1(2)*

	Amount due at maturity	Discount rate	Time	Bank discount	Proceeds
11–1.	$6,000	$3\frac{1}{2}\%$	160 days		
11–2.	$2,900	$6\frac{1}{4}\%$	180 days		

Calculate the discount period for the bank to wait to receive its money: *LU 11-2(1)*

	Date of note	Length of note	Date note discounted	Discount period
11–3.	April 12	45 days	May 2	
11–4.	March 7	120 days	June 8	

Solve for maturity value, discount period, bank discount, and proceeds (assume for Problems 11–5 and 11–6 a bank discount rate of 9%). *LU 11-2(1, 2)*

	Face value (principal)	Rate of interest	Length of note	Maturity value	Date of note	Date note discounted	Discount period	Bank discount	Proceeds
11–5.	$50,000	11%	95 days		June 10	July 18			
11–6.	$25,000	9%	60 days		June 8	July 10			

eXcel

11–7. Calculate the effective rate of interest (to the nearest hundredth percent) of the following Treasury bill. **Given:** $10,000 Treasury bill, 4% for 13 weeks. *LU 11-1(4)*

WORD PROBLEMS

Use ordinary interest as needed.

11–8. Carl Sonntag wanted to compare what proceeds he would receive with a simple interest note versus a simple discount note. Both had the same terms: \$19,500 at 8% for 2 years. Compare the proceeds. *LU 11-1(3)*

11–9. Paul and Sandy Moede signed an \$8,000 note at Citizen's Bank. Citizen's charges a $6\frac{1}{2}\%$ discount rate. If the loan is for 300 days, find **(a)** the proceeds and **(b)** the effective rate charged by the bank (to the nearest tenth percent). *LU 11-1(3)*

11–10. You were offered either a simple interest note or a simple discount note with the following terms: \$33,353 at 7% for 18 months. Based on the effective interest rate, which would you choose? *LU 11-1(3)*

11–11. On September 5, Sheffield Company discounted at Sunshine Bank a \$9,000 (maturity value), 120-day note dated June 5. Sunshine's discount rate was 9%. What proceeds did Sheffield Company receive? *LU 11-2(1)*

11–12. The Treasury Department auctioned \$21 billion in 3-month bills in denominations of \$10,000 at a discount rate of 4.965%. What would be the effective rate of interest? Round only your final answer to the nearest hundredth percent. *LU 11–1(4)*

My Money

11–13. There are some excellent free personal finance apps available: Mint.com, GoodBudget, Mvelopes, BillGuard, PocketExpense, HomeBudget, and Expensify. After using Mint.com, you realize you need to pay off one of your high interest loans to reduce your interest expense. You decide to discount a \$5,250, 345-day note at 3% to your bank at a discount rate of 4.5% on day 210. What are your proceeds? Round each answer to the nearest cent. *LU 11-2(1)*

11–14. Ron Prentice bought goods from Shelly Katz. On May 8, Shelly gave Ron a time extension on his bill by accepting a \$3,000, 8%, 180-day note. On August 16, Shelly discounted the note at Roseville Bank at 9%. What proceeds does Shelly Katz receive? *LU 11-2(1)*

11–15. Rex Corporation accepted a \$5,000, 8%, 120-day note dated August 8 from Regis Company in settlement of a past bill. On October 11, Rex discounted the note at Park Bank at 9%. What are the note's maturity value, discount period, and bank discount? What proceeds does Rex receive? *LU 11-2(1)*

11–16. On May 12, Scott Rinse accepted an \$8,000, 12%, 90-day note for a time extension of a bill for goods bought by Ron Prentice. On June 12, Scott discounted the note at Able Bank at 10%. What proceeds does Scott receive? *LU 11-2(1)*

11–17. Robinson's, an electrical supply company, sold \$4,800 of equipment to Jim Coates Wiring, Inc. Coates signed a promissory note May 12 with 4.5% interest. The due date was August 10. Short of funds, Robinson's contacted Capital One Bank on July 20; the bank agreed to take over the note at a 6.2% discount. What proceeds will Robinson's receive? *LU 11-2(1)*

My Money

11–18. At www.daveramsey.com's Financial Peace University (FPU), Dave recommends Seven Baby Steps. One of these steps is "Pay off debt using the debt snowball." After graduating from FPU, Courtney Lopez-Munoz is trying to calculate the effective interest rate she is paying for a \$1,789 simple discount note at $5\frac{1}{4}\%$ for 15 months. What rate has she been paying? Round to the nearest tenth percent. Do not round denominator calculation. *LU 11-2(1)*

11–19. Toyota Motor Company, headquartered in Nagoya, Japan, was challenged in 2017, reported Reuters. Concerns about the strength of the steel used in its vehicle manufacturing; falling quarterly sales in North America, its largest market; and where the yen is trading against the dollar all affect Toyota's bottom line. If Toyota had a ¥20,000 note at 2.5% interest for 340 days, what would Toyota's proceeds be if it discounted the note on day 215 at 4%? (Round to the nearest yen for each answer.) *LU 11-2(1)*

CHALLENGE PROBLEMS

11–20. Assume that 3-month Treasury bills totaling \$12 billion were sold in \$10,000 denominations at a discount rate of 3.605%. In addition, the Treasury Department sold 6-month bills totaling \$10 billion at a discount rate of 3.55%. **(a)** What is the discount amount for 3-month bills? **(b)** What is the discount amount for 6-month bills? **(c)** What is the effective rate for 3-month bills? **(d)** What is the effective rate for 6-month bills? Round to the nearest hundredth percent. *LU 11-1(4)*

11–21. Tina Mier must pay a \$2,000 furniture bill. A finance company will loan Tina \$2,000 for 8 months at a 9% discount rate. The finance company told Tina that if she wants to receive exactly \$2,000, she must borrow more than \$2,000. The finance company gave Tina the following formula:

$$\text{What to ask for} = \frac{\text{Amount in cash to be received}}{1 - (\text{Discount} \times \text{Time of loan})}$$

Calculate Tina's loan request and the effective rate of interest to the nearest hundredth percent. *LU 11-1(3)*

Classroom Notes

SUMMARY PRACTICE TEST

Do you need help? Connect videos have step-by-step worked-out solutions.

1. On December 12, Lowell Corporation accepted a \$160,000, 120-day, non-interest-bearing note from Able.com. What is the maturity value of the note? *LU 11-1(1)*

eXcel

2. The face value of a simple discount note is \$17,000. The discount is 4% for 160 days. Calculate the following. *LU 11-1(3)*

   a. Amount of interest charged for each note.

   b. Amount borrower would receive.

   c. Amount payee would receive at maturity.

   d. Effective rate (to the nearest tenth percent).

3. On July 14, Gracie Paul accepted a \$60,000, 6%, 160-day note from Mike Lang. On November 12, Gracie discounted the note at Lend Bank at 7%. What proceeds did Gracie receive? *LU 11-2(1)*

4. Lee.com accepted a \$70,000, $6\frac{3}{4}$%, 120-day note on July 26. Lee discounts the note on October 28 at LB Bank at 6%. What proceeds did Lee receive? *LU 11-2(1)*

5. The owner of Lease.com signed a \$60,000 note at Reese Bank. Reese charges a $7\frac{1}{4}$% discount rate. If the loan is for 210 days, find **(a)** the proceeds and **(b)** the effective rate charged by the bank (to the nearest tenth percent). *LU 11-2(1)*

6. Sam Slater buys a \$10,000, 13-week Treasury bill at $5\frac{1}{2}$%. What is the effective rate? Round to the nearest hundredth percent. *LU 11-1(4)*

INTERACTIVE VIDEO WORKSHEET

GRADING THE SUMMARY PRACTICE TEST

Go to the summary practice test video in Connect (or click on it here in the ebook). Grade your summary practice test while viewing the video.

C for Correct/I for Incorrect

1. ________	4. ________
2. ________	5. ________
3. ________	6. ________

If you achieved 100%, you are ready for your instructor's exam.

If any of the problems were incorrect, list the questions you missed and show steps to solve the problem correctly.

Replay the video to see if you have made the correct fixes to your mistakes. If you have any questions, contact your instructor asap.

Notes on Watching Videos

MY MONEY

The Right (Financial) Tools for the Job

What I need to know

Financial planning is an important tool for achieving your financial goals, but you do not have to go at it alone. There are resources available that can assist you in drafting a financial plan for your future and assistance to keep you on the right path. As with any type of goal setting, it is helpful to seek out advice and direction from sources that will provide you with the resources you need to be successful. As someone who may be new to the act of personal finance you will most likely be seeking out assistance to help point you in the right financial direction.

There are many sources of information on the topic of financial planning. It is important that you seek out assistance that matches up with the financial goals you have set for yourself. Any assistance you can find is only helpful if you are able to apply the learning to your unique financial situation. Therefore, personal financial planning is not a one-size-fits-all strategy but needs to be customized to your unique situation and finances. Even comparing your financial planning to another co-worker will not necessarily yield desired results as you are both individuals with unique financial situations and plans.

What I need to do

As a first step, while still in college, you can seek out coursework in personal finance at your institution. Financial courses, such as the one you are in, can provide you with a foundation on which to craft your personal financial plan. Many times these courses will offer you an opportunity to create some what-if scenarios based on your expected income and financial goals to determine the type of planning with which you are most comfortable.

Being new to the act of financial planning, it will benefit you to seek out advice from your parents, teachers, or others who have utilized some form of personal financial planning. In doing so you can obtain knowledge that could assist you in formalizing a plan. In addition, you will most likely learn certain things to avoid while creating and maintaining a personal financial plan. Experience is a great teacher and looking to those who have been through a similar process will be valuable to your planning.

Finally there are professional sources of assistance when it comes to personal financial planning (https://www.daveramsey.com/, https://www.suzeorman.com/). These sources allow you to learn directly from professionals within the arena of personal finance. They offer tools for you to use in drafting a personal financial plan. In addition there are many tips and tricks provided to keep you on track with your financial goals. There are some costs associated with some of these programs but it could be money well spent to get you on the right track with your finances.

Resources I can use

- Mvelopes (mobile app)
- https://www.thebalance.com/five-steps-to-an-effective-financial-plan-2386045 — effective financial planning

MY MONEY ACTIVITY

Identify what's important in your financial plan (investing, saving, reducing debt, or diversity, for example). Find sources you could use to learn more about accomplishing your goals.

PERSONAL FINANCE

A KIPLINGER APPROACH

"Lend Online and Earn up to 11%" by Elizabeth Leary from Kiplinger's, January 2017, p. 58. Used by permission of The *Kiplinger* Washington Editors, Inc.

INCOME

Lend Online and Earn Up to 11%

Peer-to-peer lending can deliver generous returns, but you'll be taking on a lot of risk. BY ELIZABETH LEARY

HERE'S A RIDDLE: HOW CAN YOU EARN annual income of up to 11% on an investment that pays cash every month and that, on average, has not lost money in any 12-month period?

The answer: peer-to-peer lending, in which ordinary people invest online in loans made to other ordinary people. The rich cash flow those loans deliver is a big reason peer-to-peer lending has exploded in recent years, attracting individuals as well as institutional investors. In 2015, the two largest players, Prosper and Lending Club, issued more than $10 billion in loans, up almost 10-fold from 2012.

It sounds risky, and it is. The loans are unsecured (meaning the lender has nothing to repossess if a borrower defaults), and in many cases the companies originating the loans don't even verify borrowers' incomes. On average, peer-to-peer loans have returned 7.1% annualized over the five-year period that ended on September 30, according to Orchard Platform, a data company that tracks the asset class. That matches the return of the average junk-bond fund over the same period, but peer-to-peer loans delivered those returns with much less volatility. For example, the average junk-bond mutual fund *lost* 4% in 2015, while the average peer-to-peer loan returned 6.9%. "You can get a robust yield compared with other fixed-income asset classes," says Sid Jajodia, chief investment officer of Lending Club.

What's the secret to those appealing returns? For starters, loans are issued at fat interest rates; the average rate on a new Lending Club loan was 12.8% during the third quarter of 2016. Borrowers are willing to pay such a high rate because it looks attractive compared with the average 16.3% rate on new credit cards (most borrowers use the loans to refinance existing debt). Also, the average-return figures encompass hundreds of thousands of loans. Many individual borrowers default, but those defaults are dwarfed by the large proportion of borrowers who repay principal and interest on time. Finally, unlike junk bonds, peer-to-peer loans lack a readily available market price, which means returns reflect only the income that lenders receive, not movements in prices.

But avoid getting carried away by the dazzling numbers. Recent years have been a good time to invest in consumer loans because both joblessness and credit card delinquency rates have been falling. Data to indicate how these loans might perform during an economic downturn is limited because Lending Club and Prosper were in the early stages of operations when the Great Recession struck.

©sanjayart/Shutterstock

BUSINESS MATH ISSUE

Lending online is the wave of the future.

1. List the key points of the article and information to support your position.
2. Write a group defense of your position using math calculations to support your view. If you are in an online course, post to a discussion board.

Classroom Notes

CHAPTER 12

Compound Interest and Present Value

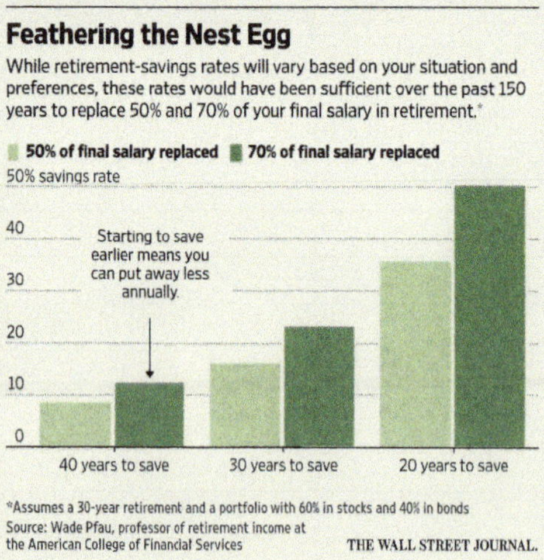

LEARNING UNIT OBJECTIVES

LU 12–1: Compound Interest (Future Value)—The Big Picture

1. Compare simple interest with compound interest.
2. Calculate the compound amount and interest manually and by table lookup.
3. Explain and compute the effective rate (APY).

LU 12–2: Present Value—The Big Picture

1. Compare compound interest (FV) with present value (PV).
2. Compute present value by table lookup.
3. Check the present value answer by compounding.

Your Guide to Successfully Completing This Chapter

Traditional book or ebook

Check box as you complete each step.

Steps

☐ Read learning unit.

  ☐ Complete practice quiz at the end of the learning unit.

☐ Grade practice quiz using provided solutions. (For more help, watch the learning unit video in Connect and have a Study Session with the authors. Then complete the additional practice quiz in Connect.)

☐ Repeat above for each of the two learning units in Chapter 12.

  ☐ Review chapter organizer.

  ☐ Complete assigned homework.

    ☐ Finish summary practice test. (Go to Connect via the ebook link and do the interactive video worksheet to grade.)

☐ Complete instructor's exam.

Wow! The chapter opening *Wall Street Journal* clip, "Feathering the Nest Egg," shows how important it is to start saving early for retirement.

In this chapter we look at the power of compounding—interest paid on earned interest. Let's begin by studying Learning Unit 12–1, which shows you how to calculate compound interest.

Learning Unit 12–1: Compound Interest (Future Value)—The Big Picture

So far we have discussed only simple interest, which is interest on the principal alone. Simple interest is either paid at the end of the loan period or deducted in advance. From the chapter introduction, you know that interest can also be compounded.

My Money

Compounding involves the calculation of interest periodically over the life of the loan (or investment). After each calculation, the interest is added to the principal. Future calculations are on the adjusted principal (old principal plus interest). **Compound interest,** then, is the interest on the principal plus the interest of prior periods. **Future value (FV),** or the **compound amount,** is the final amount of the loan or investment at the end of the last period. In the beginning of this unit, do not be concerned with how to calculate compounding but try to understand the meaning of compounding.

Figure 12.1 shows how $1 will grow if it is calculated for 4 years at 8% annually. This means that the interest is calculated on the balance once a year. In Figure 12.1, we start with $1, which is the **present value (PV).** After year 1, the dollar with interest is worth $1.08.

FIGURE 12.1

Future value of $1 at 8% for four periods

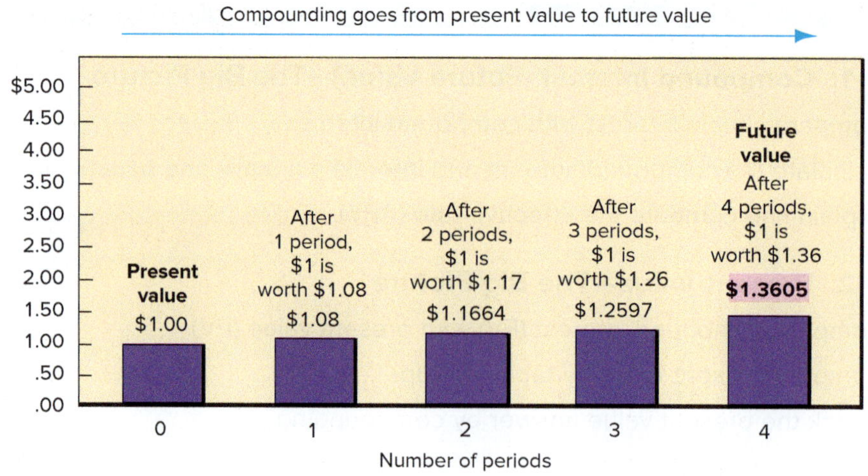

At the end of year 2, the dollar is worth $1.17. By the end of year 4, the dollar is worth $1.36. Note how we start with the present and look to see what the dollar will be worth in the future. *Compounding goes from present value to future value.*

Before you learn how to calculate compound interest and compare it to simple interest, you must understand the terms that follow. These terms are also used in Chapter 13.

- **Compounded annually:** Interest calculated on the balance once a year.
- **Compounded semiannually:** Interest calculated on the balance every 6 months or every $\frac{1}{2}$ years.
- **Compounded quarterly:** Interest calculated on the balance every 3 months or every $\frac{1}{4}$ years.
- **Compounded monthly:** Interest calculated on the balance each month.
- **Compounded daily:** Interest calculated on the balance each day.
- **Number of periods:**[1] Number of years multiplied by the number of times the interest is compounded per year. For example, if you compound $1 for 4 years at 8% annually, semiannually, or quarterly, the following periods will result:

Annually:	4 years × 1 = 4 periods
Semiannually:	4 years × 2 = 8 periods
Quarterly:	4 years × 4 = 16 periods

- **Rate for each period:**[2] Annual interest rate divided by the number of times the interest is compounded per year. Compounding changes the interest rate for annual, semiannual, and quarterly periods as follows:

Annually:	8% ÷ 1 = 8%
Semiannually:	8% ÷ 2 = 4%
Quarterly:	8% ÷ 4 = 2%

  Note that both the number of periods (4) and the rate (8%) for the annual example did not change. You will see later that rate and periods (not years) will always change unless interest is compounded yearly.

  Now you are ready to learn the difference between simple interest and compound interest.

LO 1

Simple versus Compound Interest

Did you know that money invested at 6% will double in 12 years? The following *Wall Street Journal* clip "Confused by Investing?" shows how to calculate the number of years it takes for your investment to double. Although this clip is from 2003, its information is as current today as it was then. It explains compounding and the rule of 72, so read it carefully.

The following three situations of Bill Smith will clarify the difference between simple interest and compound interest.

[1]Periods are often expressed with the letter *N* or *n* for number of periods.

[2]Rate is often expressed with the letter *i* for interest.

My Money

Confused by Investing?

If there's something about your investment portfolio that doesn't seem to add up, maybe you should check your math.

Lots of folks are perplexed by the mathematics of investing, so I thought a refresher course might help. Here's a look at some key concepts:

■ **10 Plus 10 is 21**

Imagine you invest $100, which earns 10% this year and 10% next. How much have you made? If you answered 21%, go to the head of the class.

Here's how the math works. This year's 10% gain turns your $100 into $110. Next year, you also earn 10%, but you start the year with $110. Result? You earn $11, boosting your wealth to $121.

Thus, your portfolio has earned a *cumulative* 21% return over two years, but the *annualized* return is just 10%. The fact that 21% is more than double 10% can be attributed to the effect of investment compounding, the way that you earn money each year not only on your original investment, but also on earnings from prior years that you've reinvested.

■ **The Rule of 72**

To get a feel for compounding, try the rule of 72. What's that? If you divide a particular annual return into 72, you'll find out how many years it will take to double your money. Thus, at 10% a year, an investment will double in value in a tad over seven years.

Situation 1: Calculating Simple Interest and Maturity Value

EXAMPLE Bill Smith deposited $80 in a savings account for 4 years at an annual interest rate of 8%. What is Bill's simple interest?

To calculate simple interest, we use the following simple interest formula:

$$\text{Interest } (I) = \text{Principal } (P) \times \text{Rate } (R) \times \text{Time } (T)$$

$$\$25.60 = \$80 \times .08 \times 4$$

In 4 years Bill receives a total of $105.60 ($80.00 + $25.60)—principal plus simple interest.

Now let's look at the interest Bill would earn if the bank compounded Bill's interest on his savings.

Situation 2: Calculating Compound Amount and Interest without Tables[3]

You can use the following steps to calculate the compound amount and the interest manually:

CALCULATING COMPOUND AMOUNT AND INTEREST MANUALLY

Step 1. Calculate the simple interest and add it to the principal. Use this total to figure next year's interest.

Step 2. Repeat for the total number of periods.

Step 3. Compound amount − Principal = Compound interest.

EXAMPLE Bill Smith deposited $80 in a savings account for 4 years at an annual compounded rate of 8%. What are Bill's compound amount and interest?

The following shows how the compounded rate affects Bill's interest:

	Year 1	Year 2	Year 3	Year 4
	$80.00	$86.40	$ 93.31	$100.77
	× .08	× .08	× .08	× .08
Interest	$ 6.40	$ 6.91	$ 7.46	$ 8.06
Beginning balance	+ 80.00	+ 86.40	+ 93.31	+ 100.77
Amount at year-end	$86.40	$93.31	$100.77	$108.83

[3]For simplicity of presentation, round each calculation to the nearest cent before continuing the compounding process. The compound amount will be off by 1 cent.

Note that the beginning year 2 interest is the result of the interest of year 1 added to the principal. At the end of each interest period, we add on the period's interest. This interest becomes part of the principal we use for the calculation of the next period's interest. We can determine Bill's compound interest as follows:[4]

Compound amount	$108.83	
Principal	− 80.00	
Compound interest	$ 28.83	*Note:* In Situation 1 the interest was $25.60.

We could have used the following simplified process to calculate the compound amount and interest:

Year 1	Year 2	Year 3	Year 4	
$80.00	$86.40	$ 93.31	$100.77	
× 1.08	× 1.08	× 1.08	× 1.08	
$86.40	$93.31	$100.77	$108.83	← Future value

When using this simplification, you do not have to add the new interest to the previous balance. Remember that compounding results in higher interest than simple interest. Compounding is the *sum* of principal and interest multiplied by the interest rate we use to calculate interest for the next period. So, 1.08 above is 108%, with 100% as the base and 8% as the interest.

LO 2

Situation 3: Calculating Compound Amount by Table Lookup To calculate the compound amount with a future value table, use the following steps:

CALCULATING COMPOUND AMOUNT BY TABLE LOOKUP

Step 1. Find the periods: Years multiplied by number of times interest is compounded in 1 year.

Step 2. Find the rate: Annual rate divided by number of times interest is compounded in 1 year.

Step 3. Go down the Period column of the table to the number of periods desired; look across the row to find the rate. At the intersection of the two columns is the table factor for the compound amount of $1.

Step 4. Multiply the table factor by the amount of the loan. This gives the compound amount.

In Situation 2, Bill deposited $80 into a savings account for 4 years at an interest rate of 8% compounded annually. Bill heard that he could calculate the compound amount and interest by using tables. In Situation 3, Bill learns how to do this. Again, Bill wants to know the value of $80 in 4 years at 8%. He begins by using Table 12.1.

Four Periods

1 × 4

No. of times compounded in 1 year; No. of years

Looking at Table 12.1, Bill goes down the Period column to period 4 and then across the row to the 8% column. At the intersection, Bill sees the number 1.3605. The marginal notes show how Bill arrived at the periods and rate. The 1.3605 table number means that $1 compounded at this rate will increase in value in 4 years to about $1.36. Do you recognize the $1.36? Figure 12.1 showed how $1 grew to $1.36. Since Bill wants to know the value of $80, he multiplies the dollar amount by the table factor as follows:

$80.00 × 1.3605 = $108.84*

Principal × Table factor = Compound amount (future value)

*Off 1 cent due to rounding.

[4]The formula for compounding is $A = P(1 + i)^N$, where A equals compound amount, P equals the principal, i equals interest per period, and N equals number of periods. The calculator sequence would be as follows for Bill Smith: 1 [+] .08 [y^x] 4 × 80 [=] 108.84. A Financial Calculator Guide booklet is available online that shows how to operate HP 10BII and TI BA II Plus.

TABLE 12.1 Future value of $1 at compound interest

Period	1%	$1\frac{1}{2}$%	2%	3%	4%	5%	6%	7%	8%	9%	10%
1	1.0100	1.0150	1.0200	1.0300	1.0400	1.0500	1.0600	1.0700	1.0800	1.0900	1.1000
2	1.0201	1.0302	1.0404	1.0609	1.0816	1.1025	1.1236	1.1449	1.1664	1.1881	1.2100
3	1.0303	1.0457	1.0612	1.0927	1.1249	1.1576	1.1910	1.2250	1.2597	1.2950	1.3310
4	1.0406	1.0614	1.0824	1.1255	1.1699	1.2155	1.2625	1.3108	1.3605	1.4116	1.4641
5	1.0510	1.0773	1.1041	1.1593	1.2167	1.2763	1.3382	1.4026	1.4693	1.5386	1.6105
6	1.0615	1.0934	1.1262	1.1941	1.2653	1.3401	1.4185	1.5007	1.5869	1.6771	1.7716
7	1.0721	1.1098	1.1487	1.2299	1.3159	1.4071	1.5036	1.6058	1.7138	1.8280	1.9487
8	1.0829	1.1265	1.1717	1.2668	1.3686	1.4775	1.5938	1.7182	1.8509	1.9926	2.1436
9	1.0937	1.1434	1.1951	1.3048	1.4233	1.5513	1.6895	1.8385	1.9990	2.1719	2.3579
10	1.1046	1.1605	1.2190	1.3439	1.4802	1.6289	1.7908	1.9672	2.1589	2.3674	2.5937
11	1.1157	1.1780	1.2434	1.3842	1.5395	1.7103	1.8983	2.1049	2.3316	2.5804	2.8531
12	1.1268	1.1960	1.2682	1.4258	1.6010	1.7959	2.0122	2.2522	2.5182	2.8127	3.1384
13	1.1381	1.2135	1.2936	1.4685	1.6651	1.8856	2.1329	2.4098	2.7196	3.0658	3.4523
14	1.1495	1.2318	1.3195	1.5126	1.7317	1.9799	2.2609	2.5785	2.9372	3.3417	3.7975
15	1.1610	1.2502	1.3459	1.5580	1.8009	2.0789	2.3966	2.7590	3.1722	3.6425	4.1772
16	1.1726	1.2690	1.3728	1.6047	1.8730	2.1829	2.5404	2.9522	3.4259	3.9703	4.5950
17	1.1843	1.2880	1.4002	1.6528	1.9479	2.2920	2.6928	3.1588	3.7000	4.3276	5.0545
18	1.1961	1.3073	1.4282	1.7024	2.0258	2.4066	2.8543	3.3799	3.9960	4.7171	5.5599
19	1.2081	1.3270	1.4568	1.7535	2.1068	2.5270	3.0256	3.6165	4.3157	5.1417	6.1159
20	1.2202	1.3469	1.4859	1.8061	2.1911	2.6533	3.2071	3.8697	4.6610	5.6044	6.7275
21	1.2324	1.3671	1.5157	1.8603	2.2788	2.7860	3.3996	4.1406	5.0338	6.1088	7.4002
22	1.2447	1.3876	1.5460	1.9161	2.3699	2.9253	3.6035	4.4304	5.4365	6.6586	8.1403
23	1.2572	1.4084	1.5769	1.9736	2.4647	3.0715	3.8197	4.7405	5.8715	7.2579	8.9543
24	1.2697	1.4295	1.6084	2.0328	2.5633	3.2251	4.0489	5.0724	6.3412	7.9111	9.8497
25	1.2824	1.4510	1.6406	2.0938	2.6658	3.3864	4.2919	5.4274	6.8485	8.6231	10.8347
26	1.2953	1.4727	1.6734	2.1566	2.7725	3.5557	4.5494	5.8074	7.3964	9.3992	11.9182
27	1.3082	1.4948	1.7069	2.2213	2.8834	3.7335	4.8223	6.2139	7.9881	10.2451	13.1100
28	1.3213	1.5172	1.7410	2.2879	2.9987	3.9201	5.1117	6.6488	8.6271	11.1672	14.4210
29	1.3345	1.5400	1.7758	2.3566	3.1187	4.1161	5.4184	7.1143	9.3173	12.1722	15.8631
30	1.3478	1.5631	1.8114	2.4273	3.2434	4.3219	5.7435	7.6123	10.0627	13.2677	17.4494

Note: For more detailed tables, see your reference booklet, the *Business Math Handbook.*

Note all the table factors in the future value table are greater than 1. That is because investments increase in value over time due to interest being earned.

Figure 12.2 illustrates this compounding procedure. We can say that compounding is a future value (FV) since we are looking into the future. Thus,

8% Rate

8% rate = $\frac{8\%}{1}$ → Annual rate / → No. of times compounded in 1 year

$108.84 – $80.00 = $28.84 interest for 4 years at 8% compounded annually on $80.00

Now let's look at two examples that illustrate compounding more than once a year.

FIGURE 12.2

Compounding (FV)

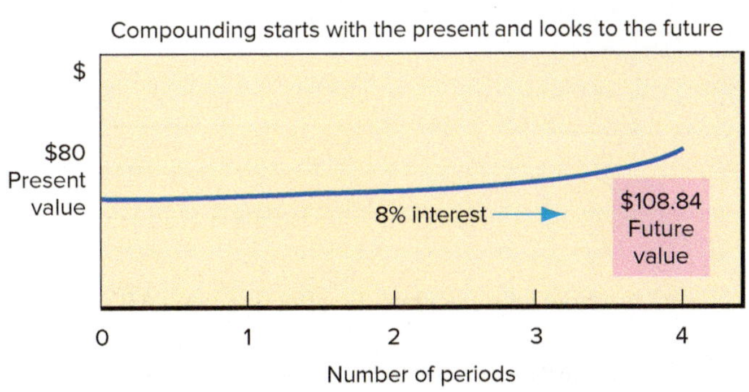

EXAMPLE Find the interest on $6,000 at 10% compounded semiannually for 5 years. We calculate the interest as follows:

Periods = 2 × 5 years = 10

Rate = 10% ÷ 2 = 5%

10 periods, 5%, in Table 12.1 = 1.6289 (table factor)

$6,000 × 1.6289 = $9,773.40
− 6,000.00
$3,773.40 interest

EXAMPLE Pam Donahue deposits $8,000 in her savings account that pays 6% interest compounded quarterly. What will be the balance of her account at the end of 5 years?

Periods = 4 × 5 years = 20

Rate = 6% ÷ 4 = $1\frac{1}{2}\%$

20 periods, $1\frac{1}{2}\%$, in Table 12.1 = 1.3469 (table factor)

8,000 × 1.3469 = **$10,775.20**

Next, let's look at bank rates and how they affect interest.

LO 3

Bank Rates—Nominal versus Effective Rates (Annual Percentage Yield, or APY)

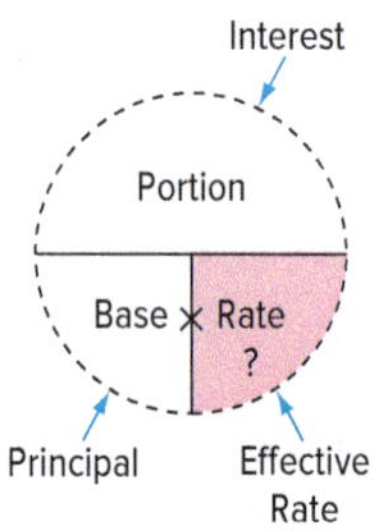

Banks often advertise their annual (nominal) interest rates and *not* their true or effective rate (annual percentage yield, or APY). This has made it difficult for investors and depositors to determine the actual rates of interest they were receiving. The Truth in Savings law forced savings institutions to reveal their actual rate of interest. The APY is defined in the Truth in Savings law as the percentage rate expressing the total amount of interest that would be received on a $100 deposit based on the annual rate and frequency of compounding for a 365-day period. As you can see from the advertisement on the left, banks now refer to the effective rate of interest as the annual percentage yield.

Let's study the rates of two banks to see which bank has the better return for the investor. Blue Bank pays 8% interest compounded quarterly on $8,000. Sun Bank offers 8% interest compounded semiannually on $8,000. The 8% rate is the **nominal rate,** or stated rate, on which the bank calculates the interest. To calculate the **effective rate (annual percentage yield,** or **APY),** however, we can use the following formula:

$$\text{Effective rate (APY)} = \frac{\text{Interest for 1 year}}{\text{Principal}}$$

Now let's calculate the effective rate (APY)[5] for Blue Bank and Sun Bank.

[5]Round to the nearest hundredth percent as needed. In practice, the rate is often rounded to the nearest thousandth.

Note the effective rates (APY) can be seen from Table 12.1 for $1:
1.0824 ← 4 periods, 2%
1.0816 ← 2 periods, 4%

Blue, 8% compounded quarterly	Sun, 8% compounded semiannually
Periods = 4 (4 × 1)	Periods = 2 (2 × 1)
Percent = $\frac{8\%}{4}$ = 2%	Percent = $\frac{8\%}{2}$ = 4%
Principal = $8,000	Principal = $8,000
Table 12.1 lookup: 4 periods, 2%	Table 12.1 lookup: 2 periods, 4%
1.0824	1.0816
× $8,000	× $8,000
Less $8,659.20	$8,652.80
principal − 8,000.00	− 8,000.00
$ 659.20	$ 652.80
Effective rate (APY) = $\frac{\$659.20}{\$8,000}$ = .0824	$\frac{\$652.80}{\$8,000}$ = .0816
= 8.24%	= 8.16%

FIGURE 12.3

Nominal and effective rates (APY) of interest compared

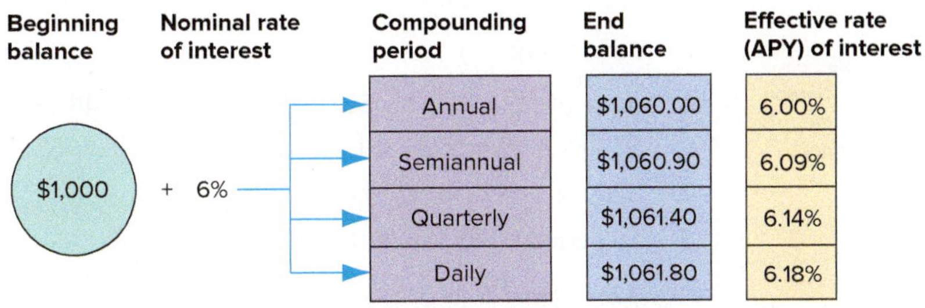

MONEY tips

Whenever possible, contribute the maximum allowed to your retirement plan(s) to defer taxes and help you prepare for a healthy financial retirement. This is especially important if your employer provides a match for your contributions.

Figure 12.3 illustrates a comparison of nominal and effective rates (APY) of interest. This comparison should make you question any advertisement of interest rates before depositing your money.

Before concluding this unit, we briefly discuss compounding interest daily.

Compounding Interest Daily

Although many banks add interest to each account quarterly, some banks pay interest that is compounded daily, and other banks use *continuous compounding*. Remember that continuous compounding sounds great, but in fact, it yields only a fraction of a percent more interest over a year than daily compounding. Today, computers perform these calculations.

Table 12.2 is a partial table showing what $1 will grow to in the future by daily compounded interest, 360-day basis. For example, we can calculate interest compounded daily on $900 at 6% per year for 25 years as follows:

$900 × 4.4811 = $4,032.99 daily compounding

TABLE 12.2 Interest on a $1 deposit compounded daily—360-day basis

Number of years	6.00%	6.50%	7.00%	7.50%	8.00%	8.50%	9.00%	9.50%	10.00%
1	1.0618	1.0672	1.0725	1.0779	1.0833	1.0887	1.0942	1.0996	1.1052
2	1.1275	1.1388	1.1503	1.1618	1.1735	1.1853	1.1972	1.2092	1.2214
3	1.1972	1.2153	1.2337	1.2523	1.2712	1.2904	1.3099	1.3297	1.3498
4	1.2712	1.2969	1.3231	1.3498	1.3771	1.4049	1.4333	1.4622	1.4917
5	1.3498	1.3840	1.4190	1.4549	1.4917	1.5295	1.5682	1.6079	1.6486
6	1.4333	1.4769	1.5219	1.5682	1.6160	1.6652	1.7159	1.7681	1.8220
7	1.5219	1.5761	1.6322	1.6904	1.7506	1.8129	1.8775	1.9443	2.0136
8	1.6160	1.6819	1.7506	1.8220	1.8963	1.9737	2.0543	2.1381	2.2253
9	1.7159	1.7949	1.8775	1.9639	2.0543	2.1488	2.2477	2.3511	2.4593
10	1.8220	1.9154	2.0136	2.1168	2.2253	2.3394	2.4593	2.5854	2.7179
15	2.4594	2.6509	2.8574	3.0799	3.3197	3.5782	3.8568	4.1571	4.4808
20	3.3198	3.6689	4.0546	4.4810	4.9522	5.4728	6.0482	6.6842	7.3870
25	4.4811	5.0777	5.7536	6.5195	7.3874	8.3708	9.4851	10.7477	12.1782
30	6.0487	7.0275	8.1645	9.4855	11.0202	12.8032	14.8747	17.2813	20.0772

Now it's time to check your progress with the following Practice Quiz.

LU 12–1 PRACTICE QUIZ

Complete this **Practice Quiz** to see how you are doing.

1. Complete the following without a table (round each calculation to the nearest cent as needed):

Principal	Time	Rate of compound interest	Compounded	Number of periods to be compounded	Total amount	Total interest
$200	1 year	8%	Quarterly	**a.**	**b.**	**c.**

2. Solve the previous problem by using compound value (FV) in Table 12.1.
3. Lionel Rodgers deposits $6,000 in Victory Bank, which pays 3% interest compounded semiannually. How much will Lionel have in his account at the end of 8 years?
4. Find the effective rate (APY) for the year: principal, $7,000; interest rate, 12%; and compounded quarterly.
5. Calculate by Table 12.2 what $1,500 compounded daily for 5 years will grow to at 7%.

*For **extra help** from your authors–Sharon and Jeff–see the videos in Connect.*

✓ Solutions

1. **a.** 4 (4 × 1)   **b.** $216.48   **c.** $16.48 ($216.48 – $200)
   $200 × 1.02 = $204 × 1.02 = $208.08 × 1.02 = $212.24 × 1.02 = $216.48
2. $200 × 1.0824 = $216.48 (4 periods, 2%)
3. 16 periods, $1\frac{1}{2}$%, $6,000 × 1.2690 = $7,614
4. 4 periods, 3%

$$\$7{,}000 \times 1.1255 = \begin{array}{r} \$7{,}878.50 \\ -\ 7{,}000.00 \\ \hline \$\ 878.50 \end{array} \qquad \frac{\$878.50}{\$7{,}000.00} = 12.55\%$$

5. $1,500 × 1.4190 = $2,128.50

Check out the plastic overlays that appear at the end of Chapter 13 to review these concepts.

Learning Unit 12–2: Present Value—The Big Picture

Figure 12.1 in Learning Unit 12–1 showed how by compounding, the *future value* of $1 became $1.36. This learning unit discusses *present value.* Before we look at specific calculations involving present value, let's look at the concept of present value.

My Money

Figure 12.4 shows that if we invested 74 cents today, compounding would cause the 74 cents to grow to $1 in the future. For example, let's assume you ask this question: "If I need $1 in 4 years in the future, how much must I put in the bank *today* (assume an 8% annual interest)?" To answer this question, you must know the present value of that $1 today.

FIGURE 12.4

Present value of $1 at 8% for four periods

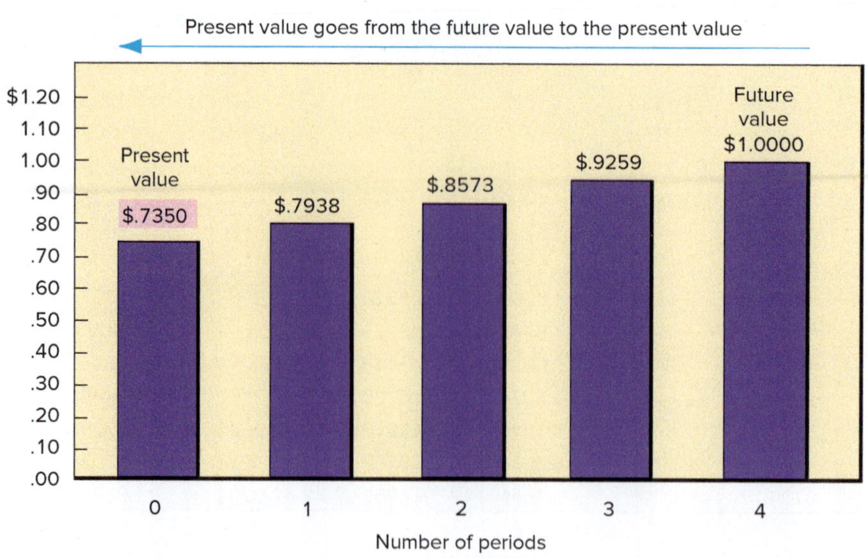

FIGURE 12.5

Present value

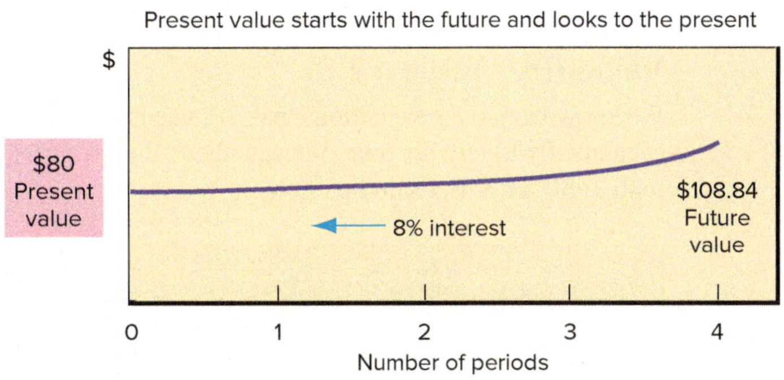

From Figure 12.4, you can see that the present value of $1 is .7350. Remember that the $1 is only worth 74 cents if you wait 4 periods to receive it. This is one reason why so many athletes get such big contracts—much of the money is paid in later years when it is not worth as much.

LO 1

Relationship of Compounding (FV) to Present Value (PV)—The Bill Smith Example Continued

In Learning Unit 12–1, our consideration of compounding started in the *present* ($80) and looked to find the *future* amount of $108.84. Present value (PV) starts with the *future* and tries to calculate its worth in the *present* ($80). For example, in Figure 12.5, we assume Bill Smith knew that in 4 years he wanted to buy a bike that cost $108.84 (future). Bill's bank pays 8% interest compounded annually. How much money must Bill put in the bank *today* (present) to have $108.84 in 4 years? To work from the future to the present, we can use a present value (PV) table. In the next section you will learn how to use this table.

LO 2

How to Use a Present Value (PV) Table[6]

To calculate present value with a present value table, use the following steps:

CALCULATING PRESENT VALUE BY TABLE LOOKUP

Step 1. Find the periods: Years multiplied by number of times interest is compounded in 1 year.

Step 2. Find the rate: Annual rate divided by number of times interest is compounded in 1 year.

Step 3. Go down the Period column of the table to the number of periods desired; look across the row to find the rate. At the intersection of the two columns is the table factor for the compound value of $1.

Step 4. Multiply the table factor times the future value. This gives the present value.

Four Periods

4 × 1 = 4

No. of years; No. of times compounded in 1 year

Table 12.3 is a present value (PV) table that tells you what $1 is worth today at different interest rates. To continue our Bill Smith example, go down the Period column in Table 12.3 to 4. Then go across to the 8% column. At 8% for 4 periods, we see a table factor of .7350. This means that $1 in the future is worth approximately 74 cents today. If Bill invested 74 cents today at 8% for 4 periods, Bill would have $1 in 4 years.

Since Bill knows the bike will cost $108.84 in the future, he completes the following calculation:

$108.84 × .7350 = $80.00

This means that $108.84 in today's dollars is worth $80.00. Now let's check this.

[6]The formula for present value is $PV = \frac{A}{(1+i)^N}$, where A equals future amount (compound amount), N equals number of compounding periods, and i equals interest rate per compounding period. The calculator sequence for Bill Smith would be as follows: 1 [+] .08 [y^x] 4 [=] [M+] 108.84 [÷] [MR] [=] 80.03.

LO 3

Comparing Compound Interest (FV) Table 12.1 with Present Value (PV) Table 12.3

We know from our calculations that Bill needs to invest $80 for 4 years at 8% compound interest annually to buy his bike. We can check this by going back to Table 12.1 and comparing it with Table 12.3. Let's do this now.

Compound value Table 12.1				Present value Table 12.3			
Table 12.1	Present value		Future value	Table 12.3	Future value		Present value
1.3605	× $80.00	=	$108.84	.7350	× $108.84	=	$80.00
(4 per., 8%)				(4 per., 8%)			
We know the present dollar amount and find what the dollar amount is worth in the future.				We know the future dollar amount and find what the dollar amount is worth in the present.			

TABLE 12.3 Present value of $1 at end of period

Period	1%	$1\frac{1}{2}$%	2%	3%	4%	5%	6%	7%	8%	9%	10%
1	.9901	.9852	.9804	.9709	.9615	.9524	.9434	.9346	.9259	.9174	.9091
2	.9803	.9707	.9612	.9426	.9246	.9070	.8900	.8734	.8573	.8417	.8264
3	.9706	.9563	.9423	.9151	.8890	.8638	.8396	.8163	.7938	.7722	.7513
4	.9610	.9422	.9238	.8885	.8548	.8227	.7921	.7629	.7350	.7084	.6830
5	.9515	.9283	.9057	.8626	.8219	.7835	.7473	.7130	.6806	.6499	.6209
6	.9420	.9145	.8880	.8375	.7903	.7462	.7050	.6663	.6302	.5963	.5645
7	.9327	.9010	.8706	.8131	.7599	.7107	.6651	.6227	.5835	.5470	.5132
8	.9235	.8877	.8535	.7894	.7307	.6768	.6274	.5820	.5403	.5019	.4665
9	.9143	.8746	.8368	.7664	.7026	.6446	.5919	.5439	.5002	.4604	.4241
10	.9053	.8617	.8203	.7441	.6756	.6139	.5584	.5083	.4632	.4224	.3855
11	.8963	.8489	.8043	.7224	.6496	.5847	.5268	.4751	.4289	.3875	.3505
12	.8874	.8364	.7885	.7014	.6246	.5568	.4970	.4440	.3971	.3555	.3186
13	.8787	.8240	.7730	.6810	.6006	.5303	.4688	.4150	.3677	.3262	.2897
14	.8700	.8119	.7579	.6611	.5775	.5051	.4423	.3878	.3405	.2992	.2633
15	.8613	.7999	.7430	.6419	.5553	.4810	.4173	.3624	.3152	.2745	.2394
16	.8528	.7880	.7284	.6232	.5339	.4581	.3936	.3387	.2919	.2519	.2176
17	.8444	.7764	.7142	.6050	.5134	.4363	.3714	.3166	.2703	.2311	.1978
18	.8360	.7649	.7002	.5874	.4936	.4155	.3503	.2959	.2502	.2120	.1799
19	.8277	.7536	.6864	.5703	.4746	.3957	.3305	.2765	.2317	.1945	.1635
20	.8195	.7425	.6730	.5537	.4564	.3769	.3118	.2584	.2145	.1784	.1486
21	.8114	.7315	.6598	.5375	.4388	.3589	.2942	.2415	.1987	.1637	.1351
22	.8034	.7207	.6468	.5219	.4220	.3418	.2775	.2257	.1839	.1502	.1228
23	.7954	.7100	.6342	.5067	.4057	.3256	.2618	.2109	.1703	.1378	.1117
24	.7876	.6995	.6217	.4919	.3901	.3101	.2470	.1971	.1577	.1264	.1015
25	.7798	.6892	.6095	.4776	.3751	.2953	.2330	.1842	.1460	.1160	.0923
26	.7720	.6790	.5976	.4637	.3607	.2812	.2198	.1722	.1352	.1064	.0839
27	.7644	.6690	.5859	.4502	.3468	.2678	.2074	.1609	.1252	.0976	.0763
28	.7568	.6591	.5744	.4371	.3335	.2551	.1956	.1504	.1159	.0895	.0693
29	.7493	.6494	.5631	.4243	.3207	.2429	.1846	.1406	.1073	.0822	.0630
30	.7419	.6398	.5521	.4120	.3083	.2314	.1741	.1314	.0994	.0754	.0573
35	.7059	.5939	.5000	.3554	.2534	.1813	.1301	.0937	.0676	.0490	.0356
40	.6717	.5513	.4529	.3066	.2083	.1420	.0972	.0668	.0460	.0318	.0221

Note: For more detailed tables, see your booklet, the *Business Math Handbook*.

FIGURE 12.6

Present value

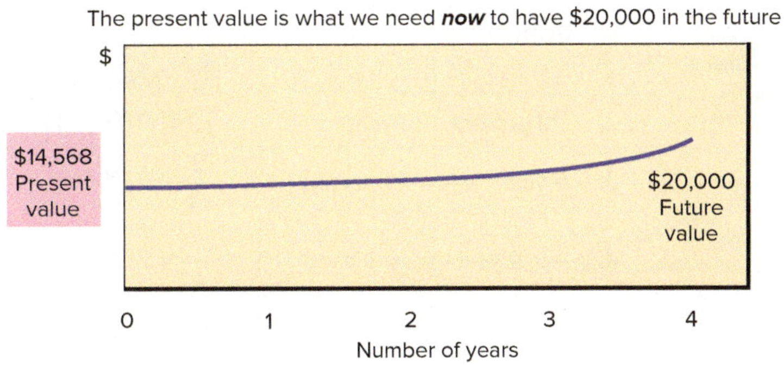

Note all the table factors in the present value table are less than 1. That is because money grows over time due to interest being earned so less money needs to be invested today to meet a future higher obligation.

Note that the table factor for compounding is over 1 (1.3605) and the table factor for present value is less than 1 (.7350). The compound value table starts with the present and goes to the future. The present value table starts with the future and goes to the present.

Let's look at another example before trying the Practice Quiz.

MONEY tips

Almost 50% of full-time workers do **not** participate in their employer's retirement plan. Many companies have dollar matching programs allowing employees to receive "free" money in addition to helping reduce their taxable income. Check with your company to see what it offers.

EXAMPLE Rene Weaver needs $20,000 for college in 4 years. She can earn 8% compounded quarterly at her bank. How much must Rene deposit at the beginning of the year to have $20,000 in 4 years?

Remember that in this example the bank compounds the interest *quarterly.* Let's first determine the period and rate on a quarterly basis:

$$\text{Periods} = 4 \times 4 \text{ years} = 16 \text{ periods} \qquad \text{Rate} = \frac{8\%}{4} = 2\%$$

Now we go to Table 12.3 and find 16 under the Period column. We then move across to the 2% column and find the .7284 table factor.

$$\$20{,}000 \times .7284 = \$14{,}568$$

(future value) (present value)

We illustrate this in Figure 12.6.

We can check the $14,568 present value by using the compound value Table 12.1:

16 periods, 2% column = 1.3728 × $14,568 = $19,998.95*

*Not quite $20,000 due to rounding of table factors.

Let's test your understanding of this unit with the Practice Quiz.

LU 12–2 PRACTICE QUIZ

Complete this **Practice Quiz** to see how you are doing.

Use the present value Table 12.3 to complete:

	Future amount desired	Length of time	Rate compounded	Table period	Rate used	PV factor	PV amount
1.	$ 7,000	6 years	6% semiannually	____	____	____	____
2.	$15,000	20 years	10% annually	____	____	____	____

3. Bill Blum needs $20,000 6 years from today to attend V.P.R. Tech. How much must Bill put in the bank today (12% quarterly) to reach his goal?

4. Bob Fry wants to buy his grandson a Ford Taurus in 4 years. The cost of a car will be $24,000. Assuming a bank rate of 8% compounded quarterly, how much must Bob put in the bank today?

*For **extra help** from your authors–Sharon and Jeff–see the videos in Connect.*

✓ Solutions

1.	12 periods (6 years × 2)	3% (6% ÷ 2)	.7014 $4,909.80 ($7,000 × .7014)
2.	20 periods (20 years × 1)	10% (10% ÷ 1)	.1486 $2,229.00 ($15,000 × .1486)
3.	6 years × 4 = 24 periods	$\frac{12\%}{4} = 3\%$	.4919 × $20,000 = $9,838
4.	4 × 4 years = 16 periods	$\frac{8\%}{4} = 2\%$	.7284 × $24,000 = $17,481.60

INTERACTIVE CHAPTER ORGANIZER

Topic/Procedure/Formula	Examples	You try it*
Calculating compound amount without tables (future value)† Determine new amount by multiplying rate times new balance (that includes interest added on). Start in present and look to future. Compound interest = Compound amount − Principal ⟼Compounding⟼ PV → FV	$100 in savings account, compounded annually for 2 years at 8%: $100 × 1.08 = $108 $108 × 1.08 = $116.64 (future value)	**Calculate compound amount (future value)** $200 for 2 years at 4%, compounded annually
Calculating compound amount (future value) by table lookup Periods = Number of times compounded per year × Years of loan $\text{Rate} = \frac{\text{Annual rate}}{\text{Number of times compounded per year}}$ Multiply table factor (intersection of period and rate) times amount of principal.	*Example:* $2,000 at 12% for 5 years and compounded quarterly: Periods = 4 × 5 years = 20 $\text{Rate} = \frac{12\%}{4} = 3\%$ 20 periods, 3% = 1.8061 (table factor) $2,000 × 1.8061 = $3,612.20 (future value)	**Calculate compound amount by table lookup** $4,000 at 6% for 6 years, compounded semiannually
Calculating effective rate (APY) $\text{Effective rate (APY)} = \frac{\text{Interest for 1 year}}{\text{Principal}}$ or Rate can be seen in Table 12.1 factor.	$1,000 at 10% compounded semiannually for 1 year. By Table 12.1: 2 periods, 5% 1.1025 means at end of year investor has earned 110.25% of original principal. Thus the interest is 10.25%. $1,000 × 1.1025 = $1,102.50 − 1,000.00 $ 102.50 $\frac{\$102.50}{\$1,000} = 10.25\%$ effective rate (APY)	**Calculate effective rate** $4,000 at 6% for 1 year, compounded semiannually
Calculating present value (PV) by table lookup‡ Start with future and calculate worth in the present. Periods and rate computed like in compound interest. ⟼Present value⟼ PV ← FV Find periods and rate. Multiply table factor (intersection of period and rate) times amount.	*Example:* Want $3,612.20 after 5 years with rate of 12% compounded quarterly: Periods = 4 × 5 = 20; % = 3% By Table 12.3: 20 periods, 3% = .5537 $3,612.20 × .5537 = $2,000.08 Invested today will yield desired amount in future	**Calculate present value by table lookup** Want $6,000 after 4 years with rate of 6%, compounded quarterly

(continues)

INTERACTIVE CHAPTER ORGANIZER

KEY TERMS			
	Annual percentage yield (APY) Compound amount Compound interest Compounded annually Compounded daily	Compounded monthly Compounded quarterly Compounded semiannually Compounding Effective rate Future value (FV)	Nominal rate Number of periods Present value (PV) Rate for each period

*Worked-out solutions are in Appendix B.

†$A = P(1 + i)^N$.

‡$\frac{A}{(1 + i)^N}$ if table not used.

Critical Thinking Discussion Questions with Chapter Concept Check

1. Explain how periods and rates are calculated in compounding problems. Compare simple interest to compound interest.
2. What are the steps to calculate the compound amount by table? Why is the compound table factor greater than $1?
3. What is the effective rate (APY)? Why can the effective rate be seen directly from the table factor?
4. Explain the difference between compounding and present value. Why is the present value table factor less than $1?
5. **Chapter Concept Check.** Create a problem using present value and compounding to show the amount you would need to put away today in a bank to have enough money to pay for a child's costs through the age of 18. Assume your own rates and periods and that the amount you put in the bank is one lump sum that will grow through compounding (without new investments).

Classroom Notes

12

END-OF-CHAPTER PROBLEMS

connect

Check figures for odd-numbered problems in Appendix B. Name ______________________ Date ________

DRILL PROBLEMS

Complete the following without using Table 12.1 (round to the nearest cent for each calculation) and then check your answer by Table 12.1 (check will be off due to rounding). *LU 12-1(2)*

	Principal	Time (years)	Rate of compound interest	Compounded	Periods	Rate	Total amount	Total interest
12–1.	$575	1	4%	Quarterly				

Complete the following using compound future value Table 12.1 or the *Business Math Handbook*: *LU 12-1(2)*

	Time	Principal	Rate	Compounded	Amount	Interest
12–2.	12 years	$15,000	$3\frac{1}{2}$%	Annually		
12–3.	6 months	$15,000	6%	Semiannually		
12–4.	2 years	$15,000	8%	Quarterly		

Calculate the effective rate (APY) of interest for 1 year. *LU 12-1(3)*

12–5. Principal: $15,500

Interest rate: 12%

Compounded quarterly

Effective rate (APY):

12–6. Using Table 12.2, calculate what $700 would grow to at $6\frac{1}{2}$% per year compounded daily for 7 years. *LU 12-1(3)*

Complete the following using present value Table 12.3 or the present value table in the *Business Math Handbook*. *LU 12-2(2)*

		Amount desired at end of period	Length of time	Rate	Compounded	On PV Table 12.3: Period used	On PV Table 12.3: Rate used	PV factor used	PV of amount desired at end of period
eXcel	**12–7.**	$6,000	8 years	3%	Semiannually				
eXcel	**12–8.**	$8,900	4 years	6%	Monthly				
eXcel	**12–9.**	$17,600	7 years	12%	Quarterly				

eXcel **12–10.** $20,000 20 years 8% Annually

12–11. Check your answer in Problem 12–9 by the compound value Table 12.1. The answer will be off due to rounding. *LU 12-2(3)*

WORD PROBLEMS

My Money

12–12. Sam Long anticipates he will need approximately $225,000 in 15 years to cover his 3-year-old daughter's college bills for a 4-year degree. How much would he have to invest today at an interest rate of 8% compounded semiannually? *LU 12-2(2)*

12–13. Lynn Ally, owner of a local Subway shop, loaned $40,000 to Pete Hall to help him open a Subway franchise. Pete plans to repay Lynn at the end of 8 years with 6% interest compounded semiannually. How much will Lynn receive at the end of 8 years? *LU 12-1(2)*

12–14. Molly Hamilton deposited $50,000 at Bank of America at 8% interest compounded quarterly. What is the effective rate (APY) to the nearest hundredth percent? *LU 12-1(3)*

eXcel

My Money

12–15. Melvin Indecision has difficulty deciding whether to put his savings in Mystic Bank or Four Rivers Bank. Mystic offers 10% interest compounded semiannually. Four Rivers offers 8% interest compounded quarterly. Melvin has $10,000 to invest. He expects to withdraw the money at the end of 4 years. Which bank gives Melvin the better deal? Check your answer. *LU 12-1(3)*

12–16. Lee Holmes deposited $15,000 in a new savings account at 9% interest compounded semiannually. At the beginning of year 4, Lee deposits an additional $40,000 at 9% interest compounded semiannually. At the end of 6 years, what is the balance in Lee's account? *LU 12-1(2)*

eXcel

12–17. Lee Wills loaned Audrey Chin $16,000 to open Snip Its Hair Salon. After 6 years, Audrey will repay Lee with 8% interest compounded quarterly. How much will Lee receive at the end of 6 years? *LU 12-1(2)*

12–18. Jazelle Momba wants to visit her family in Zimbabwe in 2025, which is 6 years from now. She knows that it will cost approximately $8,000 including flight costs, on-the-ground costs, and extra spending money to stay for 4 months. If she opens an account that compounds interest at 4% semiannually, how much does she need to deposit today to cover the total cost of her visit? *LU 12-1(2)*

My Money

12–19. After reviewing the CPI inflation calculator at inflationdata.com, Hanna Lind realized the importance of creating an investment plan for her future. She would need $10,070.34 in 2017 to have the same purchasing power her $7,000 (stored in a fireproof safe in her home since 2000) had when she put it there. To protect her savings against further inflation and to help her prepare for a healthy financial future, Hanna deposits her $7,000 in an investment account in 2017 earning 6% interest compounded quarterly. How much will Hanna have in her account in 2027? (Use tables in the *Business Math Handbook.*) *LU 12-1(2)*

12–20. The International Monetary Fund is trying to raise $500 billion in 5 years for new funds to lend to developing countries. At 6% interest compounded quarterly, how much must it invest today to reach $500 billion in 5 years? *LU 12-2(2)*

My Money

12–21. You choose to invest your $2,985 income tax refund check (rather than spend it!) in an account earning 5% compounded annually. How much will the account be worth in 30 years? *LU 12-1(2, 3)*

Imagine how much you would have in your account if you did this each year!

12–22. Jim Ryan, an owner of a Burger King restaurant, assumes that his restaurant will need a new roof in 7 years. He estimates the roof will cost him $9,000 at that time. What amount should Jim invest today at 6% compounded quarterly to be able to pay for the roof? Check your answer. *LU 12-2(2)*

eXcel

12–23. Tony Ring wants to attend Northeast College. He will need $60,000 4 years from today. Assume Tony's bank pays 12% interest compounded semiannually. What must Tony deposit today so he will have $60,000 in 4 years? *LU 12-2(2)*

12–24. Check your answer (to the nearest dollar) in Problem 12–23 by using the compound value Table 12.1. The answer will be slightly off due to rounding. *LU 12-1(3)*

My Money

12–25. Pete Air wants to buy a used Jeep in 5 years. He estimates the Jeep will cost $15,000. Assume Pete invests $10,000 now at 12% interest compounded semiannually. Will Pete have enough money to buy his Jeep at the end of 5 years? *LU 12-1(2), LU 12-2(2)*

My Money

12–26. How much could you save for retirement if you chose to invest the money you spend on Starbucks coffee in one year? Assume you buy one venti cup of caffe latte for $4.15 each weekday for 50 weeks and can invest the total amount in a mutual fund earning 5% compounded annually for 30 years. *LU 12-1(2)*

12–27. Paul Havlik promised his grandson Jamie that he would give him $6,000 8 years from today for graduating from high school. Assume money is worth 6% interest compounded semiannually. What is the present value of this $6,000? *LU 12-2(2)*

12–28. Earl Ezekiel wants to retire in San Diego when he is 65 years old. Earl is now 50. He believes he will need $300,000 to retire comfortably. To date, Earl has set aside no retirement money. Assume Earl gets 6% interest compounded semiannually. How much must Earl invest today to meet his $300,000 goal? *LU 12-2(2)*

My Money

12–29. If you saved your tax refund (Problem 12–21), quit buying vendor coffee for one year (Problem 12–26), and decided to contribute \$2,400 (you saved \$200 per month) in your Roth IRA, how much would you have for retirement if you could invest these savings at 5% compounded annually for 30 years for this one year of savings?

12–30. Treasure Mountain International School in Park City, Utah, is a public middle school interested in raising money for next year's Sundance Film Festival. If the school raises \$15,000 and invests it for 1 year at 3% interest compounded annually, what is the APY earned (round to nearest whole percent)? *LU 12-1(2, 3)*

CHALLENGE PROBLEMS

12–31. Pete's Real Estate is currently valued at \$65,000. Pete feels the value of his business will increase at a rate of 10% per year, compounded semiannually for the next 5 years. At a local fund-raiser, a competitor offered Pete \$70,000 for the business. If he sells, Pete plans to invest the money at 6% compounded quarterly. What price should Pete ask? Verify your answer. *LU 12-1(2), LU 12-2(2)*

12–32. You are the financial planner for Johnson Controls. Assume last year's profits were \$700,000. The board of directors decided to forgo dividends to stockholders and retire high-interest outstanding bonds that were issued 5 years ago at a face value of \$1,250,000. You have been asked to invest the profits in a bank. The board must know how much money you will need from the profits earned to retire the bonds in 10 years. Bank A pays 6% compounded quarterly, and Bank B pays $6\frac{1}{2}$% compounded annually. Which bank would you recommend, and how much of the company's profit should be placed in the bank? If you recommended that the remaining money not be distributed to stockholders but be placed in Bank B, how much would the remaining money be worth in 10 years? Use tables in the *Business Math Handbook.** Round final answer to nearest dollar. *LU 12-1(2, 3), LU 12-2(2)*

*Check glossary for unfamiliar terms.

Classroom Notes

SUMMARY PRACTICE TEST

Do you need help? Connect videos have step-by-step worked-out solutions.

1. Lorna Ray, owner of a Starbucks franchise, loaned $40,000 to Lee Reese to help him open a new flower shop online. Lee plans to repay Lorna at the end of 5 years with 4% interest compounded semiannually. How much will Lorna receive at the end of 5 years? *LU 12-1(2)*

2. Joe Beary wants to attend Riverside College. Eight years from today he will need $50,000. If Joe's bank pays 6% interest compounded semiannually, what must Joe deposit today to have $50,000 in 8 years? *LU 12-2(2)*

3. Shelley Katz deposited $30,000 in a savings account at 5% interest compounded semiannually. At the beginning of year 4, Shelley deposits an additional $80,000 at 5% interest compounded semiannually. At the end of 6 years, what is the balance in Shelley's account? *LU 12-1(2)*

4. Earl Miller, owner of a Papa Gino's franchise, wants to buy a new delivery truck in 6 years. He estimates the truck will cost $30,000. If Earl invests $20,000 now at 5% interest compounded semiannually, will Earl have enough money to buy his delivery truck at the end of 6 years? *LU 12-1(2), LU 12-2(2)*

5. Minnie Rose deposited $16,000 in Street Bank at 6% interest compounded quarterly. What was the effective rate (APY)? Round to the nearest hundredth percent. *LU 12-1(2, 3)*

6. Lou Ling, owner of Lou's Lube, estimates that he will need $70,000 for new equipment in 7 years. Lou decided to put aside money today so it will be available in 7 years. Reel Bank offers Lou 6% interest compounded quarterly. How much must Lou invest to have $70,000 in 7 years? *LU 12-2(2)*

7. Bernie Long wants to retire to California when she is 60 years of age. Bernie is now 40. She believes that she will need $900,000 to retire comfortably. To date, Bernie has set aside no retirement money. If Bernie gets 8% compounded semiannually, how much must Bernie invest today to meet her $900,000 goal? *LU 12-2(2)*

8. Jim Jones deposited $19,000 in a savings account at 7% interest compounded daily. At the end of 6 years, what is the balance in Jim's account? *LU 12-1(2)*

INTERACTIVE VIDEO WORKSHEET

GRADING THE SUMMARY PRACTICE TEST

Go to the summary practice test video in Connect (or click on it here in the ebook). Grade your summary practice test while viewing the video.

C for Correct/I for Incorrect

1. ________	5. ________
2. ________	6. ________
3. ________	7. ________
4. ________	8. ________

If you achieved 100%, you are ready for your instructor's exam.

If any of the problems were incorrect, list the questions you missed and show steps to solve the problem correctly.

Replay the video to see if you have made the correct fixes to your mistakes.
If you have any questions, contact your instructor asap.

Notes on Watching Videos

MY MONEY

Early Retirement Planning

What I need to know

As you embark on your chosen career field, it may seem as though retirement is a lifetime away. While this may be true, it will take proper planning today in order to achieve your retirement dreams in the future. As a soon to be college graduate, you have one of the most important advantages to reaching your financial goals—time! This **time** allows you the opportunity to grow your investments, recover from downturns in the economy, and set high financial goals for yourself. Beginning your retirement savings today will help you to achieve the goals you have for yourself and afford you the time to make adjustments along the way.

What I need to do

A great place to start this planning is through any employer sponsored retirement programs available to you. Learn as much as you can about the retirement programs offered by your employer to see how these can assist you in reaching your financial goals. Many times these plans will offer an incentive to you in the form of an employer match of the amount you are putting away towards retirement. For instance, if your employer will match your first 6% contributed to your retirement plan, it will essentially double your investment amount. While it may seem like a lot of money will be taken out of your paycheck, you may be surprised at how the tax savings positively affects your net pay—and not taking full advantage of the employer match is like saying no to the offer of free money.

You can also invest outside of your employer programs by researching individual retirement account (IRA) options. A traditional IRA allows you to make contributions up to a certain yearly limit with the possibility of receiving a tax break on your contributions by reducing your adjusted gross income. Conversely, a Roth IRA requires contributions after taxes have been taken out. When you withdraw the money in retirement, it will be taxed as ordinary income on a traditional IRA or taken tax-free on a Roth IRA.

"Don't put all your eggs into one basket" is an old saying that applies to the idea of retirement investments. It is better to have your investment monies spread across a variety of investment options to spread your risk across multiple scenarios. As shown in the previous discussion of IRAs, there are advantages and disadvantages to both a traditional and a Roth IRA. Your retirement investments should follow a similar thread. Investments can be made in stocks, bonds, CDs (certificates of deposit), and money market accounts to name a few. By spreading your investments across multiple opportunities, called diversifying, you can create a strategy to maintain your investments while weathering the ups and downs of certain markets versus the stability of others.

Resources I can use

https://www.bankrate.com/retirement/10-commandments-of-retirement-planning/#slide=1 — planning for retirement

MY MONEY ACTIVITY

Create a diversified investment portfolio for yourself. Research a variety of investment instruments: 401k, IRA, Roth IRA, CD (certificate of deposit), real estate, etc. Record rates of return versus risk. And, most importantly, begin investing for your future TODAY.

PERSONAL FINANCE

A KIPLINGER APPROACH

"Don't Step Into These Savings Traps" by Anne Kates Smith from *Kiplinger's*, February 2018, p. 42. Used by permission of The Kiplinger Washington Editors, Inc.

ANNE KATES SMITH | Your Mind and Your Money

Don't Step Into These Saving Traps

Most of us are aware that a sneaky set of psychological biases can trip us up when it comes to investing and managing our money. And some of these biases seem tailor-made to foil retirement savers. Prudential Retirement, an arm of the insurance giant that administers 401(k) plans, surveyed plan participants in 2017 to see which behavioral challenges troubled them the most. Here's what the survey found.

Longevity disconnect. Although no one is eager to meet the Grim Reaper, a lot of us think we're not going to live as long as we probably will. That's a problem when you're funding a longer-than-expected retirement. For 39% of those taking Prudential's quiz, failing to see far enough into their senior years posed the biggest behavioral challenge to achieving financial security. Counteract this bias by picturing an older version of yourself. Apps such as FaceApp can help.

Procrastination. More than one-fourth of the savers quizzed turned out to be put-it-off-until-later types. But when it comes to investing, squandering time is a huge mistake, just as the opposite, investing early and often, can work magic—the magic of compound returns. Programs that enroll employees automatically in retirement savings plans are a boon for procrastinators; many plans let you escalate contributions automatically, too.

Optimism. Some 22% of savers suffered most from rose-colored-glasses syndrome. Believing that everything will all work out in the end may be good for your blood pressure, but it's not an efficient savings strategy. Plan for inevitable bumps in the road, which can range from the catastrophic—being struck by serious illness, for example—to the merely disappointing—say, a missed promotion.

Some behavioral biases seem tailor-made to foil retirement savers. But their ill effects are smaller for those who are aware of them.

Following the crowd. Determining the best investment mix for your stage in life and your risk tolerance is a challenge. Sticking to the plan when the market is volatile requires fortitude. Prudential found that 8% of savers were most led astray by being herd-followers. Regular portfolio rebalancing can keep your assets on target; a written investment policy can help you stick to your guns.

Instant gratification. Would you wait a year for $1,000 rather than receive $500 today? How are you with impulse spending? Most people who took Prudential's quiz said that waiting, either for a higher return or to make a purchase, wasn't a big deal; only 5% revealed it as their major issue. But research published by the National Bureau of Economic Research in 2016 found that more than half (55%) of retirement savers suffered from "present bias"—a wonky term for wanting instant gratification. If you need a nudge, try a budgeting app or a commitment to cash-only spending to curb purchases you'll regret.

The NBER research found another foible that keeps people from saving as much as they should: Nearly seven in 10 retirement savers misunderstood how account balances increase over time, underestimating the exponential growth of compound returns and, therefore, investing less than they otherwise would. In other words, a lot of people aren't saving enough because they simply don't realize how much money they're leaving on the table.

"I use these biases as a conversation starter when I'm talking to clients," says ShirlyAnn Robertson, a broker and Prudential financial adviser in Schaumburg, Ill. Helping clients understand them makes it easier to navigate their financial concerns, she says. And here's the good news about these sneaky biases, according to NBER: Their ill effects are smaller for those who are aware of them. ■

ANNE KATES SMITH IS EXECUTIVE EDITOR OF *KIPLINGER'S PERSONAL FINANCE* MAGAZINE. YOU CAN CONTACT HER AT ASMITH@KIPLINGER.COM.

BUSINESS MATH ISSUE

Compounding of money avoids all savings traps.

1. List the key points of the article and information to support your position.
2. Write a group defense of your position using math calculations to support your view. If you are in an online course, post to a discussion board.

Classroom Notes

Design elements: Digital frame: (tablet) ©Radu Bercan/Shutterstock, (icons) ©12bit/Shutterstock; Globe: ©edge69/Getty Images; Money Tips: ©Pro Symbols/Shutterstock; Aha!: ©frender/Getty Images; Excel: ©McGraw-Hill Education; My Money: (tree) ©Cherkas/Shutterstock, (gold nuggets) ©JonahWong/Shutterstock; Video Worksheet: ©PureSolution/Shutterstock; author photo in circle: ©McGraw-Hill Education/Ryan McAndrews, photographer.

CHAPTER 13

Annuities and Sinking Funds

Brokerages Split on IRA Commissions

By Michael Wursthorn

Morgan Stanley will let its customers keep paying for retirement advice with commissions, setting up a split in how the brokerage industry interprets new federal regulations governing $3 trillion in assets.

A Labor Department rule taking effect in April requires investment advisers to act in the best interests of their clients when it comes to retirement accounts. But that directive has complicated implications for how clients pay for services and how brokers are compensated.

©Alexander Raths/Shutterstock

LEARNING UNIT OBJECTIVES

LU 13–1: Annuities: Ordinary Annuity and Annuity Due (Find Future Value)

1. Differentiate between contingent annuities and annuities certain.
2. Calculate the future value of an ordinary annuity and an annuity due manually and by table lookup.

LU 13–2: Present Value of an Ordinary Annuity (Find Present Value)

1. Calculate the present value of an ordinary annuity by table lookup and manually check the calculation.
2. Compare the calculation of the present value of one lump sum versus the present value of an ordinary annuity.

LU 13–3: Sinking Funds (Find Periodic Payments)

1. Calculate the payment made at the end of each period by table lookup.
2. Check table lookup by using ordinary annuity table.

Your Guide to Successfully Completing This Chapter

Traditional book or ebook

Check box as you complete each step.

Steps

- ☐ Read learning unit.
  - ☐ Complete practice quiz at the end of the learning unit.
- ☐ Grade practice quiz using provided solutions. (For more help, watch the learning unit video in Connect and have a Study Session with the authors. Then complete the additional practice quiz in Connect.)
- ☐ Repeat above for each of the three learning units in Chapter 13.
  - ☐ Review chapter organizer.
  - ☐ Complete assigned homework.
    - ☐ Finish summary practice test. (Go to Connect via the ebook link and do the interactive video worksheet to grade.)
- ☐ Complete instructor's exam.

©Boofoto/Shutterstock

The new tax act does not change the amount one can invest into retirement savings plans. The act does have new requirements for those who have plans. You should check with the IRS for details.

The *Wall Street Journal* chapter opener clip shows that many people use financial advisors to plan retirement. This chapter will show you why you need to save for retirement and not rely just on Social Security. As the following clip shows, many factors—even small ones like your daily coffee spending—could affect your retirement savings.

So, would you like to save $1,287? A *Boston Globe* article entitled "Cost of Living: A Cup a Day" (page 344) began by explaining that each month the *Globe* runs a feature on an everyday expense to see how much it costs an average person. Since many people are coffee drinkers, the *Globe* assumed that a person drank 3 cups a day of Dunkin' Donuts coffee at the cost of $1.65 a cup. For a 5-day week, the person would spend $1,287 annually

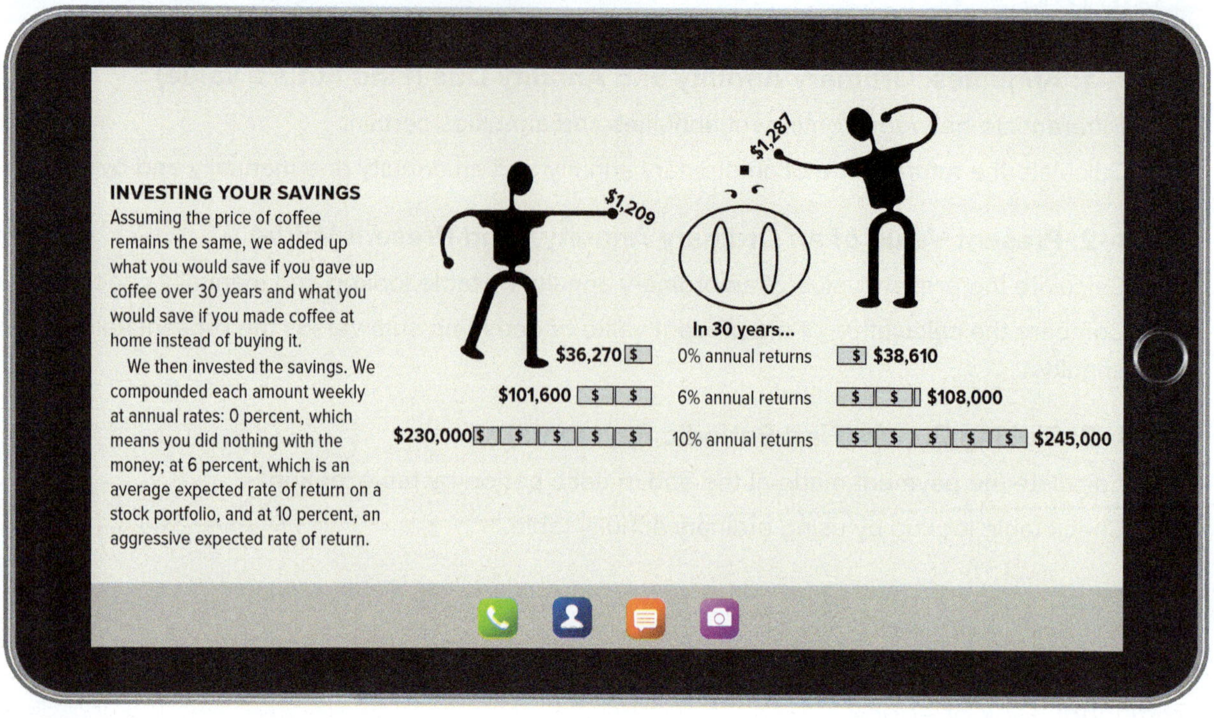

Source: Boston Sunday Globe ©2004

My Money

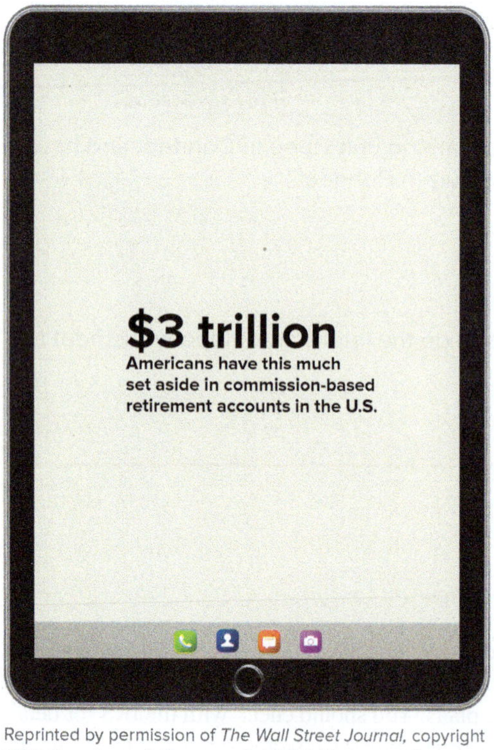

Reprinted by permission of *The Wall Street Journal*, copyright 2016 Dow Jones & Company, Inc. All rights reserved worldwide.

(52 weeks). If the person brewed the coffee at home, the cost of the beans per cup would be $0.10 with an annual expense of $78, saving $1,209 over the Dunkin' Donuts coffee. If a person gave up drinking coffee, the person would save $1,287.

The article continued with the discussion on "Investing Your Savings." Note how much you would have in 30 years if you invested your money in 0%, 6%, and 10% annual returns. Using the magic of compounding, if you saved $1,287 a year, your money could grow to a quarter of a million dollars.

This chapter shows how to compute compound interest that results from a *stream* of payments, or an annuity. Chapter 12 showed how to calculate compound interest on a lump-sum payment deposited at the beginning of a particular time. Knowing how to calculate interest compounding on a lump sum will make the calculation of interest compounding on annuities easier to understand.

We begin the chapter by explaining the difference between calculating the future value of an ordinary annuity and an annuity due. Then you learn how to find the present value of an ordinary annuity. The chapter ends with a discussion of sinking funds.

Learning Unit 13–1: Annuities: Ordinary Annuity and Annuity Due (Find Future Value)

Many parents of small children are concerned about being able to afford to pay for their children's college educations. Some parents deposit a lump sum in a financial institution when the child is in diapers. The interest on this sum is compounded until the child is 18, when the parents withdraw the money for college expenses. Parents could also fund their children's educations with annuities by depositing a series of payments for a certain time. The concept of annuities is the first topic in this learning unit.

My Money

Concept of an Annuity—The Big Picture

All of us would probably like to win $1 million in a state lottery. What happens when you have the winning ticket? You take it to the lottery headquarters. When you turn in the ticket, do you immediately receive a check for $1 million? No. Lottery payoffs are not usually made in lump sums.

FIGURE 13.1

Future value of an annuity of $1 at 8%

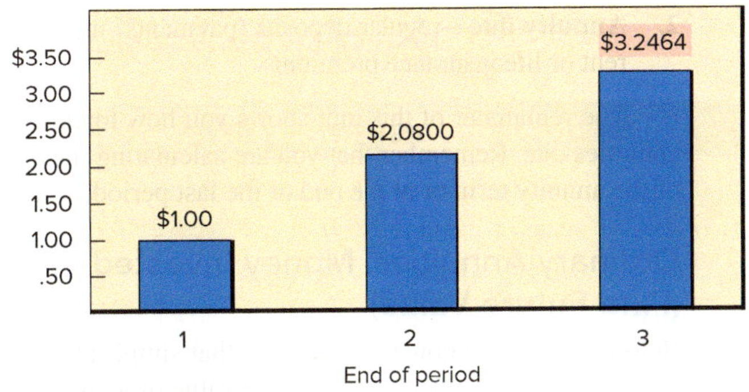

©McGraw-Hill Education/Mark Steinmetz, photographer

Used by permission of Cartoon Features Syndicate

Lottery winners receive a series of payments over a period of time—usually years. This *stream* of payments is an **annuity.** By paying the winners an annuity, lotteries do not actually spend $1 million. The lottery deposits a sum of money in a financial institution. The continual growth of this sum through compound interest provides the lottery winner with a series of payments.

When we calculated the maturity value of a lump-sum payment in Chapter 12, the maturity value was the principal and its interest. Now we are looking not at lump-sum payments but at a series of payments (usually of equal amounts over regular **payment periods**) plus the interest that accumulates. So the **future value of an annuity** is the future *dollar amount* of a series of payments plus interest.[1] The **term of the annuity** is the time from the beginning of the first payment period to the end of the last payment period.

The concept of the future value of an annuity is illustrated in Figure 13.1. Do not be concerned about the calculations (we will do them soon). Let's first focus on the big picture of annuities. In Figure 13.1 we see the following:

At end of period 1:	The $1 is still worth $1 because it was invested at the *end* of the period.
At end of period 2:	An additional $1 is invested. The $2.00 is now worth $2.08. Note the $1 from period 1 earns interest but not the $1 invested at the end of period 2.
At end of period 3:	An additional $1 is invested. The $3.00 is now worth $3.25. Remember that the last dollar invested earns no interest.

Before learning how to calculate annuities, you should understand the two classifications of annuities.

How Annuities Are Classified

Annuities have many uses in addition to lottery payoffs. Some of these uses are insurance companies' pension installments, Social Security payments, home mortgages, businesses paying off notes, bond interest, and savings for a vacation trip or college education.

Annuities are classified into two major groups: contingent annuities and annuities certain. **Contingent annuities** have no fixed number of payments but depend on an uncertain event (e.g., life insurance payments that cease when the insured dies). **Annuities certain** have a specific stated number of payments (e.g., mortgage payments on a home). Based on the time of the payment, we can divide each of these two major annuity groups into the following:

LO 1

1. **Ordinary annuity**—regular deposits (payments) made at the *end* of the period. Periods could be months, quarters, years, and so on. An ordinary annuity could be salaries, stock dividends, and so on.

[1]The term *amount of an annuity* has the same meaning as *future value of an annuity.*

2. Step 1. 5 years × 2 = 10 $\quad \frac{6\%}{2} = 3\%$

$\quad\quad\quad\quad\quad\quad\quad\quad$ + 1

$\quad\quad\quad\quad\quad\quad\quad\quad$ 11 periods

Step 2. Table factor, 12.8078

Step 3. \$4,000 × 12.8078 = \$51,231.20

Step 4. $\quad\quad\quad\quad\quad\quad$ − 4,000.00

$\quad\quad\quad\quad\quad\quad\quad\quad\quad\quad$ **\$47,231.20**

Learning Unit 13–2: Present Value of an Ordinary Annuity (Find Present Value)[4]

This unit begins by presenting the concept of present value of an ordinary annuity. Then you will learn how to use a table to calculate the present value of an ordinary annuity.

Concept of Present Value of an Ordinary Annuity—The Big Picture

Let's assume that we want to know how much money we need to invest *today* to receive a stream of payments for a given number of years in the future. This is called the **present value of an ordinary annuity.**

In Figure 13.2 you can see that if you wanted to withdraw \$1 at the end of one period, you would have to invest 93 cents *today.* If at the end of each period for three periods you wanted to withdraw \$1, you would have to put \$2.58 in the bank *today* at 8% interest. (Note that we go from the future back to the present.)

FIGURE 13.2

Present value of an annuity of \$1 at 8%

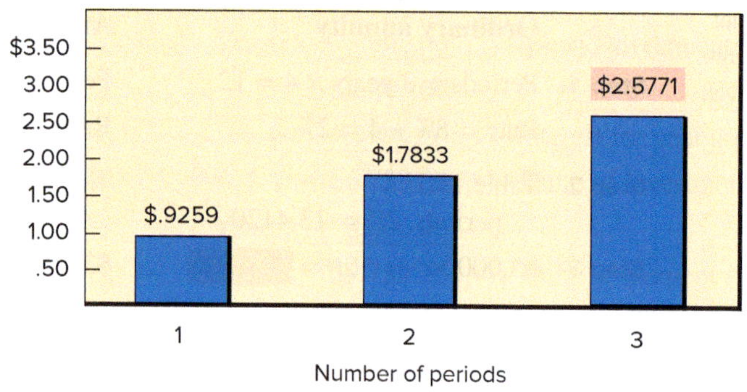

Now let's look at how we could use tables to calculate the present value of annuities and then check our answer.

LO 1

Calculating Present Value of an Ordinary Annuity by Table Lookup

Use the steps below to calculate the present value of an ordinary annuity by table lookup.[5]

CALCULATING PRESENT VALUE OF AN ORDINARY ANNUITY BY TABLE LOOKUP

Step 1. Calculate the number of periods and rate per period.

Step 2. Look up the periods and rate in the present value of an annuity table. The intersection gives the table factor for the present value of \$1.

Step 3. Multiply the withdrawal for each period by the table factor. This gives the present value of an ordinary annuity.

$$\text{Present value of ordinary annuity payment} = \text{Annuity payment} \times \text{Present value of ordinary annuity table factor}$$

[4]For simplicity we omit a discussion of present value of annuity due that would require subtracting a period and adding a 1.

[5]The formula for the present value of an ordinary annuity is $PV = PMT \times \frac{1 - 1 \div (1+i)^n}{i}$, where PV equals present value of annuity, PMT equals annuity payment, i equals interest, and n equals number of periods. The calculator sequence would be as follows for the John Fitch example that follows: 1 [+] .08 [y^x] 3 [+−] [=] [M+] 1 [−] [MR] [÷] .08 [×] 8,000 [=] 20,616.78.

TABLE 13.2 Present value of an annuity of $1

Period	2%	3%	4%	5%	6%	7%	8%	9%	10%	11%	12%	13%
1	0.9804	0.9709	0.9615	0.9524	0.9434	0.9346	0.9259	0.9174	0.9091	0.9009	0.8929	0.8850
2	1.9416	1.9135	1.8861	1.8594	1.8334	1.8080	1.7833	1.7591	1.7355	1.7125	1.6901	1.6681
3	2.8839	2.8286	2.7751	2.7232	2.6730	2.6243	2.5771	2.5313	2.4869	2.4437	2.4018	2.3612
4	3.8077	3.7171	3.6299	3.5459	3.4651	3.3872	3.3121	3.2397	3.1699	3.1024	3.0373	2.9745
5	4.7134	4.5797	4.4518	4.3295	4.2124	4.1002	3.9927	3.8897	3.7908	3.6959	3.6048	3.5172
6	5.6014	5.4172	5.2421	5.0757	4.9173	4.7665	4.6229	4.4859	4.3553	4.2305	4.1114	3.9975
7	6.4720	6.2303	6.0021	5.7864	5.5824	5.3893	5.2064	5.0330	4.8684	4.7122	4.5638	4.4226
8	7.3255	7.0197	6.7327	6.4632	6.2098	5.9713	5.7466	5.5348	5.3349	5.1461	4.9676	4.7988
9	8.1622	7.7861	7.4353	7.1078	6.8017	6.5152	6.2469	5.9952	5.7590	5.5370	5.3282	5.1317
10	8.9826	8.5302	8.1109	7.7217	7.3601	7.0236	6.7101	6.4177	6.1446	5.8892	5.6502	5.4262
11	9.7868	9.2526	8.7605	8.3064	7.8869	7.4987	7.1390	6.8052	6.4951	6.2065	5.9377	5.6869
12	10.5753	9.9540	9.3851	8.8632	8.3838	7.9427	7.5361	7.1607	6.8137	6.4924	6.1944	5.9176
13	11.3483	10.6350	9.9856	9.3936	8.8527	8.3576	7.9038	7.4869	7.1034	6.7499	6.4235	6.1218
14	12.1062	11.2961	10.5631	9.8986	9.2950	8.7455	8.2442	7.7862	7.3667	6.9819	6.6282	6.3025
15	12.8492	11.9379	11.1184	10.3796	9.7122	9.1079	8.5595	8.0607	7.6061	7.1909	6.8109	6.4624
16	13.5777	12.5611	11.6523	10.8378	10.1059	9.4466	8.8514	8.3126	7.8237	7.3792	6.9740	6.6039
17	14.2918	13.1661	12.1657	11.2741	10.4773	9.7632	9.1216	8.5436	8.0216	7.5488	7.1196	6.7291
18	14.9920	13.7535	12.6593	11.6896	10.8276	10.0591	9.3719	8.7556	8.2014	7.7016	7.2497	6.8399
19	15.6784	14.3238	13.1339	12.0853	11.1581	10.3356	9.6036	8.9501	8.3649	7.8393	7.3658	6.9380
20	16.3514	14.8775	13.5903	12.4622	11.4699	10.5940	9.8181	9.1285	8.5136	7.9633	7.4694	7.0248
25	19.5234	17.4131	15.6221	14.0939	12.7834	11.6536	10.6748	9.8226	9.0770	8.4217	7.8431	7.3300
30	22.3964	19.6004	17.2920	15.3724	13.7648	12.4090	11.2578	10.2737	9.4269	8.6938	8.0552	7.4957
40	27.3554	23.1148	19.7928	17.1591	15.0463	13.3317	11.9246	10.7574	9.7790	8.9511	8.2438	7.6344
50	31.4236	25.7298	21.4822	18.2559	15.7619	13.8007	12.2335	10.9617	9.9148	9.0417	8.3045	7.6752

EXAMPLE John Fitch wants to receive an $8,000 annuity in 3 years. Interest on the annuity is 8% annually. John will make withdrawals at the end of each year. How much must John invest today to receive a stream of payments for 3 years? Use Table 13.2. Remember that interest could be earned semiannually, quarterly, and so on, as shown in the previous unit.

Step 1. 3 years × 1 = 3 periods $\qquad \frac{8\%}{\text{Annually}} = 8\%$

Step 2. Table factor, 2.5771 (we saw this in Figure 13.2)

Step 3. 8,000 × 2.5771 = **$20,616.80**

If John wants to withdraw $8,000 at the end of each period for 3 years, he will have to deposit $20,616.80 in the bank *today.*

Amount	
$20,616.80	
+ 1,649.34	→ Interest at end of year 1 (.08 × $20,616.80)
$22,266.14	
− 8,000.00	→ First payment to John
$14,266.14	
+ 1,141.29	→ Interest at end of year 2 (.08 × $14,266.14)
$15,407.43	
− 8,000.00	→ Second payment to John
$ 7,407.43	
+ 592.59	→ Interest at end of year 3 (.08 × $7,407.43)
$ 8,000.02	
− 8,000.00	→ After end of year 3 John receives his last $8,000
.02	(off 2 cents due to rounding)

Before we leave this unit, let's work out two examples that show the relationship of Chapter 13 to Chapter 12. Use the tables in your *Business Math Handbook.*

LO 2

Lump Sum versus Annuities

EXAMPLE John Sands made deposits of \$200 semiannually to Floor Bank, which pays 8% interest compounded semiannually. After 5 years, John makes no more deposits. What will be the balance in the account 6 years after the last deposit?

Step 1. Calculate amount of annuity: (Table 13.1)

10 periods, 4% \$200 × 12.0061 = \$2,401.22

Step 2. Calculate how much the final value of the annuity will grow by the compound interest table. (Table 12.1)

12 periods, 4% \$2,401.22 × 1.6010 = **\$3,844.35**

For John, the stream of payments grows to \$2,401.22. Then this *lump sum* grows for 6 years to \$3,844.35. Now let's look at a present value example.

MONEY tips

The best rates on mortgages, loans, and credit cards are offered to those with credit scores above 740. Pay your bills on time, have one long-term credit card, keep low balances or pay off balances each month, and review your credit report regularly.

EXAMPLE Mel Rich decided to retire in 8 years to New Mexico. What amount should Mel invest today so he will be able to withdraw \$40,000 at the end of each year for 25 years *after* he retires? Assume Mel can invest money at 5% interest (compounded annually).

Step 1. Calculate the present value of the annuity: (Table 13.2)

25 periods, 5% \$40,000 × 14.0939 = \$563,756

Step 2. Find the present value of \$563,756 since Mel will not retire for 8 years: (Table 12.3)

8 periods, 5% (PV table) \$563,756 × .6768 = **\$381,550.06**

If Mel deposits \$381,550 in year 1, it will grow to \$563,756 after 8 years.

It's time to try the Practice Quiz and check your understanding of this unit.

LU 13–2 PRACTICE QUIZ

Complete this Practice Quiz to see how you are doing.

(Use tables in *Business Math Handbook*)

1. What must you invest today to receive an \$18,000 annuity for 5 years semiannually at a 10% annual rate? All withdrawals will be made at the end of each period.
2. Rase High School wants to set up a scholarship fund to provide five \$2,000 scholarships for the next 10 years. If money can be invested at an annual rate of 9%, how much should the scholarship committee invest today?
3. Joe Wood decided to retire in 5 years in Arizona. What amount should Joe invest today so he can withdraw \$60,000 at the end of each year for 30 years after he retires? Assume Joe can invest money at 6% compounded annually.

*For **extra help** from your authors–Sharon and Jeff–see the videos in Connect.*

✓ Solutions

1. **Step 1.** Periods = 5 years × 2 = 10; Rate = 10% ÷ 2 = 5%
   Step 2. Factor, 7.7217
   Step 3. \$18,000 × 7.7217 = **\$138,990.60**
2. **Step 1.** Periods = 10; Rate = 9%
   Step 2. Factor, 6.4177
   Step 3. \$10,000 × 6.4177 = **\$64,177**
3. **Step 1.** Calculate present value of annuity: 30 periods, 6%.
   \$60,000 × 13.7648 = \$825,888
   Step 2. Find present value of \$825,888 for 5 years: 5 periods, 6%.
   \$825,888 × .7473 = **\$617,186.10**

Learning Unit 13–3: Sinking Funds (Find Periodic Payments)

LO 1

A **sinking fund** is a financial arrangement that sets aside regular periodic payments of a particular amount of money. Compound interest accumulates on these payments to a specific sum at a predetermined future date. Corporations use sinking funds to discharge bonded indebtedness, to replace worn-out equipment, to purchase plant expansion, and so on.

A sinking fund is a different type of an annuity. In a sinking fund, you determine the amount of periodic payments you need to achieve a given financial goal. In the annuity, you know the amount of each payment and must determine its future value. Let's work with the following formula:

Sinking fund payment = Future value × Sinking fund table factor

EXAMPLE To retire a bond issue, Moore Company needs $60,000 in 18 years from today. The interest rate is 10% compounded annually. What payment must Moore make at the end of each year? Use Table 13.3.

We begin by looking down the Period column in Table 13.3 until we come to 18. Then we go across until we reach the 10% column. The table factor is .0219.

Now we multiply $60,000 by the factor as follows:

$60,000 × .0219 = $1,314

LO 2

This states that if Moore Company pays $1,314 at the end of each period for 18 years, then $60,000 will be available to pay off the bond issue at maturity.

We can check this by using Table 13.1 (the ordinary annuity table):

$1,314 × 45.5992 = $59,917.35 (off due to rounding)

It's time to try the following Practice Quiz.

TABLE 13.3

Sinking fund table[6] based on $1

Period	2%	3%	4%	5%	6%	8%	10%
1	1.0000	1.0000	1.0000	1.0000	1.0000	1.0000	1.0000
2	0.4951	0.4926	0.4902	0.4878	0.4854	0.4808	0.4762
3	0.3268	0.3235	0.3203	0.3172	0.3141	0.3080	0.3021
4	0.2426	0.2390	0.2355	0.2320	0.2286	0.2219	0.2155
5	0.1922	0.1884	0.1846	0.1810	0.1774	0.1705	0.1638
6	0.1585	0.1546	0.1508	0.1470	0.1434	0.1363	0.1296
7	0.1345	0.1305	0.1266	0.1228	0.1191	0.1121	0.1054
8	0.1165	0.1125	0.1085	0.1047	0.1010	0.0940	0.0874
9	0.1025	0.0984	0.0945	0.0907	0.0870	0.0801	0.0736
10	0.0913	0.0872	0.0833	0.0795	0.0759	0.0690	0.0627
11	0.0822	0.0781	0.0741	0.0704	0.0668	0.0601	0.0540
12	0.0746	0.0705	0.0666	0.0628	0.0593	0.0527	0.0468
13	0.0681	0.0640	0.0601	0.0565	0.0530	0.0465	0.0408
14	0.0626	0.0585	0.0547	0.0510	0.0476	0.0413	0.0357
15	0.0578	0.0538	0.0499	0.0463	0.0430	0.0368	0.0315
16	0.0537	0.0496	0.0458	0.0423	0.0390	0.0330	0.0278
17	0.0500	0.0460	0.0422	0.0387	0.0354	0.0296	0.0247
18	0.0467	0.0427	0.0390	0.0355	0.0324	0.0267	0.0219
19	0.0438	0.0398	0.0361	0.0327	0.0296	0.0241	0.0195
20	0.0412	0.0372	0.0336	0.0302	0.0272	0.0219	0.0175
24	0.0329	0.0290	0.0256	0.0225	0.0197	0.0150	0.0113
28	0.0270	0.0233	0.0200	0.0171	0.0146	0.0105	0.0075
32	0.0226	0.0190	0.0159	0.0133	0.0110	0.0075	0.0050
36	0.0192	0.0158	0.0129	0.0104	0.0084	0.0053	0.0033
40	0.0166	0.0133	0.0105	0.0083	0.0065	0.0039	0.0023

[6]Sinking fund table is the reciprocal of the ordinary annuity table.

MONEY tips

If you are trying to build credit by using a credit card, each time you make a purchase using the credit card, deduct that amount from your checking account. When your credit card bill is due, add up all your credit card deductions in your checking account. You will have enough to pay the credit card off in full.

LU 13–3 PRACTICE QUIZ

Complete this **Practice Quiz** to see how you are doing.

*For **extra help** from your authors–Sharon and Jeff–see the videos in Connect.*

Today, Arrow Company issued bonds that will mature to a value of \$90,000 in 10 years. Arrow's controller is planning to set up a sinking fund. Interest rates are 12% compounded semiannually. What will Arrow Company have to set aside to meet its obligation in 10 years? Check your answer. Your answer will be off due to the rounding of Table 13.3.

✓ Solution

10 years × 2 = 20 periods $\quad \frac{12\%}{2} = 6\% \quad$ \$90,000 × .0272 = **\$2,448**

Check \$2,448 × 36.7855 = \$90,050.90

INTERACTIVE CHAPTER ORGANIZER

Topic/Procedure/Formula	Examples	You try it*
Ordinary annuities (find future value) Invest money at end of each period. Find future value at maturity. Answers question of how much money accumulates. Future value of ordinary annuity = Annuity payment each period × Ordinary annuity table factor $FV = PMT\left[\frac{(1+i)^n - 1}{i}\right]$	Use Table 13.1: 2 years, \$4,000 ordinary annuity at 8% annually. Value = \$4,000 × 2.0800 = \$8,320 (2 periods, 8%) $FV = 4{,}000\left[\frac{(1+.08)^2 - 1}{.08}\right] = \$8{,}320$	**Calculate value of ordinary annuity** \$6,000, 7% annually, 4 years
Annuities due (find future value) Invest money at beginning of each period. Find future value at maturity. Should be higher than ordinary annuity since it is invested at beginning of each period. Use Table 13.1, but add one period and subtract one payment from answer. Future value of an annuity due = (Annuity payment each period × Ordinary* annuity table factor) − 1 Payment *Add 1 period. $FV_{due} = PMT\left[\frac{(1+i)^n - 1}{i}\right](1+i)$	*Example:* Same example as above but invest money at beginning of period. \$4,000 × 3.2464 = \$12,985.60 − 4,000.00 \$ 8,985.60 (3 periods, 8%) $FV_{due} = 4{,}000\left(\frac{(1+.08)^2 - 1}{.08}\right)(1+.08)$ = \$8,985.60	**Calculate value of annuity due** \$6,000, 7% annually, 4 years
Present value of an ordinary annuity (find present value) Calculate number of periods and rate per period. Use Table 13.2 to find table factor for present value of \$1. Multiply withdrawal for each period by table factor to get present value of an ordinary annuity. Present value of an ordinary annuity payment = Annuity payment × Present value of ordinary annuity table factor $PV = PMT\left[\frac{1-(1+i)^{-n}}{i}\right]$	*Example:* Receive \$10,000 for 5 years. Interest is 10% compounded annually. Table 13.2: 5 periods, 10% 3.7908 What you put in today = × \$10,000 \$37,908 $PV = 10{,}000\left[\frac{1-(1+.1)^{-5}}{.1}\right] = \$37{,}907.88$	**Calculate present value of ordinary annuity** \$20,000, 6 years, 4% interest compounded annually

INTERACTIVE CHAPTER ORGANIZER

Topic/Procedure/Formula	Examples	You try it*
Sinking funds (find periodic payment) Paying a particular amount of money for a set number of periodic payments to accumulate a specific sum. We know the future value and must calculate the periodic payments needed. Answer can be proved by ordinary annuity table. Sinking fund payment = Future value × Sinking fund table factor	*Example:* \$200,000 bond to retire 15 years from now. Interest is 6% compounded annually. By Table 13.3: \$200,000 × .0430 = \$8,600 Check by Table 13.1: \$8,600 × 23.2759 = \$200,172.74	**Calculate periodic payment** \$400,000 bond to retire 20 years from now. Interest is 5% compounded annually.

KEY TERMS			
	Annuities certain Annuity Annuity due Contingent annuities	Future value of an annuity Ordinary annuity Payment periods	Present value of an ordinary annuity Sinking fund Term of the annuity

*Worked-out solutions are in Appendix B.

Critical Thinking Discussion Questions with Chapter Concept Check

1. What is the difference between an ordinary annuity and an annuity due? If you were to save money in an annuity, which would you choose and why?
2. Explain how you would calculate ordinary annuities and annuities due by table lookup. Create an example to explain the meaning of a table factor from an ordinary annuity.
3. What is a present value of an ordinary annuity? Create an example showing how one of your relatives might plan for retirement by using the present value of an ordinary annuity. Would you ever have to use lump-sum payments in your calculation from Chapter 12?
4. What is a sinking fund? Why could an ordinary annuity table be used to check the sinking fund payment?
5. With the tight economy, more businesses are cutting back on matching the retirement contributions of their employees. Do you think this is ethical?
6. **Chapter Concept Check.** Create a retirement plan. Back up your retirement plan with calculations involving ordinary annuities as well as the present value of annuities.

Classroom Notes

END-OF-CHAPTER PROBLEMS connect 13

Check figures for odd-numbered problems in Appendix B. Name ______________________ Date ________

DRILL PROBLEMS

Complete the ordinary annuities for the following using tables in the *Business Math Handbook:* *LU 13-1(2)*

	Amount of payment	Payment payable	Years	Interest rate	Value of annuity
13–1.	$5,000	Annually	11	4%	
13–2.	$12,000	Semiannually	8	7%	

Redo Problem 13–1 as an annuity due:

13–3.

Calculate the value of the following annuity due without a table. Check your results by Table 13.1 or the *Business Math Handbook* (they will be slightly off due to rounding): *LU 13-1(2)*

	Amount of payment	Payment payable	Years	Interest rate
13–4.	$2,000	Annually	3	6%

Complete the following using Table 13.2 or the *Business Math Handbook* for the present value of an ordinary annuity: *LU 13-2(1)*

	Amount of annuity expected	Payment	Time	Interest rate	Present value (amount needed now to invest to receive annuity)
13–5.	$900	Annually	4 years	6%	
13–6.	$15,000	Quarterly	4 years	8%	

13–7. Check Problem 13–5 without the use of Table 13.2.

Using the sinking fund table, Table 13.3, or the *Business Math Handbook,* complete the following: *LU 13-3(1)*

	Required amount	Frequency of payment	Length of time	Interest rate	Payment amount end of each period
13–8.	$25,000	Quarterly	6 years	8%	
13–9.	$15,000	Annually	8 years	8%	

13–10. Check the answer in Problem 13–9 by Table 13.1. *LU 13-3(2)*

WORD PROBLEMS (Use Tables in the *Business Math Handbook*)

13–11. John Regan, an employee at Home Depot, made deposits of $800 at the end of each year for 4 years. Interest is 4% compounded annually. What is the value of Regan's annuity at the end of 4 years? *LU 13-1(2)*

13–12. Suze Orman wants to pay $1,500 semiannually to her granddaughter for 10 years for helping her around the house. If Suze can invest money at 6% compounded semiannually, how much must she invest today to meet this goal? *LU 13-2(1)*

13–13. Financial analysts recommend investing 15% to 20% of your annual income in your retirement fund to reach a replacement rate of 70% of your income by age 65. This recommendation increases to almost 30% if you start investing at 45 years old. Mallori Rouse is 25 years old and has started investing $3,000 at the end of each year in her retirement account. How much will her account be worth in 20 years at 8% interest compounded annually? How much will it be worth in 30 years? What about at 40 years? How much will it be worth in 50 years? Round to the nearest dollar. *LU 13-1(2)*

13–14. After paying off a car loan or credit card, don't remove this amount from your budget. Instead, invest in your future by applying some of it to your retirement account. How much would $450 invested at the end of each quarter be worth in 10 years at 4% interest? (Use the *Business Math Handbook* tables.) *LU 13-1(2)*

13–15. You decide to reduce the amount you spend eating out by $150 a month and invest the total saved at the end of each year in your retirement account. How much will the account be worth at 5% in 15 years? *LU 13-1(2)*

13–16. Rob Herndon, an accountant with Southwest Airlines, wants to retire 50% of Southwest Airlines bonds by 2038. Calculate the payment Rob needs to make at the end of each year at 6% compounded annually to reach his goal of paying off $300,000 in 20 years. *LU 13-1(2)*

13–17. Josef Company borrowed money that must be repaid in 20 years. The company wants to make sure the loan will be repaid at the end of year 20, so it invests $12,500 at the end of each year at 12% interest compounded annually. What was the amount of the original loan? *LU 13-1(2)*

My Money

13–18. Bankrate.com reported on a shocking statistic: only 54% of workers participate in their company's retirement plan. This means that 46% do not. With such an uncertain future for Social Security, this can leave almost 1 in 2 individuals without proper income during retirement. Jill Collins, 20, decided she needs to have \$250,000 in her retirement account upon retiring at 60. How much does she need to invest each year at 5% compounded annually to meet her goal? *Tip:* She is setting up a sinking fund. *LU 13-3(1)*

13–19. If you saved an average of \$2,900 each year from your income tax return, \$1,050 for not buying vendor coffee, and \$2,400 (saving \$200 each paycheck), how much would you have in your retirement account if you were able to invest this annual savings at the end of each year for 30 years at 5% interest compounded annually?

eXcel

13–20. Alice Longtree has decided to invest \$400 quarterly for 4 years in an ordinary annuity at 8%. As her financial adviser, calculate for Alice the total cash value of the annuity at the end of year 4. *LU 13-1(2)*

13–21. At the beginning of each period for 10 years, Merl Agnes invests \$500 semiannually at 6%. What is the cash value of this annuity due at the end of year 10? *LU 13-1(2)*

13–22. Jeff Associates needs to repay \$30,000. The company plans to set up a sinking fund that will repay the loan at the end of 8 years. Assume a 12% interest rate compounded semiannually. What must Jeff pay into the fund each period of time? Check your answer by Table 13.1. *LU 13-3(1, 2)*

eXcel

13–23. On Joe Martin's graduation from college, Joe's uncle promised him a gift of \$12,000 in cash or \$900 every quarter for the next 4 years after graduation. If money could be invested at 8% compounded quarterly, which offer is better for Joe? *LU 13-1(2), LU 13-2(1)*

eXcel

My Money

13–24. You are earning an average of \$46,500 and will retire in 10 years. If you put 20% of your gross average income in an ordinary annuity compounded at 7% annually, what will be the value of the annuity when you retire? *LU 13-1(2)*

13–25. A local Dunkin' Donuts franchise must buy a new piece of equipment in 5 years that will cost \$88,000. The company is setting up a sinking fund to finance the purchase. What will the quarterly deposit be if the fund earns 8% interest? *LU 13-3(1)*

13–26. Mike Macaro is selling a piece of land. Two offers are on the table. Morton Company offered a \$40,000 down payment and \$35,000 a year for the next 5 years. Flynn Company offered \$25,000 down and \$38,000 a year for the next 5 years. If money can be invested at 8% compounded annually, which offer is better for Mike? *LU 13-2(1)*

13–27. Al Vincent has decided to retire to Arizona in 10 years. What amount should Al invest today so that he will be able to withdraw \$28,000 at the end of each year for 15 years *after* he retires? Assume he can invest the money at 8% interest compounded annually. *LU 13-2(1)*

13–28. Victor French made deposits of \$5,000 at the end of each quarter to Book Bank, which pays 8% interest compounded quarterly. After 3 years, Victor made no more deposits. What will be the balance in the account 2 years after the last deposit? *LU 13-1(2)*

13–29. Janet Woo decided to retire to Florida in 6 years. What amount should Janet invest today so she can withdraw \$50,000 at the end of each year for 20 years after she retires? Assume Janet can invest money at 6% compounded annually. *LU 13-2(1)*

CHALLENGE PROBLEMS

13–30. Assume that you can buy a \$6,000 computer system in monthly installments for 3 years. The seller charges you 12% interest compounded monthly. What is your monthly payment? Assume your first payment is due at the end of the month. Use tables in the *Business Math Handbook*. *LU 13-2(1)*

$$\text{Monthly payment} = \frac{\text{Amount owed}}{\begin{array}{c}\text{Table factor}\\ \text{for PV of annuity}\end{array}}$$

13–31. Ajax Corporation has hired Brad O'Brien as its new president. Terms included the company's agreeing to pay retirement benefits of \$18,000 at the end of each semiannual period for 10 years. This will begin in 3,285 days. If the money can be invested at 8% compounded semiannually, what must the company deposit today to fulfill its obligation to Brad? *LU 13-2(1)*

Classroom Notes

SUMMARY PRACTICE TEST

Do you need help? Connect videos have step-by-step worked-out solutions.

(Use Tables in the *Business Math Handbook*)

1. Lin Lowe plans to deposit $1,800 at the end of every 6 months for the next 15 years at 8% interest compounded semiannually. What is the value of Lin's annuity at the end of 15 years? *LU 13-1(2)*

2. On Abby Ellen's graduation from law school, Abby's uncle, Bull Brady, promised her a gift of $24,000 or $2,400 every quarter for the next 4 years after graduating from law school. If the money could be invested at 6% compounded quarterly, which offer should Abby choose? *LU 13-2(1, 2)*

3. Sanka Blunck wants to receive $8,000 each year for 20 years. How much must Sanka invest today at 4% interest compounded annually? *LU 13-2(1)*

4. In 9 years, Rollo Company will have to repay a $100,000 loan. Assume a 6% interest rate compounded quarterly. How much must Rollo Company pay each period to have $100,000 at the end of 9 years? *LU 13-3(1)*

5. Lance Industries borrowed $130,000. The company plans to set up a sinking fund that will repay the loan at the end of 18 years. Assume a 6% interest rate compounded semiannually. What amount must Lance Industries pay into the fund each period? Check your answer by Table 13.1. *LU 13-3(1, 2)*

6. Joe Jan wants to receive $22,000 each year for the next 22 years. Assume a 6% interest rate compounded annually. How much must Joe invest today? *LU 13-2(1)*

7. Twice a year for 15 years, Warren Ford invested $1,700 compounded semiannually at 6% interest. What is the value of this annuity due? *LU 13-1(2)*

8. Scupper Molly invested $1,800 semiannually for 23 years at 8% interest compounded semiannually. What is the value of this annuity due? *LU 13-1(2)*

9. Nelson Collins decided to retire to Canada in 10 years. What amount should Nelson deposit so that he will be able to withdraw $80,000 at the end of each year for 25 years after he retires? Assume Nelson can invest money at 7% interest compounded annually. *LU 13-2(1)*

10. Bob Bryan made deposits of $10,000 at the end of each quarter to Lion Bank, which pays 8% interest compounded quarterly. After 9 years, Bob made no more deposits. What will be the account's balance 4 years after the last deposit? *LU 13-1(2)*

INTERACTIVE VIDEO WORKSHEET

GRADING THE SUMMARY PRACTICE TEST

Go to the summary practice test video in Connect (or click on it here in the ebook). Grade your summary practice test while viewing the video.

C for Correct/I for Incorrect

1. ________	5. ________	9. ________
2. ________	6. ________	10. ________
3. ________	7. ________	
4. ________	8. ________	

If you achieved 100%, you are ready for your instructor's exam.

If any of the problems were incorrect, list the questions you missed and show steps to solve the problem correctly.

Replay the video to see if you have made the correct fixes to your mistakes. If you have any questions, contact your instructor asap.

Notes on Watching Videos

MY MONEY

Retirement on a Budget

What I need to know

We know that starting early on our retirement savings is in our best interest since time is a great advantage to growing our investments toward reaching our financial goals. We also need to consider the level of finances we need to have when we reach retirement. Much of this will depend on how we view retirement and the activities that we want to pursue in retirement. How will we meet the financial obligations we have during retirement (mortgage payment, medical insurance, monthly expenses, etc.)?

Having the necessary funds as we enter into retirement will help us to enjoy retirement as we had planned. We need to be aware that our financial planning is not over simply because we are no longer employed full-time during retirement. Many of the same concepts we have been learning and applying up to this point are just as vital in retirement as they were during our working years.

What I need to do

You will need to establish and follow a budget during your retirement. For many, your retirement earnings will become more stable lacking the fluctuations you may have experienced years earlier as your income changed due to raises, for instance. This fixed income situation makes budgeting all the more important since the expenses you encounter will most likely continue to increase in price.

Getting ready for your retirement is a great time to learn from the experiences of others and avoid some of the common retirement mistakes. One big mistake is not having a plan. If you have not planned for your retirement years, it will be very difficult to determine when you should retire. Additionally, there are benefits paid out based upon your age at retirement. Retiring too early will mean you are missing out on higher amounts paid to you if you had delayed your retirement until later. The financial difference can be quite significant within a matter of only 5 years, so it is in your best interest to determine the appropriate age for you based on your financial situation and ability to work. It is also important to avoid withdrawals from your retirement accounts too early as there are significant penalties associated with such disbursements. At times during your retirement you may be tempted to tap into these reserve funds, but it will be costly to do so and you should avoid it if at all possible.

An emergency fund is needed even in retirement as surprise expenses know no age. A car repair can sometimes be in the thousands of dollars and have a significant negative impact on your budget without an emergency fund. It is best to plan on needing access to these emergency funds as we never know what the future may hold. Being prepared is the best way to address these emergencies when they occur.

Resources I can use

- https://money.usnews.com/money/retirement/slideshows/10-costs-to-include-in-your-retirement-budget — your budget in retirement

MY MONEY ACTIVITY

Describe your ideal retirement. Does it involve travel, hobbies, and part-time work? How will your ideal retirement impact your retirement budget planning?

PERSONAL FINANCE

A KIPLINGER APPROACH

A GIFT TO RETIREES

HAPPY BIRTHDAY, ROTH IRA

As the Roth turns 20, the new tax law gives it an edge for savers.

THE ROTH IRA CELEBRATES its 20th birthday this year, and like a lot of 20-year-olds, it has never looked so good.

You can't take a tax deduction for Roth contributions, but the money accumulates tax-free, and withdrawals are tax-free, too. If you contributed the maximum to a Roth for each of the past 20 years, you'd have nearly $200,000 in tax-free savings, assuming you invested it all in a fund that tracks Standard & Poor's 500-stock index and reinvested the dividends.

Ed Slott, a CPA in Rockville Centre, N.Y., recognized that Roths would become a game changer and started a newsletter in 1998 to teach advisers about the accounts. Slott calls Roth IRAs "tax insurance" because once you're invested in one, you won't have to pay taxes on contributions or earnings again.

The new tax law makes Roth contributions even more attractive, now that tax rates are the lowest they've been in years but could rise in the future. **KIMBERLY LANKFORD**

©C Squared Studios/Getty Images

Roth Versus Traditional IRA

	ROTH	TRADITIONAL IRA
Birth year	1998	1974
Maximum annual contribution	$5,500 ($6,500 if 50 or older)	$5,500 ($6,500 if 50 or older)
Income limits	$120,000 if single or $189,000 if married filing jointly; phases out at $135,000 for singles and $199,000 for couples	None
Deductible?	No	Yes, if you have no workplace plan or AGI of $63,000 or less for singles, or $101,000 or less for married couples filing jointly*
Early-withdrawal penalty	None, if limited to contributions	Ordinary income rate, plus 10% penalty (except under certain circumstances)
Required minimum distributions at 70½?	No	Yes

*2018 AGI deduction limits.

SILVER LININGS

UPSIDES OF RISING RATES

As interest rates climb, higher yields on bank accounts aren't the only bright spot. Don't overlook these perks.

Increased annuity payouts. If you're looking to buy an immediate annuity, you may get a larger monthly payout than those who invested when rates were lower, says Hersh Stern, of *Annuity Shopper Buyer's Guide*. On deferred income annuities, which delay the payout for a specified period, higher interest rates could also increase the payouts.

Relief on long-term-care insurance premiums. In recent years, premiums spiked as low interest rates hampered insurance companies' investment returns. Plus, fewer people dropped their policies before receiving payouts than insurers expected. But insurers have accounted for lower lapse rates in new policies, says Jesse Slome, of the American Association for Long-Term Care Insurance. And rising interest rates should help stabilize premiums on new policies.

A larger credit line on a reverse mortgage. The unused portion of a line of credit will grow as interest rates rise (the rate on debt you've accumulated will also rise). If you're thinking of getting a reverse mortgage, consider making the leap soon to maximize growth in a credit line over time. **LISA GERSTNER**

BUSINESS MATH ISSUE

The new tax law has a limited effect on Roth IRAs.

1. List the key points of the article and information to support your position.
2. Write a group defense of your position using math calculations to support your view. If you are in an online course, post to a discussion board.

CUMULATIVE REVIEW

A Word Problem Approach—Chapters 10, 11, 12, 13

1. Amy O'Mally graduated from high school. Her uncle promised her as a gift a check for $2,000 or $275 every quarter for 2 years. If money could be invested at 6% compounded quarterly, which offer is better for Amy? (Use the tables in the *Business Math Handbook.*) *LU 13-2(1)*

2. Alan Angel made deposits of $400 semiannually to Sag Bank, which pays 10% interest compounded semiannually. After 4 years, Alan made no more deposits. What will be the balance in the account 3 years after the last deposit? (Use the tables in the *Business Math Handbook.*) *LU 13-1(2)*

3. Roger Disney decides to retire to Florida in 12 years. What amount should Roger invest today so that he will be able to withdraw $30,000 at the end of each year for 20 years *after* he retires? Assume he can invest money at 8% interest compounded annually. (Use tables in the *Business Math Handbook.*) *LU 13-2(2)*

4. On September 15, Arthur Westering borrowed $3,000 from Vermont Bank at $10\frac{1}{2}$% interest. Arthur plans to repay the loan on January 25. Assume the loan is based on exact interest. How much will Arthur totally repay? *LU 10-1(2)*

5. Sue Cooper borrowed $6,000 on an $11\frac{3}{4}$%, 120-day note. Sue paid $300 toward the note on day 50. On day 90, Sue paid an additional $200. Using the U.S. Rule, what is Sue's adjusted balance after her first payment? *LU 10-3(1)*

6. On November 18, Northwest Company discounted an $18,000, 12%, 120-day note dated September 8. Assume a 10% discount rate. What will be the proceeds? Use ordinary interest. *LU 11-2(1)*

7. Alice Reed deposits $16,500 into Rye Bank, which pays 10% interest compounded semiannually. Using the appropriate table, what will Alice have in her account at the end of 6 years? *LU 12-1(2)*

8. Peter Regan needs $90,000 5 years from today to retire in Arizona. Peter's bank pays 10% interest compounded semiannually. What will Peter have to put in the bank today to have $90,000 in 5 years? *LU 12-2(2)*

Classroom Notes

Time-Value Relationship Appendix

One Lump Sum (Single Amount)

FIGURE 13.3 Compound (future value) of $.68 at 10% for 4 periods
FIGURE 13.4 Present value of $1.00 at 10% for 4 periods

Annuity (Stream of Payments)

FIGURE 13.5 Present value of a 4-year annuity of $1.00 at 10%
FIGURE 13.6 Future value of a 4-year annuity of $1.00 at 10%

Compare to Figure 13.4 to see relationship of compounding to present value.

FIGURE 13.3 Compound (future value) of $.68 at 10% for 4 periods

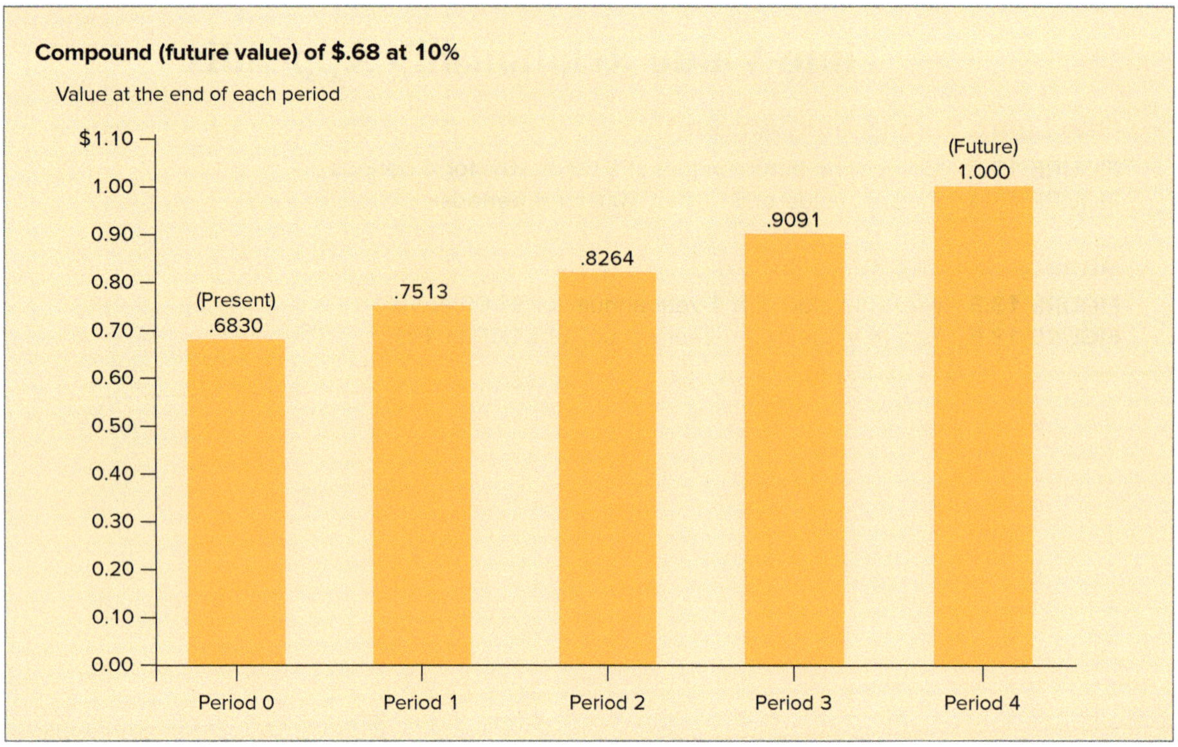

$.68 today will grow to $1.00 in the future.

What Figure 13.3 Means

If you take $.68 to a bank that pays 10%, after 4 periods you will be able to get $1.00. The $.68 is the present value, and the $1.00 is the compound value or future value. Keep in mind that the $.68 is a one lump-sum investment.

FIGURE 13.4 Present value of $1.00 at 10% for 4 periods

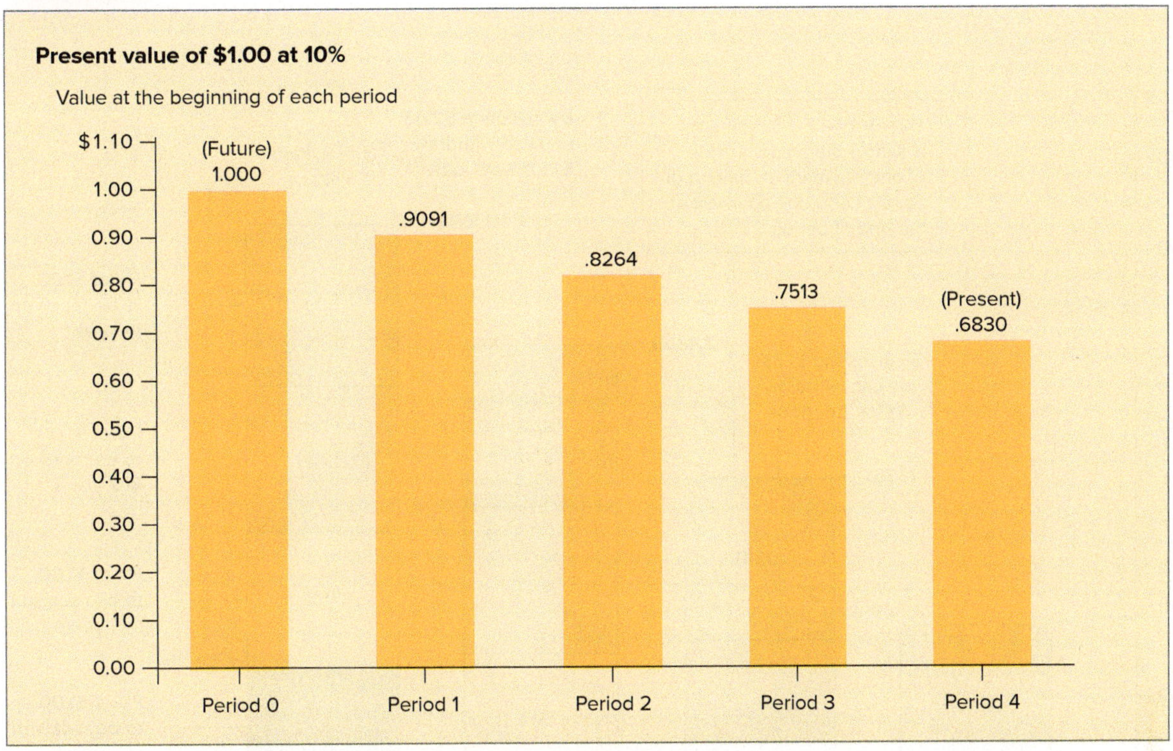

If I need $1 in four periods, I need to invest $0.68 today.

What Figure 13.4 Means

If you want to receive $1.00 at the end of 4 periods at a bank paying 10%, you will have to deposit $.68 in the bank today. The longer you have to wait for your money, the less it is worth. The $1.00 is the compound or future amount, and the $.68 is the present value of a dollar that you will not receive for 4 periods.

FIGURE 13.5 Present value of a 4-year annuity of $1.00 at 10%

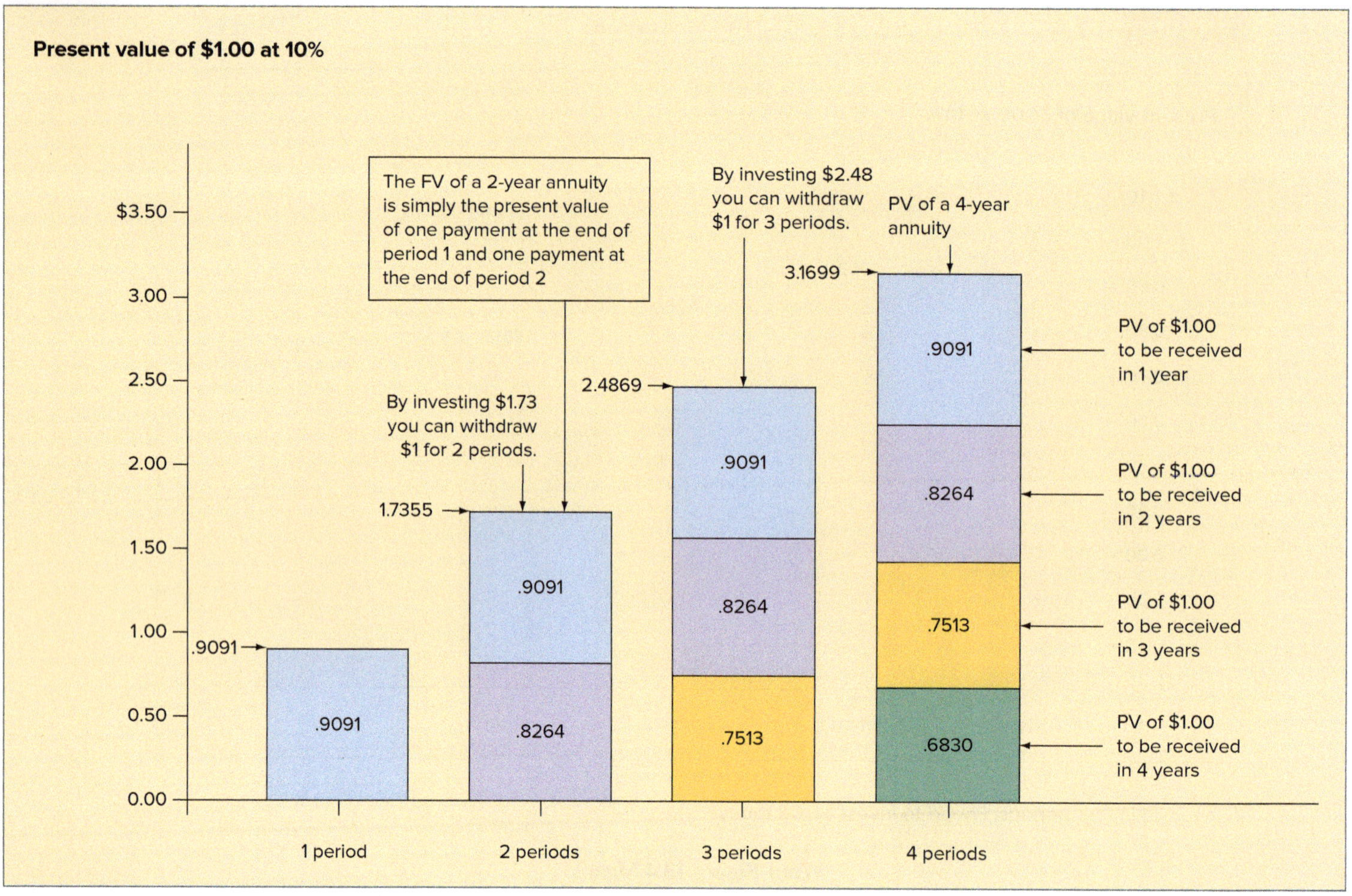

FIGURE 13.6 Future value of a 4-year annuity of $1.00 at 10%

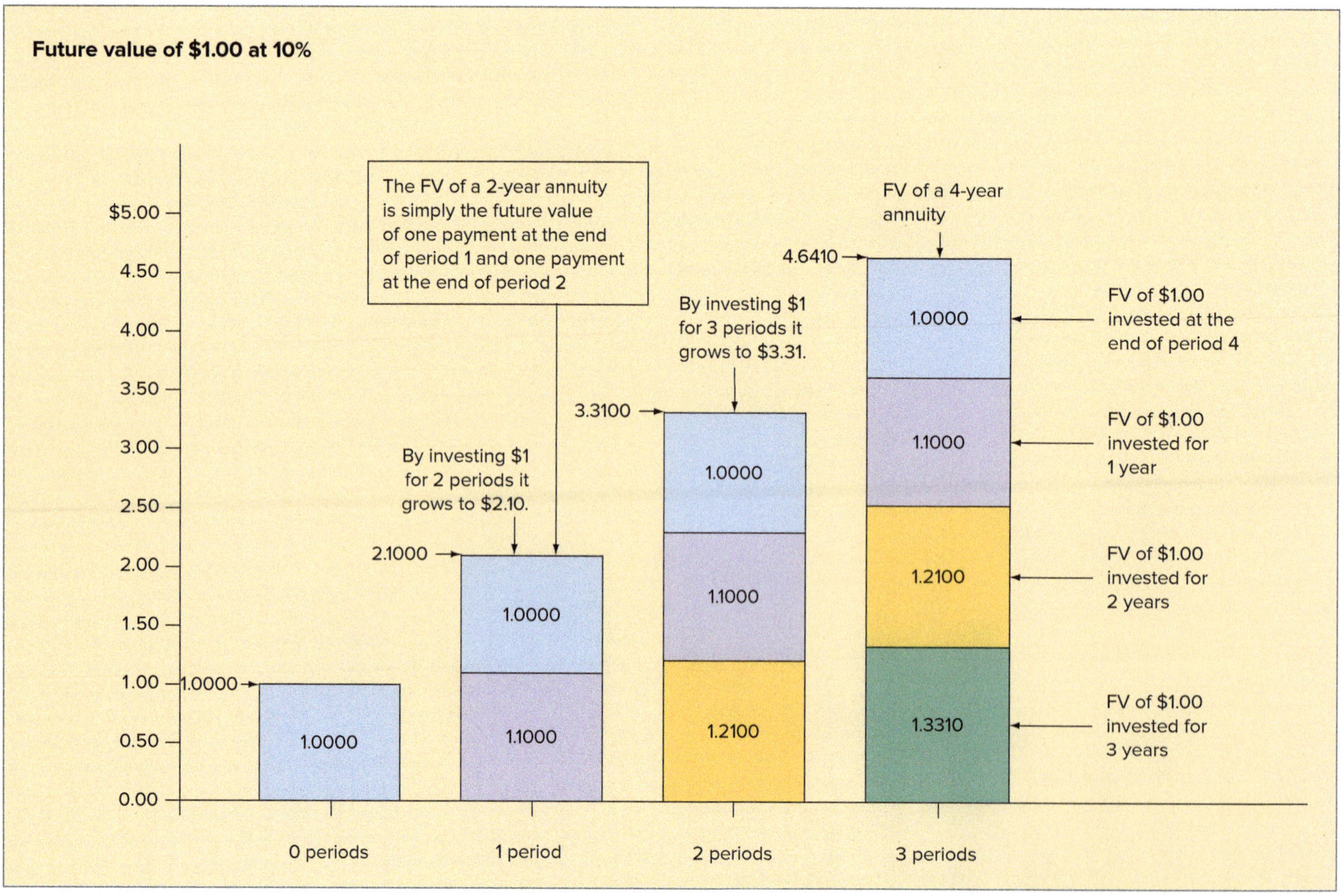

What Figure 13.5 Means*

BLUE BOX (Top box) Blue shows how to receive $1.00 after 1 period. You must put in $.91 today.

PURPLE BOX (2nd box from top) Purple shows how to receive $1.00 after 2 periods. You must put in $.83 today to get $1.00 for 2 periods. You must put in the bank today $1.74 ($.91 + $.83) to take out $1.00 for 2 periods.

YELLOW BOX (3rd box from top) Yellow shows how to receive $1.00 after 3 periods. You must put in $.75 today to get $1.00 for 3 periods. You also must put in the bank today $2.49 ($.91 + $.83 + $.75) to take out $1.00 for 3 periods.

GREEN BOX (4th box from top) Green shows how to receive $1.00 after 4 periods. You must put in $.68 today to get $1.00 after 4 periods. You also must put in the bank today $3.17 ($.91 + $.83 + $.75 + $.68) to take out $1.00 for 4 periods.

What Figure 13.6 Means*

BLUE BOX (Top box) Blue shows $1.00 invested at the end of each period. The $1.00 has no time to earn interest.

PURPLE BOX (2nd box from top) Purple shows the value of $1.00 after 2 periods. The $1.00 is now worth $1.10 due to compounding for 1 period.

YELLOW BOX (3rd box from top) Yellow shows the value of $1.00 after 3 periods. The $1.00 is now worth $1.21 due to compounding for 2 periods.

GREEN BOX (4th box from top) Green shows that the value of $1.00 after 4 periods is $1.33 due to compounding for 3 periods. If you put $1.00 in the bank at 10% for 4 years, the $4.00 grows to $4.64.

*From table in Handbook for 10%.

Periods	Amount of annuity	Present value of an annuity
1	1.0000	.9091
2	2.1000	1.7355
3	3.3100	2.4869
4	4.6410	3.1699

CHAPTER 14

Installment Buying

Higher Fed Rates Will Make 0% Financing More Painful

By Serena Ng
And Vipal Monga

The Federal Reserve is pushing interest rates higher. Don't tell that to people who have become accustomed to buying everything at 0% financing.

Years of rock-bottom interest rates have led to a proliferation of no-interest financing offers for people looking to buy everything from cars to lawn mowers, jewelry and furniture. Manufacturers and retailers have come to lean heavily on these deals, which are an inducement for shoppers considering large or discretionary purchases.

Now, with interest rates climbing, the cost of these arrangements will rise, pinching profits at companies that derive a large chunk of their sales from shoppers who prefer to pay in bite-size pieces. Most retailers will likely absorb the higher costs to stay competitive because customers may turn elsewhere if they are asked to pony up interest charges.

The cost of providing 0% financing varies from company to company, but generally retailers pay a middleman—usually a bank or finance company—a few percentage points of a product's purchase price upfront. The practice is known as "buying down the rate to zero" because retailers are in effect footing the financing costs for their customers.

©97/Getty Images

LEARNING UNIT OBJECTIVES

LU 14–1: Cost of Installment Buying

1. Calculate the amount financed, finance charge, and deferred payment.
2. Calculate the estimated APR by table lookup.
3. Calculate the monthly payment by formula and by table lookup.

LU 14–2: Revolving Charge Credit Cards

1. Calculate the finance charges on revolving charge credit card accounts.

Your Guide to Successfully Completing This Chapter

Traditional book or ebook

Check box as you complete each step.

Steps

- ☐ Read learning unit.
  - ☐ Complete practice quiz at the end of the learning unit.
- ☐ Grade practice quiz using provided solutions. (For more help, watch the learning unit video in Connect and have a Study Session with the authors. Then complete the additional practice quiz in Connect.)
- ☐ Repeat above for each of the two learning units in Chapter 14.
  - ☐ Review chapter organizer.
  - ☐ Complete assigned homework.
    - ☐ Finish summary practice test. (Go to Connect via the ebook link and do the interactive video worksheet to grade.)
- ☐ Complete instructor's exam.

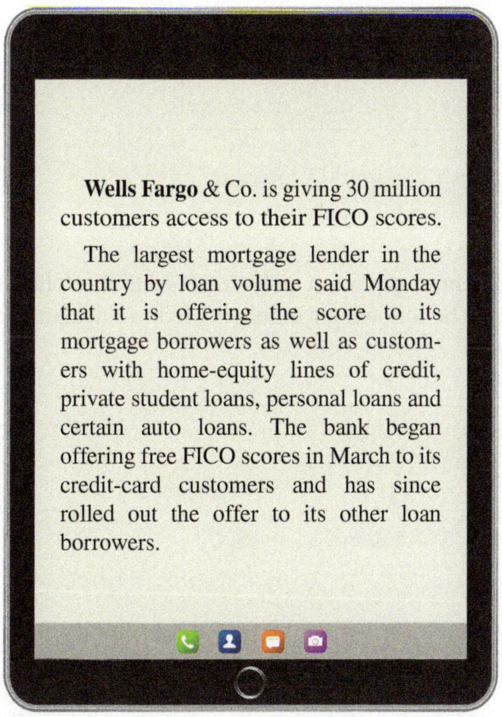

Wells Fargo & Co. is giving 30 million customers access to their FICO scores.

The largest mortgage lender in the country by loan volume said Monday that it is offering the score to its mortgage borrowers as well as customers with home-equity lines of credit, private student loans, personal loans and certain auto loans. The bank began offering free FICO scores in March to its credit-card customers and has since rolled out the offer to its other loan borrowers.

The *Wall Street Journal* chapter opener clip, "Higher Fed Rates Will Make 0% Financing More Painful," discusses the fact that interest rates are climbing. If the economy improves due to the tax act, pressure will continue for higher interest rates.

The *Wall Street Journal* clip to the left shows how Wells Fargo is providing 30 million customers with access to their FICO scores. FICO is a measure of your credit score.

This chapter discusses the cost of buying products via installments (closed-end credit) and revolving credit card (open-end credit). You will see in Learning Unit 14–1 that to buy a 4 × 4 pickup a qualified buyer must have a credit score of 720 or higher (see clip below).

Learning Unit 14–1: Cost of Installment Buying

Installment buying, a form of *closed-end credit,* can add a substantial amount to the cost of big-ticket purchases. To illustrate this, we follow the procedure of buying a pickup truck, including the amount financed, finance charge, and deferred payment price. Then we study the effect of the Truth in Lending Act.

LO 1 Amount Financed, Finance Charge, and Deferred Payment

My Money

The advertisement for the sale of a pickup truck shown on the next page appeared in a local paper. As you can see from this advertisement, after customers make a **down payment,** they can buy the truck with an **installment loan.** This loan is paid off with a series of equal periodic payments. These payments include both interest and principal. The payment process is called **amortization.** In the promissory notes of earlier chapters, the loan was paid off in one ending payment. Now let's look at the calculations involved in buying a pickup truck.

©Scott Olson/Getty Images

Checking Calculations in Pickup Advertisement

Calculating Amount Financed The **amount financed** is what you actually borrow. To calculate this amount, use the following formula:

Amount financed = Cash price − Down payment

$9,045 = $9,345 − $300

Calculating Finance Charge The words **"finance charge"** in the advertisement represent the *interest* charge. The interest charge resulting in the finance charge includes the cost of credit reports, mandatory bank fees, and so on. You can use the following formula to calculate the total interest on the loan:

Total finance charge (interest charge) = Total of all monthly payments − Amount financed

$2,617.80 = $11,662.80 − $9,045
($194.38 × 60 months)

MONEY tips

Consider the pros and cons of buying (new or used) versus leasing a vehicle. Negotiate the terms regardless of which option you choose. Note: you pay insurance and maintenance on either option.

Buy

- Build equity
- Substantial down payment
- Higher monthly payments (but cheaper long term)
- No worry about wear and tear (except for trade-in or resale value)
- Unlimited miles

Lease

- Don't own—can upgrade every few years
- Lower down payment
- Lower monthly payments
- Responsible for above-average wear and tear
- Limited miles

Calculating Deferred Payment Price The **deferred payment price** represents the total of all monthly payments plus the down payment. The following formula is used to calculate the deferred payment price:

Deferred payment price = Total of all monthly payments + Down payment

$11,962.80 = $11,662.80 + $300
($194.38 × 60)

Truth in Lending: APR Defined and Calculated

In 1969, the Federal Reserve Board established the **Truth in Lending Act** (Regulation Z). The law doesn't regulate interest charges; its purpose is to make the consumer aware of the true cost of credit.

The Truth in Lending Act requires that creditors provide certain basic information about the actual cost of buying on credit. Before buyers sign a credit agreement, creditors must inform them in writing of the amount of the finance charge and the **annual percentage rate (APR).** The APR represents the true or effective annual interest creditors charge. This is helpful to buyers who repay loans over different periods of time (1 month, 48 months, and so on).

To illustrate how the APR affects the interest rate, assume you borrow \$100 for 1 year and pay a finance charge of \$9. Your interest rate would be 9% if you waited until the end of the year to pay back the loan. Now let's say you pay off the loan and the finance charge in 12 monthly payments. Each month that you make a payment, you are losing some of the value or use of that money. So the true or effective APR is actually greater than 9%.

The APR can be calculated by formula or by tables. We will use the table method since it is more exact.

LO 2

Calculating APR Rate by Table 14.1 Note the following steps for using a table to calculate APR:

CALCULATING APR BY TABLE

Step 1. Divide the finance charge by amount financed and multiply by \$100 to get the table lookup factor.

Step 2. Go to APR Table 14.1. At the left side of the table are listed the number of payments that will be made.

Step 3. When you find the number of payments you are looking for, move to the right and look for the two numbers closest to the table lookup number. This will indicate the APR.

Now let's determine the APR for the pickup truck advertisement given earlier in the chapter.

As stated in Step 1, we begin by dividing the finance charge by the amount financed and multiply by \$100:

$$\frac{\text{Finance charge}}{\text{Amount financed}} \times \$100 = \text{Table 14.1 lookup number}$$

$$\frac{\$2{,}617.80}{\$9{,}045.00} \times \$100 = \$28.94$$

We multiply by \$100, since the table is based on \$100 of financing.

To look up \$28.94 in Table 14.1, we go down the left side of the table until we come to 60 payments (the advertisement states 60 months). Then, moving to the right, we look for \$28.94 or the two numbers closest to it. The number \$28.94 is between \$28.22 and \$28.96. So we look at the column headings and see a rate between 10.25% and 10.5%. The Truth in Lending Act requires that when creditors state the APR, it must be accurate to the nearest $\frac{1}{4}$ of 1%.[1]

LO 3

Calculating the Monthly Payment by Formula and Table 14.2 The pickup truck advertisement showed a \$194.38 monthly payment. We can check this by formula and by table lookup.

[1]If we wanted an exact reading of APR when the number is not exactly in the table, we would use the process of interpolating. We do not cover this method in this course.

TABLE 14.1 Annual percentage rate table per $100

NUMBER OF PAYMENTS	ANNUAL PERCENTAGE RATE															
	10.00%	10.25%	10.50%	10.75%	11.00%	11.25%	11.50%	11.75%	12.00%	12.25%	12.50%	12.75%	13.00%	13.25%	13.50%	13.75%
	(FINANCE CHARGE PER $100 OF AMOUNT FINANCED)															
1	0.83	0.85	0.87	0.90	0.92	0.94	0.96	0.98	1.00	1.02	1.04	1.06	1.08	1.10	1.12	1.15
2	1.25	1.28	1.31	1.35	1.38	1.41	1.44	1.47	1.50	1.53	1.57	1.60	1.63	1.66	1.69	1.72
3	1.67	1.71	1.76	1.80	1.84	1.88	1.92	1.96	2.01	2.05	2.09	2.13	2.17	2.22	2.26	2.30
4	2.09	2.14	2.20	2.25	2.30	2.35	2.41	2.46	2.51	2.57	2.62	2.67	2.72	2.78	2.83	2.88
5	2.51	2.58	2.64	2.70	2.77	2.83	2.89	2.96	3.02	3.08	3.15	3.21	3.27	3.34	3.40	3.46
6	2.94	3.01	3.08	3.16	3.23	3.31	3.38	3.45	3.53	3.60	3.68	3.75	3.83	3.90	3.97	4.05
7	3.36	3.45	3.53	3.62	3.70	3.78	3.87	3.95	4.04	4.12	4.21	4.29	4.38	4.47	4.55	4.64
8	3.79	3.88	3.98	4.07	4.17	4.26	4.36	4.46	4.55	4.65	4.74	4.84	4.94	5.03	5.13	5.22
9	4.21	4.32	4.43	4.53	4.64	4.75	4.85	4.96	5.07	5.17	5.28	5.39	5.49	5.60	5.71	5.82
10	4.64	4.76	4.88	4.99	5.11	5.23	5.35	5.46	5.58	5.70	5.82	5.94	6.05	6.17	6.29	6.41
11	5.07	5.20	5.33	5.45	5.58	5.71	5.84	5.97	6.10	6.23	6.36	6.49	6.62	6.75	6.88	7.01
12	5.50	5.64	5.78	5.92	6.06	6.20	6.34	6.48	6.62	6.76	6.90	7.04	7.18	7.32	7.46	7.60
13	5.93	6.08	6.23	6.38	6.53	6.68	6.84	6.99	7.14	7.29	7.44	7.59	7.75	7.90	8.05	8.20
14	6.36	6.52	6.69	6.85	7.01	7.17	7.34	7.50	7.66	7.82	7.99	8.15	8.31	8.48	8.64	8.81
15	6.80	6.97	7.14	7.32	7.49	7.66	7.84	8.01	8.19	8.36	8.53	8.71	8.88	9.06	9.23	9.41
16	7.23	7.41	7.60	7.78	7.97	8.15	8.34	8.53	8.71	8.90	9.08	9.27	9.46	9.64	9.83	10.02
17	7.67	7.86	8.06	8.25	8.45	8.65	8.84	9.04	9.24	9.44	9.63	9.83	10.03	10.23	10.43	10.63
18	8.10	8.31	8.52	8.73	8.93	9.14	9.35	9.56	9.77	9.98	10.19	10.40	10.61	10.82	11.03	11.24
19	8.54	8.76	8.98	9.20	9.42	9.64	9.86	10.08	10.30	10.52	10.74	10.96	11.18	11.41	11.63	11.85
20	8.98	9.21	9.44	9.67	9.90	10.13	10.37	10.60	10.83	11.06	11.30	11.53	11.76	12.00	12.23	12.46
21	9.42	9.66	9.90	10.15	10.39	10.63	10.88	11.12	11.36	11.61	11.85	12.10	12.34	12.59	12.84	13.08
22	9.86	10.12	10.37	10.62	10.88	11.13	11.39	11.64	11.90	12.16	12.41	12.67	12.93	13.19	13.44	13.70
23	10.30	10.57	10.84	11.10	11.37	11.63	11.90	12.17	12.44	12.71	12.97	13.24	13.51	13.78	14.05	14.32
24	10.75	11.02	11.30	11.58	11.86	12.14	12.42	12.70	12.98	13.26	13.54	13.82	14.10	14.38	14.66	14.95
25	11.19	11.48	11.77	12.06	12.35	12.64	12.93	13.22	13.52	13.81	14.10	14.40	14.69	14.98	15.28	15.57
26	11.64	11.94	12.24	12.54	12.85	13.15	13.45	13.75	14.06	14.36	14.67	14.97	15.28	15.59	15.89	16.20
27	12.09	12.40	12.71	13.03	13.34	13.66	13.97	14.29	14.60	14.92	15.24	15.56	15.87	16.19	16.51	16.83
28	12.53	12.86	13.18	13.51	13.84	14.16	14.49	14.82	15.15	15.48	15.81	16.14	16.47	16.80	17.13	17.46
29	12.98	13.32	13.66	14.00	14.33	14.67	15.01	15.35	15.70	16.04	16.38	16.72	17.07	17.41	17.75	18.10
30	13.43	13.78	14.13	14.48	14.83	15.19	15.54	15.89	16.24	16.60	16.95	17.31	17.66	18.02	18.38	18.74
31	13.89	14.25	14.61	14.97	15.33	15.70	16.06	16.43	16.79	17.16	17.53	17.90	18.27	18.63	19.00	19.38
32	14.34	14.71	15.09	15.46	15.84	16.21	16.59	16.97	17.35	17.73	18.11	18.49	18.87	19.25	19.63	20.02
33	14.79	15.18	15.57	15.95	16.34	16.73	17.12	17.51	17.90	18.29	18.69	19.08	19.47	19.87	20.26	20.66
34	15.25	15.65	16.05	16.44	16.85	17.25	17.65	18.05	18.46	18.86	19.27	19.67	20.08	20.49	20.90	21.31
35	15.70	16.11	16.53	16.94	17.35	17.77	18.18	18.60	19.01	19.43	19.85	20.27	20.69	21.11	21.53	21.95
36	16.16	16.58	17.01	17.43	17.86	18.29	18.71	19.14	19.57	20.00	20.43	20.87	21.30	21.73	22.17	22.60
37	16.62	17.06	17.49	17.93	18.37	18.81	19.25	19.69	20.13	20.58	21.02	21.46	21.91	22.36	22.81	23.25
38	17.08	17.53	17.98	18.43	18.88	19.33	19.78	20.24	20.69	21.15	21.61	22.07	22.52	22.99	23.45	23.91
39	17.54	18.00	18.46	18.93	19.39	19.86	20.32	20.79	21.26	21.73	22.20	22.67	23.14	23.61	24.09	24.56
40	18.00	18.48	18.95	19.43	19.90	20.38	20.86	21.34	21.82	22.30	22.79	23.27	23.76	24.25	24.73	25.22
41	18.47	18.95	19.44	19.93	20.42	20.91	21.40	21.89	22.39	22.88	23.38	23.88	24.38	24.88	25.38	25.88
42	18.93	19.43	19.93	20.43	20.93	21.44	21.94	22.45	22.96	23.47	23.98	24.49	25.00	25.51	26.03	26.55
43	19.40	19.91	20.42	20.94	21.45	21.97	22.49	23.01	23.53	24.05	24.57	25.10	25.62	26.15	26.68	27.21
44	19.86	20.39	20.91	21.44	21.97	22.50	23.03	23.57	24.10	24.64	25.17	25.71	26.25	26.79	27.33	27.88
45	20.33	20.87	21.41	21.95	22.49	23.03	23.58	24.12	24.67	25.22	25.77	26.32	26.88	27.43	27.99	28.55
46	20.80	21.35	21.90	22.46	23.01	23.57	24.13	24.69	25.25	25.81	26.37	26.94	27.51	28.08	28.65	29.22
47	21.27	21.83	22.40	22.97	23.53	24.10	24.68	25.25	25.82	26.40	26.98	27.56	28.14	28.72	29.31	29.89
48	21.74	22.32	22.90	23.48	24.06	24.64	25.23	25.81	26.40	26.99	27.58	28.18	28.77	29.37	29.97	30.57
49	22.21	22.80	23.39	23.99	24.58	25.18	25.78	26.38	26.98	27.59	28.19	28.80	29.41	30.02	30.63	31.24
50	22.69	23.29	23.89	24.50	25.11	25.72	26.33	26.95	27.56	28.18	28.80	29.42	30.04	30.67	31.29	31.92
51	23.16	23.78	24.40	25.02	25.64	26.26	26.89	27.52	28.15	28.78	29.41	30.05	30.68	31.32	31.96	32.60
52	23.64	24.27	24.90	25.53	26.17	26.81	27.45	28.09	28.73	29.38	30.02	30.67	31.32	31.98	32.63	33.29
53	24.11	24.76	25.40	26.05	26.70	27.35	28.00	28.66	29.32	29.98	30.64	31.30	31.97	32.63	33.30	33.97
54	24.59	25.25	25.91	26.57	27.23	27.90	28.56	29.23	29.91	30.58	31.25	31.93	32.61	33.29	33.98	34.66
55	25.07	25.74	26.41	27.09	27.77	28.44	29.13	29.81	30.50	31.18	31.87	32.56	33.26	33.95	34.65	35.35
56	25.55	26.23	26.92	27.61	28.30	28.99	29.69	30.39	31.09	31.79	32.49	33.20	33.91	34.62	35.33	36.04
57	26.03	26.73	27.43	28.13	28.84	29.54	30.25	30.97	31.68	32.39	33.11	33.83	34.56	35.28	36.01	36.74
58	26.51	27.23	27.94	28.66	29.37	30.10	30.82	31.55	32.27	33.00	33.74	34.47	35.21	35.95	36.69	37.43
59	27.00	27.72	28.45	29.18	29.91	30.65	31.39	32.13	32.87	33.61	34.36	35.11	35.86	36.62	37.37	38.13
60	27.48	28.22	28.96	29.71	30.45	31.20	31.96	32.71	33.47	34.23	34.99	35.75	36.52	37.29	38.06	38.83

Note: For a more detailed set of tables from 2% to 21.75%, see the reference tables in the *Business Math Handbook*.

By Formula

$$\frac{\text{Finance charge} + \text{Amount financed}}{\text{Number of payments of loan}} = \frac{\$2{,}617.80 + \$9{,}045}{60} = \$194.38$$

By Table 14.2 The **loan amortization table** (many variations of this table are available) in Table 14.2 can be used to calculate the monthly payment for the pickup truck. To calculate a monthly payment with a table, use the following steps:

TABLE 14.1 (concluded)

NUMBER OF PAYMENTS	ANNUAL PERCENTAGE RATE															
	14.00%	14.25%	14.50%	14.75%	15.00%	15.25%	15.50%	15.75%	16.00%	16.25%	16.50%	16.75%	17.00%	17.25%	17.50%	17.75%
	(FINANCE CHARGE PER $100 OF AMOUNT FINANCED)															
1	1.17	1.19	1.21	1.23	1.25	1.27	1.29	1.31	1.33	1.35	1.37	1.40	1.42	1.44	1.46	1.48
2	1.75	1.78	1.82	1.85	1.88	1.91	1.94	1.97	2.00	2.04	2.07	2.10	2.13	2.16	2.19	2.22
3	2.34	2.38	2.43	2.47	2.51	2.55	2.59	2.64	2.68	2.72	2.76	2.80	2.85	2.89	2.93	2.97
4	2.93	2.99	3.04	3.09	3.14	3.20	3.25	3.30	3.36	3.41	3.46	3.51	3.57	3.62	3.67	3.73
5	3.53	3.59	3.65	3.72	3.78	3.84	3.91	3.97	4.04	4.10	4.16	4.23	4.29	4.35	4.42	4.48
6	4.12	4.20	4.27	4.35	4.42	4.49	4.57	4.64	4.72	4.79	4.87	4.94	5.02	5.09	5.17	5.24
7	4.72	4.81	4.89	4.98	5.06	5.15	5.23	5.32	5.40	5.49	5.58	5.66	5.75	5.83	5.92	6.00
8	5.32	5.42	5.51	5.61	5.71	5.80	5.90	6.00	6.09	6.19	6.29	6.38	6.48	6.58	6.67	6.77
9	5.92	6.03	6.14	6.25	6.35	6.46	6.57	6.68	6.78	6.89	7.00	7.11	7.22	7.32	7.43	7.54
10	6.53	6.65	6.77	6.88	7.00	7.12	7.24	7.36	7.48	7.60	7.72	7.84	7.96	8.08	8.19	8.31
11	7.14	7.27	7.40	7.53	7.66	7.79	7.92	8.05	8.18	8.31	8.44	8.57	8.70	8.83	8.96	9.09
12	7.74	7.89	8.03	8.17	8.31	8.45	8.59	8.74	8.88	9.02	9.16	9.30	9.45	9.59	9.73	9.87
13	8.36	8.51	8.66	8.81	8.97	9.12	9.27	9.43	9.58	9.73	9.89	10.04	10.20	10.35	10.50	10.66
14	8.97	9.13	9.30	9.46	9.63	9.79	9.96	10.12	10.29	10.45	10.62	10.78	10.95	11.11	11.28	11.45
15	9.59	9.76	9.94	10.11	10.29	10.47	10.64	10.82	11.00	11.17	11.35	11.53	11.71	11.88	12.06	12.24
16	10.20	10.39	10.58	10.77	10.95	11.14	11.33	11.52	11.71	11.90	12.09	12.28	12.46	12.65	12.84	13.03
17	10.82	11.02	11.22	11.42	11.62	11.82	12.02	12.22	12.42	12.62	12.83	13.03	13.23	13.43	13.63	13.83
18	11.45	11.66	11.87	12.08	12.29	12.50	12.72	12.93	13.14	13.35	13.57	13.78	13.99	14.21	14.42	14.64
19	12.07	12.30	12.52	12.74	12.97	13.19	13.41	13.64	13.86	14.09	14.31	14.54	14.76	14.99	15.22	15.44
20	12.70	12.93	13.17	13.41	13.64	13.88	14.11	14.35	14.59	14.82	15.06	15.30	15.54	15.77	16.01	16.25
21	13.33	13.58	13.82	14.07	14.32	14.57	14.82	15.06	15.31	15.56	15.81	16.06	16.31	16.56	16.81	17.07
22	13.96	14.22	14.48	14.74	15.00	15.26	15.52	15.78	16.04	16.30	16.57	16.83	17.09	17.36	17.62	17.88
23	14.59	14.87	15.14	15.41	15.68	15.96	16.23	16.50	16.78	17.05	17.32	17.60	17.88	18.15	18.43	18.70
24	15.23	15.51	15.80	16.08	16.37	16.65	16.94	17.22	17.51	17.80	18.09	18.37	18.66	18.95	19.24	19.53
25	15.87	16.17	16.46	16.76	17.06	17.35	17.65	17.95	18.25	18.55	18.85	19.15	19.45	19.75	20.05	20.36
26	16.51	16.82	17.13	17.44	17.75	18.06	18.37	18.68	18.99	19.30	19.62	19.93	20.24	20.56	20.87	21.19
27	17.15	17.47	17.80	18.12	18.44	18.76	19.09	19.41	19.74	20.06	20.39	20.71	21.04	21.37	21.69	22.02
28	17.80	18.13	18.47	18.80	19.14	19.47	19.81	20.15	20.48	20.82	21.16	21.50	21.84	22.18	22.52	22.86
29	18.45	18.79	19.14	19.49	19.83	20.18	20.53	20.88	21.23	21.58	21.94	22.29	22.64	22.99	23.35	23.70
30	19.10	19.45	19.81	20.17	20.54	20.90	21.26	21.62	21.99	22.35	22.72	23.08	23.45	23.81	24.18	24.55
31	19.75	20.12	20.49	20.87	21.24	21.61	21.99	22.37	22.74	23.12	23.50	23.88	24.26	24.64	25.02	25.40
32	20.40	20.79	21.17	21.56	21.95	22.33	22.72	23.11	23.50	23.89	24.28	24.68	25.07	25.46	25.86	26.25
33	21.06	21.46	21.85	22.25	22.65	23.06	23.46	23.86	24.26	24.67	25.07	25.48	25.88	26.29	26.70	27.11
34	21.72	22.13	22.54	22.95	23.37	23.78	24.19	24.61	25.03	25.44	25.86	26.28	26.70	27.12	27.54	27.97
35	22.38	22.80	23.23	23.65	24.08	24.51	24.94	25.36	25.79	26.23	26.66	27.09	27.52	27.96	28.39	28.83
36	23.04	23.48	23.92	24.35	24.80	25.24	25.68	26.12	26.57	27.01	27.46	27.90	28.35	28.80	29.25	29.70
37	23.70	24.16	24.61	25.06	25.51	25.97	26.42	26.88	27.34	27.80	28.26	28.72	29.18	29.64	30.10	30.57
38	24.37	24.84	25.30	25.77	26.24	26.70	27.17	27.64	28.11	28.59	29.06	29.53	30.01	30.49	30.96	31.44
39	25.04	25.52	26.00	26.48	26.96	27.44	27.92	28.41	28.89	29.38	29.87	30.36	30.85	31.34	31.83	32.32
40	25.71	26.20	26.70	27.19	27.69	28.18	28.68	29.18	29.68	30.18	30.69	31.19	31.68	32.19	32.69	33.20
41	26.39	26.89	27.40	27.91	28.41	28.92	29.44	29.95	30.46	30.97	31.49	32.01	32.52	33.04	33.56	34.08
42	27.06	27.58	28.10	28.62	29.15	29.67	30.19	30.72	31.25	31.78	32.31	32.84	33.37	33.90	34.44	34.97
43	27.74	28.27	28.81	29.34	29.88	30.42	30.96	31.50	32.04	32.58	33.13	33.67	34.22	34.76	35.31	35.86
44	28.42	28.97	29.52	30.07	30.62	31.17	31.72	32.28	32.83	33.39	33.95	34.51	35.07	35.63	36.19	36.76
45	29.11	29.67	30.23	30.79	31.36	31.92	32.49	33.06	33.63	34.20	34.77	35.35	35.92	36.50	37.08	37.66
46	29.79	30.36	30.94	31.52	32.10	32.68	33.26	33.84	34.43	35.01	35.60	36.19	36.78	37.37	37.96	38.56
47	30.48	31.07	31.66	32.25	32.84	33.44	34.03	34.63	35.23	35.83	36.43	37.04	37.64	38.25	38.86	39.46
48	31.17	31.77	32.37	32.98	33.59	34.20	34.81	35.42	36.03	36.65	37.27	37.88	38.50	39.13	39.75	40.37
49	31.86	32.48	33.09	33.71	34.34	34.96	35.59	36.21	36.84	37.47	38.10	38.74	39.37	40.01	40.65	41.29
50	32.55	33.18	33.82	34.45	35.09	35.73	36.37	37.01	37.65	38.30	38.94	39.59	40.24	40.89	41.55	42.20
51	33.25	33.89	34.54	35.19	35.84	36.49	37.15	37.81	38.46	39.12	39.79	40.45	41.11	41.78	42.45	43.12
52	33.95	34.61	35.27	35.93	36.60	37.27	37.94	38.61	39.28	39.96	40.63	41.31	41.99	42.67	43.36	44.04
53	34.65	35.32	36.00	36.68	37.36	38.04	38.72	39.41	40.10	40.79	41.48	42.17	42.87	43.57	44.27	44.97
54	35.35	36.04	36.73	37.42	38.12	38.82	39.52	40.22	40.92	41.63	42.33	43.04	43.75	44.47	45.18	45.90
55	36.05	36.76	37.46	38.17	38.88	39.60	40.31	41.03	41.74	42.47	43.19	43.91	44.64	45.37	46.10	46.83
56	36.76	37.48	38.20	38.92	39.65	40.38	41.11	41.84	42.57	43.31	44.05	44.79	45.53	46.27	47.02	47.77
57	37.47	38.20	38.94	39.68	40.42	41.16	41.91	42.65	43.40	44.15	44.91	45.66	46.42	47.18	47.94	48.71
58	38.18	38.93	39.68	40.43	41.19	41.95	42.71	43.47	44.23	45.00	45.77	46.54	47.32	48.09	48.87	49.65
59	38.89	39.66	40.42	41.19	41.96	42.74	43.51	44.29	45.07	45.85	46.64	47.42	48.21	49.01	49.80	50.60
60	39.61	40.39	41.17	41.95	42.74	43.53	44.32	45.11	45.91	46.71	47.51	48.31	49.12	49.92	50.73	51.55

MONEY tips

Control your debt. Review your credit card's year-end summary to see where your money goes. Adjust your spending accordingly.

CALCULATING MONTHLY PAYMENT BY TABLE LOOKUP

Step 1. Divide the loan amount by $1,000 (since Table 14.2 is per $1,000):

$$\frac{\$9{,}045}{\$1{,}000} = 9.045$$

Step 2. Look up the rate (10.5%) and number of months (60). At the intersection is the table factor showing the monthly payment per $1,000.

Step 3. Multiply quotient in Step 1 by the table factor in Step 2:

9.045 × $21.49 = $194.38.

TABLE 14.2 Loan amortization table (monthly payment per $1,000 to pay principal and interest on installment loan)

Terms in months	7.50%	8.00%	8.50%	9.00%	10.00%	10.50%	11.00%	11.50%	12.00%
6	$170.34	$170.58	$170.83	$171.20	$171.56	$171.81	$172.05	$172.30	$172.55
12	86.76	86.99	87.22	87.46	87.92	88.15	88.38	88.62	88.85
18	58.92	59.15	59.37	59.60	60.06	60.29	60.52	60.75	60.98
24	45.00	45.23	45.46	45.69	46.14	46.38	46.61	46.84	47.07
30	36.66	36.89	37.12	37.35	37.81	38.04	38.28	38.51	38.75
36	31.11	31.34	31.57	31.80	32.27	32.50	32.74	32.98	33.21
42	27.15	27.38	27.62	27.85	28.32	28.55	28.79	29.03	29.28
48	24.18	24.42	24.65	24.77	25.36	25.60	25.85	26.09	26.33
54	21.88	22.12	22.36	22.59	23.07	23.32	23.56	23.81	24.06
60	20.04	20.28	20.52	20.76	21.25	21.49	21.74	21.99	22.24

Remember that this $194.38 fixed payment includes interest and the reduction of the balance of the loan. As the number of payments increases, interest payments get smaller and the reduction of the principal gets larger.[2]

Now let's check your progress with the Practice Quiz.

LU 14–1 PRACTICE QUIZ

Complete this Practice Quiz to see how you are doing.

©Darren Greenwood/Design Pics/Getty Images

*For **extra help** from your authors–Sharon and Jeff–see the videos in Connect.*

From the partial advertisement at the right calculate the following:

1. **a.** Amount financed.
   b. Finance charge.
   c. Deferred payment price.
   d. APR by Table 14.1.
   e. Monthly payment by formula.

$288 per month	
Sale price	$14,150
Down payment	$ 1,450
Term/Number of payments	60 months

2. Jay Miller bought a New Brunswick boat for $7,500. Jay put down $1,000 and financed the balance at 10% for 60 months. What is his monthly payment? Use Table 14.2.

✓ Solutions

1. **a.** $14,150 – $1,450 = **$12,700**
   b. $17,280 ($288 × 60) – $12,700 = **$4,580**
   c. $17,280 ($288 × 60) + $1,450 = **$18,730**
   d. $\frac{\$4,580}{\$12,700} \times \$100 = \$36.06$; between **12.75%** and **13%**
   e. $\frac{\$4,580 + \$12,700}{60} = \$288$

2. $\frac{\$6,500}{\$1,000} = 6.5 \times \$21.25 = \$138.13$ (10%, 60 months)

[2]In Chapter 15 we give an amortization schedule for home mortgages that shows how much of each fixed payment goes to interest and how much reduces the principal. This repayment schedule also gives a running balance of the loan.

TABLE 14.2 (concluded)

Terms in months	12.50%	13.00%	13.50%	14.00%	14.50%	15.00%	15.50%	16.00%
6	$172.80	$173.04	$173.29	$173.54	$173.79	$174.03	$174.28	$174.53
12	89.08	89.32	89.55	89.79	90.02	90.26	90.49	90.73
18	61.21	61.45	61.68	61.92	62.15	62.38	62.62	62.86
24	47.31	47.54	47.78	48.01	48.25	48.49	48.72	48.96
30	38.98	39.22	39.46	39.70	39.94	40.18	40.42	40.66
36	33.45	33.69	33.94	34.18	34.42	34.67	34.91	35.16
42	29.52	29.76	30.01	30.25	30.50	30.75	31.00	31.25
48	26.58	26.83	27.08	27.33	27.58	27.83	28.08	28.34
54	24.31	24.56	24.81	25.06	25.32	25.58	25.84	26.10
60	22.50	22.75	23.01	23.27	23.53	23.79	24.05	24.32

My Money

Learning Unit 14–2: Revolving Charge Credit Cards

Do you owe a balance on your credit card? Let's look at how long it will take to pay off your credit card balance by making payments for the minimum amount. Study the clipping "Pay Just the Minimum, and Get Nowhere Fast."

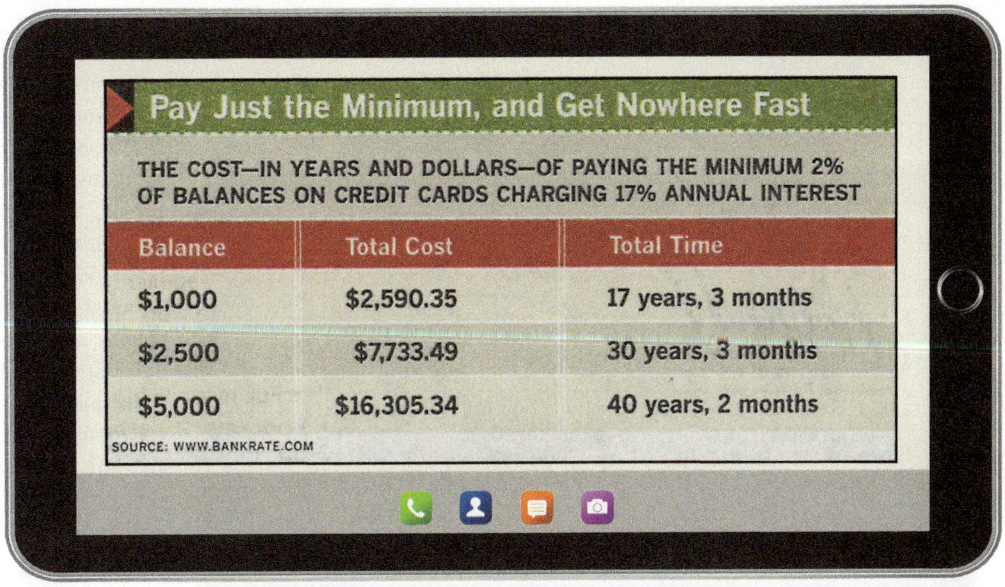

Pay Just the Minimum, and Get Nowhere Fast

THE COST—IN YEARS AND DOLLARS—OF PAYING THE MINIMUM 2% OF BALANCES ON CREDIT CARDS CHARGING 17% ANNUAL INTEREST

Balance	Total Cost	Total Time
$1,000	$2,590.35	17 years, 3 months
$2,500	$7,733.49	30 years, 3 months
$5,000	$16,305.34	40 years, 2 months

SOURCE: WWW.BANKRATE.COM

The *Wall Street Journal* clip above assumes that the minimum rate on the balance of a credit card is 2%. Note that if the annual interest cost is 17%, it will take 17 years, 3 months to pay off a balance of $1,000, and the total cost will be $2,590.35. If the balance on your revolving charge credit card is more than $1,000, you can see how fast the total cost rises. If you cannot afford the total cost of paying only the minimum, it is time for you to reconsider how you use your revolving credit card. This is why when you have financial difficulties, experts often advise you first to work on getting rid of your revolving credit card debt.

Do you know why revolving credit cards are so popular? Businesses encourage customers to use credit cards because consumers tend to buy more when they can use a credit card for their purchases. Consumers find credit cards convenient to use and valuable in establishing credit. The problem is that when consumers do not pay their balance in full each month, they do not realize how expensive it is to pay only the minimum of their balance. American Express offered a card called Serve that does not allow a customer to carry

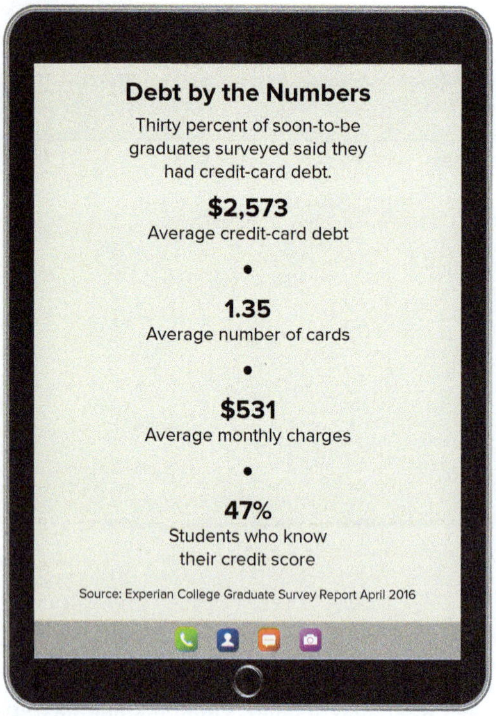

Reprinted by permission of *The Wall Street Journal*, copyright 2016 Dow Jones & Company, Inc. All rights reserved worldwide.

a balance. It is a *prepaid* card. It allows customers to use their smartphone to transfer money and pay for their purchases. The *Wall Street Journal* clip to the left, "Debt by the Numbers," shows 30% of students nearing graduation have $2,573 on average of credit card debt.

To protect consumers, Congress passed the **Fair Credit and Charge Card Disclosure Act of 1988.** This act requires that for direct-mail application or solicitation, credit card companies must provide specific details involving all fees, grace period, calculation of finance charges, and so on. In 2009 the **Credit Card Act** was passed to provide better consumer protection in dealing with credit card companies.

We begin the unit by seeing how Moe's Furniture Store calculates the finance charge on Abby Jordan's previous month's credit card balance. Then we learn how to calculate the average daily balance on the partial bill of Joan Ring.

Calculating Finance Charge on Previous Month's Balance

Abby Jordan bought a dining room set for $8,000 on credit. She has a **revolving charge account** at Moe's Furniture Store. A revolving charge account gives a buyer **open-end credit.** Abby can make as many purchases on credit as she wants until she reaches her maximum $10,000 credit limit.

Often customers do not completely pay their revolving charge accounts at the end of a billing period. When this occurs, stores add interest charges to the customers' bills. Moe's Furniture Store calculates its interest using the *unpaid balance method.* It charges $1\frac{1}{2}$% on the *previous month's balance,* or 18% per year. Moe's has no minimum monthly payment (many stores require $10 or $15, or a percent of the outstanding balance).

LO 1

Abby has no other charges on her revolving charge account. She plans to pay $500 per month until she completely pays off her dining room set. Abby realizes that when she makes a payment, Moe's Furniture Store first applies the money toward the interest and then reduces the **outstanding balance** due. (This is the U.S. Rule we discussed in Chapter 10.) For her own information, Abby worked out the first 3-month schedule of payments, shown in Table 14.3. Note how the interest payment is the rate times the outstanding balance.

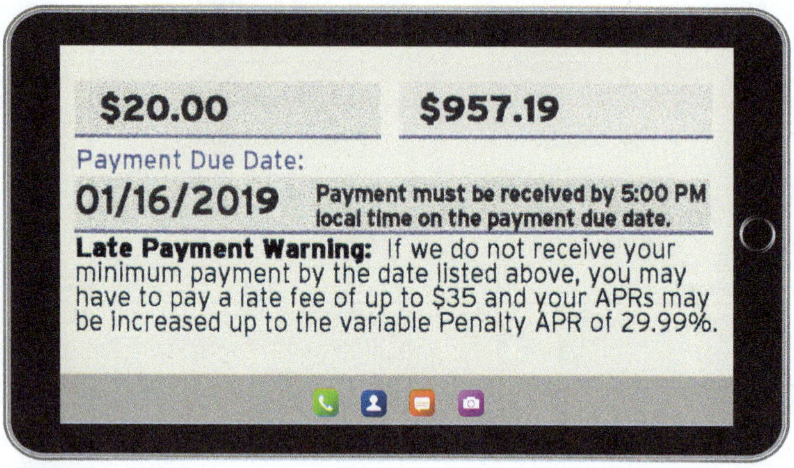

Today, most companies with credit card accounts calculate the finance charge, or interest, as a percentage of the average daily balance. Interest on credit cards can be very expensive for consumers; however, interest is a source of income for credit card companies. In the exhibit shown to the left, note the late payment warning issued by the credit card company. It states that a late payment could result in interest penalties close to 30%. The following is a letter I received from my credit card company when I questioned how my finance charge was calculated.

TABLE 14.3 Schedule of payments

Monthly payment number	Outstanding balance due	$1\frac{1}{2}$% interest payment	Amount of monthly payment	Reduction in balance due	Outstanding balance due
1	$8,000.00	$120.00	$500.00	$380.00	$7,620.00
		(.015 × $8,000.00)		($500.00 − $120.00)	($8,000.00 − $380.00)
2	$7,620.00	$114.30	$500.00	$385.70	$7,234.30
		(.015 × $7,620.00)		($500.00 − $114.30)	($7,620.00 − $385.70)
3	$7,234.30	$108.51	$500.00	$391.49	$6,842.81
		(.015 × $7,234.30)		($500.00 − $108.51)	($7,234.30 − $391.49)

How Citibank Calculates My Finance Charge

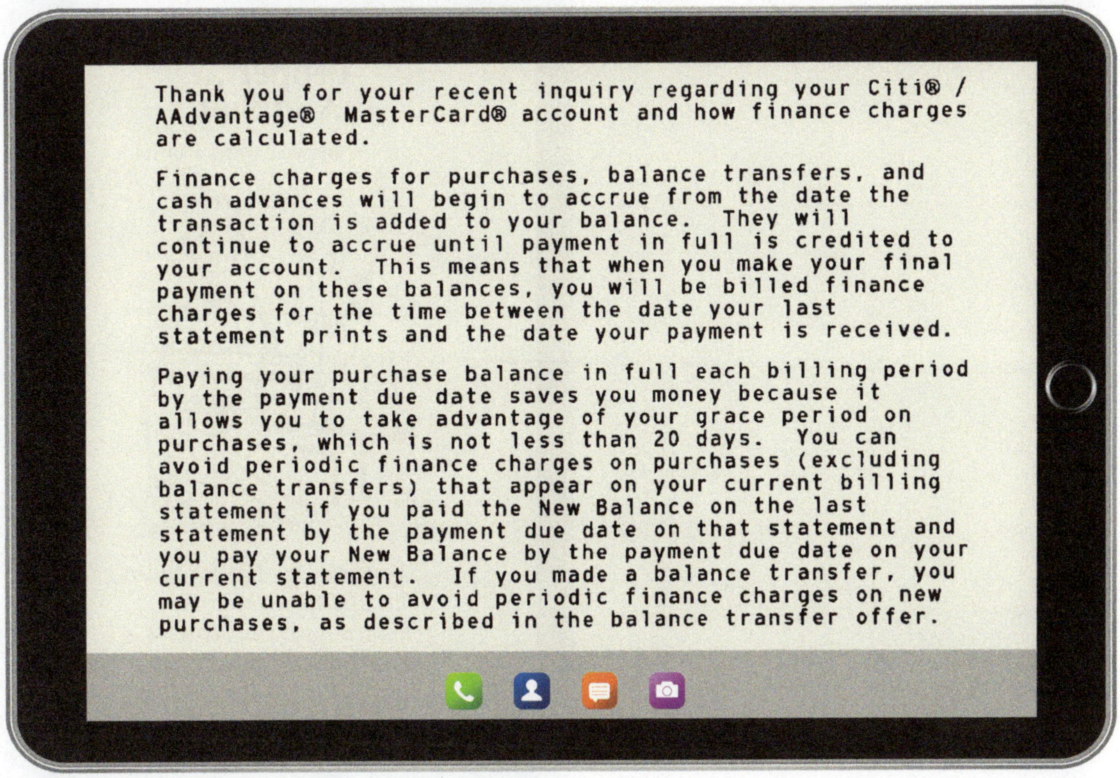

Thank you for your recent inquiry regarding your Citi® / AAdvantage® MasterCard® account and how finance charges are calculated.

Finance charges for purchases, balance transfers, and cash advances will begin to accrue from the date the transaction is added to your balance. They will continue to accrue until payment in full is credited to your account. This means that when you make your final payment on these balances, you will be billed finance charges for the time between the date your last statement prints and the date your payment is received.

Paying your purchase balance in full each billing period by the payment due date saves you money because it allows you to take advantage of your grace period on purchases, which is not less than 20 days. You can avoid periodic finance charges on purchases (excluding balance transfers) that appear on your current billing statement if you paid the New Balance on the last statement by the payment due date on that statement and you pay your New Balance by the payment due date on your current statement. If you made a balance transfer, you may be unable to avoid periodic finance charges on new purchases, as described in the balance transfer offer.

Calculating Average Daily Balance Let's look at the following steps for calculating the **average daily balance.** Remember that a **cash advance** is a cash loan from a credit card company.

CALCULATING AVERAGE DAILY BALANCE AND FINANCE CHARGE

Step 1. Calculate the daily balance or amount owed at the end of each day during the billing cycle:

$$\text{Daily balance} = \text{Previous balance} + \text{Cash advances} + \text{Purchases} - \text{Payments} - \text{Credits}$$

Step 2. When the daily balance is the same for more than 1 day, multiply it by the number of days the daily balance remained the same, or the number of days of the current balance. This gives a cumulative daily balance.

Step 3. Add the cumulative daily balances.

Step 4. Divide the sum of the cumulative daily balances by the number of days in the billing cycle.

Step 5. Finance charge = Rate per month × Average daily balance.

Step 6.* New balance = Previous balance + Cash advances + Purchases − Payments − Credits + Finance charge

*Note: There is a shortcut to this formula. Using data from the Joan Ring example (see below), take the last current daily balance and add the finance charge to it: $620 + $8.19 = $628.19.

Aha!

Always check the number of days in the billing cycle. The cycle may not be 30 or 31 days.

Following is the partial bill of Joan Ring and an explanation of how Joan's average daily balance and finance charge were calculated. Note how we calculated each **daily balance** and then multiplied each daily balance by the number of days the balance remained the same. Take a moment to study how we arrived at 8 days. The total of the cumulative daily balances

Used by permission of Cartoon Features Syndicate

was \$16,390. To get the average daily balance, we divided by the number of days in the billing cycle—30. Joan's finance charge is $1\frac{1}{2}\%$ per month on the average daily balance.

7 days had a balance of \$450
30-day cycle − 22 (7 + 3 + 9 + 3) equals 8 days left with a balance of \$620.

30-day billing cycle				
6/20	Billing date	Previous balance		\$450
6/27	Payment		\$ 50 cr.	
6/30	Charge: JCPenney		200	
7/9	Payment		40 cr.	
7/12	Cash advance		60	

MONEY tips

Understand the costs of credit. Do not spend money you do not currently have—especially if it is for entertainment. Avoid the high cost of reaching or coming close to the maximum on your credit card(s). Be careful to fully understand the terms of any credit card you use. Know your debt-to-income ratio and keep it below 10% excluding mortgage or 36% including mortgage.

	No. of days of current balance	Current daily balance	Extension	
Step 1 →	7	\$450	\$ 3,150	← Step 2
	3	400 (\$450 − \$50)	1,200	
	9	600 (\$400 + \$200)	5,400	
	3	560 (\$600 − \$40)	1,680	
	8	620 (\$560 + \$60)	4,960	
	30		\$16,390	← Step 3

Average daily balance = $\frac{\$16{,}390}{30}$ = \$546.33 ← Step 4

Step 5 → Finance charge = \$546.33 × .015 = \$8.19

Step 6 → \$450 + \$60 + \$200 − \$40 − \$50 + \$8.19 = \$628.19

Now try the following Practice Quiz to check your understanding of this unit.

LU 14–2 PRACTICE QUIZ

Complete this **Practice Quiz** to see how you are doing.

1. Calculate the balance outstanding at the end of month 2 (use U.S. Rule) given the following: purchased $600 desk at the beginning of month 1; pay back $40 per month; and charge of $2\frac{1}{2}$% interest on unpaid balance.
2. Calculate the average daily balance and finance charge from the information that follows.

31-day billing cycle				
8/20	Billing date	Previous balance		$210
8/27	Payment		$50 cr.	
8/31	Charge: Staples		30	
9/5	Payment		10 cr.	
9/10	Cash advance		60	

Rate = 2% per month on average daily balance.

*For **extra help** from your authors–Sharon and Jeff–see the videos in Connect.*

✓ Solutions

1.

Month	Balance due	Interest	Monthly payment	Reduction in balance	Balance outstanding
1	$600	$15.00 (.025 × $600)	$40	$25.00 ($40 − $15)	$575.00
2	$575	$14.38 (.025 × $575)	$40	$25.62	**$549.38**

2. Average daily balance calculated as follows:

No. of days of current balance	Current balance	Extension
7	$210	$1,470
4	160 ($210 − $50)	640
5	190 ($160 + $30)	950
5	180 ($190 − $10)	900
10	240 ($180 + $60)	2,400
31		$6,360

31 − 21(7 + 4 + 5 + 5) → 10

$$\text{Average daily balance} = \frac{\$6{,}360}{31} = \$205.16$$

Finance charge = **$4.10** ($205.16 × .02)

INTERACTIVE CHAPTER ORGANIZER

Topic/Procedure/Formula	Examples	You try it*
Amount financed Amount financed = Cash price − Down payment	60 payments of $125.67 per month; cash price $5,295 with a $95 down payment Cash price $5,295 − Down payment − 95 = Amount financed **$5,200**	**Calculate amount financed** 60 payments of $129.99 per month; Cash price $5,400 with a $100 down payment
Total finance charge (interest) Total finance charge = Total of all monthly payments − Amount financed	*(continued from above)* $125.67 per month × 60 months = $7,540.20 − Amount financed − 5,200.00 = Finance charge **$2,340.20**	**Calculate total finance charge** *(continued from above)*
Deferred payment price Deferred payment price = Total of all monthly payments + Down payment	*(continued from above)* $7,540.20 + $95 = $7,635.20	**Calculate deferred payment price** *(continued from above)*

(continues)

INTERACTIVE CHAPTER ORGANIZER

Topic/Procedure/Formula	Examples	You try it*
Calculating APR by Table 14.1 $\frac{\text{Finance charge}}{\text{Amount financed}} \times \$100 = \text{Table 14.1 lookup number}$	*(continued from above)* $\frac{\$2{,}340.20}{\$5{,}200.00} \times \$100 = \$45.004$ Search in Table 14.1 between 15.50% and 15.75% for 60 payments.	**Calculate APR by table** *(continued from above)*
Monthly payment *By formula:* $\frac{\text{Finance charge} + \text{Amount financed}}{\text{Number of payments of loan}}$ *By table:* $\frac{\text{Loan}}{\$1{,}000} \times \text{Table factor (rate, months)}$	*(continued from above)* $\frac{\$2{,}340.20 + \$5{,}200.00}{60} = \$125.67$ Given: 15.5% 60 months $5,200 loan $\frac{\$5{,}200}{\$1{,}000} = 5.2 \times \$24.05 = \$125.06$ (off due to rounding of rate)	**Calculate monthly payment** *(continued from above; use 16%)*
Open-end credit Monthly payment applied to interest first before reducing balance outstanding	$4,000 purchase $250 a month payment $2\frac{1}{2}\%$ interest on unpaid balance $4,000 × .025 = $100 interest $250 − $100 = $150 to lower balance $4,000 − $150 = $3,850 Balance outstanding after month 1.	**Calculate balance outstanding after month 1** $5,000 purchase; $275 monthly payment; $3\frac{1}{2}\%$ interest on unpaid balance
Average daily balance and finance charge Daily balance = Previous balance + Cash advances + Purchases − Payments − Credits $\text{Average daily balance} = \frac{\text{Sum of cumulative daily balances}}{\text{Number of days in billing cycle}}$ Finance charge = Monthly rate × Average daily balance	*30-day billing cycle; $1\frac{1}{2}\%$ finance charge per month* *Example:* 8/21 Balance $100 8/29 Payment $10 9/12 Charge 50 30-day billing cycle less the 8 and 14. *Average daily balance equals:* 8 days × $100 = $ 800 14 days × 90 = 1,260 8 days × 140 = 1,120 $3,180 ÷ 30 Average daily balance = $106 Finance charge = $106 × .015 = $1.59	**Calculate daily balance and finance charge** 30-day billing cycle; $2\frac{1}{2}\%$ finance charge per month Given: 9/4 bal $200 9/16 payment $80 9/20 charge $60

KEY TERMS			
	Amortization	Daily balance	Installment loan
	Amount financed	Deferred payment price	Loan amortization table
	Annual percentage rate (APR)	Down payment	Open-end credit
	Average daily balance	Fair Credit and Charge Card Disclosure Act of 1988	Outstanding balance
	Cash advance	Finance charge	Revolving charge account
	Credit Card Act		Truth in Lending Act

*Worked-out solutions are in Appendix B.

Critical Thinking Discussion Questions with Chapter Concept Check

1. Explain how to calculate the amount financed, finance charge, and APR by table lookup. Do you think the Truth in Lending Act should regulate interest charges?
2. Explain how to use the loan amortization table. Check with a person who owns a home and find out what part of each payment goes to pay interest versus the amount that reduces the loan principal.
3. What steps are used to calculate the average daily balance? Many credit card companies charge 18% annual interest. Do you think this is a justifiable rate? Defend your answer.
4. **Chapter Concept Check.** Visit the web and find information on how social networks like Facebook have had some influence on credit card companies' policies. Defend your position with the concepts learned in this chapter.

END-OF-CHAPTER PROBLEMS

Mc Graw Hill Education connect

14

Check figures for odd-numbered problems in Appendix B. Name ______________________ Date ________

DRILL PROBLEMS

Complete the following table: *LU 14-1(1)*

	Purchase price of product	Down payment	Amount financed	Number of monthly payments	Amount of monthly payments	Total of monthly payments	Total finance charge
14–1.	Landcruiser $85,000	$60,000		72	$420		
14–2.	Schwinn Mountain Bike $250	$100		12	$15.50		

Calculate **(a)** the amount financed, **(b)** the total finance charge, and **(c)** APR by table lookup. *LU 14-1(1, 2)*

	Purchase price of a used car	Down payment	Number of monthly payments	Amount financed	Total of monthly payments	Total finance charge	APR
14–3.	$5,673	$1,223	48		$5,729.76		
14–4.	$4,195	$95	60		$5,944.00		

Calculate the monthly payment for Problems 14–3 and 14–4 by table lookup and formula. (Answers will not be exact due to rounding of percents in table lookup.) *LU 14-1(3)*

14–5. (14–3) (Use 13% for table lookup.)

14–6. (14–4) (Use 15.5% for table lookup.)

14–7. Calculate the average daily balance and finance charge on the statement below. *LU 14-2(1)*

30-day billing cycle				
9/16	Billing date	Previous balance		$2,000
9/19	Payment		$ 60 cr.	
9/30	Charge: Home Depot		1,500	
10/3	Payment		60 cr.	
10/7	Cash advance		70	
Finance charge is $1\frac{1}{2}$% on average daily balance				

WORD PROBLEMS

My Money

14–8. Before purchasing a used car, Cody Lind checked www.kbb.com to learn what he should offer for the used car he wanted to buy. Then he conducted a carfax.com search on the car he found to see if the car had ever been in an accident. The Carfax was clean so he purchased the used car for $14,750. He put $2,000 down and financed the rest with a 48-month, 7.5% loan. What is his monthly car payment by table lookup? *LU 14-1(1, 2)*

14–9. Troy Juth wants to purchase new dive equipment for Underwater Connection, his retail store in Colorado Springs. He was offered a $56,000 loan at 5% for 48 months. What is his monthly payment by formula? *LU 14-1(3)*

©Performance Image/Alamy Stock Photo

14–10. Ramon Hernandez saw the following advertisement for a used Volkswagen Bug and decided to work out the numbers to be sure the ad had no errors. Please help Ramon by calculating **(a)** the amount financed, **(b)** the finance charge, **(c)** APR by table lookup, **(d)** the monthly payment by formula, and **(e)** the monthly payment by table lookup (will be off slightly). *LU 14-1(1, 2, 3)*

a. Amount financed:

b. Finance charge:

c. APR by table lookup:

d. Monthly payment by formula:

e. Monthly payment by table lookup (use 14.50%):

14–11. From this partial advertisement calculate: *LU 14-1(1, 2, 3)*

a. Amount financed.

b. Finance charge

c. Deferred payment price.

d. APR by Table 14.1.

e. Monthly payment (by formula).

eXcel

My Money

14–12. If you are trying to build credit by using a credit card, each time you make a purchase with the credit card, deduct that amount from your checking account. That way, when your credit card bill is due, you will have enough to pay the credit card off in full. Kathy Lehner is going to start doing this. She plans on paying her credit card bill in full this month. How much does she owe with a 12% APR and the following transactions? *LU 14-2(1)*

31-day billing cycle			
10/1	Previous balance		$1,168
10/3	Credit	$ 75 cr.	
10/12	Charge: King Soopers	152	
10/15	Payment	350 cr.	
10/25	Charge: Delta	325	
10/30	Charge: Holiday Fun	65	

14–13. Dallas Pierce's most recent credit card statement follows. His finance charge is 18% APR. Calculate Dallas's average daily balance, finance charge, and new balance. (Round final answers to the nearest cent.) *LU 14-2(1)*

30-day billing cycle		
9/2	Billing date	$1,200 previous balance
9/7	Payment	$ 100 cr.
9/13	Charge: Kohl's	350
9/17	Payment	200 cr.
9/28	Charge: Walmart	50

eXcel

14–14. First America Bank's monthly payment charge on a 48-month, $20,000 loan is $488.26. U.S. Bank's monthly payment fee is $497.70 for the same loan amount. What would be the APR for an auto loan for each of these banks? (Use the *Business Math Handbook.*) *LU 14-1(1, 2)* eXcel **14–15.** From the following facts, Molly Roe has requested you to calculate the average daily balance. The customer believes the average daily balance should be $877.67. Respond to the customer's concern. *LU 14-2(1)*

28-day billing cycle				
3/18	Billing date	Previous balance		$800
3/24	Payment		$ 60 cr.	
3/29	Charge: Sears		250	
4/5	Payment		20 cr.	
4/9	Charge: Macy's		200	

14–16. Jill bought a $500 rocking chair. The terms of her revolving charge are $1\frac{1}{2}\%$ on the unpaid balance from the previous month. If she pays $100 per month, complete a schedule for the first 3 months like Table 14.3. Be sure to use the U.S. Rule. *LU 14-2(1)*

Monthly payment number	Outstanding balance due	$1\frac{1}{2}\%$ interest payment	Amount of monthly payment	Reduction in balance due	Outstanding balance due

14–17. Dr. Dennis Natali plans to take advantage of a 0% interest balance transfer credit card offer to pay off a $7,250 loan he has. If his loan is at 7.5% interest for 12 months, what is his payment? How much will he save in interest?

CHALLENGE PROBLEMS

14–18. Peg Gasperoni bought a $50,000 life insurance policy for $100 per year. Ryan Life Insurance Company sent her the following billing instructions along with a premium plan example:

"Your insurance premium notice will be mailed to you in a few days. You may pay the entire premium in full without a finance charge or you may pay the premium in installments after a down payment and the balance in monthly installments of $30. The finance charge will be added to the unpaid balance. The finance charge is based on an annual percentage rate of 15%."

If the total policy premium is:	And you put down:	The balance subject to finance charge will be:	The total number of monthly installments ($30 minimum) will be:	The monthly installment before adding the finance charge will be:	The total finance charge for all installments will be:	And the total deferred payment price will be:
$100	$30.00	$ 70.00	3	$30.00	$ 1.75	$101.75
200	50.00	150.00	5	30.00	5.67	205.67
300	75.00	225.00	8	30.00	12.84	312.84

Peg feels that the finance charge of $1.75 is in error. Who is correct? Check your answer. *LU 14-2(1)*

14–19. You have a $1,100 balance on your 15% credit card. You have lost your job and been unemployed for 6 months. You have been unable to make any payments on your balance. However, you received a tax refund and want to pay off the credit card. How much will you owe on the credit card, and how much interest will have accrued? What will be the effective rate of interest after the 6 months (to the nearest hundredth percent)? *LU 14-2(1)*

SUMMARY PRACTICE TEST

Do you need help? Connect videos have step-by-step worked-out solutions.

1. Walter Lantz buys a Volvo SUV for $42,500. Walter made a down payment of $16,000 and paid $510 monthly for 60 months. What are the total amount financed and the total finance charge that Walter paid at the end of the 60 months? *LU 14-1(1)*

2. Joyce Mesnic bought an HP laptop computer at Staples for $699. Joyce made a $100 down payment and financed the balance at 10% for 12 months. What is her monthly payment? (Use the loan amortization table.) *LU 14-1(3)*

3. Lee Remick read the following partial advertisement: price, $22,500; down payment, $1,000 cash or trade; and $399.99 per month for 60 months. Calculate **(a)** the total finance charge and **(b)** the APR by Table 14.1 (or use the tables in *Business Math Handbook*) to the nearest hundredth percent. *LU 14-1(1, 2)*

4. Nancy Billows bought a $7,000 desk at Furniture.com. Based on her income, Nancy could only afford to pay back $700 per month. The charge on the unpaid balance is 3%. The U.S. Rule is used in the calculation. Calculate the balance outstanding at the end of month 2. *LU 14-2(1)*

Month	Balance due	Interest	Monthly payment	Reduction in balance	Balance outstanding

5. Calculate the average daily balance and finance charge on the statement below. *LU 14-2(1)*

	30-day billing cycle	
7/3	Balance	$400
7/18	Payment	100 cr.
7/27	Charge Walmart	250

Assume 2% finance charge on average daily balance.

INTERACTIVE VIDEO WORKSHEET

GRADING THE SUMMARY PRACTICE TEST

Go to the summary practice test video in Connect (or click on it here in the ebook). Grade your summary practice test while viewing the video.

C for Correct/I for Incorrect

1. ________
2. ________
3. ________
4. ________
5. ________

If you achieved 100%, you are ready for your instructor's exam.

If any of the problems were incorrect, list the questions you missed and show steps to solve the problem correctly.

Replay the video to see if you have made the correct fixes to your mistakes.
If you have any questions, contact your instructor asap.

Notes on Watching Videos

MY MONEY

Budgeting for My Future

What I need to know

A well-designed budget will provide you with a framework in which you can achieve your financial goals. A budget is a necessary tool to effectively manage your finances. Many people look at a budget as something that restricts your ability to use your money in a manner that you see fit. However, nothing could be further from the truth. By drafting a budget you will have an accurate picture of your income and expenses so that your financial goals have a roadmap for completion.

Good budgeting starts early and is assessed on a regular basis as things change over time. Setting a budget for yourself during your college years will help to establish the financial habits that will serve you well throughout your life. Your budget needs to be realistic, thorough, and flexible. When you begin the budgeting process look at the task as a learning opportunity to understand the big picture of your financial situation.

What I need to do

To start the budgeting process, you need to understand where you money is coming from and where it is going. This will involve taking the time to document all of your earnings as well as every expense you incur. Using your estimated salary from the Chapter 2 My Money segment, keep a log of your financial transactions for one month to determine exactly the inflows and outflows that are occurring. This process in itself can be very eye opening as you learn how you are using the money you earn. Repeat this same process for another couple of months to establish some consistency among your financial transactions and to account for earnings and expenses that may occur less frequently.

Armed with the knowledge of your financial transactions, you can now begin the process of establishing some guidelines for yourself. Examine your expenses over a few months and you will notice some patterns in how you spend your money. Maybe you notice that your expense for going out to eat is much higher than you would have guessed. Setting a limit to this particular spending is one step in establishing a budget with which you can be comfortable. You will also be able to identify expenses that remain fairly consistent over time such as rent/mortgage, utilities, and insurance. For each category of expense, set an acceptable budget amount to cover the expense within the limit of your earnings. You can use the same process to add new budget categories, such as a car payment, so that you can understand what additional expenses your budget can support.

Resources I can use

- EveryDollar Easy Budgeting App (mobile app)
- https://www.daveramsey.com/blog/the-truth-about-budgeting — budgeting tips
- What Should Your Net Worth Be By Age? http://www.moneyaftergraduation.com/2016/05/24/what-should-net-worth-be-by-age/ — your net worth by age

MY MONEY ACTIVITY

Using your estimated salary from the Chapter 2 My Money segment, create a monthly budget.

- Identify some new budget categories you are likely to encounter after college (car payment, mortgage, vacation fund, student loan payment, etc.).
- Identify some practical actions to help you stick to your budget.

A KIPLINGER APPROACH

THE BASICS

How Your Credit Score Is Calculated

Here's the formula for success no matter which model lenders use.

YOUR CREDIT SCORE—THE three-digit number that creditors use to evaluate the risk when they lend you money—helps determine which loans or interest rates you qualify for and how much you'll pay. Landlords, utilities and cell-phone companies may also check your score before doing business with you.

Dozens of credit scores may be attached to your name, including versions tailored to specific industries, such as auto lending. However, the two big consumer credit scoring models—FICO (which is used by the majority of lenders) and VantageScore (a newer model created by the three major credit bureaus)—value similar behaviors when calculating your score, even if they weight those factors differently. Both grade your creditworthiness on a scale of 300 to 850, with a score of 750 or above generally considered good enough to qualify for the best rates. You can check your VantageScore free at CreditKarma.com and your FICO score at Discover's Credit Scorecard (see kiplinger.com/links/creditcheck).

On-time payments. Both FICO and VantageScore prize on-time payments above any other factor. As long as you pay at least the minimum due each month, your payment history will stay clean (though you will rack up interest on your balance). Lenders typically don't report a late payment to the credit bureaus until it's more than 30 days overdue. If you make a late payment, it won't haunt you forever: The impact on your credit score will diminish as long as you consistently pay your bills on time.

Limits on your credit usage. Your credit utilization ratio is the amount you owe on your credit cards as a proportion of the total limit on each card, as well as the total limit for all of your cards in aggregate. VantageScore advises consumers to keep their utilization ratios below 30%, but "the lower the better," says Barry Paperno, who answers credit questions at his website, SpeakingOfCredit.com. He suggests aiming for a utilization of 1% to 9%, rather than zero, because you can pick up a few more points by showing you are managing your credit well.

You can improve your utilization ratio by spending less on your credit card and by asking your issuer to raise your limit. Applying for a new card would also increase your available credit (but having too many accounts showing balances can lower your score).

Most credit card issuers report the balance from your monthly statement to the credit bureaus. To make that balance appear lower, dole out a few mid-cycle payments or pay off your bill shortly before the closing date for your monthly statement.

A long track record. This slice of your score considers the age of your oldest account and the average age of all your accounts. Opening new cards may improve your credit utilization ratio, but it also lowers the average age of revolving accounts, which lowers your score.

Note that a closed account in good standing remains in your credit history for 10 years, so you'll benefit from your track record; however, keeping no-fee credit cards open (and using them now and then) is smart to help your utilization ratio stay low.

Other factors. A mix of revolving and installment loans also boosts your score. But don't overdo it when applying for new credit. Having "hard inquiries" on your credit report from potential lenders will temporarily shave points from your score. When you're shopping for a mortgage, student loan or auto loan, inquiries made within a certain time period, typically between two weeks and 45 days, count as one inquiry. **MIRIAM CROSS**
mcross@kiplinger.com

Adding Up the FICO Score

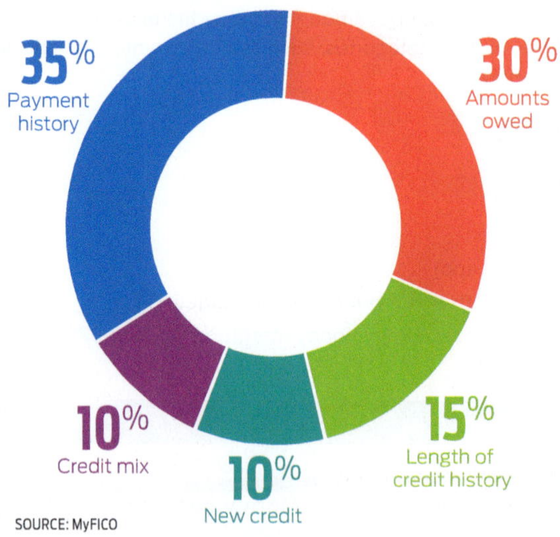

SOURCE: MyFICO

BUSINESS MATH ISSUE

A FICO score is the only factor used in the lending process.

1. List the key points of the article and information to support your position.
2. Write a group defense of your position using math calculations to support your view. If you are in an online course, post to a discussion board.

Classroom Notes

CHAPTER 15

The Cost of Home Ownership

GLOBAL

House Rules

A sampling of nations and the typical commissions paid when a home is sold.

YEAR	2002	2015
United States	6%–7%	5%–6%
Argentina	6%	5%
Australia	2.5%	2%
Canada	4.5%	3%
China	7.5%	2%
Finland	4%	2%
France	N/A	5%
Germany	4.5%	4%
Greece	4%	4%
Hong Kong	1%	2%
Israel	4%	4%
Italy	5%	5%
Japan	3%	6%
Mexico	7.5%	7.5%
Russia	7.5%	4%
Sweden	5%	1.5%
United Kingdom	1.5%	1.5%

Source: Surefield

Note: Sellers, buyers and agents may, at their discretion, pay or forfeit a portion of the commission to facilitate a sale.

©Ariel Skelley/Blend Images LLC

LEARNING UNIT OBJECTIVES

LU 15–1: Types of Mortgages and the Monthly Mortgage Payment

1. List the types of mortgages available.
2. Utilize an amortization chart to compute monthly mortgage payments.
3. Calculate the total cost of interest over the life of a mortgage.

LU 15–2: Amortization Schedule—Breaking Down the Monthly Payment

1. Calculate and identify the interest and principal portion of each monthly payment.
2. Prepare an amortization schedule.

Your Guide to Successfully Completing This Chapter

Traditional book or ebook

Check box as you complete each step.

Steps

☐ Read learning unit.

  ☐ Complete practice quiz at the end of the learning unit.

☐ Grade practice quiz using provided solutions. (For more help, watch the learning unit video in Connect and have a Study Session with the authors. Then complete the additional practice quiz in Connect.)

☐ Repeat above for each of the two learning units in Chapter 15.

  ☐ Review chapter organizer.

  ☐ Complete assigned homework.

    ☐ Finish summary practice test. (Go to Connect via the ebook link and do the interactive video worksheet to grade.)

☐ Complete instructor's exam.

Did you know the new tax act has removed the deduction for moving expenses for most individuals? For more details check with the IRS.

The *Wall Street Journal* chapter opener clip, "House Rules," shows real estate sales commissions in various countries. If property values decrease, homeowners may have to sell their properties for less than is owed. When this happens, it is called a **short sale.**

Learning Unit 15–1: Types of Mortgages and the Monthly Mortgage Payment

LO 1

Figure 15.1 lists various loan types. A type of adjustable rate mortgage called a **subprime loan** was at the root of so many foreclosures during the subprime mortgage crisis from 2007–2010. This type of home loan allowed buyers to have a very low interest rate—sometimes even a zero rate. This helped customers qualify for expensive homes that they would not otherwise have qualified for. Lenders offering subprimes assumed prices of homes would rise and most buyers would convert to a fixed rate before the rate was substantially adjusted upward. As we now know, prices of homes fell. Since that time period, the Federal Reserve has been increasing interest rates.

Purchasing a home usually involves paying a large amount of interest. Note how your author was able to save $70,121.40.

Over the life of a 30-year **fixed-rate mortgage** (see Figure 15.1) of $100,000, the interest would have cost $207,235. Monthly payments would have been $849.99. This would not include taxes, insurance, and so on.

Your author chose a **biweekly mortgage** (see Figure 15.1). This meant that every 2 weeks (26 times a year) the bank would receive $425. By paying every 2 weeks instead of once a month, the mortgage would be paid off in 23 years instead of 30—a $70,121.40 *savings* on

FIGURE 15.1 Types of mortgages available

Loan types	Advantages	Disadvantages
30-year fixed-rate mortgage	A predictable monthly payment.	If interest rates fall, you are locked in to higher rate unless you refinance. (Application and appraisal fees along with other closing costs will result.)
15-year fixed-rate mortgage	Interest rate lower than 30-year fixed (usually $\frac{1}{4}$ to $\frac{1}{2}$ of a percent). Your equity builds up faster while interest costs are cut by more than one-half.	A larger down payment is needed. Monthly payment will be higher.
Graduated-payment mortgage (GPM)	Easier to qualify for than 30- or 15-year fixed rate. Monthly payments start low and increase over time.	May have higher APR than fixed or variable rates.
Biweekly mortgage	Shortens term loan; saves substantial amount of interest; 26 biweekly payments per year. Builds equity twice as fast.	Not good for those not seeking an early loan payoff. Extra payments per year.
Adjustable rate mortgage (ARM)	Lower rate than fixed. If rates fall, could be adjusted down without refinancing. Caps available that limit how high rate could go for each adjustment period over term of loan.	Monthly payment could rise if interest rates rise. Riskier than fixed-rate mortgage in which monthly payment is stable.
Home equity loan	Cheap and reliable accessible lines of credit backed by equity in your home. Tax-deductible. Rates can be locked in. Reverse mortgages may be available to those 62 or older.	Could lose home if not paid **(foreclosure)**. No annual or interest caps.
Interest-only mortgages	Borrowers pay interest but no principal in the early years (5 to 15) of the loan.	Early years build up no equity.

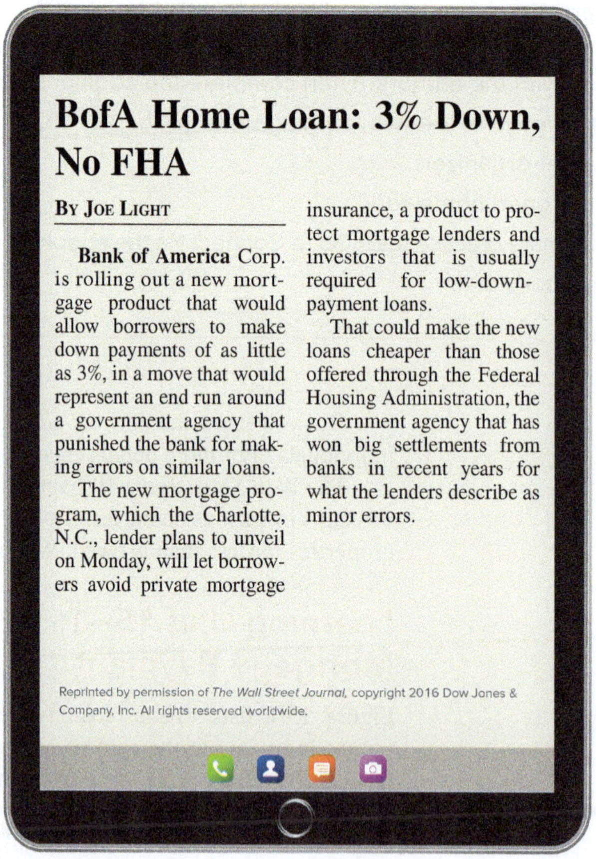

BofA Home Loan: 3% Down, No FHA

By Joe Light

Bank of America Corp. is rolling out a new mortgage product that would allow borrowers to make down payments of as little as 3%, in a move that would represent an end run around a government agency that punished the bank for making errors on similar loans.

The new mortgage program, which the Charlotte, N.C., lender plans to unveil on Monday, will let borrowers avoid private mortgage insurance, a product to protect mortgage lenders and investors that is usually required for low-down-payment loans.

That could make the new loans cheaper than those offered through the Federal Housing Administration, the government agency that has won big settlements from banks in recent years for what the lenders describe as minor errors.

MONEY tips

Should you buy or rent your home? Buying actually saves you $1,743 more, on average, per year if you stay in your home for 6 years or longer. This savings results from the allowed deduction of property taxes and mortgage interest on your personal income taxes.

interest. Why? When a payment is made every 2 weeks, the principal is reduced more quickly, which substantially reduces the interest cost.

The question facing prospective buyers concerns which type of mortgage will be best for them. Depending on how interest rates are moving when you purchase a home, you may find one type of **mortgage** to be the most advantageous for you (see Figure 15.1).

The *Wall Street Journal* article "BofA Home Loan: 3% Down, No FHA" (above) introduces a new mortgage program. Note the following *WSJ* article talks about benefits of the reverse mortgage.

Have you heard that elderly people who are house-rich and cash-poor can use their home to get cash or monthly income? The Federal Housing Administration makes it possible for older homeowners to take out a **reverse mortgage** on their homes. Under reverse mortgages,

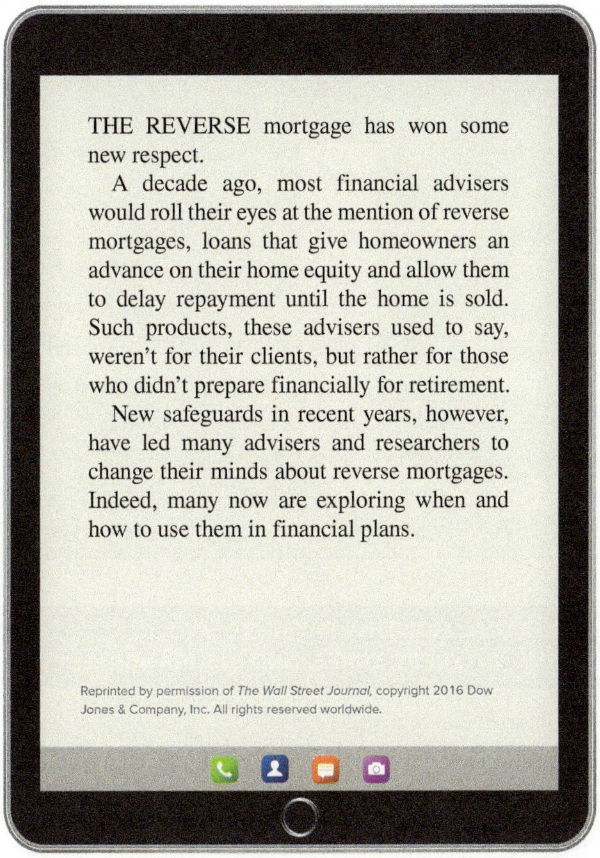

THE REVERSE mortgage has won some new respect.

A decade ago, most financial advisers would roll their eyes at the mention of reverse mortgages, loans that give homeowners an advance on their home equity and allow them to delay repayment until the home is sold. Such products, these advisers used to say, weren't for their clients, but rather for those who didn't prepare financially for retirement.

New safeguards in recent years, however, have led many advisers and researchers to change their minds about reverse mortgages. Indeed, many now are exploring when and how to use them in financial plans.

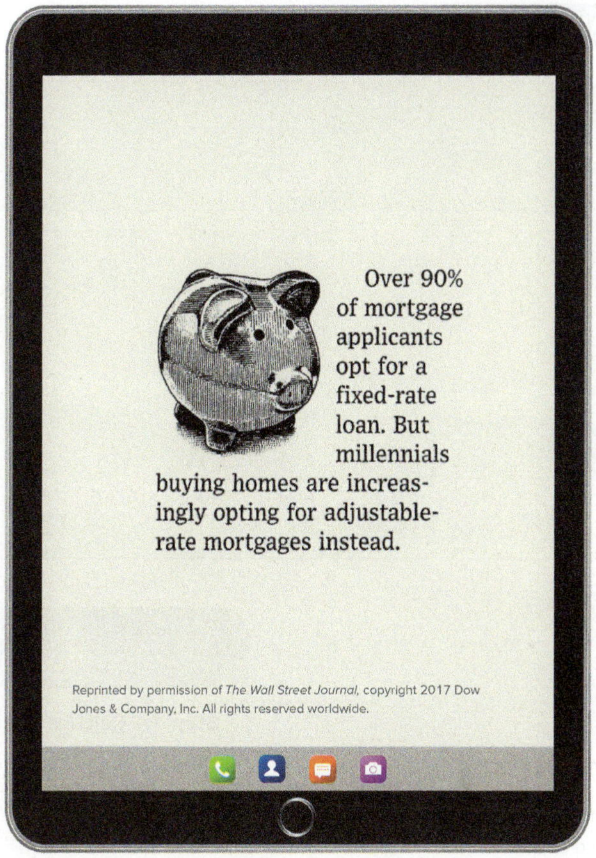

Over 90% of mortgage applicants opt for a fixed-rate loan. But millennials buying homes are increasingly opting for adjustable-rate mortgages instead.

senior homeowners borrow against the equity in their property, often getting fixed monthly checks. The debt is repaid only when the homeowners or their estate sells the home.

Note in the above clip (right) how more than 90% of mortgage applicants opt for a fixed-rate loan.

Now let's learn how to calculate a monthly mortgage payment and the total cost of loan interest over the life of a mortgage. We will use the following example in our discussion:

EXAMPLE Gary bought a home for $200,000. He made a 20% down payment. The 9% mortgage is for 30 years ($30 \times 12 = 360$ payments). What are Gary's monthly payment and total cost of interest?

LO 2

Computing the Monthly Payment for Principal and Interest

You can calculate the principal and interest of Gary's **monthly payment** using the **amortization table** shown in Table 15.1 and the following steps. (Remember that this is the same type of amortization table used in Chapter 14 for installment loans.)

MONEY tips

Research the cost of different mortgages with a variety of companies before choosing one. Compare interest rates, closing costs, mortgage length (10 year, 15 year, 20 year, 30 year) and total interest for the life of each mortgage you are considering. Ensure that there is no prepayment penalty. Look into bimonthly and biweekly mortgage payments resulting in less interest paid over the life of the mortgage, too.

COMPUTING MONTHLY PAYMENT BY USING AN AMORTIZATION TABLE

Step 1. Divide the amount of the mortgage by $1,000.

Step 2. Look up the rate and term in the amortization table. At the intersection is the table factor.

Step 3. Multiply Step 1 by Step 2.

For Gary, we calculate the following:

$$\frac{\$160{,}000 \text{ (amount of mortgage)}}{\$1{,}000} = 160 \times \$8.05 \text{ (table rate)} = \$1{,}288$$

So $160,000 is the amount of the mortgage ($200,000 less 20%). The $8.05 is the table factor of 9% for 30 years per $1,000. Since Gary is mortgaging 160 units of $1,000, the factor of $8.05 is multiplied by 160. Remember that the $1,288 payment does not include taxes, insurance, and so on.

TABLE 15.1 Amortization table (mortgage principal and interest per $1,000)

	INTEREST													
Term in years	$3\frac{1}{2}\%$	5%	$5\frac{1}{2}\%$	6%	$6\frac{1}{2}\%$	7%	$7\frac{1}{2}\%$	8%	$8\frac{1}{2}\%$	9%	$9\frac{1}{2}\%$	10%	$10\frac{1}{2}\%$	11%
10	9.89	10.61	10.86	11.11	11.36	11.62	11.88	12.14	12.40	12.67	12.94	13.22	13.50	13.78
12	8.52	9.25	9.51	9.76	10.02	10.29	10.56	10.83	11.11	11.39	11.67	11.96	12.25	12.54
15	7.15	7.91	8.18	8.44	8.72	8.99	9.28	9.56	9.85	10.15	10.45	10.75	11.06	11.37
17	6.52	7.29	7.56	7.84	8.12	8.40	8.69	8.99	9.29	9.59	9.90	10.22	10.54	10.86
20	5.80	6.60	6.88	7.17	7.46	7.76	8.06	8.37	8.68	9.00	9.33	9.66	9.99	10.33
22	5.44	6.20	6.51	6.82	7.13	7.44	7.75	8.07	8.39	8.72	9.05	9.39	9.73	10.08
25	5.01	5.85	6.15	6.45	6.76	7.07	7.39	7.72	8.06	8.40	8.74	9.09	9.45	9.81
30	4.50	5.37	5.68	6.00	6.33	6.66	7.00	7.34	7.69	8.05	8.41	8.78	9.15	9.53
35	3.99	5.05	5.38	5.71	6.05	6.39	6.75	7.11	7.47	7.84	8.22	8.60	8.99	9.37

MONEY tips

Being able to afford PI (principal and interest) may be compromised when TI (taxes and insurance) is calculated. Make certain to budget for all four mortgage expenses.

CALCULATING TOTAL MONTHLY PAYMENT—PITI: MONTHLY PRINCIPAL, INTEREST, TAXES, AND INSURANCE

Step 1. Calculate principal and interest (see above).

Step 2. Determine 1/12 of the annual property tax.

Step 3. Determine 1/12 of annual homeowner's insurance.

Step 4. Add Step 1, Step 2, and Step 3 together to get monthly PITI.

EXAMPLE $\$1{,}288 + \frac{\$2{,}345}{12} + \frac{\$1{,}578}{12}$

$$\$1{,}288 + \$195.42 + \$131.50 = \$1{,}614.92 \text{ PITI}$$

*Lenders typically require your **PITI** to not exceed 28% of your gross income. And your debt-to-income ratio should not be greater than 36%. Regardless of the loan amount you are offered, avoid being house poor. Don't buy more house than you can reasonably afford.*

LO 3

What Is the Total Cost of Interest?

We can use the following formula to calculate Gary's total interest cost over the life of the mortgage:

$$\text{Total cost of interest} = \text{Total of all monthly payments} - \text{Amount of mortgage}$$

$$\$303{,}680 = \underset{(\$1{,}288 \times 360)}{\$463{,}680} - \$160{,}000$$

My Money

Effects of Interest Rates on Monthly Payment and Total Interest Cost

Table 15.2 shows the effect that an increase in interest rates would have on Gary's monthly payment and his total cost of interest. Note that if Gary's interest rate rises to 11%, the 2% increase will result in Gary paying an additional $85,248 in total interest.

For most people, purchasing a home is a major lifetime decision. Many factors must be considered before this decision is made. Being informed about related costs and the types of available mortgages can save you thousands of dollars.

In addition to the mortgage payment, buying a home can include the following costs:

- *Closing costs:* When property passes from seller to buyer, **closing costs** may include fees for credit reports, recording costs, lawyer's fees, points, title search, and so on. A **point** is a one-time charge that is a percent of the mortgage. Two points means 2% of the mortgage. The following *Wall Street Journal* clip shows how buyers typically pay between 2% and 5% in closing costs.

TABLE 15.1 (concluded)

Term in years	INTEREST $11\frac{1}{2}\%$	$11\frac{3}{4}\%$	12%	$12\frac{1}{2}\%$	$12\frac{3}{4}\%$	13%	$13\frac{1}{2}\%$	$13\frac{3}{4}\%$	14%	$14\frac{1}{2}\%$	$14\frac{3}{4}\%$	15%	$15\frac{1}{2}\%$
10	14.06	14.21	14.35	14.64	14.79	14.94	15.23	15.38	15.53	15.83	15.99	16.14	16.45
12	12.84	12.99	13.14	13.44	13.60	13.75	14.06	14.22	14.38	14.69	14.85	15.01	15.34
15	11.69	11.85	12.01	12.33	12.49	12.66	12.99	13.15	13.32	13.66	13.83	14.00	14.34
17	11.19	11.35	11.52	11.85	12.02	12.19	12.53	12.71	12.88	13.23	13.41	13.58	13.94
20	10.67	10.84	11.02	11.37	11.54	11.72	12.08	12.26	12.44	12.80	12.99	13.17	13.54
22	10.43	10.61	10.78	11.14	11.33	11.51	11.87	12.06	12.24	12.62	12.81	12.99	13.37
25	10.17	10.35	10.54	10.91	11.10	11.28	11.66	11.85	12.04	12.43	12.62	12.81	13.20
30	9.91	10.10	10.29	10.68	10.87	11.07	11.46	11.66	11.85	12.25	12.45	12.65	13.05
35	9.77	9.96	10.16	10.56	10.76	10.96	11.36	11.56	11.76	12.17	12.37	12.57	12.98

TABLE 15.2 Effect of interest rates on monthly payments

	9%	11%	Difference
Monthly payment	$1,288 (160 × $8.05)	$1,524.80 (160 × $9.53)	$236.80 per month
Total cost of interest	$303,680 ($1,288 × 360) − $160,000	$388,928 ($1,524.80 × 360) − $160,000	$85,248 ($236.80 × 360)

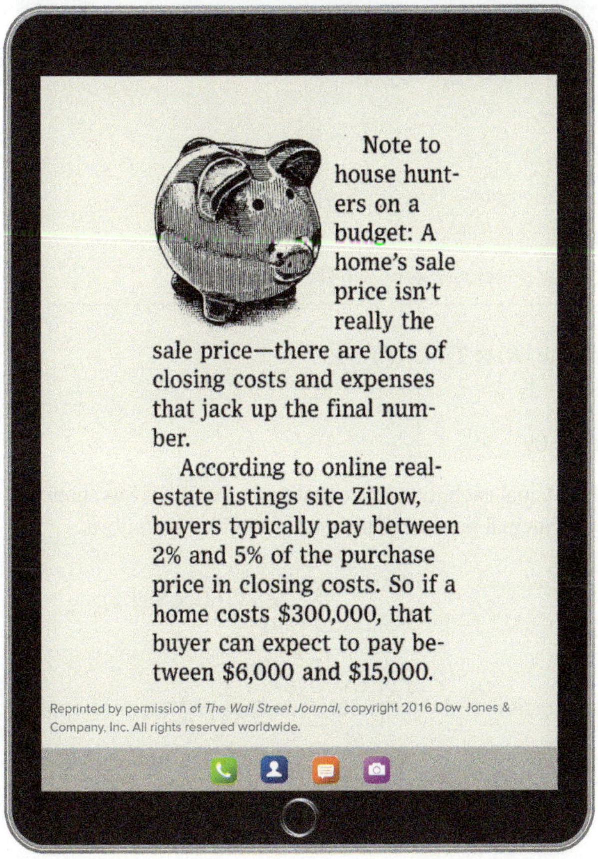

Note to house hunters on a budget: A home's sale price isn't really the sale price—there are lots of closing costs and expenses that jack up the final number.

According to online real-estate listings site Zillow, buyers typically pay between 2% and 5% of the purchase price in closing costs. So if a home costs $300,000, that buyer can expect to pay between $6,000 and $15,000.

- *Escrow amount:* Usually, the lending institution, for its protection, requires that each month 1/12 of the insurance cost and 1/12 of the real estate taxes be kept in a special account called the **escrow account.** The monthly balance in this account will change depending on the cost of the insurance and taxes. Interest is paid on escrow accounts.
- *Repairs and maintenance:* This includes paint, wallpaper, landscaping, plumbing, electrical expenses, and so on.
- *PMI insurance:* When buying a house, if you do not have 20% in cash for a down payment, lenders will require you to purchase PMI (private mortgage insurance). This can be very expensive and only benefits the lender. It is important to know that as soon as 20% equity is reached in the home (determined by an appraisal), the borrower must petition to have the PMI removed. The mortgage lender will not be tracking this and you may continue to pay PMI after it is no longer required.

As you can see, the cost of owning a home can be expensive. But remember that interest costs of your monthly payment and your real estate taxes are income tax deductible. And, your borrowing options increase if you have greater than 80% equity in your home. You can take out a **home equity loan** (a second mortgage for a fixed amount), a **home equity line of credit** (a revolving balance second mortgage), or a **cash-out refinance** (borrowing more than the amount owed) up to 80% loan to value of your home. For many, owning a home offers many advantages over renting.

Before you study Learning Unit 15–2, let's check your understanding of Learning Unit 15–1.

LU 15–1 PRACTICE QUIZ

Complete this **Practice Quiz** to see how you are doing.

Given: Price of home, \$225,000; 20% down payment; 9% interest rate; 25-year mortgage. Solve for:

1. Monthly payment and total cost of interest over 25 years.
2. If rate fell to 8%, what would be the total decrease in interest cost over the life of the mortgage?

For **extra help** from your authors–Sharon and Jeff–see the videos in Connect.

✓ Solutions

1. $\$225{,}000 - \$45{,}000 = \$180{,}000$

$$\frac{\$180{,}000}{\$1{,}000} = 180 \times \$8.40 = \$1{,}512$$

$$\$273{,}600 = \underset{(\$1{,}512 \times 300)}{\$453{,}600} - \$180{,}000$$

300 = 25 years × 12 payments per year

2. $8\% = \underset{(180 \times \$7.72)}{\$1{,}389.60}$ monthly payment

Total interest cost $\$236{,}880 = (\$1{,}389.60 \times 300) - \$180{,}000$

Savings $\$36{,}720 = (\$273{,}600 - \$236{,}880)$

Learning Unit 15–2: Amortization Schedule—Breaking Down the Monthly Payment

LO 1

In Learning Unit 15–1, we saw that over the life of Gary's \$160,000 loan, he would pay \$303,680 in interest. Now let's use the following steps to determine what portion of Gary's first monthly payment reduces the principal and what portion is interest.

CALCULATING INTEREST, PRINCIPAL, AND NEW BALANCE OF MONTHLY PAYMENT

Step 1. Calculate the interest for a month (use current principal):
Interest = Principal × Rate × Time

Step 2. Calculate the amount used to reduce the principal:
Principal reduction = Monthly payment − Interest (Step 1)

Step 3. Calculate the new principal:
Current principal − Reduction of principal (Step 2) = New principal

Step 1. Interest (I) = Principal (P) × Rate (R) × Time (T)

$$\$1{,}200 = \$160{,}000 \times .09 \times \frac{1}{12}$$

Step 2. The reduction of the \$160,000 principal each month is equal to the payment less interest. So we can calculate Gary's new principal balance at the end of month 1 as follows:

Monthly payment at 9% (from Table 15.1)	\$1,288 (160 × \$8.05)
− Interest for first month	− 1,200
= Principal reduction	\$ 88

Step 3. As the years go by, the interest portion of the payment decreases and the principal portion increases.

Principal balance	\$160,000
Principal reduction	− 88
Balance of principal	\$159,912

Let's do month 2:

Step 1. Interest = Principal × Rate × Time

$$= \$159{,}912 \times .09 \times \frac{1}{12}$$

$$= \$1{,}199.34$$

MONEY tips

Save money on interest by making 13 mortgage payments a year using one of the following methods:

1. Increase your monthly payment by 1/12.
2. Make one extra payment a year.
3. Pay half of your monthly payment every two weeks.

A $200,000 mortgage at 5% for 30 years will save you $32,699 on interest cost.

LO 2

Step 2. $1,288.00 monthly payment
− 1,199.34 interest for month 2
$ 88.66 principal reduction

Step 3. $159,912.00 principal balance
− 88.66 principal reduction
$159,823.34 balance of principal

Note that in month 2, interest costs drop 66 cents ($1,200.00 − $1,199.34). So in 2 months, Gary has reduced his mortgage balance by $176.66 ($88.00 + $88.66) After 2 months, Gary has paid a total interest of $2,399.34 ($1,200.00 + $1,199.34).

Example of an Amortization Schedule

The partial **amortization schedule** given in Table 15.3 shows the breakdown of Gary's monthly payment. Note the amount that goes toward reducing the principal and toward payment of actual interest. Also note how the outstanding balance of the loan is reduced. After 7 months, Gary still owes $159,369.97. Often when you take out a mortgage loan, you receive an amortization schedule from the company that holds your mortgage.

It's time to test your knowledge of Learning Unit 15-2 with a Practice Quiz.

TABLE 15.3 Partial amortization schedule

		MONTHLY PAYMENT, $1,288		
Payment number	**Principal (current)**	**Interest**	**Principal reduction**	**Balance of principal**
1	$160,000.00	$1,200.00	$88.00	$159,912.00
	$\left(\$160{,}000 \times .09 \times \frac{1}{12}\right)$	($1,288 − $1,200)	($160,000 − $88)	
2	$159,912.00	$1,199.34	$88.66	$159,823.34
	$\left(\$159{,}912 \times .09 \times \frac{1}{12}\right)$	($1,288 − $1,199.34)	($159,912 − $88.66)	
3	$159,823.34	$1,198.68	$89.32	$159,734.02
4	$159,734.02	$1,198.01	$89.99	$159,644.03
5	$159,644.03	$1,197.33	$90.67	$159,553.36
6	$159,553.36	$1,196.65	$91.35	$159,462.01
7	$159,462.01	$1,195.97*	$92.04	$159,369.97

*Off 1 cent due to rounding.

LU 15–2 PRACTICE QUIZ

Complete this Practice Quiz to see how you are doing.

*For **extra help** from your authors–Sharon and Jeff–see the videos in Connect.*

Prepare an amortization schedule for the first three periods for the following: mortgage, $100,000; 11%; 30 years.

✓ Solutions

$100,000 mortgage; monthly payment, $953 (100 × $9.53)

		PORTION TO—		
Payment number	**Principal (current)**	**Interest**	**Principal reduction**	**Balance of principal**
1	$100,000	$916.67	$36.33	$99,963.67
		$\left(\$100{,}000 \times .11 \times \frac{1}{12}\right)$	($953.00 − $916.67)	($100,000 − $36.33)
2	$99,963.67	$916.33	$36.67	$99,927.00
		$\left(\$99{,}963.67 \times .11 \times \frac{1}{12}\right)$	($953.00 − $916.33)	($99,963.67 − $36.67)
3	$99,927	$916.00	$37.00	$99,890.00
		$\left(\$99{,}927 \times .11 \times \frac{1}{12}\right)$	($953.00 − $916.00)	($99,927.00 − $37.00)

INTERACTIVE CHAPTER ORGANIZER

Topic/Procedure/Formula	Examples	You try it*
Computing monthly mortgage payment Based on per $1,000 (Table 15.1): $\frac{\text{Amount of mortgage}}{\$1{,}000} \times \text{Table rate}$	Use Table 15.1: 12% on $60,000 mortgage for 30 years. $\frac{\$60{,}000}{\$1{,}000} = 60 \times \$10.29$ $= \$617.40$	**Calculate monthly payment** $70,000 mortgage at 3.5% for 30 years
Calculating total interest cost Total of all monthly payments − Amount of mortgage	Using example above: 30 years = 360 (payments) ×$617.40 $222,264 − 60,000 $162,264 (mortgage interest over life of mortgage)	**Calculate total interest cost** Use the data from the problem above.
Amortization schedule $I = P \times R \times T$ $\left(I \text{ for month} = P \times R \times \frac{1}{12}\right)$ Principal reduction = Monthly payment − Interest New principal = Current principal − Reduction of principal	Using same example: Portion to— Payment number | Interest | Principal reduction | Balance of principal 1 | $600 | $17.40 | $ 59,982.60 $\left(\$60{,}000 \times .12 \times \frac{1}{12}\right)$ | $\left(\begin{array}{r}\$617.40\\-\$600.00\end{array}\right)$ | $\left(\begin{array}{r}\$60{,}000.00\\-\$17.40\end{array}\right)$ 2 | $599.83 | $17.57 | $ 59,965.03 $\left(\$59{,}982.60 \times .12 \times \frac{1}{12}\right)$ | $\left(\begin{array}{r}\$617.40\\-\$599.83\end{array}\right)$ | $\left(\begin{array}{r}\$59{,}982.60\\-\$17.57\end{array}\right)$	**Prepare amortization for first two payments** Use the data from the problem above.

KEY TERMS			
	Adjustable rate mortgage (ARM)	Escrow account	Interest-only mortgage
	Amortization schedule	Fixed-rate mortgage	Monthly payment
	Amortization table	Foreclosure	Mortgages
	Biweekly mortgage	Graduated-payment mortgages (GPM)	Points
	Cash-out refinance	Home equity line of credit	Reverse mortgage
	Closing costs	Home equity loan	Short sale
			Subprime loans

*Worked-out solutions are in Appendix B.

Critical Thinking Discussion Questions with Chapter Concept Check

1. Explain the advantages and disadvantages of the following loan types: 30-year fixed rate, 15-year fixed-rate, graduated-payment mortgage, biweekly mortgage, adjustable rate mortgage, and home equity loan. Why might a bank require a home buyer to establish an escrow account?
2. How is an amortization schedule calculated? Is there a best time to refinance a mortgage?
3. What is a point? Is paying points worth the cost?
4. Explain how rising interest rates will affect the housing market.
5. Explain a short sale.
6. Explain subprime loans and how foreclosures result.
7. **Chapter Concept Check.** Locate three mortgage options for a house you would like to buy and calculate the payment and total interest for each. Which would you choose? Why? Use concepts in the chapter to support your case.

END-OF-CHAPTER PROBLEMS

15

Check figures for odd-numbered problems in Appendix B. Name ______________________ Date ________

DRILL PROBLEMS

Complete the following amortization chart by using Table 15.1. *LU 15-1(2)*

	Selling price of home	Down payment	Principal (loan)	Rate of interest	Years	Payment per $1,000	Monthly mortgage payment
15–1.	$160,000	$20,000		$3\frac{1}{2}\%$	30		
15–2.	$90,000	$5,000		$5\frac{1}{2}\%$	30		
15–3.	$190,000	$50,000		7%	35		

15–4. What is the total cost of interest in Problem 15–2? *LU 15-1(3)*

15–5. If the interest rate rises to 7% in Problem 15–2, what is the total cost of interest? *LU 15-1(3)*

Complete the following: *LU 15-2(1)*

	Selling price	Down payment	Amount mortgage	Rate	Years	Monthly payment	First Payment Broken Down Into— Interest	Principal	Balance at end of month
15–6.	$150,000	$30,000		7%	30				
15–7.	$225,000	$45,000		5%	15				

15–8. Bob Jones bought a new log cabin for $70,000 at 11% interest for 30 years. Prepare an amortization schedule for the first three periods. *LU 15-2(2)*

Payment number	Portion to— Interest	Principal	Balance of loan outstanding

WORD PROBLEMS

15–9. CNBC.com reported mortgage applications increased 9.9% due to a decrease in the rate on 30-year fixed-rate mortgages to 4.03%. Joe Sisneros wants to purchase a vacation home for \$235,000 with 20% down. Calculate his monthly payment for a 20-year mortgage at 3.5%. Calculate total interest. *LU 15-1(3)*

My Money

15–10. If you buy a home with less than 20% down, you will pay an additional monthly fee, PMI (private mortgage insurance), until you reach 80% equity. Keep track of when you reach 80% equity so you can request to have your PMI removed. Ken Buckmiller's home recently appraised at \$290,000. His mortgage was for \$275,000 at 5% for 30 years with PMI of \$229.17 per month. What is his monthly payment plus PMI? His mortgage balance is currently \$222,990. Has he reached 80% equity? *LU 15-1(2)*

eXcel

15–11. Joe Levi bought a home in Arlington, Texas, for \$140,000. He put down 20% and obtained a mortgage for 30 years at $5\frac{1}{2}\%$. What is Joe's monthly payment? What is the total interest cost of the loan? *LU 15-1(2, 3)*

eXcel

15–12. If in Problem 15–11 the rate of interest is $7\frac{1}{2}\%$, what is the difference in interest cost? *LU 15-1(3)*

15–13. Mike Jones bought a new split-level home for \$150,000 with 20% down. He decided to use Victory Bank for his mortgage. Victory was offering $13\frac{3}{4}\%$ for 25-year mortgages. Provide Mike with an amortization schedule for the first three periods. *LU 15-2(1, 2)*

	Portion to—		
Payment number	**Interest**	**Principal**	**Balance of loan outstanding**

15–14. Harriet Marcus is concerned about the financing of a home. She saw a small cottage that sells for \$50,000. If she puts 20% down, what will her monthly payment be at **(a)** 25 years, $11\frac{1}{2}\%$; **(b)** 25 years, $12\frac{1}{2}\%$; **(c)** 25 years, $13\frac{1}{2}\%$; and **(d)** 25 years, 15%? What is the total cost of interest over the cost of the loan for each assumption? **(e)** What is the savings in interest cost between $11\frac{1}{2}\%$ and 15%? **(f)** If Harriet uses 30 years instead of 25 for both $11\frac{1}{2}\%$ and 15%, what is the difference in interest? *LU 15-1(2, 3)*

15–15. FedPrimeRate.com reported the median price of a home sold in the United States in November 2017 was $248,000. Juan Carlos Soto, Jr., wants to purchase a new home for $305,500. Juan puts 20% down and will finance the remainder of the purchase. Compare the following two mortgage options he has: 10 years at 3.5% or 15 years at 5%. Calculate Juan's monthly payment as well as his total cost of interest for both the 10- and 15-year mortgage. What is the difference in interest paid between the two options (round to nearest cent in calculations)? *LU 15-1(3)*

15–16. Daniel and Jan agreed to pay $560,000 for a four-bedroom colonial home in Waltham, Massachusetts, with a $60,000 down payment. They have a 30-year mortgage at a fixed rate of 6.00%. **(a)** How much is their monthly payment? **(b)** After the first payment, what would be the balance of the principal? *LU 15-1(2), LU 15-2(1)*

My Money

15–17. Paying 13 mortgage payments instead of 12 per year can save you thousands in mortgage interest expense. If you had a $175,000 mortgage at 6% for 30 years, how much extra would you have to pay per year to make 13 instead of 12 mortgage payments per year? How much would you pay if you paid 1/12 of it per month? *LU 15-1(1)*

My Money

15–18. Mortgage lenders base the mortgage interest rate they offer you on your credit rating. This makes it financially critical to maintain a credit score of 740 or higher. How much more interest would you pay on a $195,000 home if you put 20% down and financed the balance with a 30-year mortgage at 6% compared to a 30-year mortgage at 3 1/2%?

CHALLENGE PROBLEMS

eXcel

15–19. Rick Rueta purchased a $90,000 home at 9% for 30 years with a down payment of $20,000. His annual real estate tax is $1,800 along with an annual insurance premium of $960. Rick's bank requires that his monthly payment include an escrow deposit for the tax and insurance. What is the total payment each month for Rick? *LU 15-1(2)*

15–20. Sharon Fox decided to buy a home in Marblehead, Massachusetts, for \$275,000. Her bank requires a 30% down payment. Sue Willis, an attorney, has notified Sharon that besides the 30% down payment there will be the following additional costs:

Recording of the deed	\$ 30.00
A credit and appraisal report	155.00
Preparation of appropriate documents	48.00

In addition, there will be a transfer tax of 1.8% of the purchase price and a loan origination fee of 2.5% of the mortgage amount.

Assume a 30-year mortgage at a rate of 10%. *LU 15-1(2, 3)*

a. What is the initial amount of cash Sharon will need?

b. What is her monthly payment?

c. What is the total cost of interest over the life of the mortgage?

Classroom Notes

SUMMARY PRACTICE TEST

Do you need help? Connect videos have step-by-step worked-out solutions.

eXcel **1.** Pat Lavoie bought a home for \$180,000 with a down payment of \$10,000. Her rate of interest is 6% for 30 years. Calculate her **(a)** monthly payment; **(b)** first payment, broken down into interest and principal; and **(c)** balance of mortgage at the end of the month. *LU 15-1(2, 3)*

2. Jen Logan bought a home in Iowa for \$110,000. She put down 20% and obtained a mortgage for 30 years at $5\frac{1}{2}$%. What are Jen's monthly payment and total interest cost of the loan? *LU 15-1(2, 3)*

3. Christina Sanders is concerned about the financing of a home. She saw a small Cape Cod–style house that sells for \$90,000. If she puts 10% down, what will her monthly payment be at **(a)** 30 years, 5%; **(b)** 30 years, $5\frac{1}{2}$% **(c)** 30 years, 6%; and **(d)** 30 years, $6\frac{1}{2}$%? What is the total cost of interest over the cost of the loan for each assumption? *LU 15-1(2, 3)*

4. Loretta Scholten bought a home for \$210,000 with a down payment of \$30,000. Her rate of interest is 6% for 35 years. Calculate Loretta's payment per \$1,000 and her monthly mortgage payment. *LU 15-1(2)*

5. Using Problem 4, calculate the total cost of interest for Loretta Scholten. *LU 15-1(3)*

INTERACTIVE VIDEO WORKSHEET

GRADING THE SUMMARY PRACTICE TEST

Go to the summary practice test video in Connect (or click on it here in the ebook). Grade your summary practice test while viewing the video.

C for Correct/I for Incorrect

1. ________ 4. ________
2. ________ 5. ________
3. ________

If you achieved 100%, you are ready for your instructor's exam.

If any of the problems were incorrect, list the questions you missed and show steps to solve the problem correctly.

Replay the video to see if you have made the correct fixes to your mistakes. If you have any questions, contact your instructor asap.

Notes on Watching Videos

MY MONEY

Home Sweet Home

What I need to know

Buying a home is a big deal. Conduct research on the steps you need to take, and what you need to be aware of throughout the home buying process, and you can make an excellent investment. The knowledge you obtain will give you the confidence to effectively participate in the process and understand your responsibilities. A solid budget is an important component of making a home purchase within the limits of your income. Home ownership expenses include the mortgage payment, insurance, taxes, maintenance, and may include homeowner's association dues. Keep in mind if you have a less than 20% down payment on the appraised value of the home, you will be paying a monthly mortgage insurance premium, MIP, as well.

What I need to do

Start by getting preapproved for a mortgage. In addition, refer to your budget to determine how much you can comfortably afford to spend on your monthly house payment. Then decide the ceiling of what you want to spend on buying your home. If you are a first-time homebuyer, check out available first-time home buying programs. These programs can offer assistance with the down payment and provide other benefits as well. Shop around for a mortgage seeking the most beneficial terms: interest rate, length of the mortgage, repayment programs (bimonthly, biweekly, etc.). Avoid being house poor. Consider buying a home that is cheaper than you can afford. If you do that, you will have money left over each month for items such as entertainment.

Once you have been preapproved for your mortgage, the next step is to interview real estate agents to find one who will work with you and for you. Negotiate the commission. Then begin a search for your home. Create a list of must-haves for the home you buy and use that list as you shop. Remember the adage: location, location, location. Think about the future rental or resale value of the home. Consider any negative aspects of the location of the home such as noise, pollution, crime rates, quality of the schools, and so on.

Once you choose a home, make an educated offer. Work with your real estate agent to determine what offer to make and who will pay what at closing. When your contract has been accepted and signed by the seller, get a reputable third-party home inspection. This will provide you with data about the house and its value that you can use in the negotiation process. After you move in, be certain to do regular maintenance to maintain the value of your home.

Resources I can use

- https://www.bankrate.com/mortgages/first-time-homebuyer-grants-and-programs/ — for first-time home buyers
- https://www.usatoday.com/story/money/personalfinance/2017/06/30/25-tips-first-time-home-buyers/434449001/ — first-time home buyer tips

MY MONEY ACTIVITY

How much can you afford to spend on buying a house? Check out Zillow.com or other sites where homes are for sale and locate a house you are interested in purchasing. Apply the concepts from this chapter and determine what the down payment, PITI, and total interest cost would be based on the expected interest rate and terms your credit rating supports. Does your budget allow for this? Do you have enough money left over for home repairs and improvements? Can you afford this house?

PERSONAL FINANCE

A KIPLINGER APPROACH

"Save Money on the Sale of Your Home" by Kimberly Lankford from *Kiplinger's*, October 2016, p. 36. Used by permission of The Kiplinger Washington Editors, Inc.

KIMBERLY LANKFORD | Ask Kim

Save Money on the Sale of Your Home

I'M ABOUT TO PUT MY HOUSE on the market and I keep seeing signs for Redfin. How does it work, and would it attract as many buyers as a listing by a traditional agent?

M.S., WASHINGTON, D.C.

Redfin started in Seattle a decade ago. Since then, the real estate brokerage has expanded into 83 markets, covering most major U.S. cities.

The biggest attraction for sellers is the lower cost: Rather than paying the standard 3% commission to the selling agent, Redfin sellers pay 1.5%—or 1% in some metropolitan areas, such as Washington, D.C. (Sellers usually still pay an additional 3% commission to the buyer's agent.) A reduction of 1.5% from 3% can save the seller $6,000 on a $400,000 home. Buyers who use a Redfin agent receive a refund at closing based on the home's list price. The average refund is about $4,000, but the calculation varies by location and price. Buyers can see the amount they'd get back on a home on Redfin.com (even if the home is not listed by a Redfin agent). Redfin agents receive a salary and health benefits (the commission goes to the company), plus a bonus based on sales and customer reviews.

©Kiplinger Washington Editors, Inc.

Sellers who use Redfin cut the commission to the selling agent by half—or more.

Home listings appear on the multiple listing service (MLS) and sites such as Zillow and Trulia, as they would with a traditional agent. They also appear on Redfin's website, which includes commentary from Redfin agents about homes they have toured, details about the outcome of offers on the homes they represent, and information about how to schedule tours online. Redfin agents can notify potential buyers when your house hits the market, when the price drops or when an offer is about to be accepted.

Social Security benefits for kids. *My husband turns 66 soon and plans to claim Social Security benefits. A friend told me our 12-year-old daughter might be eligible for benefits based on his earnings history. Is this true, and if so, how does it work?*

C.C., LOS ANGELES

Unmarried children younger than 18 (or younger than 19 if full-time students through grade 12) are eligible for Social Security benefits after their parents start to receive retirement benefits. (Special rules apply for disabled children.) Children can get up to half of the parent's full retirement age (FRA) benefit. Benefits for spouses and children may be reduced if total family benefits are more than 150% to 180% of the worker's FRA benefit.

Lost tax form. *I made after-tax contributions to my traditional IRA years ago, but I can't find my Form 8606. How do I avoid getting taxed on this money again?*

D.G., SEATTLE

You should have filed Form 8606 with the IRS every year you made nondeductible IRA contributions. The form reports your basis so you avoid paying taxes twice when the money is withdrawn. Search your records for copies of Form 5498, which your IRA administrator sends to report contributions, or look for contributions on your brokerage statements. Match that information with your old tax returns to see what you deducted.

If you don't have those records, you can request a transcript from the IRS. An enrolled agent or other tax specialist can access IRS e-services for you and get the numbers, says Jeffrey Schneider, an enrolled agent in Port St. Lucie, Fla. For a list of enrolled agents, see www.naea.org.

Changing 529 beneficiaries. *I have several grandchildren and have set up a 529 college-savings account for each. Can I switch money from one grandchild's account to another?*

J.N., TAMPA

Yes. You can change the beneficiary or move money from one eligible family member's account to another without penalties or taxes. The beneficiary's spouse, parent, sibling, first cousin and a few other relatives are eligible. Check with your 529 plan for details. ■

GOT A QUESTION? ASK KIM AT ASKKIM@KIPLINGER.COM. KIMBERLY LANKFORD ANSWERS MORE QUESTIONS EACH WEEK AT KIPLINGER.COM/ASKKIM.

BUSINESS MATH ISSUE

The best way to sell a home is without a real estate broker.

1. List the key points of the article and information to support your position.
2. Write a group defense of your position using math calculations to support your view. If you are in an online course, post to a discussion board.

Classroom Notes

CHAPTER 16

How to Read, Analyze, and Interpret Financial Reports

Boeing's Unique Accounting Helped Lift Profit

By Jon Ostrower

SEATTLE—**Boeing** Co. started to make money on each 787 Dreamliner it delivers just this spring, but thanks to a unique accounting strategy the jet has been fattening the aircraft maker's bottom line for years.

Boeing is one of the few companies that uses a technique called program accounting. Rather than booking the huge costs of building the advanced 787 or other aircraft as it pays the bills, Boeing—with the blessing of its auditors and regulators and in line with accounting rules—defers those costs, spreading them out over the number of planes it expects to sell years into the future. That allows the company to include anticipated future profits in its current earnings.

Boeing calculates its Dreamliner earnings based on 10-year forecasts of supplier contracts, aircraft orders and options, productivity improvements, labor contracts and market conditions, which are reviewed by its auditors. The company, which delivered its first Dreamliner in 2011, estimates it will sell 1,300 Dreamliners over the 10 years ending in 2021. So far, it has delivered nearly 500 of them.

The problem, analysts and other critics say, is that Boeing's approach stretches its profit per plane into such a distant and uncertain future that it isn't clear if it will ever recover the nearly $30 billion it has sunk into producing the plane and validate years of projected profits.

©Mayskyphoto/Shutterstock

LEARNING UNIT OBJECTIVES

LU 16–1: Balance Sheet—Report as of a Particular Date

1. Explain the purpose and the key items on the balance sheet.
2. Explain and complete vertical and horizontal analysis.

LU 16–2: Income Statement—Report for a Specific Period of Time

1. Explain the purpose and the key items on the income statement.
2. Explain and complete vertical and horizontal analysis.

LU 16–3: Trend and Ratio Analysis

1. Explain and complete a trend analysis.
2. List, explain, and calculate key financial ratios.

Your Guide to Successfully Completing This Chapter

Traditional book or ebook

Check box as you complete each step.

Steps

- ☐ Read learning unit.
    - ☐ Complete practice quiz at the end of the learning unit.
- ☐ Grade practice quiz using provided solutions. (For more help, watch the learning unit video in Connect and have a Study Session with the authors. Then complete the additional practice quiz in Connect.)
- ☐ Repeat above for each of the three learning units in Chapter 16.
    - ☐ Review chapter organizer.
    - ☐ Complete assigned homework.
        - ☐ Finish summary practice test. (Go to Connect via the ebook link and do the interactive video worksheet to grade.)
- ☐ Complete instructor's exam.

An investor in a corporation will have to carefully read the company's financial reports. The new tax act will change how profits are reported. Check with the IRS for the latest tax law updates.

The *Wall Street Journal* chapter opener, "Boeing's Unique Accounting Helped Lift Profit," shows how a company like Boeing can use unique accounting principles to its advantage.

This chapter explains how to analyze two key financial reports: the *balance sheet* (shows a company's financial condition at a particular date) and the *income statement* (shows a company's profitability over a time period).[1] Business owners must understand their financial statements to avoid financial difficulties. This includes knowing how to read, analyze, and interpret financial reports.

[1]The third key financial report is the statement of cash flows. We do not discuss this statement. For more information on the statement of cash flows, check your accounting text.

LO 1

Learning Unit 16–1: Balance Sheet—Report as of a Particular Date

The **balance sheet** gives a financial picture of what a company is worth as of a particular date, usually at the end of a month or year. This report lists (1) how much the company owns (assets), (2) how much the company owes (liabilities), and (3) how much the owner is worth **(owner's equity).**

Note that assets and liabilities are divided into two groups: current (*short term,* usually less than 1 year); and *long term,* usually more than 1 year. The basic formula for a balance sheet is as follows:

Assets − Liabilities = Owner's equity

Like all formulas, the items on both sides of the equals sign must balance.

By reversing the above formula, we have the following common balance sheet layout:

Assets = Liabilities + Owner's equity

To introduce you to the balance sheet, let's assume that you collect baseball cards and decide to open a baseball card shop. As the owner of The Card Shop, your investment, or owner's equity, is called **capital.** Since your business is small, your balance sheet is short. After the first year of operation, The Card Shop balance sheet is shown as follows:

"Capital" does not mean "cash." It is the owner's investment in the company.

THE CARD SHOP
Balance Sheet
December 31, 2020 ← Report as of a particular date

Assets		**Liabilities**	
Cash	$ 3,000	Accounts payable	$ 2,500
Merchandise inventory (baseball cards)	4,000	**Owner's Equity**	
Equipment	3,000	E. Slott, capital	7,500
Total assets	$10,000	Total liabilities and owner's equity	$10,000

©Diego J. Robles/Getty Images

The heading gives the name of the company, title of the report, and date of the report. Note how the totals of both sides of the balance sheet are the same. This is true of all balance sheets.

We can take figures from the balance sheet of The Card Shop and use our first formula to determine how much the business is worth:

Assets − Liabilities = Owner's equity (capital)

$10,000 − $2,500 = $7,500

Since you are the single owner of The Card Shop, your business is a **sole proprietorship.** If a business has two or more owners, it is a **partnership.** A **corporation** has many owners or stockholders, and the equity of these owners is called **stockholders' equity.** *Anytime you create a balance sheet, check your accuracy by adding liabilities to owner's equity. If that total equals assets, you've got it right!*

Aha!

Elements of the Balance Sheet

The format and contents of all corporation balance sheets are similar. Figure 16.1 shows the balance sheet of Mool Company. As you can see, the formula Assets = Liabilities + Stockholders' equity (we have a corporation in this example) is also the framework of this balance sheet.

To help you understand the three main balance sheet groups (assets, liabilities, and stockholders' equity) and their elements, we have labeled them in Figure 16.1. An explanation of these groups and their elements follows this paragraph. Do not try to memorize the elements.

FIGURE 16.1 Balance sheet

Assets broken down into current assets and plant and equipment

MOOL COMPANY
Balance Sheet ← Put in heading who, what, when
December 31, 2020

Liabilities broken down into current and long-term

		Assets						**Liabilities**		
	a.	Current assets:				a.	Current liabilities:			
	b.	Cash	$ 7,000			b.	Accounts payable	$80,000		
	c.	Accounts receivable	9,000			c.	Salaries payable	12,000		
	d.	Merchandise inventory	30,000		2.	d.	Total current liabilities		$ 92,000	
	e.	Prepaid expenses	15,000			e.	Long-term liabilities:			
1.	f.	Total current assets		$ 61,000		f.	Mortgage note payable		58,000	
	g.	Plant and equipment:				g.	Total liabilities		$ 150,000	
	h.	Building (net)	$60,000				**Stockholders' Equity**			
	i.	Land	84,000			a.	Common stock	$20,000		
	j.	Total plant and equipment		144,000		b.	Retained earnings	35,000		
					3.	c.	Total stockholders' equity		55,000	
	k.	Total assets		$205,000		d.	Total liabilities and stockholders' equity		$205,000	

Total of current assets and plant and equipment

Total is double-ruled

Total of all liabilities and stockholders' equity

Just try to understand their meaning. Think of Figure 16.1 as a reference aid. You will find that the more you work with balance sheets, the easier it is for you to understand them.

1. **Assets:** Things of value *owned* by a company (economic resources of the company) that can be measured and expressed in monetary terms.
   a. **Current assets:** Assets that companies consume or convert to cash *within 1 year* or a normal operating cycle.
   b. **Cash:** Total cash in checking accounts, savings accounts, and on hand.
   c. **Accounts receivable:** Money *owed* to a company by customers from sales on account (buy now, pay later).
   d. **Merchandise inventory:** Cost of goods in stock for resale to customers.
   e. **Prepaid expenses:** The purchases of a company are assets until they expire (insurance or rent) or are consumed (supplies).
   f. **Total current assets:** Total of all assets that the company will consume or convert to cash within 1 year.
   g. **Plant and equipment:** Assets that will last longer than 1 year. These assets are used in the operation of the company.
   h. **Building (net):** The cost of the building minus the depreciation that has accumulated. Usually, balance sheets show this as "Building less accumulated depreciation." In Chapter 17 we discuss accumulated depreciation in greater detail.
   i. **Land:** This asset does not depreciate, but it can increase or decrease in value.
   j. **Total plant and equipment:** Total of building and land, including machinery and equipment.
   k. **Total assets:** Total of current assets and plant and equipment.
2. **Liabilities:** Debts or obligations of the company.
   a. **Current liabilities:** Debts or obligations of the company that are *due within 1 year.*
   b. **Accounts payable:** A current liability that shows the amount the company owes to creditors for services or items purchased.

c. **Salaries payable:** Obligations that the company must pay within 1 year for salaries earned but unpaid.

d. **Total current liabilities:** Total obligations that the company must pay within 1 year.

e. **Long-term liabilities:** Debts or obligations that the company does not have to pay within 1 year.

f. **Mortgage note payable:** Debt owed on a building that is a long-term liability; often the building is the collateral.

g. **Total liabilities:** Total of current and long-term liabilities.

3. **Stockholders' equity (owner's equity):** The rights or interest of the stockholders to assets of a corporation. If the company is not a corporation, the term *owner's equity* is used. The word *capital* follows the owner's name under the title *Owner's Equity.*

   a. **Common stock:** Amount of the initial and additional investment of corporation owners by the purchase of stock.

   b. **Retained earnings:** The amount of corporation earnings that the company retains, not necessarily in cash form.

   c. **Total stockholders' equity:** Total of stock plus retained earnings.

   d. **Total liabilities and stockholders' equity:** Total current liabilities, long-term liabilities, stock, and retained earnings. This total represents all the claims on assets—prior and present claims of creditors, owners' residual claims, and any other claims.

My Money

Aha!

Create a personal balance sheet by listing all personal assets and debts. Subtract debts from assets to calculate net worth. Do an Internet search to compare your net worth to what financial analysts recommend for your age group.

Now that you are familiar with the common balance sheet items, you are ready to analyze a balance sheet.

LO 2

Vertical and Horizontal Analyses and the Balance Sheet

Often financial statement readers want to analyze reports that contain data for two or more successive accounting periods. To make this possible, companies present a statement showing the data from these periods side by side. As you might expect, this statement is called a **comparative statement.**

Comparative reports help illustrate changes in data. Financial statement readers should compare the percents in the reports to industry percents and the percents of competitors.

Figure 16.2 shows the comparative balance sheet of Roger Company. Note that the statement analyzes each asset as a percent of total assets for a single period. The statement then analyzes each liability and equity as a percent of total liabilities and stockholders' equity. We call this type of analysis **vertical analysis.**

The following steps use the portion formula to prepare a vertical analysis of a balance sheet.

PREPARING A VERTICAL ANALYSIS OF A BALANCE SHEET

Step 1. Divide each asset (the portion) as a percent of total assets (the base). Round as indicated.

Step 2. Round each liability and stockholders' equity (the portions) as a percent of total liabilities and stockholders' equity (the base). Round as indicated.

We can also analyze balance sheets for two or more periods by using **horizontal analysis.** Horizontal analysis compares each item in 1 year by amount, percent, or both with the same item of the previous year. Note the Abby Ellen Company horizontal analysis shown in Figure 16.3 on page 424. To make a horizontal analysis, we use the portion formula and the steps that follow:

FIGURE 16.2

Comparative balance sheet: Vertical analysis

We divide each item by the total of assets.

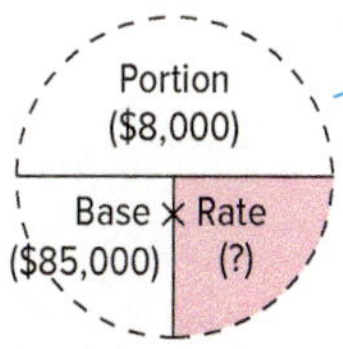

We divide each item by the total of liabilities and stockholders' equity.

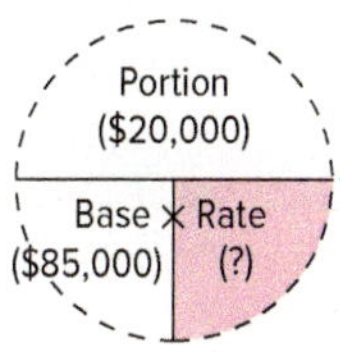

ROGER COMPANY
Comparative Balance Sheet
December 31, 2019 and 2020

	2020		2019	
	Amount	**Percent**	**Amount**	**Percent**
Assets				
Current assets:				
Cash	$22,000	25.88	$18,000	22.22
Accounts receivable	8,000	9.41	9,000	11.11
Merchandise inventory	9,000	10.59	7,000	8.64
Prepaid rent	4,000	4.71	5,000	6.17
Total current assets	$43,000	50.59	$39,000	48.15*
Plant and equipment:				
Building (net)	$18,000	21.18	$18,000	22.22
Land	24,000	28.24	24,000	29.63
Total plant and equipment	$42,000	49.41*	$42,000	51.85
Total assets	$85,000	100.00	$81,000	100.00
Liabilities				
Current liabilities:				
Accounts payable	$14,000	16.47	$ 8,000	9.88
Salaries payable	18,000	21.18	17,000	20.99
Total current liabilities	$32,000	37.65	$25,000	30.86*
Long-term liabilities:				
Mortgage note payable	12,000	14.12	20,000	24.69
Total liabilities	$44,000	51.76*	$45,000	55.56*
Stockholders' Equity				
Common stock	$20,000	23.53	$20,000	24.69
Retained earnings	21,000	24.71	16,000	19.75
Total stockholders' equity	$41,000	48.24	$36,000	44.44
Total liabilities and stockholders' equity	$85,000	100.00	$81,000	100.00

Note: All percents are rounded to the nearest hundredth percent.
*Due to rounding.

MONEY tips

Are you financially "on track"? Here is a simple calculation to help estimate what your net worth should be:

(Age × Pretax income) ÷ 10

PREPARING A HORIZONTAL ANALYSIS OF A COMPARATIVE BALANCE SHEET

Step 1. Calculate the increase or decrease (portion) in each item from the base year.

Step 2. Divide the increase or decrease in Step 1 by the old or base year.

Step 3. Round as indicated.

You can see the difference between vertical analysis and horizontal analysis by looking at the example of vertical analysis in Figure 16.2. The percent calculations in Figure 16.2 are for each item of a particular year as a percent of that year's total assets or total liabilities and stockholders' equity.

Horizontal analysis needs comparative columns because we take the difference *between* periods. In Figure 16.3, for example, the accounts receivable decreased $1,000 from 2019 to 2020. Thus, by dividing $1,000 (amount of change) by $6,000 (base year), we see that Abby's receivables decreased 16.67%.

FIGURE 16.3

Comparative balance sheet: Horizontal analysis

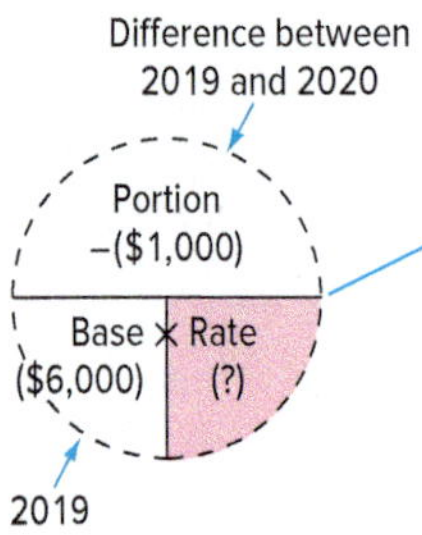

ABBY ELLEN COMPANY Comparative Balance Sheet December 31, 2019 and 2020				
			INCREASE (DECREASE)	
	2020	2019	Amount	Percent
Assets				
Current assets:				
Cash	$ 6,000	$ 4,000	$2,000	50.00*
Accounts receivable	5,000	6,000	(1,000)	−16.67
Merchandise inventory	9,000	4,000	5,000	125.00
Prepaid rent	5,000	7,000	(2,000)	−28.57
Total current assets	$25,000	$21,000	$4,000	19.05
Plant and equipment:				
Building (net)	$12,000	$12,000	–0–	–0–
Land	18,000	18,000	–0–	–0–
Total plant and equipment	$30,000	$30,000	–0–	–0–
Total assets	$55,000	$51,000	$4,000	7.84
Liabilities				
Current liabilities:				
Accounts payable	$ 3,200	$ 1,800	$1,400	77.78
Salaries payable	2,900	3,200	(300)	− 9.38
Total current liabilities	$ 6,100	$ 5,000	$1,100	22.00
Long-term liabilities:				
Mortgage note payable	17,000	15,000	2,000	13.33
Total liabilities	$23,100	$20,000	$3,100	15.50
Owner's Equity				
Abby Ellen, capital	$31,900	$31,000	$ 900	2.90
Total liabilities and owner's equity	$55,000	$51,000	$4,000	7.84

*The percents are not summed vertically in horizontal analysis.

Let's now try the following Practice Quiz.

LU 16–1 PRACTICE QUIZ

Complete this Practice Quiz to see how you are doing.

1. Complete this partial comparative balance sheet by vertical analysis. Round percents to the nearest hundredth.

	2020		2019	
	Amount	Percent	Amount	Percent
Assets				
Current assets:				
a. Cash	$42,000		$40,000	
b. Accounts receivable	18,000		17,000	
c. Merchandise inventory	15,000		12,000	
d. Prepaid expenses	17,000		14,000	
.	.		.	
.	.		.	
.	.		.	
Total current assets	$160,000		$150,000	

2. What is the amount of change in merchandise inventory and the percent increase?

For **extra help** from your authors–Sharon and Jeff–see the videos in Connect.

✓ Solutions

		2020	2019
1. a.	Cash	$\frac{\$42{,}000}{\$160{,}000} = 26.25\%$	$\frac{\$40{,}000}{\$150{,}000} = 26.67\%$
b.	Accounts receivable	$\frac{\$18{,}000}{\$160{,}000} = 11.25\%$	$\frac{\$17{,}000}{\$150{,}000} = 11.33\%$
c.	Merchandise inventory	$\frac{\$15{,}000}{\$160{,}000} = 9.38\%$	$\frac{\$12{,}000}{\$150{,}000} = 8.00\%$
d.	Prepaid expenses	$\frac{\$17{,}000}{\$160{,}000} = 10.63\%$	$\frac{\$14{,}000}{\$150{,}000} = 9.33\%$

2. $15,000
 − 12,000
Amount = $ 3,000

Percent = $\frac{\$3{,}000}{\$12{,}000} = 25\%$

Learning Unit 16–2: Income Statement—Report for a Specific Period of Time

One of the most important departments in a company is its accounting department. The job of the accounting department is to determine the financial results of the company's operations. Is the company making money or losing money?

LO 1

In this learning unit we look at the **income statement**—a financial report that tells how well a company is performing (its profitability or net profit) during a specific period of time (month, year, etc.). In general, the income statement reveals the inward flow of revenues (sales) against the outward or potential outward flow of costs and expenses.

The form of income statements varies depending on the company's type of business. However, the basic formula of the income statement is the same:

Revenues − Operating expenses = Net income

In a merchandising business like The Card Shop, we can expand on this formula:

Revenues (sales) ← After any returns, allowances, or discounts
− Cost of merchandise or goods ← Baseball cards
= Gross profit from sales
− Operating expenses
= Net income (profit)

THE CARD SHOP Income Statement For Month Ended December 31, 2020	
Revenues (sales)	$8,000
Cost of merchandise (goods) sold	3,000
Gross profit from sales	$5,000
Operating expenses	750
Net income	$4,250

Now let's look at The Card Shop's income statement to see how much profit The Card Shop made during its first year of operation. For simplicity, we assume The Card Shop sold all the cards it bought during the year. For its first year of business, The Card Shop made a profit of $4,250.

We can now go more deeply into the income statement elements as we study the income statement of a corporation.

Elements of the Corporation Income Statement

Figure 16.4 gives the format and content of the Mool Company income statement—a corporation. The five main items of an income statement are revenues, cost of merchandise (goods) sold, gross profit on sales, operating expenses, and net income. We will follow the same pattern we used in explaining the balance sheet and define the main items and the letter-coded subitems.

FIGURE 16.4 Income statement

MOOL COMPANY
Income Statement
For Month Ended December 31, 2020

		Revenues:				
1.	a.	Gross sales			$22,080	
	b.	Less: Sales returns and allowances		$ 1,082		
	c.	Sales discounts		432	1,514	
	d.	Net sales				$20,566
		Cost of merchandise (goods) sold:				
2.	a.	Merchandise inventory, December 1, 2020			$ 1,248	
	b.	Purchases		$10,512		
	c.	Less: Purchase returns and allowances	$336			
	d.	Less: Purchase discounts	204	540		
	e.	Cost of net purchases			9,972	
	f.	Cost of merchandise (goods available for sale)			$ 11,220	
	g.	Less: Merchandise inventory, December 31, 2020			1,600	
	h.	Cost of merchandise (goods) sold				9,620
3.		Gross profit from sales				$10,946
		Operating expenses:				
4.	a.	Salary		$ 2,200		
	b.	Insurance		1,300		
	c.	Utilities		400		
	d.	Plumbing		120		
	e.	Rent		410		
	f.	Depreciation		200		
	g.	Total operating expenses				4,630
5.		Net income				$ 6,316

Callouts: Report for a specific period; Put in heading who, what, when; Actual sales after discounts and returns; Inventory not yet sold; Net sales − Cost of merchandise (goods) sold; Gross profit − Operating expenses

Note: Numbers are subtotaled from left to right.

1. **Revenues:** Total earned sales (cash or credit) less any sales returns and allowances or sales discounts.
   - a. **Gross sales:** Total earned sales before sales returns and allowances or sales discounts.
   - b. **Sales returns and allowances:** Reductions in price or reductions in revenue due to goods returned because of product defects, errors, and so on. When the buyer keeps the damaged goods, an allowance results.
   - c. **Sales (not trade) discounts:** Reductions in the selling price of goods due to early customer payment. For example, a store may give a 2% discount to a customer who pays a bill within 10 days.
   - d. **Net sales:** Gross sales less sales returns and allowances less sales discounts.
2. **Cost of merchandise (goods) sold:** All the costs of getting the merchandise that the company sold. The cost of all unsold merchandise (goods) will be subtracted from this item (ending inventory).
   - a. **Merchandise inventory, December 1, 2020:** Cost of inventory in the store that was for sale to customers at the beginning of the month.
   - b. **Purchases:** Cost of additional merchandise brought into the store for resale to customers.
   - c. **Purchase returns and allowances:** Cost of merchandise returned to the store due to damage, defects, errors, and so on. Damaged goods kept by the buyer result in a cost reduction called an *allowance.*

Apple Boosts Margins, but Not Enough

If nothing else, **Apple's** latest results will serve as a sound reminder of the company's ability to watch its bills.

Granted, that isn't always enough for a company that frequently touts its own products as "revolutionary." Apple's share price slipped after hours Tuesday following the company's fiscal second-quarter results.

However, the company also managed to edge up its gross margin to 38.9%, from 38.5% in the December quarter. That is notable given the pressure of a strong dollar, along with the fact that the cost of key components keeps rising. The midpoint of Apple's gross margin projection for the June quarter is 38%, suggesting the company is confident it can keep managing those costs as production ramps up for the next iPhone.

GLOBAL

d. **Purchase discounts:** Savings received by the buyer for paying for merchandise before a certain date. These discounts can result in a substantial savings to a company.

e. **Cost of net purchases:** Cost of purchases less purchase returns and allowances less purchase discounts.

f. **Cost of merchandise (goods available for sale):** Sum of beginning inventory plus cost of net purchases.

g. **Merchandise inventory, December 31, 2020:** Cost of inventory remaining in the store to be sold.

h. **Cost of merchandise (goods) sold:** Beginning inventory plus net purchases less ending inventory. Note in the *Wall Street Journal* clip to the left the operating margin for Apple.

3. **Gross profit (gross margin) from sales:** Net sales less cost of merchandise (goods) sold.

4. **Operating expenses:** Additional costs of operating the business beyond the actual cost of inventory sold.

   a.–f. **Expenses:** Individual expenses broken down.

   g. **Total operating expenses:** Total of all the individual expenses.

5. **Net income:** Gross profit less operating expenses.

My Money

Aha!

Create a personal income statement: Salaries − Expenses = Savings/Investment.

In the next section you will learn some formulas that companies use to calculate various items on the income statement.

Calculating Net Sales, Cost of Merchandise (Goods) Sold, Gross Profit, and Net Income of an Income Statement

It is time to look closely at Figure 16.4 and see how each section is built. Use the previous vocabulary as a reference. We will study Figure 16.4 step by step.

Step 1. Calculate the net sales—what Mool earned:

Net sales = Gross sales − Sales returns and allowances − Sales discounts

$20,566 = $22,080 − $1,082 − $432

MONEY tips

Keep a record of how much you owe and to whom you owe it. Create a plan to pay each debt off, one by one. Pay off the highest interest rate balance first.

Step 2. Calculate the cost of merchandise (goods) sold:

Cost of merchandise (goods) sold = Beginning inventory + Net purchases (purchases less returns and discounts) − Ending inventory

$9,620 = $1,248 + $9,972 − $1,600

Step 3. Calculate the gross profit (gross margin) from sales—profit before operating expenses:

Gross profit from sales = Net sales − Cost of merchandise (goods) sold

$10,946 = $20,566 − $9,620

Step 4. Calculate the net income—profit after operating expenses:

Net income = Gross profit − Operating expenses

$6,316 = $10,946 − $4,630

LO 2

Analyzing Comparative Income Statements

We can apply the same procedures of vertical and horizontal analysis to the income statement that we used in analyzing the balance sheet. Let's first look at the vertical analysis for Royal Company, Figure 16.5. Then we will look at the horizontal analysis of Flint Company's 2019 and 2020 income statements shown in Figure 16.6. Note in the margin how numbers are calculated.

FIGURE 16.5

Vertical analysis

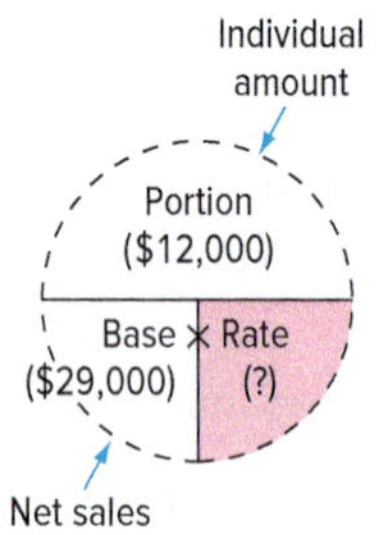

ROYAL COMPANY
Comparative Income Statement
For Years Ended December 31, 2019 and 2020

	2020	Percent of net	2019	Percent of net
Net sales	$45,000	100.00	$29,000	100.00[*]
Cost of merchandise sold	19,000	42.22	12,000	41.38
Gross profit from sales	$26,000	57.78	$17,000	58.62
Operating expenses:				
Depreciation	$ 1,000	2.22	$ 500	1.72
Selling and advertising	4,200	9.33	1,600	5.52
Research	2,900	6.44	2,000	6.90
Miscellaneous	500	1.11	200	.69
Total operating expenses	$ 8,600	19.11[†]	$ 4,300	14.83
Income before interest and taxes	$17,400	38.67	$12,700	43.79
Interest expense	6,000	13.33	3,000	10.34
Income before taxes	$ 11,400	25.33[†]	$ 9,700	33.45
Provision for taxes	5,500	12.22	3,000	10.34
Net income	$ 5,900	13.11	$ 6,700	23.10[†]

[*]Net sales = 100%
[†]Off due to rounding.

Aha!

When conducting a vertical analysis of an income statement, divide each line item by net sales.

FIGURE 16.6

Horizontal analysis

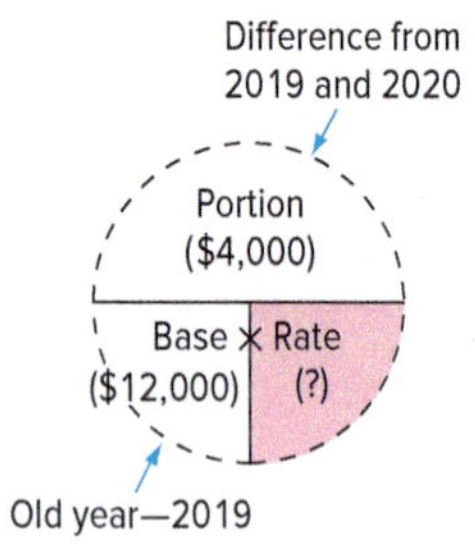

FLINT COMPANY
Comparative Income Statement
For Years Ended December 31, 2019 and 2020

			INCREASE (DECREASE)	
	2020	2019	Amount	Percent
Sales	$90,000	$80,000	$10,000	
Sales returns and allowances	2,000	2,000	–0–	
Net sales	$88,000	$78,000	$10,000	+ 12.82
Cost of merchandise (goods) sold	45,000	40,000	5,000	+ 12.50
Gross profit from sales	$43,000	$38,000	$ 5,000	+ 13.16
Operating expenses:				
Depreciation	$ 6,000	$ 5,000	$ 1,000	+ 20.00
Selling and administrative	16,000	12,000	4,000	+ 33.33
Research	600	1,000	(400)	− 40.00
Miscellaneous	1,200	500	700	+ 140.00
Total operating expenses	$23,800	$18,500	$ 5,300	+ 28.65
Income before interest and taxes	$19,200	$19,500	$ (300)	− 1.54
Interest expense	4,000	4,000	–0–	
Income before taxes	$15,200	$15,500	$ (300)	− 1.94
Provision for taxes	3,800	4,000	(200)	− 5.00
Net income	$11,400	$11,500	$ (100)	− .87

The following Practice Quiz will test your understanding of this unit.

LU 16–2 PRACTICE QUIZ

Complete this Practice Quiz to see how you are doing.

From the following information, calculate:

a. Net sales.
b. Cost of merchandise (goods) sold.
c. Gross profit from sales.
d. Net income.

Given: Gross sales, \$35,000; sales returns and allowances, \$3,000; beginning inventory, \$6,000; net purchases, \$7,000; ending inventory, \$5,500; operating expenses, \$7,900.

*For **extra help** from your authors–Sharon and Jeff–see the videos in Connect.*

✓ Solutions

a. \$35,000 − \$3,000 = \$32,000 (Gross sales − Sales returns and allowances)

b. \$6,000 + \$7,000 − \$5,500 = \$7,500 (Beginning inventory + Net purchases − Ending inventory)

c. \$32,000 − \$7,500 = \$24,500 (Net sales − Cost of merchandise sold)

d. \$24,500 − \$7,900 = \$16,600 (Gross profit from sales − Operating expenses)

Learning Unit 16–3: Trend and Ratio Analysis

Aha!

A balance sheet is like a snapshot. It reflects a company's or individual's financial position at a specific point in time. An income statement is like a video. It reflects a company's or individual's profitability over an interval of time.

Now that you understand the purpose of balance sheets and income statements, you are ready to study how experts look for various trends as they analyze the financial reports of companies. This learning unit discusses trend analysis and ratio analysis. The study of these trends is valuable to businesses, financial institutions, and consumers.

LO 1

Trend Analysis

Many tools are available to analyze financial reports. When data cover several years, we can analyze changes that occur by expressing each number as a percent of the base year. The base year is a past period of time that we use to compare sales, profits, and so on with other years. We call this **trend analysis.**

Using the data below, we complete a trend analysis with the following steps:

MONEY tips

Check out Financial Peace University at www.daveramsey.com for some excellent personal finance training and tips.

COMPLETING A TREND ANALYSIS

Step 1. Select the base year (100%).

Step 2. Express each amount as a percent of the base year amount (rounded to the nearest whole percent).

	GIVEN (BASE YEAR 2018)			
	2021	2020	2019	2018
Sales	\$621,000	\$460,000	\$340,000	\$420,000
Gross profit	182,000	141,000	112,000	124,000
Net income	48,000	41,000	22,000	38,000

	TREND ANALYSIS			
	2021	2020	2019	2018
Sales	148%	110%	81%	100%
Gross profit	147	114	90	100
Net income	126	108	58	100

How to Calculate Trend Analysis

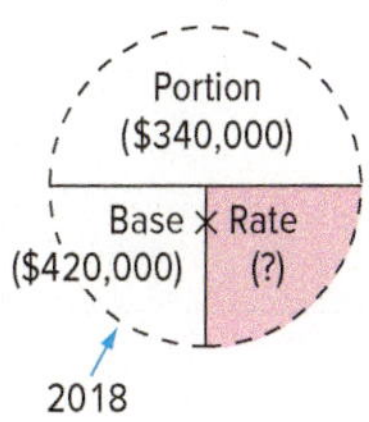

$$\frac{\text{Each item}}{\text{Base amount}} = \frac{\$340{,}000}{\$420{,}000} = 80.95\% = 81\%$$

Sales for 2019 (\$340,000); Sales for 2018 (\$420,000)

What Trend Analysis Means Sales of 2019 were 81% of the sales of 2018. Note that you would follow the same process no matter which of the three areas you were analyzing. All categories are compared to the base year—sales, gross profit, or net income.

We now will examine **ratio analysis**—another tool companies use to analyze performance.

LO 2

Ratio Analysis

A *ratio* is the relationship of one number to another. Many companies compare their ratios with those of previous years and with ratios of other companies in the industry. Companies can get ratios of the performance of other companies from their bankers, accountants, local small business centers, libraries, and newspaper articles. It is important to choose companies from similar industries when comparing ratios. For example, ratios at McDonald's will be different from ratios at Best Buy. McDonald's sells more perishable products.

Percentage ratios are used by companies to determine the following:

1. How well the company manages its assets—*asset management ratios.*
2. The company's debt situation—*debt management ratios.*
3. The company's profitability picture—*profitability ratios.*

Each company must decide the true meaning of what the three types of ratios (asset management, debt management, and profitability) are saying. Table 16.1 gives a summary of the key ratios, their calculations (rounded to the nearest hundredth), and what they mean. All calculations are from Figures 16.1 and 16.4.

Now you can check your knowledge with the Practice Quiz that follows.

MONEY tips

Keep track of the equity you have in your home. When you reach 20% or more, request to have the private mortgage insurance (PMI) removed from your mortgage. Contact your lender for details.

TABLE 16.1 Summary of key ratios: A reference guide*

Ratio	Formula	Actual calculations	What it says	Questions that could be raised
Current ratio	$\frac{\text{Current assets}}{\text{Current liabilities}}$ (Current assets include cash, accounts receivable, and marketable securities.)	$\frac{\$61{,}000}{\$92{,}000} = .66:1$ Industry average, 2 to 1	Business has 66¢ of current assets to meet each $1 of current debt.	Not enough current assets to pay off current liabilities. Industry standard is $2 for each $1 of current debt.[†]
Acid test (quick ratio) Top of fraction often referred to as ***quick assets***	$\frac{\text{Current assets} - \text{Inventory} - \text{Prepaid expenses}}{\text{Current liabilities}}$ (Inventory and prepaid expenses are excluded because it may not be easy to convert these to cash.)	$\frac{\$61{,}000 - \$30{,}000 - \$15{,}000}{\$92{,}000} = .17:1$ Industry average, 1 to 1	Business has only 17¢ to cover each $1 of current debt. This calculation excludes inventory and prepaid expenses.	Same as above but more severe.
Average day's collection	$\frac{\text{Accounts receivable}}{\left(\frac{\text{Net sales}}{360}\right)}$	$\frac{\$9{,}000}{\frac{\$20{,}566}{360}} = 158\text{ days}$ Industry average, 90–120 days	On the average, it takes 158 days to collect accounts receivable.	Could we speed up collection since industry average is 90–120 days?
Total debt to total assets	$\frac{\text{Total liabilities}}{\text{Total assets}}$	$\frac{\$150{,}000}{\$205{,}000} = 73.17\%$ Industry average, 50%–70%	For each $1 of assets, the company owes 73¢ in current and long-term debt.	73% is slightly higher than industry average.
Return on equity	$\frac{\text{Net income}}{\text{Stockholders' equity}}$	$\frac{\$6{,}316}{\$55{,}000} = 11.48\%$ Industry average, 15%–20%	For each $1 invested by the owner, a return of 11¢ results.	Could we get a higher return on money somewhere else?
Asset turnover	$\frac{\text{Net sales}}{\text{Total assets}}$	$\frac{\$20{,}566}{\$205{,}000} = 10¢$ Industry average, 3¢ to 8¢	For each $1 invested in assets, it returns 10¢ in sales.	Are assets being utilized efficiently?
Profit margin on net sales	$\frac{\text{Net income}}{\text{Net sales}}$	$\frac{\$6{,}316}{\$20{,}566} = 30.71\%$ Industry average, 25%–40%	For each $1 of sales, company produces 31¢ in profit.	Compared to competitors, are we showing enough profits versus our increased sales?

*Inventory turnover is discussed in Chapter 18.

[†]For example, Wal-Mart Stores, Inc., has a current ratio of .76:1.

LU 16–3 PRACTICE QUIZ

Complete this Practice Quiz to see how you are doing.

1. Prepare a trend analysis from the following sales, assuming a base year of 2018. Round to the nearest whole percent.

	2021	2020	2019	2018
Sales	\$29,000	\$44,000	\$48,000	\$60,000

2. **Given:** Total current assets (CA), \$15,000; accounts receivable (AR), \$6,000; total current liabilities (CL), \$10,000; inventory (Inv), \$4,000; net sales, \$36,000; total assets, \$30,000; net income (NI), \$7,500.
Calculate:
   a. Current ratio.
   b. Acid test.
   c. Average day's collection.
   d. Profit margin on sales (rounded to the nearest hundredth percent).

*For **extra help** from your authors–Sharon and Jeff–see the videos in Connect.*

✓ Solutions

1.

	2021	2020	2019	2018
Sales	48%	73%	80%	100%
	$\left(\frac{\$29{,}000}{\$60{,}000}\right)$	$\left(\frac{\$44{,}000}{\$60{,}000}\right)$	$\left(\frac{\$48{,}000}{\$60{,}000}\right)$	

2. **a.** $\frac{CA}{CL} = \frac{\$15{,}000}{\$10{,}000} = 1.5{:}1$

b. $\frac{CA - Inv}{CL} = \frac{\$15{,}000 - \$4{,}000}{\$10{,}000} = 1.1{:}1$

c. $\frac{AR}{\frac{\text{Net sales}}{360}} = \frac{\$6{,}000}{\frac{\$36{,}000}{360}} = 60 \text{ days}$

d. $\frac{NI}{\text{Net sales}} = \frac{\$7{,}500}{\$36{,}000} = 20.83\%$

INTERACTIVE CHAPTER ORGANIZER

Topic/Procedure/Formula	Examples	You try it*
Balance sheet		
Vertical analysis Process of relating each figure on a financial report (down the column) to a total figure. The denominator for a balance sheet is total assets (or total liabilities + owner's equity); for an income statement it is net sales.	Current assets \$ 520 52% Plant and equipment 480 48 Total assets \$1,000 100%	**Do vertical analysis** CA \$ 400 ? P + E 600 ? Total assets \$1,000 ?
Horizontal analysis Analyzing comparative financial reports shows rate and amount of change across columns item by item. (New line item amount – Old line item amount)/Old line item amount	2020 | 2019 | Change | % Cash, \$5,000 | \$4,000 | \$1,000 | 25% $\left(\frac{\$1{,}000}{\$4{,}000}\right)$	**Do horizontal analysis** | 2020 | 2019 | Change | % Cash | \$8,000 | \$2,000 | ? | ?
Net sales Gross sales − Sales returns and allowances − Sales discounts	\$200 gross sales − 10 sales returns and allowances − 2 sales discounts \$ 188 net sales	**Calculate net sales** Gross sales, \$400 Sales returns and allowances, \$20 Sales discount, \$5

(continues)

INTERACTIVE CHAPTER ORGANIZER

Topic/Procedure/Formula	Examples	You try it*
Cost of merchandise (goods) sold Beginning inventory + Net purchases − Ending inventory	$50 + $100 − $20 = $130 Beginning inventory + Net purchases − Ending inventory = Cost of merchandise (goods) sold	**Calculate cost of merchandise sold** Beginning inventory, $50 Net purchases, $200 Ending inventory, $20
Gross profit from sales Net sales − Cost of merchandise (goods) sold	$188 − $130 = $58 gross profit from sales Net sales − Cost of merchandise (goods) sold = Gross profit from sales	**Calculate gross profit** Net sales, $400 Cost of merchandise sold, $250
Net income Gross profit − Operating expenses	$58 − $28 = $30 Gross profit from sales − Operating expenses = Net income	**Calculate net income** Gross profit, $210 Operating expenses, $180
Trend analysis Each number expressed as a percent of the base year. $\frac{\text{Each item}}{\text{Base amount}}$	**2021** **2020** **2019** Sales $200 $300 $400 ← Base year 50% 75% 100% $\left(\frac{\$200}{\$400}\right)$ $\left(\frac{\$300}{\$400}\right)$	**Prepare a trend analysis** **2021** **2020** **2019** $1,200 $800 $1,000 ← Base year
Ratios Tools to interpret items on financial reports.	Use this example for calculating the following ratios: current assets, $30,000; accounts receivable, $12,000; total current liabilities, $20,000; inventory, $6,000; prepaid expenses, $2,000; net sales, $72,000; total assets, $60,000; net income, $15,000; total liabilities, $30,000.	**Use this example for calculating the following ratios:** Current assets, $40,000; Accounts receivable, $44,000; Total current liabilities, $160,000; Inventory, $2,000; Prepaid expenses, $3,000; Net sales, $60,000; Total assets, $70,000; Net income, $16,000; Total liabilities, $180,000.
Current ratio $\frac{\text{Current assets}}{\text{Current liabilities}}$	$\frac{\$30,000}{\$20,000} = 1.5:1$	Use the information from the example above.
Acid test (quick ratio) $\frac{\text{Current assets − Inventory − Prepaid expenses}}{\text{Current liabilities}}$ (numerator: Called quick assets)	$\frac{\$30,000 - \$6,000 - \$2,000}{\$20,000} = 1.1:1$	Use the information from the example above.
Average day's collection $\frac{\text{Accounts receivable}}{\left(\frac{\text{Net sales}}{360}\right)}$	$\frac{\$12,000}{\left(\frac{\$72,000}{360}\right)} = 60 \text{ days}$	Use the information from the example above.
Total debt to total assets $\frac{\text{Total liabilities}}{\text{Total assets}}$	$\frac{\$30,000}{\$60,000} = 50\%$	Use the information from the example above.
Return on equity $\frac{\text{Net income}}{\text{Stockholders' equity (A − L)}}$	$\frac{\$15,000}{\$30,000} = 50\%$	Use the information from the example above.

(continues)

INTERACTIVE CHAPTER ORGANIZER

Topic/Procedure/Formula	Examples	You try it*
Asset turnover $\frac{\text{Net sales}}{\text{Total assets}}$	$\frac{\$72,000}{\$60,000} = 1.2$	Use the information from the example above.
Profit margin on net sales $\frac{\text{Net income}}{\text{Net sales}}$	$\frac{\$15,000}{\$72,000} = .2083 = 20.83\%$	Use the information from the example above.

KEY TERMS

Accounts payable
Accounts receivable
Acid test
Asset turnover
Assets
Average day's collection
Balance sheet
Capital
Common stock
Comparative statement
Corporation
Cost of merchandise (goods) sold
Current assets
Current liabilities
Current ratio
Expenses
Gross profit from sales
Gross sales
Horizontal analysis
Income statement
Liabilities
Long-term liabilities
Merchandise inventory
Mortgage note payable
Net income
Net purchases
Net sales
Operating expenses
Owner's equity
Partnership
Plant and equipment
Prepaid expenses
Profit margin on net sales
Purchase discounts
Purchase returns and allowances
Purchases
Quick assets
Quick ratio
Ratio analysis
Retained earnings
Return on equity
Revenues
Salaries payable
Sales (not trade) discounts
Sales returns and allowances
Sole proprietorship
Stockholders' equity
Total debt to total assets
Trend analysis
Vertical analysis

*Worked-out solutions are in Appendix B.

Critical Thinking Discussion Questions with Chapter Concept Check

1. What is the difference between current assets and plant and equipment? Do you think land should be allowed to depreciate?
2. What items make up stockholders' equity? Why might a person form a sole proprietorship instead of a corporation?
3. Explain the steps to complete a vertical or horizontal analysis relating to balance sheets. Why are the percents not summed vertically in horizontal analysis?
4. How do you calculate net sales, cost of merchandise (goods) sold, gross profit, and net income? Why do we need two separate figures for inventory in the cost of merchandise (goods) sold section?
5. Explain how to calculate the following: current ratios, acid test, average day's collection, total debt to assets, return on equity, asset turnover, and profit margin on net sales. How often do you think ratios should be calculated?
6. What is trend analysis? Explain how the portion formula assists in preparing a trend analysis.
7. In light of the most recent economic crises, explain how companies such as GE are trying to gain market share and increase profit margins.
8. **Chapter Concept Check.** Go online and look up the financial statement of a company that interests you to see how it is doing financially. Use ratio analysis based on concepts presented in this chapter.

Classroom Notes

END-OF-CHAPTER PROBLEMS

connect

Check figures for odd-numbered problems in Appendix B. Name __________ Date ________

DRILL PROBLEMS

16–1. Prepare a December 31, 2020, balance sheet for Long Print Shop like the one for The Card Shop (LU 16–1) from the following: cash, $50,000; accounts payable, $38,000; merchandise inventory, $14,000; Joe Ryan, capital, $46,000; and equipment, $20,000. *LU 16-1(1)*

16–2. From the following, prepare a classified balance sheet for Bach Crawlers as of December 31, 2020. Ending merchandise inventory was $4,000 for the year. *LU 16-1(1)*

Cash	$6,000	Accounts payable	$1,800
Prepaid rent	1,600	Salaries payable	1,600
Prepaid insurance	4,000	Note payable (long term)	8,000
Office equipment (net)	5,000	P. Bach, capital*	9,200

*What the owner supplies to the business. Replaces common stock and retained earnings section.

eXcel **16–3.** Complete a horizontal analysis for Brown Company, rounding percents to the nearest hundredth: *LU 16-1(2)*

BROWN COMPANY Comparative Balance Sheet December 31, 2020 and 2019				
			INCREASE (DECREASE)	
	2020	2019	Amount	Percent
Assets				
Current assets:				
Cash	$ 15,750	$ 10,500		
Accounts receivable	18,000	13,500		
Merchandise inventory	18,750	22,500		
Prepaid advertising	54,000	45,000		
Total current assets	$106,500	$ 91,500		
Plant and equipment:				
Building (net)	$120,000	$126,000		
Land	90,000	90,000		
Total plant and equipment	$210,000	$216,000		
Total assets	$316,500	$307,500		
Liabilities				
Current liabilities:				
Accounts payable	$132,000	$120,000		
Salaries payable	22,500	18,000		
Total current liabilities	$154,500	$138,000		
Long-term liabilities:				
Mortgage note payable	99,000	87,000		
Total liabilities	$253,500	$225,000		
Owner's Equity				
J. Brown, capital	63,000	82,500		
Total liabilities and owner's equity	$316,500	$307,500		

16–4. Prepare an income statement for Hansen Realty for the year ended December 31, 2020. Beginning inventory was $1,248. Ending inventory was $1,600. *LU 16-2(1)*

Sales	$34,900
Sales returns and allowances	1,092
Sales discount	1,152
Purchases	10,512
Purchase discounts	540
Depreciation expense	115
Salary expense	5,200
Insurance expense	2,600
Utilities expense	210
Plumbing expense	250
Rent expense	180

16–5. Assume this is a partial list of financial highlights from a Best Buy annual report:

	2020	2019
	(dollars in millions)	
Net sales	$37,580	$33,075
Earnings before taxes	2,231	1,283
Net earnings	1,318	891

Complete a horizontal and vertical analysis from the above information. Round to the nearest hundredth percent. *LU 16-2(2)*

16–6. From the French Instrument Corporation second-quarter report ended 2020, do a vertical analysis for the second quarter of 2020. *LU 16-2(2)*

FRENCH INSTRUMENT CORPORATION AND SUBSIDIARIES
Consolidated Statements of Operation
(Unaudited) (In thousands of dollars, except share data)

	SECOND QUARTER		
	2020	2019	Percent of net
Net sales	$6,698	$6,951	
Cost of sales	4,089	4,462	
Gross margin	2,609	2,489	
Expenses:			
Selling, general and administrative	1,845	1,783	
Product development	175	165	
Interest expense	98	123	
Other (income), net	(172)	(99)	
Total expenses	1,946	1,972	
Income before income taxes	663	517	
Provision for income taxes	265	209	
Net income	$398	$308	
Net income per common share*	$.05	$.03	
Weighted-average number of common shares and equivalents	6,673,673	6,624,184	

*Income per common share reflects the deduction of the preferred stock dividend from net income.

†Off due to rounding.

16–7. Complete the comparative income statement and balance sheet for Logic Company, rounding percents to the nearest hundredth: *LU 16-1(2), LU 16-2(2)*

LOGIC COMPANY
Comparative Income Statement
For Years Ended December 31, 2019 and 2020

			INCREASE (DECREASE)	
	2020	**2019**	**Amount**	**Percent**
Gross sales	$19,000	$15,000		
Sales returns and allowances	1,000	100		
Net sales	$18,000	$14,900		
Cost of merchandise (goods) sold	12,000	9,000		
Gross profit	$ 6,000	$ 5,900		
Operating expenses:				
Depreciation	$ 700	$ 600		
Selling and administrative	2,200	2,000		
Research	550	500		
Miscellaneous	360	300		
Total operating expenses	$ 3,810	$ 3,400		
Income before interest and taxes	$ 2,190	$ 2,500		
Interest expense	560	500		
Income before taxes	$ 1,630	$ 2,000		
Provision for taxes	640	800		
Net income	$ 990	$ 1,200		

LOGIC COMPANY
Comparative Balance Sheet
December 31, 2019 and 2020

	2020		2019	
	Amount	**Percent**	**Amount**	**Percent**
Assets				
Current assets:				
Cash	$12,000		$ 9,000	
Accounts receivable	16,500		12,500	
Merchandise inventory	8,500		14,000	
Prepaid expenses	24,000		10,000	
Total current assets*	$61,000		$45,500	
Plant and equipment:				
Building (net)	$14,500		$11,000	
Land	13,500		9,000	
Total plant and equipment	$28,000		$20,000	
Total assets	$89,000		$65,500	
Liabilities				
Current liabilities:				
Accounts payable	$13,000		$ 7,000	
Salaries payable	7,000		5,000	
Total current liabilities*	$20,000		$12,000	
Long-term liabilities:				
Mortgage note payable	22,000		20,500	
Total liabilities	$42,000		$32,500	
Stockholders' Equity				
Common stock	$21,000		$21,000	
Retained earnings	26,000		12,000	
Total stockholders' equity	$47,000		$33,000	
Total liabilities and stockholders' equity	$89,000		$65,500	

*Note that the percentages for total current assets and total current liabilities may be off due to rounding.

From Problem 16–7, your supervisor has requested that you calculate the following ratios, rounded to the nearest hundredth: *LU 16-3(2)*

	2020	2019
16–8. Current ratio.		
16–9. Acid test.		
16–10. Average day's collection.		
16–11. Asset turnover.		
16–12. Total debt to total assets.		
16–13. Net income (after tax) to the net sales.		
16–14. Return on equity (after tax).		

16–8.

16–9.

16–10.

16–11.

16–12.

16–13.

16–14.

WORD PROBLEMS

eXcel **16–15.** William Burris invested $100,000 in an Australian-based franchise, Rent Your Boxes, purchasing three territories in the Washington area. After finding out the company had gone bankrupt, he rallied 10 other franchisees to join him and created a new company, Rent Our Boxes. If Rent Our Boxes had net income of $38,902 with net sales of $286,585, what was its profit margin on net sales to the nearest hundredth percent? *LU 16-3(2)*

16–16. Assume General Motors announced a quarterly profit of $119 million for 4th quarter 2019. Below is a portion of its balance sheet. Conduct a horizontal analysis of the following line items (rounding percent to nearest hundredth): *LU 16-1(2)*

	2019 (dollars in millions)	2018 (dollars in millions)	Difference	% CHG
Cash and cash equivalents	$ 15,980	$ 15,499		
Marketable securities	9,222	16,148		
Inventories	13,642	14,324		
Goodwill	—	1,278		
Total liabilities and equity	$103,249	$144,603		

eXcel

16–17. Find the following ratios for Motorola Credit Corporation's annual report: **(a)** total debt to total assets, **(b)** return on equity, **(c)** asset turnover (to nearest cent), and **(d)** profit margin on net sales. Round to the nearest hundredth percent. *LU 16-3(2)*

	(dollars in millions)
Net revenue (sales)	$ 265
Net earnings	147
Total assets	2,015
Total liabilities	1,768
Total stockholders' equity	427

16–18. Assume figures were presented for the past 5 years on merchandise sold at Chicago department and discount stores ($ million). Sales in 2022 were $3,154; in 2021, $3,414; in 2020, $3,208; in 2019, $3,152; and in 2018, $3,216. Using 2018 as the base year, complete a trend analysis. Round each percent to the nearest whole percent. *LU 16-3(1)*

16–19. Don Williams received a memo requesting that he complete a trend analysis of the following numbers using 2018 as the base year and rounding each percent to the nearest whole percent. Could you help Don with the request? *LU 16-3(1)*

	2021	2020	2019	2018
Sales	$340,000	$400,000	$420,000	$500,000
Gross profit	180,000	240,000	340,000	400,000
Net income	70,000	90,000	40,000	50,000

eXcel

16–20. If the French bank Société Générale reported its 2019 net income was 23,561 million euros and its operating expenses totaled 16,016 million euros, what was its gross profit? *LU 16-2(1)*

My Money

16–21. At age 32, you have assets of $275,658 and liabilities of $266,211. What is your net worth? Is this appropriate for your age if you have an annual household income of $69,200? See https://lifehacker.com/5859040/what-should-your-current-net-worth-be to determine if the net worth is adequate. *LU 16-1(1)*

CHALLENGE PROBLEMS

16–22. On January 1, Pete Rowe bought a ski chalet for $51,000. Pete is renting the chalet for $55 per night. He estimates he can rent the chalet for 190 nights. Pete's mortgage for principal and interest is $448 per month. Real estate tax on the chalet is $500 per year.

Pete estimates that his heating bill will run $60 per month. He expects his monthly electrical bill to be $20 per month. He pays $12 per month for cable television.

What is Pete's return on the initial investment for this year? Assume rentals drop by 30% and monthly bills for heat and electricity drop by 10% each month. What would be Pete's return on initial investment? Round to the nearest tenth percent as needed. *LU 16-3(2)*

eXcel 16–23. As the accountant for Tootsie Roll, you are asked to calculate the current ratio and the quick ratio for the following partial financial statement. Round to the nearest tenth. *LU 16-3(2)*

Assets		Liabilities	
Current assets:		Current liabilities:	
Cash and cash equivalents	$ 4,224,190	Notes payable to banks	$ 672,221
Investments	32,533,769	Accounts payable	7,004,075
Accounts receivable, less allowances of $748,000 and $744,000	16,206,648	Dividends payable	576,607
		Accrued liabilities	9,826,534
Inventories:		Income taxes payable	4,471,429
Finished goods and work in progress	12,650,955		
Raw materials and supplies	10,275,858		
Prepaid expenses	2,037,710		

SUMMARY PRACTICE TEST

Do you need help? Connect videos have step-by-step worked-out solutions.

1. Given: Gross sales, \$170,000; sales returns and allowances, \$9,000; beginning inventory, \$8,000; net purchases, \$18,000; ending inventory, \$5,000; and operating expenses, \$56,000. Calculate **(a)** net sales, **(b)** cost of merchandise (goods) sold, **(c)** gross profit from sales, and **(d)** net income. *LU 16-2(1)*

2. Complete the following partial comparative balance sheet by filling in the total current assets and percent column; assume no plant and equipment (round to the nearest hundredth percent as needed). *LU 16-1(2)*

	Amount	Percent	Amount	Percent
Assets				
Current assets:				
Cash	\$ 9,000		\$ 8,000	
Accounts receivable	5,000		7,500	
Merchandise inventory	12,000		6,900	
Prepaid expenses	7,000		8,000	
Total current assets*				

*The percentages for total current assets will be off due to rounding.

3. Calculate the amount of increase or decrease and the percent change of each item, rounding to the nearest hundredth percent as needed. *LU 16-1(2)*

	2020	**2019**	**Amount**	**Percent**
Cash	\$19,000	\$ 8,000		
Land	70,000	30,000		
Accounts payable	21,000	10,000		

4. Complete a trend analysis for sales, rounding to the nearest whole percent and using 2019 as the base year. *LU 16-3(1)*

	2022	**2021**	**2020**	**2019**
Sales	\$140,000	\$350,000	\$210,000	\$190,000

5. From the following, prepare a balance sheet for True Corporation as of December 31, 2019. *LU 16-1(1)*

Building	$40,000	Mortgage note payable	$70,000
Merchandise inventory	12,000	Common stock	10,000
Cash	15,000	Retained earnings	37,000
Land	90,000	Accounts receivable	9,000
Accounts payable	50,000	Salaries payable	8,000
Prepaid rent	9,000		

6. Solve from the following facts, rounding to the nearest hundredth. *LU 16-3(2)*

Current assets	$14,000	Net sales	$40,000
Accounts receivable	$ 5,000	Total assets	$38,000
Current liabilities	$20,000	Net income	$10,100
Inventory	$ 4,000		

a. Current ratio

b. Acid test

c. Average day's collection

d. Asset turnover

e. Profit margin on sales

INTERACTIVE VIDEO WORKSHEET

GRADING THE SUMMARY PRACTICE TEST

Go to the summary practice test video in Connect (or click on it here in the ebook). Grade your summary practice test while viewing the video.

C for Correct/I for Incorrect

1. ________	4. ________
2. ________	5. ________
3. ________	6. ________

If you achieved 100%, you are ready for your instructor's exam.

If any of the problems were incorrect, list the questions you missed and show steps to solve the problem correctly.

Replay the video to see if you have made the correct fixes to your mistakes. If you have any questions, contact your instructor asap.

Notes on Watching Videos

MY MONEY

For What It's Worth

What I need to know

Net worth is a way of providing a snapshot of your current financial standing. At its basic level, net worth subtracts what you owe from the value of what you own. As you enter the workforce after college you may not see the immediate value in determining your net worth since you are most likely in a state where what you owe vastly outweighs what you own. However, as you move through your future, understanding your net worth will be a valuable tool in assessing your financial situation and whether or not you are making progress toward your financial goals.

Another helpful tool in assessing your financial standing is the personal income statement. This statement will make a direct comparison of the inflow of money (earnings) to the outflow of money (expenses). This is a quick way to assess how you are using the earnings generated from your employment. By generating and assessing these types of reports and applying ratio analysis, over time you can gain the knowledge concerning whether your financial standing is improving or declining. With such information in hand you will be able to make the necessary changes to keep yourself on track or modify your financial position to continue making progress toward your financial goals.

What I need to do

Your budget plays a crucial role in the process of assessing your current financial standing. Having accurate information concerning the inflows and outflows of your financial resources will be paramount in assessing your financial position. Be detailed in the tracking of your expenses to ensure they are staying in line with your budget and income. As things change over time your budget will also need to adjust. Taking into account these changes over time will help to make sure you can maintain a healthy budget and begin building your net worth.

Be aware of how taking on additional debt will impact your financial position. Since what you owe is compared directly to what you own in the assessment of your financial position, you need to seriously consider any new debt you take on. Your debt-to-income ratio should be no greater than 36% and less is better. Total your monthly debt payments (credit cards, auto loans, mortgage, etc.) and divide this amount by your gross income to determine your debt-to-income ratio.

Begin building your net worth in your twenties. Most financial advisors recommend having a net worth of $1,000,000 by age 60. If you invest $10,000 per year starting at age 25 with a 5% rate of return you will net almost $1 million by your 60th birthday. To estimate what your net worth should be by age: Your age × Your annual gross income ÷ 10.

Resources I can use

- https://www.bankrate.com/calculators/smart-spending/personal-net-worth-calculator.aspx — net worth calculator
- https://www.investopedia.com/terms/p/personal-financial-statement.asp — personal financial statement

MY MONEY ACTIVITY

Use the Internet net worth calculator provided above. Experiment with changing your assets and liabilities to see their impact on your net worth.

A KIPLINGER APPROACH

"Farm Living is the Life for Them", *Kiplinger's*, April 2018, p. 24. Used by permission of The Kiplinger Washington Editors, Inc.

SUCCESS STORY

Farm Living Is the Life for Them

A former sheriff and assistant D.A. go back to their roots, raising livestock on 132 acres.

PROFILE

WHO: Kerry Dunaway, 60 (interviewed) and Robin Dunaway, 62

WHERE: Roberta, Ga.

WHAT: Owners, Greenway Farms of Georgia, LLC

How did you get into farming? We grew up on family farms, and Robin and I lived together on 18 acres. Robin pickled and made jelly all her life, and she began making goats-milk products after she and I retired. She sold them at local farmers markets, where she noticed the demand for meat and eggs. So we decided to raise livestock. We had bought an additional 22 acres early in retirement, and we rented another 100 acres after getting advice from university cooperative extension agents and other farmers.

When did you retire? At the end of 2008. I was sheriff of Crawford County for 25 years, and Robin was an assistant district attorney in Macon for 28 years. I was tired of politics, and Robin was burned out. We enjoyed sitting around, drinking coffee and watching the news. But soon we were getting on each other's last nerve and needed more to do.

Where did your capital come from? We tapped our savings for the 22 acres. When we began raising livestock, we took a line of credit from AgSouth Farm Credit, a lending cooperative. Last year, we partnered with another couple and bought a 92-acre farm. We no longer rent land. Now our assets total $520,000, and our debt is less than $90,000.

What kind of livestock do you raise? We buy calves, piglets and meat chickens. We also have 235 laying hens. We pasture-raise our animals without growth hormones or antibiotics added to their feed. When it's time, we drive them to a family-owned processor that is USDA-certified and humane. We've found that meat from an animal raised and processed humanely is more tender than meat from a stressed animal.

What do you can? In 2013, we built the cannery to make jams, jellies, pickles and sauces with local produce. Robin runs it. Her law skills were a big help in getting our kitchen federally inspected and USDA-certified.

©Matee Nuserm/Shutterstock

Where do you sell? We started at farmers markets. As we became known for our quality, we attracted some local stores and restaurants, too. We also sell via our website (www.greenwayfarmsofga.com).

What do you charge? Our chicken goes for $3.35 a pound, ground beef for $7, center-cut pork chops for $8, filet mignon for $20 and a dozen large eggs for $5. Some people say, "I can buy this hamburger or steak at Wal-Mart." I say, "You can't buy this kind of hamburger or steak at Wal-Mart."

How big are you now? In 2017, we sold about 17,000 pounds of beef, pork and chicken and about 2,500 dozen eggs, as well as 4,600 jars of our canned goods. In 2018, we expect to more than triple our meat production.

Do you make a living? We supplement our retirement income with farm income. We also benefit from agricultural tax breaks that flow back to us personally as a limited liability corporation. The work is demanding, with nonstop going and doing, but it's fun, and every day is a little different. Still, I used to wonder why my grandparents went to bed at 8 p.m., and now I know why.

PATRICIA MERTZ ESSWEIN
pesswein@kiplinger.com

BUSINESS MATH ISSUE

A small business can compete with large companies due to the new tax law.

1. List the key points of the article and information to support your position.
2. Write a group defense of your position using math calculations to support your view. If you are in an online course, post to a discussion board.

Design elements: Digital frame: (tablet) ©Radu Bercan/Shutterstock, (icons) ©12bit/Shutterstock; Globe: ©edge69/Getty Images; Money Tips: ©Pro Symbols/Shutterstock; Aha!: ©frender/Getty Images; Excel: ©McGraw-Hill Education; My Money: (tree) ©Cherkas/Shutterstock, (gold nuggets) ©JonahWong/Shutterstock; Video Worksheet: ©PureSolution/Shutterstock; author photo in circle: ©McGraw-Hill Education/Ryan McAndrews, photographer.

CHAPTER 17

Depreciation

Toyota and Mazda To Build U.S. Plant

By Adrienne Roberts

Toyota Motor Corp. and **Mazda Motor** Corp. are expected to announce plans on Friday to build a $1.6 billion assembly plant in the U.S. that would create 4,000 jobs and be up and running by 2021, according to a person briefed on the plans.

The new factory's location hasn't been decided. Once built, it would produce 300,000 vehicles a year with half being the Toyota Corolla and the rest an unspecified Mazda model, the person said.

©Bandit Chanheng/Shutterstock

LEARNING UNIT OBJECTIVES

LU 17–1: Concept of Depreciation and the Straight-Line Method

1. Explain the concept and causes of depreciation.
2. Prepare a depreciation schedule and calculate partial-year depreciation.

LU 17–2: Units-of-Production Method

1. Explain how use affects the units-of-production method.
2. Prepare a depreciation schedule.

LU 17–3: Declining-Balance Method

1. Explain the importance of residual value in the depreciation schedule.
2. Prepare a depreciation schedule.

LU 17–4: Modified Accelerated Cost Recovery System (MACRS) with Introduction to ACRS (1986, 1989, 2010, 2017)

1. Explain the goals of ACRS and MACRS and their limitations.
2. Calculate depreciation using the MACRS guidelines.

Your Guide to Successfully Completing This Chapter

Traditional book or ebook

Check box as you complete each step.

Steps

- ☐ Read learning unit.
  - ☐ Complete practice quiz at the end of the learning unit.
- ☐ Grade practice quiz using provided solutions. (For more help, watch the learning unit video in Connect and have a Study Session with the authors. Then complete the additional practice quiz in Connect.)
- ☐ Repeat above for each of the four learning units in Chapter 17.
  - ☐ Review chapter organizer.
  - ☐ Complete assigned homework.
    - ☐ Finish summary practice test. (Go to Connect via the ebook link and do the interactive video worksheet to grade.)
- ☐ Complete instructor's exam.

The new tax act has a huge impact on how a business takes depreciation charges. The new act allows a business to write off the full purchase price. In past years depreciation was taken over a period of years. The act keeps this new depreciation process in place until the year 2027. Check with the IRS for the latest details and information about which companies do not receive these new benefits.

The *Wall Street Journal* chapter opener clip, "Toyota and Mazda To Build U.S. Plant," reports that Toyota and Mazda are going to build a new assembly plant that they will be able to depreciate over time. In Learning Units 17–1 to 17–3, we discuss methods of calculating depreciation for financial reporting. In Learning Unit 17–4, we look at how tax laws force companies to report depreciation for tax purposes. Financial reporting methods and the tax-reporting methods are both legal.

Used by permission of Cartoon Features Syndicate

Learning Unit 17–1: Concept of Depreciation and the Straight-Line Method

Companies frequently buy assets such as equipment or buildings that will last longer than 1 year. As time passes, these assets depreciate, or lose some of their market value. The total cost of these assets cannot be shown in *1 year* as an expense of running the business. In a systematic and logical way, companies must estimate the asset cost they show as an expense of a particular period. This process is called **depreciation.** The next time you fly in a plane think how the airline will depreciate the cost of that plane over a number of years.

Remember that depreciation *does not* measure the amount of deterioration or decline in the market value of the asset. Depreciation is simply a means of recognizing that these assets are depreciating.

The depreciation process results in **depreciation expense** that involves three key factors: (1) **asset cost**—the amount the company paid for the asset including freight and charges relating to the asset; (2) **estimated useful life**—the number of years or time periods for which the company can use the asset; and (3) **residual value (salvage** or **trade-in value)**—the expected cash value at the end of the asset's useful life.

Depreciation expense is listed on the income statement. The **accumulated depreciation** title on the balance sheet gives the amount of the asset's depreciation taken to date. Asset cost less accumulated depreciation is the asset's book value. The **book value** shows the unused amount of the asset cost that the company may depreciate in future accounting periods. At the end of the asset's life, the asset's book value is the same as its residual value—book value cannot be less than residual value.

LO 1

Book value is important to a business owner because it provides a way to calculate the value of assets within a business.

Depending on the amount and timetable of an asset's depreciation, a company can increase or decrease its profit. If a company shows greater depreciation in earlier years, the company will have a lower reported profit and pay less in taxes. Thus, depreciation can be an indirect tax savings for the company.

Later in the chapter we will discuss the different methods of computing depreciation that spread the cost of an asset over specified periods of time. However, first let's look at some of the major causes of depreciation.

Causes of Depreciation

As assets, all machines have an estimated amount of usefulness simply because as companies use the assets, the assets gradually wear out. The cause of this depreciation is *physical deterioration.*

The growth of a company can also cause depreciation. Many companies begin on a small scale. As the companies grow, they often find their equipment and buildings inadequate. The use of depreciation enables these businesses to "write off" their old, inadequate equipment and buildings. *In fact, the depreciation method chosen by a company can be determined by when tax write-offs are needed most.* Companies cannot depreciate land. For example, a garbage dump can be depreciated but not the land.

©Helen89/Shutterstock

Another cause of depreciation is the result of advances in technology. The computers that companies bought a few years ago may be in perfect working condition but outdated. Companies may find it necessary to replace these old computers with more sophisticated, faster, and possibly more economical machines. Thus, *product obsolescence* is a key factor contributing to depreciation.

Now we are ready to begin our study of depreciation methods. The first method we will study is straight-line depreciation. It is also the most common of the three depreciation methods (straight line, units of production, and declining balance).

TABLE 17.1

Depreciation schedule for straight-line method

$$\frac{100\%}{\text{Number of years}} = \frac{100\%}{5} = 20\%$$

Thus, the company is depreciating the equipment at a 20% rate each year.

End of year	Depreciation cost of Equipment	Depreciation expense for year	Accumulated depreciation at end of year	Book value at end of year (Cost − Depreciation at end of year)
1	$2,500	$400	$ 400	$2,100 ($2,500 − $400)
2	2,500	400	800	1,700
3	2,500	400	1,200	1,300
4	2,500	400	1,600	900
5	2,500	400	2,000	500
	↑ Cost stays the same.	↑ Depreciation expense is same each year.	↑ Accumulated depreciation increases by $400 each year.	↑ Book value is lowered by $400 until residual value of $500 is reached.

LO 2

Straight-Line Method

The **straight-line method** of depreciation is used more than any other method. It tries to distribute the same amount of expense to each period of time. Most large companies, such as Gillette Corporation, Campbell's Soup, and General Mills, use the straight-line method. *Today, more than 90% of U.S. companies depreciate by straight line.* For example, let's assume Ajax Company bought equipment for $2,500. The company estimates that the equipment's period of "usefulness"—or *useful life*—will be 5 years. After 5 years the equipment will have a residual value (salvage value) of $500. The company decides to calculate its depreciation with the straight-line method and uses the following formula:

$$\text{Depreciation expense each year} = \frac{\text{Cost} - \text{Residual value}}{\text{Estimated useful life in years}}$$

$$\frac{\$2{,}500 - \$500}{5 \text{ years}} = \$400 \text{ depreciation expense taken each year}$$

MONEY tips

Do not open several credit card accounts in a short period of time if you are trying to purchase or refinance a home. Doing so negatively affects your credit rating.

Table 17.1 gives a summary of the equipment depreciation that Ajax Company will take over the next 5 years. Companies call this summary a **depreciation schedule.** Buildings for BigLot are depreciated over 40 years while equipment is depreciated from 3 to 15 years.

My Money

You may be able to deduct allowable depreciation and other business expenses for a home office.

Depreciation for Partial Years

If a company buys an asset before the 15th of the month, the company calculates the asset's depreciation for a full month. Companies do not take the full month's depreciation for assets bought after the 15th of the month. For example, assume Ajax Company (Table 17.1) bought the equipment on May 6. The company would calculate the depreciation for the first year as follows:

$$\frac{\$2{,}500 - \$500}{5 \text{ years}} = \$400 \times \frac{8}{12} = \$266.67$$

Now let's check your progress with the Practice Quiz before we look at the next depreciation method.

LU 17–1 PRACTICE QUIZ

Complete this Practice Quiz to see how you are doing.

1. Prepare a depreciation schedule using straight-line depreciation for the following:

Cost of truck	$16,000
Residual value	$ 1,000
Life	5 years

2. If the truck were bought on February 3, what would the depreciation expense be in the first year?

For **extra help** from your authors–Sharon and Jeff–see the videos in Connect.

✓ **Solutions**

1.

End of year	Cost of truck	Depreciation expense for year	Accumulated depreciation at end of year	Book value at end of year (Cost – Accumulated depreciation)
1	$16,000	$3,000	$ 3,000	$13,000 ($16,000 – $3,000)
2	16,000	3,000	6,000	10,000
3	16,000	3,000	9,000	7,000
4	16,000	3,000	12,000	4,000
5	16,000	3,000	15,000	1,000 ← Note that we are down to residual value

2. $$\frac{\$16{,}000 - \$1{,}000}{5} = \$3{,}000 \times \frac{11}{12} = \$2{,}750$$

Learning Unit 17–2: Units-of-Production Method

LO 1

Unlike in the straight-line depreciation method, in the **units-of-production method** the passage of time is not used to determine an asset's depreciation amount. Instead, the company determines the asset's depreciation according to how much the company uses the asset. This use could be miles driven, tons hauled, or units that a machine produces. For example, when a company such as Ajax Company (in Learning Unit 17–1) buys equipment, the company estimates how many units the equipment can produce. Let's assume the equipment has a useful life of 4,000 units. The following formulas are used to calculate the equipment's depreciation for the units-of-production method.

MONEY tips

Unplug electronics when they are not in use. It will extend their lifespan and reduce your energy consumption.

$$\text{Depreciation per unit} = \frac{\text{Cost} - \text{Residual value}}{\text{Total estimated units produced}} = \frac{\$2{,}500 - \$500}{4{,}000 \text{ units}} = \$.50 \text{ per unit}$$

$$\text{Depreciation amount} = \text{Unit depreciation} \times \text{Units produced} = \$.50 \text{ times actual number of units}$$

LO 2

Now we can complete Table 17.2. Note that the table gives the units produced each year.

TABLE 17.2 Depreciation schedule for units-of-production method

End of year	Cost of equipment	Units produced	Depreciation expense for year	Accumulated depreciation at end of year	Book value at end of year (Cost – Accumulated depreciation)
1	$2,500	300	$ 150 (300 × $.50)	$ 150	$2,350 ($2,500 – $150)
2	2,500	400	200	350	2,150
3	2,500	600	300	650	1,850
4	2,500	2,000	1,000	1,650	850
5	2,500	700 ↑ At the end of 5 years, the equipment produced 4,000 units. If in year 5 the equipment produced 1,500 units, only 700 could be used in the calculation, or it will go below the equipment's residual value.	350 ↑ Units produced per year times $.50 equals depreciation expense.	2,000	500 ↑ Residual value of $500 is reached. (Be sure depreciation is not taken below the residual value.)

Let's check your understanding of this unit with the Practice Quiz.

LU 17–2 PRACTICE QUIZ

Complete this Practice Quiz to see how you are doing.

$$\frac{\$20{,}000 - \$4{,}000}{16{,}000} = \$1$$

*For **extra help** from your authors–Sharon and Jeff–see the videos in Connect.*

From the following facts prepare a depreciation schedule:

Machine cost	$20,000
Residual value	$ 4,000

Expected to produce 16,000 units over its expected life

	2016	**2017**	**2018**	**2019**	**2020**
Units produced:	2,000	8,000	3,000	1,800	1,600

✓ Solutions

End of year	Cost of machine	Units produced	Depreciation expense for year	Accumulated depreciation at end of year	Book value at end of year (Cost − Accumulated depreciation)
1	$20,000	2,000	$2,000 (2,000 × $1)	$ 2,000	$18,000
2	20,000	8,000	8,000	10,000	10,000
3	20,000	3,000	3,000	13,000	7,000
4	20,000	1,800	1,800	14,800	5,200
5	20,000	1,600	1,200*	16,000	4,000

*Note that we can depreciate only 1,200 units since we cannot go below the residual value of $4,000.

Learning Unit 17–3: Declining-Balance Method

In the declining-balance method, we cannot depreciate below the residual value.

LO 1, 2

The **declining-balance method** is another type of **accelerated depreciation** that takes larger amounts of depreciation expense in the earlier years of the asset. The straight-line method, you recall, estimates the life of the asset and distributes the same amount of depreciation expense to each period. To take larger amounts of depreciation expense in the asset's earlier years, the declining-balance method uses up to *twice* the **straight-line rate** in the first year of depreciation. A key point to remember is that the declining-balance method does not deduct the residual value in calculating the depreciation expense. Today, the declining-balance method is the basis of current tax depreciation.

For all problems, we will use double the straight-line rate unless we indicate otherwise. Today, the rate is often 1.5 or 1.25 times the straight-line rate. Again we use our $2,500 equipment with its estimated useful life of 5 years. As we build the depreciation schedule in Table 17.3 on page 454, note the following steps:

MONEY tips

Do you have enough money? Before purchasing or adopting a pet, estimate the annual cost of caring for it to determine whether you can afford it.

Step 1. Rate is equal to $\frac{100\%}{5 \text{ years}} \times 2 = 40\%$.

Or another way to look at it is that the straight-line rate is $\frac{1}{5} \times 2 = \frac{2}{5} = 40\%$.

Step 2.

$$\frac{\text{Depreciation expense}}{\text{each year}} = \frac{\text{Book value of equipment}}{\text{at beginning of year}} \times \frac{\text{Depreciation}}{\text{rate}}$$

Step 3. We cannot depreciate the equipment below its residual value ($500). The straight-line method automatically reduced the asset's book value to the residual value. This is not true with the declining-balance method. So you must be careful when you prepare the depreciation schedule.

TABLE 17.3 Depreciation schedule for declining-balance method

End of year	Cost of equipment	Accumulated depreciation at beginning of year	Book value at beginning of year (Cost − Accumulated depreciation)	Depreciation (Book value at beginning of year × Rate)	Accumulated depreciation at end of year	Book value at end of year (Cost − Accumulated depreciation)
1	$2,500	—	$2,500	$1,000 ($2,500 × .40)	$1,000	$1,500 ($2,500 − $1,000)
2	2,500	$1,000	1,500	600 ($1,500 × .40)	1,600	900
3	2,500	1,600	900	360 ($900 × .40)	1,960	540
4	2,500	1,960	540	40	2,000	500
5	2,500	2,000	500	—	2,000	500
	↑ Original cost of $2,500 does not change. Residual value was not subtracted.	↑ Ending accumulated depreciation of 1 year becomes next year's beginning.	↑ Cost less accumulated depreciation	↑ *Note:* In year 4, only $40 is taken since we cannot depreciate below residual value of $500. In year 5, no depreciation is taken.	↑ Accumulated depreciation balance plus depreciation expense this year.	↑ Book value now equals residual value.

Now let's check your progress again with another Practice Quiz.

LU 17–3 PRACTICE QUIZ

Complete this **Practice Quiz** to see how you are doing.

*For **extra help** from your authors–Sharon and Jeff–see the videos in Connect.*

Prepare a depreciation schedule from the following:

Cost of machine: $16,000 Estimated life: 5 years
Rate: 40% (this is twice the straight-line rate) Residual value: $1,000

✓ Solutions

End of year	Cost of machine	Accumulated depreciation at beginning of year	Book value at beginning of year (Cost − Accumulated depreciation)	Depreciation (Book value at beginning of year × Rate)	Accumulated depreciation at end of year	Book value at end of year (Cost − Accumulated depreciation)
1	$16,000	$ -0-	$16,000.00	$6,400.00	$ 6,400.00	$9,600.00
2	16,000	6,400.00	9,600.00	3,840.00	10,240.00	5,760.00
3	16,000	10,240.00	5,760.00	2,304.00	12,544.00	3,456.00
4	16,000	12,544.00	3,456.00	1,382.40	13,926.40	2,073.60
5	16,000	13,926.40	2,073.60	829.44*	14,755.84	1,244.16

*Since we do not reach the residual value of $1,000, another $244.16 could have been taken as depreciation expense to bring it to the estimated residual value of $1,000.

Learning Unit 17–4: Modified Accelerated Cost Recovery System (MACRS) with Introduction to ACRS (1986, 1989, 2010, 2017)

In Learning Units 17–1 to 17–3, we discussed the depreciation methods used for financial reporting. Since 1981, federal tax laws have been passed that state how depreciation must be taken for income tax purposes. Assets put in service from 1981 through 1986 fell under the federal **Accelerated Cost Recovery System (ACRS)** tax law enacted in 1981. The Tax Reform Act of 1986 established the **Modified Accelerated Cost Recovery System (MACRS)** for all property placed into service after December 31, 1986. This system, used by businesses to calculate depreciation for tax purposes based on the tax laws of 1986, 1989, and 2010, is also known as the **General Depreciation System (GDS).** For the latest updates, check with the IRS regarding the tax overhaul act of 2017. Airplanes for commercial use are usually depreciated by MACRS for 7 years or ADS (Alternative Depreciation System) for 12 years.

LO 1

Depreciation for Tax Purposes Based on the Tax Reform Act of 1986 (MACRS)

Tables 17.4 and 17.5 give the classes of recovery and annual depreciation percentages that MACRS established in 1986. The key points of MACRS are

1. It calculates depreciation for tax purposes.
2. It ignores residual value.
3. Depreciation in the first year (for personal property) is based on the assumption that the asset was purchased halfway through the year. (A new law adds a midquarter convention for all personal property if more than 40% is placed in service during the last 3 months of the taxable year.)
4. Classes 3, 5, 7, and 10 use a 200% declining-balance method for a period of years before switching to straight-line depreciation. You do not have to determine the year in which to switch since Table 17.5 builds this into the calculation.
5. Classes 15 and 20 use a 150% declining-balance method before switching to straight-line depreciation.
6. Classes 27.5 and 31.5 use straight-line depreciation.

©Cheryl Ann Quigley/Shutterstock

Racehorses, when put in training, can be depreciated by MACRS.

TABLE 17.4

Modified Accelerated Cost Recovery System (MACRS) for assets placed in service after December 31, 1986

Class recovery period (life)	Asset types
3-year*	Racehorses more than 2 years old or any horse other than a racehorse that is more than 12 years old at the time placed into service; special tools of certain industries.
5-year*	Automobiles (not luxury); taxis; light general-purpose trucks; semiconductor manufacturing equipment; computer-based telephone central-office switching equipment; qualified technological equipment; property used in connection with research and experimentation.
7-year*	Railroad track; single-purpose agricultural (pigpens) or horticultural structures; fixtures; equipment; furniture.
10-year*	New law doesn't add any specific property under this class.
15-year†	Municipal wastewater treatment plants; telephone distribution plants and comparable equipment used for two-way exchange of voice and data communications.
20-year†	Municipal sewers.
27.5-year‡	Only residential rental property.
31.5-year‡	Only nonresidential real property.

*These classes use a 200% declining-balance method before switching to the straight-line method.

†These classes use a 150% declining-balance method before switching to the straight-line method.

‡These classes use a straight-line method.

TABLE 17.5 Annual recovery for MACRS

Recovery year	3-year class (200% D.B.)	5-year class (200% D.B.)	7-year class (200% D.B.)	10-year class (200% D.B.)	15-year class (150% D.B.)	20-year class (150% D.B.)
1	33.00	20.00	14.28	10.00	5.00	3.75
2	45.00	32.00	24.49	18.00	9.50	7.22
3	15.00*	19.20	17.49	14.40	8.55	6.68
4	7.00	11.52*	12.49	11.52	7.69	6.18
5		11.52	8.93*	9.22	6.93	5.71
6		5.76	8.93	7.37	6.23	5.28
7			8.93	6.55*	5.90*	4.89
8			4.46	6.55	5.90	4.52
9				6.55	5.90	4.46*
10				6.55	5.90	4.46
11				3.29	5.90	4.46
12					5.90	4.46
13					5.90	4.46
14					5.90	4.46
15					5.90	4.46
16					3.00	4.46

*Identifies when switch is made to straight line.

LO 2

EXAMPLE Using the same equipment cost of $2,500 for Ajax, prepare a depreciation schedule under MACRS assuming the equipment is a 5-year class and not part of the tax bill of 1989. Use Table 17.5. Note that percent figures from Table 17.5 have been converted to decimals.

End of year	Cost	Depreciation expense	Accumulated depreciation	Book value at end of year
1	$2,500	$500 (.20 × $2,500)	$ 500	$2,000
2	2,500	800 (.32 × $2,500)	1,300	1,200
3	2,500	480 (.1920 × $2,500)	1,780	720
4	2,500	288 (.1152 × $2,500)	2,068	432
5	2,500	288 (.1152 × $2,500)	2,356	144
6	2,500	144 (.0576 × $2,500)	2,500	–0–

MONEY tips

Consider refinancing your home to obtain a lower fixed interest rate. Determine whether your savings offsets the refinance costs. A rate reduction of 1.5% is generally worth the cost in the long run.

Check your understanding of this learning unit with the below Practice Quiz.

LU 17–4 PRACTICE QUIZ

Complete this Practice Quiz to see how you are doing.

*For **extra help** from your authors–Sharon and Jeff–see the videos in Connect.*

1. In 2019, Rancho Corporation bought semiconductor equipment for $80,000. Using MACRS, what is the depreciation expense in year 3?
2. What would depreciation be the first year for a wastewater treatment plant that cost $800,000?

✓ Solutions

1. $80,000 × .1920 = **$15,360**
2. $800,000 × .05 = **$40,000**

INTERACTIVE CHAPTER ORGANIZER

Topic/Procedure/Formula	Examples	You try it*
Straight-line method $\text{Depreciation expense each year} = \dfrac{\text{Cost} - \text{Residual value}}{\text{Estimated useful life in years}}$ For partial years if purchased before 15th of month depreciation is taken.	Truck, $25,000; $5,000 residual value, 4-year life. $\text{Depreciation expense} = \dfrac{\$25{,}000 - \$5{,}000}{4}$ $= \$5{,}000$ per year	**Calculate depreciation expense** Truck, $50,000; $10,000 residual value; 4-year life.
Units-of-production method $\text{Depreciation per unit} = \dfrac{\text{Cost} - \text{Residual value}}{\text{Total estimated units produced}}$ Do not depreciate below residual value even if actual units are greater than estimate.	Machine, $5,000; estimated life in units, 900; residual value, $500. Assume first year produced 175 units. $\text{Depreciation expense} = \dfrac{\$5{,}000 - \$500}{900}$ $= \dfrac{\$4{,}500}{900}$ = $5 depreciation per unit 175 units × $5 = $875 depreciation expense	**Calculate depreciation expense** Machine, $4,000; estimated life in units, 700; residual value, $500. Assume first year produced 150 units.
Declining-balance method An accelerated method. Residual value not subtracted from cost in depreciation schedule. Do not depreciate below residual value. $\text{Depreciation expense each year} = \text{Book value of equipment at beginning of year} \times \text{Depreciation rate}$	Truck, $50,000; estimated life, 5 years; residual value, $10,000. $\frac{1}{5} = 20\% \times 2 = 40\%$ (assume double the straight-line rate) **Year** / **Cost** / **Depreciation expense** / **Book value at end of year** 1 / $50,000 / $20,000 ($50,000 × .40) / $30,000 ($50,000 − $20,000) 2 / $50,000 / $12,000 ($30,000 × .40) / $18,000 ($50,000 − $32,000)	**Calculate depreciation expense and book value for 2 years** Truck, $40,000; estimated life, 4 years; residual value, $5,000.
MACRS/Tax Bill of 1989, 2010, 2017 After December 31, 1986, depreciation calculation is modified. Tax Act of 1989, 2010, modifies way to depreciate equipment. The new tax act of 2017 speeds up depreciation.	Auto: $8,000, 5 years. First year, .20 × $8,000 = $1,600 depreciation expense	Auto: $7,000, 5 years. Second year = ? depreciation expense

KEY TERMS			
Accelerated Cost Recovery System (ACRS)	Depreciation expense	Residual value	
Accelerated depreciation	Depreciation schedule	Salvage value	
Accumulated depreciation	Estimated useful life	Straight-line method	
Asset cost	General Depreciation System (GDS)	Straight-line rate	
Book value	Modified Accelerated Cost Recovery System (MACRS)	Trade-in value	
Declining-balance method		Units-of-production method	
Depreciation			

*Worked-out solutions are in Appendix B.

Critical Thinking Discussion Questions with Chapter Concept Check

1. What is the difference between depreciation expense and accumulated depreciation? Why does the book value of an asset never go below the residual value?
2. Compare the straight-line method to the units-of-production method. Should both methods be based on the passage of time?
3. Why is it possible in the declining-balance method for a person to depreciate below the residual value by mistake?
4. Explain the Modified Accelerated Cost Recovery System. Do you think this system will be eliminated in the future?
5. **Chapter Concept Check.** Search the web for a car of your choice and use concepts from this chapter to provide a depreciation schedule for the car.

Classroom Notes

17

END-OF-CHAPTER PROBLEMS

Connect

Check figures for odd-numbered problems in Appendix B. Name ____________________ Date ________

DRILL PROBLEMS

From the following facts, complete a depreciation schedule by using the straight-line method: *LU 17-1(2)*

Given		
	Cost of Honda Accord Hybrid	$40,000
	Residual value	$10,000
	Estimated life	6 years

	End of year	**Cost of Accord**	**Depreciation expense for year**	**Accumulated depreciation at end of year**	**Book value at end of year**
17–1.					
17–2.					
17–3.					
17–4.					
17–5.					
17–6.					

From the following facts, prepare a depreciation schedule using the declining-balance method (twice the straight-line rate): *LU 17-3(2)*

Given		
	Chevrolet Colorado	$25,000
	Residual value	$ 5,000
	Estimated life	5 years

	End of year	**Cost of Chevy truck**	**Accumulated depreciation at beginning of year**	**Book value at beginning of year**	**Depreciation expense for year**	**Accumulated depreciation at end of year**	**Book value at end of year**
17–7.							
17–8.							
17–9.							
17–10.							

For the first 2 years, calculate the depreciation expense for a $7,000 car under MACRS. This is a nonluxury car. *LU 17-4(2)*

		MACRS			**MACRS**
17–11.	Year 1		**17–12.**	Year 2	

Complete the following table given this information:

Cost of machine	$94,000	Estimated units machine will produce	100,000	
Residual value	$ 4,000	Actual production:	**Year 1**	**Year 2**
Useful life	5 years		60,000	15,000

	Method	**Depreciation Expense**	
		Year 1	**Year 2**
17–13.	Straight line *LU 17-1(2)*		
17–14.	Units of production *LU 17-2(2)*		
17–15.	Declining balance *LU 17-3(2)*		
17–16.	MACRS (5-year class) *LU 17-4(2)*		

WORD PROBLEMS

17–17. Shearer's Foods, part of the $374 billion global snack food industry, employs 3,300 people in Brewster, Ohio. If Shearer's purchased a packaging unit for $185,000 with a life expectancy of 695,000 units and a residual value of $46,000, what is the depreciation expense for year 1 if 75,000 units were produced? *LU 17-2(2)*

eXcel

17–18. Lena Horn bought a Toyota Tundra on January 1 for $30,000 with an estimated life of 5 years. The residual value of the truck is $5,000. Assume a straight-line method of depreciation. **(a)** What will be the book value of the truck at the end of year 4? **(b)** If the Tundra was bought the first year on April 12, how much depreciation would be taken the first year? *LU 17-1(2)*

eXcel

17–19. Jim Company bought a machine for $36,000 with an estimated life of 5 years. The residual value of the machine is $6,000. Calculate **(a)** the annual depreciation and **(b)** the book value at the end of year 3. Assume straight-line depreciation. *LU 17-1(2)*

eXcel

17–20. Using Problem 17–19, calculate the first 2 years' depreciation, assuming the units-of-production method. This machine is expected to produce 120,000 units. In year 1, it produced 19,000 units, and in year 2, 38,000 units. *LU 17-2(2)*

My Money

17–21. CNBC reported that one in five consumers who purchase bitcoin do so using their credit card. Melissa Gamez purchased a used RV with 19,000 miles for $46,900. Originally the RV sold for $70,000 with a residual value of $20,000. After subtracting the residual value, depreciation allowance per mile was $.86. How much was Melissa's purchase price over or below the book value? Does she have any equity that might assist her with purchasing bitcoin? *LU 17-2(1)*

17–22. Volkswagen car sales hit record high sales of $6.23 million in 2017. If one of Volkswagen's assembly lines purchased a new quality control computer verifying VIN numbers for $7,985 with a 5-year life and residual value of $1,100, what is the depreciation expense in year 2 for the QC computer? Use the straight-line method. *LU 17-1(2)*

My Money

17–23. CNN reported in early 2018 that higher-income shoppers are the biggest bargain shoppers searching Amazon, clicking on their mobile devices and even querying Alexa looking for the best deals. If Rebecca Johnson purchased a state-of-the-art Biologique Recherche facial machine for her home business for $108,000, with a useful life of 3 years and a residual value of $35,000, what would be the book value of the machine after the first year using the straight-line depreciation method? Round your answers to the nearest dollar. *LU 17-1(2)*

17–24. If corporate headquarters for UPS in Atlanta is considering adding to its 96,000+ fleet of delivery vans, what is year 5's depreciation expense using MACRS if one van costs $78,500? *LU 17-4(2)*

CHALLENGE PROBLEMS

17–25. A delivered price (including attachments) of a crawler dozer tractor is $135,000 with a residual value of 35%. The useful life of the tractor is 7,700 hours. *LU 17-2(2)*

a. What is the total amount of depreciation allowed?

b. What is the amount of depreciation per hour?

c. If the tractor is operated five days a week for an average of $7\frac{1}{4}$ hours a day, what would be the depreciation for the first year?

d. If the hours of operation were the same each year, what would be the total number of years of useful life for the tractor? Round years to the nearest whole number.

17–26. Assume a piece of equipment was purchased July 26, 2019, at a cost of $72,000. The estimated residual value is $5,400 with a useful life of 5 years. Assume a production life of 60,000 units. Compute the depreciation for years 2019 and 2020 using **(a)** straight-line and **(b)** units-of-production (in 2019, 5,000 units were produced and in 2020, 18,000 units were produced). *LU 17-1(2), LU 17-2(2)*

SUMMARY PRACTICE TEST

Do you need help? Connect videos have step-by-step worked-out solutions.

1. Leo Lucky, owner of a Pizza Hut franchise, bought a delivery truck for $30,000. The truck has an estimated life of 5 years with a residual value of $10,000. Leo wants to know which depreciation method will be the best for his truck. He asks you to prepare a depreciation schedule using the declining-balance method at twice the straight-line rate. *LU 17-3(2)*

2. Using MACRS, what is the depreciation for the first year on furniture costing $12,000? *LU 17-4(2)*

eXcel 3. Abby Matthew bought a new Jeep Commander for $30,000. The Jeep Commander has a life expectancy of 5 years with a residual value of $10,000. Prepare a depreciation schedule for the straight-line method. *LU 17-1(2)*

eXcel 4. Car.com bought a Toyota for $28,000. The Toyota has a life expectancy of 10 years with a residual value of $3,000. After 3 years, the Toyota was sold for $19,000. What was the difference between the book value and the amount received from selling the car if Car.com used the straight-line method of depreciation? *LU 17-1(2)*

5. A machine cost $70,200; it had an estimated residual value of $6,000 and an expected life of 300,000 units. What would be the depreciation in year 3 if 60,000 units were produced? (Round to nearest cent.) *LU 17-2(2)*

INTERACTIVE VIDEO WORKSHEET

GRADING THE SUMMARY PRACTICE TEST

Go to the summary practice test video in Connect (or click on it here in the ebook). Grade your summary practice test while viewing the video.

C for Correct/I for Incorrect

1. ________
2. ________
3. ________
4. ________
5. ________

If you achieved 100%, you are ready for your instructor's exam.

If any of the problems were incorrect, list the questions you missed and show steps to solve the problem correctly.

Replay the video to see if you have made the correct fixes to your mistakes. If you have any questions, contact your instructor asap.

Notes on Watching Videos

MY MONEY

Work@Home

What I need to know

As you think about your life after college you may be considering opening a small business or possibly running a home-based business. Consider the pros and cons of being an entrepreneur. The flexibility and freedom that comes with doing what you love often outweighs the challenges of administrative details, no regular salary, and learning how to be competitive. Having a home office provides one of the best known tax advantages for a home-based business: a percentage of your household expenses (mortgage, utilities, property taxes, insurance, Internet, vehicle usage, etc.) may be tax deductible. Consult with a tax accountant for details.

What I need to do

Give thought to your financial goals and whether or not opening a small/home business is in line with these goals. Part of this consideration should take into account the rationale behind a decision to start a business. Are you seeking to satisfy a particular niche in the market? Are you interested in supplementing your income with this business venture on the side? Is there a clear demand for the type of product or service you are offering as part of your business? These can be complex questions, but your answers can help you decide if there is an opportunity for success. Consulting with your local Small Business Development Center (SBDC) and/or with SCORE (Service Corps of Retired Executives) can assist you with this important assessment process.

Determining the amount of time and resources required for running your own business is also necessary. In the beginning there will be a significant amount of time needed to get your business idea off the ground. You need to determine if you can work another job in addition to the work on your small/home business or if it makes sense to devote 100% of your time toward this business venture. You should also consider the financial investment needed to launch a business and the sources from which you could generate these funds. For instance, you may need to take on partners to help finance the venture or possibly take out loans to cover your expenses. The financing option you choose will have a significant impact on achieving the business goals you have set for yourself.

Resources I can use

- https://www.thebalance.com/taxes-and-deductions-for-home-based-business-398592 — tax considerations for home-based businesses
- https://www.thebalance.com/the-pros-and-cons-of-working-from-home-2951766 — pros and cons of a home-based business
- https://www.sba.gov/tools/local-assistance/sbdc — small business assistance
- https://www.score.org/ — experts helping small businesses

MY MONEY ACTIVITY

Assume you want to open a small/home business. What would it be? Identify reasons behind your decision to start your own business. Make a list of the resources you will need and begin analyzing the market, your capabilities, and financial resources to help determine if you should pursue the business idea.

PERSONAL FINANCE

A KIPLINGER APPROACH

"Is It Taxable? Take Our Quiz", *Kiplinger's*, January 2017, p. 72. Used by permission of The Kiplinger Washington Editors, Inc.

Form 1040 | TAKEAWAY | IRS Use Only–Do not write or staple in this space.

Is It Taxable? Take Our Quiz

You may be surprised at the sources of income and good fortune that are taxable. Which of the following do you think are **true**, and which are **false**?

©Quarta/Shutterstock

1. You received a $100 cash rebate after you purchased your new laptop. It isn't taxable.
☐ T ☐ F

2. You found buried treasure in your backyard. You don't have to tell the IRS.
☐ T ☐ F

3. You received a college scholarship that covers tuition, room and board. All of it is tax-free.
☐ T ☐ F

4. Your ex pays you alimony each month. You owe taxes on that income.
☐ T ☐ F

5. Your boss gave you a gift certificate for the holidays. It's not taxable.
☐ T ☐ F

6. You were out of work for a few months last year. Your unemployment benefits aren't taxable.
☐ T ☐ F

7. A settlement to cover your medical costs after a car accident is tax-free.
☐ T ☐ F

8. You sold the home you've lived in for the past eight years for $100,000 more than you paid for it. All of the proceeds are tax-free.
☐ T ☐ F

9. You hit it big in Vegas. Now you must pay taxes on your jackpot.
☐ T ☐ F

10. You used bitcoin that you bought several years ago to buy a first-class airline ticket. You don't have to worry about taxes.
☐ T ☐ F

Answers:
1. True. The IRS considers a cash rebate a discount.
2. False. Property you find that was lost or abandoned is taxed at its fair market value.
3. False. But only the portion of the award that covers room and board is taxable.
4. True. But under the new tax law, alimony paid through divorce agreements reached after December 31, 2018, will be tax-free. Child-support payments continue to be tax-free.
5. False. Gifts of cash or gift certificates from your employer—but not from relatives or friends—are taxable income.
6. False.
7. True. However, if you received punitive damages, a portion of that money is taxable.
8. True. If you lived in the home for at least two of the past five years, up to $250,000 of your profits ($500,000 for a married couple) is tax-free.
9. True. Gambling winnings are taxable.
10. False. If the bitcoin you use to buy goods or services is worth more than you paid for it, you're expected to pay taxes on your profits at capital gains rates—just as you do with stocks and bonds. You may also be able to take a deduction if it's worth less than you paid for it.

Form 1040

BUSINESS MATH ISSUE

The new tax law means all equipment will never be depreciated.

1. List the key points of the article and information to support your position.
2. Write a group defense of your position using math calculations to support your view. If you are in an online course, post to a discussion board.

Classroom Notes

CHAPTER 18

Inventory and Overhead

Retailers Embrace Barer Shelves

Home Depot, others trim inventories amid a rethink prompted by online shopping

By Paul Ziobro

CONROE, Texas—**Home Depot** Inc. is bringing a new philosophy to its cavernous stores: Less is more.

HOW WE SHOP

Instead of filling its warehouse-style racks to the ceiling with Makita drills, rolls of **Owens Corning** insulation and cans of Rust-Oleum paint, Home Depot wants fewer items on its shelves and it wants them to be within customers' reach.

"Get comfortable with days of inventory, not weeks," Tom Shortt, Home Depot's senior vice president of supply chain, says is the message going out to stores. The retailer is targeting sales growth of nearly 15% by 2018, but wants to keep inventory levels flat or slightly down.

It is a shift happening across the retail sector as companies try to figure out ways to profitably serve the growing needs of online shoppers while making their network of stores less of a financial burden. Chains must predict whether demand will come from the internet or a store visit, and whether they'll ship online orders from a distribution center or a store. Every move of inventory is an added cost that eats away at already thin margins.

Online shopping "has forced the industry to rethink not only the math and science behind the inventory pool, but also the strategy," said Scott Fenwick, a senior director at Manhattan Associates Inc., which makes supply-chain software.

Inventory Management

By boosting sales and stocking less items, Home Depot and other retailers have increased the percentage of cash they get back from the amount they invest in inventory.

*Fiscal year ends Aug. 31 †Excludes credit card earnings in 2010
Source: Customer Growth Partners

THE WALL STREET JOURNAL.

LEARNING UNIT OBJECTIVES

LU 18–1: Assigning Costs to Ending Inventory—Specific Identification; Weighted Average; FIFO; LIFO

1. List the key assumptions of each inventory method.
2. Calculate the cost of ending inventory and cost of goods sold for each inventory method.

LU 18–2: Retail Method; Gross Profit Method; Inventory Turnover; Distribution of Overhead

1. Calculate the cost ratio and ending inventory at cost for the retail method.
2. Calculate the estimated inventory using the gross profit method.
3. Explain and calculate inventory turnover.
4. Explain overhead; allocate overhead according to floor space and sales.

Your Guide to Successfully Completing This Chapter

Traditional book or ebook

Check box as you complete each step.

Steps

☐ Read learning unit.

  ☐ Complete practice quiz at the end of the learning unit.

☐ Grade practice quiz using provided solutions. (For more help, watch the learning unit video in Connect and have a Study Session with the authors. Then complete the additional practice quiz in Connect.)

☐ Repeat above for each of the two learning units in Chapter 18.

  ☐ Review chapter organizer.

  ☐ Complete assigned homework.

    ☐ Finish summary practice test. (Go to Connect via the ebook link and do the interactive video worksheet to grade.)

☐ Complete instructor's exam.

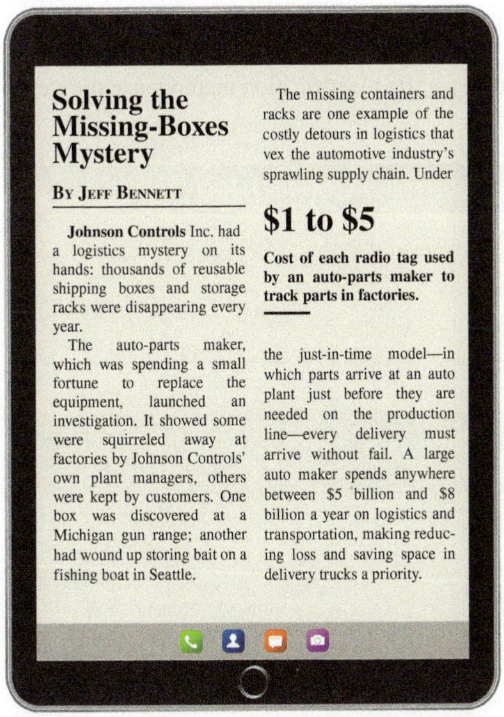

Solving the Missing-Boxes Mystery

By Jeff Bennett

Johnson Controls Inc. had a logistics mystery on its hands: thousands of reusable shipping boxes and storage racks were disappearing every year.

The auto-parts maker, which was spending a small fortune to replace the equipment, launched an investigation. It showed some were squirreled away at factories by Johnson Controls' own plant managers, others were kept by customers. One box was discovered at a Michigan gun range; another had wound up storing bait on a fishing boat in Seattle.

The missing containers and racks are one example of the costly detours in logistics that vex the automotive industry's sprawling supply chain. Under the just-in-time model—in which parts arrive at an auto plant just before they are needed on the production line—every delivery must arrive without fail. A large auto maker spends anywhere between $5 billion and $8 billion a year on logistics and transportation, making reducing loss and saving space in delivery trucks a priority.

$1 to $5

Cost of each radio tag used by an auto-parts maker to track parts in factories.

The new tax act does not limit car dealers from writing off interest on loans that are used to get inventory into the dealership. Other companies get this benefit based on income. Check with the IRS for the latest updates.

The *Wall Street Journal* chapter opener clip, "Retailers Embrace Barer Shelves," shows how difficult it is for Home Depot to have enough inventory while keeping costs down. Note in the *Wall Street Journal* clip to the left how radio tags are solving inventory concerns. The two methods that a company can use to monitor its inventory are the *perpetual* method and the *periodic* method.

The perpetual inventory system should be familiar to most consumers. Today, it is common for cashiers to run scanners across the product code of each item sold. These scanners read pertinent information into a computer terminal, such as the item's number, department, and price. The computer then uses the **perpetual inventory system** as it subtracts outgoing merchandise from inventory and adds incoming merchandise to inventory. However, as you probably know, the computer cannot be completely relied on to maintain an accurate count of merchandise in stock. Since some products may be stolen or lost, periodically a physical count is necessary to verify the computer count.

With the increased use of computers, many companies are changing to a perpetual inventory system of maintaining inventory records. Some small stores, however, still use the **periodic inventory system.** This system usually does not keep a running account of a store's inventory but relies only on a physical inventory count taken at least once a year. The store then uses various accounting methods to value the cost of its merchandise. In this chapter we discuss the periodic method of inventory.

©Peter Yates/Getty Images

You may wonder why a company should know the status of its inventory. In Chapter 16 we introduced you to the balance sheet and the income statement. Companies cannot accurately prepare these statements unless they have placed the correct value on their inventory. To do this, a company must know (1) the cost of its ending inventory (found on the balance sheet) and (2) the cost of the goods (merchandise) sold (found on the income statement).

No longer do retailers get a few seasonal deliveries; they now receive new items often, maybe even weekly, to keep their store looking fresh. Frequently, the same type of merchandise flows into a company at different costs. The value assumptions a company makes about the merchandise it sells affect the cost assigned to its ending inventory. Remember that different costs result in different levels of profit on a firm's financial reports.

This chapter begins by using the Blue Company to discuss four common methods (specific identification, weighted average, FIFO, and LIFO) that companies use to calculate the cost of ending inventory and the cost of goods sold. In these methods, the flow of costs does not always match the flow of goods. The chapter continues with a discussion of two methods of estimating ending inventory (retail and gross profit methods), inventory turnover, and the distribution of overhead.

Aha!

A company must declare on its financial statements the inventory method used.

Learning Unit 18–1: Assigning Costs to Ending Inventory—Specific Identification; Weighted Average; FIFO; LIFO

LO 1, 2

Blue Company is a small artist supply store. Its beginning inventory is 40 tubes of art paint that cost $320 (at $8 a tube) to bring into the store. As shown in Figure 18.1, Blue made additional purchases in April, May, October, and December. Note that because of inflation and other competitive factors, the cost of the paint rose from $8 to $13 per tube. At the end of December, Blue had 48 unsold paint tubes. During the year, Blue had 120 paint tubes to sell. Blue wants to calculate (1) the cost of ending inventory (not sold) and (2) the cost of goods sold.

Specific Identification Method

Companies use the **specific identification method** when they can identify the original purchase cost of an item with the item. For example, Blue Company color codes its paint tubes as they come into the store. Blue can then attach a specific invoice price to each paint tube. This makes the flow of goods and flow of costs the same. Then, when Blue computes its ending inventory and cost of goods sold, it can associate the actual invoice cost with each item sold and in inventory.

Companies that sell high-cost items such as autos, jewelry, antiques, and so on, usually use the specific identification method.

To help Blue calculate its inventory with the specific identification method, use the steps that follow.

MONEY tips

Keep an inventory of your personal property and the cost of each item. This will help you calculate your net worth as well as assist you with itemizing in the event of a loss.

CALCULATING THE SPECIFIC IDENTIFICATION METHOD

Step 1. Calculate the cost of goods (merchandise available for sale).

Step 2. Calculate the cost of the ending inventory.

Step 3. Calculate the cost of goods sold (Step 1 – Step 2).

FIGURE 18.1

Blue Company—a case study

	Number of units purchased	Cost per unit	Total cost	
Beginning inventory	40	$ 8	$ 320	
First purchase (April 1)	20	9	180	
Second purchase (May 1)	20	10	200	
Third purchase (October 1)	20	12	240	
Fourth purchase (December 1)	20	13	260	
Goods (merchandise) available for sale	120		$1,200	← Step 1
Units sold	72			
Units in ending inventory	48			

First, Blue must actually count the tubes of paint on hand. Since Blue coded these paint tubes, it can identify the tubes with their purchase cost and multiply them by this cost to arrive at a total cost of ending inventory. Let's do this now.

	Cost per unit	Total cost	
20 units from April 1	$ 9	$180	
20 units from October 1	12	240	
8 units from December 1	13	104	
Cost of ending inventory		$524	← **Step 2**

Blue uses the following cost of goods sold formula to determine its cost of goods sold:

Cost of goods available for sale	−	Cost of ending inventory	=	Cost of goods sold	← **Step 3**
↑		↑		↑	
$1,200 (Figure 18.1)	−	$524	=	$676	

Note that the $1,200 for cost of goods available for sale comes from Figure 18.1. *Remember we are focusing our attention on Blue's purchase costs. Blue's actual selling price does not concern us here.*

Now let's look at how Blue would use the weighted-average method.

Weighted-Average Method[1]

The **weighted-average method** prices the ending inventory by using an average unit cost. Let's replay Blue Company and use the weighted-average method to find the average unit cost of its ending inventory and its cost of goods sold. Blue would use the steps that follow.

CALCULATING THE WEIGHTED-AVERAGE METHOD

Step 1. Calculate the average unit cost.

Step 2. Calculate the cost of the ending inventory.

Step 3. Calculate the cost of goods sold.

In the table that follows, Blue makes the calculation using the above steps.

	Number of units purchased	Cost per unit	Total cost
Beginning inventory	40	$ 8	$ 320
First purchase (April 1)	20	9	180
Second purchase (May 1)	20	10	200
Third purchase (October 1)	20	12	240
Fourth purchase (December 1)	20	13	260
Goods (merchandise) available for sale	120		$1,200
Units sold	72		
Units in ending inventory	48		

$$\text{Weighted-average unit cost} = \frac{\text{Total cost of goods available for sale}}{\text{Total number of units available for sale}} = \frac{\$1{,}200}{120 \text{ units}} = \$10 \text{ average unit cost}$$ ← **Step 1**

Average cost of ending inventory: 48 units at $10 = $480 ← **Step 2**

Cost of goods available for sale	−	Cost of ending inventory	=	Cost of goods sold	
↑		↑		↑	
$1,200	−	$480	=	$720	← **Step 3**

[1]Virtually all countries permit the use of the weighted-average method.

Remember that some of the costs we used to determine the average unit cost were higher and others were lower. The weighted-average method, then, calculates an *average unit price* for goods. Companies with similar units of goods, such as rolls of wallpaper, often use the weighted-average method. Also, companies with homogeneous products such as fuels and grains may use the weighted-average method.

Now let's see how Blue Company would value its inventory with the FIFO method.

FIFO—First-In, First-Out Method

The **first-in, first-out (FIFO)** inventory valuation method assumes that the first goods (paint tubes for Blue) brought into the store are the first goods sold. Thus, FIFO assumes that each sale is from the oldest goods in inventory. FIFO also assumes that the inventory remaining in the store at the end of the period is the most recently acquired goods. This cost flow assumption may or may not hold in the actual physical flow of the goods. An example of a corporation using the FIFO method is Gillette Corporation.

Use the following steps to calculate inventory with the FIFO method.

CALCULATING THE FIFO INVENTORY

Step 1. List the units to be included in the ending inventory and their costs.

Step 2. Calculate the cost of the ending inventory.

Step 3. Calculate the cost of goods sold.

In the table that follows, we show how to calculate FIFO for Blue using the above steps.

FIFO (bottom up)	Number of units purchased	Cost per unit	Total cost
Beginning inventory	40	$ 8	$ 320
First purchase (April 1)	20	9	180
Second purchase (May 1)	20	10	200
Third purchase (October 1)	20	12	240
Fourth purchase (December 1)	20	13	260
Goods (merchandise) available for sale	120		$1,200
Units sold	72		
Units in ending inventory	48		

20 units from December 1 purchased at $13		$260
20 units from October 1 purchased at $12	← Step 1 →	240
8 units from May 1 purchased at $10		80
48 units result in an ending inventory cost of		$580 ← Step 2

Cost of goods available for sale − Cost of ending inventory = Cost of goods sold

$1,200 − $580 = $620 ← Step 3

In FIFO, the cost flow of goods tends to follow the physical flow. For example, a fish market could use FIFO because it wants to sell its old inventory first. Note that during inflation, FIFO produces a higher income than other methods. So companies using FIFO during this time must pay more taxes.

We conclude this unit by using the LIFO method to value Blue Company's inventory.

LIFO—Last-In, First-Out Method

If Blue Company chooses the **last-in, first-out (LIFO)** method of inventory valuation, then the goods sold by Blue will be the last goods brought into the store. The ending inventory would consist of the old goods that Blue bought earlier.

You can calculate inventory with the LIFO method by using the steps that follow.

CALCULATING THE LIFO INVENTORY

Step 1. List the units to be included in the ending inventory and their costs.

Step 2. Calculate the cost of the ending inventory.

Step 3. Calculate the cost of goods sold.

Now we use the above steps to calculate LIFO for Blue.

LIFO (top down)	Number of units purchased	Cost per unit	Total cost
Beginning inventory	40	$ 8	$ 320
First purchase (April 1)	20	9	180
Second purchase (May 1)	20	10	200
Third purchase (October 1)	20	12	240
Fourth purchase (December 1)	20	13	260
Goods (merchandise) available for sale	120		$1,200
Units sold	72		
Units in ending inventory	48		

40 units of beginning inventory at $8		$320	
8 units from April at $9	← **Step 1** →	72	
48 units result in an ending inventory cost of		$392 ← **Step 2**	

Cost of goods available for sale − Cost of ending inventory = Cost of goods sold

$1,200 − $392 = $808 ← **Step 3**

Although LIFO doesn't always match the physical flow of goods, companies do still use it to calculate the flow of costs for products such as DVDs and computers, which have declining replacement costs. Also, during inflation, LIFO produces less income than other methods. This results in lower taxes for companies using LIFO.

Before concluding this unit, we will make a summary for the cost of ending inventory and cost of goods sold under the weighted-average, FIFO, and LIFO methods. From this summary, you can see that in times of rising prices, LIFO gives the highest cost of goods sold ($808). This results in a tax savings for Blue. The weighted-average method tends to smooth out the fluctuations between LIFO and FIFO and falls in the middle.

The key to this discussion of inventory valuation is that different costing methods produce different results. So management, investors, and potential investors should understand the different inventory costing methods and should know which method a particular company uses. For example, Fruit of the Loom, Inc., changed its inventories from LIFO to FIFO due to cost reductions.

MONEY tips

Take inventory of your assets and record all serial numbers. Save receipts as proof of purchase for high dollar items. Take a video of assets and store the file in a fireproof container offsite in the event of a loss to assist with filing an insurance claim.

Inventory method	Cost of goods available for sale	Cost of ending inventory	Cost of goods sold
Weighted average	\$1,200	\$480 **Step 1:** Total goods, \$1,200 Total units, 120 $\frac{\$1,200}{120} = \$10$ **Step 2:** \$10 × 48 = \$480	\$1,200 − \$480 = \$720
FIFO	\$1,200	Bottom up to inventory level (48) 20 × \$13 = \$260 20 × \$12 = 240 8 × \$10 = 80 \$580	\$1,200 − \$580 = \$620
LIFO	\$1,200	Top down to inventory level (48) 40 × \$8 = \$320 8 × \$9 = 72 \$ 392	\$1,200 − \$392 = \$808

Let's check your understanding of this unit with a Practice Quiz.

LU 18–1 PRACTICE QUIZ

Complete this **Practice Quiz** to see how you are doing.

From the following, calculate **(a)** the cost of ending inventory and **(b)** the cost of goods sold under the assumption of (1) weighted-average method, (2) FIFO, and (3) LIFO (ending inventory shows 72 units):

	Number of books purchased for resale	Cost per unit	Total
January 1 inventory	30	\$3	\$ 90
March 1	50	2	100
April 1	20	4	80
November 1	60	6	360

*For **extra help** from your authors–Sharon and Jeff–see the videos in Connect.*

✓ Solutions

1. a. 72 units of ending inventory × \$3.94 = \$283.68 cost of ending inventory
($630 ÷ 160)

b. Cost of goods available for sale − Cost of ending inventory = Cost of goods sold

\$630 − \$283.68 = \$346.32

2. a.

60 units from November 1 purchased at \$6	\$360
12 units from April 1 purchased at \$4	48
72 units Cost of ending inventory	\$408

b. Cost of goods available for sale − Cost of ending inventory = Cost of goods sold

\$630 − \$408 = \$222

3. a.

30 units from January 1 purchased at \$3	\$ 90
42 units from March 1 purchased at \$2	84
72 Cost of ending inventory	\$174

b. Cost of goods available for sale − Cost of ending inventory = Cost of goods sold

\$630 − \$174 = \$456

Learning Unit 18–2: Retail Method; Gross Profit Method; Inventory Turnover; Distribution of Overhead

Customers want stores to have products available for sale as soon as possible. This has led to outsourced warehouses offshore where tens of thousands of products can be stored ready to be quickly shipped to various stores.

When retailers receive their products, they go into one of their most important assets—their inventory. When the product is sold, it must be removed from inventory so it can be replaced or discontinued. Often these transactions occur electronically at the registers that customers use to pay for products. How is inventory controlled when the register of the store cannot perform the task of adding and subtracting products from inventory?

©DreamPictures/Shannon Faulk/Blend Images LLC

Convenience stores often try to control their inventory by taking physical inventories. This can be time-consuming and expensive. Some stores draw up monthly financial reports but do not want to spend the time or money to take a monthly physical inventory.

Many stores estimate the amount of inventory on hand. Stores may also have to estimate their inventories when they have a loss of goods due to fire, theft, flood, and the like. This unit begins with two methods of estimating the value of ending inventory—the *retail method* and the *gross profit method.*

Retail Method

Many companies use the **retail method** to estimate their inventory. As shown in Figure 18.2, this method does not require that a company calculate an inventory cost for each item. To calculate the $3,500 ending inventory in Figure 18.2, Green Company used the steps that follow:

LO 1

CALCULATING THE RETAIL METHOD

Step 1. Calculate the cost of goods available for sale at cost and retail: $6,300; $9,000.

Step 2. Calculate a cost ratio using the following formula:

$$\frac{\text{Cost of goods available for sale at cost}}{\text{Cost of goods available for sale at retail}} = \frac{\$6,300}{\$9,000} = .70$$

Step 3. Deduct net sales from cost of goods available for sale at retail: $9,000 – $4,000.

Step 4. Multiply the cost ratio by the ending inventory at retail: .70 × $5,000.

Now let's look at the gross profit method.

Gross Profit Method

To use the **gross profit method** to estimate inventory, the company must keep track of (1) average gross profit rate, (2) net sales at retail, (3) beginning inventory, and (4) net purchases. You can use the following steps to calculate the gross profit method:

LO 2

CALCULATING THE GROSS PROFIT METHOD

Step 1. Calculate the cost of goods available for sale (Beginning inventory + Net purchases).

Step 2. Multiply the net sales at retail by the complement of the gross profit rate. This is the estimated cost of goods sold.

Step 3. Calculate the cost of estimated ending inventory (Step 1 – Step 2).

FIGURE 18.2

Estimating inventory with the retail method

	Cost	Retail
Beginning inventory	$4,000	$6,000
Net purchases during month	2,300	3,000
Cost of goods available for sale **(Step 1)**	$6,300	$9,000
Less net sales for month		4,000 **(Step 3)**
Ending inventory at retail		$5,000
Cost ratio ($6,300 ÷ $9,000) **(Step 2)**		70%
Ending inventory at cost (.70 × $5,000) **(Step 4)**		$3,500

EXAMPLE Assume Radar Company has the following information in its records:

Gross profit on sales	30%
Beginning inventory, January 1, 2019	$20,000
Net purchases	$ 8,000
Net sales at retail for January	$12,000

If you use the gross profit method, what is the company's estimated inventory?

The gross profit method calculates Radar's estimated cost of ending inventory at the end of January as follows:

Goods available for sale				
Beginning inventory, January 1, 2019			$20,000	
Net purchases			8,000	
Cost of goods available for sale			$28,000	← Step 1
Less estimated cost of goods sold:				
Net sales at retail		$12,000		
Cost percentage (100% − 30%)	Step 2 →	.70		
Estimated cost of goods sold			8,400	
Estimated ending inventory, January 31, 2019			$19,600	← Step 3

Note that the cost of goods available for sale less the estimated cost of goods sold gives the estimated cost of ending inventory.

Since this chapter has looked at inventory flow, let's discuss inventory turnover—a key business ratio.

LO 3

Inventory Turnover

Inventory turnover is the number of times the company replaces inventory during a specific time. Companies use the following two formulas to calculate inventory turnover:

$$\text{Inventory turnover at retail} = \frac{\text{Net sales}}{\text{Average inventory at retail}}$$

$$\text{Inventory turnover at cost} = \frac{\text{Cost of goods sold}}{\text{Average inventory at cost}}$$

Aha!

Note that inventory turnover at retail is usually lower than inventory turnover at cost. This is due to theft, markdowns, spoilage, and so on. Also, retail outlets and grocery stores usually have a high turnover, but jewelry and appliance stores have a low turnover.

Now let's use an example to calculate the inventory turnover at retail and at cost.

EXAMPLE The following facts are for Abby Company, a local sporting goods store (rounded to the nearest hundredth):

Net sales	$32,000	Cost of goods sold	$22,000
Beginning inventory at retail	$11,000	Beginning inventory at cost	$ 7,500
Ending inventory at retail	$ 8,900	Ending inventory at cost	$ 5,600

With these facts, we can make the following calculations to determine **average inventory:**

$$\text{Average inventory} = \frac{\text{Beginning inventory} + \text{Ending inventory}}{2}$$

$$\text{At retail: } \frac{\$32,000}{\frac{\$11,000 + \$8,900}{2}} = \frac{\$32,000}{\$9,950} = 3.22$$

$$\text{At cost: } \frac{\$22,000}{\frac{\$7,500 + \$5,600}{2}} = \frac{\$22,000}{\$6,550} = 3.36$$

What Turnover Means Inventory is often a company's most expensive asset. The turnover of inventory can have important implications. Too much inventory results in the use of needed space, extra insurance coverage, and so on. A low inventory turnover could indicate customer dissatisfaction, too much tied-up capital, and possible product obsolescence. A high inventory turnover might mean insufficient amounts of inventory causing stockouts that may lead to future lost sales. If inventory is moving out quickly, perhaps the company's selling price is too low compared to that of its competitors.

In recent years the **just-in-time (JIT) inventory system** from Japan has been introduced in the United States. Under ideal conditions, manufacturers must have suppliers that will provide materials daily as the manufacturing company needs them, thus eliminating inventories. The companies that are using this system, however, have often not been able to completely eliminate the need to maintain some inventory.

LO 4

Distribution of Overhead

In Chapter 16 we studied the cost of goods sold and operating expenses shown on the income statement. The operating expenses included **overhead expenses**—expenses that are *not* directly associated with a specific department or product but that contribute indirectly to the running of the business. Examples of such overhead expenses are rent, taxes, and insurance.

Companies must allocate their overhead expenses to the various departments in the company. The two common methods of calculating the **distribution of overhead** are by (1) floor space (square feet) or (2) sales volume.

Calculations by Floor Space To calculate the distribution of overhead by floor space, use the steps that follow:

CALCULATING THE DISTRIBUTION OF OVERHEAD BY FLOOR SPACE

Step 1. Calculate the total square feet in all departments.

Step 2. Calculate the ratio for each department based on floor space.

Step 3. Multiply each department's floor space ratio by the total overhead.

EXAMPLE Roy Company has three departments with the following floor space:

Department A	6,000 square feet
Department B	3,000 square feet
Department C	1,000 square feet

The accountant's job is to allocate $90,000 of overhead expenses to the three departments.

To allocate this overhead by floor space:

	Floor space in square feet	Ratio	
Department A	6,000	$\frac{6,000}{10,000} = 60\%$	
Department B	3,000	$\frac{3,000}{10,000} = 30\%$	← **Steps 1 and 2**
Department C	1,000	$\frac{1,000}{10,000} = 10\%$	
	10,000 total square feet		

Department A	.60 × $90,000 =	$54,000	
Department B	.30 × $90,000 =	27,000	← **Step 3**
Department C	.10 × $90,000 =	9,000	
		$90,000	

MONEY tips

Track your overhead expenses such as electricity, gas, water, sewage, etc., to determine trends. Make adjustments where needed.

Calculations by Sales To calculate the distribution of overhead by sales, use the steps that follow:

CALCULATING THE DISTRIBUTION OF OVERHEAD BY SALES

Step 1. Calculate the total sales in all departments.

Step 2. Calculate the ratio for each department based on sales.

Step 3. Multiply each department's sales ratio by the total overhead.

EXAMPLE Morse Company distributes its overhead expenses based on the sales of its departments. For example, last year Morse's overhead expenses were $60,000. Sales of its two departments were as follows, along with its ratio calculation.

Since Department A makes 80% of the sales, it is allocated 80% of the overhead expenses.

	Sales	Ratio	
Department A	$ 80,000	$\frac{\$80{,}000}{\$100{,}000} = .80$	← Steps 1 and 2
Department B	20,000	$\frac{\$20{,}000}{\$100{,}000} = .20$	
Total sales	$100,000		

These ratios are then multiplied by the overhead expense to be allocated.

Department A	.80 × $60,000 =	$48,000
Department B	.20 × $60,000 =	12,000 ← Step 3
		$60,000

It's time to try another Practice Quiz.

LU 18–2 PRACTICE QUIZ

Complete this Practice Quiz to see how you are doing.

1. From the following facts, calculate the cost of ending inventory using the retail method (round the cost ratio to the nearest tenth percent):

January 1—inventory at cost	$ 18,000
January 1—inventory at retail	58,000
Net purchases at cost	220,000
Net purchases at retail	376,000
Net sales at retail	364,000

2. Given the following, calculate the estimated cost of ending inventory using the gross profit method:

Gross profit on sales	40%
Beginning inventory, January 1, 2019	$27,000
Net purchases	$ 7,500
Net sales at retail for January	$15,000

3. Calculate the inventory turnover at cost and at retail from the following (round the turnover to the nearest hundredth):

Average inventory at cost	Average inventory at retail	Net sales	Cost of goods sold
$10,590	$19,180	$109,890	$60,990

4. From the following, calculate the distribution of overhead to Departments A and B based on floor space.

Amount of overhead expense to be allocated	Square footage
$70,000	10,000 Department A
	30,000 Department B

*For **extra help** from your authors–Sharon and Jeff–see the videos in Connect.*

✓ Solutions

		Cost	Retail
1.	Beginning inventory	$ 18,000	$ 58,000
	Net purchases during the month	220,000	376,000
	Cost of goods available for sale	$238,000	$434,000
	Less net sales for the month		364,000
	Ending inventory at retail		$ 70,000
	Cost ratio ($238,000 ÷ $434,000)		54.8%
	Ending inventory at cost (.548 × $70,000)		$ 38,360

2.	**Goods available for sale**		
	Beginning inventory, January 1, 2019		$ 27,000
	Net purchases		7,500
	Cost of goods available for sale		$ 34,500
	Less estimated cost of goods sold:		
	Net sales at retail	$ 15,000	
	Cost percentage (100% – 40%)	.60	
	Estimated cost of goods sold		9,000
	Estimated ending inventory, January 31, 2019		$ 25,500

3. $$\text{Inventory turnover at cost} = \frac{\text{Cost of goods sold}}{\text{Average inventory at cost}} = \frac{\$60{,}900}{\$10{,}590} = 5.75$$

$$\text{Inventory turnover at retail} = \frac{\text{Net sales}}{\text{Average inventory at retail}} = \frac{\$109{,}890}{\$19{,}180} = 5.73$$

4.

		Ratio
Department A	10,000	$\frac{10{,}000}{40{,}000} = .25 \times \$70{,}000 = \$17{,}500$
Department B	30,000	$\frac{30{,}000}{40{,}000} = .75 \times \$70{,}000 = 52{,}500$
		$70,000

INTERACTIVE CHAPTER ORGANIZER

Topic/Procedure/Formula	Examples	You try it*
Specific identification method Identification could be by serial number, physical description, or coding. The flow of goods and flow of costs are the same.		**Cost per unit** / **Total cost** April 1, 3 units at $7 — $21 May 5, 4 units at 8 — 32 $53 If 1 unit from each group is left, ending inventory is: 1 × $7 = $ 7 + 1 × 8 = 8 $15 Cost of goods available for sale – Cost of ending inventory = Cost of goods sold $53 – $15 = $38	**Calculate ending inventory and cost of goods sold** **Cost per unit** / **Total cost** May 1, 4 units at $ 9 June 6, 3 units at 10 Assume one unit from each group is left.

(continues)

INTERACTIVE CHAPTER ORGANIZER

Topic/Procedure/Formula	Examples	You try it*

Weighted-average method

Topic/Procedure/Formula

$$\text{Weighted-average unit cost} = \frac{\text{Total cost of goods available for sale}}{\text{Total number of units available for sale}}$$

Examples

	Cost per unit	Total cost
1/XX, 4 units at	$4	$16
5/XX, 2 units at	5	10
8/XX, 3 units at	6	18
		$44

$$\text{Unit cost} = \frac{\$44}{9} = \$4.89$$

If 5 units left, cost of ending inventory is 5 units × $4.89 = $24.45

You try it*

Calculate unit cost and cost of ending inventory

	Cost per unit	Total cost
1/XX, 6 units at	$5	$30
5/XX, 4 units at	6	24
8/XX, 5 units at	7	35
		$89

4 units left

FIFO—first-in, first-out method

Topic/Procedure/Formula

Sell old inventory first. Ending inventory is made up of last merchandise brought into store.

Examples

Using example above:
5 units left:

(Last into store)	3 units at $6	$18
	2 units at $5	10
Cost of ending inventory		$28

You try it*

Calculate cost of inventory by FIFO

Use weighted-average example.

LIFO—last-in, first-out method

Topic/Procedure/Formula

Sell last inventory brought into store first. Ending inventory is made up of oldest merchandise in store.

Examples

Using weighted-average example:
5 units left:

(First into store)	4 units at $4	$16
	1 unit at $5	5
Cost of ending inventory		$21

You try it*

Calculate cost of inventory by LIFO

Use weighted-average example.

Retail method

Topic/Procedure/Formula

Ending inventory at cost equals:

$$\frac{\text{Cost of goods available at cost}}{\text{Cost of goods available at retail}} \times \text{Ending inventory at retail}$$

(This is cost ratio.)

Examples

	Cost	Retail
Beginning inventory	$52,000	$ 83,000
Net purchases	28,000	37,000
Cost of goods available for sale	$80,000	$120,000
Less net sales for month		80,000
Ending inventory at retail		$ 40,000

$$\text{Cost ratio} = \frac{\$80,000}{\$120,000} = .67 = 67\%$$

Rounded to nearest percent.
Ending inventory at cost, $26,800
(.67 × $40,000)

You try it*

Calculate cost of ending inventory at cost and at retail

	Cost	Retail
Beginning inventory	$60,000	$80,000
Net purchases	28,000	37,000

Assume net sales of $90,000.

(Round ratio to nearest percent.)

Gross profit method

Topic/Procedure/Formula

$$\text{Beg. inv.} + \text{Net purchases} - \text{Estimated cost of goods sold} = \text{Estimated ending inventory}$$

Examples

Goods available for sale

Beginning inventory		$30,000
Net purchases		3,000
Cost of goods available for sale		$33,000
Less: Estimated cost of goods sold:		
Net sales at retail	$18,000	
Cost percentage (100% – 30%)	.70	
Estimated cost of goods sold		12,600
Estimated ending inventory		$ 20,400

You try it*

Calculate estimated ending inventory

Given: Net sales at retail of $20,000 and a 75% gross profit.

Goods available for sale

Beginning inventory	$40,000
Net purchases	2,000

(continues)

INTERACTIVE CHAPTER ORGANIZER

Topic/Procedure/Formula	Examples	You try it*
Inventory turnover at retail and at cost $\frac{\text{Net sales}}{\text{Average inventory at retail}}$ or $\frac{\text{Cost of goods sold}}{\text{Average inventory at cost}}$	Inventory, January 1 at cost $20,000 Inventory, December 31 at cost 48,000 Cost of goods sold 62,000 At cost: $\frac{\$62{,}000}{\frac{\$20{,}000 + \$48{,}000}{2}} = 1.82$ (inventory turnover at cost)	**Calculate inventory turnover at cost** Jan 1 inventory at cost $40,000 Dec 31 inventory at cost $60,000 Cost of goods sold $90,000
Distribution of overhead Based on floor space or sales volume, calculate: **1.** Ratios of department floor space or sales to the total. **2.** Multiply ratios by total amount of overhead to be distributed.	Total overhead to be distributed, $10,000 **Floor space** Department A 6,000 sq. ft. Department B 2,000 sq. ft. 8,000 sq. ft. $\text{Ratio A} = \frac{6{,}000}{8{,}000} = .75$ $\text{Ratio B} = \frac{2{,}000}{8{,}000} = .25$ $\text{Dept. A} = .75 \times \$10{,}000 = \$7{,}500$ $\text{Dept. B} = .25 \times \$10{,}000 = \$2{,}500$	**Calculate overhead cost to each department** Total overhead to be distributed, $30,000 **Floor space** Department A 4,000 sq. ft. Department B 6,000 sq. ft.

KEY TERMS

- Average inventory
- Distribution of overhead
- First-in, first-out (FIFO) method
- Gross profit method
- Inventory turnover
- Just-in-time (JIT) inventory system
- Last-in, first-out (LIFO) method
- Overhead expenses
- Periodic inventory system
- Perpetual inventory system
- Retail method
- Specific identification method
- Weighted-average method

*Worked-out solutions are in Appendix B.

Critical Thinking Discussion Questions with Chapter Concept Check

1. Explain how you would calculate the cost of ending inventory and cost of goods sold for specific identification, FIFO, LIFO, and weighted-average methods. Explain why during inflation LIFO results in a tax savings for a business.
2. Explain the cost ratio in the retail method of calculating inventory. What effect will the increased use of computers have on the retail method?
3. What is inventory turnover? Explain the effect of a high inventory turnover during the Christmas shopping season.
4. How is the distribution of overhead calculated by floor space or sales? Give an example of why a store in your area cut back one department to expand another. Did it work?
5. Discuss how levels of inventory have been affected by the economic crises at your local mall.
6. **Chapter Concept Check.** Search the web to find the latest techniques used by stores to control their inventory. Use all the concepts in the chapter to discuss the privacy issue as well.

Classroom Notes

18

END-OF-CHAPTER PROBLEMS

connect

Check figures for odd-numbered problems in Appendix B. Name ______________________ Date ________

DRILL PROBLEMS

18–1. Using the specific identification method, calculate **(a)** the cost of ending inventory and **(b)** the cost of goods sold given the following: *LU 18-1(2)*

Date	Units purchased	Cost per unit	Ending inventory
June 1	15 Echo Show's 360	$275	2 Echo Show's from June
July 1	45 Echo Show's 360	250	15 Echo Show's from July
August 1	60 Echo Show's 360	240	12 Echo Show's from August

From the following, calculate the **(a)** cost of ending inventory (round the average unit cost to the nearest cent) and **(b)** cost of goods sold using the weighted-average method, FIFO, and LIFO (ending inventory shows 61 units). *LU 18-1(2)*

	Number purchased	Cost per unit	Total
January 1 inventory	40	$4	$160
April 1	60	7	420
June 1	50	8	400
November 1	55	9	495

18–2. Use weighted average:

18–3. Use FIFO:

18–4. Use LIFO:

From the following (18–5 to 18–12), calculate the cost of ending inventory and cost of goods sold for the LIFO (18–13), FIFO (18–14), and weighted-average (18–15) methods (make sure to first find total cost to complete the table); ending inventory is 49 units: *LU 18-1(2)*

	Beginning inventory and purchases	Units	Unit cost	Total dollar cost
18–5.	Beginning inventory, January 1	5	$2.00	
18–6.	April 10	10	2.50	
18–7.	May 15	12	3.00	
18–8.	July 22	15	3.25	
18–9.	August 19	18	4.00	
18–10.	September 30	20	4.20	
18–11.	November 10	32	4.40	
18–12.	December 15	16	4.80	

18–13. LIFO:

Cost of ending inventory **Cost of goods sold**

18–14. FIFO:

Cost of ending inventory **Cost of goods sold**

18–15. Weighted average:

Cost of ending inventory **Cost of goods sold**

18–16. From the following, calculate the cost ratio (round to the nearest hundredth percent) and the cost of ending inventory to the nearest cent under the retail method. *LU 18-2(1)*

Net sales at retail for year	$40,000	Purchases—cost	$14,000
Beginning inventory—cost	$27,000	Purchases—retail	$19,000
Beginning inventory—retail	$49,000		

18–17. Complete the following (round answers to the nearest hundredth): *LU 18-2(3)*

a. Average inventory at cost	b. Average inventory at retail	c. Net sales	d. Cost of goods sold	e. Inventory turnover at cost	f. Inventory turnover at retail
$14,000	$21,540	$70,000	$49,800		

Complete the following (assume $90,000 of overhead to be distributed): *LU 18-2(4)*

		Square feet	Ratio	Amount of overhead allocated
18–18.	Department A	10,000		
18–19.	Department B	30,000		

18–20. Given the following, calculate the estimated cost of ending inventory using the gross profit method. *LU 18-2(2)*

Gross profit on sales	55%	Net purchases	$ 3,900
Beginning inventory	$29,000	Net sales at retail	$17,000

WORD PROBLEMS

eXcel **18–21.** If Exxon uses FIFO for its inventory valuation, calculate the cost of ending inventory and cost of goods sold if ending inventory is 110 barrels of crude oil. *LU 18-1(2)*

Beginning inventory and purchases	Barrels	Barrel cost	Total cost
Beginning inventory: Jan 1	125	$ 95	$11,875
March 1	50	101	5,050
June 1	65	98	6,370
September 1	75	90	6,750
December 1	50	103	5,150

18–22. Marvin Company has a beginning inventory of 12 sets of paints at a cost of $1.50 each. During the year, the store purchased 4 sets at $1.60, 6 sets at $2.20, 6 sets at $2.50, and 10 sets at $3.00. By the end of the year, 25 sets were sold. Calculate **(a)** the number of paint sets in ending inventory and **(b)** the cost of ending inventory under the LIFO, FIFO, and weighted-average methods. Round to nearest cent for the weighted average. *LU 18-1(2)*

18–23. Better Finance (previously BillFloat), based in San Francisco, California, provides leasing and credit solutions to consumers and small businesses. If Better Finance wants to distribute $45,000 worth of overhead by sales, calculate the overhead expense for each department: *LU 18-2(4)*

New customer sales (NCS)	$ 5,120,000
Current customer new sales (CCNS)	4,480,000
Current customer loan extension sales (CCLES)	3,200,000

18–24. If Comcast is upgrading its cable boxes and has 500 obsolete boxes in ending inventory, what is the cost of ending inventory using FIFO, LIFO, and the weighted-average method? *LU 18-1(2)*

Beginning inventory and purchases	Boxes	Box cost	Total cost
Beginning inventory: January 1	15,500	$15	$232,500
March 1	6,500	16	104,000
June 1	2,500	20	50,000
September 1	1,500	23	34,500
December 1	1,000	32	32,000

18–25. May's Dress Shop's inventory at cost on January 1 was $39,000. Its retail value was $59,000. During the year, May purchased additional merchandise at a cost of $195,000 with a retail value of $395,000. The net sales at retail for the year were $348,000. Calculate May's inventory at cost by the retail method. Round the cost ratio to the nearest whole percent. *LU 18-2(1)*

18–26. A sneaker outlet has made the following wholesale purchases of new running shoes: 12 pairs at $45, 18 pairs at $40, and 20 pairs at $50. An inventory taken last week indicates that 23 pairs are still in stock. Calculate the cost of this inventory by FIFO. *LU 18-1(2)*

18–27. Over the past 3 years, the gross profit rate for Jini Company was 35%. Last week a fire destroyed all Jini's inventory. Using the gross profit method, estimate the cost of inventory destroyed in the fire, given the following facts that were recorded in a fireproof safe: *LU 18-2(2)*

Beginning inventory	$ 6,000
Net purchases	64,000
Net sales at retail	49,000

18–28. Calculate cost of goods sold and ending inventory for Emergicare's bandages orders using FIFO, LIFO and average cost. There are 35 units in ending inventory. *LU 18-1(2)*

Date	Units purchased	Cost per unit	Total cost
January 1	50	$7.50	$ 375.00
April 1	45	6.75	303.75
June 1	60	6.50	390.00
September 1	55	7.00	385.00
Total	210		$1,453.75

My Money

18–29. Your home business uses 350 square feet of your 1,750 square foot home. If household expenses for the year were $17,558, how much was alloted to your business?

CHALLENGE PROBLEMS

18–30. Monroe Company had a beginning inventory of 350 cans of paint at $12 each on January 1 at a cost of $4,200. During the year, the following purchases were made:

February 15	280 cans at $14.00
April 30	110 cans at $14.50
July 1	100 cans at $15.00

Monroe marks up its goods at 40% on cost. At the end of the year, ending inventory showed 105 units remaining. Calculate the amount of sales assuming a FIFO flow of inventory. *LU 18-1(2)*

18–31. Logan Company uses a perpetual inventory system on a FIFO basis. Assuming inventory on January 1 was 800 units at $8 each, what is the cost of ending inventory at the end of October 5? *LU 18-1(2)*

Received			Sold	
Date	**Quantity**	**Cost per unit**	**Date**	**Quantity**
Apr. 15	220	$5	Mar. 8	500
Nov. 12	1,900	9	Oct. 5	200

SUMMARY PRACTICE TEST

Do you need help? Connect videos have step-by-step worked-out solutions.

1. Writing.com has a beginning inventory of 16 sets of pens at a cost of $2.12 each. During the year, Writing.com purchased 8 sets at $2.15, 9 sets at $2.25, 14 sets at $3.05, and 13 sets at $3.20. By the end of the year, 29 sets were sold. Calculate **(a)** the number of pen sets in stock and **(b)** the cost of ending inventory under LIFO, FIFO, and weighted-average methods. *LU 18-1(2)*

2. Lee Company allocates overhead expenses to all departments on the basis of floor space (square feet) occupied by each department. The total overhead expenses for a recent year were $200,000. Department A occupied 8,000 square feet; Department B, 20,000 square feet; and Department C, 7,000 square feet. What is the overhead allocated to Department C? In your calculations, round to the nearest whole percent. *LU 18-2(4)*

3. A local college bookstore has a beginning inventory costing $80,000 and an ending inventory costing $84,000. Sales for the year were $300,000. Assume the bookstore markup rate on selling price is 70%. Based on the selling price, what is the inventory turnover at cost? Round to the nearest hundredth. *LU 18-2(3)*

4. Dollar Dress Shop's inventory at cost on January 1 was $82,800. Its retail value was $87,500. During the year, Dollar purchased additional merchandise at a cost of $300,000 with a retail value of $325,000. The net sales at retail for the year were $295,000. Calculate Dollar's inventory at cost by the retail method. Round the cost ratio to the nearest whole percent. *LU 18-2(1)*

5. On January 1, Randy Company had an inventory costing $95,000. During January, Randy had net purchases of $118,900. Over recent years, Randy's gross profit in January has averaged 45% on sales. The company's net sales in January were $210,800. Calculate the estimated cost of ending inventory using the gross profit method. *LU 18-2(2)*

INTERACTIVE VIDEO WORKSHEET

GRADING THE SUMMARY PRACTICE TEST

Go to the summary practice test video in Connect (or click on it here in the ebook). Grade your summary practice test while viewing the video.

C for Correct/I for Incorrect

1. ________
2. ________
3. ________
4. ________
5. ________

If you achieved 100%, you are ready for your instructor's exam.

If any of the problems were incorrect, list the questions you missed and show steps to solve the problem correctly.

Replay the video to see if you have made the correct fixes to your mistakes. If you have any questions, contact your instructor asap.

Notes on Watching Videos

MY MONEY

Personal Property Protection (Say That Three Times Fast!)

What I need to know

As you enter the workforce after college and start working in your chosen career field, you will most likely earn far more income than you have experienced to date. With this increase in earnings you may find yourself making purchases of higher cost items (furniture, televisions, cars, etc.). As you begin to make these types of purchases, you will no doubt wonder what safeguards should be in place to protect your valuables. Proper documentation, such as sales receipts, serial numbers, videos, etc., of your personal property is crucial to establishing the proof you need in the event of a loss. Not only will this documentation be important for an insurance claim, but it will also be incredibly helpful to your local authorities in the event of a loss due to theft.

What I need to do

Get organized! Develop a method for storing the receipts from your purchases. This can be done by keeping documentation within a fireproof file cabinet, safety deposit box, or by storing electronic copies of your receipts. Either way, you will be able to document the purchases you have made should you ever be in a situation in which this documentation may be required (such as for an insurance claim for a personal property loss). Having a storage method in place for this documentation will assist in locating it quickly, which makes the process of filing a claim a bit easier.

In addition to the receipt documentation, it will be helpful to have some additional proof of the personal property you have purchased. This can be done by creating a video recording of the physical items in your possession. Not only will this serve as proof of the items you have, it will also be helpful to you in remembering the property you had in the event of a property loss such as burglary or fire. The experience of personal property loss can be quite upsetting and the video will assist you to ensure all personal property is accounted for when filing your insurance claim.

It will also be helpful for you to understand the limitations of your insurance coverage (renters, homeowners, etc.). When selecting your insurance coverage be sure to note the items that will be covered by the policy and those that are not. In addition, familiarize yourself with the types of losses that will be covered. Many policies will cover theft and loss due to fire but may not cover loss due to flooding, etc. Understanding your coverage will assist you should you need to file a claim.

Resources I can use

- Foreceipt Receipt Tracker App (mobile app)
- https://www.allstate.com/tools-and-resources/insurance-basics/personal-property-coverage.aspx — personal property coverage explanation

MY MONEY ACTIVITY

Assume you needed to produce documentation for all of your school supplies (books, laptop, notebooks, pens/pencils, etc.).

- Which items would be easy to locate?
- Which items would be difficult to locate?
- What changes could you make to improve your purchase documentation?

A KIPLINGER APPROACH

"Cops Rescue Hometown Bakery", *Kiplinger's*, January 2018, p. 22. Used by permission of The Kiplinger Washington Editors, Inc.

SUCCESS STORY

Cops Rescue Hometown Bakery

Nine policemen buy a doughnut shop and help revive their hometown's main street.

PROFILE

WHO: Alan "Bubba" White, 54 (interviewed), and Greg "Ryno" Rynearson, 55

WHERE: Clare, Mich.

WHAT: Owners, Cops & Doughnuts

You're both policemen? Retired. Ryno and I grew up together and served in Clare, our hometown of 3,100 people in central Michigan.

And you bought a bakery? We did, with seven other Clare cops. The joke around town is that we were having lunch one day and heard that the doughnut shop was closing, so the police department sprang into action. But the truth is, we were worried about what would happen to our little town when the bakery closed. At the time, in 2009, our downtown already had 11 empty storefronts. Ryno said, "Why don't we buy the bakery?" Everyone laughed. Then we realized we had an opportunity to play up the stereotype of cops and doughnuts.

How did you finance the business? The first year, we leased the bakery for $12,000. Each owner put in $1,500. We had an option to buy the building, equipment, inventory and recipes for $160,000. By year-end, it was obvious we should buy, so we took a loan from a local bank.

Word spread quickly? A reporter with AP interviewed us. The story went national, then international. Our web hits went from seven to 70,000 in one day. Pretty soon, the phone rang off the hook. You couldn't find a place to park downtown, and we couldn't make product fast enough.

What do you make? Doughnuts and other baked goods, made from scratch with the best ingredients. We make them big. Our apple fritters are the size of a puppy, and our Long Johns have twice the usual amount of filling. Plus, the Clare location serves breakfast and lunch.

How do you manage the business? We are all equal partners, and we have a monthly owners meeting. It's like nine brothers working together—sometimes sniping but always looking out for one another. Two people were elected managing members to make decisions for the group. And we hired a chief operating officer, who hires general managers for the bakeries.

You've expanded? We've bought five other bakeries in Michigan and Indiana. To buy three of the locations, we borrowed $400,000. We paid for the next two from cash flow. We have 167 year-round employees. In 2016, we had gross sales of more than $4.3 million. We expect gross sales of $5.7 million in 2017, with a payroll of $1.9 million.

Your marketing is fun. Yeah, our best ideas come from sitting around a fire with a six-pack. Ryno and I do videos, and we've learned that goofy and funny works [to see videos and order online, go to www.copsdoughnuts.com].

How's Clare doing? For the first time in my life, Clare has no empty storefronts.

Do you think you'll ever retire again? Driving around, making appearances, giving talks, shaking hands, giving out free Bubba hugs—I can do it forever.

PATRICIA MERTZ ESSWEIN
pesswein@kiplinger.com

BUSINESS MATH ISSUE

A small business should always use FIFO in valuating ending inventory.

1. List the key points of the article and information to support your position.
2. Write a group defense of your position using math calculations to support your view. If you are in an online course, post to a discussion board.

Classroom Notes

CHAPTER 19

Sales, Excise, and Property Taxes

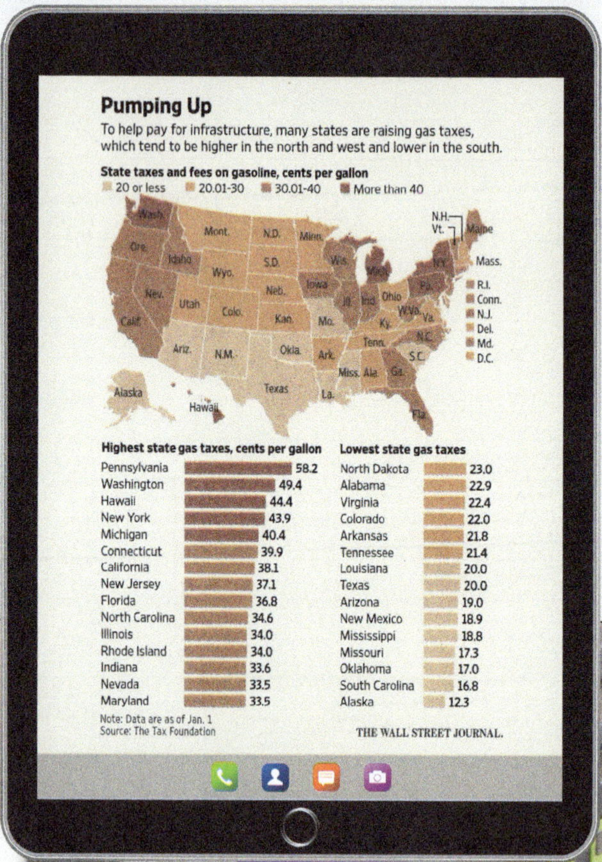

©Kaspars Grinvalds/Shutterstock

LEARNING UNIT OBJECTIVES

LU 19–1: Sales and Excise Taxes

1. Compute sales tax on goods sold involving trade and cash discounts and shipping charges.
2. Explain and calculate excise tax.

LU 19–2: Property Tax

1. Calculate the tax rate in decimal.
2. Convert tax rate in decimal to percent, per $100 of assessed value, per $1,000 of assessed value, and in mills.
3. Compute property tax due.

Your Guide to Successfully Completing This Chapter

Traditional book or ebook

Check box as you complete each step.

Steps

☐ Read learning unit.

☐ Complete practice quiz at the end of the learning unit.

☐ Grade practice quiz using provided solutions. (For more help, watch the learning unit video in Connect and have a Study Session with the authors. Then complete the additional practice quiz in Connect.)

☐ Repeat above for each of the two learning units in Chapter 19.

☐ Review chapter organizer.

☐ Complete assigned homework.

☐ Finish summary practice test. (Go to Connect via the ebook link and do the interactive video worksheet to grade.)

☐ Complete instructor's exam.

In 2018 the new tax law allows a deduction cap of only $10,000 per tax return for state and local income and property tax. Check with the IRS for the latest details.

The *Wall Street Journal* chapter opener, "Pumping Up," shows many states are raising gas taxes.

In Learning Unit 19–1 you will learn how sales taxes are calculated. Learning Unit 19–1 discusses the excise tax that is collected in addition to the sales tax. Learning Unit 19–2 explains the use of property tax.

Learning Unit 19–1: Sales and Excise Taxes

Today, many states have been raising their sales tax and excise tax.

LO 1

Sales Tax

Effective June 2018 the Supreme Court ruled that states could collect sales tax from online sales. The change will happen slowly so you should check for latest updates.

In many cities, counties, and states, the sellers of certain goods and services collect **sales tax** and forward it to the appropriate government agency. Forty-five states have a sales tax. Of the 45 states, 28 states and the District of Columbia exempt food; 44 states and the District of Columbia exempt prescription drugs. The Tax Foundation map, page 496, shows sales tax rates by state.

My Money

Sales tax is usually computed electronically by the new cash register systems and scanners. However, it is important to know how sellers calculate sales tax manually. The example of a car battery will show you how to manually calculate sales tax.

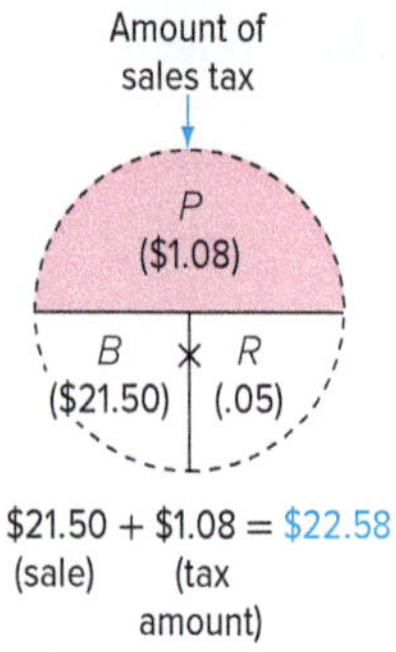

EXAMPLE

Selling price of a Sears battery	$32.00	Shipping charge	$3.50
Trade discount to local garage	$10.50	Sales tax	5%

Manual calculation

$32.00 − $10.50 = $21.50 taxable
× .05
$ 1.08 tax
+ 21.50 taxable
+ 3.50 shipping
$26.08 total price with tax and shipping

Check

100% is base + 5% is tax = 105%
1.05 × $21.50 = $22.58
+ 3.50 shipping
$26.08

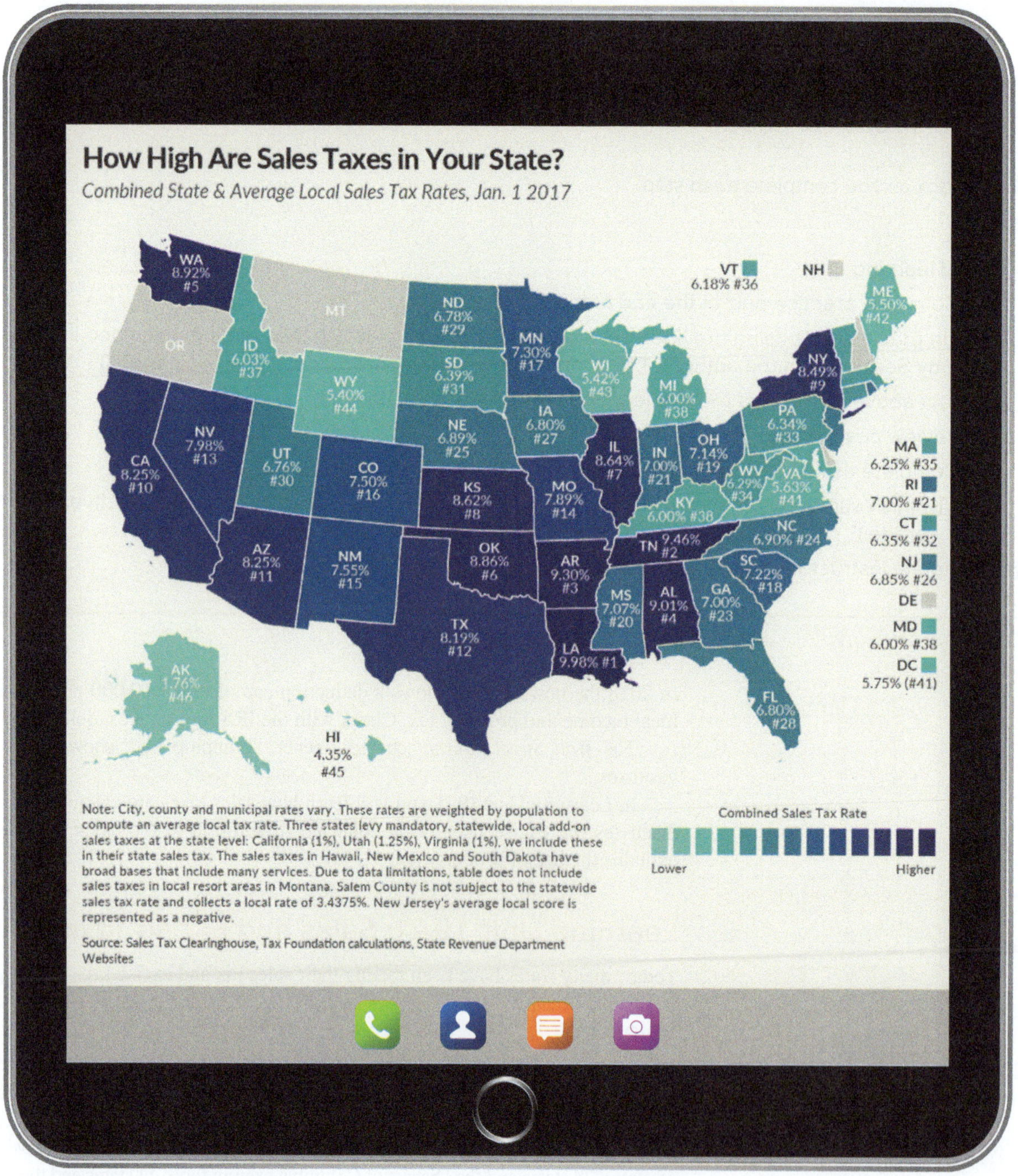

Source: Sales Tax Clearinghouse. Tax Foundation calculations, State Revenue Department Websites

Aha!

In this example, note how the trade discount is subtracted from the selling price before any cash discounts are taken. If the buyer is entitled to a 6% cash discount, it is calculated as follows:

$.06 \times \$21.50 = \$1.29$

Remember cash discounts are not taken on sales tax or shipping charges.

Calculating Actual Sales Managers often use the cash register to get a summary of their total sales for the day. The total sales figure includes the sales tax. So the sales tax must be deducted from the total sales. To illustrate this, let's assume the total sales for the day were $40,000, which included a 7% sales tax. What were the actual sales?

Hint: $40,000 is 107% of actual sales.

$$\text{Actual sales} = \frac{\text{Total sales}}{1 + \text{Tax rate}}$$

$$\text{Actual sales} = \frac{\$40{,}000}{1.07} = \$37{,}383.18$$

Total sales

100%	sales
+ 7%	tax
107%	→ 1.07

Thus, the store's actual sales were $37,383.18. The actual sales plus the tax equals $40,000.

Check

$37,383.18 × .07 =	$ 2,616.82	sales tax
	+ 37,383.18	actual sales
	$40,000.00	total sales including sales tax

LO 2

Excise Tax

Governments (local, federal, and state) levy **excise tax** on particular products and services. This can be a sizable source of revenue for these governments.

Consumers pay the excise tax in addition to the sales tax. The excise tax is based on a percent of the *retail* price of a product or service. This tax, which varies in different states, is imposed on luxury items or nonessentials. Examples of products or services subject to the excise tax include airline travel, telephone service, alcoholic beverages, jewelry, furs, fishing rods, tobacco products, and motor vehicles. Although excise tax is often calculated as a percent of the selling price, the tax can be stated as a fixed amount per item sold. The following example calculates excise tax as a percent of the selling price.[1]

My Money

MONEY tips

When buying online at sites such as Amazon.com, choose an out-of-state vendor and save on sales tax at point-of-purchase.

EXAMPLE On June 1, Angel Rowe bought a fur coat for a retail price of $5,000. Sales tax is 7% with an excise tax of 8%. Her total cost is as follows:

	$5,000	
+	350	sales tax (.07 × $5,000)
+	400	excise tax (.08 × $5,000)
	$5,750	

Let's check your progress with a Practice Quiz.

LU 19–1 PRACTICE QUIZ

Complete this Practice Quiz to see how you are doing.

From the following shopping list, calculate the total sales tax. Food items are excluded from sales tax, which is 8%.

Chicken	$6.10	Orange juice	$1.29	Shampoo	$4.10
Lettuce	$ .75	Laundry detergent	$3.65		

*For **extra help** from your authors–Sharon and Jeff–see the videos in Connect.*

✓ Solutions

Shampoo	$4.10
Laundry detergent	+ 3.65
	$7.75 × .08 = $.62

[1]If excise tax were a stated fixed amount per item, it would have to be added to the cost of goods or services before any sales tax was taken. For example, a $100 truck tire with a $4 excise tax would be $104 before the sales tax was calculated.

Learning Unit 19–2: Property Tax

My Money

When you own property, you must pay property tax. In this unit we listen in on a conversation between a property owner and a tax assessor.

Defining Assessed Value

Property Can Have Two Meanings

Both subject to property tax:
1. **Real property**—land, buildings, etc.
2. **Personal property**—possessions like jewelry, autos, furniture, etc.

Bill Adams was concerned when he read in the local paper that the property tax rate had been increased. Bill knows that the revenue the town receives from the tax helps pay for fire and police protection, schools, and other public services. However, Bill wants to know how the town set the new rate and the amount of the new property tax.

Bill went to the town assessor's office to get specific details. The assessor is a local official who estimates the fair market value of a house. Before you read the summary of Bill's discussion, note the following formula:

$$\text{Assessed value} = \text{Assessment rate} \times \text{Market value}$$

Bill: What does *assessed value* mean?

Assessor: **Assessed value** is the value of the property for purposes of computing property taxes. We estimated the market value of your home at $210,000. In our town, we assess property at 30% of the market value. Thus, your home has an assessed value of $63,000 ($210,000 × .30). Usually, assessed value is rounded to the nearest dollar.

Bill: I know that the **tax rate** multiplied by my assessed value ($63,000) determines the amount of my property tax. What I would like to know is how did you set the new tax rate?

LO 1

Determining the Tax Rate

Assessor: In our town first we estimate the total amount of revenue needed to meet our budget. Then we divide the total of all assessed property into this figure to get the *tax rate*. The formula looks like this:[2]

$$\text{Tax rate} = \frac{\text{Budget needed}}{\text{Total assessed value}}$$

Our town budget is $125,000, and we have a total assessed property value of $1,930,000. Using the formula, we have the following:

$$\frac{\$125{,}000}{\$1{,}930{,}000} = \$.0647668 = .0648 \text{ tax rate per dollar}$$

Note that the rate should be rounded up to the indicated digit, *even if the digit is less than 5.* Here we rounded to the nearest ten thousandth.

LO 2

How the Tax Rate Is Expressed

Assessor: We can express the .0648 tax rate per dollar in the following forms:

By percent	Per $100 of assessed value	Per $1,000 of assessed value	In mills
6.48%	$6.48	$64.80	64.80
(Move decimal two places to right.)	(.0648 × 100)	(.0648 × 1,000)	$\left(\frac{.0648}{.001}\right)$

MONEY tips

Review your property tax assessment annually and appeal it if it is too high. Property values may have fallen, leaving you at risk for paying too much for your property taxes.

A **mill** is $\frac{1}{10}$ of a cent or $\frac{1}{1{,}000}$ of a dollar (.001). To represent the number of mills as a tax rate per dollar, we divide the tax rate in decimal by .001. Rounding practices vary from state to state. Colorado tax bills are now rounded to the thousandth mill. An alternative to finding the rate in mills is to multiply the rate per dollar by 1,000, since a dollar has 1,000 mills. In the problems in this text, we round the mills per dollar to the nearest hundredth.

[2]Remember that exemptions to total assessed value include land and buildings used for educational and religious purposes and the like.

LO 3

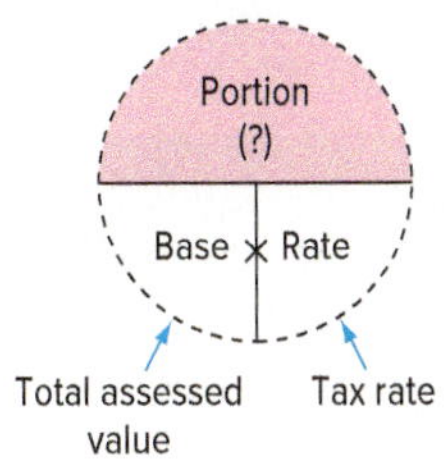

How to Calculate Property Tax Due[3]

Assessor: The following formula will show you how we arrive at your **property tax:**

$$\text{Total property tax due (Portion)} = \text{Tax rate (Rate)} \times \text{Total assessed value (Base)}$$

$$\$4{,}082.40 = .0648 \times \$63{,}000$$

We can use the other forms of the decimal tax rate to show you how the property tax will not change even when expressed in various forms:

By percent	Per $100	Per $1,000	Mills
6.48% × $63,000 = $4,082.40	$\frac{\$63{,}000}{\$100} = 630$ 630 × $6.48 = $4,082.40	$\frac{\$63{,}000}{\$1{,}000} = 63$ 63 × $64.80 = $4,082.40	Property tax due = Mills × .001 × Assessed value = 64.80 × .001 × $63,000 = $4,082.40

MONEY tips

Track the estimated value of your property using websites such as Zillow.com.

Keep in mind you always round up when calculating the tax rate—even if the digit being rounded is less than 5.

Now it's time to try the Practice Quiz.

LU 19–2 PRACTICE QUIZ

Complete this Practice Quiz to see how you are doing.

From the following facts: (1) calculate the assessed value of Bill's home; (2) calculate the tax rate for the community in decimal (to the nearest ten thousandth); (3) convert the decimal to **(a)** %, **(b)** per $100 of assessed value, **(c)** per $1,000 of assessed value, and **(d)** mills (to the nearest hundredth); and (4) calculate the property tax due on Bill's home **(a)** in decimal, **(b)** per $100, **(c)** per $1,000, and **(d)** in mills.

Given

Assessed market value	40%	Total budget needed	$ 176,000
Market value of Bill's home	$210,000	Total assessed value	$1,910,000

✓ Solutions

*For **extra help** from your authors–Sharon and Jeff–see the videos in Connect.*

1. $.40 \times \$210{,}000 = \$84{,}000$

2. $\frac{\$176{,}000}{\$1{,}910{,}000} = .0922$ per dollar

3. a. $.0922 = 9.22\%$

b. $.0922 \times 100 = \$9.22$

c. $.0922 \times 1{,}000 = \$92.20$

d. $\frac{.0922}{.001} = 92.2$ mills (or $.0922 \times 1{,}000$)

4. a. $.0922 \times \$84{,}000 = \$7{,}744.80$

b. $\$9.22 \times 840 = \$7{,}744.80$

c. $\$92.20 \times 84 = \$7{,}744.80$

d. $92.20 \times .001 \times \$84{,}000 = \$7{,}744.80$

[3]Some states have credits available to reduce what the homeowner actually pays. For example, 42 out of 50 states give tax breaks to people over age 65. In Alaska, the state's homestead exemption reduces the property tax of a $168,000 house from $1,512 to $253.

INTERACTIVE CHAPTER ORGANIZER

Topic/Procedure/Formula	Examples	You try it*
Sales tax Sales tax is not calculated on trade discounts. Shipping charges, etc., also are not subject to sales tax. $\text{Actual sales} = \dfrac{\text{Total sales}}{1 + \text{Tax rate}}$ Cash discounts are calculated on sale price before sales tax is added on.	Calculate sales tax: Purchased 12 bags of mulch at \$59.40; 10% trade discount; 5% sales tax. \$59.40 − \$5.94 = \$53.46 \$53.46 × .05 **\$2.67 sales tax** Any cash discount would be calculated on \$53.46.	**Calculate sales tax** 14 bags of mulch at \$62.80; 8% trade discount; 6% sales tax
Excise tax Excise tax is calculated separately from sales tax and is an additional tax. It is based as a percent of the selling price. It could be stated as a fixed amount per item sold. In that case, the excise tax would be added to the cost of the item before any sales tax calculations. Rates for excise tax vary.	Jewelry \$4,000 retail price Sales tax 7% Excise tax 10% \$4,000 + 280 sales tax + 400 excise tax **\$4,680**	**Calculate cost of jewelry** \$6,000 retail price Sales tax 5% Excise tax 10%
Assessed value Assessment rate × Market value	\$100,000 house; rate, 30%; **\$30,000** assessed value.	**Calculate assessed value** \$200,000 house; rate, 40%.
Tax rate $\dfrac{\text{Budget needed}}{\text{Total assessed value}} = \text{Tax rate}$ (Round rate up to indicated digit even if less than 5.)	$\dfrac{\$800{,}000}{\$9{,}200{,}000} = .08695 =$ **.0870** tax rate per \$1	**Calculate tax rate** Budget needed, \$700,000; Total assessed value, \$8,400,000. (Round up to 4 digits.)
Expressing tax rate in other forms **1.** Percent: Move decimal two places to right. Add % sign. **2.** Per \$100: Multiply by 100. **3.** Per \$1,000: Multiply by 1,000. **4.** Mills: Divide by .001.	**1.** .0870 = **8.7%** **2.** .0870 × 100 = **\$8.70** **3.** .0870 × 1,000 = **\$87** **4.** $\dfrac{.0870}{.001} =$ **87 mills**	**Using the above tax rate, calculate tax rate in:** **1.** Percent **2.** Per \$100 **3.** Per \$1,000 **4.** Mills
Calculating property tax $\text{Total property tax due} = \text{Tax rate} \times \text{Total assessed value}$ Various forms: **1.** Percent × Assessed value **2.** Per \$100: $\dfrac{\text{Assessed value}}{\$100} \times \text{Rate}$ **3.** Per \$1,000: $\dfrac{\text{Assessed value}}{\$1{,}000} \times \text{Rate}$ **4.** Mills: Mills × .001 × Assessed value	*Example:* Rate, .0870 per \$1; \$30,000 assessed value **1.** (.087)8.7% × \$30,000 = **\$2,610** **2.** $\dfrac{\$30{,}000}{\$100} = 300 \times \$8.70 =$ **\$2,610** **3.** $\dfrac{\$30{,}000}{\$1{,}000} = 30 \times \$87 =$ **\$2,610** **4.** $\dfrac{.0870}{.001} = 87$ mills 87 mills × .001 × \$30,000 = **\$2,610**	**Calculate property tax for various forms given:** \$.0950 per \$1; \$40,000 assessed value

KEY TERMS			
	Assessed value Excise tax Mill	Personal property Property tax Real property	Sales tax Tax rate

*Worked-out solutions are in Appendix B.

Critical Thinking Discussion Questions with Chapter Concept Check

1. Explain sales and excise taxes. Should all states have the same tax rate for sales tax?
2. Explain how to calculate actual sales when the sales tax was included in the sales figure. Is a sales tax necessary?
3. How is assessed value calculated? If you think your value is unfair, what could you do?
4. What is a mill? When we calculate property tax in mills, why do we use .001 in the calculation?
5. **Chapter Concept Check.** Search the web to find the latest information on taxing online sales. Do you think it is fair? Defend your position using concepts learned in this chapter.

19

END-OF-CHAPTER PROBLEMS

Check figures for odd-numbered problems in Appendix B. Name ______________________ Date ________

DRILL PROBLEMS

Calculate the following: *LU 19-1(1, 2)*

	Retail selling price	Sales tax (6%)	Excise tax (10%)	Total price including taxes
19–1.	$800			
19–2.	$1,200			

Calculate the actual sales since the sales and sales tax were rung up together. Assume a 6% sales tax and round your answer to the nearest cent. *LU 19-1(1)*

19–3. $88,000

19–4. $26,000

Calculate the assessed value of the following pieces of property: *LU 19-2(2)*

	Assessment rate	Market value	Assessed value
19–5.	30%	$130,000	
19–6.	80%	$210,000	

Calculate the tax rate in decimal form to the nearest ten thousandth: *LU 19-2(2)*

	Required budget	Total assessed value	Tax rate per dollar
19–7.	$920,000	$39,500,000	

Complete the following:

	Tax rate per dollar	In percent	Per $100	Per $1,000	Mills
19–8.	.0956				
19–9.	.0699				

Complete the amount of property tax due to the nearest cent for each situation: *LU 19-2(3)*

	Tax rate	Assessed value	Amount of property tax due
19–10.	40 mills	$ 65,000	
19–11.	$42.50 per $1,000	105,000	
19–12.	$8.75 per $100	125,000	
19–13.	$94.10 per $1,000	180,500	

WORD PROBLEMS

eXcel My Money

19–14. Be careful when signing a work-for-hire agreement if you are a songwriter. You may lose all rights to your song in the copyright law world. If your song sells thousands on iTunes, you do not get to share in any of the publishing income. If you live in New Jersey and iTunes sold five of your songs for a total of $65,000, what is the tax owed at 7.0%? *LU 19-1(1)*

eXcel

19–15. Don Chather bought a new Dell computer for $1,995. This included a 6% sales tax. What is the amount of sales tax and the selling price before the tax? *LU 19-1(1)*

My Money

19–16. Homeowners enjoy many benefits, including a federal tax deduction for state and local property taxes paid. Fishers, Indiana, was voted one of the top 100 best places to live in 2017 by *Money* magazine. With a population of 86,357, a median home price of $236,167, and estimated property taxes at 10.6 mills, how much does the average homeowner pay in property taxes? *LU 19-2(3)*

19–17. The median home price in Arlington, Virginia, is $634,000. If the assessment rate is 100%, what is the assessed value? *LU 19-2(2)*

eXcel

19–18. Bemidji, Minnesota, needs $3,850,000 for its 2019 budget. If total assessed value of property in Bemidji is $353,211,009, what is the tax rate expressed as a percent, per $100, per $1,000, and in mills? *LU 19-2(2)*

19–19. Lois Clark bought a ring for $6,000. She must still pay a 5% sales tax and a 10% excise tax. The jeweler is shipping the ring, so Lois must also pay a $40 shipping charge. What is the total purchase price of Lois's ring? *LU 19-1(1, 2)*

19–20. Blunt County needs $700,000 from property tax to meet its budget. The total value of assessed property in Blunt is $110,000,000. What is the tax rate of Blunt? Round to the nearest ten thousandth. Express the rate in mills. *LU 19-2(1, 2)*

19–21. Bill Shass pays a property tax of $3,200. In his community, the tax rate is 50 mills. What is Bill's assessed value? *LU 19-2(2)*

eXcel

19–22. The home of Bill Burton is assessed at $80,000. The tax rate is 18.50 mills. What is the tax on Bill's home? *LU 19-2(3)*

19–23. New Hampshire ranks as the #1 most expensive state for property taxes. The median 2017 property tax was $4,636. If the 2019 rate is $1.25 per $100 of assessed value, how much does a homeowner owe for a property assessed at $378,150? *LU 19-2(3)*

19–24. Bill Blake pays a property tax of $2,500. In his community, the tax rate is 55 mills. What is Bill's assessed value? Round to the nearest dollar. *LU 19-2(2)*

19–25. Assume the property tax rate for Minneapolis is $8.73 per square foot, and the Denver rate is $2.14 a square foot. If 3,500 square feet is occupied at each location, what is the difference paid in property taxes? *LU 19-2(3)*

CHALLENGE PROBLEMS

19–26. Ginny Fieg expanded her beauty salon by increasing her space by 20%. Ginny paid property taxes of $2,800 at 22 mills. The new rate is now 24 mills. As Ginny's accountant, estimate what she may have to pay for property taxes this year. Round the final answer to the nearest dollar. In the calculation, round assessed value to the nearest dollar. *LU 19-2(2)*

eXcel **19–27.** Art Neuner, an investor in real estate, bought an office condominium. The market value of the condo was $250,000 with a 70% assessment rate. Art feels that his return should be 12% per month on his investment after all expenses. The tax rate is $31.50 per $1,000. Art estimates it will cost $275 per month to cover general repairs, insurance, and so on. He pays a $140 condo fee per month. All utilities and heat are the responsibility of the tenant. Calculate the monthly rent for Art. Round your answer to the nearest dollar (at intermediate stages). *LU 19-2(2)*

SUMMARY PRACTICE TEST

Do you need help? Connect videos have step-by-step worked-out solutions.

1. Carol Shan bought a new Apple iPod at Best Buy for $299. The price included a 5% sales tax. What are the sales tax and the selling price before the tax? *LU 19-1(1)*

2. Jeff Jones bought a ring for $4,000 from Zales. He must pay a 7% sales tax and 10% excise tax. Since the jeweler is shipping the ring, Jeff must also pay a $30 shipping charge. What is the total purchase price of Jeff's ring? *LU 19-1(1, 2)*

3. The market value of a home in Boston, Massachusetts, is $365,000. The assessment rate is 40%. What is the assessed value? *LU 19-2(1)*

4. Jan County needs $910,000 from its property tax to meet the budget. The total value of assessed property in Jan is $180,000,000. What is Jan's tax rate? Round to the nearest ten thousandth. Express the rate in mills (to the nearest tenth). *LU 19-2(2)*

5. The home of Nancy Billows is assessed at $250,000. The tax rate is 4.95 mills. What is the tax on Nancy's home? *LU 19-2(3)*

6. V's Warehouse has a market value of $880,000. The property in V's area is assessed at 35% of the market value. The tax rate is $58.90 per $1,000 of assessed value. What is V's property tax? *LU 19-2(3)*

INTERACTIVE VIDEO WORKSHEET

GRADING THE SUMMARY PRACTICE TEST

Go to the summary practice test video in Connect (or click on it here in the ebook). Grade your summary practice test while viewing the video.

C for Correct/I for Incorrect

1. ________ 4. ________
2. ________ 5. ________
3. ________ 6. ________

If you achieved 100%, you are ready for your instructor's exam.

If any of the problems were incorrect, list the questions you missed and show steps to solve the problem correctly.

Replay the video to see if you have made the correct fixes to your mistakes. If you have any questions, contact your instructor asap.

Notes on Watching Videos

MY MONEY

Financial Fitness—It's Exhilarating and Exhausting!

What I need to know

Personal financial planning is also referred to as financial fitness. Staying fit requires a regimen for both health and personal finances. Personal financial planning is the process of assessing what your current financial status is, determining where you want to be, and taking the steps to get you there. Four good financial habits worth getting addicted to are **saving money, avoiding debt, investing for the future, and protecting what you have**. Personal financial plans help you stay financially fit. Going through the process of analyzing where you currently are financially, where you want to be, what you need to do to get you there, and how to protect it is all part of it.

What I need to do

Save money: Saving money can be accomplished in many ways. Many of us don't really even know where all of our money is going. Change that habit. Know what you spend and where you spend it. Cut back wherever you can to provide you with more discretionary income. Create and stick to a budget. Spend your money wisely. An emergency fund consisting of six months of monthly expenses should be funded and kept in a liquid account (savings account, money market account, or certificates of deposit with staggered maturity dates).

Avoid debt: Avoiding debt is critical to a healthy financial position. Many North Americans were raised on spending rather than saving. Relatively easy access to credit has taken its toll on personal savings, too. To help turn this trend around and enhance your personal financial health, eliminate late fees and pay off your credit card balances monthly so you do not pay finance charges. Check your credit score at least three times per year. A high credit score provides you with lower interest rates charged by lenders on credit cards, mortgages, car loans, etc. An improved score can save you thousands of dollars in interest costs.

Invest for the future: Investing for the future has become a necessity. Make your money work for you and pay yourself first. There are many different types of investments with varying returns. The average stock market return over the past 30 years, for example, has been 10.4% (http://www.stockpickssystem.com/historical-rate-of-return/). Create a diversified portfolio and watch your money grow. Money is worth much more invested than spent. Opportunity costs run rampant when money is spent frivolously.

Protect what you have: Protecting what you have is critical. Knowing what types of insurance you need depends on your age and family situation. Do an evaluation of your insurance needs and then compare rates through online sources such as www.insurance.com. Don't be caught underinsured. The costs can be overwhelming. Write a will and prepare your dependents and those surviving you for a healthy financial future. If you have no dependents, at the minimum set aside money for your burial. Leave instructions for those surviving you on what your wishes are. Consider writing letters to your loved ones and keeping these and your will in a safe deposit box or fireproof safe.

Resources I can use

- https://www.wikihow.com/Write-a-Personal-Financial-Plan
- https://www.fidelity.com/growing-managing-wealth/estate-planning/overview

MY MONEY ACTIVITY

Make a list of how you can save more money, avoid additional debt, and/or pay off debt, invest for the future in a diversified portfolio, and protect what you have through insurance and estate planning. Choose two items on the list each month to focus on and create a plan allowing you to work toward completion of each by a specified due date.

PERSONAL FINANCE

A KIPLINGER APPROACH

"My Property tax bill has skyrocketed, how can I reduce it?", *Kiplinger's*, May 2017, p. 37. Used by permission of The Kiplinger Washington Editors, Inc.

GAME PLAN

Q My property tax bill has skyrocketed. How can I reduce it?

START BY MAKING SURE you're taking advantage of all the property tax breaks available to you. Many jurisdictions will exclude a portion of a home's value from property taxes if you're a senior or a veteran, or if you're disabled. In Florida, all homeowners are eligible for a homestead exemption of up to $50,000; those 65 and older who meet certain income limits can claim up to an additional $50,000. Other jurisdictions reduce your tax bill by a certain percentage if you meet specific criteria.

These tax breaks are valuable, but they're often overlooked. For example, when Chicago increased property taxes by an average of 13% last year, it included a rebate program for low- and middle-income homeowners. The rebates were worth up to $200, but only 11% of eligible homeowners claimed them.

Rebates and other property tax breaks aren't automatic; you usually have to apply for them and show proof of eligibility. Go to your tax assessor's website for details.

©karen roach/Shutterstock

Fighting city hall. You can score an even bigger tax cut by challenging the assessed value of your home, which is used to calculate your tax bill. The National Taxpayers Union, an advocacy group, estimates that 30% to 60% of property in the U.S. is assessed for more than it's worth.

See how often your jurisdiction assesses property. If it's not every year, there's a greater chance your home's value has changed since the last assessment. Check how market value is determined. An appraiser might compare your property with similar, recently sold properties to determine its market value, then multiply that by a set fraction, known as the assessment ratio. So if a property's market value is determined to be $100,000 and the assessment ratio is 80%, the assessed value for property tax purposes would be $80,000. Property tax bills are typically calculated by multiplying the home's assessed value by the local tax rate. You can find this information on the tax assessor's website.

Next, review the assessor's data on your home. You'll find this on your property's record card, which should be on file at your assessor's office and may be available online. Look for errors, such as an incorrect number of bathrooms or inflated lot size. If you can't find a glaring mistake but believe your home is still being overassessed, check out the property cards of similar homes in your neighborhood to see how their assessments compare with yours. If the assessments on those properties are lower than yours, you can make the case that your property's assessment is too high. Recent sales of homes in your neighborhood could also help you demonstrate that your property is overvalued.

Armed with this information, request an informal meeting with your assessor. He or she may agree to adjust your assessment on the spot, says Aaron Terrazas, senior economist at Zillow, an online real estate marketplace. If that doesn't work, request a formal review. Procedures vary, but a typical review takes one to three months, and you'll usually receive the results in writing. You can find the procedures for your jurisdiction on the tax assessor's website. Pay attention to deadlines. Most jurisdictions give you 90 days to challenge a new assessment, but some give you only 30 days to appeal.

If that doesn't work, most jurisdictions allow you to appeal to an independent board. The burden of proof is usually on the property owner, so come prepared. Zillow (www.zillow.com) offers a tool you can use to check recent sales of properties in your neighborhood. Alternatively, ask a real estate agent to point out three to five comparable homes that have sold in the past 60 to 90 days, or get a professional appraisal (expect to spend several hundred dollars for one). **SANDRA BLOCK**

BUSINESS MATH ISSUE

The new tax law has little effect on how much you pay for property tax.

1. List the key points of the article and information to support your position.
2. Write a group defense of your position using math calculations to support your view. If you are in an online course, post to a discussion board.

Classroom Notes

Life, Fire, and Auto Insurance

Smartphone Use Lifts Car-Insurance Rates

By Leslie Scism
And Nicole Friedman

Distracted by their smartphones, America's drivers are becoming more dangerous by the day. That is pushing auto-insurance rates higher as insurers struggle to keep up.

Costs associated with crashes are outpacing premium increases for some companies, according to insurers, and they say the use of smartphones to talk, text and access the internet while on the road is a new and important factor behind the wrecks.

It is "an epidemic issue for this country," said Michael LaRocco, chief executive of State Auto Financial Corp., at an insurance-industry conference last month.

State Farm Mutual Automobile Insurance Co., the largest U.S. auto insurer by market share, said 36% of the people it surveyed in 2015 admitted to texting while driving, and 29% said they access the internet, compared with 31% and 13%, respectively, in 2009. Among drivers aged 18 to 29, 64% said in the 2015 survey that they text while driving, and 54% said they use the internet behind the wheel.

State Farm's survey found that 52% of respondents in 2011 owned a smartphone, and 88% owned one in 2015.

The connection between phones and collisions is surfacing in insurers' earnings. Fourth-quarter underwriting results for personal auto insurance worsened at Travelers Cos., Hartford Financial Services Group Inc., and Horace Mann Educators Corp., and all three said distracted driving was partly to blame. The three companies insure millions of vehicles across the U.S.

©Daxiao Productions/Shutterstock

LEARNING UNIT OBJECTIVES

LU 20–1: Life Insurance

1. Explain the types of life insurance; calculate life insurance premiums.
2. Explain and calculate cash value and other nonforfeiture options.

LU 20–2: Fire Insurance

1. Explain and calculate premiums for fire insurance of buildings and their contents.
2. Calculate refunds when the insured and the insurance company cancel fire insurance.
3. Explain and calculate insurance loss when coinsurance is not met.

LU 20–3: Auto Insurance

1. Explain and calculate the cost of auto insurance.
2. Determine the amount paid by the insurance carrier and the insured after an auto accident.

Your Guide to Successfully Completing This Chapter

Traditional book or ebook

Check box as you complete each step.

Steps

☐ Read learning unit.
  ☐ Complete practice quiz at the end of the learning unit.
☐ Grade practice quiz using provided solutions. (For more help, watch the learning unit video in Connect and have a Study Session with the authors. Then complete the additional practice quiz in Connect.)
☐ Repeat above for each of the three learning units in Chapter 20.
  ☐ Review chapter organizer.
  ☐ Complete assigned homework.
    ☐ Finish summary practice test. (Go to Connect via the ebook link and do the interactive video worksheet to grade.)
☐ Complete instructor's exam.

My Money

The chapter opener clip, "Smartphone Use Lifts Car-Insurance Rates," shows insurance companies are raising premiums due to accidents caused by smartphone use.

Regardless of the type of insurance you buy—life, auto, nursing home, property, or fire—be sure to read and understand the policy before you buy the insurance. It has been reported that half of the people in the United States who have property insurance have not read their policy and 60% do not understand their policy. If you do not understand your life, fire, or auto insurance policies, this chapter should answer many of your questions. We begin by studying life insurance.

Learning Unit 20–1: Life Insurance

The following *Wall Street Journal* clip shows 30% of U.S. households have no life-insurance coverage.

Bob Brady owns Bob's Deli. He is 40 years of age, married, and has three children. Bob wants to know what type of life insurance protection will best meet his needs. Following is a discussion between an insurance agent, Rick Jones, and Bob.

Bob: I would like to buy a life insurance policy that will pay my wife $200,000 in the event of my death. My problem is that I do not have much cash. You know, bills, bills, bills. Can you explain some types of life insurance and their costs?

Reprinted by permission of *The Wall Street Journal*, copyright 2017 Dow Jones & Company, Inc. All rights reserved worldwide.

Rick: Let's begin by explaining some life insurance terminology. The **insured** is you—the **policyholder** receiving coverage. The **insurer** is the company selling the insurance policy. Your wife is the **beneficiary.** As the beneficiary, she is the person named in the policy to receive the insurance proceeds at the death of the insured (that's you, Bob). The amount stated in the policy, say, $200,000, is the **face amount** of the policy. The **premium** (determined by **statisticians** called *actuaries*) is the periodic payments you agree to make for the cost of the insurance policy. You can pay premiums annually, semiannually, quarterly, or monthly. The more frequent the payment, the higher the total cost due to increased paperwork, billing, and so on. Now we look at the different types of insurance.

Types of Insurance

In this section Rick explains term insurance, straight life (ordinary life), 20-payment life, 20-year endowment, and universal life insurance.

Term Insurance[1]

Rick: The cheapest type of life insurance is **term insurance,** but it only provides *temporary* protection. Term insurance pays the face amount to your wife (beneficiary) only if you die within the period of the insurance (1, 5, 10 years, and so on).

For example, let's say you take out a 5-year term policy. The insurance company automatically allows you to renew the policy at increased rates until age 70. A new policy called **level premium term** may be less expensive than an annual term policy since each year for, say, 50 years, the premium will be fixed.

LO 1

The policy of my company lets you convert to other insurance types without a medical examination. To determine your rates under 5-year term insurance, check this table (Table 20.1). The annual premium at 40 years per $1,000 of insurance is $3.52. We use the following steps to calculate the total yearly premium.

CALCULATING ANNUAL LIFE INSURANCE PREMIUMS

Step 1. Look up the age of the insured (for females, subtract 3 years) and the type of insurance in Table 20.1. This gives the premium cost per $1,000.

Step 2. Divide the amount of coverage by $1,000 and multiply the answer by the premium cost per $1,000.

$$\frac{\$200{,}000\,(\text{coverage})}{\$1{,}000} = 200 \times \$3.52 = \$704$$

Number of thousands (200); Cost per $1,000 for age 40 ($3.52); Annual premium ($704)

Airport flight insurance is a type of term insurance.

From this formula you can see that for $704 per year for the next 5 years, we, your insurance company, offer to pay your wife $200,000 in the event of your death. At the end of the 5th year, you are not entitled to any cash from your paid premiums. If you do not renew your policy (at a higher rate) and die in the 6th year, we will not pay your wife anything. Term insurance provides protection for only a specific period of time.

Bob: Are you telling me that my premium does not build up any cash savings that you call **cash value?**

Rick: The term insurance policy does not build up cash savings. Let me show you a policy that does build up cash value. This policy is straight life.

[1]A new term policy is available that covers policyholders until their expected retirement age.

TABLE 20.1

Life insurance rates for males (for females, subtract 3 years from the age)[2]

Age	Five-year term	Age	Straight life	Age	Twenty-payment life	Age	Twenty-year endowment
20	1.85	20	5.90	20	8.28	20	13.85
21	1.85	21	6.13	21	8.61	21	14.35
22	1.85	22	6.35	22	8.91	22	14.92
23	1.85	23	6.60	23	9.23	23	15.54
24	1.85	24	6.85	24	9.56	24	16.05
25	1.85	25	7.13	25	9.91	25	17.55
26	1.85	26	7.43	26	10.29	26	17.66
27	1.86	27	7.75	27	10.70	27	18.33
28	1.86	28	8.08	28	11.12	28	19.12
29	1.87	29	8.46	29	11.58	29	20.00
30	1.87	30	8.85	30	12.05	30	20.90
31	1.87	31	9.27	31	12.57	31	21.88
32	1.88	32	9.71	32	13.10	32	22.89
33	1.95	33	10.20	33	13.67	33	23.98
34	2.08	34	10.71	34	14.28	34	25.13
35	2.23	35	11.26	35	14.92	35	26.35
36	2.44	36	11.84	36	15.60	36	27.64
37	2.67	37	12.46	37	16.30	37	28.97
38	2.95	38	13.12	38	17.04	38	30.38
39	3.24	39	13.81	39	17.81	39	31.84
40	3.52	40	14.54	40	18.61	40	33.36
41	3.79	41	15.30	41	19.44	41	34.94
42	4.04	42	16.11	42	20.31	42	36.59
43	4.26	43	16.96	43	21.21	43	38.29
44	4.50	44	17.86	44	22.15	44	40.09

Straight Life (Ordinary Life)

Rick: Straight-life insurance provides *permanent* protection rather than the temporary protection provided by term insurance. The insured pays the same premium each year or until death.[3] The premium for straight life is higher than that for term insurance because straight life provides both protection and a built-in cash savings feature. According to our table (Table 20.1), your annual premium, Bob, would be

Face value is usually the amount paid to the beneficiary at the time of the insured's death.

$$\frac{\$200{,}000}{\$1{,}000} = 200 \times \$14.54 = \$2{,}908 \text{ annual premium}$$

Bob: Compared to term, straight life is quite expensive.

Rick: Remember that term insurance has no cash value accumulating, as straight life does. Let me show you another type of insurance—20-payment life—that builds up cash value.

Twenty-Payment Life

Rick: A **20-payment life** policy is similar to straight life in that 20-payment life provides permanent protection and cash value, but you (the insured) pay premiums for only the first 20 years. After 20 years you own **paid-up insurance.** According to my table (Table 20.1), your annual premium would be

$$\frac{\$200{,}000}{\$1{,}000} = 200 \times \$18.61 = \$3{,}722 \text{ annual premium}$$

[2]The life insurance tables in this chapter show premiums for a sampling of age groups, options, and coverage available to those under 45 years of age.

[3]In the following section on nonforfeiture values, we show how a policyholder in later years can stop making payments and still be covered by using the accumulated cash value built up.

Bob: The 20-payment life policy is more expensive than straight life.

Rick: This is because you are only paying for 20 years. The shorter period of time does result in increased yearly costs. Remember that in straight life you pay premiums over your entire life. Let me show you another alternative that we call 20-year endowment.

Twenty-Year Endowment

Rick: The **20-year endowment** insurance policy is the most expensive. It is a combination of term insurance and cash value. For example, from age 40 to 60, you receive term insurance protection in that your wife would receive $200,000 should you die. At age 60, your protection *ends* and you receive the face value of the policy that equals the $200,000 cash value. Let's use my table again (Table 20.1) to see how expensive the 20-year endowment is:

$$\frac{\$200{,}000}{\$1{,}000} = 200 \times \$33.36 = \$6{,}672 \text{ annual premium}$$

In summary, Bob, following is a review of the costs for the various types of insurance we have talked about:

	5-year term	Straight life	20-payment life	20-year endowment
Premium cost per year	$704	$2,908	$3,722	$6,672

Before we proceed, I have another policy that may interest you—universal life.

LO 2

Universal Life Insurance

Rick: **Universal life** is basically a **whole-life** insurance plan with flexible premium schedules and death benefits. Under whole life, the premiums and death benefits are fixed. Universal has limited guarantees with greater risk on the holder of the policy. For example, if interest rates fall, the policyholder must pay higher premiums, increase the number of payments, or switch to smaller death benefits in the future.

Bob: That policy is not for me—too much risk. I'd prefer fixed premiums and death benefits.

Rick: OK, let's look at how straight life, 20-payment life, and 20-year endowment can build up cash value and provide an opportunity for insurance coverage without requiring additional premiums. We call these options **nonforfeiture values.**

Nonforfeiture Values

Rick: Except for term insurance, the other types of life insurance build up cash value as you pay premiums. These policies provide three options should you, the policyholder, ever want to cancel your policy, stop paying premiums, or collect the cash value. As shown in Figure 20.1, these options are cash value; **reduced paid-up insurance;** and **extended term insurance.**

FIGURE 20.1

Nonforfeiture options

Option 1: Cash value (cash surrender value)
a. Receive cash value of policy.
b. Policy is terminated.
The longer the policy has been in effect, the higher the cash value because more premiums have been paid in.
Option 2: Reduced paid-up insurance
a. Cash value buys protection without paying new premiums.
b. Face amount of policy is related to cash value buildup and age of insured. The **face amount is less than original policy.**
c. Policy continues for life (at a reduced face amount).
Option 3: Extended term insurance
a. Original face amount of policy continues for a certain period of time.
b. Length of policy depends on cash value built up and on insured's age.
c. This option results automatically if policyholder doesn't pay premiums and fails to elect another option.

For example, Bob, let's assume that at age 40 we sell you a $200,000 straight-life policy. Assume that at age 55, after the policy has been in force for 15 years, you want to stop paying premiums. From this table (Table 20.2), I can show you the options that are available.

TABLE 20.2 Nonforfeiture options based on $1,000 face value

Years insurance policy in force	STRAIGHT LIFE				20-PAYMENT LIFE				20-YEAR ENDOWMENT			
	Cash value	Amount of paid-up insurance	EXTENDED TERM Years	EXTENDED TERM Days	Cash value	Amount of paid-up insurance	EXTENDED TERM Years	EXTENDED TERM Days	Cash value	Amount of paid-up insurance	EXTENDED TERM Years	EXTENDED TERM Days
5	29	86	9	91	71	220	19	190	92	229	23	140
10	96	259	18	76	186	521	28	195	319	520	30	160
15	148	371	20	165	317	781	32	176	610	790	35	300
20	265	550	21	300	475	1,000		Life	1,000	1,000		Life

Option 1: Cash value

$\frac{\$200{,}000}{\$1{,}000} = 200 \times \$148 = \$29{,}600$

Option 2: Reduced paid-up insurance

$\frac{\$200{,}000}{\$1{,}000} = 200 \times \$371 = \$74{,}200$

Option 3: Extended term insurance

Bob could continue this $200,000 policy for 20 years and 165 days.

MONEY tips

Review your insurance policies, coverages, and deductibles annually. Consider disability, renter's, life, auto, burial, and home. Use the same insurance company for all your insurance needs to take advantage of discounts.

Insight into Health and Business Insurance Often people who interview for a new job are more concerned with the salary offered than the whole health care package such as eye care, dental care, hospital and doctor care, and so on. Be sure you know exactly what the new job offers in health insurance. For employees, company health insurance and life insurance benefits can be an important job consideration.

Some of the key types of business insurance that you may need as a business owner include fire insurance, business interruption insurance (business loss until physical damages are fixed), casualty insurance (insurance against a customer's suing your business due to an accident on company property), workers' compensation (insurance against injuries or sickness from being on the job), and group insurance (life, health, and accident).

Although group health insurance costs have soared recently, many companies still pay the major portion of the cost. Some companies also provide health insurance benefits for retirees. As health costs continue to rise, we can expect to see some changes in this employee benefit.

Companies vary in the type of life insurance benefits they provide to their employees. This insurance can be a percent of the employee's salary with the employee naming the beneficiary; or in the case of key employees, the company can be the beneficiary.

If as an employer you need any of the types of insurance mentioned in this section, be sure to shop around for the best price. If you are in the job market, consider the benefits offered by a company as part of your salary and make your decisions accordingly.

Aha!

Don't forget to subtract 3 years from a female's age when using the life insurance tables. In the next unit, we look specifically at fire insurance. Now let's check your understanding of this unit with a Practice Quiz.

LU 20–1 PRACTICE QUIZ

Complete this **Practice Quiz** to see how you are doing.

1. Bill Boot, age 39, purchased a $60,000, 5-year term life insurance policy. Calculate his annual premium from Table 20.1. After 4 years, what is his cash value?

2. Ginny Katz, age 32, purchased a $78,000, straight-life policy. Calculate her annual premium. If after 10 years she wants to surrender her policy, what options and what amounts are available to her?

For **extra help** from your authors–Sharon and Jeff–see the videos in Connect.

✓ Solutions

1. $\frac{\$60,000}{\$1,000} = 60 \times \$3.24 = \$194.40$ No cash value in term insurance.

2. $\frac{\$78,000}{\$1,000} = 78 \times \$8.46^* = \$659.88$

Option 1: Cash value	78 × $96 = $7,488
Option 2: Paid up	78 × $259 = $20,202
Option 3: Extended term	18 years and 76 days

*For females we subtract 3 years.

LO 1

My Money

Learning Unit 20–2: Fire Insurance

Periodically, some areas of the United States, especially California, have experienced drought followed by devastating fires. These fires spread quickly and destroy wooded areas and homes. When the fires occur, the first thought of the owners is the adequacy of their **fire insurance.** Homeowners are made more aware of the importance of fire insurance that provides for the replacement value of their home. Out-of-date fire insurance policies can result in great financial loss.

In this unit, Alice Swan meets with her insurance agent, Bob Jones, to discuss fire insurance needs for her new dress shop at 4 Park Plaza. (Alice owns the building.)

Alice: What is *extended coverage?*

Bob: Your basic fire insurance policy provides financial protection if fire or lightning damages your property. However, the extended coverage protects you from smoke, chemicals, water, or other damages that firefighters may cause to control the fire. We have many options available.

Alice: What is the cost of a fire insurance policy?

Bob: Years ago, if you bought a policy for 2, 3, 5, or more years, reduced rates were available. Today, with rising costs of reimbursing losses from fires, most insurance companies write policies for 1 to 3 years. The cost of a 3-year policy premium is 3 times the annual premium. Because of rising insurance premiums, your total costs are cheaper if you buy one 3-year policy than three 1-year policies.

©Pixtal/AgeFotostock

Alice: For my purpose, I will need coverage for 1 year. Before you give me the premium rates, what factors affect the cost of my premium?

Bob: In your case, you have several factors in your favor that will result in a lower premium. For example, (1) your building is brick, (2) the roof is fire-resistant, (3) the building is located next to a fire hydrant, (4) the building is in a good location (not next to a gas station) with easy access for the fire department, and (5) the goods within your store are not as flammable as, say, those of a paint store. I have a table here (Table 20.3) that gives an example of typical fire insurance rates for buildings and contents (furniture, fixtures, etc.).

TABLE 20.3

Fire insurance rates per $100 of coverage for buildings and contents

	CLASSIFICATION OF BUILDING			
	CLASS A		CLASS B	
Rating of area	Building	Contents	Building	Contents
1	.28	.35	.41	.54
2	.33	.47	.50	.60
3	.41	.50	.61	.65

Fire insurance premium equals premium for building and premium for contents.

Let's assume your building has an insured value of $190,000, is rated Class B, and has an area rating of 2. We insure your contents for $80,000. Using the rates shown in Table 20.3, we would calculate your total annual premium for building and contents as follows:

$$\text{Premium} = \frac{\text{Insured value}}{\$100} \times \text{Rate}$$

Building

$$\frac{\$190{,}000}{\$100} = 1{,}900 \times \$.50 = \$950$$

Contents

$$\frac{\$80{,}000}{\$100} = 800 \times \$.60 = \$480$$

Total premium = $950 + $480 = $1,430

For our purpose, we round all premiums to the nearest cent. In practice, the premium is rounded to the nearest dollar.

LO 2

Canceling Fire Insurance

Alice: What if my business fails in 7 months? Do I get back any portion of my premium when I cancel?

Bob: If the insured—that's you, Alice—cancels or wants a policy for less than 1 year, we use this **short-rate table** (Table 20.4). These rates are higher because it is more expensive to process a policy for a short time. For example, if you cancel at the end of 7 months, the premium cost is 67% of the annual premium. We would calculate your refund as follows:

MONEY tips

Make a video of the contents of your home, garage, sheds, etc., to inventory your belongings. Keep the video in a safety deposit box or fireproof box offsite. In the event of a loss, you will have a recording of your belongings. You may be amazed how much can be forgotten without documentation.

Short-rate premium = Annual premium × Short rate

$958.10 = $1,430 × .67

Refund = Annual premium − Short-rate premium

$471.90 = $1,430 − $958.10

Alice: Let's say that I don't pay my premium or follow the fire codes. What happens if your insurance company cancels me?

Bob: If the insurance company cancels you, the company is *not* allowed to use the short-rate table. To calculate what part of the premium the company may keep,[4] you can prorate the premium based on the actual days that have elapsed. We can illustrate the amount of your refund by assuming you are canceled after 7 months:

Note that when the insurance company cancels the policy, the refund ($595.83) is greater than if the insured cancels ($471.90).

For insurance company:

$$\text{Charge} = \$1{,}430 \text{ annual premium} \times \frac{7 \text{ months elapsed}}{12}$$

Charge = $834.17

For insured:

Refund = $1,430 annual premium − $834.17 charge

Refund = $595.83

TABLE 20.4

Fire insurance short-rate and cancellation table

Time policy is in force		Percent of annual rate to be charged	Time policy is in force		Percent of annual rate to be charged
Days:	5	8%	Months:	5	52%
	10	10		6	61
	20	15		7	67
	25	17		8	74
Months:	1	19		9	81
	2	27		10	87
	3	35		11	96
	4	44		12	100

[4]Many companies use $\frac{\text{Days}}{365}$.

LO 3

Coinsurance

Alice: My friend tells me that I should meet the coinsurance clause. What is coinsurance?

Bob: Usually, fire does not destroy the entire property. **Coinsurance** means that you and the insurance company *share* the risk. The reason for this coinsurance clause[5] is to encourage property owners to purchase adequate coverage.

Alice: What is adequate coverage?

Bob: In the fire insurance industry, the usual rate for coinsurance is 80% of the current replacement cost. This cost equals the value to replace what was destroyed. If your insurance coverage is 80% of the current value, the insurance company will pay all damages up to the face value of the policy.

Alice: Hold it, Bob! Will you please show me how this coinsurance is figured?

Bob: Yes, Alice, I'll be happy to show you how we figure coinsurance. Let's begin by looking at the following steps so you can see what amount of the insurance the company will pay.

MONEY tips

Request quotes from your insurance provider using different deductibles for each policy. Choose the one that meets your financial needs. Keep the amount of each deductible in an interest-earning account for easy access if and when it is needed.

CALCULATING WHAT INSURANCE COMPANY PAYS WITH COINSURANCE CLAUSE

Step 1. Set up a fraction. The numerator is the actual amount of the insurance carried on the property. The denominator is the amount of insurance you should be carrying on the property to meet coinsurance (80% times the replacement value).

Step 2. Multiply the fraction by the amount of loss (up to the face value of the policy).

Let's assume for this example that you carry $60,000 fire insurance on property that will cost $100,000 to replace. If the coinsurance clause in your policy is 80% and you suffer a loss of $20,000, your insurance company will pay the following:

Although there are many types of property and homeowner's insurance policies, they usually include fire protection.

Insurance coverage → $60,000 (numerator); Loss → $20,000

Step 1 $$\frac{\$60{,}000}{\$80{,}000} \times \$20{,}000 = \$15{,}000$$ **Step 2**

What you should have carried → $80,000 ($100,000 × .80)

This kind of limited insurance payment for a loss is often called an **indemnity.** If you had had actual insurance coverage of $80,000, then the insurance company would have paid $20,000. Remember that if the coinsurance clause is met, the most an insurance company will pay is the face value of the policy.

You are now ready for the following Practice Quiz.

LU 20–2 PRACTICE QUIZ

Complete this **Practice Quiz** to see how you are doing.

1. Calculate the total annual premium of a warehouse that has an area rating of 2 with a building classification of B. The value of the warehouse is $90,000 with contents valued at $30,000.
2. If the insured in problem 1 cancels at the end of month 9, what are the costs of the premium and the refund?
3. Jones insures a building for $120,000 with an 80% coinsurance clause. The replacement value is $200,000. Assume a loss of $60,000 from fire. What will the insurance company pay? If the loss was $160,000 and coinsurance *was* met, what would the insurance company pay?

[5]In some states (including Wisconsin), the clause is not in effect for losses under $1,000.

*For **extra help** from your authors–Sharon and Jeff–see the videos in Connect.*

✓ Solutions

1. $\frac{\$90{,}000}{\$100} = 900 \times \$.50 = \$450$

$\frac{\$30{,}000}{\$100} = 300 \times \$.60 = \frac{180}{\$630}$ ← total premium

2. $\$630 \times .81 = \$510.30$ $\quad\quad$ $\$630 - \$510.30 = \$119.70$

3. $\frac{\$120{,}000}{\$160{,}000} = \frac{3}{4} \times \$60{,}000 = \$45{,}000$

↑

$(.80 \times \$200{,}000)$ $\quad\quad$ $\$160{,}000$ never more than face value

Learning Unit 20–3: Auto Insurance

My Money

If you own an auto, you have had some experience purchasing auto insurance. Often first-time auto owners do not realize that auto insurance can be a substantial expense. Insurance rates often increase when a driver is involved in an accident. Some insurance companies give reduced rates to accident-free drivers—a practice that has encouraged drivers to be more safety conscious. For example, State Farm Insurance offers a discount to drivers who maintain a safety record. An important factor in safe driving is the use of a seat belt. Make it a habit to always put on your seat belt. Note in the following *Wall Street Journal* clip, "How's My Driving?" the difference between two groups of people as to who files the most claims.

How's My Driving?

States with the biggest gap in the percentage of auto-insurance claims filed by homeowners and renters.

	% of homeowners who file claims	% of renters who file claims	Differential
Utah	11.93%	16.16%	4.23%
Oregon	14.52%	20.17%	5.65%
Maryland	15.88%	21.12%	5.25%
South Carolina	16.99%	21.34%	4.35%
Nebraska	15.17%	22.65%	7.48%

Source: Insurance.com

Source: *The Wall Street Journal*, 1/2/15.

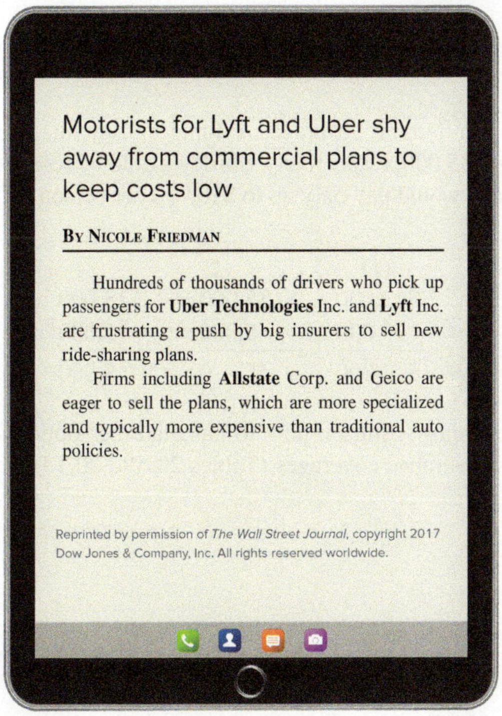

Motorists for Lyft and Uber shy away from commercial plans to keep costs low

By Nicole Friedman

Hundreds of thousands of drivers who pick up passengers for **Uber Technologies** Inc. and **Lyft** Inc. are frustrating a push by big insurers to sell new ride-sharing plans.

Firms including **Allstate** Corp. and Geico are eager to sell the plans, which are more specialized and typically more expensive than traditional auto policies.

The *Wall Street Journal* clip, "Motorists for Lyft and Uber shy away from commercial plans to keep costs low," reflects drivers are not interested in ride-sharing plans because of the expense.

In this unit we follow Shirley as she learns about auto insurance. Shirley, who just bought a new auto, has never purchased auto insurance. So she called her insurance agent, Bob Long, who agreed to meet her for lunch. We will listen in on their conversation.

Shirley: Bob, where do I start?

Bob: Our state has two kinds of **liability insurance,** or **compulsory insurance,** that by law you must buy (regulations and requirements vary among states). Liability insurance covers any physical damages that you inflict on others or their property. You must buy liability insurance for the following:

1. **Bodily injury** to others: 10/20. This means that the insurance company will pay damages to people injured or killed by your auto up to $10,000 for injury to one person per accident or a total of $20,000 for injuries to two or more people per accident.

Liability insurance includes
1. **Bodily injury**—injury or death to people in passenger car or other cars, etc.
2. **Property damage**—injury to other people's autos, trees, buildings, hydrants, etc.

2. **Property damage** to someone else's property: 5. The insurance company will pay up to $5,000 for damages that you have caused to the property of others.

Now we leave Shirley and Bob for a few moments as we calculate Shirley's premium for compulsory insurance.

TABLE 20.5 Compulsory insurance (based on class of driver)

BODILY INJURY TO OTHERS		DAMAGE TO SOMEONE ELSE'S PROPERTY	
Class	**10/20**	**Class**	**5M***
10	$ 55	10	$129
17	98	17	160
18	80	18	160
20	116	20	186

Explanation of 10/20 and 5

10	20	5
Maximum paid to one person per accident for bodily injury	Maximum paid for total bodily injury per accident	Maximum paid for property damage per accident

*M means thousands.

LO 1

Calculating Premium for Compulsory Insurance[6]

The tables we use in this unit are for Territory 5. Other tables are available for different territories.

Insurance companies base auto insurance rates on the territory you live in, the class of driver (class 10 is experienced driver with driver training), whether the auto is for business use, how much you drive the car, the age of the car, and the make of the car (symbol). Shirley lives in Territory 5 (suburbia). She is classified as 17 because she is an inexperienced operator licensed for less than 6 years. Her car is age 3 and symbol 4 (make of car). We use Table 20.5 to calculate Shirley's compulsory insurance. Note that the table rates in this unit are not representative of all areas of the country. In case of lawsuits, the minimum coverage may not be adequate. Some states add surcharges to the premium if the person has a poor driving record. The tables are designed to show how rates are calculated. From Table 20.5, we can determine Shirley's premium for compulsory insurance as follows:

Bodily injury	$ 98
+ Property damage	160
	$258

Remember that the $258 premium represents minimum coverage. Assume Shirley hits two people and the courts award them $13,000 and $5,000, respectively. Shirley would be responsible for $3,000 because the insurance company would pay only up to $10,000 per person and a total of $20,000 per accident.

Although total damages of $18,000 are less than $20,000, the insurance company pays only $15,000.

	(1)		(2)		
	$13,000	+	$5,000 =		$18,000
Paid by insurance company →	− 10,000	−	5,000 =		− 15,000
Paid by Shirley →	$ 3,000	+	$ 0 =		$ 3,000

We return to Shirley and Bob. Bob now shows Shirley how to calculate her optional insurance coverage. Remember that optional insurance coverages (Tables 20.6 to 20.10) are added to the costs in Table 20.5.

[6]Some states may offer medical payment insurance (a supplement to policyholders' health and accident insurance) as well as personal injury protection against uninsured or underinsured motorists.

TABLE 20.6

Bodily injury

Class	15/30	20/40	20/50	25/50	25/60	50/100	100/300	250/500	500/1,000
10	27	37	40	44	47	69	94	144	187
17	37	52	58	63	69	104	146	228	298
18	33	46	50	55	60	89	124	193	251
20	41	59	65	72	78	119	168	263	344

TABLE 20.7

Damage to someone else's property

Class	10M	25M	50M	100M
10	132	134	135	136
17	164	166	168	169
18	164	166	168	169
20	191	193	195	197

Calculating Optional Insurance Coverage

Bob: In our state, you can add optional bodily injury to the compulsory amount. If you finance your car, the lender may require specific amounts of optional insurance to protect its investment. I have two tables (Tables 20.6 and 20.7) here that we use to calculate the option of 250/500/50. This means that in an accident the insurance company will pay $250,000 per person, up to $500,000 per accident, and up to $50,000 for property damage.

Bob then explains the tables to Shirley. By studying the tables, you can see how insurance companies figure bodily injury and damage to someone else's property. Shirley is Class 17:

Bodily injury
250/500 = $228
Property damage
50M = +168
$396 premium for optional bodily injury and property damage

Note: These are additional amounts to compulsory.

Collision and comprehensive are optional insurance types that pay only the insured. Note that Tables 20.8 and 20.9 are based on territory, age, and car symbol. The higher the symbol, the more expensive the car.

Shirley: Is that all I need?

Bob: No, I would recommend two more types of optional coverage: **collision** and **comprehensive.** Collision provides protection against damages to your car caused by a moving vehicle. It covers the cost of repairs less **deductibles** (amount of repair you cover first before the insurance company pays the rest) and depreciation.[7] In collision, insurance companies pay the resale or book value. So as the car gets older, after 5 or more years, it might make sense to drop the collision. The decision depends on how much risk you are willing to assume. Comprehensive covers damages resulting from theft, fire, falling objects, and so on. Now let's calculate the cost of these two types of coverage—assuming a $100 deductible for collision and a $200 deductible for comprehensive—with some more of my tables (Tables 20.8 and 20.9, page 522).

	Class	Age	Symbol	Premium
Collision	17	3	4	$191 ($148 + $43)
Comprehensive	17	3	4	+ 56 ($52 + $4)
				$247

Cost to reduce deductibles

Total premium for collision and comprehensive

Shirley: Anything else?

Bob: I would also recommend that you buy towing and substitute transportation coverage. The insurance company will pay up to $25 for each tow. Under substitute transportation, the

[7]In some states, repair to glass has no deductible and many insurance companies now use a $500 deductible instead of $300.

TABLE 20.8 Collision

Classes	Age group	Symbols 1–3 $300 ded.	Symbol 4 $300 ded.	Symbol 5 $300 ded.	Symbol 6 $300 ded.	Symbol 7 $300 ded.	Symbol 8 $300 ded.	Symbol 10 $300 ded.
10–20	1	180	180	187	194	214	264	279
	2	160	160	166	172	190	233	246
	3	148	148	154	166	183	221	233
	4	136	136	142	160	176	208	221
	5	124	124	130	154	169	196	208

These classes would use all this information.

To find the premium, use the age and symbol only.

Additional cost to reduce deductible

Class	From $300 to $200	From $300 to $100
10	13	27
17	20	43
18	16	33
20	26	55

TABLE 20.9 Comprehensive

Classes	Age group	Symbols 1–3 $300 ded.	Symbol 4 $300 ded.	Symbol 5 $300 ded.	Symbol 6 $300 ded.	Symbol 7 $300 ded.	Symbol 8 $300 ded.	Symbol 10 $300 ded.
10–25	1	61	61	65	85	123	157	211
	2	55	55	58	75	108	138	185
	3	52	52	55	73	104	131	178
	4	49	49	52	70	99	124	170
	5	47	47	49	67	94	116	163

Additional cost to reduce deductible: From $300 to $200 add $4

TABLE 20.10

Transportation and towing

Substitute transportation	$16
Towing and labor	4

MONEY tips

Keep a memory card with photos of both the inside and outside of all the vehicles, structures, and property you own in a safety deposit box. Photographing VIN and serial numbers is a wise idea as well. This will help the claim process in the event of a loss.

insurance company will pay you $12 a day for renting a car, up to $300 total. Again, from another table (Table 20.10), we find the additional premium for towing and substitute transportation is $20 ($16 + $4).

We leave Shirley and Bob now as we make a summary of Shirley's total auto premium in Table 20.11.

No-Fault Insurance Some states have **no-fault insurance,** a type of auto insurance that was intended to reduce premium costs on bodily injury. With no fault, one forfeits the right to sue for *small* claims involving medical expense, loss of wages, and so on. Each person collects the bodily injury from his or her insurance company no matter who is at fault. In reality, no-fault insurance has not reduced premium costs, due to large lawsuits, fraud, and operating costs of insurance companies. Many states that were once considering no fault are no longer pursuing its adoption. Note that states with no-fault insurance require the purchase of *personal-injury protection (PIP).* The most successful no-fault law seems to be in Michigan, since it has tough restrictions on the right to sue along with unlimited medical and rehabilitation benefits.

Premiums for collision, property damage, and comprehensive are not reduced by no fault.

TABLE 20.11

Worksheet for calculating Shirley's auto premium

Compulsory insurance	Limits	Deductible	Premium
Bodily injury to others	$10,000 per person $20,000 per accident	None	$ 98 (Table 20.5)
Damage to someone else's property	$5,000 per accident	None	$160 (Table 20.5)
Options			
Optional bodily injury to others	$250,000 per person $500,000 per accident	None	$228 (Table 20.6)
Optional property damage	$50,000 per accident	None	$168 (Table 20.7)
Collision	Actual cash value	$100	$191 (Table 20.8) ($148 + $43)
Comprehensive	Actual cash value	$200	$ 56 (Table 20.9) ($52 + $4)
Substitute transportation	Up to $12 per day or $300 total	None	$ 16 (Table 20.10)
Towing and labor	$25 per tow	None	$ 4 (Table 20.10)
			$921 Total premium

LO 2

Calculating What the Insurance Company and the Insured Pay after an Auto Accident

My Money

When an automobile accident occurs, the insurance company pays up to the maximum of insurance coverage. The insured pays whatever is left. Because you will be financially responsible for damages not covered by your insurance, you should always get quotes from various companies for different coverage limits as well as deductibles. The cost to change to the next level of coverage may not be significant but lack of better coverage can be very costly.

Let's look at a typical example.

MONEY tips

Drive carefully. Tickets and accidents significantly affect the cost of your automobile insurance premiums. When paying your premiums, consider how important it is to have insurance when you need it and what a gift it is to have insurance and not need it.

EXAMPLE Mario Andreety was at fault in an auto accident. He destroyed a fence and tree in a yard, causing damages of $2,500. He injured three passengers in the car he hit, which resulted in the following medical expenses: passenger 1, $25,250, passenger 2, $17,589, passenger 3, $12,567. The damage to the BMW he hit amounted to $45,888. Finally, the damage to his own vehicle came to $9,772.

a. If Mario has 15/30/10 coverage with a $500 deductible for collision and $100 deductible for comprehensive, how much will the insurance company pay? How much will Mario pay?

Insurance company pays	Mario pays
$2,500	$0 Property damage (He has $10,000 property damage.)
$15,000	$25,250 − $15,000 = $10,250 Passenger 1 (His coverage is $15,000 per person bodily injury with an accident maximum of $30,000.)
$15,000	$17,589 − $15,000 = $2,589 Passenger 2 (The $30,000 [$15,000 + $15,000] maximum bodily injury coverage has been met.)
$0 max	$12,567 − $0 = $12,567 Passenger 3 ($30,000 per accident has been reached.)
$7,500	$45,888 − $7,500 = $38,388 (His coverage is $10,000 property damage with $2,500 used.)
$9,272	$500 ($9,772 personal vehicle − $500 collision deductible)
$49,272	$64,294

b. If Mario has 50/100/50 with a $500 deductible for collision and $100 deductible for comprehensive, how much will the insurance company pay? How much will Mario pay?

Insurance company pays	Mario pays
$2,500	$0 Property damage
$25,250	$0 Passenger 1
$17,589	$0 Passenger 2 ($25,250 + $17,589 = $42,839)
$12,567	$0 Passenger 3 ($42,839 + $12,567 = $55,406 < $100,000)
$45,888	$0 ($2,500 + $45,888 = $48,388 < $50,000)
$9,272	$500 ($9,772 − $500)
$113,066	$500

Clearly, Mario would be better off paying higher premiums for better coverage. His premiums won't amount to anywhere near $64,000 a year but, after only one accident with insufficient coverage, his out-of-pocket costs could easily exceed that amount.

It's time to take your final Practice Quiz in this chapter.

LU 20–3 PRACTICE QUIZ

Complete this **Practice Quiz** to see how you are doing.

1. Calculate the annual auto premium for Mel Jones who lives in Territory 5, is a driver classified 18, and has a car with age 4 and symbol 7. His state has compulsory insurance, and Mel wants to add the following options:
   - **a.** Bodily injury, 100/300.
   - **b.** Damage to someone else's property, 10M.
   - **c.** Collision, $200 deductible.
   - **d.** Comprehensive, $200 deductible.
   - **e.** Towing.
2. Calculate how much the insurance company and Carl Burns, the insured, pay if Carl carries 10/20/5 with $500 deductible for collision and $100 deductible for comprehensive and is at fault in an auto accident causing the following damage: $7,981, personal property; $6,454, injury to passenger 1; $4,239, injury to passenger 2; $25,250, injury to passenger 3; and $12,120 damage to Carl's car.

*For **extra help** from your authors–Sharon and Jeff–see the videos in Connect.*

✓ Solutions

1.

Compulsory			
Bodily	$ 80		(Table 20.5)
Property	160		(Table 20.5)
Options			
Bodily	124		(Table 20.6)
Property	164		(Table 20.7)
Collision	192	($176 + $16)	(Table 20.8)
Comprehensive	103	($99 + $4)	(Table 20.9)
Towing	4		(Table 20.10)
Total annual premium	$827		

2. Carl's coverage is $10,000 per person bodily injury with an accident maximum of $20,000 bodily injury and $5,000 property damage.

Insurance company pays	Carl pays
$ 5,000	$ 2,981 Property damage
6,454	0 Passenger 1
4,239	0 Passenger 2 ($6,454 + $4,239 = $10,693)
9,307 ($20,000 − $10,693)	15,943 Passenger 3 ($25,250 − $9,307)
11,620	500
$36,620	$19,424

INTERACTIVE CHAPTER ORGANIZER

Topic/Procedure/Formula	Examples	You try it*
Life insurance Using Table 20.1, per $1,000: $\frac{\text{Coverage desired}}{\$1,000} \times \text{Rate}$ For females, subtract 3 years.	**Given** $80,000 of insurance desired; age 34; male. **1.** 5-year term: $\frac{\$80,000}{\$1,000} = 80 \times \$2.08 = \$166.40$ **2.** Straight life: $\frac{\$80,000}{\$1,000} = 80 \times \$10.71 = \$856.80$ **3.** 20-payment life: $\frac{\$80,000}{\$1,000} = 80 \times \$14.28 = \$1,142.40$ **4.** 20-year endowment: $\frac{\$80,000}{\$1,000} = 80 \times \$25.13 = \$2,010.40$	**Given** $90,000 of insurance desired; age 36; male. **Calculate these premiums:** 1. 5-year term 2. Straight life 3. 20-payment life 4. 20-year endowment
Nonforfeiture values **By Table 20.2** Option 1: Cash surrender value. Option 2: Reduced paid-up insurance policy continues for life at reduced face amount. Option 3: Extended term—original face policy continued for a certain period of time.	A $50,000 straight-life policy was issued to Jim Rose at age 28. At age 48 Jim wants to stop paying premiums. What are his nonforfeiture options? Option 1: $\frac{\$50,000}{\$1,000} = 50 \times \$265 = \$13,250$ Option 2: $50 \times \$550 = \$27,500$ Option 3: 21 years 300 days	**Given** $60,000 straight-life policy issued to Ron Lee at age 30. At age 50, Ron wants to stop paying premium. **Calculate his nonforfeiture options**
Fire insurance Per $100 $\text{Premium} = \frac{\text{Insurance value}}{\$100} \times \text{Rate}$ Rate can be for buildings or contents.	**Given** Area 3; Class B; building insured for $90,000; contents, $30,000. Building: $\frac{\$90,000}{\$100} = 900 \times \$.61 = \$549$ Contents: $\frac{\$30,000}{\$100} = 300 \times \$.65 = \$195$ Total: $549 + $195 = $744	**Calculate fire insurance premium** Area 3; Class B; insurance for $80,000; contents, $20,000.
Canceling fire insurance—short-rate Table 20.4 (canceling by policyholder) $\text{Short-rate premium} = \text{Annual premium} \times \text{Short rate}$ $\text{Refund} = \text{Annual premium} - \text{Short-rate premium}$ If insurance company cancels, do not use Table 20.4.	Annual premium is $400. Short rate is .35 (cancel end of 3 months). $400 × .35 = $140 Refund = $400 − $140 = $260	**Calculate refund** Annual premium is $600; insurance cancels after 4 months.
Canceling by insurance company $\text{Annual premium} \times \frac{\text{Months elapsed}}{12}$ (Refund is higher since company cancels.)	Using example above, assume the insurance company cancels at end of 3 months. $\$400 \times \frac{1}{4} = \$100$ Refund = $400 − $100 = $300	**Calculate refund from example above if insurance company cancels**

(continues)

INTERACTIVE CHAPTER ORGANIZER

Topic/Procedure/Formula	Examples	You try it*
Coinsurance Amount insurance company pays: $\frac{\text{Actual insurance} \rightarrow \text{Insurance carried (Face value)}}{\text{What coverage should have been} \rightarrow \text{Insurance required to meet coinsurance (Rate} \times \text{Replacement value)}} \times \text{Loss}$ Insurance company never pays more than the face value.	**Given** Face value, \$30,000; replacement value, \$50,000; coinsurance rate, 80%; loss, \$10,000; insurance to meet required coinsurance, \$40,000. $\frac{\$30,000}{\$40,000} \times \$10,000 = \$7,500$ paid by insurance company (\$40,000 = \$50,000 × .80)	**Calculate coinsurance** **Given** Face value, \$40,000; replacement, \$60,000; rate, 80%; loss, \$9,000.
Auto insurance **Compulsory** Required insurance. **Optional** Added to cost of compulsory. Bodily injury—pays for injury to person caused by insured. Property damage—pays for property damage (not for insured auto). Collision—pays for damages to insured auto. Comprehensive—pays for damage to insured auto for fire, theft, etc. Towing. Substitute transportation.	Calculate the annual premium. Driver class 10; compulsory 10/20/5. **Optional** Bodily—100/300 Property—10M Collision—age 3, symbol 10, \$100 deductible Comprehensive—\$300 deductible 10/20/5 \$184 Table 20.5 (\$55 + \$129) Bodily 94 Table 20.6 Property 132 Table 20.7 Collision 260 Table 20.8 (\$233 + \$27) Comprehensive 178 Table 20.9 Total premium \$848	**Calculate annual premium** Driver class 10; compulsory 10/20/5. **Optional** Bodily—100/300 Property—10M Collision—age 5, symbol 8, \$100 deductible Comprehensive—\$300 deductible

KEY TERMS

Beneficiary
Bodily injury
Cash value
Coinsurance
Collision insurance
Comprehensive insurance
Compulsory insurance
Deductibles
Extended term insurance
Face amount
Face value
Fire insurance
Indemnity
Insured
Insurer
Level premium term
Liability insurance
No-fault insurance
Nonforfeiture values
Paid-up insurance
Policyholder
Premium
Property damage
Reduced paid-up insurance
Short-rate table
Statisticians
Straight-life insurance
Term insurance
20-payment life
20-year endowment
Universal life
Whole life

*Worked-out solutions are in Appendix B.

Critical Thinking Discussion Questions with Chapter Concept Check

1. Compare and contrast term insurance versus whole-life insurance. At what age do you think people should take out life insurance?
2. What is meant by *nonforfeiture values*? If you take the cash value option, should it be paid in a lump sum or over a number of years?
3. How do you use a short-rate table? Explain why an insurance company gets less in premiums if it cancels a policy than if the insured cancels.
4. What is coinsurance? Do you feel that an insurance company should pay more than the face value of a policy in the event of a catastrophe?
5. Explain compulsory auto insurance, collision, and comprehensive. If your car is stolen, explain the steps you might take with your insurance company.
6. "Health insurance is not that important. It would not be worth the premiums." Please take a stand.
7. **Chapter Concept Check.** Based on concepts in the chapter, what would it cost you to set up a life insurance policy that would fit your needs?

20

END-OF-CHAPTER PROBLEMS

connect

Check figures for odd-numbered problems in Appendix B. Name ________________ Date ________

DRILL PROBLEMS

Calculate the annual premium for the following policies using Table 20.1 (for females subtract 3 years from the table). *LU 20-1(1)*

	Amount of coverage (face value of policy)	Age and sex of insured	Type of insurance policy	Annual premium
20–1.	$200,000	42 F	Straight life	
20–2.	$200,000	42 M	20-payment life	
20–3.	$75,000	29 F	5-year term	
20–4.	$50,000	27 F	20-year endowment	

Calculate the following nonforfeiture options for Lee Chin, age 42, who purchased a $200,000 straight-life policy. At the end of year 20, Lee stopped paying premiums. *LU 20-1(2)*

20–5. Option 1: Cash surrender value

20–6. Option 2: Reduced paid-up insurance

20–7. Option 3: Extended term insurance

Calculate the total cost of a fire insurance premium (rounded to nearest cent) for a building and its contents given the following: *LU 20-2(1)*

		Rating of area	Class	Building	Contents	Total premium cost
eXcel	**20–8.**	3	B	$90,000	$40,000	

Calculate the short-rate premium and refund of the following: *LU 20-2(2)*

		Annual premium	Canceled after	Short-rate premium	Refund
eXcel	**20–9.**	$700	8 months by insured		
eXcel	**20–10.**	$360	4 months by insurance company		

Complete the following: *LU 20-2(3)*

		Replacement value of property	Amount of insurance	Kind of policy	Actual fire loss	Amount insurance company will pay
eXcel	**20–11.**	$100,000	$60,000	80% coinsurance	$22,000	
	20–12.	$60,000	$40,000	80% coinsurance	$42,000	

Calculate the annual auto insurance premium for the following: *LU 20-3(1)*

20–13. Britney Sper, Territory 5
Class 17 operator
Compulsory, 10/20/5 ______

Optional

a. Bodily injury, 500/1,000 ______

b. Property damage, 25M ______

c. Collision, $100 deductible ______
Age of car is 2; symbol of car is 7

d. Comprehensive, $200 deductible ______

Total annual premium ______

WORD PROBLEMS

My Money

20–14. The average Roman's lifespan 2,000 years ago was 22 years. In 1900, a person was expected to live 47.3 years. In 2017, life expectancy was 78.5. If you are a 47-year-old female, what annual premium would you pay for a $200,000, 5-year term life insurance policy? What will be the cash value after 3 years? *LU 20-1(1, 2)*

My Money

20–15. CBS News reported four ways to cut down on the cost of life insurance: (1) Shop around for the best rates from reputable companies, (2) improve your life expectancy by quitting smoking, etc., (3) buy life insurance when you are young, and (4) negotiate for lower premiums. Warren Kawano, age 34, was quoted $20 per month for a $100,000 term life insurance policy. Compare this to the rates in Table 20.1. *LU 20-1(1)*

eXcel

20–16. Kathleen Osness, a 38-year-old massage therapist, decided to take out a limited-payment life policy. She chose this since she expects her income to decline in future years. Kathleen decided to take out a 20-year payment life policy with a coverage amount of $90,000. Could you advise Kathleen about what her annual premium will be? If she decides to stop paying premiums after 15 years, what will be her cash value? *LU 20-1(1)*

My Money

20–17. Life insurance for a single parent is critical. Protect your children. The good news is life insurance for most single parents is very inexpensive. Buy term life insurance with a 20-year term, multiply your annual income by 7 to 10 times to determine how much to buy but get what you can afford now. Compare premiums. Choose a beneficiary. Janette Raffa, a single mom, has two young children and wants to take out an additional $300,000 of 5-year term insurance. Janette is 40 years old. What will be her additional annual premium? In 3 years, what cash value will have been built up? *LU 20-1(1)*

20–18. Roger's office building has a $320,000 value, a 2 rating, and a B building classification. The contents in the building are valued at $105,000. Could you help Roger calculate his total annual premium? *LU 20-2(1)*

eXcel

20–19. Abby Ellen's toy store is worth $400,000 and is insured for $200,000. Assume an 80% coinsurance clause and that a fire caused $190,000 damage. What is the liability of the insurance company? *LU 20-2(3)*

My Money

20–20. To an insurer, you are a statistic. Your premiums are based on your risk factors, including your credit rating. Bad credit increases the amount you pay for your premiums. Make certain to check your credit report annually for accuracy. Calculate the premium for someone in class 20 for 10/20/5. Then determine how much the premium will be for 50/100/50. What is the difference between the two? *LU 20-3(1)*

20–21. As given via the Internet, auto insurance quotes gathered online could vary from $947 to $1,558. A class 18 operator carries compulsory 10/20/5 insurance. He has the following optional coverage: bodily injury, 500/1,000; property damage, 50M; and collision, $200 deductible. His car is 1 year old, and the symbol of the car is 8. He has comprehensive insurance with a $200 deductible. Using your text, what is the total annual premium? *LU 20-3(1)*

eXcel

20–22. Earl Miller insured his pizza shop for $100,000 for fire insurance at an annual rate per $100 of $.66. At the end of 11 months, Earl canceled the policy since his pizza shop went out of business. What was the cost of Earl's premium and his refund? *LU 20-2(2)*

eXcel

20–23. Warren Ford insured his real estate office with a fire insurance policy for $95,000 at a cost of $.59 per $100. Eight months later the insurance company canceled his policy because of a failure to correct a fire hazard. What did Warren have to pay for the 8 months of coverage? Round to the nearest cent. *LU 20-2(2)*

My Money

20–24. If you had 10/20/5 coverage and were in a car accident causing injury to three people with injuries totaling $15,000, $9,000, and $5,000, how much would you have to pay out of pocket? If you had 50/100/50 coverage for the same scenario, how much would you have to pay out of pocket? What is the difference? *LU 20-3(1)*

20–25. Tina Grey bought a new Honda Civic and insured it with only 10/20/5 compulsory insurance. Driving up to her ski chalet one snowy evening, Tina hit a parked van and injured the couple inside. Tina's car had damage of $4,200, and the van she struck had damage of $5,500. After a lengthy court suit, the injured persons were awarded personal injury judgments of $16,000 and $7,900, respectively. What will the insurance company pay for this accident, and what is Tina's responsibility? *LU 20-3(1)*

20–26. Rusty Reft, who lives in Territory 5, carries 10/20/5 compulsory liability insurance along with optional collision that has a $300 deductible. Rusty was at fault in an accident that caused $3,600 damage to the other auto and $900 damage to his own. Also, the courts awarded $15,000 and $7,000, respectively, to the two passengers in the other car for personal injuries. How much will the insurance company pay, and what is Rusty's share of the responsibility? *LU 20-3(1)*

20–27. Marika Katz bought a new Blazer and insured it with only compulsory insurance 10/20/5. Driving up to her summer home one evening, Marika hit a parked car and injured the couple inside. Marika's car had damage of $7,500, and the car she struck had damage of $5,800. After a lengthy court suit, the couple struck were awarded personal injury judgments of $18,000 and $9,000, respectively. What will the insurance company pay for this accident, and what is Marika's responsibility? *LU 20-3(1)*

20–28. In Problem 20–27, what will the insurance company pay and what is Marika's responsibility if Marika has 25/50/25 coverage with $200 deductible for collision instead of 10/20/5? *LU 20-3(2)*

CHALLENGE PROBLEMS

20–29. Money.cnn.com states, "The single most important reason to own life insurance is to provide support for your dependents." Insurance4usa.com states, "Professionals suggest you have 8 to 12 times your income in life insurance." Pat and Bonnie Marsh are calculating how much life insurance they need. They have two young children with no college fund set up. Bonnie is a 35-year-old stay-at-home mom. Pat, 39, earns $68,000 per year. How much life insurance do you recommend each person have? (Note that a spouse who stays at home to raise the family generates the equivalent of a salary that needs to be taken into account. Assume Bonnie's salary is $25,000.) What will be the cost of straight-life insurance for both policies if the lowest recommended amount is used? What is the monthly premium owed? *LU 20-1(1)*

20–30. Lou Ralls insured a building and contents (area 2, class B) for $150,000. After 1 month, he canceled the policy. The next day he received a cancellation notice by the company. It stated that he was being canceled due to his previous record. How does Lou save by this insurance cancellation versus his planned cancellation? *LU 20-2(1, 2)*

SUMMARY PRACTICE TEST

Do you need help? Connect videos have step-by-step worked-out solutions.

1. Howard Slater, age 44, an actor, expects his income to decline in future years. He decided to take out a 20-year payment life policy with a $90,000 coverage. What will be Howard's annual premium? If he decides to stop paying premiums after 15 years, what will be his cash value? *LU 20-1(1, 2)*

2. J.C. Monahan, age 40, bought a straight-life insurance policy for $210,000. Calculate her annual premium. If after 20 years J.C. no longer pays her premiums, what nonforfeiture options will be available to her? *LU 20-1(1, 2)*

3. The property of Pote's Garage is worth $900,000. Pote has a $375,000 fire insurance policy that contains an 80% coinsurance clause. What will the insurance company pay on a fire that causes $450,000 damage? If Pote meets the coinsurance, how much will the insurance company pay? *LU 20-2(3)*

4. Lee Collins insured her pizza shop with a $90,000 fire insurance policy at a $1.10 annual rate per $100. At the end of 7 months, Lee's pizza shop went out of business so she canceled the policy. What is the cost of Lee's premium and her refund? *LU 20-2(2)*

5. Charles Prose insured his real estate office with a $300,000 fire insurance policy at $.78 annual rate per $100. Nine months later the insurance company canceled his policy because Charles failed to correct a fire hazard. What was Charles's cost for the 9-month coverage? Round to the nearest cent. *LU 20-2(2)*

6. Roger Laut, who lives in Territory 5, carries 10/20/5 compulsory liability insurance along with optional collision that has a $1,000 deductible. Roger was at fault in an accident that caused $4,800 damage to the other car and $8,800 damage to his own car. Also, the courts awarded $19,000 and $9,000, respectively, to the two passengers in the other car for personal injuries. How much does the insurance company pay, and what is Roger's share of the responsibility? *LU 20-3(1)*

INTERACTIVE VIDEO WORKSHEET

GRADING THE SUMMARY PRACTICE TEST

Go to the summary practice test video in Connect (or click on it here in the ebook). Grade your summary practice test while viewing the video.

C for Correct/I for Incorrect

1. ________
2. ________
3. ________
4. ________
5. ________
6. ________

If you achieved 100%, you are ready for your instructor's exam.

If any of the problems were incorrect, list the questions you missed and show steps to solve the problem correctly.

Replay the video to see if you have made the correct fixes to your mistakes. If you have any questions, contact your instructor asap.

Notes on Watching Videos

MY MONEY

Insuring Your Future

What I need to know

Insurance can be an intimidating subject for many consumers. Premiums paid but not needed might initially seem like a waste, but they provide the assurance of minimizing the financial impact incurred in the event of a loss (such as accident, theft, death, or fire). The process of finding and purchasing the insurance coverage needed in the amounts that will be sufficient may initially seem like a daunting task. Determining the value of your personal property as well as identifying your current financial obligations will assist you in determining the amount of insurance coverage appropriate to your unique situation. Remember that insurance is what protects you from the risks associated with loss. Determining an accurate assessment of your potential financial risks will assist you in locating the type and coverage amounts of insurance you need.

What I need to do

Determine your coverage amounts. For your personal property (such as car, home, possessions, and so on) you will want to assess the fair market value of these items and what they would cost to replace. Documenting your purchases with receipts is a great way to account for the value of your personal property. For the purchase of life insurance you should take into account all of your current debt, such as student loans, home mortgage, and car loan, and calculate what amount would be needed to pay off these debts in their entirety. Determine the value of all paid and unpaid work you do (such as childcare, mowing the yard, cleaning the house, and cooking meals). Add this to the amount needed to pay off all debts. (A rule of thumb is to purchase 10 times your income plus $100,000 per child for college expenses.) In addition you would want to consider the funeral and burial costs associated with a loss of life as part of these calculations. By determining all of these financial obligations you can purchase a life insurance policy that would cover all of your current expenses and avoid leaving a financial burden on your family.

Shop around. There are many insurance providers able to assist you with your insurance needs. Not all of these providers will charge the same amount for the coverage provided. And not all insurance providers are reputable. Check the Better Business Bureau and customer feedback sites for each company you are considering. It is in your best interest to obtain insurance quotes from a few providers to determine which offers you the best value. Make sure these quotes are all representing the same levels of insurance coverage so that direct comparisons can be made on the price of each insurance policy. Understanding what you receive in return for your insurance premiums paid will help you make the best decision. Over time insurance rates and coverage options will change as will your need for different levels of insurance coverage. For instance, the value of your home will most likely increase and the amount of insurance coverage you initially purchased for your home will need to be adjusted accordingly. Therefore, you will want to conduct an annual insurance review to make sure your insurance coverage and premiums are still in line with your budget and insurance needs. Keep in mind financial planners recommend everyone have the following types of insurance: health, disability, life, auto, homeowners/renters, and burial.

Resources I can use

- https://www.daveramsey.com/blog/types-insurance-cant-go-without — types of insurance you need
- https://www.daveramsey.com/blog/how-much-car-insurance — determine how much car insurance you need

MY MONEY ACTIVITY

- Request quotes from three different car insurance providers for your vehicle. Note the difference in insurance premiums from each. Repeat this activity for homeowners/renters insurance.
- What is the value in combining your car and home/renters policies?

PERSONAL FINANCE

A KIPLINGER APPROACH

"How to Get a Great Deal on a Lease" by David Muhlbaum from *Kiplinger's*, January 2017, p. 38. Used by permission of The Kiplinger Washington Editors, Inc.

DAVID MUHLBAUM | Drive Time

How to Get a Great Deal on a Lease

Leasing has spread beyond luxury brands and their affluent buyers to deals involving family sedans and crossovers, compacts, and even pickups. That reflects a mutual enthusiasm among car shoppers looking to lower monthly payments and carmakers, who often subsidize the cost of leases to prop up sales. Leasing hit an all-time high in 2016, accounting for nearly one in three new-car transactions.

The biggest reason people of all generations choose leasing is an old one: It allows you to put a more expensive vehicle in your driveway. Because you're paying for a car's depreciation only over the term of the lease, your payments are lower than if you financed the entire cost. Another plus: With most lease terms, you're always under warranty.

If you tend to keep your car until it quits, leasing isn't for you. But if you finance new cars and trade them in often enough that you always have a new-car payment, a lease could be a good deal.

The biggest reason people choose leasing is an old one: A lease allows you to put a more expensive vehicle in your driveway.

How to negotiate a lease. Lessees often fixate on the monthly payment rather than the overall cost of a lease, and that's a mistake. In their defense, leases are complex agreements that seem even more daunting because they have their own special jargon. In the leasing world, for example, the price of the vehicle is called the **capitalized cost**. You should haggle over that cost just as you would over the sticker price of a car you're buying. "A lot of people don't even think to do that," says Jack Nerad, executive market analyst for Kelley Blue Book.

The **capital cost reduction** is basically the down payment. The more you put down, the lower your payments—but you should put down as little as you can. One of the benefits of leasing is that you are tying up as little money as possible in a depreciating asset. Plus, if your car is stolen (and not recovered) or totaled early in the lease, it's unlikely you will get that money back after insurance pays off the car.

A couple of other terms to focus on as you compare leases: **Residual value** (or resale value) is what the vehicle is expected to be worth at the end of the lease and is the same as, or close to, the purchase price you'd pay then. An inflated residual value can lower your monthly payments, but it can also make it more difficult to sell the lease, trade in your vehicle in the middle of the lease or buy the vehicle at the end of the lease. The **money factor** is the interest you'll pay (the lower the number, the better). You multiply that number by 2,400 to get an estimate of the annual percentage rate.

You know those TV commercials you see for lease deals around the holidays? Lowering payments by pushing up residual values is one of the ways carmakers meet end-of-year sales goals, especially with slower-selling models. That can spell savings for you. But carmakers have also learned that subsidizing the lease business can burn them. They're the ones on the hook after lessees return all those cars and the glut drives down used-car prices. Recently, lessors have been trying to prevent prices from falling by lowering the mileage limits that come with leases to make the used vehicles more valuable. Some current lease deals offer as little as 10,000 miles a year. You can buy more miles for about 25 cents a mile, but don't add on more mileage than the warranty covers, lest you undermine one of leasing's advantages.

Comparison shopping is the only way to be sure you're getting a good deal. Shop for your lease at the dealer as well as banks and credit unions, focusing on the money factor and the residual value. (No matter who writes your lease, you'll have to haggle with the dealer over the capitalized cost.) Check out Edmunds.com's new tool for finding lease deals (www.edmunds.com/lease-deals). LeaseHackr has a calculator (www.leasehackr.com/calculator) that helps you compare the cost of leases. Or consider LeaseWise.org. For $350, the service will shop at least five dealers in your area and guarantee the deals. ■

YOU CAN FOLLOW DAVID MUHLBAUM'S AUTOMOTIVE MUSINGS ON TWITTER AT WWW.TWITTER.COM/DAVEYDOG.

BUSINESS MATH ISSUE

Leasing always beats buying a vehicle.

1. List the key points of the article and information to support your position.
2. Write a group defense of your position using math calculations to support your view. If you are in an online course, post to a discussion board.

Classroom Notes

CHAPTER 21

Stocks, Bonds, and Mutual Funds

Walt Disney on Thursday locked in the lowest long-term borrowing costs of any U.S. company ever.

LEARNING UNIT OBJECTIVES

LU 21–1: Stocks

1. Read, calculate, and explain stock quotations.
2. Calculate dividends of preferred and common stocks; calculate return on investment.

LU 21–2: Bonds

1. Read, calculate, and explain bond quotations.
2. Compare bond yields to bond premiums and discounts.

LU 21–3: Mutual Funds

1. Explain and calculate net asset value and mutual fund commissions.
2. Read and explain mutual fund quotations.

Your Guide to Successfully Completing This Chapter

Traditional book or ebook

Check box as you complete each step.
Steps

- ☐ Read learning unit.
  - ☐ Complete practice quiz at the end of the learning unit.
- ☐ Grade practice quiz using provided solutions. (For more help, watch the learning unit video in Connect and have a Study Session with the authors. Then complete the additional practice quiz in Connect.)
- ☐ Repeat above for each of the three learning units in Chapter 21.
  - ☐ Review chapter organizer.
  - ☐ Complete assigned homework.
    - ☐ Finish summary practice test. (Go to Connect via the ebook link and do the interactive video worksheet to grade.)
- ☐ Complete instructor's exam.

When you make financial investments there is always some degree of risk. Should you invest in Disney stock, or just keep your money in cash? The *Wall Street Journal* clip in the chapter opener offers some information about long-term borrowing costs for Disney.

Before we explain the concept of stock, consider the following general investor principles: (1) know your risk tolerance and the risk of the investments you are considering—determine whether you are a low-risk conservative investor or a high-risk speculative investor; (2) know your time frame—how soon you need your money; (3) know the liquidity of the investments you are considering—how easy it is to get your money; (4) know the return you can expect on your money—how much your money should earn; and (5) do not put "all your eggs in one basket"—diversify with a mixture of stocks, bonds, and cash equivalents. It is most important that before you seek financial advice from others, you go to the library and/or the Internet for information. When you do your own research first, you can judge the advice you receive from others.

This chapter introduces you to the major types of investments—stocks, bonds, and mutual funds. These investments indicate the performance of the companies they represent and the economy of the country at home and abroad.

Learning Unit 21–1: Stocks

We begin this unit with an introduction to the basic stock terms. Then we explain the reason why people buy stocks, newspaper stock quotations, dividends on preferred and common stocks, and return on investment.

LO 1

Introduction to Basic Stock Terms

Companies sell shares of ownership in their company to raise money to finance operations, plan expansion, and so on. These ownership shares are called **stocks.** The buyers of the stock **(stockholders)** receive **stock certificates** verifying the number of shares of stock they own.

If you own 50 shares of common stock, you are entitled to 50 votes in company elections. Preferred stockholders do not have this right.

The two basic types of stock are **common stock** and **preferred stock.** Common stockholders have voting rights. Preferred stockholders do not have voting rights, but they receive preference over common stockholders in **dividends** (payments from profit) and in the company's assets if the company goes bankrupt. **Cumulative preferred stock** entitles its owners to a specific amount of dividends in 1 year. Should the company fail to pay these dividends, the **dividends in arrears** accumulate. The company pays no dividends to common stockholders until the company brings the preferred dividend payments up to date.

Why Buy Stocks?

My Money

Some investors own stock because they think the stock will become more valuable, for example, if the company makes more profit, new discoveries, and the like. Other investors own stock to share in the profit distributed by the company in dividends (cash or stock).

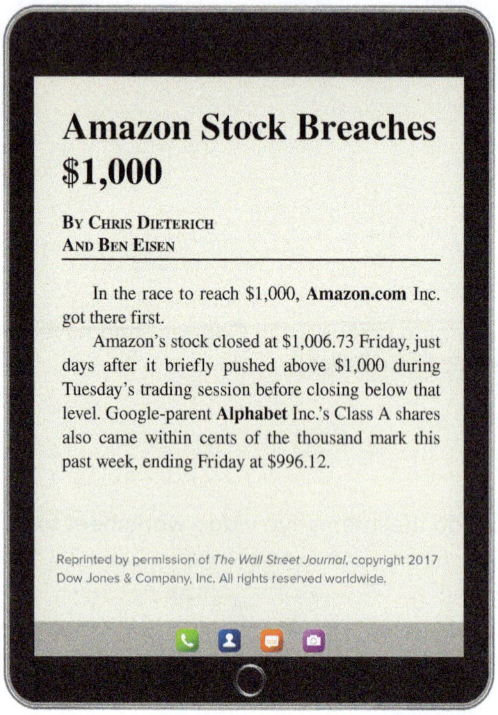
Amazon Stock Breaches $1,000

By Chris Dieterich and Ben Eisen

In the race to reach $1,000, **Amazon.com** Inc. got there first.

Amazon's stock closed at $1,006.73 Friday, just days after it briefly pushed above $1,000 during Tuesday's trading session before closing below that level. Google-parent **Alphabet** Inc.'s Class A shares also came within cents of the thousand mark this past week, ending Friday at $996.12.

For various reasons, investors at different times want to sell their stock or buy more stock. Strikes, inflation, or technological changes may cause some investors to think their stock will decline in value. These investors may decide to sell. Then the law of supply and demand takes over. As more people want to sell, the stock price goes down. Should more people want to buy, the stock price would go up.

How Are Stocks Traded?

Stock exchanges provide an orderly trading place for stock. You can think of these exchanges as an auction place. Only **stockbrokers** and their representatives are allowed to trade on the floor of the exchange. Stockbrokers charge commissions for stock trading—buying and selling stock for investors. As you might expect, in this age of the Internet, stock trades can also be made on the Internet. Electronic trading is growing each day. Note in the *Wall Street Journal* clip Amazon stock broke $1,000 per share. We will now look at how to read stock quotations.

How to Read Stock Quotations in the Newspaper's Financial Section[1]

We will use AT&T stock to learn how to read the stock quotations found in your newspaper. Note the following newspaper listing of AT&T stock:

52 WEEKS HI	LO	STOCK (SYM)	YLD %	PE	LAST	NET CHG
43.48	35.81	AT&T (T)	5.1	18	38.43	0.15

New York Stock Exchange

Stockholders
↓
elect
↓
board of directors
↓
elect
↓
officers of corporation

The highest price at which AT&T stock traded during the past 52 weeks was $43.48 per share. This means that during the year someone was willing to pay $43.48 for a share of stock.

The lowest price at which AT&T stock traded during the year was $35.81 per share. The newspaper lists the company name. The symbol that AT&T uses for trading is T.

The **stock yield** percent tells stockholders that the dividend per share is returning a rate of 5.1% to investors. This 5.1% is based on the closing price. AT&T declared a dividend of $1.96. The calculation is

$$\text{Stock yield} = \frac{\text{Annual dividend per share}}{\text{Today's last price per share}} = \frac{\$1.96}{\$38.43} = 5.1\% \quad \text{(rounded to nearest tenth percent)}$$

The 5.1% return may seem low to people who could earn a better return on their money elsewhere. Remember that if the stock price rises and you sell, your investment may result in a high rate of return.

[1]For centuries, stocks were traded and reported in fraction form. In 2001 the New York Stock Exchange and NASDAQ began the conversion to decimals, which is how stocks are reported today.

The AT&T stock is selling at \$38.43; it is selling at 18 times its **earnings per share (EPS).** Earnings per share are not listed on the stock quote.

Earnings per share	=	Last price	÷	Price-earnings ratio
(\$2.14)	=	(\$38.43)	÷	(18)

The **price-earnings ratio,** or **PE ratio,** measures the relationship between the closing price per share of stock and the annual earnings per share. For AT&T we calculate the following price-earnings ratio. (This is not listed in the newspaper.)

Round PE to the nearest whole number.

LO 2

$$\text{PE ratio} = \frac{\text{Last price per share of stock}}{\text{Annual earnings per share}} = \frac{\$38.43}{\$2.14} = 18$$

Used by permission of Cartoon Features Syndicate

If the PE ratio column shows ". . . ," this means the company has no earnings. The PE ratio will often vary depending on quality of stock, future expectations, economic conditions, and so on.

The last trade of the day, called the closing price, was at \$38.43 per share. On the *previous day,* the closing price was \$38.28 (not given). The result is that the last price is up \$0.15 from the *previous day.*

Dividends on Preferred and Common Stocks

If you own stock in a company, the company may pay out dividends. (Not all companies pay dividends.) The amount of the dividend is determined by the net earnings of the company listed in its financial report.

Earlier we stated that cumulative preferred stockholders must be paid all past and present dividends before common stockholders can receive any dividends. Following is an example to illustrate the calculation of dividends on preferred and common stocks for 2020 and 2021.

EXAMPLE The stock records of Jason Corporation show the following:

Preferred stock issued: 20,000 shares.

Preferred stock cumulative at \$.80 per share.

Common stock issued: 400,000 shares.

In 2020, Jason paid no dividends.

In 2021, Jason paid \$512,000 in dividends.

Remember that common stockholders do not have the cumulative feature of preferred stockholders.

Since Jason declared no dividends in 2020, the company has \$16,000 (20,000 shares × \$.80 = \$16,000) dividends in arrears to preferred stockholders. The dividend of \$512,000 in 2021 is divided between preferred and common stockholders as follows:

	2020	2021	
Dividends paid	0	\$512,000	
Preferred stockholders*	Paid: 0	Paid for 2020 (20,000 shares × \$.80)	\$ 16,000
	Owe: Preferred, \$16,000 (20,000 shares × \$.80)	Paid for 2021	16,000
			\$ 32,000
Common stockholders	0	Total dividend	\$512,000
		Paid preferred for 2020 and 2021	− 32,000
		To common	\$480,000
		$\frac{\$480{,}000}{400{,}000 \text{ shares}} = \$1.20$ per share	

*For a discussion of par value (arbitrary value placed on stock for accounting purposes) and cash and stock dividend distribution, check your accounting text.

MONEY tips

An effective long-term strategy for investing in stocks is to continue to invest even when stock prices are falling. As stock prices fall, you purchase more shares with the same amount of money, giving you additional future earning potential.

Shares are typically traded in groups of 100 called ***round lots.*** *Purchases of fewer than 100 shares of stock are called* ***odd lots.***

Calculating Return on Investment

Now let's learn how to calculate a return on your investment if you bought a different stock than AT&T. Let's assume you decided to buy stock of General Mills given the following:

Bought 200 shares at $39.09.

Sold at end of 1 year 200 shares at $41.10.

1% commission rate on buying and selling stock.

Current $1.21 dividend per share in effect.

MONEY tips

Set financial and personal goals. Each goal should be specific, quantifiable, realistic, and given a due date. Review and update each goal annually.

Bought		**Sold**	
200 shares at $39.09	$7,818.00	200 shares at $41.10	$8,220.00
+ Broker's commission (.01 × $7,818)	+ 78.18	− Broker's commission (.01 × $8,220.00)	− 82.20
Total cost	$7,896.18	Total receipt	$8,137.80

Note: A commission is charged on both the buying and selling of stock.

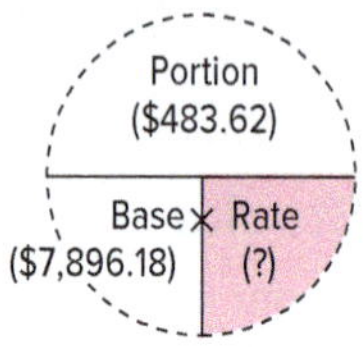

Total receipt	$8,137.80
Total cost	− 7,896.18
Net gain	$ 241.62
Dividends	+ 242.00 (200 shares × $1.21)
Total gain	$ 483.62

Portion → $\frac{\$483.62}{\$7,896.18}$ = 6.12% rate of return (to nearest hundredth percent)

↑ Base

It's time for another Practice Quiz.

LU 21–1 PRACTICE QUIZ

Complete this Practice Quiz to see how you are doing.

1. From the following Texaco stock quotation **(a)** explain the letters, **(b)** estimate the company's earnings per share, and **(c)** show how "YLD %" was calculated.

52 WEEKS HI	LO	STOCK (SYM)	YLD %	PE	LAST	NET CHG
73.90	48.25	Texaco TX	2.5	14	72.25	+0.46
(A)	(B)	(C)	(D)	(E)	(F)	(G)

2. Given: 30,000 shares of preferred cumulative stock at $.70 per share; 200,000 shares of common; 2017, no dividend; 2018, $109,000. How much is paid to each class of stock in 2018?

*For **extra help** from your authors–Sharon and Jeff–see the videos in Connect.*

✓ Solutions

1. a. (A) Highest price traded in last 52 weeks.
(B) Lowest price traded in past 52 weeks.
(C) Name of corporation is Texaco (symbol TX).
(D) Yield for year is 2.5%.
(E) Texaco stock sells at 14 times its earnings.
(F) The last price (closing price for the day) is $72.25.
(G) Stock is up $.46 from closing price yesterday.

b. $\text{EPS} = \frac{\$72.25}{14} = \$5.16$ per share

c. $\frac{?}{\$72.25} = 2.5\%$ $\quad \$72.25 \times 2.5\% = \$1.80$*

*Rounding difference

2. Preferred: 30,000 × \$.70 = \$21,000 Arrears 2017
+ 21,000 2018
\$42,000

Common: **\$67,000** (\$109,000 – \$42,000)

Learning Unit 21–2: Bonds

My Money

Have you heard of the Rule of 115? This rule is used as a rough measure to show how quickly an investment will triple in value. To use the rule, divide 115 by the rate of return your money earns. For example, if a bond earns 5% interest, divide 115 by 5. This measure estimates that your money in the bond will triple in 23 years.

This unit begins by explaining the difference between bonds and stocks. Then you will learn how to read bond quotations and calculate bond yields.

When you own stock, you own a share of a company. When you own a bond, you are lending the company money, similar to how banks lend money.

LO 1

Reading Bond Quotations

Bond quotes are stated in percents of the face value of the bond and not in dollars as stock is. Interest is paid semiannually.

Sometimes companies raise money by selling bonds instead of stock. When you buy stock, you become a part owner in the company. To raise money, companies may not want to sell more stock and thus dilute the ownership of their current stock owners, so they sell bonds. **Bonds** represent a promise from the company to pay the face amount to the bond owner at a future date, along with interest payments at a stated rate. See the *Wall Street Journal* clip in the margin about J.Crew and its debt load.

Once a company issues bonds, they are traded as stock is. If a company goes bankrupt, bondholders have the first claim to the assets of the corporation—before stockholders. As with stock, changes in bond prices vary according to supply and demand. Brokers also charge commissions on bond trading. These commissions vary.

LO 2

How to Read the Bond Section of the Newspaper

The bond section of the newspaper shows the bonds that are traded that day. The information given on bonds differs from the information given on stocks. The newspaper states bond prices in *percents of face amount, not in dollar amounts* as stock prices are stated. Also, bonds are usually in denominations of \$1,000 (the face amount).

When a bond sells at a price below its face value, the bond is sold at a discount. Why? The interest that the bond pays may not be as high as the current market rate. When this happens, the bond is not as attractive to investors, and it sells for a **discount.** The opposite could, of course, also occur. The bond may sell at a **premium,** which means that the bond sells for more than its face value or the bond interest is higher than the current market rate.

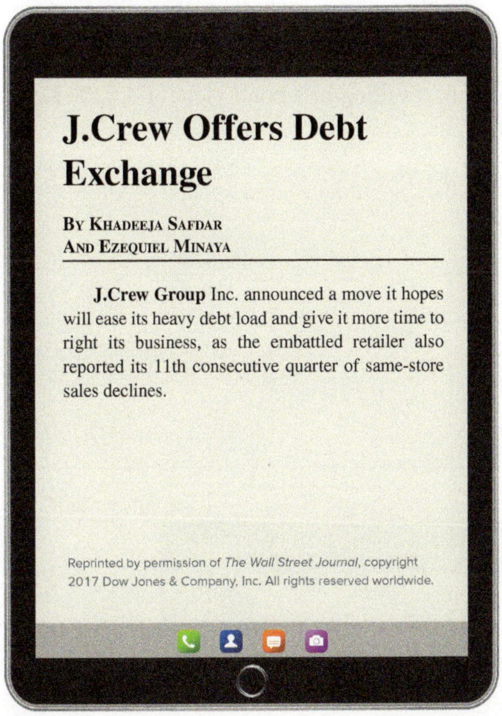

J.Crew Offers Debt Exchange

BY KHADEEJA SAFDAR AND EZEQUIEL MINAYA

J.Crew Group Inc. announced a move it hopes will ease its heavy debt load and give it more time to right its business, as the embattled retailer also reported its 11th consecutive quarter of same-store sales declines.

Let's look at this newspaper information given for Aflac bonds:

Bonds	Current yield	Vol.	Close	Net change
Aflac 4 23	4.02%	214,587	99.50	−1

Note: Bond prices are stated as a percent of face amount.

The name of the company is Aflac. It produces a wide range of insurance coverage. The interest on the bond is 4%. The company pays the interest semiannually. The bond matures (comes due) in 2023. The total interest for the year is \$40 (.04 × \$1,000). Remember that the face value of the bond is \$1,000. Now let's show this with the following formula:

Yearly interest = Face value of bond × Stated yearly interest rate

\$40.00 = \$1,000 × .04

Note this bond is selling for more than $1,000 since its interest is very attractive compared to other new offerings.

We calculate the 4.02% yield by dividing the total annual interest of the bond by the total cost of the bond. (For our purposes, we will omit the commission cost.) We will calculate more bond yields in a moment.

$$\frac{\text{Yearly interest}}{\text{Cost of bond at closing}} = \frac{\$40\ (.04 \times \$1{,}000)}{\$995\ (.9950 \times \$1{,}000)} = 4.02\%$$

This is the same as 99.50%.

Used by permission of Cartoon Features Syndicate

On this day, $214,587 worth of bonds were traded. Note that we do *not* add two zeros as we did to the sales volume of stock.

The last bond traded on this day was 99.50% of face value, or in dollars, $.9950.

The last trade of the day was down 1% of the face value from the last trade of yesterday. In dollars this is 1% = $10.

$$1\% = .01 \times \$1{,}000 = \$10$$

Thus, the closing price on this day, 99.50% + 1%, equals yesterday's close of 100.50% ($1,005). Note that *yesterday's close is not listed in today's quotations.*

Aha!

Remember: Bond prices are quoted as a percent of $1,000 but without the percent sign. A bond quote of 99 means 99% of $1,000, or $990.

Calculating Bond Yields

The Aflac bond (selling at a discount) pays 4% interest when it is yielding investors 4.02%.

$$\text{Bond yield} = \frac{\text{Total annual interest of bond}}{\text{Total current cost of bond at closing}^*}$$

*We assume this to be the buyer's purchase price.

The following example will show us how to calculate **bond yields.**

MONEY tips

Spend your money wisely. Cut back on frivolous spending to provide more discretionary income for an emergency fund and retirement savings. Spend less money than you earn. Make your money make more money through investing and save for the future.

EXAMPLE Jim Smith bought 5 bonds of Aflac at the closing price of 99.50 (remember that in dollars 99.50% is $995). Jim's total cost excluding commission is:

$$5 \times \$995 = \$4{,}975$$

What is Jim's interest?

No matter what Jim pays for the bonds, he will still receive interest of $40 per bond (.04 × $1,000). Jim bought the bonds at $995 each, resulting in a bond yield of 4.02%. Let's calculate Jim's yield to the nearest tenth percent:

(5 bonds × $40 interest per bond per year)

$$\frac{\$200.00}{\$4{,}975} = 4.02\%$$

Now let's try another Practice Quiz.

LU 21–2 PRACTICE QUIZ

Complete this **Practice Quiz** to see how you are doing.

Bonds	Yield	Sales	Close	Net change
Aetna 6.375% 23	6.4	20	100.375	.875

From the above bond quotation, **(1)** calculate the cost of 5 bonds at closing (disregard commissions) and **(2)** check the current yield of 6.4%.

*For **extra help** from your authors–Sharon and Jeff–see the videos in Connect.*

✓ Solutions

1. $100.375\% = 1.00375 \times \$1{,}000 = \$1{,}003.75 \times 5 =$ **\$5,018.75**
2. $6.375\% = .06375 \times \$1{,}000 = \$63.75$ annual interest

$$\frac{\$63.75}{\$1{,}003.75} = 6.35\% = 6.4\%$$

©Darren McCollester/Getty Images

Learning Unit 21–3: Mutual Funds

In recent years, mutual funds have increased dramatically and people in the United States have invested billions in mutual funds. Investors can choose from several fund types—stock funds, bond funds, international funds, balanced funds (stocks and bonds), and so on. This learning unit tells you why investors choose mutual funds and discusses the net asset value of mutual funds, mutual fund commissions, and how to read a mutual fund quotation.

Why Investors Choose Mutual Funds

The main reasons investors choose mutual funds are the following:

My Money

1. **Diversification.** When you invest in a mutual fund, you own a small portion of many different companies. This protects you against the poor performance of a single company but not against a sell-off in the market (stock and bond exchanges) or fluctuations in the interest rate.
2. **Professional management.** You are hiring a professional manager to look after your money when you own shares in mutual funds. The success of a particular fund is often due to the person(s) managing the fund.

   Some investors invest in a mutual fund simply because of who the mutual fund manager is.
3. **Liquidity.** Most funds will buy back your fund shares whenever you decide to sell.
4. **Low fund expenses.** Competition forces funds to keep their expenses low to maximize their performance. Because stocks and bonds in a mutual fund represent thousands of shareholders, funds can trade in large blocks, reducing transaction costs.
5. **Access to foreign markets.** Through mutual funds, investors can conveniently and inexpensively invest in foreign markets.

Aha!

Net Asset Value

LO 1

Investing in a **mutual fund** means that you buy shares in the fund's portfolio (group of stocks and/or bonds). The value of your mutual fund share is expressed in the share's **net asset value (NAV),** which is the dollar value of one mutual fund share. You calculate the NAV by subtracting the fund's current liabilities from the current market value of the fund's investments and dividing this by the number of shares outstanding.

$$\text{NAV} = \frac{\text{Current market value of fund's investments} - \text{Current liabilities}}{\text{Number of shares outstanding}}$$

The NAV helps investors track the value of their fund investment. After the market closes on each business day, the fund uses the closing prices of the investments it owns to find the dollar value of one fund share, or NAV. This is the price investors receive if they sell fund shares on that day or pay if they buy fund shares on that day.

Commissions When Buying Mutual Funds

The following table is a quick reference for the cost of buying mutual fund shares. Commissions vary from 0% to $8\frac{1}{2}\%$ depending on how the mutual fund is classified.

Classification	Commission charge*	Offer price to buy
No-load (NL) fund	No sales charge	NAV (buy directly from investment company)
Low-load (LL) fund	3% or less	NAV + commission % (buy directly from investment company or from a broker)
Load fund	$8\frac{1}{2}\%$ or less	NAV + commission % (buy from a broker)

*On a front-end load, you pay a commission when you purchase the fund shares, while on a back-end load, you pay when you redeem or sell. In general, if you hold the shares for more than 5 years, you pay no commission charge.

The offer price to buy a share for a low-load or load fund is the NAV plus the commission. Now let's look at how to read a mutual fund quotation.

LO 2

How to Read a Mutual Fund Quotation

MONEY tips

A will provides peace of mind to those surviving. Review your will annually to ensure it is up to date. If you do not have a will, write one. There are many free resources online. Put your will in a safe deposit box or fireproof safe and ensure survivors can locate it.

We will be studying the American Funds. Cindy Joelson has invested in the Growth Fund with the hope that over the years this will provide her with financial security when she retires. On May 27, Cindy turns to the *Wall Street Journal* and looks up the American Funds Growth quotation.

The name of the fund is Growth, which has the investment objective of growth securities as set forth in the fund's prospectus (document giving information about the fund). Note that this is only one fund in the American Funds family of funds.

- The $48.70 figure is the NAV plus the sales commission.
- The fund has increased $0.04 from the NAV quotation of the previous day.
- The fund has a 15.8% return this year (January through December). This assumes reinvestments of all distributions. Sales charges are not reflected.

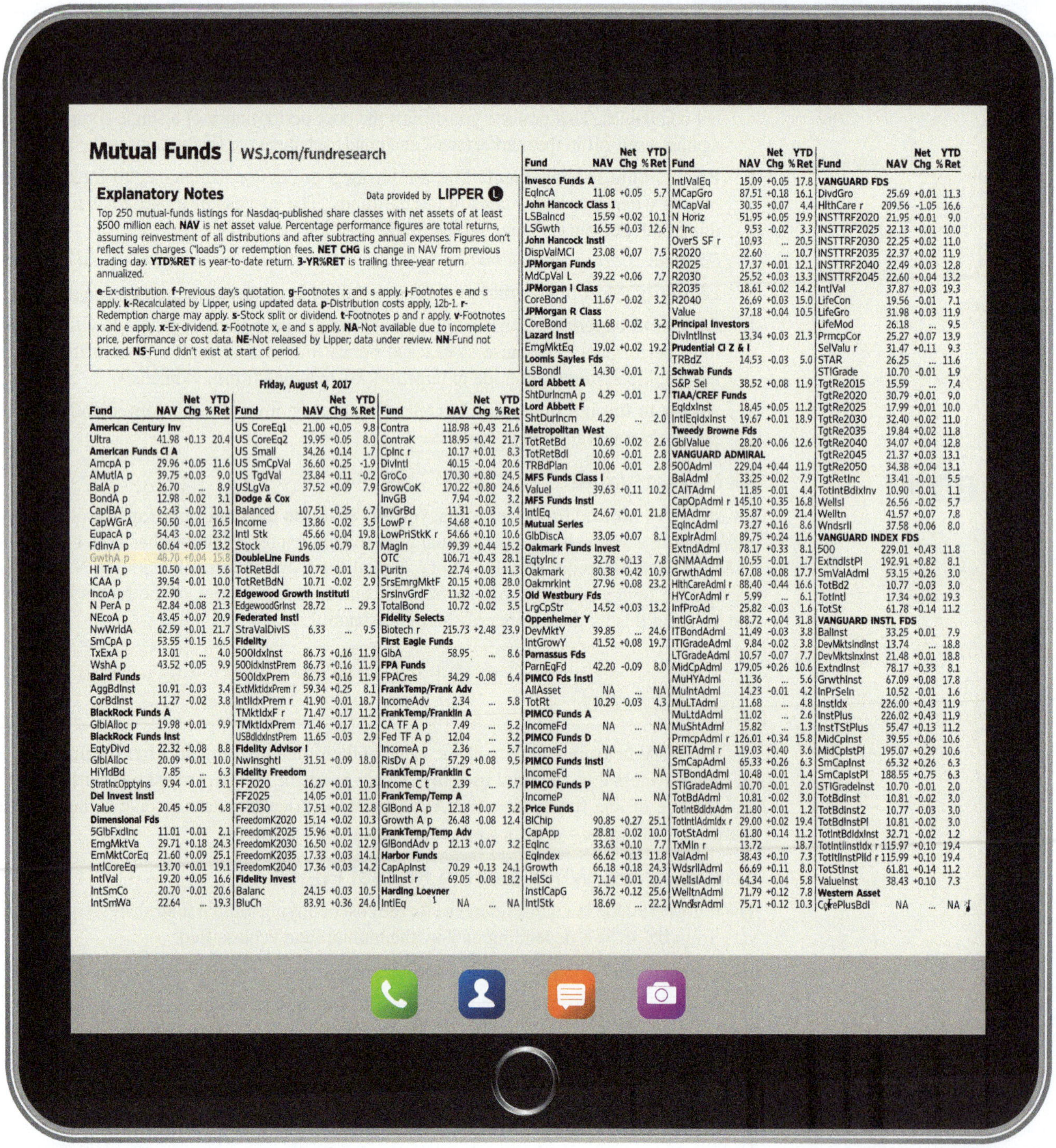

Mutual Funds | WSJ.com/fundresearch

Explanatory Notes Data provided by LIPPER

Top 250 mutual-funds listings for Nasdaq-published share classes with net assets of at least $500 million each. **NAV** is net asset value. Percentage performance figures are total returns, assuming reinvestment of all distributions and after subtracting annual expenses. Figures don't reflect sales charges ("loads") or redemption fees. **NET CHG** is change in NAV from previous trading day. **YTD%RET** is year-to-date return. **3-YR%RET** is trailing three-year return annualized.

e-Ex-distribution. **f**-Previous day's quotation. **g**-Footnotes x and s apply. **j**-Footnotes e and s apply. **k**-Recalculated by Lipper, using updated data. **p**-Distribution costs apply, 12b-1. **r**-Redemption charge may apply. **s**-Stock split or dividend. **t**-Footnotes p and r apply. **v**-Footnotes x and e apply. **x**-Ex-dividend. **z**-Footnote x, e and s apply. **NA**-Not available due to incomplete price, performance or cost data. **NE**-Not released by Lipper; data under review. **NN**-Fund not tracked. **NS**-Fund didn't exist at start of period.

Friday, August 4, 2017

Fund	NAV	Net Chg	YTD %Ret
American Century Inv			
Ultra	41.98	+0.13	20.4
American Funds Cl A			
AmcpA p	29.96	+0.05	11.6
AMutlA p	39.75	+0.03	9.0
BalA p	26.70	...	8.9
BondA p	12.98	-0.02	3.1
CapIBA p	62.43	-0.02	10.1
CapWGrA	50.50	-0.01	16.5
EupacA p	54.43	-0.02	23.2
FdInvA p	60.64	+0.05	13.2
GwthA p	48.70	+0.04	15.8
HI TrA p	10.50	+0.01	5.6
ICAA p	39.54	-0.01	10.0
IncoA p	22.90	...	7.2
N PerA p	42.84	+0.08	21.3
NEcoA p	43.45	+0.07	20.9
NwWrldA	62.59	+0.01	21.7
SmCpA p	53.55	+0.15	16.5
TxExA p	13.01	...	4.0
WshA p	43.52	+0.05	9.9
Baird Funds			
AggBdInst	10.91	-0.03	3.4
CorBdInst	11.27	-0.02	3.8
BlackRock Funds A			
GlblAlloc p	19.98	+0.01	9.9
BlackRock Funds Inst			
EqtyDivd	22.32	+0.08	8.8
GlblAlloc	20.09	+0.01	10.0
HiYldBd	7.85	...	6.3
StratIncOpptyIns	9.94	-0.01	3.1
Del Invest Instl			
Value	20.45	+0.05	4.8
Dimensional Fds			
5GlbFxdInc	11.01	-0.01	2.1
EmgMktVa	29.71	+0.18	24.3
EmMktCorEq	21.60	+0.09	25.1
IntlCoreEq	13.70	+0.01	19.1
IntlVal	19.20	+0.05	16.6
IntSmCo	20.70	-0.01	20.6
IntSmWa	22.64	...	19.3
US CoreEq1	21.00	+0.05	9.8
US CoreEq2	19.95	+0.05	8.0
US Small	34.26	+0.14	1.7
US SmCpVal	36.60	+0.25	-1.9
US TgdVal	23.84	+0.11	-0.2
USLgVa	37.52	+0.09	7.9
Dodge & Cox			
Balanced	107.51	+0.25	6.7
Income	13.86	-0.02	3.5
Intl Stk	45.66	+0.04	19.8
Stock	196.05	+0.79	8.7
DoubleLine Funds			
TotRetBdI	10.72	-0.01	3.1
TotRetBdN	10.71	-0.02	2.9
Edgewood Growth Instituti			
EdgewoodGrInst	28.72	...	29.3
Federated Instl			
StraValDivIS	6.33	...	9.5
Fidelity			
500IdxInst	86.73	+0.16	11.9
500IdxInstPrem	86.73	+0.16	11.9
500IdxPrem	86.73	+0.16	11.9
ExtMktIdxPrem r	59.34	+0.25	8.1
IntlIdxPrem r	41.90	-0.01	18.7
TMktIdxF r	71.47	+0.17	11.2
TMktIdxPrem	71.46	+0.17	11.2
USBdIdxInstPrem	11.65	-0.03	2.9
Fidelity Advisor I			
NwInsghtI	31.51	+0.09	18.0
Fidelity Freedom			
FF2020	16.27	+0.01	10.3
FF2025	14.05	+0.01	11.0
FF2030	17.51	+0.02	12.8
FreedomK2020	15.14	+0.02	10.3
FreedomK2025	15.96	+0.01	11.0
FreedomK2030	16.50	+0.02	12.9
FreedomK2035	17.33	+0.03	14.1
FreedomK2040	17.36	+0.03	14.2
Fidelity Invest			
Balanc	24.15	+0.03	10.5
BluCh	83.91	+0.36	24.6
Contra	118.98	+0.43	21.6
ContraK	118.95	+0.42	21.7
CpInc r	10.17	+0.01	8.3
DivIntl	40.15	-0.04	20.6
GroCo	170.30	+0.80	24.5
GrowCoK	170.22	+0.80	24.6
InvGB	7.94	-0.02	3.2
InvGrBd	11.31	-0.03	3.4
LowP r	54.68	+0.10	10.5
LowPriStkK r	54.66	+0.10	10.6
MagIn	99.39	+0.44	15.2
OTC	106.71	+0.43	28.1
Puritn	22.74	+0.03	11.3
SrsEmrgMktF	20.15	+0.08	28.0
SrsInvGrdF	11.32	-0.02	3.5
TotalBond	10.72	-0.02	3.5
Fidelity Selects			
Biotech r	215.73	+2.48	23.9
First Eagle Funds			
GlbA	58.95	...	8.6
FPA Funds			
FPACres	34.29	-0.08	6.4
FrankTemp/Frank Adv			
IncomeAdv	2.34	...	5.8
FrankTemp/Franklin A			
CA TF A p	7.49	...	5.2
Fed TF A p	12.04	...	3.2
IncomeA p	2.36	...	5.7
RisDv A p	57.29	+0.08	9.5
FrankTemp/Franklin C			
Income C t	2.39	...	5.7
FrankTemp/Temp A			
GlBond A p	12.18	+0.07	3.2
Growth A p	26.48	-0.08	12.4
FrankTemp/Temp Adv			
GlBondAdv p	12.13	+0.07	3.2
Harbor Funds			
CapApInst	70.29	+0.13	24.1
IntlInst r	69.05	-0.08	18.2
Harding Loevner			
IntlEq	NA	...	NA
Invesco Funds A			
EqIncA	11.08	+0.05	5.7
John Hancock Class 1			
LSBalncd	15.59	+0.02	10.1
LSGwth	16.55	+0.03	12.6
John Hancock Instl			
DispValMCI	23.08	+0.07	7.5
JPMorgan Funds			
MdCpVal L	39.22	+0.06	7.7
JPMorgan I Class			
CoreBond	11.67	-0.02	3.2
JPMorgan R Class			
CoreBond	11.68	-0.02	3.2
Lazard Instl			
EmgMktEq	19.02	+0.02	19.2
Loomis Sayles Fds			
LSBondI	14.30	-0.01	7.1
Lord Abbett A			
ShtDurIncmA p	4.29	-0.01	1.7
Lord Abbett F			
ShtDurIncm	4.29	...	2.0
Metropolitan West			
TotRetBd	10.69	-0.02	2.6
TotRetBdI	10.69	-0.01	2.8
TRBdPlan	10.06	-0.01	2.8
MFS Funds Class I			
ValueI	39.63	+0.11	10.2
MFS Funds Instl			
IntlEq	24.67	+0.01	21.8
Mutual Series			
GlbDiscA	33.05	+0.07	8.1
Oakmark Funds Invest			
EqtyInc r	32.78	+0.13	7.8
Oakmark	80.38	+0.42	10.9
OakmrkInt	27.96	+0.08	23.2
Old Westbury Fds			
LrgCpStr	14.52	+0.03	13.2
Oppenheimer Y			
DevMktY	39.85	...	24.6
IntGrowY	41.52	+0.08	19.7
Parnassus Fds			
ParnEqFd	42.20	-0.09	8.0
PIMCO Fds Instl			
AllAsset	NA	...	NA
TotRt	10.29	-0.03	4.3
PIMCO Funds A			
IncomeFd	NA	...	NA
PIMCO Funds D			
IncomeFd	NA	...	NA
PIMCO Funds Instl			
IncomeFd	NA	...	NA
PIMCO Funds P			
IncomeP	NA	...	NA
Price Funds			
BlChip	90.85	+0.27	25.1
CapApp	28.81	-0.02	10.0
EqInc	33.63	+0.10	7.7
EqIndex	66.62	+0.13	11.8
Growth	66.18	+0.18	24.3
HelSci	71.14	+0.01	20.4
InstlCapG	36.72	+0.12	25.6
IntlStk	18.69	...	22.2
IntlValEq	15.09	+0.05	17.8
MCapGro	87.51	+0.18	16.1
MCapVal	30.35	+0.07	4.4
N Horiz	51.95	+0.05	19.9
N Inc	9.53	-0.02	3.3
OverS SF r	10.93	...	20.5
R2020	22.60	...	10.7
R2025	17.37	+0.01	12.1
R2030	25.52	+0.03	13.3
R2035	18.61	+0.02	14.2
R2040	26.69	+0.03	15.0
Value	37.18	+0.04	10.5
Principal Investors			
DivIntlInst	13.34	+0.03	21.3
Prudential Cl Z & I			
TRBdZ	14.53	-0.03	5.0
Schwab Funds			
S&P Sel	38.52	+0.08	11.9
TIAA/CREF Funds			
EqIdxInst	18.45	+0.05	11.2
IntlEqIdxInst	19.67	+0.01	18.9
Tweedy Browne Fds			
GblValue	28.20	+0.06	12.6
VANGUARD ADMIRAL			
500Adml	229.04	+0.44	11.9
BalAdml	33.25	+0.02	7.9
CAITAdml	11.85	-0.01	4.4
CapOpAdml r	145.10	+0.35	16.8
EMAdmr	35.87	+0.09	21.4
EqIncAdml	73.27	+0.16	8.6
ExplrAdml	89.75	+0.24	11.6
ExtndAdml	78.17	+0.33	8.1
GNMAAdml	10.55	-0.01	1.7
GrwthAdml	67.08	+0.08	17.7
HlthCareAdml r	88.40	-0.44	16.6
HYCorAdml r	5.99	...	6.1
InfProAd	25.82	-0.03	1.6
IntlGrAdml	88.72	+0.04	31.8
ITBondAdml	11.49	-0.03	3.8
ITIGradeAdml	9.84	-0.02	3.8
LTGradeAdml	10.57	-0.07	7.7
MidCpAdml	179.05	+0.26	10.6
MuHYAdml	11.36	...	5.6
MuIntAdml	14.23	-0.01	4.2
MuLTAdml	11.68	...	4.8
MuLtdAdml	11.02	...	2.6
MuShtAdml	15.82	...	1.3
PrmcpAdml r	126.01	+0.34	15.8
REITAdml r	119.03	+0.40	3.6
SmCapAdml	65.33	+0.26	6.3
STBondAdml	10.48	-0.01	1.4
STIGradeAdml	10.70	-0.01	2.0
TotBdAdml	10.81	-0.02	3.0
TotIntBdIdxAdm	21.80	-0.01	1.2
TotIntlAdmIdx r	29.00	+0.02	19.4
TotStAdml	61.80	+0.14	11.2
TxMIn r	13.72	...	18.7
ValAdml	38.43	+0.10	7.3
WdsrIIAdml	66.69	+0.11	8.0
WellsIAdml	64.34	-0.04	5.8
WelltnAdml	71.79	+0.12	7.8
WndsrAdml	75.71	+0.12	10.3
VANGUARD FDS			
DivdGro	25.69	+0.01	11.3
HlthCare r	209.56	-1.05	16.6
INSTTRF2020	21.95	+0.01	9.0
INSTTRF2025	22.13	+0.01	10.0
INSTTRF2030	22.25	+0.02	11.0
INSTTRF2035	22.37	+0.02	11.9
INSTTRF2040	22.49	+0.03	12.8
INSTTRF2045	22.60	+0.04	13.2
IntlVal	37.87	+0.03	19.3
LifeCon	19.56	-0.01	7.1
LifeGro	31.98	+0.03	11.9
LifeMod	26.18	...	9.5
PrmcpCor	25.27	+0.07	13.9
SelValu r	31.47	+0.11	9.3
STAR	26.25	...	11.6
STIGrade	10.70	-0.01	1.9
TgtRe2015	15.59	...	7.4
TgtRe2020	30.79	+0.01	9.0
TgtRe2025	17.99	+0.01	10.0
TgtRe2030	32.40	+0.02	11.0
TgtRe2035	19.84	+0.02	11.8
TgtRe2040	34.07	+0.04	12.8
TgtRe2045	21.37	+0.03	13.1
TgtRe2050	34.38	+0.04	13.1
TgtRetInc	13.41	-0.01	5.5
TotIntBdIxInv	10.90	-0.01	1.1
WellsI	26.56	-0.02	5.7
Welltn	41.57	+0.07	7.8
WndsrII	37.58	+0.06	8.0
VANGUARD INDEX FDS			
500	229.01	+0.43	11.8
ExtndIstPl	192.91	+0.82	8.1
SmValAdml	53.15	+0.26	3.0
TotBd2	10.77	-0.03	3.0
TotIntl	17.34	+0.02	19.3
TotSt	61.78	+0.14	11.2
VANGUARD INSTL FDS			
BalInst	33.25	+0.01	7.9
DevMktsIndInst	13.74	...	18.8
DevMktsInxInst	21.48	+0.01	18.8
ExtndInst	78.17	+0.33	8.1
GrwthInst	67.09	+0.08	17.8
InPrSeIn	10.52	-0.01	1.6
InstIdx	226.00	+0.43	11.9
InstPlus	226.02	+0.43	11.9
InstTStPlus	55.47	+0.13	11.2
MidCpInst	39.55	+0.06	10.6
MidCpIstPl	195.07	+0.29	10.6
SmCapInst	65.32	+0.26	6.3
SmCapIstPl	188.55	+0.75	6.3
STIGradeInst	10.70	-0.01	2.0
TotBdInst	10.81	-0.02	3.0
TotBdInst2	10.77	-0.03	3.0
TotBdInstPl	10.81	-0.02	3.0
TotIntBdIdxInst	32.71	-0.02	1.2
TotIntlInstIdx r	115.97	+0.10	19.4
TotItlInstPlId r	115.99	+0.10	19.4
TotStInst	61.81	+0.14	11.2
ValueInst	38.43	+0.10	7.3
Western Asset			
CorePlusBdI	NA	...	NA

My Money

Financial analysts recommend that individual retirement accounts contain some mixture of stocks and bonds. Retirement accounts should be heavily invested in stocks while the investor is young and gradually shift holdings to bonds as retirement approaches. Mutual funds can invest in a variety of securities such as stocks, bonds, money market instruments, real estate, and similar assets. Investors may use a bond fund for income. Consider the following example:

EXAMPLE Bonnie and Pat Meyer are in their retirement years. They just received $250,000 after taxes from the sale of their vacation home and decided to invest the money in a bond mutual fund. They chose a no-load mutual fund that yields 4.5%. How much will they receive each year? How much would they need to invest if they want to earn $15,000 per year?

Step 1. $I = PRT = \$250{,}000 \times .045 \times 1 = \$11{,}250$

Step 2. $P = \dfrac{I}{RT} = \dfrac{\$15{,}000}{.045 \times 1} = \$333{,}333.33$

If Bonnie and Pat invest $250,000, they will receive $11,250 in interest each year. If they need to earn $15,000 in interest each year, they must invest an additional $83,333.33: $333,333.33 − $250,000 = $83,333.33.

Now let's check your understanding of this unit with a Practice Quiz.

LU 21–3 PRACTICE QUIZ

Complete this **Practice Quiz** to see how you are doing.

From the following mutual fund quotation of the Franklin Temp Growth A complete the following:

1. NAV ________
2. NAV change ________
3. Total return, YTD ________
4. You are interested in earning $6,500 each year on a no-load 5% yield mutual fund. How much must you invest?

Fund	NAV	Net Chg	YTD % Ret
GrowCoK	170.20	-0.34	24.6
InvGB	7.94	...	3.2
InvGrBd	11.32	+0.01	3.6
LowP r	54.53	-0.23	10.2
LowPriStkK r	54.51	-0.23	10.3
MagIn	99.23	-0.12	15.0
OTC	106.60	-0.41	27.9
Puritn	22.75	-0.01	11.4
SrsEmrgMktF	20.18	-0.16	28.2
SrsInvGrdF	11.32	+0.01	3.5
TotalBond	10.71	...	3.4
Fidelity Selects			
Biotech r	212.99	-1.63	22.4
First Eagle Funds			
GlbA	58.68	...	8.1
FPA Funds			
FPACres	34.09	-0.09	5.8
FrankTemp/Frank Adv			
IncomeAdv	2.33	-0.01	5.4
FrankTemp/Franklin A			
CA TF A p	7.50	+0.01	5.3
Fed TF A p	12.05	+0.01	3.3
IncomeA p	2.35	-0.01	5.3
RisDv A p	56.96	-0.06	8.9
FrankTemp/Franklin C			
Income C t	2.38	-0.01	5.3
FrankTemp/Temp A			
GlBond A p	12.10	-0.07	2.5
Growth A p	26.30	-0.12	11.6
FrankTemp/Temp Adv			
GlBondAdv p	12.06	-0.06	2.7
Harbor Funds			
CapApInst	70.35	-0.22	24.2
IntlInst r	68.30	-0.33	16.9
Harding Loevner			
IntlEq	NA	...	NA
Invesco Funds A			
EqIncA	11.06	-0.02	5.5
John Hancock Class 1			
LSBalncd	15.55	-0.03	9.8
LSGwth	16.49	-0.05	12.2
John Hancock Instl			
DispValMCI	22.96	-0.05	6.9
JPMorgan Funds			
MdCpVal L	39.10	-0.13	7.4
JPMorgan I Class			
CoreBond	11.67	+0.01	3.2
JPMorgan R Class			

*For **extra help** from your authors–Sharon and Jeff–see the videos in Connect.*

✓ Solutions

1. 26.30
2. −0.12
3. 11.6%
4. $P = \dfrac{I}{RT} = \dfrac{\$6{,}500}{.05 \times 1} = \$130{,}000$

INTERACTIVE CHAPTER ORGANIZER

Topic/Procedure/Formula	Examples	You try it*
Stock yield $\frac{\text{Annual dividend per share}}{\text{Today's last price per share}}$ (Round yield to nearest hundredth percent.)	Annual dividend, \$.72 Today's last price, \$42.375 $\frac{\$.72}{\$42.375} = 1.70\%$	**Calculate stock yield to nearest hundredth percent** Annual dividend, \$.88 Today's closing price, \$53.88
Price-earnings ratio $\text{PE} = \frac{\text{Last price per share of stock}}{\text{Annual earnings per share}}$ (Round answer to nearest whole number.)	From previous example: Last price, \$42.375 Annual earnings per share, \$4.24 $\frac{\$42.375}{\$4.24} = 9.99 = 10$	**Calculate PE ratio** From previous example: Closing price, \$53.88 Annual earnings per share, \$3.70
Dividends with cumulative preferred stock Cumulative preferred stock is entitled to all dividends in arrears before common stock receives dividend	2019 dividend omitted; in 2020, \$400,000 in dividends paid out. Preferred is cumulative at \$.90 per share; 20,000 shares of preferred issued and 100,000 shares of common issued. To preferred: 20,000 shares × \$.90 = \$18,000 In arrears 2019: 20,000 shares × .90 = 18,000 Dividend to preferred \$36,000 To common: \$364,000 (\$400,000 − \$36,000) $\frac{\$364{,}000}{100{,}000 \text{ shares}}$ = \$3.64 dividend to common per share	**Calculate dividends to preferred and common stock** 2019, no dividend; 2020, \$300,000 Preferred—\$.80 cumulative, 30,000 shares issued Common—60,000 shares issued
Cost of a bond Bond prices are stated as a percent of the face value. Bonds selling for less than face value result in bond discounts. Bonds selling for more than face value result in bond premiums.	Bill purchases 5 \$1,000, 12% bonds at closing price of $103\frac{1}{4}$. What is his cost (omitting commissions)? $103\frac{1}{4}\% = 103.25\% = 1.0325$ in decimal 1.0325 × \$1,000 bond = \$1,032.50 per bond 5 bonds × \$1,032.50 = \$5,162.50	**Calculate cost of bonds** 6 \$1,000, 3% bonds at 102.25
Bond yield $\frac{\text{Total annual interest of bond}}{\text{Total current cost of bond at closing}}$ (Round to nearest tenth percent.)	Calculate bond yield from last example on one bond. (\$1,000 × .12) $\frac{\$120}{\$1{,}032.50} = 11.6\%$	**Calculate bond yield** 4% bond selling for \$1,011.20
Mutual fund $\text{NAV} = \frac{\text{Current market value of fund's investment} - \text{Current liabilities}}{\text{Number of shares outstanding}}$	The NAV of the Scudder Income Bond Fund was \$12.84. The NAV change was 0.01. What was the NAV yesterday? \$12.83	**Calculate yesterday's NAV** Today—\$12.44 Change—.05

(continues)

INTERACTIVE CHAPTER ORGANIZER

Topic/Procedure/Formula	Examples		You try it*
KEY TERMS	Bond yield Bonds Common stocks Cumulative preferred stock Discount Dividends Dividends in arrears	Earnings per share (EPS) Mutual fund Net asset value (NAV) Odd lot PE ratio Preferred stock Premium	Price-earnings ratio Round lot Stock certificate Stock yield Stockbrokers Stockholders Stocks

*Worked-out solutions are in Appendix B.

Critical Thinking Discussion Questions with Chapter Concept Check

1. Explain how to read a stock quotation. What are some of the red flags of buying stock?
2. What is the difference between odd and round lots? Explain why the commission on odd lots could be quite expensive.
3. Explain how to read a bond quote. What could be a drawback of investing in bonds?
4. Compare and contrast stock yields and bond yields. As a conservative investor, which option might be better? Defend your answer.
5. Explain what NAV means. What is the difference between a load and a no-load fund? How safe are mutual funds?
6. **Chapter Concept Check.** Determine what you would invest in today if you were building a portfolio. Keep in mind your age, marital status, and the financial goals you want to achieve. Use the concepts in this chapter to develop your investment strategy.

Classroom Notes

21

END-OF-CHAPTER PROBLEMS connect

Check figures for odd-numbered problems in Appendix B. Name ______________________ Date ________

DRILL PROBLEMS

Calculate the cost (omit commission) of buying the following shares of stock: *LU 21-1(1)*

21–1. 500 shares of Delta Air at $50.61

21–2. 1,200 shares of Apple at $125.50

Calculate the yield of each of the following stocks (rounded to the nearest tenth percent): *LU 21-1(1)*

	Company	Yearly dividend	Closing price per share	Yield
21–3.	Boeing	$.68	$64.63	____
21–4.	Best Buy	$.07	$9.56	___

Calculate the price-earnings ratio (21-5) and closing price per share (21-6) (to nearest whole number) or stock price as needed: *LU 21-1(1)*

	Company	Earnings per share	Closing price per share	Price-earnings ratio
21–5.	BellSouth	$3.15	$40.13	__
21–6.	American Express	$3.85	______	26

21–7. Calculate the total cost of buying 400 shares of CVS at $102.90. Assume a 2% commission. *LU 21-1(1)*

21–8. If in Problem 21–1 the 500 shares of Delta Air stock were sold at $50, what would be the loss? Commission is omitted. *LU 21-1(1)*

21–9. Given: 20,000 shares cumulative preferred stock ($2.25 dividend per share); 40,000 shares common stock. Dividends paid: 2018, $8,000; 2019, 0; and 2020, $160,000. How much will preferred and common stockholders receive each year? *LU 21-1(2)*

For each of these bonds, calculate the total dollar amount you would pay at the quoted price (disregard commission or any interest that may have accrued): *LU 21-2(1)*

	Company	Bond price	Number of bonds purchased	Dollar amount of purchase price
21–10.	Petro	87.75	3	_______
21–11.	Wang	114	2	_______

For the following bonds, calculate the total annual interest, total cost, and current yield (to the nearest tenth percent): *LU 21-2(2)*

	Bond	Number of bonds purchased	Selling price	Total annual interest	Total cost	Current yield
21–12.	Sharn $11\frac{3}{4}$ 33	2	115	______	______	______
21–13.	Wang $6\frac{1}{2}$ 26	4	68.125	______	______	______

21–14. From the following calculate the net asset values. Round to the nearest cent. *LU 21-3(1)*

	Current market value of fund investment	Current liabilities	Number of shares outstanding	NAV
a.	$5,550,000	$770,000	600,000	______
b.	$13,560,000	$780,000	840,000	______

21–15. From the following mutual fund quotation, complete the blanks: *LU 21-3(2)*

				TOTAL RETURN		
	Inv. obj.	NAV	NAV chg.	YTD	4 wks.	1 yr.
EuGr	ITL	12.04	−0.06	+8.2	+0.9	+9.6

NAV ______ NAV change ______

Total return, 1 year ______

WORD PROBLEMS

eXcel **21–16.** Ryan Neal bought 1,200 shares of Ford at $15.98 per share. Assume a commission of 2% of the purchase price. What is the total cost to Ryan? *LU 21-1(1)*

eXcel **21–17.** Assume in Problem 21–16 that Ryan sells the stock for $22.25 with the same 2% commission rate. What is the bottom line for Ryan? *LU 21-1(1)*

eXcel **21–18.** Jim Corporation pays its cumulative preferred stockholders $1.60 per share. Jim has 30,000 shares of preferred and 75,000 shares of common. In 2018, 2019, and 2020, due to slowdowns in the economy, Jim paid no dividends. Now in 2021, the board of directors decided to pay out $500,000 in dividends. How much of the $500,000 does each class of stock receive as dividends? *LU 21-1(2)*

21–19. Maytag Company earns $4.80 per share. Today the stock is trading at $59.25. The company pays an annual dividend of $1.40. Calculate **(a)** the price-earnings ratio (rounded to the nearest whole number) and **(b)** the yield on the stock (to the nearest tenth percent). *LU 21-1(1)*

21–20. Jimmy Comfort was interested in pursuing a second career after retiring from the military. He signed up with Twitter to help network with individuals in his field. Within 1 week, he received an offer from a colleague to join her start-up business in Atlanta, Georgia. Along with his salary, he receives 100 shares of stock each month. If the stock is worth $4.50 a share, what is the value of the 100 shares he receives each month? *LU 21-1(1, 2)*

21–21. The following bond was quoted in *The Wall Street Journal:* *LU 21-2(1)*

Bonds	Curr. yld.	Vol.	Close	Net chg.
NJ 4.125 35	3.5	5	96.875	$+1\frac{1}{2}$

Five bonds were purchased yesterday, and 5 bonds were purchased today. How much more did the 5 bonds cost today (in dollars)?

21–22. DailyFinance.com reported one $40 share of Coca-Cola's stock bought in 1919, with dividends reinvested, would be worth $9.8 million today. If the price-earnings ratio was 28.42 at that time, what were the annual earnings per share? Round to the nearest cent. *LU 21-1(1)*

21–23. Dairy Queen, as part of Warren Buffet's Berkshire Hathaway (BRKA) with 6,400 locations in the USA, gave away free ice cream cones to celebrate its 75th anniversary. If Warren Buffet has a bond bought at 105.25 at $4\frac{3}{4}$ 25, what is the current yield to the nearest percent? *LU 21-2(2)*

21–24. Abby Sane decided to buy corporate bonds instead of stock. She desired to have the fixed-interest payments. She purchased 5 bonds of Meg Corporation $11\frac{3}{4}$ 24 at 88.25. As the stockbroker for Abby (assume you charge her a $5 commission per bond), please provide her with the following: **(a)** the total cost of the purchase, **(b)** total annual interest to be received, and **(c)** current yield (to nearest tenth percent). *LU 21-2(1)*

21–25. Mary Blake is considering whether to buy stocks or bonds. She has a good understanding of the pros and cons of both. The stock she is looking at is trading at $59.25, with an annual dividend of $3.99. Meanwhile, the bond is trading at 96.25, with an annual interest rate of $11\frac{1}{2}$%. Calculate for Mary her yield (to the nearest tenth percent) for the stock and the bond. *LU 21-1(1), LU 21-2(1)*

My Money

21–26. Wall Street performs a sort of "financial alchemy" enabling individuals to benefit from institutions lending money to them, according to Adam Davidson, cofounder of NPR's "Planet Money." Individuals can invest small amounts of their money in a 401(k), pooling their capital and spreading the risk. If you invested in Fidelity New Millennium, FMILX, how much would you pay for 80 shares if the 52-week high is $32.26, the 52-week low is $26.38, and the NAV is $31.88? *LU 21-3(1)*

21–27. Louis Hall read in the paper that Fidelity Growth Fund has an NAV of $16.02. He called Fidelity and asked how the NAV was calculated. Fidelity gave him the following information:

Current market value of fund investment	$8,550,000
Current liabilities	$ 860,000
Number of shares outstanding	480,000

Did Fidelity provide Louis with the correct information? *LU 21-3(1)*

21–28. Lee Ray bought 130 shares of a mutual fund with an NAV of $13.10. This fund also has a load charge of $8\frac{1}{2}$%. **(a)** What is the offer price and **(b)** what did Lee pay for his investment? *LU 21-3(1)*

21–29. Ron and Madeleine Couple received their 2019 Form 1099-DIV (dividends received) in the amount of $1,585. Ron and Madeleine are in the 28% bracket. What would be their tax liability on the dividends received? *LU 21-1(2)*

My Money

21–30. Bob Eberhart wants to retire in 10 years. He's heard that he needs either $1 million or 10 to 12 times his current income of $60,000 saved. He doesn't have either. His goal now is to save, save, save. Saving can double his nest egg if the stock market continues to deliver 7% annually in this decade. He sold his second car and is saving $500 per month. How many mutual fund shares can he purchase monthly with an NAV of $22.74?

CHALLENGE PROBLEMS

21–31. Here's an example of how breakpoint discounts on sales commissions for mutual fund investors work:
Sales charge

Less than $25,000, 5.75%
$25,000 to $49,999, 5.50%
$50,000 to 99,999, 4.75%
$100,000 to $249,999, 3.75%

Nancy Dolan is interested in the T Rowe Price Mid Cap Fund. Assume the NAV is 19.43. **(a)** What minimum amount of shares must Nancy purchase to have a sales charge of 5.50%? **(b)** What are the minimum shares Nancy must purchase to have a sales charge of 4.75%? **(c)** What are the minimum shares Nancy must purchase to have a sales charge of 3.75%? **(d)** What would be the total purchase price for **(a), (b),** or **(c)?** Round up to the nearest share even if it is less than 5. *LU 21-3(1)*

21–32. On September 6, Irene Westing purchased one bond of Mick Corporation at 98.50. The bond pays $8\frac{3}{4}$ interest on June 1 and December 1. The stockbroker told Irene that she would have to pay the accrued interest and the market price of the bond and a $6 brokerage fee. What was the total purchase price for Irene? Assume a 360-day year (each month is 30 days) in calculating the accrued interest. (*Hint:* Final cost = Cost of bond + Accrued interest + Brokerage fee. Calculate time for accrued interest.) *LU 21-2(2)*

SUMMARY PRACTICE TEST

Do you need help? Connect videos have step-by-step worked-out solutions.

1. Russell Slater bought 700 shares of Disney stock at $106.50 per share. Assume a commission of 4% of the purchase price. What is the total cost to Russell? *LU 21-1(1)*

2. HM Company earns $2.50 per share. Today, the stock is trading at $18.99. The company pays an annual dividend of $.25. Calculate **(a)** the price-earnings ratio (to the nearest whole number) and **(b)** the yield on the stock (to the nearest tenth percent). *LU 21-1(1)*

3. The stock of Aware is trading at $4.90. The price-earnings ratio is 4 times earnings. Calculate the earnings per share (to the nearest cent) for Aware. *LU 21-1(1)*

4. Tom Fox bought 8 bonds of UXY Company $3\frac{1}{2}$ 25 at 84 and 4 bonds of Foot Company $4\frac{1}{8}$ 26 at 93. Assume the commission on the bonds is $3 per bond. What was the total cost of all the purchases? *LU 21-2(1)*

5. Leah Long bought one bond of Vick Company for 147. The original bond was 8.25 30. Leah wants to know the current yield to the nearest tenth percent. Help Leah with the calculation. *LU 21-2(2)*

6. Cumulative preferred stockholders of Rale Company receive $.80 per share. The company has 70,000 shares outstanding. For the last 9 years, Rale paid no dividends. This year, Rale paid $400,000 in dividends. What is the amount of dividends in arrears that is still owed to preferred stockholders? *LU 21-1(2)*

7. Bill Roundy bought 800 shares of a mutual fund with an NAV of $14.10. This fund has a load charge of 3%. **(a)** What is the offer price and **(b)** what did Bill pay for the investment? *LU 21-3(1)*

INTERACTIVE VIDEO WORKSHEET

GRADING THE SUMMARY PRACTICE TEST

Go to the summary practice test video in Connect (or click on it here in the ebook). Grade your summary practice test while viewing the video.

C for Correct/I for Incorrect

1. ________	5. ________
2. ________	6. ________
3. ________	7. ________
4. ________	

If you achieved 100%, you are ready for your instructor's exam.

If any of the problems were incorrect, list the questions you missed and show steps to solve the problem correctly.

Replay the video to see if you have made the correct fixes to your mistakes. If you have any questions, contact your instructor asap.

Notes on Watching Videos

MY MONEY

Planning Your Estate Today

What I need to know

One of the most important parts of financial planning is to have a will. It may feel like you have years to plan your estate. However, the best time to plan for how you want your estate to be handled is today. When the time comes to carry out your wishes, your family and friends will be glad you took the time to plan and discuss your wishes with them. Recording your wishes in detail will make the process easier for everyone involved. Know that if you die without a will in place, your state's laws will determine who gets what of your assets as well as who will raise your minor children. You should understand how your estate will be taxed, too.

What I need to do

Make the necessary plans to take care of your family *today*. Write a will. Consider a living trust. A living trust is a legal document that can assist you in sharing your wishes and allows you to specify exactly how you would like to distribute your assets through your chosen executor/s, an individual or individuals whom you identify as your representative/s after your death. As part of this process you will need to decide who is best suited to make decisions on your behalf. If you are no longer able, who should make decisions concerning your health care or finances? The person(s) to whom you give your power of attorney and medical power of attorney should be considered carefully to ensure they possess the capacity to make these difficult decisions. Power of attorney allows someone to make legal and financial decisions on your behalf, whereas the medical power of attorney concentrates on decisions related to your health care.

Create a letter of instruction identifying any situations needing attention after your death. These situations can be of a personal or financial nature. In drafting such a letter you are alerting those who need to know about situations requiring their attention and providing the necessary information for them to effectively handle these matters. This could include burial or cremation plans, your obituary, organ or body donation, and desired funeral or memorial service requests. In addition you can document accounts becoming payable on your death such as bank accounts and life insurance policies. Basically, you want to use your letter of instruction as a tool to assist your executor/s in locating the information needed to carry out your wishes and to do so in an organized and efficient manner.

As with any effective planning it is important to review the documents you are creating as part of your estate planning on an annual basis to account for any changes you may want to make. For instance, you may have accounted for the guardianship of your children when they were minors but when they are older this will no longer be necessary. A good practice as part of this annual review is to ensure your executor/s know where to access your estate plans through a simple reminder. This way you are able to confirm they know how to access your plans and are aware of any changes you have made based on your annual review.

Resources I can use

- https://money.usnews.com/money/personal-finance/family-finance/articles/2017-07-14/10-essential-estate-planning-tips-everyone-should-know — tips for estate planning
- https://www.thebalance.com/living-will-vs-living-trust-3505198 — living trust versus a will

MY MONEY ACTIVITY

- Interview an older friend or family member concerning their estate planning. What did you learn from your interview? How can you use what you've learned?
- Conduct an Internet search on estate laws of your state. Consider meeting with an estate attorney for assistance with writing your will and trusts.

PERSONAL FINANCE

A KIPLINGER APPROACH

"Surviving Amazon", *Kiplinger's*, July 2017, p. 60. Used by permission of The Kiplinger Washington Editors, Inc.

■ AT A TIME WHEN MANY RETAILERS ARE UNDER PRESSURE FROM THE INTERNET, COSTCO IS THRIVING. ONE REASON: LOYAL CUSTOMERS.

STOCKS

Surviving Amazon

Warehouse giant Costco continues to prosper despite the growth of internet retailing. **BY DAREN FONDA**

TAKING ON AMAZON.COM (SYMBOL AMZN) seems foolhardy these days. The company is expanding so fast that it could gobble up more than half of online retail sales in the U.S. by 2021, up from one-third in 2016, according to brokerage firm Needham & Co. Yet even with Amazon raking in sales, some big retail chains are flourishing. One of the most successful is **COSTCO WHOLESALE (COST)**. Appealing to millennials and baby boomers alike, the warehouse club has homed in on a surprisingly effective way to keep Amazon at bay: selling food.

Although Costco's lineup spans everything from jewelry to appliances, more than 80% of its members go specifically for the groceries, according to a recent survey by investment firm Barclays. Customers pile in to buy household goods in bulk, at prices that are tough to beat online. Costco also sells fresh meats, produce and baked goods. Add it all up and the warehouse club's "mousetrap" looks well protected against Amazon, Barclays says.

COSTCO WHOLESALE

Share price: $178
Market value: $78.1 billion
Annual sales*: $121.2 billion
Annual profit*: $2.4 billion
Price-earnings ratio†: 30
Yield: 1.1%

Share price and related figures are as of April 30. *For the past 12 months. †Based on estimated earnings for the next year. SOURCES: Yahoo, Zacks Investment Research.

Groceries are just the bait, though. Many customers also fuel their cars with Costco's discounted gasoline or purchase other items that catch their eye. "You go in for tires and wind up buying a generator," says David Marcus, manager of Evermore Global Value Fund, who views Costco as one of the few retailers that can fight off online competitors.

Granted, Amazon is pushing into the food business. Its grocery delivery service, Fresh, offers more than 500,000 items, including perishables and prepared meals. Amazon is testing robots and automation in Fresh warehouses to lower labor costs, says Barclays, and it could deploy "robot vans" for home deliveries. A decade from now, Barclays estimates, Fresh could reach 15% of U.S. households.

Yet Amazon and Costco should be able to coexist. Costco members shop on Amazon mainly for electronics, books and clothing, with groceries accounting for just 25% of their purchases, says Barclays. That could change as Fresh expands. But Costco may still beat or match Amazon's prices because shoppers who buy in bulk already get deep discounts. Costco offers a grocery delivery service through Instacart.com, available in 24 states and Washington, D.C., and is testing another delivery service, Shipt, in Florida. Moreover, Fresh costs $15 a month, on top of the $99 annual membership for Amazon Prime. One other advantage for Costco is that it sells things you still can't find on Amazon, such as prescription drugs, vacation packages and car rental services.

Profit boost. Costco members tend to be loyal, too, renewing their annual memberships at a roughly 90% rate. An increase in annual fees was slated to take effect in June, generating cash that will go straight to the bottom line. Membership fees, in fact, account for 72% of pretax profits, and the increase in fees will add 42 cents to earnings per share over the next two years, about a 7% lift, estimates UBS. Another avenue for growth: Costco continues to expand worldwide, opening new stores this year in Korea, Taiwan and other locales.

For investors, the major drawback is Costco's pricey stock. It trades at 30 times estimated year-ahead earnings, well above the 10-year median price-earnings ratio of 22 and the stock market's overall P/E of 18.

But analysts argue that Costco deserves a rich valuation because it's one of the few retailers that can thrive in the age of Amazon. Costco is also doling out hefty cash payouts to try to keep shareholders happy. The company was set to pay a "special dividend" of $7 per share on May 26, on top of the regular quarterly dividend, which it recently hiked from 45 to 50 cents per share. ■

BUSINESS MATH ISSUE

Costco really cannot compete with Amazon.

1. List the key points of the article and information to support your position.
2. Write a group defense of your position using math calculations to support your view. If you are in an online course, post to a discussion board.

Classroom Notes

CHAPTER 22

Business Statistics

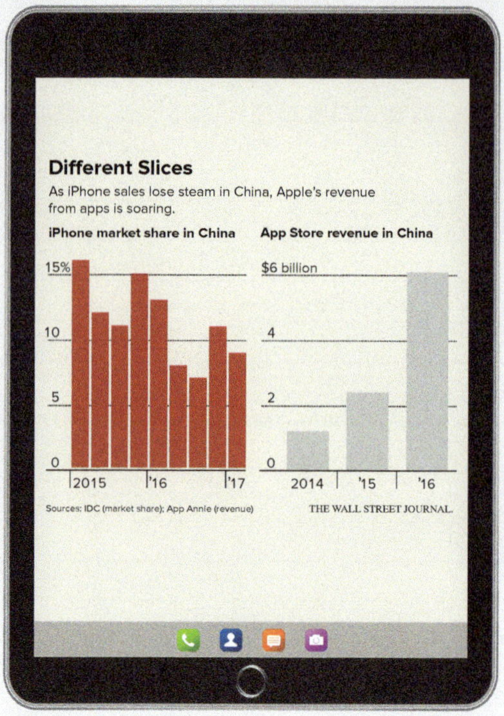

LEARNING UNIT OBJECTIVES

LU 22–1: Mean, Median, and Mode

1. Define and calculate the mean.
2. Explain and calculate a weighted mean.
3. Define and calculate the median.
4. Define and identify the mode.

LU 22–2: Frequency Distributions and Graphs

1. Prepare a frequency distribution.
2. Prepare bar, line, and circle graphs.
3. Calculate price relatives and cost comparisons.

LU 22–3: Measures of Dispersion (Optional)

1. Explain and calculate the range.
2. Define and calculate the standard deviation.
3. Estimate percentage of data by using standard deviations.

Your Guide to Successfully Completing This Chapter

Traditional book or ebook

Check box as you complete each step.

Steps

- ☐ Read learning unit.
  - ☐ Complete practice quiz at the end of the learning unit.
- ☐ Grade practice quiz using provided solutions. (For more help, watch the learning unit video in Connect and have a Study Session with the authors. Then complete the additional practice quiz in Connect.)
- ☐ Repeat above for each of the three learning units in Chapter 22.
  - ☐ Review chapter organizer.
  - ☐ Complete assigned homework.
    - ☐ Finish summary practice test. (Go to Connect via the ebook link and do the interactive video worksheet to grade.)
- ☐ Complete instructor's exam.

In this chapter we look at various techniques that analyze and graphically represent business statistics. For example, in the *Wall Street Journal* chapter opener clip, "Different Slices," we see that Apple's revenue from apps is rising. Learning Unit 22–1 discusses the mean, median, and mode. Learning Unit 22–2 explains how to gather data by using frequency distributions and express these data visually in graphs. Emphasis is placed on whether graphs are indeed giving accurate information. The chapter concludes with an introduction to index numbers—an application of statistics—and an optional learning unit on measures of dispersion.

Learning Unit 22–1: Mean, Median, and Mode

Companies frequently use averages and measurements to guide their business decisions. The mean and median are the two most common averages used to indicate a single value that represents an entire group of numbers. The mode can also be used to describe a set of data.

LO 1

Mean

The accountant of Bill's Sport Shop told Bill, the owner, that the average daily sales for the week were \$150.14. The accountant stressed that \$150.14 was an average and did not represent specific daily sales. Bill wanted to know how the accountant arrived at \$150.14.

Used by permission of Cartoon Features Syndicate

The accountant went on to explain that he used an arithmetic average, or **mean** (a measurement), to arrive at \$150.14 (rounded to the nearest hundredth). He showed Bill the following formula:

$$\text{Mean} = \frac{\text{Sum of all values}}{\text{Number of values}}$$

The accountant used the following data:

	Sun.	Mon.	Tues.	Wed.	Thur.	Fri.	Sat.
Sport Shop sales	\$400	\$100	\$68	\$115	\$120	\$68	\$180

To compute the mean, the accountant used these data:

$$\text{Mean} = \frac{\$400 + \$100 + \$68 + \$115 + \$120 + \$68 + \$180}{7} = \$150.14$$

When values appear more than once, businesses often look for a **weighted mean.** The format for the weighted mean is slightly different from that for the mean. The concept, however, is the same except that you weight each value by how often it occurs (its frequency). Thus, considering the frequency of the occurrence of each value allows a weighting of each day's sales in proper importance. To calculate the weighted mean, use the following formula:

LO 2

$$\text{Weighted mean} = \frac{\text{Sum of products}}{\text{Sum of frequencies}}$$

Let's change the sales data for Bill's Sport Shop and see how to calculate a weighted mean:

	Sun.	Mon.	Tues.	Wed.	Thur.	Fri.	Sat.
Sport Shop sales	\$400	\$100	\$100	\$80	\$80	\$100	\$400

Value	Frequency	Product
\$400	2	\$ 800
100	3	300
80	2	160
		\$1,260

$$\text{The weighted mean is } \frac{\$1,260}{7} = \$180$$

Note how we multiply each value by its frequency of occurrence to arrive at the product. Then we divide the sum of the products by the sum of the frequencies.

When you calculate your grade point average (GPA), you are using a weighted average. The following formula is used to calculate GPA:

$$\text{GPA} = \frac{\text{Total points}}{\text{Total credits}}$$

Now let's show how Jill Rivers calculated her GPA to the nearest tenth.

Given A = 4; B = 3; C = 2; D = 1; F = 0

Courses	Credits attempted	Grade received	Points (Credits × Grade)	
Introduction to Computers	4	A	16 (4 × 4)	
Psychology	3	B	9 (3 × 3)	
English Composition	3	B	9 (3 × 3)	
Business Law	3	C	6 (2 × 3)	
Business Math	3	B	9 (3 × 3)	
	16		49	$\frac{49}{16} = 3.1$

When high or low numbers do not significantly affect a list of numbers, the mean is a good indicator of the center of the data. If high or low numbers do have an effect, the median may be a better indicator to use.

LO 3

Median

The **median** is another measurement that indicates the center of the data. An average that has one or more extreme values is not distorted by the median. For example, let's look at the following yearly salaries of the employees of Rusty's Clothing Shop.

Alice Knight	$95,000	Jane Wang	$67,000
Jane Hess	27,000	Bill Joy	40,000
Joel Floyd	32,000		

Note how Alice's salary of $95,000 will distort an average calculated by the mean.

$$\frac{\$95{,}000 + \$27{,}000 + \$32{,}000 + \$67{,}000 + \$40{,}000}{5} = \$52{,}200$$

The $52,200 average salary is considerably more than the salary of three of the employees. So it is not a good representation of the store's average salary. The following *Wall Street Journal* clip, "Filling Open Slots," shows how educators are trying to better match their programs with new job opportunities. We use the following steps to find the median of a group of numbers.

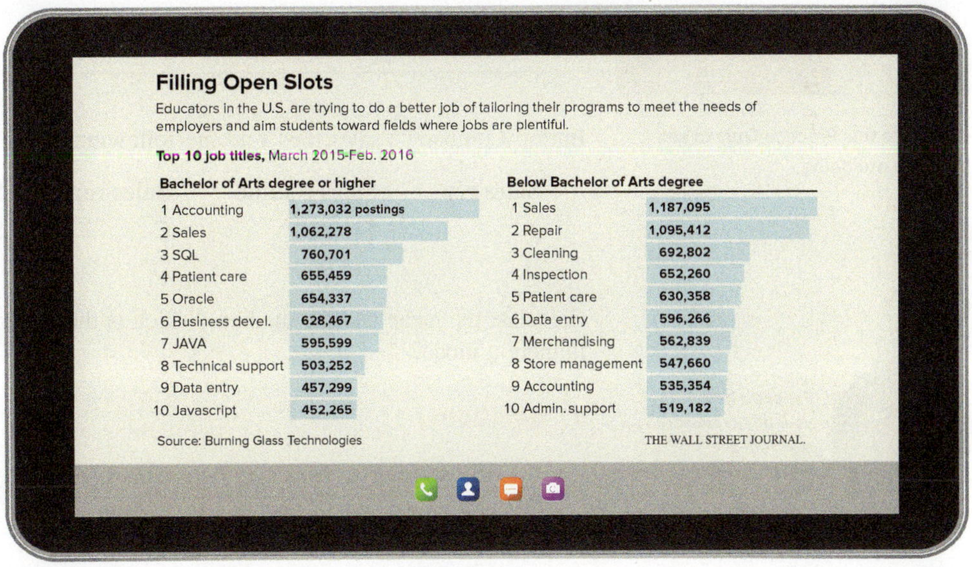

Filling Open Slots

Educators in the U.S. are trying to do a better job of tailoring their programs to meet the needs of employers and aim students toward fields where jobs are plentiful.

Top 10 job titles, March 2015-Feb. 2016

Bachelor of Arts degree or higher		Below Bachelor of Arts degree	
1 Accounting	1,273,032 postings	1 Sales	1,187,095
2 Sales	1,062,278	2 Repair	1,095,412
3 SQL	760,701	3 Cleaning	692,802
4 Patient care	655,459	4 Inspection	652,260
5 Oracle	654,337	5 Patient care	630,358
6 Business devel.	628,467	6 Data entry	596,266
7 JAVA	595,599	7 Merchandising	562,839
8 Technical support	503,252	8 Store management	547,660
9 Data entry	457,299	9 Accounting	535,354
10 Javascript	452,265	10 Admin. support	519,182

Source: Burning Glass Technologies

THE WALL STREET JOURNAL.

FINDING THE MEDIAN OF A GROUP OF VALUES

Step 1. Orderly arrange values from the smallest to the largest.

Step 2. Find the middle value.

a. *Odd number of values:* Median is the middle value. You find this by first dividing the total number of numbers by 2. The next-higher number is the median.

b. *Even number of values:* Median is the average of the two middle values.

MONEY tips

If you are the victim of identity theft: (1) Check all of your credit reports: Equifax, Experian, and TransUnion. (2) Monitor your credit card and bank accounts for unauthorized activity. (3) Put a freeze on your credit file. (4) Place a fraud alert on your credit file. (5) File your taxes early to prevent a scammer from collecting your refund. Be aware that your child's information may be stolen, too.

For Rusty's Clothing Shop, we find the median as follows:

1. Arrange values from smallest to largest:
   $27,000; $32,000; **$40,000;** $67,000; $95,000
2. Since we have a total number of five values and 5 is an odd number, we divide 5 by 2 to get $2\frac{1}{2}$. The next-higher number is 3, so our median is the third-listed number, **$40,000.**

   If Jane Hess ($27,000) were not on the payroll, we would find the median as follows:

1. Arrange values from smallest to largest:
   $32,000; $40,000; $67,000; $95,000
2. Average the two middle values:

$$\frac{\$40{,}000 + \$67{,}000}{2} = \$53{,}500$$

Note that the median results in two salaries below and two salaries above the average.
Now we'll look at another measurement tool—the mode.

LO 4

Mode

The **mode** is a measurement that also records values. In a series of numbers, the value that occurs most often is the mode. If all the values are different, there is no mode. If two or more numbers appear most often, you may have two or more modes. Note that we do not have to arrange the numbers in the lowest-to-highest order, although this could make it easier to find the mode.

EXAMPLE 3, 4, 5, 6, 3, 8, 9, 3, 5, 3

3 is the mode since it is listed 4 times.

Use a bar graph to find the mode if you do not have a list of the data set.

Now let's check your progress with a Practice Quiz.

LU 22–1 PRACTICE QUIZ

Complete this Practice Quiz to see how you are doing.

Barton Company's sales reps sold the following last month:

Sales rep	Sales volume	Sales rep	Sales volume
A	$16,500	C	$12,000
B	15,000	D	48,900

Calculate the mean and the median. Which is the better indicator of the center of the data? Is there a mode?

*For **extra help** from your authors–Sharon and Jeff–see the videos in Connect.*

✓ Solutions

$$\text{Mean} = \frac{\$16{,}500 + \$15{,}000 + \$12{,}000 + \$48{,}900}{4} = \$23{,}100$$

$$\text{Median} = \frac{\$15{,}000 + \$16{,}500}{2} = \$15{,}750$$

$12,000, $15,000, $16,500, $48,900. Note how we arrange numbers from smallest to highest to calculate median.

Median is the better indicator since in calculating the mean, the $48,900 puts the average of $23,100 much too high. There is no mode.

Learning Unit 22–2: Frequency Distributions and Graphs

My Money

In this unit you will learn how to gather data and illustrate these data. Today, computer software programs can make beautiful color graphics. But how accurate are these graphics? The *Wall Street Journal* clip below, "What's Wrong With this Picture?" gives an example of graphics that do not agree with the numbers beneath them. The clip reminds all readers to check the numbers illustrated by the graphics. This is an old clip that is still relevant today.

What's Wrong With this Picture? Utility's Glasses Are Never Empty

By Kathleen Deveny
Staff Reporter of The Wall Street Journal

When Les Waas, an investor in Philadelphia Suburban Corp., paged through the company's 1994 annual report, he was impressed by what he saw.

The water utility had used a series of charts to represent its revenues, net income and book value per share, among other results. Each figure was represented by the level of water in a glass. Each chart showed strong growth.

Then Mr. Waas looked a little more carefully. The bars in the chart seemed to indicate far more impressive growth than the numbers beneath them. A chart showing the growth in the number of Philadelphia Suburban's water customers, for example, seemed to indicate the company's customer base had more than tripled since 1990. But the numbers actually increased only 6.4%.

The reason for the disparity: The charts don't begin at zero. Even an empty glass in the accompanying chart would represent a customer base of 230,000.

LO 1

Collecting raw data and organizing the data is a prerequisite to presenting statistics graphically. Let's illustrate this by looking at the following example.

A computer industry consultant wants to know how much college freshmen are willing to spend to set up a computer in their dormitory rooms. After visiting a local college dorm, the consultant gathered the following data on the amount of money 20 students spent on computers:

\$1,000	\$7,000	\$4,000	\$1,000	\$ 5,000	\$1,000	\$3,000
5,000	2,000	3,000	3,000	3,000	8,000	9,000
3,000	6,000	6,000	1,000	10,000	1,000	

Price of computer	Tally	Frequency
\$ 1,000	卌	5
2,000	|	1
3,000	卌	5
4,000	|	1
5,000	||	2
6,000	||	2
7,000	|	1
8,000	|	1
9,000	|	1
10,000	|	1

Note that these raw data are not arranged in any order. To make the data more meaningful, the consultant made the **frequency distribution** table shown on the left. Think of this distribution table as a way to organize a list of numbers to show the patterns that may exist.

As you can see, 25% ($\frac{5}{20}=\frac{1}{4}=25\%$) of the students spent \$1,000 and another 25% spent \$3,000. Only four students spent \$7,000 or more.

Aha!

Typically between 5 and 20 classes are used in a frequency distribution for ease in analyzing the data.

Now let's see how we can use bar graphs.

LO 2

Bar Graphs

Bar graphs help readers see the changes that have occurred over a period of time. This is especially true when the same type of data is repeatedly studied.

The following *Wall Street Journal* clip, "One Hundred Candles," uses bar graphs to show the number of Americans age 100-plus up through 2050.

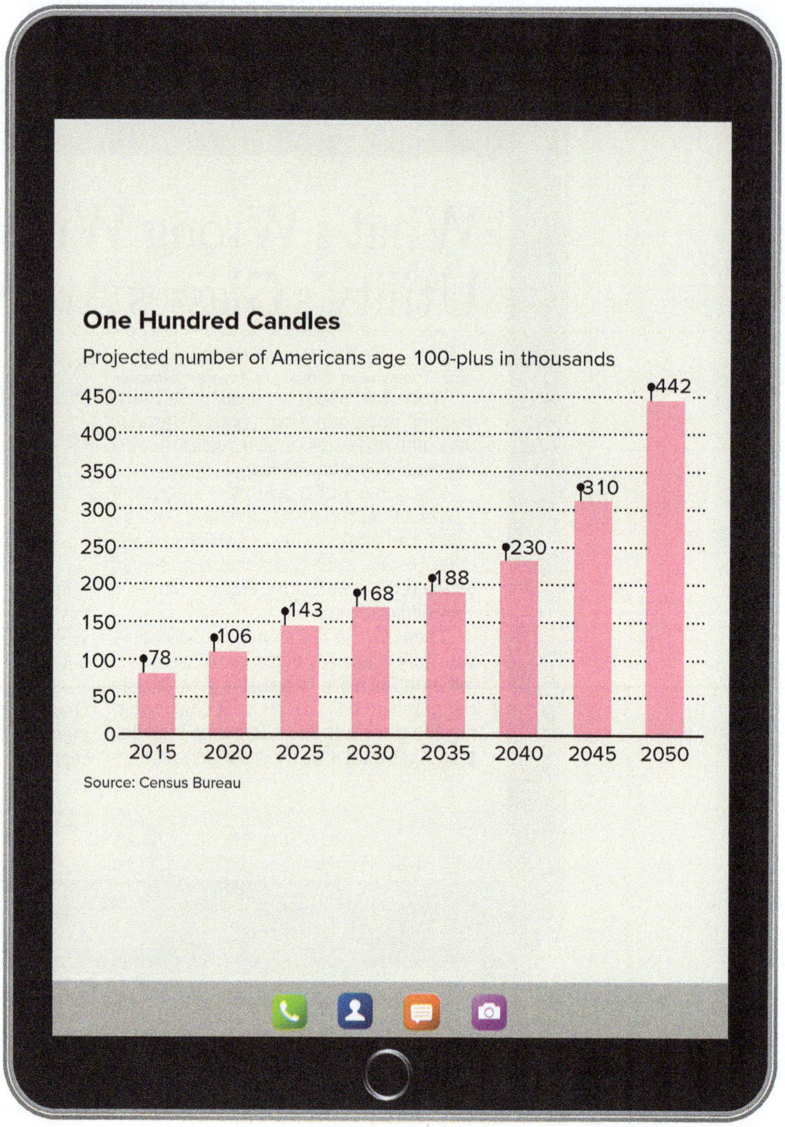

Source: *The Wall Street Journal.*

Let's return to our computer consultant example and make a bar graph of the computer purchases data collected by the consultant. Note that the height of the bar represents the frequency of each purchase. Bar graphs can be vertical or horizontal. In the following *Wall Street Journal* clip, "3-D Appeal," the bar chart shows how much is spent on 3-D printing worldwide.

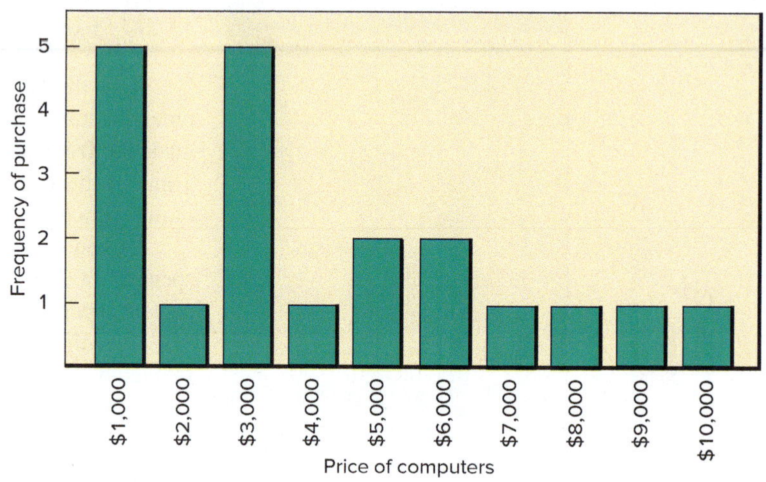

GLOBAL

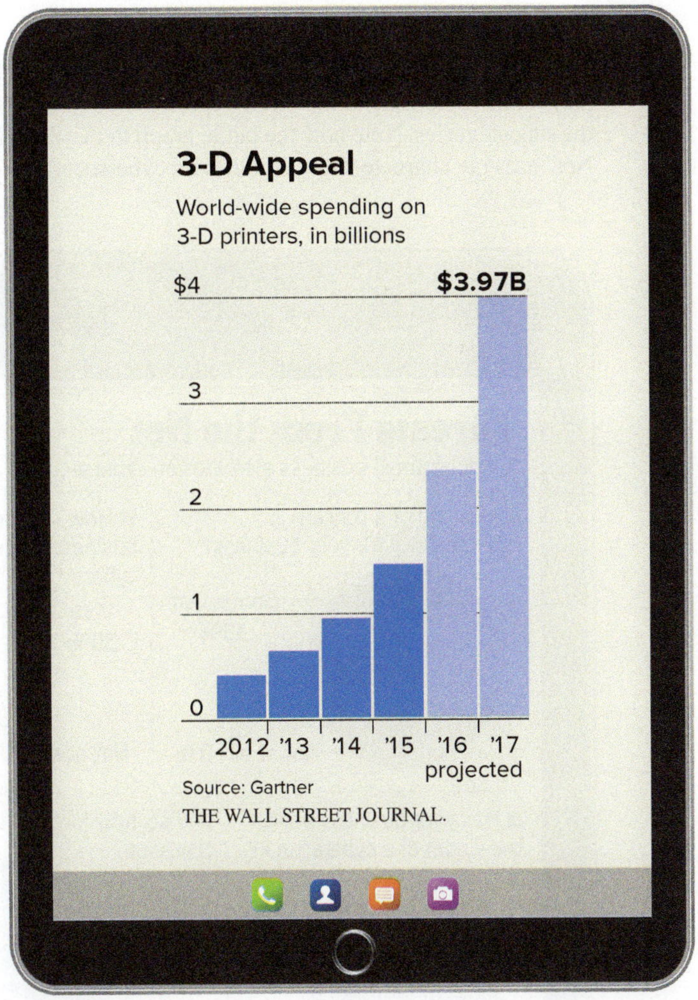

We can simplify this bar graph by grouping the prices of the computers. The grouping, or *intervals,* should be of equal sizes.

A bar graph for the grouped data follows.

Class	Frequency
$1,000– $3,000.99	11
3,001– 5,000.99	3
5,001– 7,000.99	3
7,001– 9,000.99	2
9,001– 11,000.99	1

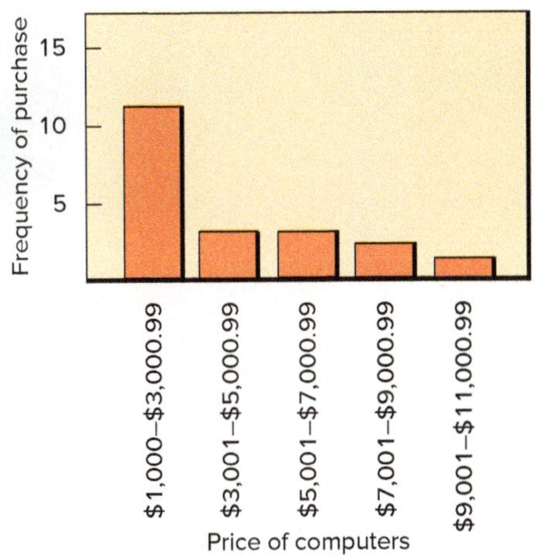

Next, let's see how we can use line graphs.

Line Graphs

A **line graph** shows trends over a period of time. Often separate lines are drawn to show the comparison between two or more trends.

The *Wall Street Journal* clip, "Aging America," shows the growth in the percentage of Americans 65 and older.

We conclude our discussion of graphics with the use of the circle graph.

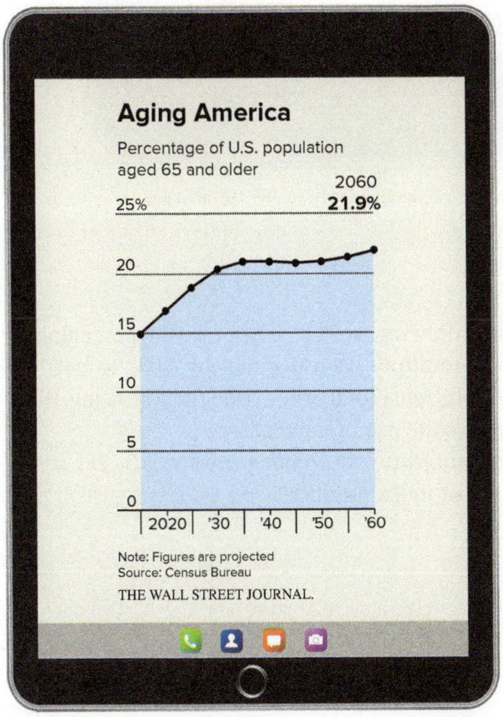

$.15 \times 360° = 54.0$
$.11 \times 360° = 39.6$
$.36 \times 360° = 129.6$
$.38 \times 360° = 136.8$
360.0

Circle Graphs

Circle graphs, often called *pie charts,* are especially helpful for showing the relationship of parts to a whole. The entire circle represents 100%, or 360°; the pie-shaped pieces represent the subcategories. Note how the circle graph in the *Wall Street Journal* clip "Threats From the Net" uses pie charts to show attitudes on cybersecurity.

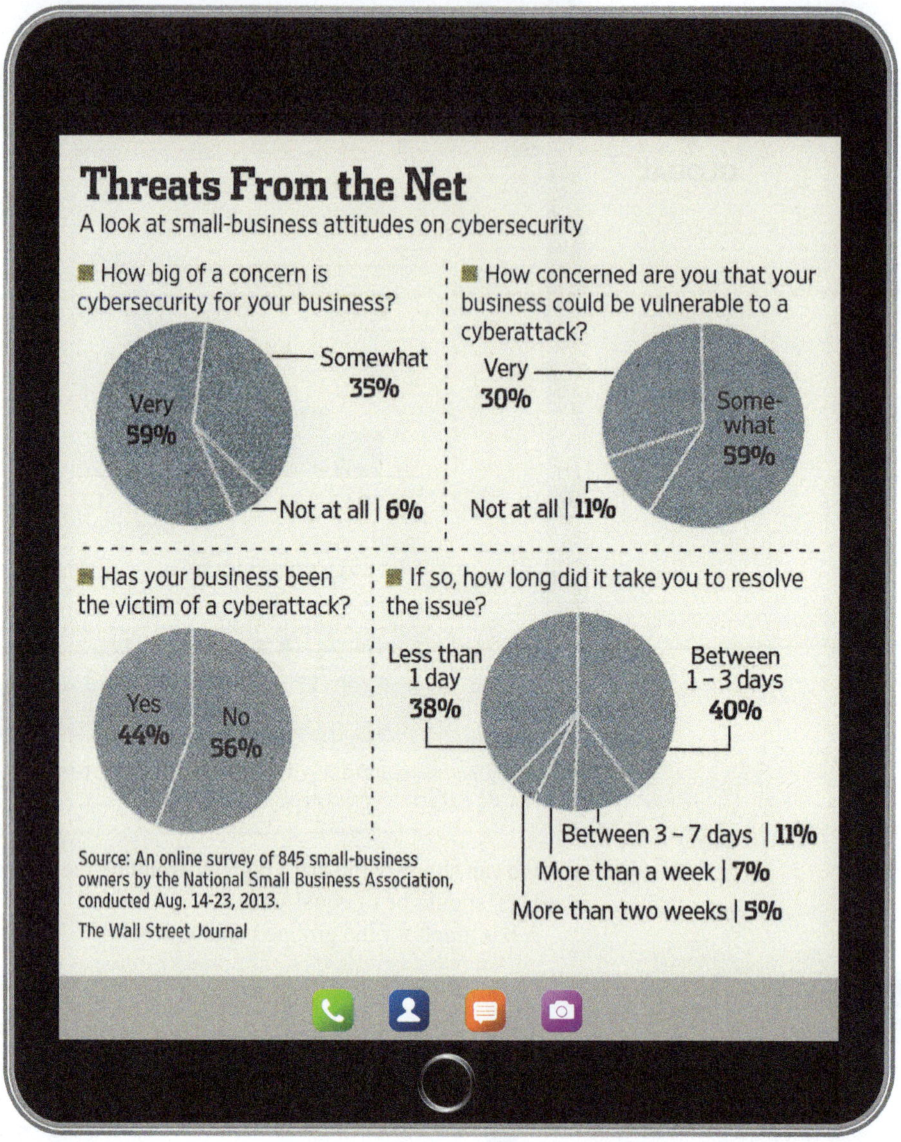

To draw a circle graph (or pie chart), begin by drawing a circle. Then take the percentages and convert each percentage to a decimal. Next multiply each decimal by 360° to get the degrees represented by the percentage. Circle graphs must total 360°. Note the following *Wall Street Journal* clip "An Enduring Challenge" shows the use of a pie chart.

You can use Excel, and many other software programs, to create graphs easily. Try it!

We conclude this unit with a brief discussion of index numbers.

GLOBAL

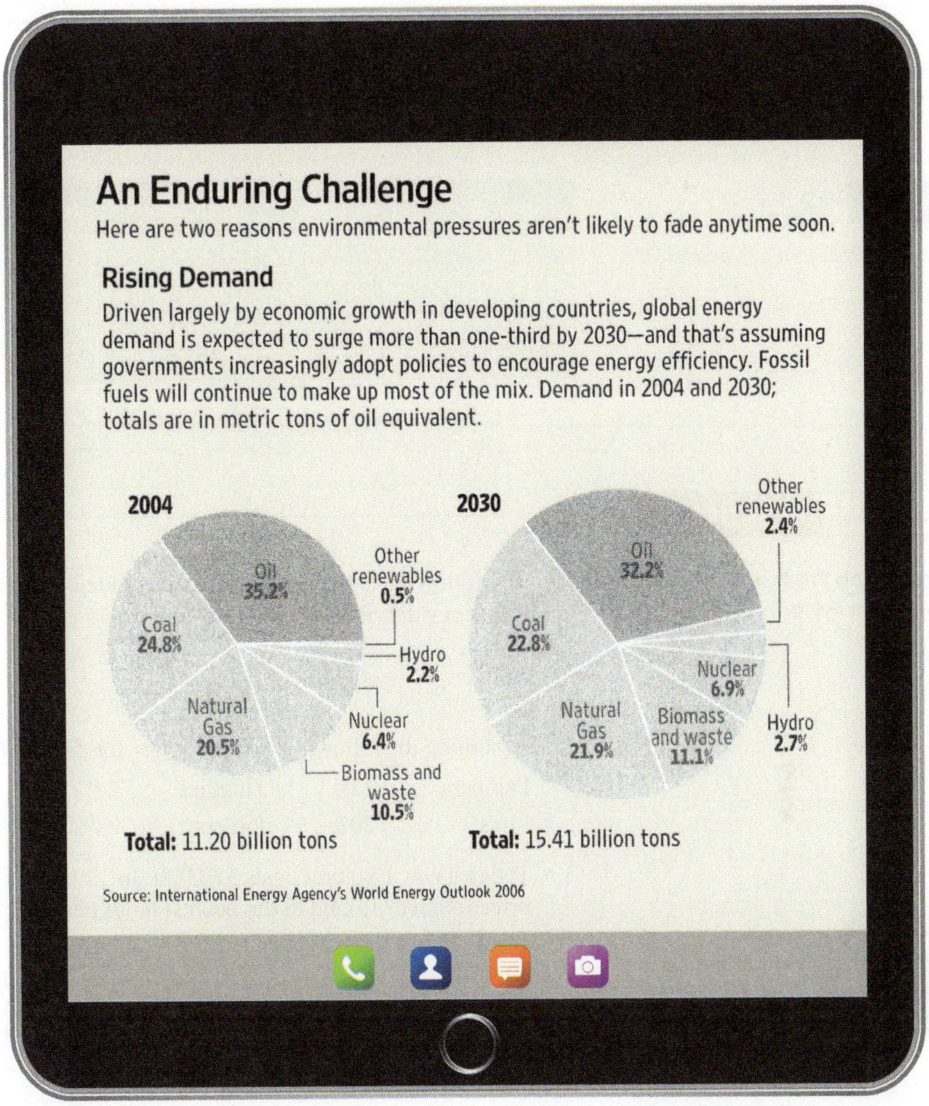

LO 3

An Application of Statistics: Index Numbers

The financial section of a newspaper often gives different index numbers describing the changes in business. These **index numbers** express the relative changes in a variable compared with some base, which is taken as 100. The changes may be measured from time to time or from place to place. Index numbers function as percents and are calculated like percents.

Frequently, a business will use index numbers to make comparisons of a current price relative to a given year. For example, a calculator may cost \$9 today relative to a cost of \$75 some 30 years ago. The **price relative** of the calculator is $\frac{\$9}{\$75} \times 100 = 12\%$. The calculator now costs 12% of what it cost some 30 years ago. A price relative, then, is the current price divided by some previous year's price—the base year—multiplied by 100.

$$\text{Price relative} = \frac{\text{Current price}}{\text{Base year's price}} \times 100$$

MONEY tips

All too often, the incorrect beneficiary is listed on life insurance policies, retirement accounts, and bank accounts. Create a list and review your accounts annually to verify the correct beneficiary is listed. Ensure a survivor knows where to locate the list and your estate documents, in the event of death.

Index numbers can also be used to estimate current prices at various geographic locations. The frequently quoted Consumer Price Index (CPI), calculated and published monthly by the U.S. Bureau of Labor Statistics, records the price relative percentage cost of many goods and services nationwide compared to a base period. Table 22.1 gives a portion of the CPI that uses 1982–84 as its base period. Note that the table shows, for example, that the price relative for housing in Los Angeles is 139.3% of what it cost in 1982–84. Thus, Los Angeles housing costs amounting to \$100.00 in 1982–84 now cost \$139.30. So if you built a \$90,000 house in 1982–84, it is worth \$125,370 today. (Convert 139.3% to the decimal 1.393; multiply \$90,000 by 1.393 = \$125,370.)

Once again, we complete the unit with a Practice Quiz.

TABLE 22.1

Consumer Price Index (in percent)

Expense	Atlanta	Chicago	New York	Los Angeles
Food	131.9	130.3	139.6	130.9
Housing	128.8	131.4	139.3	139.3
Clothing	133.8	124.3	121.8	126.4
Medical care	177.6	163.0	172.4	163.3

LU 22–2 PRACTICE QUIZ

Complete this Practice Quiz to see how you are doing.

1. The following is the number of sales made by 20 salespeople on a given day. Prepare a frequency distribution and a bar graph. Do not use intervals for this example.

5	8	9	1	4	4	0	3	2	8
8	9	5	1	9	6	7	5	9	10

2. Assuming the following market shares for diapers 5 years ago, prepare a circle graph:

Pampers	32%	Huggies	24%
Luvs	20%	Others	24%

3. Today a new Explorer costs \$30,000. In 1991 the Explorer cost \$19,000. What is the price relative? Round to the nearest tenth percent.

*For **extra help** from your authors–Sharon and Jeff–see the videos in Connect.*

✓ Solutions

1.

Number of sales	Tally	Frequency
0	I	1
1	II	2
2	I	1
3	I	1
4	II	2
5	III	3
6	I	1
7	I	1
8	III	3
9	IIII	4
10	I	1

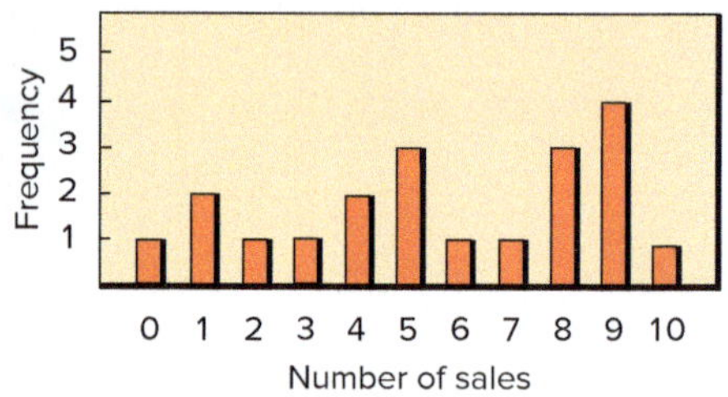

2.

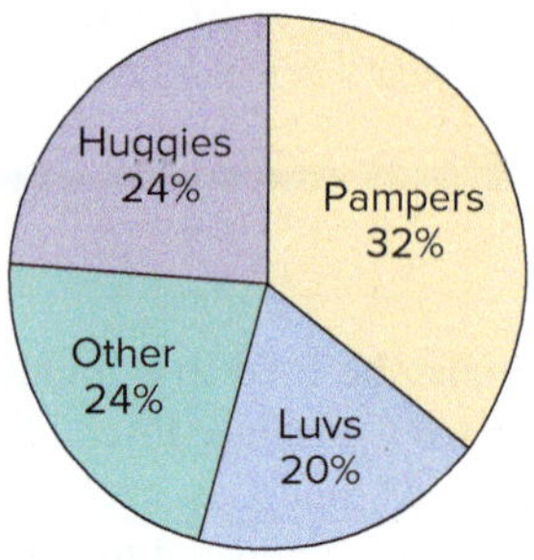

$.32 \times 360° = 115.20°$

$.20 \times 360° = 72.00°$

$.24 \times 360° = 86.40°$

$.24 \times 360° = 86.40°$

3. $\frac{\$30{,}000}{\$19{,}000} \times 100 = 157.9$

Learning Unit 22–3: Measures of Dispersion (Optional)

In Learning Unit 22–1 you learned how companies use the mean, median, and mode to indicate a single value, or number, that represents an entire group of numbers, or data. Often it is valuable to know how the information is scattered (spread or dispersed) within a data set. A **measure of dispersion** is a number that describes how the numbers of a set of data are spread out or dispersed.

This learning unit discusses three measures of dispersion—range, standard deviation, and normal distribution. We begin with the range—the simplest measure of dispersion.

LO 1

Range

The **range** is the difference between the two extreme values (highest and lowest) in a group of values or a set of data. For example, often the actual extreme values of hourly temperature readings during the past 24 hours are given but not the range or difference between the high and low readings. To find the range in a group of data, subtract the lowest value from the highest value.

Range = Highest value − Lowest value

Thus, if the high temperature reading during the past 24 hours was 90° and the low temperature reading was 60° the range is 90° − 60°, or 30°. The range is limited in its application because it gives only a general idea of the spread of values in a data set.

EXAMPLE Find the range of the following values: 83.6, 77.3, 69.2, 93.1, 85.4, 71.6.

Range = 93.1 − 69.2 = 23.9

LO 2

Standard Deviation

Since the **standard deviation** is intended to measure the spread of data around the mean, you must first determine the mean of a set of data. The following diagram shows two sets of data—A and B. In the diagram, the means of A and B are equal. Now look at how the data in these two sets are spread or dispersed.

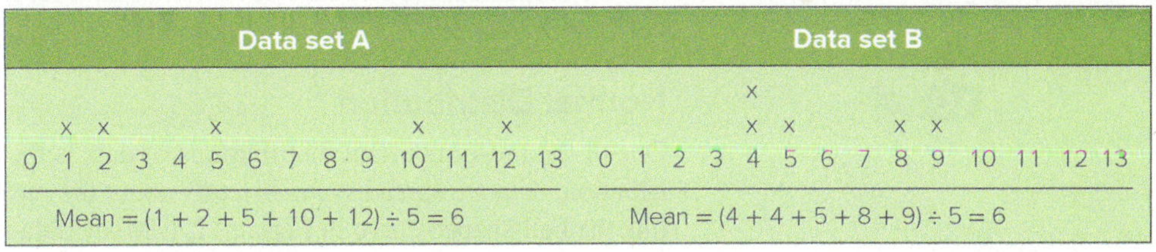

Note that although the means of data sets A and B are equal, A is more widely dispersed, which means B will have a smaller standard deviation than A.

A statistics calculator will allow you to calculate the mean, variance, standard deviation, and many more basic and advanced statistics calculations easily.

To find the standard deviation of an ungrouped set of data, use the following steps:

FINDING THE STANDARD DEVIATION

Step 1. Find the mean of the set of data.

Step 2. Subtract the mean from each piece of data to find each deviation.

Step 3. Square each deviation (multiply the deviation by itself).

Step 4. Sum all squared deviations.

Step 5. Divide the sum of the squared deviations by $n - 1$, where n equals the number of pieces of data.

Step 6. Find the square root ($\sqrt{\ }$) of the number obtained in Step 5 (use a calculator). This is the standard deviation. (The square root is a number that when multiplied by itself equals the amount shown inside the square root symbol.)

Two additional points should be made. First, Step 2 sometimes results in negative numbers. Since the sum of the deviations obtained in Step 2 should always be zero, we would not be able to find the average deviation. This is why we square each deviation—to generate positive quantities only. Second, the standard deviation we refer to is used with *sample* sets of data, that is, a collection of data from a population. The population is the *entire* collection of data. When the standard deviation for a population is calculated, the sum of the squared deviations is divided by n instead of by $n - 1$. In all problems that follow, sample sets of data are being examined.

EXAMPLE Calculate the standard deviations for the sample data sets A and B given in the previous diagram. Round the final answer to the nearest tenth. Note that Step 1—find the mean—is given in the diagram.

Standard deviation of data sets A and B: The table on the left uses Steps 2 through 6 to find the standard deviation of data set A, and the table on the right uses Steps 2 through 6 to find the standard deviation of data set B.

Data	Step 2 Data − Mean	Step 3 (Data − Mean)^2
1	1 − 6 = −5	25
2	2 − 6 = −4	16
5	5 − 6 = −1	1
10	10 − 6 = 4	16
12	12 − 6 = 6	36
	Total 0	94 **(Step 4)**

Step 5: Divide by $n - 1$: $\frac{94}{5-1} = \frac{94}{4} = 23.5$

Step 6: The square root of $\sqrt{23.5}$ is 4.8 (rounded).

The standard deviation of data set A is 4.8.

Data	Step 2 Data − Mean	Step 3 (Data − Mean)^2
4	4 − 6 = −2	4
4	4 − 6 = −2	4
5	5 − 6 = −1	1
8	8 − 6 = 2	4
9	9 − 6 = 3	9
	Total 0	22 **(Step 4)**

Step 5: Divide by $n - 1$: $\frac{22}{5-1} = \frac{22}{4} = 5.5$

Step 6: The square root of $\sqrt{5.5}$ is 2.3.

The standard deviation of data set B is 2.3.

As suspected, the standard deviation of data set B is less than that of set A. The standard deviation value reinforces what we see in the diagram.

LO 3

Normal Distribution

One of the most important distributions of data is the **normal distribution.** In a normal distribution, data are spread *symmetrically* about the mean. A graph of such a distribution looks like the bell-shaped curve in Figure 22.1. Many data sets are normally distributed. Examples are the life span of automobile engines, women's heights, and intelligence quotients.

FIGURE 22.1

Standard deviation and the normal distribution

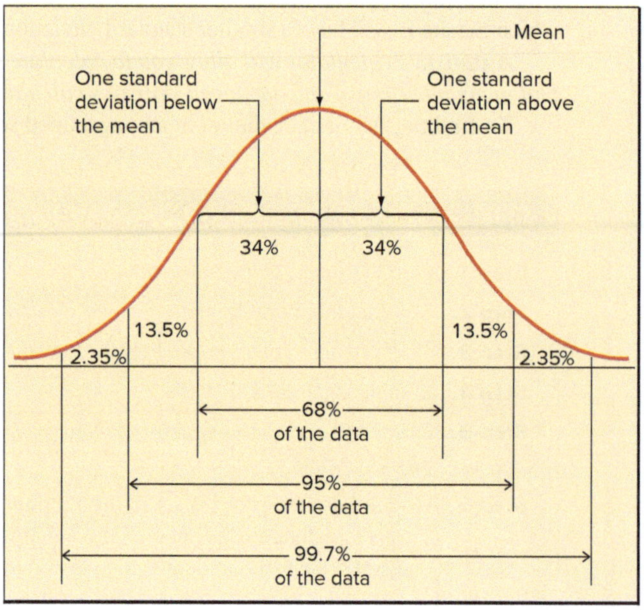

In a normal distribution, the mean, median, and mode are all equal. Additionally, when the data are normally distributed, the **Empirical Rule** (Three Sigma Rule) applies, stating that for a normal distribution, also known as a bell curve, approximately

- 68% of the observations will fall within ± 1 standard deviation of the mean,
- 95% will fall within ± 2 standard deviations of the mean, and
- 99.7% will fall within ± 3 standard deviations of the mean.

Figure 22.1 above illustrates these facts.

EXAMPLE Assume that the mean useful life of a particular lightbulb is 2,000 hours and is normally distributed with a standard deviation of 300 hours. Calculate the useful life of the lightbulb with **(a)** one standard deviation of the mean and **(b)** two standard deviations of the mean; also **(c)** calculate the percent of lightbulbs that will last 2,300 hours or longer.

MONEY tips

Start achieving financial security today! Reduce spending, pay off debt, and start saving. Set small milestones, increasing them as you reach each goal. It's worth the effort and will make tomorrow financially brighter.

a. The useful life of the lightbulb one standard deviation from the mean is one standard deviation above *and* below the mean.

$2{,}000 \pm 300 = 1{,}700$ and $2{,}300$ hours

The useful life is somewhere between 1,700 and 2,300 hours.

b. The useful life of the lightbulb within two standard deviations of the mean is within two standard deviations above *and* below the mean.

$2{,}000 \pm 2(300) = 1{,}400$ and $2{,}600$ hours

c. Since 50% of the data in a normal distribution lie below the mean and 34% represent the amount of data one standard deviation above the mean, we must calculate the percent of data that lie beyond one standard deviation above the mean.

$100\% - (50\% + 34\%) = 16\%$

So 16% of the bulbs should last 2,300 hours or longer.

It's time for another Practice Quiz.

LU 22–3 PRACTICE QUIZ

Complete this Practice Quiz to see how you are doing.

1. Calculate the range for the following data: 58, 13, 17, 26, 5, 41.
2. Calculate the variance and the standard deviation for the following sample set of data: 113, 92, 77, 125, 110, 93, 111. Round answers to the nearest tenth.
3. If the mean tax refund for the year is $3,000 with a $300 standard deviation, what is the refund range within three standard deviations above and below the mean?

*For **extra help** from your authors–Sharon and Jeff–see the videos in Connect.*

✓ Solutions

1. $58 - 5 = 53$ range
2.

Data	Data – Mean	(Data – Mean)²
113	113 – 103 = 10	100
92	92 – 103 = –11	121
77	77 – 103 = –26	676
125	125 – 103 = 22	484
110	110 – 103 = 7	49
93	93 – 103 = –10	100
111	111 – 103 = 8	64
	Total	1,594

$1{,}594 \div (7 - 1) = 265.6666667$ variance

$\sqrt{265.6666667} = 16.3$ standard deviation

3. $3,000 +/– ($300 × 3) = Between $2,100 and $3,900

INTERACTIVE CHAPTER ORGANIZER

Topic/Procedure/Formula	Examples	You try it*
Mean $\frac{\text{Sum of all values}}{\text{Number of values}}$	Age of team players: 22, 28, 31, 19, 15 $\text{Mean} = \frac{22 + 28 + 31 + 19 + 15}{5}$ $= 23$	**Calculate mean** 41, 29, 16, 15, 18
Weighted mean $\frac{\text{Sum of products}}{\text{Sum of frequencies}}$	**S. M. T. W. Th. F. S.** Sales $90 $75 $80 $75 $80 $90 $90 **Value** / **Frequency** / **Product** $90 / 3 / $270 75 / 2 / 150 80 / 2 / 160 7 / $580 $\text{Mean} = \frac{\$580}{7} = \$82.86$	**Calculate weighted mean** **S. M. T. W. Th. Fr. S.** Sales 80 90 100 80 80 90 90
Median **1.** Arrange values from smallest to largest. **2.** Find the middle value. **a. Odd number of values:** median is middle value. $\left(\frac{\text{Total number of numbers}}{2}\right)$ Next-higher number is median. **b. Even number of values:** average of two middle values.	12, 15, 8, 6, 3 **1.** 3 6 8 12 15 **2.** $\frac{5}{2} = 2.5$ Median is third number, 8.	**Calculate median** 14, 16, 9, 7, 4
Mode Value that occurs most often in a set of numbers	6, 6, 8, 5, 6 Mode is 6	**Find mode** 7, 7, 4, 3, 2, 7
Frequency distribution Method of listing numbers or amounts not arranged in any particular way by columns for numbers (amounts), tally, and frequency	Number of sodas consumed in one day: 1, 5, 4, 3, 4, 2, 2, 3, 2, 0 **Number of sodas** / **Tally** / **Frequency** 0 / I / 1 1 / I / 1 2 / III / 3 3 / II / 2 4 / II / 2 5 / I / 1	**Prepare frequency distribution** Number of coffees consumed in one day: 1, 4, 5, 8, 2, 2, 3, 0
Bar graphs Height of bar represents frequency. Bar graph used for grouped data. Bar graphs can be vertical or horizontal.	From soda example above: 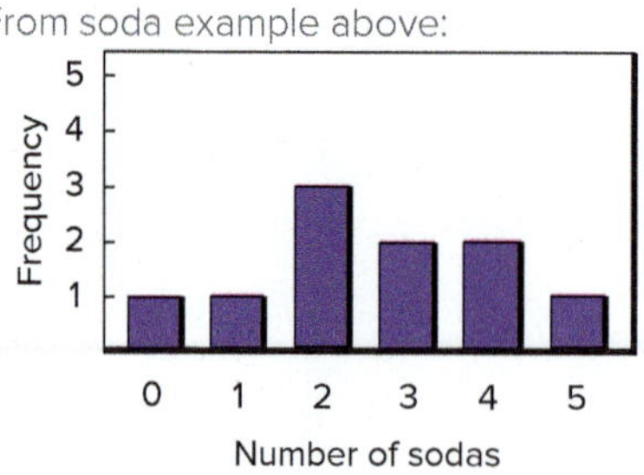	**From coffee example above, prepare bar graph**
Line graphs Shows trend. Helps to put numbers in order.	**Sales** 2018 $1,000 2019 2,000 2020 3,000 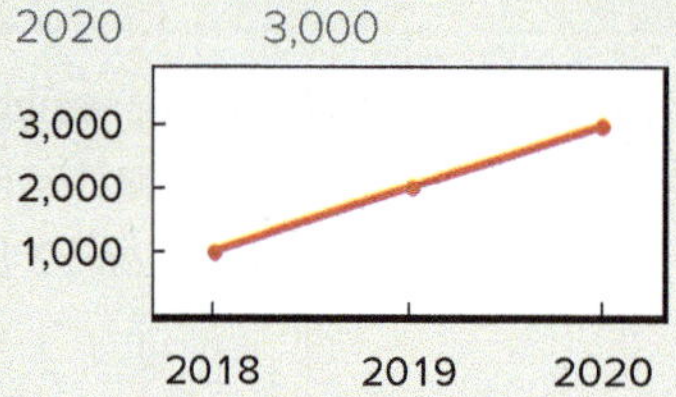	**Prepare line graph** **Sales** 2018 $5,000 2019 3,000 2020 2,000

(continues)

INTERACTIVE CHAPTER ORGANIZER

Topic/Procedure/Formula	Examples	You try it*
Circle graphs Circle = 360° % × 360° = Degrees of pie to represent percent Total should = 360°	60% favor diet soda 40% favor sugared soda 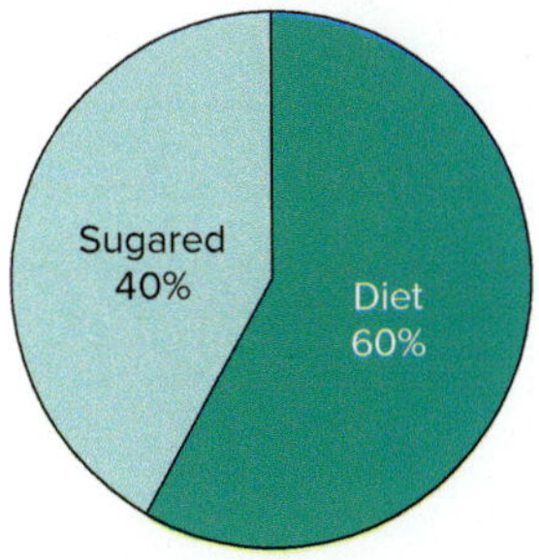 $.60 \times 360° = 216°$ $.40 \times 360° = \underline{144°}$ $360°$	**Create circle graph** 70% coffee drinkers 30% non-coffee-drinkers
Price relative $\text{Price relative} = \dfrac{\text{Current price}}{\text{Base year's price}} \times 100$	A station wagon's sticker price was \$8,799 in 1982. Today it is \$14,900. $\text{Price relative} = \dfrac{\$14,900}{\$8,799} \times 100 = 169.3$ (rounded to nearest tenth percent)	**Calculate price relative** Old price, \$ 9,000 Today's price, 12,000
Range (optional) Range = Highest value − Lowest value	Calculate range of the data set consisting of 5, 9, 13, 2, 8 Range = 13 − 2 = 11	**Calculate range** 6, 8, 14, 2, 9
Standard deviation (optional) **1.** Calculate mean. **2.** Subtract mean from each piece of data. **3.** Square each deviation. **4.** Sum squares. **5.** Divide sum of squares by $n - 1$, where n = number of pieces of data. **6.** Take square root of number obtained in Step 5, to find the standard deviation.	Calculate the standard deviation of this set of data: 7, 2, 5, 3, 3. **1.** $\text{Mean} = \dfrac{20}{5} = 4$ **2.** $7 - 4 = 3$ $2 - 4 = -2$ $5 - 4 = 1$ $3 - 4 = -1$ $3 - 4 = -1$ **3.** $(3)^2 = 9$ $(-2)^2 = 4$ $(1)^2 = 1$ $(-1)^2 = 1$ $\underline{(-1)^2 = 1}$ **4.** 16 **5.** $16 \div 4 = 4$ **6.** Standard deviation = 2	**Calculate standard deviation** 8, 1, 6, 2, 2

KEY TERMS			
	Bar graph Circle graph Empirical Rule Frequency distribution Index numbers	Line graph Mean Measure of dispersion Median Mode	Normal distribution Price relative Range Standard deviation Weighted mean

*Worked-out solutions are in Appendix B.

Critical Thinking Discussion Questions with Chapter Concept Check

1. Explain the mean, median, and mode. Give an example that shows you must be careful when you read statistics in an article.
2. Explain frequency distributions and the types of graphs. Locate a company annual report and explain how the company shows graphs to highlight its performance. Does the company need more or fewer of these visuals? Could price relatives be used?
3. Explain the statement, "Standard deviations are not accurate."
4. **Chapter Concept Check.** Visit the Apple website. Gather new statistics on the iPad, Apple Watch, and/or iPhone. Use concepts in this chapter to create a presentation.

22

END-OF-CHAPTER PROBLEMS

connect

Check figures for odd-numbered problems in Appendix B. Name ____________________ Date ________

DRILL PROBLEMS (*Note:* Problems for optional Learning Unit 22–3 follow the Challenge Problem 22–24)

Calculate the mean (to the nearest hundredth): *LU 22-1(1)*

22–1. 9, 3, 2, 11

22–2. 5, 4, 8, 12, 15

22–3. \$55.83, \$66.92, \$108.93

22–4. \$1,001, \$68.50, \$33.82, \$581.95

eXcel **22–5.** Calculate the grade point average: A = 4, B = 3, C = 2, D = 1, F = 0 (to nearest tenth). *LU 22-1(2)*

Courses	Credits	Grade
Computer Principles	3	B
Business Law	3	C
Logic	3	D
Biology	4	A
Marketing	3	B

22–6. Find the weighted mean (to the nearest tenth): *LU 22-1(2)*

Value	Frequency	Product
4	7	
8	3	
2	9	
4	2	

Find the median: *LU 22-1(3)*

22–7. 55, 10, 19, 38, 100, 25

22–8. 95, 103, 98, 62, 31, 15, 82

Find the mode: *LU 22-1(4)*

22–9. 8, 9, 3, 4, 12, 8, 8, 9

22–10. 22, 19, 15, 16, 18, 18, 5, 18

22–11. **Given:**

Truck cost	2012	\$30,000
Truck cost	2008	\$21,000

Calculate the price relative (rounded to the nearest tenth percent). *LU 22-2(3)*

eXcel **22–12.** Given the following sales of Lowe Corporation, prepare a line graph (run sales from \$5,000 to \$20,000). *LU 22-2(2)*

2017	\$ 8,000
2018	11,000
2019	13,000
2020	18,000

22–13. Prepare a frequency distribution from the following weekly salaries of teachers at Pikes Peak Community College. Use the following intervals: *LU 22-2(1)*

$200–$299.99
$300–$399.99
$400–$499.99
$500–$599.99

$210	$505	$310	$380	$275
290	480	550	490	200
286	410	305	444	368

22–14. Prepare a bar graph from the frequency distribution in Problem 22–13. *LU 22-2(2)*

22–15. How many degrees on a circle graph would be given to each of the following? *LU 22-2(2)*

Wear digital watch	42%
Wear traditional watch	51
Wear no watch	7

WORD PROBLEMS

eXcel **22–16.** The first Super Bowl on January 15, 1967, charged $42,000 for a 30-second commercial. Create a line graph for the following Super Bowl 30-second commercial costs: 2011, $3,100,000; 2012, $3,500,000; 2013 and 2014, $4,000,000; 2015, $4,500,000; 2016, $5,000,000; and 2017 and 2018, $5,020,000. *LU 22-2(2)*

eXcel **22–17.** The American Kennel Club announced the "Most Popular Dogs in the U.S." Labrador retrievers remained number one for the 26th consecutive year. German shepherds came in second followed by Yorkshire terriers, golden retrievers, and beagles. Create a circle graph for Dogs for Life Kennel Club with the following members: 52 Labrador retrievers, 33 German shepherds, 22 golden retrievers, 15 bulldogs, and 10 beagles. *LU 22-2(2)*

My Money

22–18. Despite tuition skyrocketing, a college education is still valuable. Recent calculations by the Federal Reserve Bank in San Francisco demonstrate a college degree is worth $830,000 in lifetime earnings compared to the average high school education. If graduates in 2019 earn $40,632, $35,554, $42,192, $33,432, $69,479 and $43,589, what is the standard deviation for this sample? Round to a whole number for each calculation. *LU 22-3(2)*

22–19. Costcotravel.com provided a member with the following information regarding her upcoming travel. Construct a circle graph for the member. *LU 22-2(2)*

Transportation	35%
Hotel	28
Food and entertainment	20
Miscellaneous	17

22–20. Jim Smith, a marketing student, observed how much each customer spent in a local convenience store. Based on the following results, prepare **(a)** a frequency distribution and **(b)** a bar graph. Use intervals of $0–$5.99, $6.00–$11.99, $12.00–$17.99, and $18.00–$23.99. *LU 22-2(2)*

$18.50	$18.24	$ 6.88	$9.95
16.10	3.55	14.10	6.80
12.11	3.82	2.10	
15.88	3.95	5.50	

22–21. Angie's Bakery bakes bagels. Find the weighted mean (to the nearest whole bagel) given the following daily production for June: *LU 22-1(2)*

200	150	200	150	200
150	190	360	360	150
190	190	190	200	150
360	400	400	150	200
400	360	150	400	360
400	400	200	150	150

22–22. The United Nations states the gender pay gap will not close for 70 years. Women across the world earn $0.76 for every $1.00 of what men earn. Construct a bar graph reflecting the following Harvard University study on pay for women based on $1.00 for men: financial specialists, $0.66; physicians, $0.71; aircraft pilots, $0.71; accountants, $0.76; lawyers, $0.82; and nurses, $0.89. *LU 22-2(2)*

CHALLENGE PROBLEMS

22–23. Listed below are annual revenues for a few travel agencies:

AAA Travel Agency	$86,700,000
Riser Group	63,200,000
Casto Travel	62,900,000
Balboa Travel	36,200,000
Hunter Travel Managers	36,000,000

(a) What would be the mean and the median? **(b)** What is the total revenue percent of each agency? **(c)** Prepare a circle graph depicting the percents. *LU 22-1(1, 2), LU 22-2(2)*

My Money

22–24. Review the two circle graphs for recommendations on how to budget. Then look at the circle graph budget for Ron Rye and his family for a month. Ron would like you to calculate the percent (to the hundredth) for each part of the circle graph along with the appropriate number of degrees. Should Ron adjust his spending? If yes, how? *LU 22-2(2)*

Recommended percentage of overall spending: A

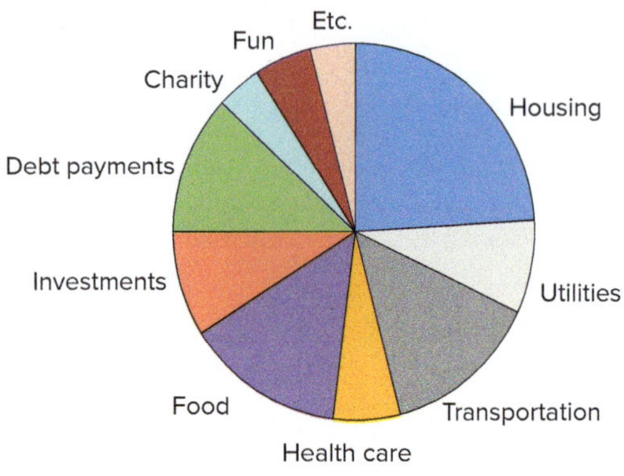

Recommended percentage of overall spending: B

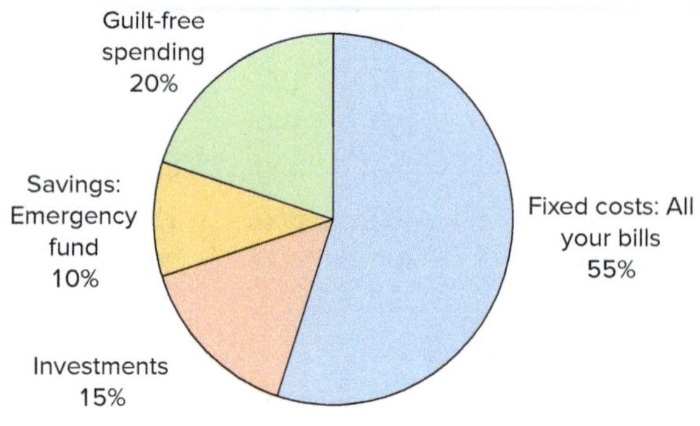

Ron Rye Family Budget

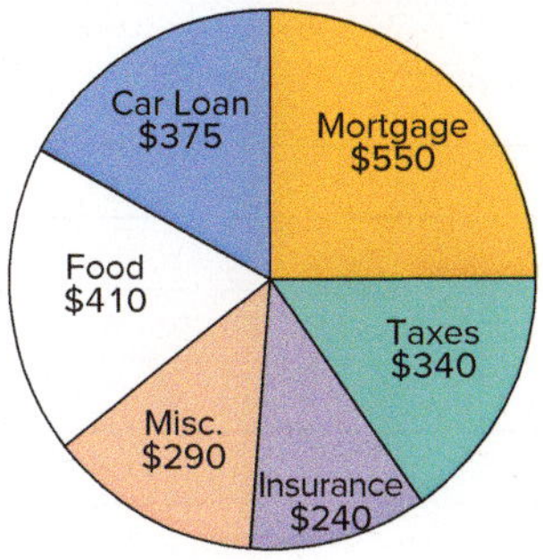

Classroom Notes

LEARNING UNIT 22–3 OPTIONAL ASSIGNMENTS connect

Check figures for odd-numbered problems in Appendix B. Name ______________________ Date ________

DRILL PROBLEMS

1. Calculate the range for the following set of data: 117, 98, 133, 52, 114, 35. *LU 22-3(1)*

Calculate the standard deviation for the following sample sets of data. Round the final answers to the nearest tenth. *LU 22-3(2)*

2. 83.6, 92.3, 56.5, 43.8, 77.1, 66.7

3. 7, 3, 12, 17, 5, 8, 9, 9, 13, 15, 6, 6, 4, 5

4. 41, 41, 38, 27, 53, 56, 28, 45, 47, 49, 55, 60

WORD PROBLEMS

eXcel

5. The mean useful life of car batteries is 48 months. They have a standard deviation of 3. If the useful life of batteries is normally distributed, calculate **(a)** the percent of batteries with a useful life of less than 45 months and **(b)** the percent of batteries that will last longer than 54 months. *LU 22-3(2)*

6. The average weight of a particular box of crackers is 24.5 ounces with a standard deviation of 0.8 ounce. The weights of the boxes are normally distributed. What percent of the boxes **(a)** weigh more than 22.9 ounces and **(b)** weigh less than 23.7 ounces? *LU 22-3(2)*

7. An examination is normally distributed with a mean score of 77 and a standard deviation of 6. Find the percent of individuals scoring as indicated below. *LU 22-3(2)*

a. Between 71 and 83

b. Between 83 and 65

c. Above 89

d. Less than 65

e. Between 77 and 65

8. Listed below are the sales figures in thousands of dollars for a group of insurance salespeople. Calculate the mean sales figure and the standard deviation. *LU 22-3(1, 2)*

\$117	\$350	\$400	\$245	\$420
223	275	516	265	135
486	320	285	374	190

9. The time in seconds it takes for 20 individual sewing machines to stitch a border onto a particular garment is listed below. Calculate the mean stitching time and the standard deviation to the nearest hundredth. *LU 22-3(1, 2)*

67	69	64	71	73
58	71	64	62	67
62	57	67	60	65
60	63	72	56	64

Classroom Notes

SUMMARY PRACTICE TEST

Do you need help? Connect videos have step-by-step worked-out solutions.

eXcel **1.** In July, Lee Realty sold 10 homes at the following prices: \$140,000; \$166,000; \$80,000; \$98,000; \$185,000; \$150,000; \$108,000; \$114,000; \$142,000; and \$250,000. Calculate the mean and median. *LU 22-1(1, 3)*

2. Lowes counted the number of customers entering the store for a week. The results were 1,100; 950; 1,100; 1,700; 880; 920; and 1,100. What is the mode? *LU 22-1(4)*

3. This semester Hung Lee took four 3-credit courses at Riverside Community College. She received an A in accounting and C's in history, psychology, and algebra. What is her cumulative grade point average (assume A = 4 and C = 2) to the nearest hundredth? *LU 22-1(2)*

4. Pete's Variety Shop reported the following sales for the first 20 days of May. Prepare a frequency distribution for Pete's. *LU 22-2(1)*

\$100	\$400	\$600	\$400	\$600
100	600	300	500	700
200	600	700	500	200
100	600	100	700	700

eXcel **5.** Leeds Company produced the following number of maps during the first 5 weeks of last year. Prepare a bar graph. *LU 22-2(2)*

Week	Maps
1	800
2	600
3	400
4	700
5	300

6. Laser Corporation reported record profits of 30%. It stated in the report that the cost of sales was 40% with expenses of 30%. Prepare a circle graph for Laser. *LU 22-2(2)*

7. Today a new Explorer costs $39,900. In 1990, Explorers cost $24,000. What is the price relative to the nearest tenth percent? *LU 22-2(3)*

***8.** Calculate the standard deviation for the following set of data: 7, 2, 5, 3, 3, 10. Round the final answer to the nearest tenth. *LU 22-3(2)*

*Optional problem.

INTERACTIVE VIDEO WORKSHEET

GRADING THE SUMMARY PRACTICE TEST

Go to the summary practice test video in Connect (or click on it here in the ebook). Grade your summary practice test while viewing the video.

C for Correct/I for Incorrect

1.	________	5.	________
2.	________	6.	________
3.	________	7.	________
4.	________	*8.	________

If you achieved 100%, you are ready for your instructor's exam.

If any of the problems were incorrect, list the questions you missed and show steps to solve the problem correctly.

Replay the video to see if you have made the correct fixes to your mistakes. If you have any questions, contact your instructor asap.

*Optional problem.

MY MONEY

My Health

What I need to know

Throughout the My Money segments in this book you have no doubt learned that managing your finances is an important step to achieving your financial goals. The same can be said about managing your health. The hard work you put into reaching your financial goals won't matter much if you are not around to reap the rewards of your diligence. A healthy lifestyle is just as important, if not more so, than a healthy financial position. Creating good health habits early on in life will make these changes more of a lifestyle than just a fad or temporary activity. As with the habits you are learning about managing your finances, managing your health will demand a concerted effort if you hope to achieve and maintain success.

What I need to do

Take an active approach to life. Engage in a regular form of exercise. It can be as simple as walking or maybe joining a competitive sports league. Find ways in which you can exercise throughout your day such as taking the steps versus the escalator or elevator. Join a fitness center and commit to exercising there at least 3 times per week for 30 minutes each session. As you become more accustomed to these workouts, you can increase the time per session or the number of days per week in which you are exercising. Over time you will find the act of exercising becomes easier and actually more enjoyable as you establish good health habits. Explore the use of a fitness tracker to help you stay on track with your exercising goals and track your progress. There are many easy-to-use fitness trackers available for your smartphone or computer.

A balanced diet is another part of creating and maintaining a healthy lifestyle. Changes to your eating habits don't have to be drastic. Finding ways to cut the total calories you consume on a daily basis is a great way to make positive strides in your diet. For instance, you could reduce your sugar intake by switching from soda to water or opting for a handful of almonds versus a cookie. Also, selecting fresh fruits and vegetables versus prepackaged processed foods will leave you feeling fuller and provide you with key nutrients your body needs. If you consume alcohol, do so in moderation.

Stay on track by maintaining your healthy lifestyle. Regular checkups with your medical and dental professionals will also keep you on the right path to a healthy life. These professionals will also provide you with helpful ways in which you can continue to maintain a healthy lifestyle based on your unique health situation. If you are experiencing stress, seek out ways in which you can reduce the stressful situations in your life. Conduct an Internet search or ask your medical professional to assist you in finding ways to effectively deal with the stress in your life and keep it from negatively impacting your overall health. Getting the right amount of restful sleep will not only help with your stress management but will also allow your body to reenergize itself.

Resources I can use

- Lose It!—Calorie Counter (mobile app)
- https://www.webmd.com/fitness-exercise/ss/twelve-habits-super-healthy-people — healthy habits

MY MONEY ACTIVITY

For three weeks, commit to exercising three days a week for 30 minutes per day. After three weeks this should become more of a habit rather than just an "assignment."

PERSONAL FINANCE

A KIPLINGER APPROACH

"Retire a Millionaire, Time Is On Your Side", *Kiplinger's*, September 2017, p. 33. Used by permission of The Kiplinger Washington Editors, Inc.

Road to Riches

Retire a Millionaire: Time Is on Your Side

If you're starting out and want to save $1 million by the time you retire, time is on your side. Thanks to the miracle of compounding, even modest monthly contributions can grow into a seven-figure nest egg by the time you turn 65. Late bloomers can still reach the magic number, but it will take a bigger chunk of your paycheck. We're assuming an all-stock portfolio with an 8% annual return—less than the stock market's historical average but slightly more than some analysts expect the market to return over the next decade.

©Image Source/Getty Images

©langstrup/123RF

©Caia Image/Glow Images

©Don Mason/Blend Images

	At Age 25	At Age 35	At Age 45	At Age 55
YOU'VE SAVED	$0	$45,000*	$63,000*	$117,000*
WHAT YOU NEED TO SAVE PER MONTH	$310	$400	$1,250	$4,200
YOU'VE SAVED	$16,000*	$65,000	$100,000	$250,000
WHAT YOU NEED TO SAVE PER MONTH	$210	$250	$960	$2,750

TIP: You may qualify for a tax credit of 10% to 50% of the amount you contribute to an IRA, 401(k) or other retirement account. The credit can reduce your tax bill by up to $2,000. To qualify, your income must be $31,000 or less if you're single, $46,500 or less if you're a head of household, or $62,000 or less if you're married and file jointly.

TIP: Despite other demands on your paycheck (such as a mortgage, contributions to college savings and car loans), resolve to contribute at least enough to your 401(k) to capture your employer's matching contribution—that's free money. Try to save 15% of your gross income for retirement, including your employer match.

TIP: Use extra income to supercharge your savings. You can contribute up to $18,000 to your 401(k) or similar employer-sponsored plan this year, plus an additional $6,000 if you're 50 or older. You can also contribute up to $5,500 to a traditional or Roth IRA, plus an additional $1,000 if you're 50 or older.

TIP: Planning to work a few years longer can boost your savings. And because you won't be taking withdrawals, your money will have more time to compound and grow.

*MEDIAN AMOUNT SAVED FOR THE AGE GROUP.
SOURCE: TRANSAMERICA CENTER FOR RETIREMENT STUDIES 16TH ANNUAL RETIREMENT SURVEY OF WORKERS

BUSINESS MATH ISSUE

Most people cannot retire as a millionaire because the cost of living is so high.

1. List the key points of the article and information to support your position.
2. Write a group defense of your position using math calculations to support your view. If you are in an online course, post to a discussion board.

Classroom Notes

APPENDIX A Additional Homework by Learning Unit

Learning Unit 1–1: Reading, Writing, and Rounding Whole Numbers

DRILL PROBLEMS

1. Express the following numbers in verbal form:

a. 8,821 ______________________

b. 160,501 ______________________

c. 2,098,767 ______________________

d. 58,003 ______________________

e. 50,025,212,015 ______________________

2. Write in numeric form:

a. Eighty thousand, two hundred eighty-one ________

b. Fifty-eight thousand, three ________

c. Two hundred eighty thousand, five ________

d. Three million, ten ________

e. Sixty-seven thousand, seven hundred sixty ________

3. Round the following numbers:

a. To the nearest ten:

76 ________ 379 ________ 855 ________ 5,981 ________ 206 ________

b. To the nearest hundred:

9,664 ________ 2,074 ________ 888 ________ 271 ________ 75 ________

c. To the nearest thousand:

21,486 ________ 621 ________ 3,504 ________ 9,735 ________

4. Round off each number to the nearest ten, nearest hundred, nearest thousand, and round all the way. (Remember that you are rounding the original number each time.)

	Nearest ten	**Nearest hundred**	**Nearest thousand**	**Round all the way**
a. 4,752	________	________	________	________
b. 70,351	________	________	________	________
c. 9,386	________	________	________	________
d. 4,983	________	________	________	________
e. 408,119	________	________	________	________
f. 30,051	________	________	________	________

5. Name the place position (place value) of the underlined digit.

a. 8,<u>3</u>48 ______________________

b. <u>9</u>,734 ______________________

c. 3<u>4</u>7,107 ______________________

d. 7<u>2</u>3 ______________________

e. 2<u>8</u>,200,000,121 ______________________

f. 7<u>0</u>6,359,005 ______________________

g. 27,<u>5</u>63,530 ______________________

WORD PROBLEMS

6. Gim Smith was shopping for an Apple computer. He went to three different websites and found the computer he wanted at three different prices. At website A the price was $2,018, at website B the price was $1,985, and at website C the price was $2,030. What is the approximate price Gim will have to pay for the computer? Round to the nearest thousand. (Just one price.)

7. Amy Parker had to write a check at the bookstore when she purchased her books for the new semester. The total cost of the books was $564. How will she write this amount in verbal form on her check?

8. Matt Schaeffer was listening to the news and heard that steel production last week was one million, five hundred eighty-seven thousand tons. Express this amount in numeric form.

9. Jackie Martin is the city clerk and must go to the aldermen's meetings and take notes on what is discussed. At last night's meeting, they were discussing repairs for the public library, which will cost three hundred seventy-five thousand, nine hundred eighty-five dollars. Write this in numeric form as Jackie would.

10. A government survey revealed that 25,963,400 people are employed as office workers. To show the approximate number of office workers, round the number all the way.

11. Bob Donaldson wished to present his top student with a certificate of achievement at the end of the school year in 2019. To make it appear more official, he wanted to write the year in verbal form. How did he write the year?

12. Nancy Morrissey has a problem reading large numbers and determining place value. She asked her brother to name the place value of the 4 in the number 13,542,966. Can you tell Nancy the place value of the 4? What is the place value of the 3?

    The 4 is in the ____________________ place.

    The 3 is in the ____________________ place.

Learning Unit 1–2: Adding and Subtracting Whole Numbers

DRILL PROBLEMS

1. Add by totaling each separate column:

a.	b.	c.	d.	e.	f.	g.	h.
668	43	493	36	716	535	751	75,730
338	58	826	76	458	107	378	48,531
	96		43	397	778	135	15,797
			24	139	215	747	
				478	391	368	

2. Estimate by rounding all the way, and then add the actual numbers:

a.	b.	c.
580	1,470	475
971	7,631	837
548	4,383	213
430		775
506		432

d.	e.	f.
442	2,571	10,928
609	3,625	9,321
766	4,091	12,654
410	928	15,492
128		

3. Estimate by rounding all the way, and then subtract the actual numbers:

a.	b.	c.
90 − 38	91 − 33	68 − 59

d.	e.	f.
981 − 283	622 − 328	1,125 − 913

4. Subtract and check:

a.	b.	c.
4,947 − 4,362	3,724 − 2,138	474,820 − 85,847

d.	e.	f.
50,000 − 21,762	65,003 − 24,987	15,715 − 3,503

5. In the following sales report, total the rows and the columns, and then check that the grand total is the same both horizontally and vertically.

Salesperson	Region 1	Region 2	Region 3	Total
a. Becker	$ 5,692	$ 7,403	$ 3,591	
b. Edwards	7,652	7,590	3,021	
c. Graff	6,545	6,738	4,545	
d. Jackson	6,937	6,950	4,913	
e. Total				

WORD PROBLEMS

6. June Long owes $8,600 on her car loan for a new Chevy Volt, plus interest of $620. How much will it cost her to pay off this loan?

7. Sales at Rich's Convenience Store were $3,587 on Monday, $3,944 on Tuesday, $4,007 on Wednesday, $3,890 on Thursday, and $4,545 on Friday. What were the total sales for the week?

8. Poor's Variety Store sold $5,000 worth of lottery tickets in the first week of August; it sold $289 less in the second week. How much were the lottery ticket sales in the second week of August?

9. A truck weighed 9,550 pounds when it was empty. After being filled with rubbish, it was driven to the dump where it weighed in at 22,347 pounds. How much did the rubbish weigh?

10. Joanne Hoster had $610 in her checking account when she went to the bookstore. Joanne purchased an accounting book for $140, the working papers for $30, and a study guide for $35. After Joanne writes a check for the entire purchase, how much money will remain in her checking account?

11. A used Ford truck is advertised with a base price of $6,986 delivered. However, the window sticker on the truck reads as follows: tinted glass, $210; automatic transmission, $650; power steering, $210; power brakes, $215; safety locks, $95; air conditioning, $1,056. Estimate the total price, including the accessories, by rounding all the way and *then* calculating the exact price.

12. Four different stores are offering the same make and model of a Panasonic Smart television:

Store A	Store B	Store C	Store D
$1,285	$1,380	$1,440	$1,355

Find the difference between the highest price and the lowest price. Check your answer.

13. A Xerox copy machine has a suggested retail price of $1,395. The net price is $649. How much is the discount on the copy machine?

Learning Unit 1–3: Multiplying and Dividing Whole Numbers

DRILL PROBLEMS

1. In the following problems, first estimate by rounding all the way, and then work the actual problems and check:

	Actual	Estimate	Check
a.	160×15		
b.	$4,216 \times 45$		
c.	$52,376 \times 309$		
d.	$3,106 \times 28$		

2. Multiply; use the shortcut when applicable:

a. $4{,}072 \times 100$

b. $5{,}100 \times 40$

c. $76{,}000 \times 1{,}200$

d. $93 \times 100{,}000$

3. Divide by rounding all the way; then do the actual calculation and check showing the remainder as a whole number.

Actual **Estimate** **Check**

a. $8\overline{)7{,}709}$

b. $26\overline{)5{,}910}$

c. $151\overline{)3{,}783}$

d. $46\overline{)19{,}550}$

4. Divide by the shortcut method:

a. $200\overline{)5{,}400}$

b. $50\overline{)5{,}650}$

c. $1{,}200\overline{)43{,}200}$

d. $17{,}000\overline{)510{,}000}$

WORD PROBLEMS

5. Mia Kaminsky sells state lottery tickets in her variety store. If Mia's Variety Store sells 720 lottery tickets per day, how many tickets will be sold in a 7-day period?

6. Arlex Oil Company employs 100 people who are eligible for profit sharing. The financial manager has announced that the profits to be shared amount to $64,000. How much will each employee receive?

7. John Duncan's employer withheld $4,056 in federal taxes from his pay for the year. If equal deductions are made each week, what is John's weekly deduction?

8. Anne Domingoes drives a Volvo that gets 32 miles per gallon of gasoline. How many miles can she travel on 25 gallons of gas?

9. How many 8-inch pieces of yellow ribbon can be cut from a spool of ribbon that contains 6 yards (1 yard = 36 inches)?

10. The number of commercials aired per day on a local television station is 672. How many commercials are aired in 1 year?

11. The computer department at City College purchased 18 computers at a cost of $2,400 each. What was the total price for the computer purchase?

12. Net income for Goodwin's Partnership was $64,500. The five partners share profits and losses equally. What was each partner's share?

13. Ben Krenshaw's supervisor at the construction site told Ben to divide a load of 1,423 bricks into stacks containing 35 bricks each. How many stacks will there be when Ben has finished the job? How many "extra" bricks will there be?

Learning Unit 2–1: Types of Fractions and Conversion Procedures

DRILL PROBLEMS

1. Identify the type of fraction—proper, improper, or mixed number:

a. $\frac{9}{8}$ **b.** $\frac{7}{9}$ **c.** $\frac{29}{27}$

d. $9\frac{3}{11}$ **e.** $\frac{18}{5}$ **f.** $9\frac{1}{8}$

2. Convert to a mixed number:

a. $\frac{29}{4}$ **b.** $\frac{137}{8}$ **c.** $\frac{27}{5}$

d. $\frac{29}{9}$ **e.** $\frac{71}{8}$ **f.** $\frac{43}{6}$

3. Convert the mixed number to an improper fraction:

a. $9\frac{1}{5}$ **b.** $12\frac{3}{11}$ **c.** $4\frac{3}{7}$

d. $20\frac{4}{9}$ **e.** $10\frac{11}{12}$ **f.** $17\frac{2}{3}$

4. Tell whether the fractions in each pair are equivalent or not:

a. $\frac{3}{4}$ $\frac{9}{12}$ ________ b. $\frac{2}{3}$ $\frac{12}{18}$ ________ c. $\frac{7}{8}$ $\frac{15}{16}$ ________

d. $\frac{4}{5}$ $\frac{12}{15}$ ________ e. $\frac{3}{2}$ $\frac{9}{4}$ ________ f. $\frac{5}{8}$ $\frac{7}{11}$ ________

g. $\frac{7}{12}$ $\frac{7}{24}$ ________ h. $\frac{5}{4}$ $\frac{30}{24}$ ________ i. $\frac{10}{26}$ $\frac{12}{26}$ ________

5. Find the greatest common divisor by the step approach and reduce to lowest terms:

a. $\frac{36}{42}$

b. $\frac{30}{75}$

c. $\frac{74}{148}$

d. $\frac{15}{600}$

e. $\frac{96}{132}$

f. $\frac{84}{154}$

6. Convert to higher terms:

a. $\frac{9}{10} = \frac{\quad}{70}$

b. $\frac{2}{15} = \frac{\quad}{30}$

c. $\frac{6}{11} = \frac{\quad}{132}$

d. $\frac{4}{9} = \frac{\quad}{36}$

e. $\frac{7}{20} = \frac{\quad}{100}$

f. $\frac{7}{8} = \frac{\quad}{560}$

WORD PROBLEMS

7. Ken drove to college in $3\frac{1}{4}$ hours. How many quarter-hours is that? Show your answer as an improper fraction.

8. Mary looked in the refrigerator for a dozen eggs. When she found the box, only 5 eggs were left. What fractional part of the box of eggs was left?

9. At a recent meeting of a local Boosters Club, 17 of the 25 members attending were men. What fraction of those in attendance were men?

10. By weight, water is two parts out of three parts of the human body. What fraction of the body is water?

11. Three out of 5 students who begin college will continue until they receive their degree. Show in fractional form how many out of 100 beginning students will graduate.

12. Tina and her friends came in late to a party and found only $\frac{3}{4}$ of a pizza remaining. In order for everyone to get some pizza, she wanted to divide it into smaller pieces. If she divides the pizza into twelfths, how many pieces will she have? Show your answer in fractional form.

13. Sharon and Spunky noted that it took them 35 minutes to do their exercise routine. What fractional part of an hour is that? Show your answer in lowest terms.

14. Norman and his friend ordered several pizzas, which were all cut into eighths. The group ate 43 pieces of pizza. How many pizzas did they eat? Show your answer as a mixed number.

Learning Unit 2–2: Adding and Subtracting Fractions

DRILL PROBLEMS

1. Find the least common denominator (LCD) for each of the following groups of denominators using the prime numbers:

   a. 8, 16, 32 **b.** 9, 15, 20

   c. 12, 15, 32 **d.** 7, 9, 14, 28

2. Add and reduce to lowest terms or change to a mixed number if needed:

   a. $\frac{1}{9}+\frac{4}{9}$ **b.** $\frac{5}{12}+\frac{8}{15}$

c. $\frac{7}{8}+\frac{5}{12}$

d. $7\frac{2}{3}+5\frac{1}{4}$

e. $\frac{2}{3}+\frac{4}{9}+\frac{1}{4}$

3. Subtract and reduce to lowest terms:

a. $\frac{5}{9}-\frac{2}{9}$

b. $\frac{14}{15}-\frac{4}{15}$

c. $\frac{8}{9}-\frac{5}{6}$

d. $\frac{7}{12}-\frac{9}{16}$

e. $33\frac{5}{8}-27\frac{1}{2}$

f. $9-2\frac{3}{7}$

g. $15\frac{1}{3}-9\frac{7}{12}$

h. $92\frac{3}{10}-35\frac{7}{15}$

i. $93-57\frac{5}{12}$

j. $22\frac{5}{8}-17\frac{1}{4}$

WORD PROBLEMS

4. Dan Lund took a cross-country trip. He drove $5\frac{3}{8}$ hours on Monday, $6\frac{1}{2}$ hours on Tuesday, $9\frac{3}{4}$ hours on Wednesday, $6\frac{3}{8}$ hours on Thursday, and $10\frac{1}{4}$ hours on Friday. Find the total number of hours Dan drove in the first 5 days of his trip.

5. Sharon Parker bought 20 yards of material to make curtains. She used $4\frac{1}{2}$ yards for one bedroom window, $8\frac{3}{5}$ yards for another bedroom window, and $3\frac{7}{8}$ yards for a hall window. How much material did she have left?

6. Molly Ring visited a local gym and lost $2\frac{1}{4}$ pounds the first weekend and $6\frac{1}{8}$ pounds in week 2. What is Molly's total weight loss?

7. Bill Williams had to drive $46\frac{1}{4}$ miles to work. After driving $28\frac{5}{6}$ miles he noticed he was low on gas and had to decide whether he should stop to fill the gas tank. How many more miles does Bill have to drive to get to work?

8. Albert's Lumber Yard purchased $52\frac{1}{2}$ cords of lumber on Monday and $48\frac{3}{4}$ cords on Tuesday. It sold $21\frac{3}{8}$ cords on Friday. How many cords of lumber remain at Albert's Lumber Yard?

9. At Arlen Oil Company, where Dave Bursett is the service manager, it took $42\frac{1}{3}$ hours to clean five boilers. After a new cleaning tool was purchased, the time for cleaning five boilers was reduced to $37\frac{4}{9}$ hours. How much time was saved?

Learning Unit 2–3: Multiplying and Dividing Fractions

DRILL PROBLEMS

1. Multiply; use the cancellation technique:

a. $\frac{6}{13} \times \frac{26}{12}$

b. $\frac{3}{8} \times \frac{2}{3}$

c. $\frac{5}{7} \times \frac{9}{10}$

d. $\frac{3}{4} \times \frac{9}{13} \times \frac{26}{27}$

e. $6\frac{2}{5} \times 3\frac{1}{8}$

f. $2\frac{2}{3} \times 2\frac{7}{10}$

g. $45 \times \frac{7}{9}$

h. $3\frac{1}{9} \times 1\frac{2}{7} \times \frac{3}{4}$

i. $\frac{3}{4} \times \frac{7}{9} \times 3\frac{1}{3}$

j. $\frac{1}{8} \times 6\frac{2}{3} \times \frac{1}{10}$

2. Multiply; do not use canceling but reduce by finding the greatest common divisor:

a. $\frac{3}{4} \times \frac{8}{9}$

b. $\frac{7}{16} \times \frac{8}{13}$

3. Multiply or divide as indicated:

a. $\frac{25}{36} \div \frac{5}{9}$

b. $\frac{18}{8} \div \frac{12}{16}$

c. $2\frac{6}{7} \div 2\frac{2}{5}$

d. $3\frac{1}{4} \div 16$

e. $24 \div 1\frac{1}{3}$

f. $6 \times \frac{3}{2}$

g. $3\frac{1}{5} \times 7\frac{1}{2}$

h. $\frac{3}{8} \div \frac{7}{4}$

i. $9 \div 3\frac{3}{4}$

j. $\frac{11}{24} \times \frac{24}{33}$

k. $\frac{12}{14} \div 27$

l. $\frac{3}{5} \times \frac{2}{7} \times \frac{3}{10}$

WORD PROBLEMS

4. Mary Smith plans to make 12 meatloafs to store in her freezer. Each meatloaf requires $2\frac{1}{4}$ pounds of ground beef. How much ground beef does Mary need?

5. Judy Carter purchased a real estate lot for $24,000. She sold it 2 years later for $1\frac{5}{8}$ times as much as she had paid for it. What was the selling price?

6. Lynn Clarkson saw an ad for a camcorder that cost $980. She knew of a discount store that would sell it to her for a markdown of $\frac{3}{20}$ off the advertised price. How much is the discount she can get?

7. To raise money for their club, the members of the Marketing Club purchased 68 bushels of popcorn to resell. They plan to repackage the popcorn in bags that hold $\frac{2}{21}$ of a bushel each. How many bags of popcorn will they be able to fill?

8. Richard Tracy paid a total of $375 for lumber costing $\$9\frac{3}{8}$ per foot. How many feet did he purchase?

9. While training for a marathon, Kristin Woods jogged $7\frac{3}{4}$ miles per hour for $2\frac{2}{3}$ hours. How many miles did Kristin jog?

10. On a map, 1 inch represents 240 miles. How many miles are represented by $\frac{3}{8}$ of an inch?

11. In Massachusetts, the governor wants to allot $\frac{1}{6}$ of the total sales tax collections to public education. The total sales tax collected is $2,472,000; how much will go to education?

Learning Unit 3–1: Rounding Decimals; Fraction and Decimal Conversions

DRILL PROBLEMS

1. Write in decimal:

a. Sixty-two hundredths ______

b. Six tenths ______

c. Nine hundred fifty-three thousandths ______

d. Four hundred one thousandths ______

e. Six hundredths ______

2. Round each decimal to the place indicated:

a. .8624 to the nearest thousandth ______

b. .051 to the nearest tenth ______

c. 8.207 to the nearest hundredth ______

d. 2.094 to the nearest hundredth ______

e. .511172 to the nearest ten thousandth ______

3. Name the place position of the underlined digit:

a. $.8\underline{2}6$ ____________________

b. $.91\underline{4}$ ____________________

c. $3.\underline{1}169$ ____________________

d. $53.17\underline{5}$ ____________________

e. $1.017\underline{4}$ ____________________

4. Convert to fractions (do not reduce):

a. .91 ______ **b.** .426 ______ **c.** 2.516 ______

d. $.62\frac{1}{2}$ ______ **e.** 13.007 ______ **f.** $5.03\frac{1}{4}$ ______

5. Convert to fractions and reduce to lowest terms:

a. .4 **b.** .44 **c.** .53

d. .336 **e.** .096 **f.** .125

g. .3125 **h.** .008 **i.** 2.625

j. 5.75 **k.** 3.375 **l.** 9.04

6. Convert the following fractions to decimals and round your answer to the nearest hundredth:

a. $\frac{1}{8}$ **b.** $\frac{7}{16}$

c. $\frac{2}{3}$ **d.** $\frac{3}{4}$

e. $\frac{9}{16}$ **f.** $\frac{5}{6}$

g. $\frac{7}{9}$ **h.** $\frac{38}{79}$

i. $2\frac{3}{8}$ **j.** $9\frac{1}{3}$

k. $11\frac{19}{50}$ **l.** $6\frac{21}{32}$

m. $4\frac{83}{97}$ **n.** $1\frac{2}{5}$

o. $2\frac{2}{11}$ **p.** $13\frac{30}{42}$

WORD PROBLEMS

7. Alan Angel got 2 hits in his first 7 times at bat. What is his average to the nearest thousandths place?

8. Bill Breen earned $1,555, and his employer calculated that Bill's total FICA deduction should be $118.9575. Round this deduction to the nearest cent.

9. At the local college, .566 of the students are men. Convert to a fraction. Do not reduce.

10. The average television set is watched 2,400 hours a year. If there are 8,760 hours in a year, what fractional part of the year is spent watching television? Reduce to lowest terms.

11. On Saturday, the employees at the Empire Fish Company work only $\frac{1}{3}$ of a day. How could this be expressed as a decimal to the nearest thousandth?

12. The North Shore Cinema has 610 seats. At a recent film screening there were 55 vacant seats. Show as a fraction the number of filled seats. Reduce as needed.

13. Michael Sullivan was planning his marketing strategy for a new product his company had produced. He was fascinated to discover that Rhode Island, the smallest state in the United States, was only twenty thousand, five hundred seven ten millionths the size of the largest state, Alaska. Write this number in decimal.

14. Bull Moose Company purchased a new manufacturing plant, located on an acre of land, for a total price of $2,250,000. The accountant determined that $\frac{3}{7}$ of the total price should be allocated as the price of the building. What decimal portion is the price of the building? Round to the nearest thousandth.

Learning Unit 3–2: Adding, Subtracting, Multiplying, and Dividing Decimals

DRILL PROBLEMS

1. Add:

a. 7.57 + 6.2 + 13.008 + 4.83

b. 1.0625 + 4.0881 + .0775

c. .903 + .078 + .17 + .1 + .96

d. 3.38 + .175 + .0186 + .2

2. Rearrange and subtract:

a. .96 – .43

b. .885 – .069

c. 11.67 – .935

d. 261.2 – 8.08

3. Multiply and round to the nearest tenth:

a. 13.6 × .02 **b.** 1.73 × .069 **c.** 400 × 3.7 **d.** 0.025 × 5.6

4. Divide and round to the nearest hundredth:

a. 13.869 ÷ .6 **b.** 1.0088 ÷ .14 **c.** 18.7 ÷ 2.16 **d.** 15.64 ÷ .34

5. Complete by the shortcut method:

a. 6.87 × 1,000 = **b.** 927,530 ÷ 100 = **c.** 27.2 ÷ 1,000 =

d. .21 × 1,000 = **e.** 347 × 100 = **f.** 347 ÷ 100 =

g. .0021 ÷ 10 = **h.** 85.44 × 10,000 = **i.** 83.298 × 100 =

j. 23.0109 ÷ 100 =

WORD PROBLEMS (Use *Business Math Handbook* Tables as Needed.)

6. Andy Hay noted his Ford Explorer odometer reading of 18,969.4 at the beginning of his vacation. At the end of his vacation the reading was 21,510.4. How many miles did he drive during his vacation?

7. Jeanne Allyn purchased 12.25 yards of ribbon for a craft project. The ribbon cost 37¢ per yard. What was the total cost of the ribbon?

8. Leo Green wanted to find out the gas mileage for his company truck. When he filled the gas tank, he wrote down the odometer reading of 9,650.7. The next time he filled the gas tank the odometer reading was 10,112.2. He looked at the gas pump and saw that he had taken 18.5 gallons of gas. Find the gas mileage per gallon for Leo's truck. Round to the nearest tenth.

9. At Halley's Rent-a-Car, the cost per day to rent a medium-size car is \$35.25 plus 37¢ a mile. What would be the charge to rent this car for 1 day if you drove 205.4 miles?

10. A trip to Mexico costs 6,000 pesos. What is this in U.S. dollars? Check your answer.

11. If a commemorative gold coin weighs 7.842 grams, find the number of coins that can be produced from 116 grams of gold. Round to the nearest whole number.

Learning Unit 4–1: The Checking Account

DRILL PROBLEMS

1. The following is a deposit slip made out by Fred Young of the F. W. Young Company.

a. How much cash did Young deposit? ________

b. How many checks did Young deposit? ________

c. What was the total amount deposited? ________

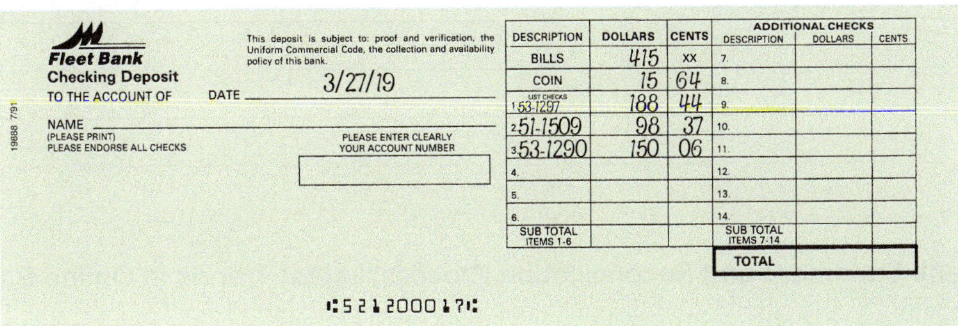
Fleet Bank
Checking Deposit
TO THE ACCOUNT OF
DATE 3/27/19

This deposit is subject to: proof and verification, the Uniform Commercial Code, the collection and availability policy of this bank.

NAME
(PLEASE PRINT)
PLEASE ENDORSE ALL CHECKS

PLEASE ENTER CLEARLY YOUR ACCOUNT NUMBER

19888 7/91

DESCRIPTION	DOLLARS	CENTS	ADDITIONAL CHECKS DESCRIPTION	DOLLARS	CENTS
BILLS	415	XX	7.		
COIN	15	64	8.		
LIST CHECKS 1 53-1297	188	44	9.		
2 51-1509	98	37	10.		
3 53-1290	150	06	11.		
4.			12.		
5.			13.		
6.			14.		
SUB TOTAL ITEMS 1-6			SUB TOTAL ITEMS 7-14		
			TOTAL		

⑆521200017⑆

2. Blackstone Company had a balance of \$2,173.18 in its checking account. Henry James, Blackstone's accountant, made a deposit that consisted of 2 fifty-dollar bills, 120 ten-dollar bills, 6 five-dollar bills, 14 one-dollar bills, \$9.54 in change, plus two checks the company had accepted, one for \$16.38 and the other for \$102.50. Find the amount of the deposit and the new balance in Blackstone's checking account.

3. Answer the following questions using the illustration:

No. 113 $ 750 00/100
October 4 20 XX
To Neuner Realty
For real estate

	DOLLARS	CENTS
BALANCE	1,020	93
AMT. DEPOSITED	2,756	80
TOTAL	3,777	73
AMT. THIS CHECK	750	00
BALANCE FORWARD	3,027	73

Jones Company
22 Aster Road
Salem, MA 01970

No. 113

October 4 20 XX 5-13/110

PAY TO THE ORDER OF Neuner Realty Company $ 750 00/100

Seven Hundred Fifty and 00/100 DOLLARS

Fleet Bank FLEET BANK OF MASSACHUSETTS, NATIONAL ASSOCIATION BOSTON, MASSACHUSETTS

Kevin Jones

MEMO real estate

A011000138A 14 0380 113

a. Who is the payee? ______

b. Who is the drawer? ______

c. Who is the drawee? ______

d. What is the bank's identification number ______

e. What is Jones Company's account number? ______

f. What was the balance in the account on September 30? ______

g. For how much did Jones write Check No. 113? ______

h. How much was deposited on October 1? ______

i. How much was left after Check No. 113 was written? ______

4. Write each of the following amounts in verbal form as you would on a check:

a. $40 ______

b. $245.75 ______

c. $3.98 ______

d. $1,205.05 ______

e. $3,013 ______

f. $510.10 ______

Learning Unit 4–2: Bank Statement and Reconciliation Process; Latest Trends in Online Banking

WORD PROBLEMS

1. Find the bank balance on January 31.

Date	Checks and payments			Deposits	Balance
January 1					401.17
January 2	108.64				______
January 5	116.50			432.16	______
January 6	14.92	150.00	10.00		______
January 11	12.29			633.89	______
January 18	108.64	18.60			______
January 25	43.91	23.77		657.22	______
January 26	75.00				______
January 31	6.75 sc				______

2. Joe Madruga, of Madruga's Taxi Service, received a bank statement for the month of May showing a balance of $932.36. His records show that the bank had not yet recorded two of his deposits, one for $521.50 and the other for $98.46. There are outstanding checks in the amounts of $41.67, $135.18, and $25.30. The statement also shows a service charge of $3.38. The balance in the check register is $1,353.55. Prepare a bank reconciliation for Madruga's as of May 31.

3. In reconciling the checking account for Nasser Enterprises, Beth Accomando found that the bank had collected a $3,000 promissory note on the company's behalf and had charged a $15 collection fee. There was also a service charge of $7.25. What amount should be added/subtracted from the checkbook balance to bring it up to date?

   Add: ________ Deduct: ________

4. In reconciling the checking account for Colonial Cleaners, Steve Papa found that a check for $34.50 had been recorded in the check register as $43.50. The bank returned an NSF check in the amount of $62.55. Interest income of $8.25 was earned and a service charge of $10.32 was assessed. What amount should be added/subtracted from the checkbook balance to bring it up to date?

   Add: ________ Deduct: ________

5. Matthew Stokes was completing the bank reconciliation for Parker's Tool and Die Company. The check register balance was $1,503.67. Matthew found that a $76.00 check had been recorded in the check register as $67.00; that a note for $1,500 had been collected by the bank for Parker's and the collection fee was $12.00; that $15.60 interest was earned on the account; and that an $8.35 service charge had been assessed. What should the check register balance be after Matthew updates it with the bank reconciliation information?

6. Consumers, community activists, and politicians are decrying the new line of accounts because several include a $3 service charge for some customers who use bank tellers for transactions that can be done through an automated teller machine. Bill Wade banks at a local bank that charges this fee. He was having difficulty balancing his checkbook because he did not notice this fee on his bank statement. His bank statement showed a balance of $822.18. Bill's checkbook had a balance of $206.48. Check No. 406 for $116.08 and Check No. 407 for $12.50 were outstanding. A $521 deposit was not on the statement. Bill has his payroll check electronically deposited to his checking account—the payroll check was for $1,015.12. (Bill's payroll checks vary each month.) There are also a $1 service fee and a teller fee of $6. Complete Bill's bank reconciliation.

7. At First National Bank in San Diego, some customers have to pay \$25 each year as an ATM card fee. John Levi banks at First National Bank and just received his bank statement showing a balance of \$829.25; his checkbook balance is \$467.40. The bank statement shows an ATM card fee of \$25.00, teller fee of \$9.00, interest of \$1.80, and John's \$880 IRS refund check, which was processed by the IRS and deposited to his account. John has two checks that have not cleared—No. 112 for \$620.10 and No. 113 for \$206.05. There is also a deposit in transit for \$1,312.10. Prepare John's bank reconciliation.

Learning Unit 5–1: Solving Equations for the Unknown

DRILL PROBLEMS

1. Write equations for the following situations. Use N for the unknown number. Do not solve the equations.
   - **a.** Four times a number is 180.
   - **b.** A number increased by 13 equals 25.
   - **c.** Seven less than a number is 5.
   - **d.** Fifty-seven decreased by 3 times a number is 21.
   - **e.** Fourteen added to one-third of a number is 18.
   - **f.** Twice the sum of a number and 4 is 32.
   - **g.** Three-fourths of a number is 9.
   - **h.** Two times a number plus 3 times the same number plus 8 is 68.

2. Solve for the unknown number:
   - **a.** $C + 40 = 90$
   - **b.** $29 + M = 44$
   - **c.** $D - 77 = 98$
   - **d.** $7N = 63$
   - **e.** $\frac{X}{12} = 11$
   - **f.** $3Q + 4Q + 2Q = 108$
   - **g.** $H + 5H + 3 = 57$
   - **h.** $2(N - 3) = 62$
   - **i.** $\frac{3R}{4} = 27$
   - **j.** $E - 32 = 41$
   - **k.** $5(2T - 2) = 120$
   - **l.** $12W - 5W = 98$

m. $49 - X = 37$ **n.** $12(V + 2) = 84$ **o.** $7D + 4 = 5D + 14$

p. $7(T - 2) = 2T - 9$

Learning Unit 5–2: Solving Word Problems for the Unknown

WORD PROBLEMS

1. A sweater at the Gap was marked down $30. The sale price was $50. What was the original price?

Unknown(s)	Variable(s)	Relationship

2. Goodwin's Corporation found that $\frac{2}{3}$ of its employees were vested in their retirement plan. If 124 employees are vested, what is the total number of employees at Goodwin's?

Unknown(s)	Variable(s)	Relationship

3. Eileen Haskin's utility and telephone bills for the month totaled $180. The utility bill was 3 times as much as the telephone bill. How much was each bill?

Unknown(s)	Variable(s)	Relationship

4. Ryan and his friends went to the golf course to hunt for golf balls. Ryan found 15 more than $\frac{1}{3}$ of the total number of golf balls that were found. How many golf balls were found if Ryan found 75 golf balls?

Unknown(s)	Variable(s)	Relationship

5. Linda Mills and Sherry Somers sold 459 tickets for the Advertising Club's raffle. If Linda sold 8 times as many tickets as Sherry, how many tickets did each one sell?

Unknown(s)	Variable(s)	Relationship

6. Jason Mazzola wanted to buy a suit at Giblee's. Jason did not have enough money with him, so Mr. Giblee told him he would hold the suit if Jason gave him a deposit of $\frac{1}{5}$ of the cost of the suit. Jason agreed and gave Mr. Giblee \$79. What was the price of the suit?

Unknown(s)	Variable(s)	Relationship

7. Peter sold watches (\$7) and necklaces (\$4) at a flea market. Total sales were \$300. People bought 3 times as many watches as necklaces. How many of each did Peter sell? What were the total dollar sales of each?

Unknown(s)	Variable(s)	Price	Relationship

8. Peter sold watches (\$7) and necklaces (\$4) at a flea market. Total sales for 48 watches and necklaces were \$300. How many of each did Peter sell? What were the total dollar sales of each?

Unknown(s)	Variable(s)	Price	Relationship

9. A 3,000 piece of direct mailing cost \$1,435. Printing cost is \$550, about $3\frac{1}{2}$ times the cost of typesetting. How much did the typesetting cost? Round to the nearest cent.

Unknown(s)	Variable(s)	Relationship

10. In 2019, Tony Rigato, owner of MRM, saw an increase in sales to \$13.5 million. Rigato states that since 2018, sales have more than tripled. What were his sales in 2018?

Unknown(s)	Variable(s)	Relationship

Learning Unit 6–1: Conversions

DRILL PROBLEMS

1. Convert the following to percents; round to the nearest tenth of a percent if needed:

a. .08 ________ % **b.** .729 ________ % **c.** .009 ________ %

d. 8.3 ________ % **e.** 5.26 ________ % **f.** 6 ________ %

g. .0105 ________ % **h.** .1180 ________ % **i.** 5.0375 ________ %

j. .862 ________ % **k.** .2615 ________ % **l.** .8 ________ %

m. .025 ________ % **n.** .06 ________ %

2. Convert the following to decimals; do not round:

a. 68% ________ **b.** .09% ________ **c.** 4.7% ________

d. 9.67% ________ **e.** .2% ________ **f.** $\frac{1}{4}\%$ ________

g. .76% ________ **h.** 110% ________ **i.** $12\frac{1}{2}\%$ ________

j. 5% ________ **k.** .004% ________ **l.** $7\frac{5}{10}\%$ ________

m. $\frac{3}{4}\%$ ________ **n.** 1% ________

3. Convert the following to percents; round to the nearest tenth of a percent if needed:

a. $\frac{7}{10}$ ________ % **b.** $\frac{1}{5}$ ________ % **c.** $1\frac{5}{8}$ ________ %

d. $\frac{2}{7}$ ________ % **e.** 2 ________ % **f.** $\frac{14}{100}$ ________ %

g. $\frac{1}{6}$ ________ % **h.** $\frac{1}{2}$ ________ % **i.** $\frac{3}{5}$ ________ %

j. $\frac{3}{25}$ ________ % **k.** $\frac{5}{16}$ ________ % **l.** $\frac{11}{50}$ ________ %

m. $4\frac{3}{4}$ ________ % **n.** $\frac{3}{200}$ ________ %

4. Convert the following to fractions in simplest form:

a. 40% ________ **b.** 15% ________ **c.** 50% ________

d. 75% ________ **e.** 35% ________ **f.** 85% ________

g. $12\frac{1}{2}\%$ ________ **h.** $37\frac{1}{2}\%$ ________ **i.** $33\frac{1}{3}\%$ ________

j. 3% ________ **k.** 8.5% ________ **l.** $5\frac{3}{4}\%$ ________

m. 100% ________ **n.** 10% ________

5. Complete the following table by finding the missing fraction, decimal, or percent equivalent:

	Fraction	Decimal	Percent		Fraction	Decimal	Percent
a.	______	.25	25%	**h.**	$\frac{1}{6}$	$.16\overline{6}$	______
b.	$\frac{3}{8}$	______	$37\frac{1}{2}\%$	**i.**	______	$.083\overline{3}$	$8\frac{1}{3}\%$
c.	$\frac{1}{2}$	.5	______	**j.**	$\frac{1}{9}$	______	$11\frac{1}{9}\%$
d.	$\frac{2}{3}$	______	$66\frac{2}{3}\%$	**k.**	______	.3125	$31\frac{1}{4}\%$
e.	______	.4	40%	**l.**	$\frac{3}{40}$	.075	______
f.	$\frac{3}{5}$	.6	______	**m.**	$\frac{1}{5}$	______	20%
g.	$\frac{7}{10}$	______	70%	**n.**	______	1.125	$112\frac{1}{2}\%$

WORD PROBLEMS

6. If in 2019 Mutual of New York reported that 80% of its new sales came from existing clients, what fractional part of its new sales came from existing clients? Reduce to simplest form.

7. Six hundred ninety corporations and design firms competed for the Industrial Design Excellence Award (IDEA). Twenty were selected as the year's best and received gold awards. Show the gold award winners as a fraction; then show what percent of the entrants received gold awards. Round to the nearest tenth of a percent.

8. If in the first half of 2018 stock prices in the Standard & Poor's 500 stock index rose 4.1%, show the increase as a decimal.

9. In the recent banking crisis, many banks were unable to cover their bad loans. Citicorp, the nation's largest real estate lender, was reported as having only enough reserves to cover 39% of its bad loans. What fractional part of its loan losses was covered?

10. Dave Mattera spent his vacation in Las Vegas. He ordered breakfast in his room, and when he went downstairs to the coffee shop, he discovered that the same breakfast was much less expensive. He had paid 1.884 times as much for the breakfast in his room. What was the percent of increase for the breakfast in his room?

11. Putnam Management Company of Boston recently increased its management fee by .09%. What is the increase as a decimal? What is the same increase as a fraction?

12. Joel Black and Karen Whyte formed a partnership and drew up a partnership agreement, with profits and losses to be divided equally after each partner receives a $7\frac{1}{2}\%$ return on his or her capital contribution. Show their return on investment as a decimal and as a fraction. Reduce.

Learning Unit 6–2: Application of Percents—Portion Formula

DRILL PROBLEMS

1. Fill in the amount of the base, rate, and portion in each of the following statements:
   a. The Logans spend $4,000 a month on food, which is 30% of their monthly income of $20,000.
   Base ________ Rate ________ Portion ________
   b. Rocky Norman got a $15 discount when he purchased a new camera. This was 20% off the sticker price of $75.
   Base ________ Rate ________ Portion ________
   c. Mary Burns got a 12% senior citizens discount when she bought a $7.00 movie ticket. She saved $0.84.
   Base ________ Rate ________ Portion ________
   d. Arthur Bogey received a commission of $13,500 when he sold the Brown's house for $225,000. His commission rate is 6%.
   Base ________ Rate ________ Portion ________
   e. Leo Davis deposited $5,000 in a certificate of deposit (CD). A year later he received an interest payment of $450, which was a yield of 9%.
   Base ________ Rate ________ Portion ________
   f. Grace Tremblay is on a diet that allows her to eat 1,600 calories per day. For breakfast she had 600 calories, which is $37\frac{1}{2}$% of her allowance.
   Base ________ Rate ________ Portion ________
2. Find the portion; round to the nearest hundredth if necessary:
   a. 7% of 74 ________ b. 12% of 205 ________ c. 16% of 630 ________
   d. 7.5% of 920 ________ e. 25% of 1,004 ________ f. 10% of 79 ________
   g. 103% of 44 ________ h. 30% of 78 ________ i. .2% of 50 ________
   j. 1% of 5,622 ________ k. $6\frac{1}{4}$% of 480 ________ l. 150% of 10 ________
   m. 100% of 34 ________ n. $\frac{1}{2}$% of 27 ________
3. Find the rate; round to the nearest tenth of a percent as needed:
   a. 30 is what percent of 90? ________ b. 6 is what percent of 200? ________
   c. 275 is what percent of 1,000? ________ d. .8 is what percent of 44? ________
   e. 67 is what percent of 2,010? ________ f. 550 is what percent of 250? ________
   g. 13 is what percent of 650? ________ h. $15 is what percent of $455? ________
   i. .05 is what percent of 100? ________ j. $6.25 is what percent of $10? ________
4. Find the base; round to the nearest tenth as needed:
   a. 63 is 30% of ________ b. 60 is 33% of ________ c. 150 is 25% of ________
   d. 47 is 1% of ________ e. $21 is 120% of ________ f. 2.26 is 40% of ________
   g. 75 is $12\frac{1}{2}$% of ________ h. 18 is 22.2% of ________ i. $37.50 is 50% of ________
   j. 250 is 100% of ________
5. Find the percent of increase or decrease; round to the nearest tenth percent as needed:

	Last year	This year	Amount of change	Percent of change
a.	5,962	4,378	________	________
b.	$10,995	$12,250	________	________
c.	120,000	140,000	________	________
d.	120,000	100,000	________	________

WORD PROBLEMS

6. A machine that originally cost $8,000 was sold for $800 at the end of 5 years. What percent of the original cost is the selling price?

7. Joanne Byrne invested $75,000 in a candy shop and is making 12% per year on her investment. How much money per year is she making on her investment?

8. There was a fire in Bill Porper's store that caused 2,780 inventory items to be destroyed. Before the fire, 9,565 inventory items were in the store. What percent of inventory was destroyed? Round to nearest tenth percent.

9. Elyse's Dress Shoppe makes 25% of its sales for cash. If the cash receipts on January 21 were $799, what were the total sales for the day?

10. The YMCA is holding a fund-raiser to collect money for a new gym floor. So far it has collected $7,875, which is 63% of the goal. What is the amount of the goal? How much more money must the YMCA collect?

11. Leslie Tracey purchased her home for $51,500. She sold it last year for $221,200. What percent profit did she make on the sale? Round to nearest tenth percent.

12. Maplewood Park Tool & Die had an annual production of 375,165 units this year. This is 140% of the annual production last year. What was last year's annual production?

Learning Unit 7–1: Trade Discounts—Single and Chain*

DRILL PROBLEMS

1. Calculate the trade discount amount for each of the following items:

Item	List price	Trade discount	Trade discount amount
a. iPhone	$ 400	30%	______
b. Flat-screen TV	$1,200	30%	______
c. Suit	$ 500	10%	______
d. Bicycle	$ 800	$12\frac{1}{2}\%$	______
e. David Yurman bracelet	$ 950	40%	______

2. Calculate the net price for each of the following items:

Item	List price	Trade discount amount	Net price
a. Home Depot table	$600	$250	______
b. Bookcase	$525	$129	______
c. Rocking chair	$480	$95	______

3. Fill in the missing amount for each of the following items:

Item	List price	Trade discount amount	Net price
a. Sears electric saw	______	$19	$56.00
b. Electric drill	$90	______	$68.50
c. Ladder	$56	$15.25	______

4. For each of the following, find the percent paid (complement of trade discount) and the net price:

List price	Trade discount	Percent paid	Net price
a. $45	15%	______	______
b. $195	12.2%	______	______
c. $325	50%	______	______
d. $120	18%	______	______

5. In each of the following examples, find the net price equivalent rate and the single equivalent discount rate:

Chain discount	Net price equivalent rate	Single equivalent discount rate
a. 25/5	______	______
b. 15/15	______	______
c. 15/10/5	______	______
d. 12/12/6	______	______

*Freight problems to be shown in LU 7–2 material.

6. In each of the following examples, find the net price and the trade discount:

	List price	Chain discount	Net price	Trade discount
a.	$5,000	10/10/5	________	________
b.	$7,500	9/6/3	________	________
c.	$898	20/7/2	________	________
d.	$1,500	25/10	________	________

7. The list price of a handheld calculator is $19.50, and the trade discount is 18%. Find the trade discount amount.

8. The list price of a silver picture frame is $29.95, and the trade discount is 15%. Find the trade discount amount and the net price.

9. The net price of a set of pots and pans is $65, and the trade discount is 20%. What is the list price?

10. Jennie's Variety Store has the opportunity to purchase candy from three different wholesalers; each of the wholesalers offers a different chain discount. Company A offers 25/5/5, Company B offers 20/10/5, and Company C offers 15/20. Which company should Jennie deal with? *Hint:* Choose the company with the highest single equivalent discount rate.

11. The list price of a television set is $625. Find the net price after a series discount of 30/20/10.

12. Mandy's Accessories Shop purchased 12 purses with a total list price of $726. What was the net price of each purse if the wholesaler offered a chain discount of 25/20?

13. Kransberg Furniture Store purchased a bedroom set for $1,097.25 from Furniture Wholesalers. The list price of the set was $1,995. What trade discount rate did Kransberg receive?

14. Susan Monk teaches second grade and receives a discount at the local art supply store. Recently she paid $47.25 for art supplies after receiving a chain discount of 30/10. What was the regular price of the art supplies?

Learning Unit 7–2: Cash Discounts, Credit Terms, and Partial Payments

DRILL PROBLEMS

1. Complete the following table:

	Date of invoice	Date goods received	Terms	Last day of discount period	End of credit period
a.	February 8		2/10, n/30	______	______
b.	August 26		2/10, n/30	______	______
c.	October 17		3/10, n/60	______	______
d.	March 11	May 10	3/10, n/30, ROG	______	______
e.	September 14		2/10, EOM	______	______
f.	May 31		2/10, EOM	______	______

2. Calculate the cash discount and the net amount paid.

	Invoice amount	Cash discount rate	Discount amount	Net amount paid
a.	$75	3%	______	______
b.	$1,559	2%	______	______
c.	$546.25	2%	______	______
d.	$9,788.75	1%	______	______

3. Use the complement of the cash discount to calculate the net amount paid. Assume all invoices are paid within the discount period.

	Terms of invoice	Amount of invoice	Complement	Net amount paid
a.	3/10, n/30	$1,400	______	______
b.	3/10, n/30 ROG	$4,500	______	______
c.	2/10, EOM	$375.50	______	______
d.	1/15, n/45	$3,998	______	______

4. Calculate the amount of cash discount and the net amount paid.

	Date of invoice	Terms of invoice	Amount of invoice	Date paid	Cash discount	Amount paid
a.	January 12	2/10, n/30	$5,320	January 22	______	______
b.	May 28	2/10, n/30	$975	June 7	______	______
c.	August 15	2/10, n/30	$7,700	August 26	______	______
d.	March 8	2/10, EOM	$480	April 10	______	______
e.	January 24	3/10, n/60	$1,225	February 3	______	______

5. Complete the following table:

	Total invoice	Freight charges included in invoice total	Date of invoice	Terms of invoice	Date of payment	Cash discount	Amount paid
a.	$852	$12.50	3/19	2/10, n/30	3/29	______	______
b.	$669.57	$15.63	7/28	3/10, EOM	9/10	______	______
c.	$500	$11.50	4/25	2/10, n/60	6/5	______	______
d.	$188	$9.70	1/12	2/10, EOM	2/10	______	______

6. In the following table, assume that all the partial payments were made within the discount period.

	Amount of invoice	Terms of invoice	Partial payment	Amount to be credited	Balance outstanding
a.	$481.90	2/10, n/30	$90.00	________	________
b.	$1,000	2/10, EOM	$500.00	________	________
c.	$782.88	3/10, n/30, ROG	$275.00	________	________
d.	$318.80	2/15, n/60	$200.00	________	________

WORD PROBLEMS

7. Ray Chemical Company received an invoice for $16,500, dated March 14, with terms of 2/10, n/30. If the invoice was paid March 22, what was the amount due?

8. On May 27, Trotter Hardware Store received an invoice for trash barrels purchased for $13,650 with terms of 3/10, EOM; the freight charge, which is included in the price, is $412. What are **(a)** the last day of the discount period and **(b)** the amount of the payment due on this date?

9. The Glass Sailboat received an invoice for $930.50 with terms 2/10, n/30 on April 19. On April 29, it sent a payment of $430.50. **(a)** How much credit will be given on the total due? **(b)** What is the new balance due?

10. Dallas Ductworks offers cash discounts of 2/10, 1/15, n/30 on all purchases. If an invoice for $544 dated July 18 is paid on August 2, what is the amount due?

11. The list price of a Luminox watch is $299.90 with trade discounts of 10/20 and terms of 3/10, n/30. If a retailer pays the invoice within the discount period, what amount must the retailer pay?

12. The invoice of a sneakers supplier totaled $2,488.50, was dated February 7, and offered terms 2/10, ROG. The shipment of sneakers was received on March 7. What are **(a)** the last date of the discount period and **(b)** the amount of the discount that will be lost if the invoice is paid after that date?

13. Starburst Toy Company receives an invoice amounting to $1,152.30 with terms of 2/10, EOM and dated November 6. If a partial payment of $750 is made on December 8, what are **(a)** the credit given for the partial payment and **(b)** the balance due on the invoice?

14. Todd's Sporting Goods received an invoice for soccer equipment dated July 26 with terms 3/10, 1/15, n/30 in the amount of $3,225.83, which included shipping charges of $375.50. If this bill is paid on August 5, what amount must be paid?

Learning Unit 8–1: Markups Based on Cost (100%)

DRILL PROBLEMS

1. Fill in the missing numbers:

	Cost	Dollar markup	Selling price
a.	$16.10	$3.80	_______
b.	$8.32	_______	$11.04
c.	$25.27	_______	$29.62
d.	_______	$75.00	$165.00
e.	$86.54	$29.77	

2. Calculate the markup based on cost; round to the nearest cent.

	Cost	Markup (percent of cost)	Dollar markup
a.	$425.00	30%	_________
b.	$1.52	20%	_________
c.	$9.90	$12\frac{1}{2}\%$	_________
d.	$298.10	50%	_________
e.	$74.25	38%	_________
f.	$552.25	100%	_________

3. Calculate the dollar markup and rate of the markup as a percent of cost, rounding percents to nearest tenth percent. Verify your result, which may be slightly off due to rounding.

	Cost	Selling price	Dollar markup	Markup (percent of cost)	Verify
a.	$2.50	$4.50	_______	_______	
b.	$12.50	$19.00	_______	_______	
c.	$0.97	$1.25	_______	_______	
d.	$132.25	$175.00	_______	_______	
e.	$65.00	$89.99	_______	_______	

4. Calculate the dollar markup and the selling price.

	Cost	Markup (percent of cost)	Dollar markup	Selling price
a.	$2.20	40%	_______	_______
b.	$2.80	16%	_______	_______
c.	$840.00	$12\frac{1}{2}\%$	_______	_______
d.	$24.36	30%	_______	_______

5. Calculate the cost, rounding to the nearest cent.

	Selling price	Rate of markup based on cost	Cost
a.	$1.98	30%	________
b.	$360.00	60%	________
c.	$447.50	20%	________
d.	$1,250.00	100%	________

6. Find the missing numbers. Round money to the nearest cent and percents to the nearest tenth percent.

	Cost	Dollar markup	Percent markup on cost	Selling price
a.	$72.00	________	40%	________
b.	________	$7.00	________	$35.00
c.	$8.80	$1.10	________	________
d.	________	________	28%	$19.84
e.	$175.00	________	________	$236.25

WORD PROBLEMS

7. If the cost of a Pottery Barn chair is $499 and the markup rate is 40% of the cost, what are **(a)** the dollar markup and **(b)** the selling price?

8. If Barry's Furniture Store purchased a floor lamp for $120 and plans to add a markup of $90, **(a)** what will the selling price be and **(b)** what is the markup as a percent of cost?

9. If Lesjardin's Jewelry Store is selling a gold bracelet for $349, which includes a markup of 35% on cost, what are **(a)** Lesjardin's cost and **(b)** the amount of the dollar markup?

10. Toll's Variety Store sells an alarm clock for $14.75. The alarm clock cost Toll's $9.90. What is the markup amount as a percent of cost? Round to the nearest whole percent.

11. Swanson's Audio Supply marks up its merchandise by 40% on cost. If the markup on a cassette player is $85, what are **(a)** the cost of the cassette player and **(b)** the selling price?

12. Brown's Department Store is selling a shirt for $55. If the markup is 70% on cost, what is Brown's cost (to the nearest cent)?

13. Ward's Greenhouse purchased tomato flats for $5.75 each. Ward's has decided to use a markup of 42% on cost. Find the selling price.

Learning Unit 8–2: Markups Based on Selling Price (100%)

DRILL PROBLEMS

1. Calculate the markup based on the selling price.

Selling price	Markup (percent of selling price)	Dollar markup
a. $25.00	40%	________
b. $230.00	25%	________
c. $81.00	42.5%	________
d. $72.88	$37\frac{1}{2}\%$	________
e. $1.98	$7\frac{1}{2}\%$	________

2. Calculate the dollar markup and the markup as a percent of selling price (to the nearest tenth percent). Verify your answer, which may be slightly off due to rounding.

Cost	Selling price	Dollar markup	Markup (percent of selling price)	Verify
a. $2.50	$4.25	________	________	
b. $16.00	$24.00	________	________	
c. $45.25	$85.00	________	________	
d. $0.19	$0.25	________	________	
e. $5.50	$8.98	________	________	

3. Given the *cost* and the markup as a percent of *selling price,* calculate the selling price.

Cost	Markup (percent of selling price)	Selling price
a. $5.90	15%	________
b. $600	32%	________
c. $15	50%	________
d. $120	30%	________
e. $0.29	20%	________

4. Given the selling price and the percent markup on selling price, calculate the cost.

Cost	Markup (percent of selling price)	Selling price
a. ________	40%	$6.25
b. ________	20%	$16.25
c. ________	19%	$63.89
d. ________	$62\frac{1}{2}\%$	$44.00

5. Calculate the equivalent rate of markup, rounding to the nearest hundredth percent.

	Markup on cost	Markup on selling price		Markup on cost	Markup on selling price
a.	40%	_______	b.	50%	_______
c.	_______	50%	d.	_______	35%
e.	_______	40%			

WORD PROBLEMS

6. Fisher Equipment is selling a Wet/Dry Shop Vac for $49.97. If Fisher's markup is 40% of the selling price, what is the cost of the Shop Vac?

7. Gove Lumber Company purchased a 10-inch table saw for $225 and will mark up the price 35% on the selling price. What will the selling price be?

8. To realize a sufficient gross margin, City Paint and Supply Company marks up its paint 27% on the selling price. If a gallon of Latex Semi-Gloss Enamel has a markup of $4.02, find **(a)** the selling price and **(b)** the cost.

9. A Magnavox 20-inch color TV cost $180 and sells for $297. What is the markup based on the selling price? Round to the nearest hundredth percent.

10. Bargain Furniture sells a five-piece country maple bedroom set for $1,299. The cost of this set is $700. What are **(a)** the markup on the bedroom set, **(b)** the markup percent on cost, and **(c)** the markup percent on the selling price? Round to the nearest hundredth percent.

11. Robert's Department Store marks up its sundries by 28% on the selling price. If a 6.4-ounce tube of toothpaste costs $1.65, what will the selling price be?

12. To be competitive, Tinker Toys must sell the DS software for $89.99. To meet expenses and make a sufficient profit, Tinker Toys must add a markup on the selling price of 23%. What is the maximum amount that Tinker Toys can afford to pay a wholesaler for the DS software?

13. Nicole's Restaurant charges $7.50 for a linguini dinner that costs $2.75 for the ingredients. What rate of markup is earned on the selling price? Round to the nearest hundredth percent.

Learning Unit 8–3: Markdowns and Perishables

DRILL PROBLEMS

1. Find the dollar markdown and the sale price.

	Original selling price	Markdown percent	Dollar markdown	Sale price
a.	$200	40%	______	______
b.	$2,099.98	25%	______	______
c.	$729	30%	______	______

2. Find the dollar markdown and the markdown percent on original selling price.

	Original selling price	Sale price	Dollar markdown	Markdown percent
a.	$19.50	$9.75	______	______
b.	$250	$175	______	______
c.	$39.95	$29.96	______	______

3. Find the original selling price.

	Sale price	Markdown percent	Original selling price
a.	$328	20%	______
b.	$15.85	15%	______

4. Calculate the final selling price.

	Original selling price	First markdown	Second markdown	Final markup	Final selling price
a.	$4.96	25%	8%	5%	______
b.	$130	30%	10%	20%	______

5. Find the missing amounts.

	Number of units	Unit cost	Total cost	Estimated* spoilage	Desired markup (percent of cost)	Total selling price	Selling price per unit
a.	72	$3	_______	12%	50%	_______	_______
b.	50	$0.90	_______	16%	42%	_______	_______

*Round to the nearest whole unit as needed.

WORD PROBLEMS

6. Speedy King is having a 30%-off sale on its box springs and mattresses. A queen-size, back-supporter mattress is priced at $325. What is the sale price of the mattress?

7. Murray and Sons sells a Dell computer for $602.27. It is having a sale, and the computer is marked down to $499.88. What is the percent of the markdown?

8. Coleman's is having a clearance sale. A lamp with an original selling price of $249 is now selling for $198. Find the percent of the markdown. Round to the nearest hundredth percent.

9. Johnny's Sports Shop has advertised markdowns on certain items of 22%. A soccer ball is marked with a sale price of $16.50. What was the original price of the soccer ball?

10. Sam Grillo sells seasonal furnishings. Near the end of the summer a five-piece patio set that was priced $349.99 had not been sold, so he marked it down by 12%. As Labor Day approached, he still had not sold the patio set, so he marked it down an additional 18%. What was the final selling price of the patio set?

11. Calsey's Department Store sells its down comforters for a regular price of $325. During its white sale the comforters were marked down 22%. Then, at the end of the sale, Calsey's held a special promotion and gave a second markdown of 10%. When the sale was over, the remaining comforters were marked up 20%. What was the final selling price of the remaining comforters?

12. The New Howard Bakery wants to make a 60% profit on the cost of its pies. To calculate the price of the pies, it estimated that the usual amount of spoilage is five pies. Calculate the selling price for each pie if the number of pies baked each day is 24 and the cost of the ingredients for each pie is $1.80.

13. Sunshine Bakery bakes 660 loaves of bread each day and estimates that 10% of the bread will go stale before it is sold and thus will have to be discarded. The owner of the bakery wishes to realize a 55% markup on cost on the bread. If the cost to make a loaf of bread is $0.46, what should the owner sell each loaf for?

Learning Unit 8–4: Breakeven Analysis

DRILL PROBLEMS

1. Calculate the contribution margin.

	Selling price per unit	Variable cost per unit	Contribution margin
a.	$14.00	$8.00	
b.	$15.99	$4.88	
c.	$18.99	$4.99	
d.	$251.86	$110.00	
e.	$510.99	$310.00	
f.	$1,000.10	$410.00	

2. Calculate the selling price per unit.

	Selling price per unit	Variable cost per unit	Contribution margin
a.		$12.18	$4.10
b.		$19.19	$5.18
c.		$21.00	$13.00
d.		$41.00	$14.88
e.		$128.10	$79.50
f.		$99.99	$60.00

3. Calculate the breakeven point, rounding to the nearest whole unit.

	Breakeven point	Fixed cost	Selling price per unit	Variable cost per unit
a.		$50,000	$4.00	$1.00
b.		$30,000	$6.00	$2.00
c.		$20,000	$9.00	$3.00
d.		$100,000	$12.00	$4.00
e.		$120,000	$14.00	$5.00
f.		$90,000	$26.00	$8.00

WORD PROBLEMS

4. Jones Co. produces bars of candy. Each bar sells for $3.99. The variable cost per unit is $2.85. What is the contribution margin for Jones Co.?

5. Logan Co. produces stuffed animals. It has $40,000 in fixed costs. Logan sells each animal for $19.99 with a $12.10 cost per unit. What is the breakeven point for Logan? Round to the nearest whole number.

6. Ranyo Company produces lawn mowers. It has a breakeven point of 6,000 lawn mowers. If its contribution margin is $150, what is Ranyo's fixed cost?

7. Moore company has $100,000 in fixed costs. Its contribution margin is $4.50. Calculate the breakeven point for Moore to the nearest whole number.

Learning Unit 9–1: Calculating Various Types of Employees' Gross Pay

DRILL PROBLEMS

1. Fill in the missing amounts for each of the following employees. Do not round the overtime rate in your calculations and round your final answers to the nearest cent.

	Employee	Total hours	Rate per hour	Regular pay	Overtime pay	Gross pay
a.	Mel Jones	38	$11.25	______	______	______
b.	Casey Guitare	43	$9.00	______	______	______
c.	Norma Harris	37	$7.50	______	______	______
d.	Ed Jackson	45	$12.25	______	______	______

2. Calculate each employee's gross from the following data. Do not round the overtime rate in your calculation but round your final answers to the nearest cent.

	Employee	S	M	Tu	W	Th	F	S	Total hours	Rate per hour	Regular pay	Overtime pay	Gross pay
a.	L. Adams	0	8	8	8	8	8	0	____	$8.10	______	______	______
b.	M. Card	0	9	8	9	8	8	4	____	$11.35	______	______	______
c.	P. Kline	2	$7\frac{1}{2}$	$8\frac{1}{4}$	8	$10\frac{3}{4}$	9	2	$47\frac{1}{2}$	$10.60	______	______	______
d.	J. Mack	0	$9\frac{1}{2}$	$9\frac{3}{4}$	$9\frac{1}{2}$	10	10	4	$52\frac{3}{4}$	$9.95	______	______	______

3. Calculate the gross wages of the following production workers.

	Employee	Rate per unit	No. of units produced	Gross pay
a.	A. Bossie	$0.67	655	______
b.	J. Carson	$0.87\frac{1}{2}$	703	______

4. Using the given differential scale, calculate the gross wages of the following production workers.

Units produced	Amount per unit
From 1–50	$.55
From 51–100	.65
From 101–200	.72
More than 200	.95

	Employee	Units produced	Gross pay
a.	F. Burns	190	______
b.	B. English	210	______
c.	E. Jackson	200	______

5. Calculate the following salespersons' gross wages.

a. Straight commission:

Employee	Net sales	Commission	Gross pay
M. Salley	$40,000	13%	______

b. Straight commission with draw:

Employee	Net sales	Commission	Draw	Commission minus draw
G. Gorsbeck	$38,000	12%	$600	________

c. Variable commission scale:

Up to $25,000	8%
Excess of $25,000 to $40,000	10%
More than $40,000	12%

Employee	Net sales	Gross pay
H. Lloyd	$42,000	________

d. Salary plus commission:

Employee	Salary	Commission	Quota	Net sales	Gross pay
P. Floyd	$2,500	3%	$400,000	$475,000	________

WORD PROBLEMS

For all problems with overtime, be sure to round only the final answer.

6. In the first week of December, Dana Robinson worked 52 hours. His regular rate of pay is $11.25 per hour. What was Dana's gross pay for the week?

7. Davis Fisheries pays its workers for each box of fish they pack. Sunny Melanson receives $.30 per box. During the third week of July, Sunny packed 2,410 boxes of fish. What was Sunny's gross pay?

8. Maye George is a real estate broker who receives a straight commission of 6%. What would her commission be for a house that sold for $197,500?

9. Devon Company pays Eileen Haskins a straight commission of $12\frac{1}{2}$% on net sales. In January, Devon gave Eileen a draw of $600. She had net sales that month of $35,570. What was Eileen's commission minus draw?

10. Parker and Company pays Selma Stokes on a variable commission scale. In a month when Selma had net sales of $155,000, what was her gross pay based on the following schedule?

Net sales	Commission rate
Up to $40,000	5%
Excess of $40,000 to $75,000	5.5%
Excess of $75,000 to $100,000	6%
More than $100,000	7%

11. Marsh Furniture Company pays Joshua Charles a monthly salary of $1,900 plus a commission of $2\frac{1}{2}$% on sales over $12,500. Last month, Joshua had net sales of $17,799. What was Joshua's gross pay for the month?

12. Amy McWha works at Lamplighter Bookstore where she earns $7.75 per hour plus a commission of 2% on her weekly sales in excess of $1,500. Last week, Amy worked 39 hours and had total sales of $2,250. What was Amy's gross pay for the week?

Learning Unit 9–2: Computing Payroll Deductions for Employees' Pay; Employers' Responsibilities

DRILL PROBLEMS

Use tables in the *Business Math Handbook* (assume FICA rates in text).

	Employee	Allowances and marital status	Cumulative earnings	Salary per week	Taxable earnings S.S.	Taxable earnings Medicare
1.	Pete Small	M—3	$127,900	$2,300	**a.** ______	**b.** ______
2.	Alice Hall	M—1	$128,400	$1,100	**c.** ______	**d.** ______
3.	Jean Rose	M—2	$150,000	$2,000	**e.** ______	**f.** ______

4. What is the tax for Social Security and Medicare for Pete in Problem 1?

5. Calculate Pete's FIT by the percentage method.

6. What would the employer contribute for this week's payroll for SUTA and FUTA?

WORD PROBLEMS

7. Cynthia Pratt has earned $126,900 thus far this year. This week she earned $3,500. Find her total FICA tax deduction (Social Security and Medicare).

8. If Cynthia (Problem 7) earns $1,050 the following week, what will be her new total FICA tax deduction?

9. Roger Alley, a service dispatcher, has weekly earnings of $750. He claimed four allowances on his W-4 form and is married. Besides his FIT and FICA deductions, he has deductions of $35.16 for medical insurance and $17.25 for union dues. Calculate his net earnings for the third week in February. Use the percentage method.

10. Nicole Mariotte is unmarried and claimed one withholding allowance on her W-4 form. In the second week of February, she earned $707.35. Deductions from her pay included federal withholding, Social Security, Medicare, health insurance for $47.75, and $30.00 for the company meal plan. What is Nicole's net pay for the week? Use the percentage method.

11. Gerald Knowlton had total gross earnings of $128,100 in the last week of November. His earnings for the first week in December were $804.70. His employer uses the percentage method to calculate federal withholding. If Gerald is married, claims two allowances, and has medical insurance of $52.25 deducted each week from his pay, what is his net pay for the week?

Learning Unit 10–1: Calculation of Simple Interest and Maturity Value

DRILL PROBLEMS

1. Find the simple interest for each of the following loans:

	Principal	Rate	Time	Interest
a.	$12,000	2%	1 year	________
b.	$3,000	12%	3 years	________
c.	$18,000	$8\frac{1}{2}\%$	10 months	________

2. Find the simple interest for each of the following loans; use the exact interest method. Use the days-in-a-year calendar in the text when needed.

	Principal	Rate	Time	Interest
a.	$900	4%	30 days	________
b.	$4,290	8%	250 days	________
c.	$1,500	8%	Made March 11 Due July 11	________

3. Find the simple interest for each of the following loans using the ordinary interest method (Banker's Rule).

	Principal	Rate	Time	Interest
a.	$5,250	$7\frac{1}{2}\%$	120 days	________
b.	$700	3%	70 days	________
c.	$2,600	11%	Made on June 15 Due October 17	________

WORD PROBLEMS

4. On October 17, Gill Iowa borrowed $6,000 at a rate of 4%. She promised to repay the loan in 7 months. What are **(a)** the amount of the simple interest and **(b)** the total amount owed upon maturity?

5. Marjorie Folsom borrowed $5,500 to purchase a computer. The loan was for 9 months at an annual interest rate of $12\frac{1}{2}\%$. What are **(a)** the amount of interest Marjorie must pay and **(b)** the maturity value of the loan?

6. Eric has a loan for $1,200 at an ordinary interest rate of 9.5% for 80 days. Julie has a loan for $1,200 at an exact interest rate of 9.5% for 80 days. Calculate **(a)** the total amount due on Eric's loan and **(b)** the total amount due on Julie's loan.

7. Roger Lee borrowed $5,280 at $13\frac{1}{2}$% on May 24 and agreed to repay the loan on August 24. The lender calculates interest using the exact interest method. How much will Roger be required to pay on August 24?

8. On March 8, Jack Faltin borrowed $10,225 at $9\frac{3}{4}$%. He signed a note agreeing to repay the loan and interest on November 8. If the lender calculates interest using the ordinary interest method, what will Jack's repayment be?

9. Dianne Smith's real estate taxes of $641.49 were due on November 1, 2018. Due to financial difficulties, Dianne was unable to pay her tax bill until January 15, 2019. The penalty for late payment is $13\frac{3}{8}$% ordinary interest. What is the penalty Dianne will have to pay, and what is Dianne's total payment on January 15?

10. On August 8, Rex Eason had a credit card balance of $550, but he was unable to pay his bill. The credit card company charges interest of $18\frac{1}{2}$% annually on late payments. What amount will Rex have to pay if he pays his bill 1 month late?

11. An issue of *Your Money* discussed average consumers who carry a balance of $2,000 on one credit card. If the yearly rate of interest is 18%, how much are consumers paying in interest per year?

12. AFBA Industrial Bank of Colorado Springs, Colorado, charges a credit card interest rate of 11% per year. If you had a credit card debt of $1,500, what would your interest amount be after 3 months?

Learning Unit 10–2: Finding Unknown in Simple Interest Formula

DRILL PROBLEMS

1. Find the principal in each of the following. Round to the nearest cent. Assume 360 days. *Calculator hint:* Do denominator calculation first, and do not round; when answer is displayed, save it in memory by pressing [M+]. Now key in the numerator (interest amount), [÷], [MR], [=] for the answer. Be sure to clear memory after each problem by pressing [MR] again so that the M is no longer in the display.

	Rate	Time	Interest	Principal
a.	8%	70 days	$68	________
b.	11%	90 days	$125	________
c.	9%	120 days	$103	________
d.	$8\frac{1}{2}$%	60 days	$150	________

2. Find the rate in each of the following. Round to the nearest tenth of a percent. Assume 360 days.

	Principal	Time	Interest	Rate
a.	$7,500	120 days	$350	________
b.	$975	60 days	$25	________
c.	$20,800	220 days	$910	________
d.	$150	30 days	$2.10	________

3. Find the time (to the nearest day) in each of the following. Assuming ordinary interest, use 360 days.

	Principal	Rate	Interest	Time (days)	Time (years) (Round to nearest hundredth)
a.	$400	11%	$7.33	________	________
b.	$7,000	12.5%	$292	________	________
c.	$1,550	9.2%	$106.95	________	________
d.	$157,000	10.75%	$6,797.88	________	________

4. Complete the following. Assume 360 days for all examples.

	Principal	Rate (nearest tenth percent)	Time (nearest day)	Simple interest
a.	$345	________	150 days	$14.38
b.	________	12.5%	90 days	$46.88
c.	$750	12.2%	________	$19.06
d.	$20,260	16.7%	110 days	________

WORD PROBLEMS

Use 360 days.

5. In June, Becky opened a $20,000 bank CD paying 1% interest, but she had to withdraw the money in a few days to cover one child's college tuition. The bank charged her $1,000 in penalties for the withdrawal. What percent of the $20,000 was she charged?

6. Dr. Vaccarro invested his money at $12\frac{1}{2}$% for 175 days and earned interest of $760. How much money did Dr. Vaccarro invest?

7. If you invested $10,000 at 5% interest in a 6-month CD compounding interest daily, you would earn $252.43 in interest. How much would the same $10,000 invested in a bank paying simple interest earn?

8. Thomas Kyrouz opened a savings account and deposited $750 in a bank that was paying 2.5% simple interest. How much were his savings worth in 200 days?

9. Mary Millitello paid the bank $53.90 in interest on a 66-day loan at 9.8%. How much money did Mary borrow? Round to the nearest dollar.

10. If Anthony Lucido deposits $2,400 for 66 days and makes $60.72 in interest, what interest rate is he receiving?

11. Find how long in days David Wong must invest $23,500 of his company's cash at 8.4% in order to earn $652.50 in interest.

Learning Unit 10–3: U.S. Rule—Making Partial Note Payments before Due Date

DRILL PROBLEMS

1. A merchant borrowed $3,000 for 320 days at 11% (assume a 360-day year). Use the U.S. Rule to complete the following table:

Payment number	Payment day	Amount paid	Interest to date	Principal payment	Adjusted balance
					$3,000
1	75	$500	______	______	______
2	160	$750	______	______	______
3	220	$1,000	______	______	______
4	320	______	______	______	______

2. Use the U.S. Rule to solve for total interest costs, balances, and final payments; use ordinary interest.

Given

Principal, \$6,000, 5%, 100 days
Partial payments on 30th day, \$2,000
on 70th day, \$1,000

WORD PROBLEMS

3. John Joseph borrowed \$10,800 for 1 year at 14%. After 60 days, he paid \$2,500 on the note. On the 200th day, he paid an additional \$5,000. Use the U.S. Rule and ordinary interest to find the final balance due.

4. Doris Davis borrowed \$8,200 on March 5 for 90 days at $8\frac{3}{4}$%. After 32 days, Doris made a payment on the loan of \$2,700. On the 65th day, she made another payment of \$2,500. What is her final payment if you use the U.S. Rule with ordinary interest?

5. David Ring borrowed $6,000 on a 13%, 60-day note. After 10 days, David paid $500 on the note. On day 40, David paid $900 on the note. What are the total interest and ending balance due by the U.S. Rule? Use ordinary interest.

Learning Unit 11–1: Structure of Promissory Notes; the Simple Discount Note

DRILL PROBLEMS

1. Identify each of the following characteristics of promissory notes with an **I** for simple interest note, a **D** for simple discount note, or a **B** if it is true for both.

___ Interest is computed on face value, or what is actually borrowed.
___ A promissory note for a loan usually less than 1 year.
___ Borrower receives proceeds = Face value − Bank discount.
___ Maturity value = Face value + Interest.
___ Maturity value = Face value.
___ Borrower receives the face value.
___ Paid back by one payment at maturity.
___ Interest computed on maturity value, or what will be repaid, and not on actual amount borrowed.

2. Find the bank discount and the proceeds for the following; assume 360 days:

	Maturity value	Discount rate	Time (days)	Bank discount	Proceeds
a.	$8,000	3%	120	________	________
b.	$4,550	8.1%	110	________	________
c.	$19,350	12.7%	55	________	________
d.	$63,400	10%	90	________	________
e.	$13,490	7.9%	200	________	________
f.	$780	$12\frac{1}{2}\%$	65	________	________

3. Find the effective rate of interest for each of the loans in Problem 2. Use the answers you calculated in Problem 2 to solve these problems; round to the nearest tenth percent.

	Maturity value	Discount rate	Time (days)	Effective rate
a.	$8,000	3%	120	________
b.	$4,550	8.1%	110	________
c.	$19,350	12.7%	55	________
d.	$63,400	10%	90	________
e.	$13,490	7.9%	200	________
f.	$780	$12\frac{1}{2}\%$	65	________

WORD PROBLEMS

Assume 360 days.

4. Kaylee Putty signed an $8,000 note for 140 days at a discount rate of 5%. Find the discount and the proceeds Kaylee received.

5. The Salem Cooperative Bank charges an $8\frac{3}{4}\%$ discount rate. What are the discount and the proceeds for a $16,200 note for 60 days?

6. Bill Jackson is planning to buy a used car. He went to City Credit Union to take out a loan for $6,400 for 300 days. If the credit union charges a discount rate of $11\frac{1}{2}\%$, what will the proceeds of this loan be?

7. Mike Drislane goes to the bank and signs a note for $9,700. The bank charges a 15% discount rate. Find the discount and the proceeds if the loan is for 210 days.

8. Flora Foley plans to have a deck built on the back of her house. She decides to take out a loan at the bank for $14,300. She signs a note promising to pay back the loan in 280 days. If the note was discounted at 9.2%, how much money will Flora receive from the bank?

9. At the end of 280 days, Flora (Problem 8) must pay back the loan. What is the maturity value of the loan?

10. Dave Cassidy signed a $7,855 note at a bank that charges a 14.2% discount rate. If the loan is for 190 days, find **(a)** the proceeds and **(b)** the effective rate charged by the bank (to the nearest tenth percent).

11. How much money must Dave (Problem 10) pay back to the bank?

Learning Unit 11–2: Discounting an Interest-Bearing Note before Maturity

DRILL PROBLEMS

1. Calculate the maturity value for each of the following promissory notes; use 360 days:

	Date of note	Principal of note	Length of note (days)	Interest rate	Maturity value
a.	June 9	$5,000	180	3%	________
b.	August 23	$15,990	85	13%	________
c.	December 10	$985	30	11.5%	________

2. Find the maturity date and the discount period for the following; assume no leap years. *Hint:* See Exact Days-in-a-Year Calendar, Chapter 7.

	Date of note	Length of note (days)	Date of discount	Maturity date	Discount period
a.	March 11	200	June 28	________	________
b.	January 22	60	March 2	________	________
c.	April 19	85	June 6	________	________
d.	November 17	120	February 15	________	________

3. Find the bank discount for each of the following; use 360 days:

	Date of note	Principal of note	Length of note	Interest rate	Bank discount rate	Date of discount	Bank discount
a.	October 5	$2,475	88 days	11%	9.5%	December 10	________
b.	June 13	$9,055	112 days	15%	16%	August 11	________
c.	March 20	$1,065	75 days	12%	11.5%	May 24	________

4. Find the proceeds for each of the discounted notes in Problem 3.

a. __________

b. __________

c. __________

WORD PROBLEMS

5. Connors Company received a $4,000, 90-day, 10% note dated April 6 from one of its customers. Connors Company held the note until May 16, when the company discounted it at a bank at a discount rate of 12%. What were the proceeds that Connors Company received?

6. Souza & Sons accepted a 9%, $22,000, 120-day note from one of its customers on July 22. On October 2, the company discounted the note at Cooperative Bank. The discount rate was 12%. What were **(a)** the bank discount and **(b)** the proceeds?

7. The Fargate Store accepted an $8,250, 75-day, 9% note from one of its customers on March 18. Fargate discounted the note at Parkside National Bank at $9\frac{1}{2}\%$ on March 29. What proceeds did Fargate receive?

8. On November 1, Marjorie's Clothing Store accepted a $5,200, $8\frac{1}{2}\%$, 90-day note from Mary Rose in granting her a time extension on her bill. On January 13, Marjorie's discounted the note at Seawater Bank, which charged a 10% discount rate. What were the proceeds that Marjorie's received?

9. On December 3, Duncan's Company accepted a $5,000, 90-day, 12% note from Al Finney in exchange for a $5,000 bill that was past due. On January 29, Duncan's discounted the note at The Sidwell Bank at 13.1%. What were the proceeds from the note?

10. On February 26, Sullivan Company accepted a 60-day, 10% note in exchange for a $1,500 past-due bill from Tabot Company. On March 28, Sullivan Company discounted at National Bank the note received from Tabot Company. The bank discount rate was 12%. What are **(a)** the bank discount and **(b)** the proceeds?

11. On June 4, Johnson Company received from Marty Russo a 30-day, 11% note for $720 to settle Russo's debt. On June 17, Johnson discounted the note at Eastern Bank at 15%. What proceeds did Johnson receive?

12. On December 15, Lawlers Company went to the bank and discounted a 10%, 90-day, $14,000 note dated October 21. The bank charged a discount rate of 12%. What were the proceeds of the note?

Learning Unit 12–1: Compound Interest (Future Value)—The Big Picture

DRILL PROBLEMS

1. In the following examples, calculate manually the amount at year-end for each of the deposits, assuming that interest is compounded annually. Round to the nearest cent each year.

Principal	Rate	Number of years	Year 1	Year 2	Year 3	Year 4
a. $530	4%	2	______	______		
b. $1,980	12%	4	______	______	______	______

2. In the following examples, calculate the simple interest, the compound interest, and the difference between the two. Round to the nearest cent; do not use tables.

Principal	Rate	Number of years	Simple interest	Compound interest	Difference
a. $4,600	10%	2	______	______	______
b. $18,400	9%	4	______	______	______
c. $855	$7\frac{1}{5}\%$	3	______	______	______

3. Find the future value and the compound interest using the Future Value of $1 at Compound Interest table or the Compound Daily table. Round to the nearest cent.

Principal	Investment terms	Future value	Compound interest
a. $20,000	6 years at 4% compounded annually	______	______
b. $10,000	6 years at 8% compounded quarterly	______	______
c. $8,400	7 years at 12% compounded semiannually	______	______
d. $2,500	15 years at 10% compounded daily	______	______
e. $9,600	5 years at 6% compounded quarterly	______	______
f. $20,000	2 years at 6% compounded monthly	______	______

4. Calculate the effective rate (APY) of interest using the Future Value of $1 at Compound Interest table.

Investment terms	Effective rate (annual percentage yield)
a. 12% compounded quarterly	________
b. 12% compounded semiannually	________
c. 6% compounded quarterly	________

WORD PROBLEMS

5. John Mackey deposited $7,000 in his savings account at Salem Savings Bank. If the bank pays 2% interest compounded semiannually, what will be the balance of his account at the end of 3 years?

6. Pine Valley Savings Bank offers a certificate of deposit at 12% interest compounded quarterly. What is the effective rate (APY) of interest?

7. Jack Billings loaned $6,000 to his brother-in-law Dan, who was opening a new business. Dan promised to repay the loan at the end of 5 years, with interest of 8% compounded semiannually. How much will Dan pay Jack at the end of 5 years?

8. Eileen Hogarty deposits $5,630 in City Bank, which pays 12% interest compounded quarterly. How much money will Eileen have in her account at the end of 7 years?

9. If Kevin Bassage deposits $3,500 in Scarsdale Savings Bank, which pays 8% interest compounded quarterly, what will be in his account at the end of 6 years? How much interest will he have earned at that time?

10. Arlington Trust pays 6% compounded semiannually. How much interest would be earned on $7,200 for 1 year?

11. Paladium Savings Bank pays 9% compounded quarterly. Find the amount and the interest on $3,000 after three quarters. Do not use a table.

12. David Siderski bought an $8,000 bank certificate paying 4% compounded semiannually. How much money did he obtain upon cashing in the certificate 3 years later?

13. An issue of *Your Money* showed that the more frequently the bank compounds your money, the better. Consider a $10,000 investment earning 6% interest in a 5-year certificate of deposit at the following three banks. What would be the interest earned at each bank?

a. Bank A (simple interest, no compounding)
b. Bank B (quarterly compounding)
c. Bank C (daily compounding)

Learning Unit 12–2: Present Value—The Big Picture

DRILL PROBLEMS

1. Use the *Business Math Handbook* to find the table factor for each of the following:

	Future value	Rate	Number of years	Compounded	Table value
a.	$1.00	2%	5	Annually	______
b.	$1.00	12%	8	Semiannually	______
c.	$1.00	6%	10	Quarterly	______
d.	$1.00	12%	2	Monthly	______
e.	$1.00	8%	15	Semiannually	______

2. Use the *Business Math Handbook* to find the table factor and the present value for each of the following:

	Future value	Rate	Number of years	Compounded	Table value	Present value
a.	$1,000	2%	6	Semiannually	______	______
b.	$1,000	16%	7	Quarterly	______	______
c.	$1,000	8%	7	Quarterly	______	______
d.	$1,000	8%	7	Semiannually	______	______
e.	$1,000	8%	7	Annually	______	______

3. Find the present value and the interest earned for the following:

	Future value	Number of years	Rate	Compounded	Present value	Interest earned
a.	$2,500	6	8%	Annually	______	______
b.	$4,600	10	6%	Semiannually	______	______
c.	$12,800	8	10%	Semiannually	______	______
d.	$28,400	7	8%	Quarterly	______	______
e.	$53,050	1	12%	Monthly	______	______

4. Find the missing amount (present value or future value) for each of the following:

	Present value	Investment terms	Future value
a.	$3,500	5 years at 8% compounded annually	______
b.	______	6 years at 12% compounded semiannually	$9,000
c.	$4,700	9 years at 14% compounded semiannually	______

WORD PROBLEMS

Solve for future value or present value.

5. Paul Palumbo assumes that he will need to have a new roof put on his house in 4 years. He estimates that the roof will cost him $17,000 at that time. What amount of money should Paul invest today at 2%, compounded semiannually, to be able to pay for the roof?

6. Tilton, a pharmacist, rents his store and has signed a lease that will expire in 3 years. When the lease expires, Tilton wants to buy his own store. He wants to have a down payment of $35,000 at that time. How much money should Tilton invest today at 6%, compounded quarterly, to yield $35,000?

7. Brad Morrissey loans $8,200 to his brother-in-law. He will be repaid at the end of 5 years, with interest at 10% compounded semiannually. Find out how much he will be repaid.

8. The owner of Waverly Sheet Metal Company plans to buy some new machinery in 6 years. He estimates that the machines he wishes to purchase will cost $39,700 at that time. What must he invest today at 8%, compounded semiannually, to have sufficient money to purchase the new machines?

9. Paul Stevens's grandparents want to buy him a car when he graduates from college in 4 years. They feel that they should have $27,000 in the bank at that time. How much should they invest at 12%, compounded quarterly, to reach their goal?

10. Gilda Nardi deposits $5,325 in a bank that pays 12% interest compounded quarterly. Find the amount she will have at the end of 7 years.

11. Mary Wilson wants to buy a new set of golf clubs in 2 years. They will cost $775. How much money should she invest today at 9%, compounded annually, so that she will have enough money to buy the new clubs?

12. Jack Beggs plans to invest $30,000 at 10%, compounded semiannually, for 5 years. What is the future value of the investment?

13. Ron Thrift expects his Honda Pilot will last 3 more years. Ron does not like to finance his purchases. He went to First National Bank to find out how much money he should put in the bank to purchase a $20,300 car in 3 years. The bank's 3-year CD is compounded quarterly with a 4% rate. How much should Ron invest in the CD?

14. The Downers Grove YMCA had a fund-raising campaign to build a swimming pool in 6 years. Members raised $825,000; the pool is estimated to cost $1,230,000. The money will be placed in Downers Grove Bank, which pays daily interest at 6%. Will the YMCA have enough money to pay for the pool in 6 years?

Learning Unit 13–1: Annuities: Ordinary Annuity and Annuity Due (Find Future Value)

DRILL PROBLEMS

1. Find the value of the following ordinary annuities; calculate manually:

	Amount of each annual deposit	Interest rate	Value at end of year 1	Value at end of year 2	Value at end of year 3
a.	$1,000	8%	______	______	______
b.	$2,500	12%	______	______	______
c.	$7,200	10%	______	______	______

2. Use the Ordinary Annuity Table: Compound Sum of an Annuity of $1 to find the value of the following ordinary annuities:

	Annuity payment	Payment period	Term of annuity	Interest rate	Value of annuity
a.	$650	Semiannually	5 years	6%	________
b.	$3,790	Annually	13 years	12%	________
c.	$500	Quarterly	1 year	8%	________

3. Find the annuity due (deposits are made at beginning of period) for each of the following using the Ordinary Annuity Table:

	Amount of payment	Payment period	Interest rate	Time (years)	Amount of annuity
a.	$900	Annually	7%	6	________
b.	$1,200	Annually	11%	4	________
c.	$550	Semiannually	10%	9	________

4. Find the amount of each annuity:

	Amount of payment	Payment period	Interest rate	Time (years)	Type of annuity	Amount of annuity
a.	$600	Semiannually	12%	8	Ordinary	________
b.	$600	Semiannually	12%	8	Due	________
c.	$1,100	Annually	9%	7	Ordinary	________

WORD PROBLEMS

5. At the end of each year for the next 9 years, D'Aldo Company will deposit $25,000 in an ordinary annuity account paying 9% interest compounded annually. Find the value of the annuity at the end of the 9 years.

6. David McCarthy is a professional baseball player who expects to play in the major leagues for 10 years. To save for the future, he will deposit $50,000 at the beginning of each year into an account that pays 11% interest compounded annually. How much will he have in this account at the end of 10 years?

7. Tom and Sue plan to get married. Because they hope to have a large wedding, they are going to deposit $1,000 at the end of each month into an account that pays 24% compounded monthly. How much will they have in this account at the end of 1 year?

8. Chris Dennen deposits $15,000 at the end of each year for 13 years into an account paying 7% interest compounded annually. What is the value of her annuity at the end of 13 years? How much interest will she have earned?

9. Amanda Blinn is 52 years old today and has just opened an IRA. She plans to deposit $500 at the end of each quarter into her account. If Amanda retires on her 62nd birthday, what amount will she have in her account if the account pays 8% interest compounded quarterly?

10. Jerry Davis won the citywide sweepstakes and will receive a check for $2,000 at the beginning of each 6 months for the next 5 years. If Jerry deposits each check in an account that pays 8% compounded semiannually, how much will he have at the end of 5 years?

11. Mary Hynes purchased an ordinary annuity from an investment broker at 8% interest compounded semiannually. If her semiannual deposit is $600, what will be the value of the annuity at the end of 15 years?

Learning Unit 13–2: Present Value of an Ordinary Annuity (Find Present Value)

DRILL PROBLEMS

1. Use the Present Value of an Annuity of $1 table to find the amount to be invested today to receive a stream of payments for a given number of years in the future. Show the manual check of your answer. (Check may be a few pennies off due to rounding.)

	Amount of expected payments	Payment period	Interest rate	Term of annuity	Present value of annuity
a.	$1,500	Yearly	9%	2 years	________
b.	$2,700	Yearly	13%	3 years	________
c.	$2,700	Yearly	6%	3 years	________

2. Find the present value of the following annuities. Use the Present Value of an Annuity of $1 table.

	Amount of each payment	Payment period	Interest rate	Time (years)	Compounded	Present value of annuity
a.	$2,000	Year	7%	25	Annually	________
b.	$7,000	Year	11%	12	Annually	________
c.	$850	6 months	12%	5	Semiannually	________
d.	$1,950	6 months	14%	9	Semiannually	________
e.	$500	Quarter	12%	10	Quarterly	________

WORD PROBLEMS

3. Tom Hanson would like to receive $200 each quarter for the 4 years he is in college. If his bank account pays 8% compounded quarterly, how much must he have in his account when he begins college?

4. Jean Reith has just retired and will receive a $12,500 retirement check every 6 months for the next 20 years. If her employer can invest money at 12% compounded semiannually, what amount must be invested today to make the semiannual payments to Jean?

5. Tom Herrick will pay $4,500 at the end of each year for the next 7 years to pay the balance of his college loans. If Tom can invest his money at 7% compounded annually, how much must he invest today to make the annual payments?

6. Helen Grahan is planning an extended sabbatical for the next 3 years. She would like to invest a lump sum of money at 10% interest so that she can withdraw $6,000 every 6 months while on sabbatical. What is the amount of the lump sum that Helen must invest?

7. Linda Rudd has signed a rental contract for office equipment, agreeing to pay $3,200 at the end of each quarter for the next 5 years. If Linda can invest money at 12% compounded quarterly, find the lump sum she can deposit today to make the payments for the length of the contract.

8. Sam Adams is considering lending his brother John $6,000. John said that he would repay Sam $775 every 6 months for 4 years. If money can be invested at 8%, calculate the equivalent cash value of the offer today. Should Sam go ahead with the loan?

9. The State Lotto Game offers a grand prize of $1,000,000 paid in 20 yearly payments of $50,000. If the state treasurer can invest money at 9% compounded annually, how much must she invest today to make the payments to the grand prize winner?

10. Thomas Martin's uncle has promised him upon graduation a gift of $20,000 in cash or $2,000 every quarter for the next 3 years. If money can be invested at 8%, which offer will Thomas accept? (Thomas is a business major.)

11. Paul Sasso is selling a piece of land. He has received two solid offers. Jason Smith has offered a $60,000 down payment and $50,000 a year for the next 5 years. Kevin Bassage offered $35,000 down and $55,000 a year for the next 5 years. If money can be invested at 7% compounded annually, which offer should Paul accept? (To make the comparison, find the equivalent cash price of each offer.)

12. Abe Hoster decided to retire to Spain in 10 years. What amount should Abe invest today so that he will be able to withdraw $30,000 at the end of each year for 20 years after he retires? Assume he can invest money at 8% interest compounded annually.

Learning Unit 13–3: Sinking Funds (Find Periodic Payments)

DRILL PROBLEMS

1. Given the number of years and the interest rate, use the Sinking Fund Table based on $1 to calculate the amount of the periodic payment.

	Frequency of payment	Length of time	Interest rate	Future amount	Sinking fund payment
a.	Annually	19 years	5%	$125,000	________
b.	Annually	7 years	10%	$205,000	________
c.	Semiannually	10 years	6%	$37,500	________
d.	Quarterly	9 years	12%	$12,750	________
e.	Quarterly	6 years	8%	$25,600	________

2. Find the amount of each payment into the sinking fund and the amount of interest earned.

	Maturity value	Interest rate	Term (years)	Frequency of payment	Sinking fund payment	Interest earned
a.	$45,500	5%	13	Annually	________	________
b.	$8,500	10%	20	Semiannually	________	________
c.	$11,000	8%	5	Quarterly	________	________
d.	$66,600	12%	$7\frac{1}{2}$	Semiannually	________	________

WORD PROBLEMS

3. To finance a new police station, the town of Pine Valley issued bonds totaling $600,000. The town treasurer set up a sinking fund at 8% compounded quarterly in order to redeem the bonds in 7 years. What is the quarterly payment that must be deposited into the fund?

4. Arlex Oil Corporation plans to build a new garage in 6 years. To finance the project, the financial manager established a $250,000 sinking fund at 6% compounded semianually. Find the semiannual payment required for the fund.

5. The City Fisheries Corporation sold $300,000 worth of bonds that must be redeemed in 9 years. The corporation agreed to set up a sinking fund to accumulate the $300,000. Find the amount of the periodic payments made into the fund if payments are made annually and the fund earns 8% compounded annually.

6. Gregory Mines Corporation wishes to purchase a new piece of equipment in 4 years. The estimated price of the equipment is $100,000. If the corporation makes periodic payments into a sinking fund with 12% interest compounded quarterly, find the amount of the periodic payments.

7. The Best Corporation must buy a new piece of machinery in $4\frac{1}{2}$ years that will cost $350,000. If the firm sets up a sinking fund to finance this new machine, what will the quarterly deposits be assuming the fund earns 8% interest compounded quarterly?

8. The Lowest-Price-in-Town Company needs $75,500 in 6 years to pay off a debt. The company makes a decision to set up a sinking fund and make semiannual deposits. What will its payments be if the fund pays 10% interest compounded semiannually?

9. The WIR Company plans to renovate its offices in 5 years. It estimates that the cost will be $235,000. If the company sets up a sinking fund that pays 12% quarterly, what will its quarterly payments be?

Learning Unit 14–1: Cost of Installment Buying

DRILL PROBLEMS

1. For the following installment problems, find the amount financed and the finance charge.

	Sale price	Down payment	Number of monthly payments	Monthly payment	Amount financed	Finance charge
a.	$1,500	$300	24	$58	______	______
b.	$12,000	$3,000	30	$340	______	______
c.	$62,500	$4,700	48	$1,500	______	______
d.	$4,975	$620	18	$272	______	______
e.	$825	$82.50	12	$67.45	______	______

2. For each of the above purchases, find the deferred payment price.

	Sale price	Down payment	Number of monthly payments	Monthly payment	Deferred payment price
a.	$1,500	$300	24	$58	______
b.	$12,000	$3,000	30	$340	______
c.	$62,500	$4,700	48	$1,500	______
d.	$4,975	$620	18	$272	______
e.	$825	$82.50	12	$67.45	______

3. Use the Annual Percentage Rate Table per $100 to calculate the estimated APR for each of the previous purchases.

	Sale price	Down payment	Number of monthly payments	Monthly payment	Annual percentage rate
a.	$1,500	$300	24	$58	_______
b.	$12,000	$3,000	30	$340	_______
c.	$62,500	$4,700	48	$1,500	_______
d.	$4,975	$620	18	$272	_______
e.	$825	$82.50	12	$67.45	_______

4. Given the following information, calculate the monthly payment by the loan amortization table.

	Amount financed	Interest rate	Number of months of loan	Monthly payment
a.	$12,000	10%	18	_______
b.	$18,000	11%	36	_______
c.	$25,500	13.50%	54	_______

WORD PROBLEMS

5. Jill Walsh purchases a bedroom set for a cash price of $3,920. The down payment is $392, and the monthly installment payment is $176 for 24 months. Find **(a)** the amount financed, **(b)** the finance charge, and **(c)** the deferred payment price.

6. An automaker promotion loan on a $20,000 automobile and a down payment of 20% are being financed for 48 months. The monthly payments will be $367.74. What will be the APR for this auto loan? Use the table in the *Business Math Handbook*.

7. David Nason purchased a recreational vehicle for $25,000. David went to City Bank to finance the purchase. The bank required that David make a 10% down payment and monthly payments of $571.50 for 4 years. Find **(a)** the amount financed, **(b)** the finance charge, and **(c)** the deferred payment that David paid.

8. Calculate the estimated APR that David (Problem 7) was charged per $100 using the Annual Percentage Rate Table.

9. Young's Motors advertised a new car for $16,720. Young's offered an installment plan of 5% down and 42 monthly payments of $470. What are **(a)** the deferred payment price and **(b)** the estimated APR for this car? Use the table.

10. Angie French bought a used car for $9,000. Angie put down $2,000 and financed the balance at 11.50% for 36 months. What is her monthly payment? Use the loan amortization table.

Learning Unit 14–2: Revolving Charge Credit Cards

DRILL PROBLEMS

1. Use the U.S. Rule to calculate the outstanding balance due for each of the following independent situations:

	Monthly payment number	Outstanding balance due	$1\frac{1}{2}\%$ interest payment	Amount of monthly payment	Reduction in balance due	Outstanding balance due
a.	1	$9,000.00	_______	$600	_______	_______
b.	5	$5,625.00	_______	$1,000	_______	_______
c.	4	$926.50	_______	$250	_______	_______
d.	12	$62,391.28	_______	$1,200	_______	_______
e.	8	$3,255.19	_______	$325	_______	_______

2. Complete the missing data for a $6,500 purchase made on credit. The annual interest charge on this revolving charge account is 18%, or $1\frac{1}{2}\%$ interest on previous month's balance. Use the U.S. Rule.

Monthly payment number	Outstanding balance due	$1\frac{1}{2}\%$ interest payment	Amount of monthly payment	Reduction in balance due	Outstanding balance due
1	$6,500	______	$700	______	______
2	______	______	$700	______	______
3	______	______	$700	______	______

3. Calculate the average billing daily balance for each of the monthly statements for the following revolving credit accounts; assume a 30-day billing cycle:

	Billing date	Previous balance	Payment date	Payment amount	Charge date(s)	Charge amount(s)	Average daily balance
a.	4/10	$329	4/25	$35	4/29	$56	______
b.	6/15	$573	6/25	$60	6/26 6/30	$25 $72	______
c.	9/15	$335.50	9/20	$33.55	9/25 9/26	$12.50 $108	______

4. Find the finance charge for each monthly statement (Problem 3) if the annual percentage rate is 15%.

a. ______ **b.** ______ **c.** ______

WORD PROBLEMS

5. Niki Marshall is going to buy a new bedroom set at Scottie's Furniture Store, where she has a revolving charge account. The cost of the bedroom set is $5,500. Niki does not plan to charge anything else to her account until she has completely paid for the bedroom set. Scottie's Furniture Store charges an annual percentage rate of 18%, or $1\frac{1}{2}\%$ per month. Niki plans to pay $1,000 per month until she has paid for the bedroom set. Set up a schedule for Niki to show her outstanding balance at the end of each month after her $1,000 payment and also the amount of her final payment. Use the U.S. Rule.

6. Frances Dollof received her monthly statement from Brown's Department Store. The following is part of the information contained on that statement. Finance charge is calculated on the average daily balance.

Date	Reference	Department	Description	Amount
Dec. 15	5921	359	Petite sportswear	84.98
Dec. 15	9612	432	Footwear	55.99
Dec. 15	2600	126	Women's fragrance	35.18
Dec. 23	6247	61	Ralph Lauren towels	20.99
Dec. 24	0129	998	Payment received—thank you	100.00CR

Previous balance		Annual percentage rate	Billing date
719.04	12/13	18%	JAN 13

Brown's Charge Account Terms

Payment is required in monthly installments upon receipt of monthly statement in accordance with Brown's payment terms.

When my new balance is:	My minimum required payment is:	When my new balance is:	My minimum required payment is:
Up to $20.00	New Balance	$350.01 to $400.00	$40.00
$ 20.01 to $200.00	$20.00	$400.01 to $450.00	$45.00
$200.01 to $250.00	$25.00	$450.01 to $500.00	$50.00
$250.01 to $300.00	$30.00	More than $500.00	$50.00 plus $10.00 for each $50.00 (or fraction thereof) of New Balance over $500.00
$300.01 to $350.00	$35.00		

a. Calculate the average daily balance for the month.

b. What is Ms. Dollof's finance charge?

c. What is the new balance for Ms. Dollof's account?

d. What is the minimum payment Frances is required to pay according to Brown's payment terms?

7. What is the finance charge for a Brown's customer who has an average daily balance of $3,422.67?
8. What is the minimum payment for a Brown's customer with a new balance of $522.00?
9. What is the minimum payment for a Brown's customer with a new balance of $325.01?
10. What is the new balance for a Brown's customer with a previous balance of $309.35 whose purchases totaled $213.00, given that the customer made a payment of $75.00 and the finance charge was $4.65?

RECAP OF WORD PROBLEMS IN LU 14–1

11. A home equity loan on a $20,000 automobile with a down payment of 20% is being financed for 48 months. The interest is tax deductible. The monthly payments will be $401.97. What is the APR on this loan? Use the table in the *Business Math Handbook.* If the person is in the 28% income tax bracket, what will be the tax savings with this type of a loan?

12. An automobile with a total transaction price of $20,000 with a down payment of 20% is being financed for 48 months. Banks and credit unions require a monthly payment of $400.36. What is the APR for this auto loan? Use the table in the *Business Math Handbook.*

13. Assume you received a $2,000 rebate that brought the price of a car down to $20,000; the financing rate was for 48 months, and your total interest was $3,279. Using the table in the *Business Math Handbook,* what was your APR?

Learning Unit 15–1: Types of Mortgages and the Monthly Mortgage Payment

DRILL PROBLEMS

1. Use the table in the *Business Math Handbook* to calculate the monthly payment for principal and interest for the following mortgages:

Price of home	Down payment	Interest rate	Term in years	Monthly payment
a. $200,000	15%	6%	25	______
b. $200,000	15%	$5\frac{1}{2}\%$	30	______
c. $450,000	10%	$11\frac{3}{4}\%$	30	______
d. $450,000	10%	11%	30	______

2. For each of the mortgages, calculate the amount of interest that will be paid over the life of the loan.

Price of home	Down payment	Interest rate	Term in years	Total interest paid
a. $200,000	15%	$6\frac{1}{2}\%$	25	______
b. $200,000	15%	$10\frac{1}{2}\%$	30	______
c. $450,000	10%	$11\frac{3}{4}\%$	30	______
d. $450,000	10%	11%	30	______

3. Calculate the increase in the monthly mortgage payments for each of the rate increases in the following mortgages. Then calculate what percent of change the increase represents, rounded to the nearest tenth percent.

	Mortgage amount	Term in years	Interest rate	Increase in interest rate	Increase in monthly payment	Percent change
a.	$175,000	22	9%	1%	________	______
b.	$300,000	30	$11\frac{3}{4}\%$	$\frac{3}{4}\%$	________	______

4. Calculate the increase in total interest paid for the increase in interest rates in Problem 3.

	Mortgage amount	Term in years	Interest rate	Increase in interest rate	Increase in total interest paid
a.	$175,000	22	9%	1%	________
b.	$300,000	30	$11\frac{3}{4}\%$	$\frac{3}{4}\%$	________

WORD PROBLEMS

5. The Counties are planning to purchase a new home that costs $150,000. The bank is charging them 6% interest and requires a 20% down payment. The Counties are planning to take a 25-year mortgage. How much will their monthly payment be for principal and interest?

6. The MacEacherns wish to buy a new house that costs $299,000. The bank requires a 15% down payment and charges $11\frac{1}{2}\%$ interest. If the MacEacherns take out a 15-year mortgage, what will their monthly payment for principal and interest be?

7. Because the monthly payments are so high, the MacEacherns (Problem 6) want to know what the monthly payments would be for **(a)** a 25-year mortgage and **(b)** a 30-year mortgage. Calculate these two payments.

8. If the MacEacherns choose a 30-year mortgage instead of a 15-year mortgage, **(a)** how much money will they "save" monthly and **(b)** how much more interest will they pay over the life of the loan?

9. If the MacEacherns choose the 25-year mortgage instead of the 30-year mortgage, **(a)** how much more will they pay monthly and **(b)** how much less interest will they pay over the life of the loan?

10. Larry and Doris Davis plan to purchase a new home that costs $415,000. The bank that they are dealing with requires a 20% down payment and charges $12\frac{3}{4}\%$. The Davises are planning to take a 25-year mortgage. What will the monthly payment be?

11. How much interest will the Davises (Problem 10) pay over the life of the loan?

Learning Unit 15–2: Amortization Schedule—Breaking Down the Monthly Payment

DRILL PROBLEMS

1. In the following, calculate the monthly payment for each mortgage, the portion of the first monthly payment that goes to interest, and the portion of the payment that goes toward the principal.

	Amount of mortgage	Interest rate	Term in years	Monthly payment	Portion to interest	Portion to principal
a.	$170,000	8%	22	________	________	________
b.	$222,000	$11\frac{3}{4}\%$	30	________	________	________
c.	$167,000	$10\frac{1}{2}\%$	25	________	________	________
d.	$307,000	13%	15	________	________	________
e.	$409,500	$12\frac{1}{2}\%$	20	________	________	________

2. Prepare an amortization schedule for the first 3 months of a 25-year, 12% mortgage on $265,000.

Payment number	Monthly payment	Portion to interest	Portion to principal	Balance of loan outstanding
1	________	________	________	________
2	________	________	________	________
3	________	________	________	________

3. Prepare an amortization schedule for the first 4 months of a 30-year, $10\frac{1}{2}$% mortgage on $195,500.

Payment number	Monthly payment	Portion to interest	Portion to principal	Balance of loan outstanding
1	________	________	________	________
2	________	________	________	________
3	________	________	________	________
4	________	________	________	________

WORD PROBLEMS

4. Jim and Janice Hurst are buying a new home for $235,000. The bank that is financing the home requires a 20% down payment and charges a $13\frac{1}{2}$% interest rate. Janice wants to know **(a)** what the monthly payment for the principal and interest will be if they take out a 30-year mortgage and **(b)** how much of the first payment will be for interest on the loan.

5. The Hursts (Problem 4) thought that a lot of their money was going to interest. They asked the banker just how much they would be paying for interest over the life of the loan. Calculate the total amount of interest that the Hursts will pay.

6. The banker told the Hursts (Problem 4) that they could, of course, save on the interest payments if they took out a loan for a shorter period of time. Jim and Janice decided to see if they could afford a 15-year mortgage. Calculate how much more the Hursts would have to pay each month for principal and interest if they took a 15-year mortgage for their loan.

7. The Hursts (Problem 4) thought that they might be able to afford the 15-year mortgage (Problem 6), but first they wanted to see **(a)** how much of the first payment would go to the principal and **(b)** how much total interest they would be paying with a 15-year mortgage.

8.

	1980	2019
Cost of median-priced new home	$44,200	$136,600
10% down payment	$4,420	
Fixed-rate, 30-year mortgage		
Interest rate	8.9%	$7\frac{1}{2}$%
Total monthly principal and interest	$316	

Complete the 2019 year.

9. You can't count on your home mortgage lender to keep you from getting in debt over your head. The old standards of allowing 28% of your income for mortgage debt (including taxes and insurance) usually still apply. If your total monthly payment is $1,033, what should be your annual income to buy a home?

10. Assume that a 30-year fixed-rate mortgage for $100,000 was 9% at one date as opposed to 7% the previous year. What is the difference in monthly payments for these 2 years?

11. If you had a \$100,000 mortgage with $7\frac{1}{2}\%$ interest for 25 years and wanted a $7\frac{1}{2}\%$ loan for 35 years, what would be the change in monthly payments? How much more would you pay in interest?

Learning Unit 16–1: Balance Sheet—Report as of a Particular Date

DRILL PROBLEMS

1. Complete the balance sheet for David Harrison, Attorney, on December 31, 2019, and show that

   Assets = Liabilities + Owner's equity

   Account totals are as follows: accounts receivable, \$4,800; office supplies, \$375; building (net), \$130,000; accounts payable, \$1,200; notes payable, \$137,200; cash, \$2,250; prepaid insurance, \$1,050; office equipment (net), \$11,250; land, \$75,000; capital, \$85,900; and salaries payable, \$425.

DAVID HARRISON, ATTORNEY Balance Sheet December 31, 2019			
Assets			
Current assets:			
Cash	______		
Accounts receivable	______		
Prepaid insurance	______		
Office supplies	______		
Total current assets		______	
Plant and equipment:			
Office equipment (net)	______		
Building (net)	______		
Land	______		
Total plant and equipment		______	
Total assets			______
Liabilities			
Current liabilities:			
Accounts payable	______		
Salaries payable	______		
Total current liabilities		______	
Long-term liabilities:			
Notes payable		______	
Total liabilities			______
Owner's Equity			
David Harrison, capital, December 31, 2019			______
Total liabilities and owner's equity			______

2. Given the amounts in each of the accounts of Fisher-George Electric Corporation, fill in these amounts on the balance sheet for December 31, 2019, to show that

   Assets = Liabilities + Stockholders' equity

   Account totals are as follows: cash, \$2,500; merchandise inventory, \$1,325; automobiles (net), \$9,250; common stock, \$10,000; accounts payable, \$275; office equipment (net), \$5,065; accounts receivable, \$300; retained earnings, \$6,895; prepaid insurance, \$1,075; salaries payable, \$175; and mortgage payable, \$2,170.

FISHER-GEORGE ELECTRIC CORPORATION Balance Sheet December 31, 2019			
Assets			
Current assets:			
Cash	____		
Accounts receivable	____		
Merchandise inventory	____		
Prepaid insurance	____		
Total current assets		____	
Plant and equipment:			
Office equipment (net)	____		
Automobiles (net)	____		
Total plant and equipment		____	
Total assets			____
Liabilities			
Current liabilities:			
Accounts payable	____		
Salaries payable	____		
Total current liabilities		____	
Long-term liabilities:			
Mortgage payable		____	
Total liabilities			____
Stockholders' Equity			
Common stock		____	
Retained earnings		____	
Total stockholders' equity			____
Total liabilities and stockholders' equity			____

3. Complete a vertical analysis of the following partial balance sheet; round all percents to the nearest hundredth percent.

THREEMAX, INC. Comparative Balance Sheet Vertical Analysis At December 31, 2018 and 2019	2019		2018	
	Amount	Percent	Amount	Percent
Assets				
Cash	$ 8,500	____	$ 10,200	____
Accounts receivable (net)	11,750	____	15,300	____
Merchandise inventory	55,430	____	54,370	____
Store supplies	700	____	532	____
Office supplies	650	____	640	____
Prepaid insurance	2,450	____	2,675	____
Office equipment (net)	12,000	____	14,300	____
Store equipment (net)	32,000	____	31,000	____
Building (net)	75,400	____	80,500	____
Land	200,000	____	150,000	____
Total assets	$398,880	____	$359,517	____

4. Complete a horizontal analysis of the following partial balance sheet; round all percents to the nearest hundredth percent.

THREEMAX, INC. Comparative Balance Sheet Horizontal Analysis At December 31, 2018 and 2019	2019	2018	Change	Percent
Assets				
Cash	$ 8,500	$ 10,200	______	______
Accounts receivable (net)	11,750	15,300	______	______
Merchandise inventory	55,430	54,370	______	______
Store supplies	700	532	______	______
Office supplies	650	640	______	______
Prepaid insurance	2,450	2,675	______	______
Office equipment (net)	12,000	14,300	______	______
Store equipment (net)	32,000	31,000	______	______
Building (net)	75,400	80,500	______	______
Land	200,000	150,000	______	______
Total assets	$398,880	$359,517		

Learning Unit 16–2: Income Statement—Report for a Specific Period of Time

DRILL PROBLEMS

1. Complete the income statement for the year ended December 31, 2019, for Foley Realty, doing all the necessary addition. Account totals are as follows: office salaries expense, $15,255; advertising expense, $2,400; rent expense, $18,000; telephone expense, $650; insurance expense, $1,550; office supplies, $980; depreciation expense, office equipment, $990; depreciation expense, automobile, $2,100; sales commissions earned, $98,400; and management fees earned, $1,260.

FOLEY REALTY Income Statement For the Year Ended December 31, 2019		
Revenues:		
Sales commissions earned	______	
Management fees earned	______	
Total revenues		______
Operating expenses:		
Office salaries expense	______	
Advertising expense	______	
Rent expense	______	
Telephone expense	______	
Insurance expense	______	
Office supplies expense	______	
Depreciation expense, office equipment	______	
Depreciation expense, automobile	______	
Total operating expenses		______
Net income		______

2. Complete the income statement for Toll's, Inc., a merchandising concern, doing all the necessary addition and subtraction. Sales were $250,000; sales returns and allowances were $1,400; sales discounts were $2,100; merchandise inventory, December 31, 2018, was $42,000; purchases were $156,000; purchases returns and allowances were $1,100; purchases discounts were $3,000; merchandise inventory, December 31, 2019, was $47,000; selling expenses were $37,000; and general and administrative expenses were $29,000.

TOLL'S, INC. Income Statement For the Year Ended December 31, 2019				
Revenues:				
Sales				______
Less: Sales return and allowances			______	
Sales discounts			______	______
Net sales				______
Cost of goods sold:				
Merchandise inventory, December 31, 2018			______	
Purchases		______		
Less: Purchases returns and allowances	______			
Purchase discounts	______	______		
Cost of net purchases			______	
Goods available for sale			______	
Merchandise inventory, December 31, 2019			______	
Total cost of goods sold				______
Gross profit from sales				______
Operating expenses:				
Selling expenses			______	
General and administrative expenses			______	
Total operating expenses				______
Net income				______

3. Complete a vertical analysis of the following partial income statement; round all percents to the nearest hundredth percent. Note net sales are 100%.

THREEMAX, INC. Comparative Income Statement Vertical Analysis For Years Ended December 31, 2018 and 2019				
	2019		2018	
	Amount	Percent	Amount	Percent
Sales	$795,450		$665,532	
Sales returns and allowances	−6,250		−5,340	
Sales discounts	−6,470		−5,125	
Net sales	$782,730		$655,067	
Cost of goods sold:				
Beginning inventory	$ 75,394		$ 81,083	
Purchases	575,980		467,920	
Purchase discounts	−4,976		−2,290	
Goods available for sale	$646,398		$546,713	
Less ending inventory	−66,254		− 65,712	
Total costs of goods sold	$580,144		$481,001	
Gross profit	$202,586		$174,066	

4. Complete a horizontal analysis of the following partial income statement. Round all percents to the nearest hundredth percent.

THREEMAX, INC. Comparative Income Statement Horizontal Analysis For Years Ended December 31, 2018 and 2019				
	2019	2018	Change	Percent
Sales	$795,450	$665,532	____	____
Sales returns and allowances	−6,250	− 5,340	____	____
Sales discounts	−6,470	− 5,125	____	____
Net sales	$782,730	$655,067	____	____
Cost of goods sold:				
Beginning inventory	$ 75,394	$ 81,083	____	____
Purchases	575,980	467,920	____	____
Purchase discounts	−4,976	−2,290	____	____
Goods available for sale	$646,398	$546,713	____	____
Less ending inventory	−66,254	− 65,712	____	____
Total cost of goods sold	$580,144	$481,001	____	____
Gross profit	$202,586	$ 174,066	____	____

Learning Unit 16–3: Trend and Ratio Analysis

DRILL PROBLEMS

1. Express each amount as a percent of the base-year (2017) amount. Round to the nearest tenth percent.

	2020	2019	2018	2017
Sales	$562,791	$560,776	$588,096	$601,982
Percent	____	____	____	____
Gross profit	$168,837	$196,271	$235,238	$270,891
Percent	____	____	____	____
Net income	$67,934	$65,927	$56,737	$62,762
Percent	____	____	____	____

2. If current assets = $42,500 and current liabilities = $56,400, what is the current ratio (to the nearest hundredth)?

3. In Problem 2, if inventory = $20,500 and prepaid expenses = $9,750, what is the quick ratio, or acid test (to the nearest hundredth)?

4. If accounts receivable = $36,720 and net sales = $249,700, what is the average day's collection (to the nearest whole day)?

5. If total liabilities = $243,000 and total assets = $409,870, what is the ratio of total debt to total assets (to the nearest hundredth percent)?

6. If net income = $55,970 and total stockholders' equity = $440,780, what is the return on equity (to the nearest hundredth percent)?

7. If net sales = $900,000 and total assets = $1,090,000, what is the asset turnover (to the nearest hundredth)?

8. In Problem 7, if the net income is $36,600, what is the profit margin on net sales (to the nearest hundredth percent)?

WORD PROBLEMS

9. Calculate trend percentages for the following items using 2018 as the base year. Round to the nearest hundredth percent.

	2021	2020	2019	2018
Sales	$298,000	$280,000	$264,000	$249,250
Cost of goods sold	187,085	175,227	164,687	156,785
Accounts receivable	29,820	28,850	27,300	26,250

10. According to the balance sheet for Ralph's Market, current assets = $165,500 and current liabilities = $70,500. Find the current ratio (to the nearest hundredth).

11. On the balance sheet for Ralph's Market (Problem 10), merchandise inventory = $102,000. Find the quick ratio (acid test).

12. The balance sheet of Moses Contractors shows cash of $5,500, accounts receivable of $64,500, an inventory of $42,500, and current liabilities of $57,500. Find Moses' current ratio and acid test ratio (both to the nearest hundredth).

13. Moses' income statement shows gross sales of $413,000, sales returns of $8,600, and net income of $22,300. Find the profit margin on net sales (to the nearest hundredth percent).

14. Given:

Cash	$ 39,000	Retained earnings	$194,000
Accounts receivable	109,000	Net sales	825,000
Inventory	150,000	Cost of goods sold	528,000
Prepaid expenses	48,000	Operating expenses	209,300
Plant and equipment (net)	487,000	Interest expense	13,500
Accounts payable	46,000	Income taxes	32,400
Other current liabilities	43,000	Net income	41,800
Long-term liabilities	225,000		
Common stock	325,000		

Calculate (to nearest hundredth or hundredth percent as needed):

a. Current ratio.

b. Quick ratio.

c. Average day's collection.

d. Total debt to total assets.

e. Return on equity.

f. Asset turnover.

g. Profit margin on net sales.

15. The Vale Group lost $18.4 million in profits for the year 2019 as sales dropped to $401 million. Sales in 2018 were $450.6 million. What percent is the decrease in Vale's sales? Round to the nearest hundredth percent.

Learning Unit 17–1: Concept of Depreciation and the Straight-Line Method

DRILL PROBLEMS

1. Find the annual straight-line rate of depreciation, given the following estimated lives.

Life	Annual rate	Life	Annual rate
a. 25 years	______	**b.** 4 years	______
c. 10 years	______	**d.** 5 years	______
e. 8 years	______	**f.** 30 years	______

2. Find the annual depreciation using the straight-line depreciation method. Round to the nearest whole dollar.

	Cost of asset	Residual value	Useful life	Annual depreciation
a.	$2,460	$400	4 years	______
b.	$24,300	$2,000	6 years	______
c.	$350,000	$42,500	12 years	______
d.	$17,325	$5,000	5 years	______
e.	$2,550,000	$75,000	30 years	______

3. Find the annual depreciation and ending book value for the first year using the straight-line depreciation method. Round to the nearest dollar.

	Cost	Residual value	Useful life	Annual depreciation	Ending book value
a.	$6,700	$600	3 years	______	______
b.	$11,600	$500	6 years	______	______
c.	$9,980	–0–	5 years	______	______
d.	$36,950	$2,500	12 years	______	______
e.	$101,690	$3,600	27 years	______	______

4. Find the first-year depreciation to the nearest dollar for the following assets, which were only owned for part of a year. Round to the nearest whole dollar the annual depreciation for in-between calculations.

	Date of purchase	Cost of asset	Residual value	Useful life	First year depreciation
a.	April 8	$10,500	$1,200	4 years	______
b.	July 12	$23,900	$3,200	6 years	______
c.	June 19	$8,880	$800	3 years	______
d.	November 2	$125,675	$6,000	17 years	______
e.	May 25	$44,050	–0–	9 years	______

WORD PROBLEMS

5. North Shore Grinding purchased a lathe for $37,500. This machine has a residual value of $3,000 and an expected useful life of 4 years. Prepare a depreciation schedule for the lathe using the straight-line depreciation method.

6. Colby Wayne paid $7,750 for a photocopy machine with an estimated life of 6 years and a residual value of $900. Prepare a depreciation schedule using the straight-line depreciation method. Round to the nearest whole dollar. (Last year's depreciation may have to be adjusted due to rounding.)

7. The Leo Brothers purchased a machine for $8,400 that has an estimated life of 3 years. At the end of 3 years the machine will have no value. Prepare a depreciation schedule using the straight-line depreciation method for this machine.

8. Fox Realty bought a computer table for $1,700. The estimated useful life of the table is 7 years. The residual value at the end of 7 years is $370. Find **(a)** the annual rate of depreciation to the nearest hundredth percent, **(b)** the annual amount of depreciation, and **(c)** the book value of the table at the end of the *third* year using the straight-line depreciation method.

9. Cashman, Inc., purchased an overhead projector for $560. It has an estimated useful life of 6 years, at which time it will have no remaining value. Find the book value at the end of 5 years using the straight-line depreciation method. Round the annual depreciation to the nearest whole dollar.

10. Shelley Corporation purchased a new machine for $15,000. The estimated life of the machine is 12 years with a residual value of $2,400. Find **(a)** the annual rate of depreciation by the straight-line method to the nearest hundredth percent, **(b)** the annual amount of depreciation, **(c)** the accumulated depreciation at the end of 7 years, and **(d)** the book value at the end of 9 years.

11. Wolfe Ltd. purchased a supercomputer for $75,000 on July 7, 2017. The computer has an estimated life of 5 years and will have a residual value of $15,000. Find **(a)** the annual depreciation amount by the straight-line method, **(b)** the depreciation amount for 2017, **(c)** the accumulated depreciation at the end of 2018, and **(d)** the book value at the end of 2019.

Learning Unit 17–2: Units-of-Production Method

DRILL PROBLEMS

1. Find the depreciation per unit for each of the following assets. Round to three decimal places.

	Cost of asset	Residual value	Estimated production	Depreciation per unit
a.	$3,500	$800	9,000 units	_______
b.	$309,560	$22,000	1,500,000 units	_______
c.	$54,890	$6,500	275,000 units	_______

2. Find the annual depreciation expense for each of the assets in Problem 1.

	Cost of asset	Residual value	Estimated production	Depreciation per unit	Units produced	Amount of depreciation
a.	$3,500	$800	9,000 units	_______	3,000	_______
b.	$309,560	$22,000	1,500,000 units	_______	45,500	_______
c.	$54,890	$6,500	275,000 units	_______	4,788	_______

3. Find the book value at the end of the first year for each of the assets in Problems 1 and 2.

	Cost of asset	Residual value	Estimated production	Depreciation per unit	Units produced	Book value
a.	$3,500	$800	9,000 units	_______	3,000	_______
b.	$309,560	$22,000	1,500,000 units	_______	45,500	_______
c.	$54,890	$6,500	275,000 units	_______	4,788	_______

4. Calculate the accumulated depreciation at the end of year 2 for each of the following machines. Carry out the unit depreciation to three decimal places.

	Cost of machine	Residual value	Estimated life	Hours used during year 1	Hours used during year 2	Accumulated depreciation
a.	$67,900	$4,300	19,000 hours	5,430	4,856	_______
b.	$3,810	$600	33,000 hours	10,500	9,330	_______
c.	$25,000	$4,900	80,000 hours	7,000	12,600	_______

WORD PROBLEMS

5. Prepare a depreciation schedule for the following machine: The machine cost $63,400; it has an estimated residual value of $5,300 and expected life of 290,500 units. The units produced were:

Year 1	95,000 units
Year 2	80,000 units
Year 3	50,000 units
Year 4	35,500 units
Year 5	30,000 units

6. Forsmann & Smythe purchased a new machine that cost $46,030. The machine has a residual value of $2,200 and estimated output of 430,000 hours. Prepare a units-of-production depreciation schedule for this machine, rounding the unit depreciation to three decimal places. The hours of use were:

Year 1	90,000 hours
Year 2	150,000 hours
Year 3	105,000 hours
Year 4	90,000 hours

7. Young Electrical Company depreciates its vans using the units-of-production method. The cost of its new van was $24,600, the useful life is 125,000 miles, and the trade-in value is $5,250. What are **(a)** the depreciation expense per mile (to three decimal places) and **(b)** the book value at the end of the first year if it is driven 29,667 miles?

8. Tremblay Manufacturing Company purchased a new machine for $52,000. The machine has an estimated useful life of 185,000 hours and a residual value of $10,000. The machine was used for 51,200 hours the first year. Find **(a)** the depreciation rate per hour, rounded to three decimal places, **(b)** the depreciation expense for the first year, and **(c)** the book value of the machine at the end of the first year.

Learning Unit 17–3: Declining-Balance Method

DRILL PROBLEMS

1. Find the declining-balance rate of depreciation, given the following estimated lives.

Life	Declining rate
a. 25 years	________
b. 10 years	________
c. 8 years	________

2. Find the first year depreciation amount for the following assets using the declining-balance depreciation method. Round to the nearest whole dollar.

Cost of asset	Residual value	Useful life	First year depreciation
a. $2,460	$400	4 years	________
b. $24,300	$2,000	6 years	________
c. $350,000	$42,500	12 years	________
d. $17,325	$5,000	5 years	________
e. $2,550,000	$75,000	30 years	________

3. Find the depreciation expense and ending book value for the first year, using the declining-balance depreciation method. Round to the nearest dollar.

Cost	Residual value	Useful life	First year depreciation	Ending book value
a. $6,700	$600	3 years	________	________
b. $11,600	$500	6 years	________	________
c. $9,980	-0-	5 years	________	________
d. $36,950	$2,500	12 years	________	________
e. $101,690	$3,600	27 years	________	________

WORD PROBLEMS

4. North Shore Grinding purchased a lathe for $37,500. This machine has a residual value of $3,000 and an expected useful life of 4 years. Prepare a depreciation schedule for the lathe using the declining-balance depreciation method. Round to the nearest whole dollar.

5. Colby Wayne paid $7,750 for a photocopy machine with an estimated life of 6 years and a residual value of $900. Prepare a depreciation schedule using the declining-balance depreciation method. Round to the nearest whole dollar.

6. The Leo Brothers purchased a machine for $8,400 that has an estimated life of 3 years. At the end of 3 years, the machine will have no value. Prepare a depreciation schedule for this machine using the declining-balance depreciation method. Round to the nearest whole dollar.

7. Fox Realty bought a computer table for $1,700. The estimated useful life of the table is 7 years. The residual value at the end of 7 years is $370. Find **(a)** the declining depreciation rate to the nearest hundredth percent, **(b)** the amount of depreciation at the end of the *third* year, and **(c)** the book value of the table at the end of the *third* year using the declining-balance depreciation method. Round to the nearest whole dollar.

8. Cashman, Inc., purchased an overhead projector for $560. It has an estimated useful life of 6 years, at which time it will have no remaining value. Find the book value at the end of 5 years using the declining-balance depreciation method. Round to the nearest whole dollar.

9. Shelley Corporation purchased a new machine for $15,000. The estimated life of the machine is 12 years with a residual value of $2,400. Find **(a)** the declining-balance depreciation rate as a fraction and as a percent (hundredth percent), **(b)** the amount of depreciation at the end of the first year, **(c)** the accumulated depreciation at the end of 7 years, and **(d)** the book value at the end of 9 years. Round to the nearest dollar.

Learning Unit 17–4: Modified Accelerated Cost Recovery System (MACRS) with Introduction to ACRS

DRILL PROBLEMS

1. Using the MACRS method of depreciation, find the recovery rate, first-year depreciation expense, and book value of the asset at the end of the first year. Round to the nearest whole dollar.

	Cost of asset	Recovery period	Recovery rate	Depreciation expense	End-of-year book value
a.	$2,500	3 years	_____	_______	_______
b.	$52,980	3 years	_____	_______	_______
c.	$4,250	5 years	_____	_______	_______
d.	$128,950	10 years	_____	_______	_______
e.	$13,775	5 years	_____	_______	_______

2. Find the accumulated depreciation at the end of the second year for each of the following assets. Round to the nearest whole dollar.

	Cost of asset	Recovery period	Accumulated depreciation at end of 2nd year using MACRS	Book value at end of 2nd year using MACRS
a.	$2,500	3 years	_______	_______
b.	$52,980	3 years	_______	_______
c.	$4,250	5 years	_______	_______
d.	$128,950	10 years	_______	_______
e.	$13,775	5 years	_______	_______

WORD PROBLEMS

3. Colby Wayne paid $7,750 for a photocopy machine that is classified as equipment and has a residual value of $900. Prepare a depreciation schedule using the MACRS depreciation method. Round all calculations to the nearest whole dollar.

4. Fox Realty bought a computer table for $1,700. The table is classified as furniture. The residual value at the end of the table's useful life is $370. Using the MACRS depreciation method, find **(a)** the amount of depreciation at the end of the *third* year, **(b)** the total accumulated depreciation at the end of year 3, and **(c)** the book value of the table at the end of the *third* year. Round all calculations to the nearest dollar.

5. Cashman, Inc., purchased an overhead projector for $560. It is classified as office equipment and will have no residual value. Find the book value at the end of 5 years using the MACRS depreciation method. Round to the nearest whole dollar.

6. Shelley Corporation purchased a new machine for $15,000. The machine is comparable to equipment used for two-way exchange of voice and data with a residual value of $2,400. Find **(a)** the amount of depreciation at the end of the first year, **(b)** the accumulated depreciation at the end of 7 years, and **(c)** the book value at the end of 9 years. Round to the nearest dollar.

7.* Wolfe Ltd. purchased a supercomputer for $75,000 at the beginning of 1996. The computer is classified as a 5-year asset and will have a residual value of $15,000. Using MACRS, find **(a)** the depreciation amount for 1996, **(b)** the accumulated depreciation at the end of 1997, **(c)** the book value at the end of 1998, and **(d)** the last year that the asset will be depreciated.

*These problems are placed here for a quick review.

8.* Cummins Engine Company uses a straight-line depreciation method to calculate the cost of an asset of $1,200,000 with a $200,000 residual value and a life expectancy of 15 years. How much would Cummins have for depreciation expense for each of the first 2 years? Round to the nearest dollar for each year.

9.* An article in an issue of *Management Accounting* stated that Cummins Engine Company changed its depreciation. The cost of its asset was $1,200,000 with a $200,000 residual value (with a life expectancy of 15 years) and an estimated productive capacity of 864,000 products. Cummins produced 59,000 products this year. What would it write off for depreciation using the units-of-production method?

*These problems are placed here for a quick review.

Learning Unit 18–1: Assigning Costs to Ending Inventory—Specific Identification; Weighted Average; FIFO; LIFO

DRILL PROBLEMS

1. Given the value of the beginning inventory, purchases for the year, and ending inventory, find the cost of goods available for sale and the cost of goods sold.

	Beginning inventory	Purchases	Ending inventory	Cost of goods available for sale	Cost of goods sold
a.	$1,000	$4,120	$2,100	________	________
b.	$52,400	$270,846	$49,700	________	________
c.	$205	$48,445	$376	________	________
d.	$78,470	$2,788,560	$100,600	________	________
e.	$965	$53,799	$2,876	________	________

2. Find the missing amounts; then calculate the number of units available for sale and the cost of the goods available for sale.

Date	Category	Quantity	Unit cost	Total cost
January 1	Beginning inventory	1,207	$45	________
February 7	Purchase	850	$46	________
April 19	Purchase	700	$47	________
July 5	Purchase	1,050	$49	________
November 2	Purchase	450	$52	________
Goods available for sale		____		________

3. Using the *specific identification* method, find the ending inventory and cost of goods sold for the merchandising concern in Problem 2.

Remaining inventory	Unit cost	Total cost
20 units from beginning inventory	____	________
35 units from February 7	____	________
257 units from July 5	____	________
400 units from November 2	____	________
Cost of ending inventory		________
Cost of goods sold		________

4. Using the *weighted-average* method, find the average cost per unit (to the nearest cent) and the cost of ending inventory.

	Units available for sale	Cost of goods available for sale	Units in ending inventory	Weighted-average unit cost	Cost of ending inventory
a.	2,350	$120,320	1,265	______	________
b.	7,090	$151,017	1,876	______	________
c.	855	$12,790	989	______	________
d.	12,964	$125,970	9,542	______	________
e.	235,780	$507,398	239,013	______	________

5. Use the *FIFO* method of inventory valuation to determine the value of ending inventory, which consists of 40 units, and the cost of goods sold.

Date	Category	Quantity	Unit cost	Total cost
January 1	Beginning inventory	37	$219.00	________
March 5	Purchases	18	230.60	________
June 17	Purchases	22	255.70	________
October 18	Purchases	34	264.00	________
Goods available for sale		___		________

Ending inventory = ________ Cost of goods sold = ________

6. Use the *LIFO* method of inventory valuation to determine the value of the ending inventory, which consists of 40 units, and the cost of goods sold.

Date	Category	Quantity	Unit cost	Total cost
January 1	Beginning inventory	37	$219.00	________
March 5	Purchases	18	230.60	________
June 17	Purchases	22	255.70	________
October 18	Purchases	34	264.00	________
Goods available for sale		___		________

Ending inventory = ________ Cost of goods sold = ________

WORD PROBLEMS

7. At the beginning of September, Green's of Gloucester had 13 yellow raincoats in stock. These raincoats cost $36.80 each. During the month, Green's purchased 14 raincoats for $37.50 each and 16 raincoats for $38.40 each, and it sold 26 raincoats. Calculate **(a)** the average unit cost rounded to the nearest cent and **(b)** the ending inventory value using the weighted-average method.

8. If Green's of Gloucester (Problem 7) used the FIFO method, what would the value of the ending inventory be?

9. If Green's of Gloucester (Problem 7) used the LIFO method, what would the value of the ending inventory be?

10. Hobby Caterers purchased recycled-paper sketch pads during the year as follows:

January	350 pads for $.27 each
March	400 pads for $.31 each
July	200 pads for $.36 each
October	850 pads for $.26 each
November	400 pads for $.31 each

At the end of the year, the company had 775 of these sketch pads in stock. Find the ending inventory value using **(a)** the weighted-average method (round to the nearest cent), **(b)** the FIFO method, and **(c)** the LIFO method.

11. On March 1, Sandler's Shoe Store had the following sports shoes in stock:

13 pairs running shoes for $33 a pair
22 pairs walking shoes for $29 a pair
35 pairs aerobic shoes for $26 a pair
21 pairs cross-trainers for $52 a pair

During the month Sandler's sold 10 pairs of running shoes, 15 pairs of walking shoes, 28 pairs of aerobic shoes, and 12 pairs of cross-trainers. Use the specific identification method to find **(a)** the cost of the goods available for sale, **(b)** the value of the ending inventory, and **(c)** the cost of goods sold.

Learning Unit 18–2: Retail Method; Gross Profit Method; Inventory Turnover; Distribution of Overhead

DRILL PROBLEMS

1. Given the following information, calculate **(a)** the goods available for sale at cost and retail, **(b)** the cost ratio (to the nearest thousandth), **(c)** the ending inventory at retail, and **(d)** the cost of the March 31 inventory (to the nearest dollar) by the retail inventory method.

	Cost	Retail
Beginning inventory, March 1	$57,300	$95,500
Purchases during March	$28,400	$48,000
Sales during March		$79,000

2. Given the following information, use the gross profit method to calculate **(a)** the cost of goods available for sale, **(b)** the cost percentage, **(c)** the estimated cost of goods sold, and **(d)** the estimated cost of the inventory as of April 30.

Beginning inventory, April 1	$30,000
Net purchases during April	81,800
Sales during April	98,000
Average gross profit on sales	40%

3. Given the following information, find the average inventory.

Merchandise inventory, January 1, 20XX	$82,000
Merchandise inventory, December 31, 20XX	$88,000

4. Given the following information, find the inventory turnover for the company in Problem 3 to the nearest hundredth.

Cost of goods sold (12/31/XX)	$625,000

5. Given the following information, calculate the **(a)** average inventory at retail, **(b)** average inventory at cost, **(c)** inventory turnover at retail, and **(d)** inventory turnover at cost. Round to the nearest hundredth.

	Cost	Retail
Merchandise inventory, January 1	$ 250,000	$ 355,000
Merchandise inventory, December 31	$ 235,000	$ 329,000
Cost of goods sold	$1,525,000	
Sales		$2,001,000

6. Given the floor space for the following departments, find the entire floor space and the percent each department represents.

		Percent of floor space
Department A	15,000 square feet	_____
Department B	25,000 square feet	_____
Department C	10,000 square feet	_____
Total floor space	50,000 square feet	_____

7. If the total overhead for all the departments (Problem 6) is $200,000, how much of the overhead expense should be allocated to each department?

	Overhead/department
Department A	________
Department B	________
Department C	________

WORD PROBLEMS

8. During the accounting period, Ward's Greenery sold $290,000 of merchandise at marked retail prices. At the end of the period, the following information was available from Ward's records:

	Cost	**Retail**
Beginning inventory	$ 53,000	$ 79,000
Net purchases	$204,000	$280,000

Use the retail method to estimate Ward's ending inventory at cost. Round the cost ratio to the nearest thousandth.

9. On January 1, Benny's Retail Mart had a $49,000 inventory at cost. During the first quarter of the year, Benny's made net purchases of $199,900. Benny's records show that during the past several years, the store's gross profit on sales has averaged 35%. If Benny's records show $275,000 in sales for the quarter, estimate the ending inventory for the first quarter, using the gross profit method.

10. On April 4, there was a big fire and the entire inventory of R. W. Wilson Company was destroyed. The company records were salvaged. They showed the following information:

Sales (January 1 through April 4)	$127,000
Merchandise inventory, January 1	16,000
Net purchases	71,250

On January 1, the inventory was priced to sell for $38,000 and additional items bought during the period were priced to sell for $102,000. Using the retail method, calculate the cost of the inventory that was destroyed by the fire. Round the cost ratio to the nearest thousandth.

11. During the past 4 years, the average gross margin on sales for R. W. Wilson Company was 36% of net sales. Using the data in Problem 10 and the gross profit method, calculate the cost of the ending inventory destroyed by fire.

12. Chase Bank has to make a decision on whether to grant a loan to Sally's Furniture Store. The lending officer is interested in how often Sally's inventory turns over. Using selected information from Sally's income statement, calculate the inventory turnover for Sally's Furniture Store (to the nearest hundredth).

Merchandise inventory, January 1	$ 43,000
Merchandise inventory, December 31	55,000
Cost of goods sold	128,000

13. Wanting to know more about a business he was considering buying, Jake Paige studied the business's books. He found that beginning inventory for the previous year was $51,000 at cost and $91,800 at retail, ending inventory was $44,000 at cost and $72,600 at retail, sales were $251,000, and cost of goods sold was $154,000. Using this information, calculate for Jake the inventory turnover at cost and the inventory turnover at retail.

14. Ralph's Retail Outlet has calculated its expenses for the year. Total overhead expenses are $147,000. Ralph's accountant must allocate this overhead to four different departments. Given the following information regarding the floor space occupied by each department, calculate how much overhead expense should be allocated to each department.

Department W	12,000 square feet	Department Y	14,000 square feet
Department X	9,000 square feet	Department Z	7,000 square feet

15. How much overhead would be allocated to each department of Ralph's Retail Outlet (Problem 14) if the basis of allocation were the sales of each department? Sales for each of the departments were:

Department W	$110,000	Department Y	$170,000
Department X	$120,000	Department Z	$100,000

Learning Unit 19–1: Sales and Excise Taxes

DRILL PROBLEMS

1. Calculate the sales tax and the total amount due for each of the following:

	Total sales	Sales tax rate	Sales tax	Total amount due
a.	$536	5%	______	______
b.	$11,980	6%	______	______
c.	$3,090	$8\frac{1}{4}\%$	______	______
d.	$17.65	$5\frac{1}{2}\%$	______	______
e.	$294	7.42%	______	______

2. Find the amount of actual sales and amount of sales tax on the following total receipts:

	Total receipts	Sales tax rate	Actual sales	Sales tax
a.	$27,932.15	5.5%	______	______
b.	$35,911.53	7%	______	______
c.	$115,677.06	$6\frac{1}{2}\%$	______	______
d.	$142.96	$5\frac{1}{4}\%$	______	______
e.	$5,799.24	4.75%	______	______

3. Find the sales tax, excise tax, and total cost for each of the following items:

	Retail price	Sales tax, 5.2%	Excise tax, 11%	Total cost
a.	$399	______	______	______
b.	$22,684	______	______	______
c.	$7,703	______	______	______

4. Calculate the amount, subtotal, sales tax, and total amount due of the following:

Quantity	Description	Unit price	Amount
3	Taxable item	$4.30	______
2	Taxable item	$5.23	______
4	Taxable item	$1.20	______
		Subtotal	______
		5% sales tax	______
		Total	______

5. Given the sales tax rate and the amount of the sales tax, calculate the price of the following purchases (before tax was added):

	Tax rate	Tax amount	Price of purchase
a.	7%	$71.61	______
b.	$5\frac{1}{2}\%$	$3.22	______

6. Given the sales tax rate and the total price (including tax), calculate the price of the following purchases (before the tax was added):

	Tax rate	Total price	Price of purchase
a.	5%	$340.20	______
b.	6%	$1,224.30	______

WORD PROBLEMS

7. In a state with a 4.75% sales tax, what will be the sales tax and the total price of a video game marked $110?

8. Browning's invoice included a sales tax of $38.15. If the sales tax rate is 6%, what was the total cost of the taxable goods on the invoice?

9. David Bowan paid a total of $2,763 for a new computer. If this includes a sales tax of 5.3%, what was the marked price of the computer?

10. After a 5% sales tax and a 12% excise tax, the total cost of a leather jacket was $972. What was the selling price of the jacket?

11. A customer at the RDM Discount Store purchased four tubes of toothpaste priced at $1.88 each, six toothbrushes for $1.69 each, and three bottles of shampoo for $2.39 each. What did the customer have to pay if the sales tax is $5\frac{1}{2}$%?

12. Bill Harrington purchased a mountain bike for $875. Bill had to pay a sales tax of 6% and an excise tax of 11%. What was the total amount Bill had to pay for his mountain bike?

13. Donna DeCoff received a bill for $754 for a new chair she had purchased. The bill included a 6.2% sales tax and a delivery charge of $26. What was the selling price of the chair?

Learning Unit 19–2: Property Tax

DRILL PROBLEMS

1. Find the assessed value of the following properties, rounding to the nearest whole dollar:

	Market value	Assessment rate	Assessed value		Market value	Assessment rate	Assessed value
a.	$195,000	35%	________	d.	$2,585,400	65%	________
b.	$1,550,900	50%	________	e.	$349,500	85%	________
c.	$75,000	75%	________				

2. Find the tax rate for each of the following municipalities, rounding to the nearest tenth of a percent:

	Budget needed	Total assessed value	Tax rate		Budget needed	Total assessed value	Tax rate
a.	$2,594,000	$44,392,000	______	d.	$13,540,000	$143,555,500	______
b.	$17,989,000	$221,900,000	______	e.	$1,099,000	$12,687,000	______
c.	$6,750,000	$47,635,000	______				

3. Express each of the following tax rates in all the indicated forms:

	By percent	Per $100 of assessed value	Per $1,000 of assessed value	In mills
a.	7.45%	______	______	______
b.	______	$14.24	______	______
c.	______	______	______	90.8
d.	______	______	$62.00	______

4. Calculate the property tax due for each of the following:

	Total assessed value	Tax rate	Total property tax due		Total assessed value	Tax rate	Total property tax due
a.	$12,900	$6.60 per $100	________	e.	$78,900	59 mills	________
b.	$175,400	43 mills	________	f.	$225,550	$11.39 per $1,000	________
c.	$320,500	2.7%	________	g.	$198,750	$2.63 per $100	________
d.	$2,480,000	$17.85 per $1,000	________				

WORD PROBLEMS

5. The county of Chelsea approved a budget of $3,450,000, which had to be raised through property taxation. If the total assessed value of properties in the county of Chelsea was $37,923,854, what will the tax rate be? The tax rate is stated per $100 of assessed valuation.

6. Linda Tawse lives in Camden and her home has a market value of $235,000. Property in Camden is assessed at 55% of its market value, and the tax rate for the current year is $64.75 per $1,000. What is the assessed valuation of Linda's home?

7. Using the information in Problem 6, find the amount of property tax that Linda will have to pay.

8. Mary Faye Souza has property with a fair market value of $219,500. Property in Mary Faye's city is assessed at 65% of its market value and the tax rate is $3.64 per $100. How much is Mary Faye's property tax due?

9. Cagney's Greenhouse has a fair market value of $1,880,000. Property is assessed at 35% by the city. The tax rate is 6.4%. What is the property tax due for Cagney's Greenhouse?

10. In Chester County, property is assessed at 40% of its market value, the residential tax rate is $12.30 per $1,000, and the commercial tax rate is $13.85 per $1,000. What is the property tax due on a home that has a market value of $205,000?

11. Using the information in Problem 10, find the property tax due on a grocery store with a market value of $5,875,000.

12. Bob Rose's home is assessed at $195,900. Last year the tax rate was 11.8 mills, and this year the rate was raised to 13.2 mills. How much more will Bob have to pay in taxes this year?

Learning Unit 20–1: Life Insurance

DRILL PROBLEMS

1. Use the table in the *Business Math Handbook* to find the annual premium per $1,000 of life insurance and calculate the annual premiums for each policy listed. Assume the insureds are males.

	Face value of policy	Type of insurance	Age at issue	Annual premium per $1,000	Number of $1,000s in face value	Annual premium
a.	$25,000	Straight life	31	______	______	______
b.	$40,500	20-year endowment	40	______	______	______
c.	$200,000	Straight life	44	______	______	______
d.	$62,500	20-payment life	25	______	______	______
e.	$12,250	5-year term	35	______	______	______
f.	$42,500	20-year endowment	42	______	______	______

2. Use Table 20.1 to find the annual premium for each of the following life insurance policies. Assume the insured is a 30-year-old male.

	Face value of policy	Five-year term policy	Straight life policy	Twenty-payment life policy	Twenty-year endowment
a.	$50,000	______	______	______	______
b.	$1,000,000	______	______	______	______
c.	$250,000	______	______	______	______
d.	$72,500	______	______	______	______

3. Use the table in the *Business Math Handbook* to find the annual premium for each of the following life insurance policies. Assume the insured is a 30-year-old female.

	Face value of policy	Five-year term policy	Straight life policy	Twenty-payment life policy	Twenty-year endowment
a.	$50,000	______	______	______	______
b.	$1,000,000	______	______	______	______
c.	$250,000	______	______	______	______
d.	$72,500	______	______	______	______

4. Use the table in the *Business Math Handbook* to find the nonforfeiture options for the following policies:

	Years policy in force	Type of policy	Face value	Cash value	Amount of paid-up insurance	Extended term
a.	10	Straight life	$25,000	_______	_______	_______
b.	20	20-year endowment	$500,000	_______	_______	_______
c.	5	20-payment life	$2,000,000	_______	_______	_______
d.	15	Straight life	$750,000	_______	_______	_______
e.	5	20-year endowment	$93,500	_______	_______	_______

WORD PROBLEMS

5. If Mr. Davis, aged 39, buys a $90,000 straight-life policy, what is the amount of his annual premium?

6. If Miss Jennie McDonald, age 27, takes out a $65,000 20-year endowment policy, what premium amount will she pay each year?

7. If Gary Thomas decides to cash in his $45,000 20-payment life insurance policy after 15 years, what cash surrender value will he receive?

8. Mary Allyn purchased a $70,000 20-year endowment policy when she was 26 years old. Ten years later, she decided that she could no longer afford the premiums. If Mary decides to convert her policy to paid-up insurance, what amount of paid-up insurance coverage will she have?

9. Peter and Jane Rizzo are both 28 years old and are both planning to take out $50,000 straight-life insurance policies. What is the difference in the annual premiums they will have to pay?

10. Paul Nasser purchased a $125,000 straight-life policy when he was 30 years old. He is now 50 years old. Two months ago, he slipped in the bathtub and injured his back; he will not be able to return to his regular job for several months. Due to a lack of income, he feels that he can no longer continue to pay the premiums on his life insurance policy. If Paul decides to surrender his policy for cash, how much cash will he receive?

11. If Paul Nasser (Problem 10) chooses to convert his policy to paid-up insurance, what will the face value of his new policy be?

Learning Unit 20–2: Fire Insurance

DRILL PROBLEMS

1. Use the tables in the *Business Math Handbook* to find the premium for each of the following:

	Rating of area	Building class	Building value	Value of contents	Total annual premium
a.	3	A	$80,000	$32,000	_______
b.	2	B	$340,000	$202,000	_______
c.	2	A	$221,700	$190,000	_______
d.	1	B	$96,400	$23,400	_______
e.	3	B	$65,780	$62,000	_______

2. Use the tables in the *Business Math Handbook* to find the short-term premium and the amount of refund due if the insured cancels.

	Annual premium	Months of coverage	Short-term premium	Refund due
a.	$1,860	3	_______	_______
b.	$650	7	_______	_______
c.	$1,200	10	_______	_______
d.	$341	12	_______	_______
e.	$1,051	4	_______	_______

3. Find the amount to be paid for each of the following losses:

	Property value	Coinsurance clause	Insurance required	Insurance carried	Amount of loss	Insurance company pays (indemnity)
a.	$85,000	80%	________	$70,000	$60,000	________
b.	$52,000	80%	________	$45,000	$50,000	________
c.	$44,000	80%	________	$33,000	$33,000	________
d.	$182,000	80%	________	$127,400	$61,000	________

WORD PROBLEMS

4. Mary Rose wants to purchase fire insurance for her building, which is rated as Class B; the rating of the area is 2. If her building is worth $225,000 and the contents are worth $70,000, what will her annual premium be?

5. Janet Ambrose owns a Class A building valued at $180,000. The contents of the building are valued at $145,000. The territory rating is 3. What is her annual fire insurance premium?

6. Jack Altshuler owns a building worth $355,500. The contents are worth $120,000. The classification of the building is B, and the rating of the area is 1. What annual premium must Jack pay for his fire insurance?

7. Jay Viola owns a store valued at $460,000. His fire insurance policy (which has an 80% coinsurance clause) has a face value of $345,000. A recent fire resulted in a loss of $125,000. How much will the insurance company pay?

8. The building that is owned by Tally's Garage is valued at $275,000 and is insured for $225,000. The policy has an 80% coinsurance clause. If there is a fire in the building and the damages amount to $220,000, how much of the loss will be paid for by the insurance company?

9. Michael Dannon owns a building worth $420,000. He has a fire insurance policy with a face value of $336,000 (there is an 80% coinsurance clause). There was recently a fire that resulted in a $400,000 loss. How much money will he receive from the insurance company?

10. Rice's Rent-A-Center business is worth $375,000. He has purchased a $250,000 fire insurance policy. The policy has an 80% coinsurance clause. What will Rice's reimbursement be **(a)** after a $150,000 fire and **(b)** after a $330,000 fire?

11. If Maria's Pizza Shop is valued at $210,000 and is insured for $147,000 with a policy that contains an 80% coinsurance clause, what settlement is due after a fire that causes **(a)** $150,000 in damages and **(b)** $175,000 in damages?

Learning Unit 20–3: Auto Insurance

DRILL PROBLEMS

1. Calculate the annual premium for compulsory coverage for each of the following.

	Driver classification	Bodily	Property	Total premium
a.	17	______	______	______
b.	20	______	______	______
c.	10	______	______	______

2. Calculate the amount of money the insurance company and the driver should pay for each of the following accidents, assuming the driver carries compulsory insurance only.

	Accident and court award	Insurance company pays	Driver pays
a.	Driver hit one person and court awarded $15,000.	________	________
b.	Driver hit one person and court awarded $12,000 for personal injury.	________	________
c.	Driver hit two people; court awarded first person $9,000 and the second person $12,000.	________	________

3. Calculate the additional premium payment for each of the following options.

Optional insurance coverage	Addition to premium
a. Bodily injury 50/100/25, driver class 20	______
b. Bodily injury 25/60/10, driver class 17	______
c. Collision insurance, driver class 10, age group 3, symbol 5, deductible $100	______
d. Comprehensive insurance, driver class 10, age group 3, symbol 5, deductible $200	______
e. Substitute transportation, towing, and labor; driver class 10, age group 3, symbol 5	______

4. Compute the annual premium for compulsory insurance with optional liability coverage for bodily injury and damage to someone else's property.

Driver classification	Bodily coverage	Premium
a. 17	50/100/25	______
b. 20	100/300/10	______
c. 10	25/60/25	______
d. 18	250/500/50	______
e. 20	25/50/10	______

5. Calculate the annual premium for each of the following drivers with the indicated options. All drivers must carry compulsory insurance.

Driver classification	Car age	Car symbol	Bodily injury	Collision	Comprehensive	Transportation and towing	Annual premium
a. 10	2	4	50/100/10	$100 deductible	$300 deductible	Yes	______
b. 18	3	2	25/60/25	$200 deductible	$200 deductible	Yes	______

WORD PROBLEMS

6. Ann Centerino's driver classification is 10. She carries only compulsory insurance coverage. What annual insurance premium must she pay?

7. Gary Hines is a class 18 driver. He wants to add optional bodily injury and property damage of 250/500/50 to his compulsory insurance coverage. What will be Gary's total annual premium?

8. Sara Goldberg wants optional bodily injury coverage of 50/100/25 and collision coverage with a deductible of $300 in addition to the compulsory coverage her state requires. Sara is a class 17 driver and has a symbol 4 car that is 2 years old. What annual premium must Sara pay?

9. Karen Babson has just purchased a new car with a symbol of 8. She wants bodily injury and property liability of 500/1,000/100, comprehensive and collision insurance with a $200 deductible, and transportation and towing coverage. If Karen is a class 10 driver, what will be her annual insurance premium? There is no compulsory insurance requirement in her state. Assume age group 1.

10. Craig Haberland is a class 18 driver. He has a 5-year-old car with a symbol of 4. His state requires compulsory insurance coverage. In addition, he wishes to purchase collision and comprehensive coverage with the maximum deductible. He also wants towing insurance. What will Craig's annual insurance premium be?

11. Nancy Poland has an insurance policy with limits of 10/20. If Nancy injures a pedestrian and the judge awards damages of $18,000, **(a)** how much will the insurance company pay and **(b)** how much will Nancy pay?

12. Peter Bell carries insurance with bodily injury limits of 25/60. Peter is in an accident and is charged with injuring four people. The judge awards damages of $10,000 to each of the injured parties. How much will the insurance company pay? How much will Peter pay?

13. Jerry Greeley carries an insurance policy with bodily injury limits of 25/60. Jerry is in an accident and is charged with injuring four people. If the judge awards damages of $20,000 to each of the injured parties, **(a)** how much will the insurance company pay and **(b)** how much will Jerry pay?

14. An issue of *Your Money* reported that the Illinois Department of Insurance gave a typical premium for a brick house in Chicago built in 1950, assuming no policy discounts and a replacement cost estimated at $100,000. With a $100 deductible, the annual premium will be $653. Using the rate in your textbook, with a rating area 3 and class B, what would be the annual premium? (This problem reviews fire insurance.)

15. An issue of *Money* ran a story on cutting car insurance premiums. Raising the car insurance deductible to $500 will cut the collision premium 15%. Theresa Mendex insures her car; her age group is 5 and symbol is 5. What would be her reduction if she changed her policy to a $500 deductible? What would the collision insurance now cost?

16. Robert Stuono lost his life insurance when he was downsized from an investment banking company early this year. Stuono, age 44, enlisted the help of an independent agent who works with several insurance companies. His goal is $350,000 in term coverage with a level premium for 5 years. What will Robert's annual premium be for term insurance? (This problem reviews life insurance.)

Learning Unit 21–1: Stocks

DRILL PROBLEMS

52 weeks Hi	Lo	Stocks	SYM	Div	Yld %	PE	Vol 100s	High	Low	Close	Net chg
43.88	25.51	Disney	DIS	.21	.8	49	49633	27.69	26.50	27.69	+0.63

1. From the listed information for Disney, complete the following:

a. __________ was the highest price at which Disney stock traded during the year.
b. __________ was the lowest price at which Disney stock traded during the year.
c. __________ was the amount of the dividend Disney paid to shareholders last year.
d. __________ is the dividend amount a shareholder with 100 shares would receive.
e. __________ is the rate of return the stock yielded to its stockholders.
f. __________ is how many times its earnings per share the stock is selling for.
g. __________ is the number of shares traded on the day of this stock quote.
h. __________ is the highest price paid for Disney stock on this day.
i. __________ is the lowest price paid for Disney stock on this day.
j. is the change in price from yesterday's closing price.

2. Use the Disney information to show how the yield percent was calculated.

3. What was the price of the last trade of Disney stock yesterday?

WORD PROBLEMS

4. Assume a stockbroker's commission of 2%. What will it cost to purchase 200 shares of Saplent Corporation at $10.75?

5. In Problem 4, the stockbroker's commission for selling stock is the same as that for buying stock. If the customer who purchased 200 shares at $10.75 sells the 200 shares of stock at the end of the year at $18.12, what will be the gain on investment?

6. Holtz Corporation's records show 80,000 shares of preferred stock issued. The preferred dividend is $2 per share, which is cumulative. The records show 750,000 shares of common stock issued. In 2018, no dividends were paid. In 2019, the board of directors declared a dividend of $582,500. What are **(a)** the total amount of dividends paid to preferred stockholders, **(b)** the total amount of dividends paid to common stockholders, and **(c)** the amount of the common dividend per share?

7. Melissa Tucker bought 300 shares of Delta Air Lines stock listed at $61.22 per share. What is the total amount she paid if the stockbroker's commission is 2.5%?

8. A year later, Melissa (Problem 7) sold the stock she had purchased. The market price of the stock at this time was $72.43. Delta Air Lines had paid its shareholders a dividend of $1.20 per share. If the stockbroker's commission to sell stock is 2.5%, what gain did Melissa realize?

9. The board of directors of Parker Electronics, Inc., declared a $539,000 dividend. If the corporation has 70,000 shares of common stock outstanding, what is the dividend per share?

Learning Unit 21–2: Bonds

DRILL PROBLEMS

Complete problems 1-5 using the following information:

Bond	Current yield	Sales	Close	Net change
IBM $10\frac{1}{4}$ 25	10.0	11	102.5	+.125

1. From the bond listing above complete the following:

a. ________ is the name of the company.

b. ________ is the percent of interest paid on the bond.

c. ________ is the year in which the bond matures.

d. ________ is the total interest for the year.

e. ________ was the previous day's closing price on the IBM bond.

2. Show how to calculate the current yield of 10.0% for IBM. (Trade commissions have been omitted.)

3. Use the information for the IBM bond to calculate **(a)** the amount the last bond traded for on this day and **(b)** the amount the last bond traded for yesterday.

4. What will be the annual interest payment **(a)** to the bondholder assuming he paid $101\frac{3}{4}$ and **(b)** to the bondholder who purchased the bond for $102\frac{1}{2}$?

5. If Terry Gambol purchased three IBM bonds at this day's closing price, **(a)** what will be her total cost excluding commission and **(b)** how much interest will she receive for the year?

6. Calculate the bond yield (to the nearest tenth percent) for each of the following:

Bond interest rate	Purchase price	Bond yield
a. 7%	97	______
b. $9\frac{1}{2}$%	101.625	______
c. $13\frac{1}{4}$%	104.25	______

7. For each of the following, state whether the bond sold at a premium or a discount and give the amount of the premium or discount.

	Bond interest rate	Purchase price	Premium or discount
a.	7%	97	________
b.	$9\frac{1}{2}$%	101.625	________
c.	$13\frac{1}{4}$%	104.25	________

WORD PROBLEMS

8. Rob Morrisey purchased a $1,000 bond that was quoted at 102.25 and paying $8\frac{7}{8}$% interest. **(a)** How much did Rob pay for the bond? **(b)** What was the premium or discount? **(c)** How much annual interest will he receive?

9. Jackie Anderson purchased a bond that was quoted at 62.50 and paying interest of $10\frac{1}{2}$%. **(a)** How much did Jackie pay for the bond? **(b)** What was the premium or discount? **(c)** What interest will Jackie receive annually? **(d)** What is the bond's current annual yield (to the nearest tenth percent)?

10. Swartz Company issued bonds totaling $2,000,000 in order to purchase updated equipment. If the bonds pay interest of 11%, what is the total amount of interest the Swartz Company must pay semiannually?

11. The RJR and ACyan companies have both issued bonds that are paying $7\frac{3}{8}$% interest. The quoted price of the RJR bond is 94.125, and the quoted price of the ACyan bond is $102\frac{7}{8}$. Find the current annual yield on each (to the nearest tenth percent).

12. Mary Rowe purchased 25 of Chrysler Corporation $8\frac{3}{8}$% bonds of 2035. The bonds closed at 93.25. Find **(a)** the total purchase price and **(b)** the amount of the first semiannual interest payment Mary will receive.

13. What is the annual yield (to the nearest hundredth percent) of the bonds Mary Rowe purchased (see Problem 12)?

14. Mary Rowe purchased a $1,000 bond listed as ARch $10\frac{7}{8}$ 28 for 122.75. What is the annual yield of this bond (to the nearest tenth percent)?

Learning Unit 21–3: Mutual Funds

DRILL PROBLEMS

From the following, calculate the NAV. Round to the nearest cent.

	Current market value of fund investments	Current liabilities	Number of shares outstanding	NAV
1.	$6,800,000	$850,000	500,000	______
2.	$11,425,000	$690,000	810,000	______
3.	$22,580,000	$1,300,000	1,400,000	______

Complete the following using this information:

NAV	Net change	Fund name	Inv. obj.	YTD %Ret	Total return 1 Yr R
$23.48	+.14	EuroA	Eu	+37.3	+7.6 E

4. NAV ________
5. NAV change ________
6. Total return year to date ________
7. Return for the last 12 months ________
8. What does an E rating mean? ________

Calculate the commission (load) charge and the offer to buy.

	NAV	% commission (load) charge	Dollar amount of commission (load) charge	Offer price
9.	\$17.00	$8\frac{1}{2}\%$	________	________
10.	\$21.55	6%	________	________
11.	\$14.10	4%	________	________

WORD PROBLEMS

12. Paul wanted to know how his Fidelity mutual fund \$14.33 NAV in the newspaper was calculated. He called Fidelity, and he received the following information:

Current market value of fund investment	\$7,500,000
Current liabilities	\$910,000
Number of shares outstanding	460,000

Please calculate the NAV for Paul. Was the NAV in the newspaper correct?

13. Jeff Jones bought 150 shares of Putnam Vista Fund. The NAV of the fund was \$9.88. The offer price was \$10.49. What did Jeff pay for these 150 shares?

14. Pam Long purchased 300 shares of the no-load Scudder's European Growth Company Fund. The NAV is \$12.61. What did Pam pay for the 300 shares?

15. Assume in Problem 14 that 8 years later Pam sells her 300 shares. The NAV at the time of sale was \$12.20. What is the amount of her profit or loss on the sale?

16. Financial planner J. Michael Martin recommended that Jim Kelly choose a long-term bond because it gives high income while Kelly waits for better stock market opportunities down the road. The bond Martin recommended matures in 2030 and was originally issued at $8\frac{1}{2}\%$ interest and the current yield is 7.9%. What would be the current selling price for this bond and how would that price appear in the bond quotations?

17.

Bonds	Vol.	Close	Net chg.
Comp USA $9\frac{1}{2}$ 25	70	102.375	−.125
GMA 7 27	5	101.625	−1.25

From the above information, compare the two bonds for:

a. When the bonds expire.
b. The yield of each bond.
c. The current selling price.
d. Whether the bond is selling at a discount or premium.
e. Yesterday's bond close.

Learning Unit 22–1: Mean, Median, and Mode

Note: Optional problems for LU 22–3 are found after the Challenge Problems in Chapter 22.

DRILL PROBLEMS

1. Find the mean for the following lists of numbers. Round to the nearest hundredth.

a. 12, 16, 20, 25, 29 Mean _____

b. 80, 91, 98, 82, 68, 82, 79, 90 Mean _____

c. 9.5, 12.3, 10.5, 7.5, 10.1, 18.4, 9.8, 6.2, 11.1, 4.8, 10.6 Mean _____

2. Find the weighted mean for the following. Round to the nearest hundredth.

a. 4, 4, 6, 8, 8, 13, 4, 6, 8 Weighted mean _____

b. 82, 85, 87, 82, 82, 90, 87, 63, 100, 85, 87 Weighted mean _____

3. Find the median for the following:

a. 56, 89, 47, 36, 90, 63, 55, 82, 46, 81 Median _____

b. 59, 22, 39, 47, 33, 98, 50, 73, 54, 46, 99 Median _____

4. Find the mode for the following:

24, 35, 49, 35, 52, 35, 52 Mode _____

5. Find the mean, median, and mode for each of the following:

a. 72, 48, 62, 54, 73, 62, 75, 57, 62, 58, 78
Mean _______ Median _______ Mode _______

b. \$0.50, \$1.19, \$0.58, \$1.19, \$2.83, \$1.71, \$2.21, \$0.58, \$1.29, \$0.58
Mean _______ Median _______ Mode _______

c. \$92, \$113, \$99, \$117, \$99, \$105, \$119, \$112, \$95, \$116, \$102, \$120
Mean _______ Median _______ Mode _______

d. 88, 105, 120, 119, 105, 128, 160, 151, 90, 153, 107, 119, 105
Mean _______ Median _______ Mode _______

WORD PROBLEMS

6. The sales for the year at the 8 Bed and Linen Stores were \$1,442,897, \$1,556,793, \$1,703,767, \$1,093,320, \$1,443,984, \$1,665,308, \$1,197,692, and \$1,880,443. Find the mean earnings for a Bed and Linen Store for the year.

7. To avoid having an extreme number affect the average, the manager of Bed and Linen Stores (Problem 6) would like you to find the median earnings for the 8 stores.

8. The Bed and Linen Store in Salem sells many different towels. Following are the prices of all the towels that were sold on Wednesday: \$7.98, \$9.98, \$9.98, \$11.49, \$11.98, \$7.98, \$12.49, \$12.49, \$11.49, \$9.98, \$9.98, \$16.00, and \$7.98. Find the mean price of a towel.

9. Looking at the towel prices, the Salem manager (Problem 8) decided that he should have calculated a weighted mean. Find the weighted mean price of a towel.

10. The manager of the Salem Bed and Linen Store (Problem 8) would like to find another measure of the central tendency called the *median.* Find the median price for the towels sold.

11. The manager at the Salem Bed and Linen Store (Problem 8) would like to know the most popular towel among the group of towels sold on Wednesday. Find the mode for the towel prices for Wednesday.

Learning Unit 22–2: Frequency Distributions and Graphs

DRILL PROBLEMS

1. A local dairy distributor wants to know how many containers of yogurt health club members consume in a month. The distributor gathered the following data:

17	17	22	14	26	23	23	15	18	16
18	15	23	18	29	20	24	17	12	15
18	19	18	20	28	21	25	21	26	14
16	18	15	19	27	15	22	19	19	13
20	17	13	24	28	18	28	20	17	16

Construct a frequency distribution table to organize these data.

2. Construct a bar graph for the Problem 1 data. The height of each bar should represent the frequency of each amount consumed.

3. To simplify the amount of data concerning yogurt consumption, construct a relative frequency distribution table for the data in Problem 1. The range will be from 1 to 30 with five class intervals: 1–6, 7–12, 13–18, 19–24, and 25–30.

4. Construct a bar graph for the grouped data in Problem 1.

5. Prepare a pie chart to represent the data in Problem 1.

WORD PROBLEMS

6. The women's department of a local department store lists its total sales for the year: January, $39,800; February, $22,400; March, $32,500; April, $33,000; May, $30,000; June, $29,200; July, $26,400; August, $24,800; September, $34,000; October, $34,200; November, $38,400; December, $41,100. Draw a line graph to represent the monthly sales of the women's department for the year. The vertical axis should represent the dollar amount of the sales.

7. The following list shows the number of television sets sold in a year by the sales associates at Souza's TV and Appliance Store.

115	125	139	127	142	153	169	126	141
130	137	150	169	157	146	173	168	156
140	146	134	123	142	129	141	122	141

Construct a relative frequency distribution table to represent the data. The range will be from 115 to 174 with intervals of 10.

8. Use the data in the distribution table for Problem 7 to construct a bar graph for the grouped data.

9. Expenses for Flora Foley Real Estate Agency for the month of June were as follows: salaries expense, $2,790; utilities expense, $280; rent expense, $2,000; commissions expense, $4,800; and other expenses, $340. Present these data in a circle graph. (First calculate the percent relationship between each item and the total; then determine the number of degrees that represents each item.)

10. Today a new Jeep costs $25,000. In 1970, the Jeep cost $4,500. What is the price relative? (Round to the nearest tenth percent.)

Check Figures

APPENDIX B

Worked-Out Solutions to You Try It Problems

Odd-Numbered Drill and Word Problems for End-of-Chapter Problems.

Challenge Problems (all).

Cumulative Reviews (all).

Odd-Numbered Additional Assignments by Learning Unit from Appendix A.

Worked-Out Solutions to You Try It Problems, Check Figures to Drill and Word Problems (Odds), Challenge Problems, and Cumulative Reviews

Chapter 1

You Try It

1. 571 → Five hundred seventy-one
7,943 → Seven thousand, nine hundred forty-three

2. $691 = 691 = 690$
(Identify digit; Less than 5)

3. 429,685 → 429,685 → 400,000
(Identify digit; Less than 5)

4. $\begin{array}{r} \overset{1}{7}6 \\ +38 \\ \hline 114 \end{array}$

5. $\begin{array}{r} \overset{512}{\not{6}\not{2}9} \\ -134 \\ \hline 495 \end{array}$

6. $\begin{array}{r} 491 \\ \times 28 \\ \hline 3928 \\ 982 \\ \hline 13{,}748 \end{array}$

$13 \times 10 = 130$ (attach 1 zero)
$13 \times 1{,}000 = 13{,}000$ (attach 3 zeros)

7. $16\overline{)95}$ = 5 R15
$\begin{array}{r} 95 \\ 80 \\ \hline 5 \end{array}$

$4{,}000 \div 100 = 40$ (drop 2 zeros)
$4{,}000 \div 1{,}000 = 4$ (drop 3 zeros)

End-of-Chapter Problems

1–1. 105
1–3. 154
1–5. 13,580
1–7. 113,690
1–9. 38
1–11. 3,600
1–13. 1,074
1–15. 31,110
1–17. 340,531
1–19. 126,000
1–21. 90
1–23. 86 R4
1–25. 405
1–27. 1,616
1–29. 24,876
1–31. 17,989; 18,000
1–33. 80
1–35. 133
1–37. 216
1–39. 19 R21
1–41. 7,690; 6,990
1–43. 70,470; 72,000
1–45. 700
1–47. $27,738
1–49. $240; $200; $1,200; $1,080
1–51. $2,436; $3,056; $620 more
1–53. 905,600
1–55. 1,080
1–57. **a.** $4,569
b. $4,600
c. $31
1–59. $212
1–61. $1,872,000
1–63. $4,815; $250,380
1–65. $64,180
1–67. 200,000; 10,400,000
1–69. $1,486
1–71. Average $33; no concern
1–73. $40 per sq yard
1–75. $7,680 difference between drugstore and bakery
1–76. $12,000 difference

Chapter 2

You Try It

1. $\frac{3}{10}$ proper, $\frac{9}{8}$ improper, $1\frac{4}{5}$ mixed

2. $\frac{18}{7} = 2\frac{4}{7}$ $\quad 5\frac{1}{7} = \frac{35+1}{7} = \frac{36}{7}$

3. $\frac{16 \div 8}{24 \div 8} = \frac{2}{3}$

4. $\frac{20}{50} = 20\overline{)50}$ = 2; $50 - 40 = 10$
$10\overline{)20}$ = 2; $20 - 20 = 0$
10 is greatest common denominator

5. $\frac{16}{31} = \frac{}{310}$
$310 \div 31 = 10 \qquad 10 \times 16 = 160$

6. $\frac{3}{7} + \frac{2}{7} = \frac{5}{7} \quad \frac{5}{7} - \frac{2}{7} = \frac{3}{7}$

$\begin{array}{r} \frac{5}{8} = \frac{25}{40} \\ +\frac{3}{40} = \frac{3}{40} \\ \hline \frac{28}{40} = \frac{7}{10} \end{array}$

7. Prime numbers 2, 3, 5, 7, 11, 13, 17

8. $\frac{1}{2} + \frac{1}{4} + \frac{1}{5} = \quad \begin{array}{r|rrr} 2 & 2 & 4 & 5 \\ \hline & 1 & 2 & 5 \end{array}$
$2 \times 1 \times 2 \times 5 = 20\text{ LCD}$

9. $\begin{array}{r} 2\frac{1}{4} \\ +3\frac{3}{4} \\ \hline 5\frac{4}{4} = 6 \end{array}$

10. $\begin{array}{r} 11\frac{1}{3} \\ -2\frac{2}{3} \\ \hline \end{array} \quad \begin{array}{r} 10\frac{4}{3} \\ -2\frac{2}{3} \\ \hline 8\frac{2}{3} \end{array}$

11. $\frac{4}{5} \times \frac{25}{26} = \frac{\overset{2}{\not{4}}}{\underset{1}{\not{5}}} \times \frac{\overset{5}{\not{25}}}{\underset{13}{\not{26}}} = \frac{10}{13}$

12. $2\frac{1}{4} \times 3\frac{1}{4} = \frac{9}{4} \times \frac{13}{4} = \frac{117}{16} = 7\frac{5}{16}$

13. $\frac{1}{8} \div \frac{1}{4} = \frac{1}{\underset{2}{\not{8}}} \times \overset{1}{\not{4}} = \frac{1}{2}$

14. $3\frac{1}{4} \div 1\frac{4}{5} = \frac{13}{4} \div \frac{9}{5} = \frac{13}{4} \times \frac{5}{9} = \frac{65}{36}$

End-of-Chapter Problems

2–1. Proper
2–3. Improper

2–5. $61\frac{2}{5}$

2–7. $\frac{59}{3}$

2–9. $\frac{11}{13}$

2–11. 60 $(2 \times 2 \times 3 \times 5)$

2–13. 96 $(2 \times 2 \times 2 \times 2 \times 2 \times 3)$

2–15. $\frac{13}{21}$

2–17. $15\frac{5}{12}$

2–19. $\frac{5}{6}$

2–21. $7\frac{4}{9}$

2–23. $\frac{5}{16}$

2–25. $\frac{3}{25}$

2–27. $\frac{1}{3}$

2–29. $\frac{7}{18}$

2–31. $215,658

2–33. 486,000

2–35. $35\frac{1}{4}$ hours

2–37. $10\frac{3}{4}$ hours

2–39. $6\frac{1}{2}$ gallons

2–41. $875

2–43. $\frac{23}{36}$

2–45. $7,560

2–47. $3\frac{3}{4}$ lbs apple; $8\frac{1}{8}$ cups flour; $\frac{5}{8}$ cup marg.; $5\frac{15}{16}$ cups sugar; 5 teaspoons cin.

2–49. 400 people

2–51. 275 gloves

2–53. $450

2–55. $45\frac{3}{16}$

2–57. $62,500,000; $37,500,000

2–59. $\frac{3}{8}$

2–61. $2\frac{3}{5}$ hours

2–63. $8\frac{31}{48}$ feet; Yes

2–64. **a.** 400 homes **b.** $320,000 **c.** 3,000 people **d.** 2,500 people **e.** $112.50 **f.** $8,800,000

Chapter 3

You Try It

1. .8256 → Ten thousandths place

2. .841 = .8 (↑ Less than 5)

3. $\frac{9}{1,000} = .009$

$\frac{3}{10,000} = .0003$

4. $\frac{1}{7} = .142 = .1$

5. $5\frac{4}{5} = \frac{4}{5} = .80 + 5 = 5.80$

6. $.865 \quad \frac{865}{} \quad \frac{865}{1} \quad \frac{865}{1,000}$ (attach 3 zeros)

7.
$$\begin{array}{r} 1.7 \\ 3.0 \\ .8 \\ \hline 5.5 \end{array} \qquad \begin{array}{r} {}^{5\,10} \\ \not{6}.\not{0}0 \\ -4.10 \\ \hline 1.90 \end{array}$$

8.
$$\begin{array}{rl} 3.49 & \text{(2 places)} \\ .015 & \text{(3 places)} \\ \hline 1745 & \\ 349 & \\ \hline .05235 & \end{array}$$

9.
$$\begin{array}{r} 1.5 \\ 33\overline{)49.5} \\ \underline{33} \\ 165 \\ \underline{165} \\ 0 \end{array}$$

10.
$$\begin{array}{r} .46 = .5 \\ 3.2\overline{)1.480} \\ \underline{128} \\ 200 \\ 192 \end{array}$$

11. $6.92 \times 100 = 692$ (move 2 places to right)
$6.92 \div 100 = .0692$ (move 2 places to left)

End-of-Chapter Problems

3–1. Hundredths

3–3. .7; .74; .739

3–5. 5.8; 5.83; 5.831

3–7. 6.6; 6.56; 6.556

3–9. $4,822.78

3–11. .08

3–13. .06

3–15. .91

3–17. 16.61

3–19. $\frac{71}{100}$

3–21. $\frac{125}{10,000}$

3–23. $\frac{825}{1,000}$

3–25. $\frac{7,065}{10,000}$

3–27. $28\frac{48}{100}$

3–29. .005

3–31. .0085

3–33. 818.1279

3–35. 3.4

3–37. 2.32

3–39. 1.2; 1.26791

3–41. 4; 4.0425

3–43. 24,526.67

3–45. 161.29

3–47. 6.82

3–49. .04

3–51. .63

3–53. 2.585

3–55. .0086

3–57. 486

3–59. 3.950

3–61. 7,913.2

3–63. .583

3–65. $19.57

3–67. $0.75

3–69. $119.47

3–71. $29.00

3–73. 91 million

3–75. $423.16

3–77. $105.08

3–79. $25,001.16

3–81. 1.8%

3–83. $1.58; $3,713

3–85. $6,465.60

3–86. Yes, $16,200

3–87. $560.45

Cumulative Review 1, 2, 3

1. $216
2. $200,000
3. $50,560,000
4. $25.50
5. $225,000
6. $750
7. $369.56
8. $130,000,000
9. $63.64

Chapter 4

You Try It

Sample

1. Pete Co. 24-111-9

Pay to the order of Reel Bank Pete Co. 24-111-9

Pay to the order of Reel Bank for deposit only Pete Co. 24-111-9

Checkbook

	Beg. balance		$300
2.	*Less:* NSF	$50	
	ATM service charge	20	70
	Ending balance		$230

End-of-Chapter Problems

4–1. $4,720.33

4–3. $4,705.33

4–5. $753

4–7. $540.82

4–9. $577.95

4–11. $998.86
4–13. $1,530
4–14. $1,862.13
4–15. $3,061.67

Chapter 5

You Try It

1. $E + 15 = 14$; $-15 \quad -15$; $E = -1$

2. $B - 40 = 80$; $+40 \quad +40$; $B = 120$

3. $\frac{3C}{3} = 75$; $C = 15$

4. $\frac{A}{6} = 60 \quad (6)\frac{A}{6} = 6(60)$; $A = 360$

5. $\frac{C}{4} + 10 = 17$; $-10 \quad -10$; $\frac{C}{4} = 7$; $(4)\frac{C}{4} = 7(4)$; $C = 28$

6. $7(B - 10) = 35$; $7B - 70 = 35$; $+70 \quad +70$; $\frac{7B}{7} = \frac{105}{7}$; $B = 15$

7. $5B + 3B = 16$; $\frac{8B}{8} = \frac{16}{8}$; $B = 2$

Sit. 1. $P - \$53 = \$110$; $+53 \quad +53$; $P = \$163$

Sit. 2. $\frac{1}{7}B = \$6{,}000$; $7\left(\frac{B}{7}\right) = 6{,}000(7)$; $B = \$42{,}000$

Sit. 3. $9S - S = 640$; $\frac{8S}{8} = \frac{640}{8}$; $S = 80 \qquad 9S = 720$

Sit. 4. $9S + S = 640$; $\frac{10S}{10} = \frac{640}{10}$; $S = 64 \qquad 9S = 576$

Sit. 5. $400(3N) + 300N = 15{,}000$; $\frac{1{,}500N}{1{,}500} = \frac{15{,}000}{1{,}500}$; $N = 10$; $3N = 30$

Sit. 6. $400S + 300(40 - S) = 15{,}000$; $400S + 12{,}000 - 300S = 15{,}000$; $100S + 12{,}000 = 15{,}000$; $-12{,}000 \quad -12{,}000$; $\frac{100S}{100} = \frac{3{,}000}{100}$; $S = 30$; $40 - S = 10$

End-of-Chapter Problems

5–1. $X = 440$
5–3. $Q = 300$
5–5. $Y = 15$
5–7. $Y = 12$
5–9. $P = 25$
5–11. Fred 25; Lee 35
5–13. Josh, 16; Jessica, 240
5–15. 50 shorts; 200 T-shirts
5–17. $B = 70$
5–19. $N = 63$
5–21. $Y = 7$
5–23. $P = \$610.99$
5–25. Pete = 90; Bill = 450
5–27. 48 boxes pens; 240 batteries
5–29. $A = 135$
5–31. $M = 60$
5–33. 211 Boston; 253 Colorado Springs
5–35. $W = 129$
5–37. Shift 1: 3,360; shift 2: 2,240
5–39. 22 boxes of hammers
18 boxes of wrenches
5–41. 135,797 lenders
5–42. **a.** 2.5
b. 15 miles
c. 6 hours
5–43. $B = 4$

Chapter 6

You Try It

1. $.92 = 92\%$
$.009 = .9\%$
$5.46 = 546\%$

2. $\frac{2}{9} = 22.222\% = 22.22\%$

3. $78\% = .0078$ (2 places to left)
$96\% = .96$ (2 places to left)
$246\% = 2.46$ (2 places to left)
$7\frac{3}{4}\% = 7.75\% = .0775$
$\frac{3}{4}\% = .75\% = .0075$
$\frac{1}{2}\% = .50\% = .0050$

4. $\frac{3}{5} = .60 = 60\%$

5. $74\% \rightarrow 74 \times \frac{1}{100} = \frac{74}{100} = \frac{37}{50}$
$\frac{1}{5}\% \rightarrow \frac{1}{5} \times \frac{1}{100} = \frac{1}{500}$
$121\% \rightarrow 121 \times \frac{1}{100} = \frac{121}{100} = 1\frac{21}{100}$
$17\frac{1}{5}\% \rightarrow \frac{86}{5} \times \frac{1}{100} = \frac{86}{500} = \frac{43}{250}$
$17.75\% \rightarrow 17\frac{3}{4}\% = \frac{71}{4} \times \frac{1}{100} = \frac{71}{400}$

6. Portion ($1,600) = Base ($2,000) × Rate (.80)

7. Rate (25%) = $\frac{\text{Portion (\$500)}}{\text{Base (\$2,000)}}$

8. Base ($1,000) = $\frac{\text{Portion (\$200)}}{\text{Rate (.20)}}$

9. $\frac{\text{Difference in price (\$100)}}{\text{Base (orig. \$500)}} = 20\%$

End-of-Chapter Problems

6–1. 88%
6–3. 40%
6–5. 356.1%
6–7. .04
6–9. .643
6–11. 1.19
6–13. 8.3%
6–15. 87.5%
6–17. $\frac{1}{25}$
6–19. $\frac{19}{60}$
6–21. $\frac{27}{400}$
6–23. 10.5
6–25. 102.5
6–27. 156.6
6–29. 114.88
6–31. 16.2
6–33. 141.67
6–35. 10,000
6–37. 17,777.78
6–39. 108.2%
6–41. 110%
6–43. 400%
6–45. 59.40
6–47. 1,100
6–49. 40%
6–51. +20%
6–53. 80%
6–55. $10,000
6–57. $640 per month
6–59. 677.78%
6–61. 6%
6–63. $2,434.50
6–65. Yes, $15,480
6–67. 900
6–69. $742,500
6–71. $220,000
6–73. 33.3%

6–75. .38%
6–77. $39,063.83
6–79. $138.89
6–81. $1,900
6–83. $102.50
6–85. 3.7%
6–87. $2,571
6–89. $41,176
6–91. 40%
6–93. 585,000
6–94. **a.** 68%
b. 125%
c. $749,028
d. $20
e. 7 people
6–95. $55,429

Chapter 7

You Try It

1. $\$700 \times .20 = \$140$

2. $\begin{array}{r} 1.00 \\ -\ .20 \\ \hline .80 \end{array}$ $\quad \$700 \times .80 = \$560$

3. Seller will pay the freight

4. $\dfrac{\$240}{.40} = \$600$
(100% – 60%)

5. $\begin{array}{r} \$200 \\ \times\ .06 \\ \hline \$12.00 \end{array} \quad \begin{array}{r} \$188 \\ \times\ .08 \\ \hline \$15.04 \end{array}$

$\begin{array}{r} \$188.00 \\ -\ 15.04 \\ \hline \$172.96 \end{array} \quad .94 \times .92 = \begin{array}{r} .8648 \text{ NPER} \\ \times\ \$200 \\ \hline \$172.96 \end{array}$

6. $.94 \times .92 \times \$2,000 = \$1,729.60$

7. $\begin{array}{r} 1.0000 \\ -\ .8648 \\ \hline .1352 \end{array}$ $(.94 \times .92)$
$.1352 \times \$2,000 = \$270.40$

8. $\begin{array}{r} \$2,000 \\ -\ \ 80 \\ \hline \$1,920 \end{array}$ (Freight and returns)
$\$1,920 \times .02 = \$38.40$

9. April 12, May 2

10. $\begin{array}{r} \$700 \\ -100 \\ \hline \$600 \end{array} \times .98 = \begin{array}{r} \$588 \\ +100 \\ \hline \$688 \end{array}$

11. No discount; pay full $700
12. November 10; November 30
13. $\$300/.98 = \$306.12$

End-of-Chapter Problems

7–1. .9504; .0496; $59.52; $1,140.48
7–3. .893079; .106921; $28.76; $240.24
7–5. $369.70; $80.30
7–7. $1,392.59; $457.41
7–9. June 28; July 18
7–11. June 15; July 5
7–13. July 10; July 30
7–15. $138; $6,862
7–17. $2; $198
7–19. $408.16; $291.84
7–21. $59.80; $239.20
7–23. .648; .352; $54.56; $100.44
7–25. $576.06; $48.94
7–27. $5,100; $5,250
7–29. $5,850
7–31. $1,357.03
7–33. $8,173.20
7–35. $8,333.33; $11,666.67
7–37. $99.99
7–39. $489.90; $711.10
7–41. $4,658.97
7–43. $1,083.46; $116.54
7–45. $5,008.45
7–47. $781.80 paid
7–48. **a.** $1,500
b. 8.34%
c. $164.95
d. $16,330.05
e. $1,664.95
7–49. $4,794.99

Chapter 8

You Try It

1. $S = C + M$
$S = \$400 + \$200$
$S = \$600$

2. $\dfrac{\$50}{\$200} = 25\%$

$\dfrac{\$50}{.25} = \$200$

3. $S = C + M$
$S = \$8 + .10(\$8)$
$S = \$8 + \$.80$
$S = \$8.80$

4. $S = C + M$
$\$200 = C + .60\,C$
$\dfrac{\$200}{1.60} = \dfrac{\cancel{1.60}C}{\cancel{1.60}}$
$\$125 = C$

5. $M = S - C$
$(\$2,500) = (\$4,500) - (\$2,000)$

6. $\dfrac{\$700}{\$2,800} = 25\%$

$\dfrac{\$700}{.50} = \$1,400$

7. $S = C + M$
$S = \$800 + .40(S)$
$\underline{-.40} \qquad \underline{-.40}$
$\dfrac{\cancel{.60}S}{\cancel{.60}} = \dfrac{\$800}{.60}$
$S = \$1333.33$

8. $S = C + M$
$\$2,000 = C + .70(\$2,000)$
$\$2,000 = C + \$1,400$
$\dfrac{-1,400}{\$600} = \dfrac{-1,400}{C}$

9. $\dfrac{.47}{1 + .47} = \dfrac{.47}{1.47} = 32\%$ rounded

10. $\begin{array}{r} \$50 \\ \times .20 \\ \hline \$10 \end{array} \quad \dfrac{\$10}{\$50} = 20\%$

11. $TS = TC + TM$
$TS = \$9 + .30(\$9)$
$TS = \$9 + \$2.7$
$TS = \$11.70$
$\dfrac{\$11.70}{45} = \$.26$

12. $\dfrac{\$70,000}{\$20} = 3,500$ units

End-of-Chapter Problems

8–1. $600; $2,600
8–3. $4,285.71
8–5. $6.90; 45.70%
8–7. $450; $550
8–9. $110.83
8–11. $34.20; 69.8%
8–13. 11%
8–15. $3,830.40; $1,169.60; 23.39%
8–17. $16,250; $4.00
8–19. $166.67
8–21. $14.29
8–23. $600; $262.50
8–25. $84
8–27. 42.86%
8–29. $320
8–31. 20,000
8–33. $44; 56%
8–35. $195
8–37. $.59
8–39. $2.31
8–41. 12,000
8–42. $266
8–43. $94.98; $20.36; loss

Cumulative Review 6, 7, 8

1. 650,000
2. $296.35
3. $133
4. $2,562.14
5. $48.75
6. $259.26
7. $1.96; $1.89

Chapter 9

You Try It

1. $38 \text{ hrs} \times \$9.25 = \$351.50$

2. **Reg $** **Overtime $**
$(40 \times \$7) + (3 \times \$10.50)$
$\$280 + \$31.50 = \$311.50$
gross pay

3. $2,250 \times \$.79 = \$1,777.50$

4. $600 \times \$.79 = \$474$
$300 \times \$.88 = \underline{+264}$
$\$738$

5. $\$175,000 \times .07 = \$12,286.96$

6. $\$6,000 \times .05 = \$300$
$\$2,000 \times .09 = 180$
$\$4,000 \times .12 = \underline{480}$
$\$960$

7. $\$600 + (\$6{,}000 \times .04)$
$\$600 + 240 = \$840$

8. Gross \$490.00   \$490.00
Less: FIT 18.82   − 79.80
SS 30.38   \$410.20
Med. 7.11   −222.00
\$433.69 $\$188.20 \times .10 = \$18.82$

9. Social Security $= \$128{,}400 \times .062 = \$7{,}960.80$
Medicare $= \$150{,}000 \times .0145 = \$2{,}175$

10. \$1,400.00   (\$79.80 × 3)
239.40   \$36.60 + .12(\$572.60)
1,160.60   \$36.60 + \$68.71
− 588.00   \$105.31 FIT
\$ 572.60

11. FUTA $\$200 \times .006 = \$1.20$
SUTA $\$200 \times .054 = \$10.80$

End-of-Chapter Problems

9–1. 37; \$331.15
9–3. \$12.00; \$452
9–5. \$1,680
9–7. \$60
9–9. \$13,000
9–11. \$4,500
9–13. \$11,900; \$6,900; \$138; \$388
9–15. \$378.20; \$101.50
9–17. \$199.83; \$124.00; \$29.00; \$1,647.17
9–19. \$752.60; \$85.20
9–21. \$1,311.06
9–23. \$2,247.21
9–25. \$825
9–27. \$1,116.26
9–29. \$357; \$7,960.80
9–31. \$1,096.44
9–32. **a.** \$280
b. \$196.91
c. \$36.47
9–33. \$1,653.60, \$193.13 understated; \$52

Chapter 10

You Try It

1. $\$4{,}000 \times .03 \times \frac{18}{12} = \$180$

2. $\$3{,}000 \times .04 \times \frac{45}{365} = \$14.79$ Feb 22   53; Jan 8   − 8; 45

3. $\$3{,}000 \times .04 \times \frac{45}{360} = \$15.00$

4. $\$2{,}000 \times .04 \times \frac{90}{360} = \$20$

5. $\dfrac{\$20}{.04 \times \frac{90}{360}} = \$2{,}000$

6. $\dfrac{\$20}{\$2{,}000 \times \frac{90}{360}} = 4\%$

7. $\dfrac{20}{\$2{,}000 \times .04} = .25 \times 360 = 90 \text{ days}$

8. $\$4{,}000 \times .04 \times \frac{30}{360} = \$13.33$
\$400.00
− 13.33
\$386.67
$\$4{,}000 - 386.67 = \$3{,}613.33$
$\$3{,}613.33 \times .04 \times \frac{40}{360} = \$16.06$
$\$300 - \$16.06 = \$283.94$
$\$3{,}613.33 - \$283.94 = \$3{,}329.39$
$\$3{,}329.39 \times .04 \times \frac{20}{360} = \$7.40$
$\$3{,}329.39 + \$7.40 = \$3{,}336.79$
Total interest $= \$13.33 + \$16.06 + \$7.40 = \$36.79$

End-of-Chapter Problems

10–1. \$303.75; \$9,303.75
10–3. \$1,012.50; \$21,012.50
10–5. \$28.23; \$613.23
10–7. \$20.38; \$1,020.38
10–9. \$73.78; \$1,273.78
10–11. \$1,904.76
10–13. \$4,390.61 balance due
10–15. \$618.75; \$15,618.75
10–17. \$2,377.70; Save \$1.08
10–19. 4.7 years
10–21. \$21,596.11
10–23. \$714.87; \$44.87
10–25. 266 days
10–27. \$2,608.65
10–29. \$18,666.85
10–31. 12.37%
10–33. 15 days
10–35. 5.6%
10–36. **a.** \$1,000
b. 8%
c. \$280; \$1,400
10–37. \$7.82; \$275.33

Chapter 11

You Try It

1. $\$4{,}000 \times .02 \times \frac{30}{360} = \$6.67$
\$4,000.00
− 6.67
\$3,993.33 Proceeds

2. $\$15{,}000 \times .04 \times \frac{40}{360} = \$66.67$
\$15,000.00
− 66.67 → $\dfrac{\$66.67}{\$14{,}933.33 \times \frac{40}{360}} = 4.02\%$

3. Dec 15   349
Nov 5   −309
40 days
$\$2{,}000 \times .03 \times \frac{60}{360} = \$10$
$MV = \$2{,}010$ (Left to go) → 20
$\$2{,}010 \times .05 \times \frac{20}{360} = \$5.58$
$\$2{,}010 - \$5.58 = \$2{,}004.42$ Proceeds

End-of-Chapter Problems

11–1. \$93.33; \$5,906.67
11–3. 25 days
11–5. \$51,451.39; 57; \$733.18; \$50,718.21
11–7. 4.04%
11–9. \$7,566.67; 6.9%
11–11. \$8,937
11–13. \$5,309.80
11–15. \$5,133.33; 56; \$71.87; \$5,061.46
11–17. \$4,836.44
11–19. ¥20,188
11–20. **a.** \$90.13
b. \$177.50
c. 3.64%
d. 3.61%
11–21. \$2,127.66; 9.57%

Chapter 12

You Try It

1. $\begin{array}{rr} \$200 & \$\ \ 208 \\ \times 1.04 & \times\ \ 1.04 \\ \hline \$208 & \$216.32 \end{array}$

2. $\$4,000 \times 1.4258$ (3% 12 periods) $= \$5,703.20$

3. Table 1.0609 (3% 2 periods)
↓
6.09%

$$\begin{array}{rr} \$4,000 \times 1.0609 = & \$4,243.60 \\ & -4,000.00 \\ \hline & \$\ \ 243.60 \end{array}$$

$$\frac{\$243.60}{\$4,000.00} = 6.09\%$$

4. Table $\begin{array}{r} .7880 \\ \times \$6,000 \\ \hline \$4,728 \end{array}$ (1.5% 16 periods)

End-of-Chapter Problems

12–1. 4; 1%; \$598.35; \$23.35
12–3. \$15,450; \$450
12–5. 12.55%
12–7. 16; $1\frac{1}{2}\%$; .7880; \$4,728
12–9. 28; 3%; .4371; \$7,692.96
12–11. \$17,600.72
12–13. \$64,188
12–15. Mystic \$4,775, Four Rivers \$3,728
12–17. \$25,734.40
12–19. \$12,698
12–21. \$12,900.87
12–23. \$37,644
12–25. Yes, \$17,908 (compounding) or \$8,376 (p. v.)
12–27. \$3,739.20
12–29. \$27,757.40
12–31. \$105,878.50
12–32. \$689,125; \$34,125 Bank B

Chapter 13

You Try It

1. $\begin{array}{r} 4.4399 \\ \times\ \ \ 6,000 \\ \hline \$26,639.40 \end{array}$ (7% 4 periods)

2. $\begin{array}{rr} \$6,000 \times 5.7507 = & \$34,504.20 \\ & -\ \ 6000.00 \\ \hline & \$28,504.20 \end{array}$ (7% 5 periods)

3. $\begin{array}{r} 5.2421 \\ \times\ \$\ 20,000 \\ \hline \$104,842 \end{array}$ (4% 6 periods)

4. $\begin{array}{r} .0302 \\ \times \$400,000 \\ \hline \$\ 12,080 \end{array}$ (5% 20 periods)

End-of-Chapter Problems

13–1. \$67,431.50
13–3. \$53,135.10
13–5. \$3,118.59
13–7. End of first year \$2,405.71
13–9. \$1,410
13–11. \$3,397.20
13–13. \$137,286; \$1,721,313
13–15. \$38,841.30
13–17. \$900,655
13–19. \$421,885.11
13–21. \$13,838.25
13–23. Annuity \$12,219.11 or \$12,219.93
13–25. \$3,625.60
13–27. \$111,013.29
13–29. \$404,313.97
13–30. \$199.29
13–31. \$120,747.09

Cumulative Review 10, 11, 12, 13

1. Annuity \$2,058.62 or \$2,058.59
2. \$5,118.70
3. \$116,963.02
4. \$3,113.92
5. \$5,797.92
6. \$18,465.20
7. \$29,632.35
8. \$55,251

Chapter 14

You Try It

1. $\begin{array}{r} \$5,400 \\ -\ \ \ 100 \\ \hline \$5,300 \end{array}$ amount financed

2. $\begin{array}{rr} & \$7,799.40 \\ \$129.99 \times 60 = & -\ 5,300.00 \\ \hline & \$2,499.40\text{FC} \end{array}$

3. $\$7,799.40 + \$100 = \$7,899.40$

4. $\frac{\$2,499.40}{\$5,300.00} \times \$100 = 47.16$ (between 16.25% and 16.50%)

5. $\frac{\$2,499.40 + \$5,300}{60} = \$129.99$

$\frac{\$5,300}{1,000} = 5.3 \times 24.32 = \$128.9$
(off due to using 16% instead of using between 16.25% and 16.50%)

6. $\$5,000 \times .035 = \$175$
$\$275 - \$175 = \$100$
$\$5,000 - \$100 = \$4,900$

7. 12 days × \$200 = \$2,400
4 days × \$120 [\$200 − \$80] = \$480
14 days × \$180 [\$120 + \$60] = \$2,520
Total = \$5,400
\$5,400/30 = \$180 daily balance
Finance charge = \$180 × 2.5% = \$4.50

End-of-Chapter Problems

14–1. Finance charge \$5,240
14–3. Finance charge \$1,279.76; 12.75%–13%
14–5. \$119.39; \$119.37
14–7. \$2,741; \$41.12
14–9. \$1,400
14–11. **a.** \$4,050
b. \$1,656
c. \$5,756
d. \$40.89, falls between 14.25% and 14.50%
e. \$95.10
14–13. \$1,245; \$18.68; \$1,318.68
14–15. \$940.36
14–17. \$298.12
14–18. Peg is correct
14–19. 15.48%

Chapter 15

You Try It

1. $\frac{\$70{,}000}{\$1{,}000} = 70 \times \$4.50 = \$315$

2. 30 years = × \$ 360 payments / 315
\$113,400 − \$70,000 = \$43,400 interest

3.

Payment	Interest	Principal reduction	Balance
1	\$204.17	\$110.83	\$69,889.17
	$\left(\$70{,}000 \times .035 \times \frac{1}{12} = \$204.17\right)$	(\$315 − \$204.17)	(\$70,000 − \$110.83)
2	\$203.84	\$111.16	\$69,778.01
	$\left(\$69{,}889.17 \times .035 \times \frac{1}{12}\right)$	(\$315 − \$203.84)	(\$69,889.17 − \$111.16)

End-of-Chapter Problems

15–1. \$630
15–3. \$894.60
15–5. \$118,796
15–7. \$1,423.80; \$179,326.20
15–9. \$73,696
15–11. \$636.16; \$117,017.60
15–13. Payment 3, \$119,857.38
15–15. \$57,921.60
15–17. \$87.50
15–19. \$793.50
15–20. **a.** \$92,495.50
b. \$1,690.15
c. \$415,954

Chapter 16

You Try It

1.

\$ 400	40%
+ 600	60%
\$1,000	100%

2.

	2020	2019	Change	%	
Cash	\$8,000	\$2,000	\$6,000	300%	$\frac{\$6{,}000}{\$2{,}000}$

3. \$400 − \$20 − \$5 = \$375 net sales

4. \$50 + \$200 − \$20 = \$230

5. \$400 − \$250 = \$150 gross profit

6. \$210 − \$180 = \$30 net income

7.

2021	2020	2019
1,200	800	1,000
120%	80%	100%

$\left(\frac{1{,}200}{1{,}000}\right) \frac{200}{1{,}000}$

8. $\frac{\$40{,}000}{\$160{,}000} = .25$

9. $\frac{\$40{,}000 - \$2{,}000 - \$3{,}000}{\$160{,}000} = \frac{\$35{,}000}{\$160{,}000} = .22$

10. $\frac{\$4{,}000}{\left(\frac{\$60{,}000}{360}\right)} = 24 \text{ days}$

11. $\frac{\$180{,}000}{\$70{,}000} = 257.14$

12. $\frac{\$16{,}000}{-\$110{,}000} = -14.55\%$

13. $\frac{\$60{,}000}{70{,}000} = .86$

14. $\frac{\$16{,}000}{\$60{,}000} = .27$

End-of-Chapter Problems

16–1. Total assets \$84,000
16–3. Inventory −16.67%; mortgage note +13.79%
16–5. Net sales 13.62%; Net earnings 2020 47.92%
16–7. Depreciation \$100; + 16.67%
16–9. 1.43; 1.79
16–11. .20; .23
16–13. .06; .08
16–15. 13.57%
16–17. 87.74%; 34.43%; .13; 55.47%
16–19. 2021 68% sales
16–21. \$9,447 net worth
16–22. \$3,470; 6.8%; \$431; .8%
16–23. 3.5; 2.3

Chapter 17

You Try It

1. $\frac{\$50{,}000 - \$10{,}000}{4} = \frac{\$40{,}000}{4} = \$10{,}000 \text{ per year}$

2. $\frac{\$4{,}000 - \$500}{700} = \frac{\$3{,}500}{700} = \$5 \text{ depreciation per unit}$
150 × \$5 = \$750

3.

Year	Cost	Depreciation expense		Book value at end of year
1	$40,000	$20,000 ($40,000 × .50)		$20,000
2	$20,000	$10,000 ($20,000 × .50)	=	$10,000

4. .20 × $7,000 = $1,400 depreciation expense

End-of-Chapter Problems

17–1. Book value (end of year) $35,000
17–3. Book value (end of year) $25,000
17–5. Book value (end of year) $15,000
17–7. Book value (end of year) $15,000
17–9. Book value (end of year) $5,400
17–11. $2,240
17–13. $18,000
17–15. $22,560
17–17. $15,000
17–19. $6,000; $18,000
17–21. $6,760 below
17–23. $83,667
17–25. **a.** $87,750
b. $11.40
c. $21,489
d. 4 years
17–26. $13,320; $1.11

Chapter 18

You Try It

1. $4 \times 9 = 36$
$3 \times 10 = \underline{30}$
66 total cost
$1 \times \$9 = \$9$
$1 \times \$10 = \underline{\$10}$
$19
$66 – 19 = $47 Cost of goods sold

2. $\frac{89}{15} = \$5.93$ unit cost
$4 \times \$5.93 = \$23.72$

3. FIFO $4 \times \$7 = \$28$
4. LIFO $4 \times \$5 = \$20$

5.

	Cost	Retail
Cost of goods available for sale	$88,000	$117,000
		– 90,000
Net sales		$27,000

Cost ratio: $\frac{\$88,000}{\$117,000} = 75\%$

$.75 \times \$27,000 = \$20,250$

6.

Cost of goods available for sale		$42,000
Net sales at retail	$20,000	
	× .25	
COGS at retail		5,000
Ending inventory		$37,000

7. $$\frac{\$90,000}{\left(\frac{\$40,000 + \$60,000}{2}\right)} = \frac{\$90,000}{\$50,000} = 1.8$$

8. Total sq. ft. for dept. 10,000
.40 to Dept A $30,000 × .40 = $12,000
.60 to Dept B 30,000 × .60 = 18,000

End-of-Chapter Problems

18–1. $7,180; $22,635
18–3. $543; $932
18–5. $10
18–7. $36
18–9. $72
18–11. $140.80
18–13. $147.75; $345.60
18–15. $188.65; $304.70
18–17. 3.56; 3.25
18–19. .75; $67,500
18–21. $10,550; $24,645
18–23. $45,000
18–25. $55,120
18–27. $38,150
18–29. $3,511.16
18–30. $13,499.50
18–31. $1,900

Chapter 19

You Try It

1. $62.80 – $5.02 = $57.78
× .06
$3.47 sales tax

2. $6,000 + $300 + $600 = $6,900
3. $200,000 × .40 = $80,000 assessed value
4. $\frac{\$700,000}{\$8,400,000} = .0833$
5. 1. 8.33% 2. $8.33
3. $83.3 4. $\frac{.0833}{.001} = 83.3 = 83$ mills
6. 1. 9.5% × $40,000 = $3,800
2. $\frac{\$40,000}{\$100} = 400 \times \$9.50 = \$3,800$
3. $\frac{\$40,000}{\$1,000} = 40 \times \$95 = \$3,800$
4. $\frac{\$.0950}{.001} = 95 \times .001 \times \$40,000 = \$3,800$

End-of-Chapter Problems

19–1. $928
19–3. $83,018.87
19–5. $39,000
19–7. $.0233
19–9. 6.99%; $6.99; $69.90; 69.90
19–11. $4,462.50
19–13. $16,985.05
19–15. $112.92
19–17. $634,000
19–19. $6,940
19–21. $64,000
19–23. $4,726.88
19–25. $23,065 more in Minn.
19–26. $3,665
19–27. $979

Chapter 20

You Try It

1. 1. $\frac{\$90{,}000}{\$1{,}000} = 90 \times \$2.44 = \$219.60$

2. $\frac{\$90{,}000}{\$1{,}000} = 90 \times \$11.84 = \$1{,}065.60$

3. $\frac{\$90{,}000}{\$1{,}000} = 90 \times \$15.60 = \$1{,}404.00$

4. $\frac{\$90{,}000}{\$1{,}000} = 90 \times \$27.64 = \$2{,}487.60$

2. Option 1: $\frac{\$60{,}000}{\$1{,}000} = 60 \times \$265 = \$15{,}900$

Option 2: $60 \times \$550 = \$33{,}000$

Option 3: 21 yr 300 days

3. $\frac{\$80{,}000}{\$100} = 800 \times \$.61 = \$488$

$\frac{\$20{,}000}{\$100} = 200 \times \$.65 = \underline{\$130}$

Total $\underline{\underline{\$618}}$

4. $600 \times \$.44 = \$264$

Refund $\$600 - \$264 = \$336$

5. $\$600 \times \frac{1}{3} = \$200$

$\$600 - \$200 = \$400$

6. $\frac{\$40{,}000}{\$60{,}000} \times \$9{,}000 = \$6{,}000$

7.

10/20/5	\$184 (\$55 + \$129)
Bodily	94
Property	132
Collision	196
Comprehensive	178
Total premium	\$784

End-of-Chapter Problems

20–1. \$2,762
20–3. \$138.75
20–5. \$53,000
20–7. 21 years, 300 days
20–9. \$518; \$182
20–11. \$16,500
20–13. \$1,067
20–15. \$208 vs \$240
20–17. \$801 No cash value
20–19. \$118,750
20–21. \$1,100
20–23. \$373.67
20–25. \$22,900; \$10,700
20–27. \$24,000; \$16,300
20–29. \$7,512.64; \$1,942.00; \$787.89
20–30. \$176.00

Chapter 21

You Try It

1. $\frac{\$.88}{\$53.88} = 1.63\%$

2. $\frac{\$53.88}{\$3.70} = 14.56 = 15$

3. $30{,}000 \times \$.80 = 24{,}000$
$30{,}000 \times \$.80 = \underline{24{,}000}$
48,000 to preferred

$\$300{,}000$
$\underline{-\ 48{,}000}$
$\$252{,}000 \div 60{,}000 = \$4.20$ to common

4. $\$1{,}022.25 \times 6 = \$6{,}133.50$

5. $\frac{\$40}{1{,}011.20} = 3.96\%$

6. $\$12.44 + \$.05 = \$12.49$

End-of-Chapter Problems

21–1. \$25,305
21–3. 1.1%
21–5. 13
21–7. \$41,983.20
21–9. 2018 preferred \$8,000
2019 0
2020 preferred \$127,000
common \$33,000
21–11. \$2,280
21–13. \$260; \$2,725; 9.5%
21–15. \$12.04; –\$.06; 9.6%
21–17. Gain \$6,606.48
21–19. 12; 2.4%
21–21. \$4,843.75; \$75
21–23. 4.5%
21–25. Stock 6.7%; bond 11.9%
21–27. Yes, \$16.02
21–29. \$443.80
21–31. **a.** 1,287 shares
b. 2,574 shares
c. 5,147 shares
d. \$26,381.76 for (a); \$52,388.43 for (b); \$103,756.44 for (c)
21–32. \$1,014.33

Chapter 22

You Try It

1. $\frac{41 + 29 + 16 + 15 + 18}{5} = 23.8$

2.

Value	Frequency	Product
80	2	160
90	3	270
100	1	100
	6	690

Mean $= \frac{690}{6} = 115$

3. 4 7 (9) 14 16

4. 7

5.

Coffees consumed	Tally	Frequency
0	|	1
1	|	1
2	||	2
3	|	1
4	|	1
5	|	1
6		0
7		0
8	|	1

6.

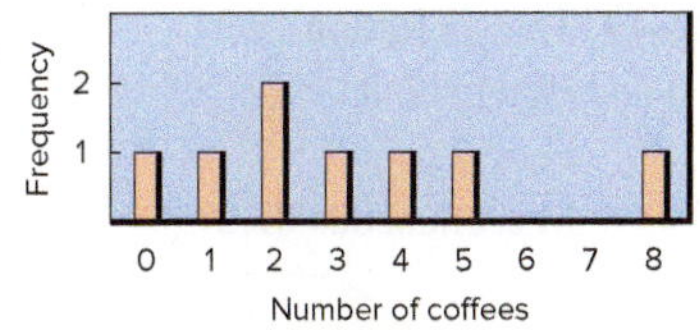

7.

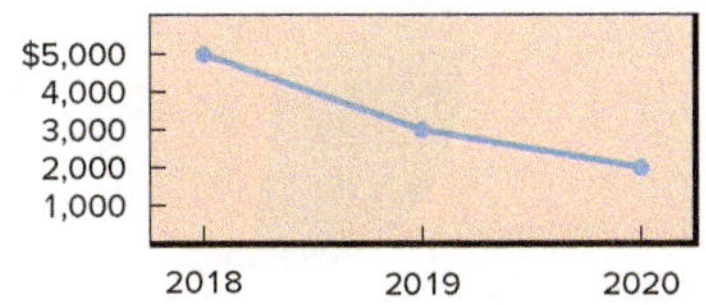

8.

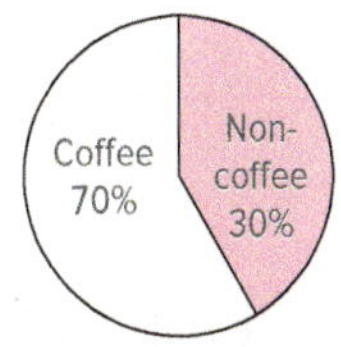

$.70 \times 360° = 252°$
$.30 \times 360° = 108°$

9. $\frac{\$12{,}000}{\$9{,}000} \times 100 = 133.3$

10. Range $= 14 - 2 = 12$

11. 1. Mean $= \frac{19}{5} = 3.8$

2. $8 - 3.8 = 4.2$
$1 - 3.8 = -2.8$
$6 - 3.8 = 2.2$
$2 - 3.8 = -1.8$
$2 - 3.8 = -1.8$

3. $(4.2)^2 = 17.64$
$(-2.8)^2 = 7.84$
$(2.2)^2 = 4.84$
$(-1.8)^2 = 3.24$
$(-1.8)^2 = 3.24$

4. 36.8

5. $36.8 \div 4 = 9.2$

6. Standard deviation = 3.03

End-of-Chapter Problems

22–1. 6.25
22–3. \$77.23
22–5. 2.7
22–7. 31.5
22–9. 8
22–11. 142.9
22–13. \$200–\$299.99 𝍸
22–15. Traditional watch 183.6°
22–17.

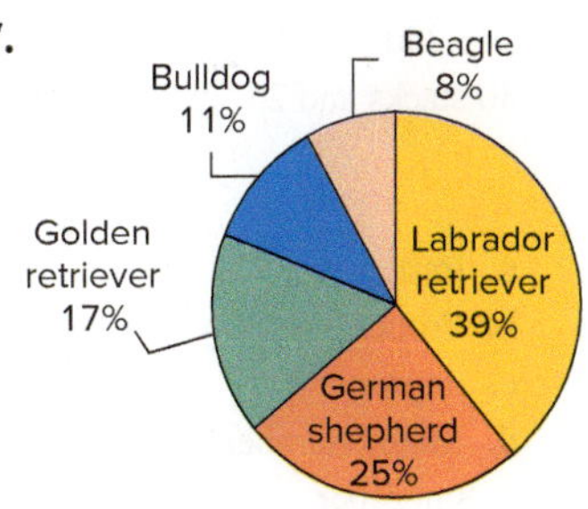

22–19. Transportation 126°
Hotel 100.8°
Food 72°
Miscellaneous 61.2°

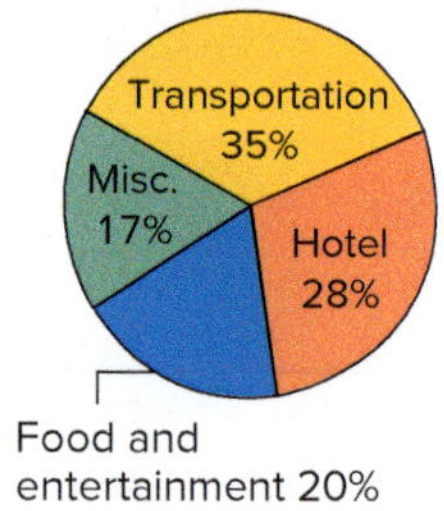

22–21. 250

22–23. a. 57,000,000 mean
62,900,000 median
b. AAA = 30.42%
Riser = 22.18%
Casto = 22.07%
Balbon = 12.70%
Hunter = 12.63%
c. 109.51°, 79.85°, 79.45°, 45.72°, 45.47°

22–24. 24.94%; 15.42%; 10.88%; 13.15%; 18.59%; 17.01%
89.78°, 55.51°, 39.17°, 47.34°, 66.92°, 61.24°

Optional Assignment

1. 98
3. 4.3
5. 16%; 2.5%
7. 68%; 81.5%; 2.5%; 2.5%; 47.5%
9. 5.02

Check Figures (Odds) to Additional Assignments by Learning Unit from Appendix A

LU 1–1

1. a. Eight thousand, eight hundred twenty-one
d. Fifty-eight thousand, three
3. a. 80; 380; 860; 5,980; 210
c. 21,000; 1,000; 4,000; 10,000
5. a. Hundreds place
c. Ten thousands place
e. Billions place
7. Five hundred sixty-five
9. $375,985
11. Two thousand, nineteen

LU 1–2

1. a. 1,006
c. 1,319
d. 179
3. a. Estimated 50; 52
c. Estimated 10; 9
5. $71,577
7. $19,973
9. 12,797 lbs
11. Estimated $9,400; $9,422
13. $746 discount

LU 1–3

1. a. Estimated 4,000; actual 2,400
c. Estimated 15,000,000; actual 16,184,184
3. a. Estimated 1,000; actual 963 R5
c. Estimated 20; actual 25 R8
5. 5,040
7. $78
9. 27
11. $43,200
13. 40 stacks and 23 "extra" bricks

LU 2–1

1. a. Improper
b. Proper
c. Improper
d. Mixed number
e. Improper
f. Mixed number

3. a. $\frac{46}{5}$ **c.** $\frac{31}{7}$ **f.** $\frac{53}{3}$
5. a. $6;\frac{6}{7}$ **b.** $15;\frac{2}{5}$ **e.** $12;\frac{8}{11}$
7. $\frac{13}{4}$
9. $\frac{17}{25}$
11. $\frac{60}{100}$
13. $\frac{7}{12}$

LU 2–2

1. a. 32 **b.** 180 **c.** 480 **d.** 252
3. a. $\frac{1}{3}$ **b.** $\frac{2}{3}$ **e.** $6\frac{1}{8}$ **h.** $56\frac{5}{6}$
5. $3\frac{1}{40}$ yards
7. $17\frac{5}{12}$ miles
9. $4\frac{8}{9}$ hours

LU 2–3

1. a. 1 **b.** $\frac{1}{4}$ **g.** 35 **i.** $1\frac{17}{18}$
3. a. $1\frac{1}{4}$ **b.** 3 **g.** 24 **l.** $\frac{4}{7}$
5. $39,000
7. 714
9. $20\frac{2}{3}$ miles
11. $412,000

LU 3–1

1. a. .62 **b.** .6 **c.** .953
d. .401 **e.** .06
3. a. Hundredths place
d. Thousandths place
5. a. $\frac{2}{5}$ **b.** $\frac{11}{25}$
g. $\frac{5}{16}$ **l.** $9\frac{1}{25}$
7. .286
9. $\frac{566}{1,000}$
11. .333
13. .0020507

LU 3–2

1. a. 31.608 **b.** 5.2281 **d.** 3.7736
3. a. .3 **b.** .1 **c.** 1,480.0 **d.** .1
5. a. 6,870 **c.** .0272
e. 34,700 **i.** 8,329.8
7. $4.53
9. $111.25
11. 15

LU 4–1

1. a. $430.64 **b.** 3 **c.** $867.51
3. a. Neuner Realty Co.
b. Kevin Jones
h. $2,756.80

LU 4–2

1. $1,435.42
3. Add $3,000; deduct $22.25
5. $2,989.92
7. $1,315.20

LU 5–1

1. a. $4N = 180$ **e.** $14 + \frac{N}{3} = 18$
h. $2N + 3N + 8 = 68$

LU 5–2

1. $80
3. $45 telephone; $135 utility
5. 51 tickets—Sherry; 408 tickets—Linda
7. 12 necklaces ($48); 36 watches ($252)
9. $157.14

LU 6–1

1. a. 8% **b.** 72.9%
i. 503.8% **l.** 80%
3. a. 70% **c.** 162.5%
h. 50% **n.** 1.5%

5. a. $\frac{1}{4}$ b. .375 c. 50%
   d. $.66\overline{6}$ n. $1\frac{1}{8}$
7. 2.9%
9. $\frac{39}{100}$
11. $\frac{9}{10,000}$

LU 6–2

1. a. $20,000; 30%; $4,000
   c. $7.00; 12%; $.84
3. a. 33.3% b. 3% c. 27.5%
5. a. −1,584; −26.6%
   d. −20,000; −16.7%
7. $9,000
9. $3,196
11. 329.5%

LU 7–1

1. a. $120 b. $360 c. $50
   d. $100 e. $380
3. a. $75 b. $21.50; $40.75
5. a. .7125; .2875 b. .7225; .2775
7. $3.51
9. $81.25
11. $315
13. 45%

LU 7–2

1. a. February 18; March 10
   d. May 20; June 9
   e. October 10; October 30
3. a. .97; $1,358 c. .98; $367.99
5. a. $16.79; $835.21
7. $16,170
9. a. $439.29 b. $491.21
11. $209.45
13. a. $765.31 b. $386.99

LU 8–1

1. a. $19.90 b. $2.72
   c. $4.35 d. $90 e. $116.31
3. a. $2; 80% b. $6.50; 52%
   c. $.28; 28.9%
5. a. $1.52 b. $225
   c. $372.92 d. $625
7. a. $199.60 b. $698.60
9. a. $258.52 b. $90.48
11. a. $212.50 b. $297.50
13. $8.17

LU 8–2

1. a. $10.00 b. $57.50
   c. $34.43 d. $27.33 e. $.15
3. a. $6.94 b. $882.35 c. $30
   d. $171.43 e. $0.36
5. a. 28.57% b. 33.33%
   d. 53.85%
7. $346.15
9. 39.39%
11. $2.29
13. 63.33%

LU 8–3

1. a. $80; $120 b. $525; $1,574.98
3. a. $410 b. $18.65
5. a. $216; $324; $5.14
   b. $45; $63.90; $1.52
7. 17%
9. $21.15
11. $273.78
13. $.79

LU 8–4

1. a. $6.00 b. $11.11
3. a. 16,667 b. 7,500
5. 5,070
7. 22,222

LU 9–1

1. a. $427.50; 0; $427.50
   b. $360; $40.50; $400.50
3. a. $438.85 b. $615.13
5. a. $5,200 b. $3,960
   c. $3,740 d. $4,750
7. $723.00
9. $3,846.25
11. $2,032.48

LU 9–2

1. a. $500; $2,300
3. $0; $2,000
5. $248.27
7. $143.75
9. $619.33
11. $678.73

LU 10–1

1. a. $240 b. $1,080 c. $1,275
3. a. $131.25 b. $4.08 c. $98.51
5. a. $515.63 b. $6,015.63
7. a. $5,459.66
9. $659.36
11. $360

LU 10–2

1. a. $4,371.44 b. $4,545.45
   c. $3,433.33
3. a. 60; .17 b. 120; .33
   c. 270; .75 d. 145; .40
5. 5%
7. $250
9. $3,000
11. 119 days

LU 10–3

1. a. $2,568.75; $1,885.47; $920.04;$0
3. $4,267.59
5. $4,715.30; $115.30

LU 11–1

1. I; B; D; I; D; I; B; D
3. a. 2% c. 13%
5. $15,963.75
7. $848.75; $8,851.25
9. $14,300
11. $7,855

LU 11–2

1. a. $5,075.00
   b. $16,480.80
   c. $994.44
3. a. $14.76
   b. $223.25
   c. $3.49
5. $4,031.67
7. $8,262.74
9. $5,088.16
11. $721.45

LU 12–1

1. a. $573.25 year 2
   b. $3,115.57 year 4
3. a. $25,306; $5,306
   b. $16,084; $6,084
5. $7,430.50
7. $8,881.20
9. $2,129.40
11. $3,207.09; $207.09
13. $3,000; $3,469; $3,498

LU 12–2

1. a. .9804 b. .3936
   c. .5513
3. a. $1,575.50; $924.50
   b. $2,547.02; $2,052.98
5. $14,509.50
7. $13,356.98
9. $16,826.40
11. $652.32
13. $18,014.22

LU 13–1

1. a. $1,000; $2,080; $3,246.40
3. a. $6,888.60 b. $6,273.36
5. $325,525
7. $13,412
9. $30,200.85
11. $33,650.94

LU 13–2

1. a. $2,638.65
   b. $6,375.24; $7,217.10
3. $2,715.54
5. $24,251.85
7. $47,608
9. $456,425
11. Accept Jason $265,010

LU 13–3

1. **a.** $4,087.50 **b.** $21,607
   c. $1,395 **d.** $201.45
   e. $842.24
3. $16,200
5. $24,030
7. $16,345
9. $8,742

LU 14–1

1. **a.** $1,200; $192
   b. $9,000; $1,200
3. **a.** 14.75% **b.** 10% **c.** 11.25%
5. **a.** $3,528 **b.** $696 **c.** $4,616
7. **a.** $22,500 **b.** $4,932 **c.** $29,932
9. **a.** $20,576 **b.** 12.75%

LU 14–2

1. **a.** $465; $8,535
   b. $915.62; $4,709.38
3. **a.** $332.03 **b.** $584.83
   c. $384.28
5. Final payment $784.39
7. $51.34
9. $35
11. $922.48
13. 7.50% to 7.75%

LU 15–1

1. **a.** $1,096.50
   b. $965.60; $4,090.50; $3,859.65
3. **a.** $117.25, 7.7% **b.** $174, 5.7%
5. $774
7. $2,584.71; $2,518.63
9. **a.** $66.08 **b.** $131,293.80
11. $773,560

LU 15–2

1. **a.** $1,371.90; $1,133.33; $238.57
3. #4 balance outstanding $195,183.05
5. $587,612.80
7. $327.12; $251,581.60
9. $44,271.43
11. $61,800

LU 16–1

1. Total assets $224,725
3. Merch. inventory 13.90%; 15.12%

LU 16–2

1. Net income $57,765
3. Purchases 73.59%; 71.43%

LU 16–3

1. Sales 2017, 93.5%; 2016, 93.2%
3. .22
5. 59.29%
7. .83
9. COGS 119.33%; 111.76%; 105.04%
11. .90

13. 5.51%
15. 11.01%

LU 17–1

1. **a.** 4% **b.** 25%
   c. 10% **d.** 20%
3. **a.** $2,033; $4,667
   b. $1,850; $9,750
5. $8,625 depreciation per year
7. $2,800 depreciation per year
9. $95
11. **a.** $12,000 **b.** $6,000
    c. $18,000 **d.** $45,000

LU 17–2

1. **a.** $.300 **b.** $.192 **c.** $.176
3. **a.** $.300, $2,600
   b. $.192, $300,824
5. $5,300 book value end of year 5
7. **a.** $.155 **b.** $20,001.61

LU 17–3

1. **a.** 8% **b.** 20% **c.** 25%
3. **a.** $4,467; $2,233
   b. $3,867; $7,733
5. $121, year 6
7. **a.** 28.57% **b.** $248 **c.** $619
9. **a.** 16.67% **b.** $2,500
   c. $10,814 **d.** $2,907

LU 17–4

1. **a.** 33%; $825; $1,675
3. Depreciation year 8, $346
5. $125
7. **a.** $15,000 **b.** $39,000
   c. $21,600 **d.** 2001
9. $68,440

LU 18–1

1. **a.** $5,120; $3,020
   b. $323,246; $273,546
3. $35,903; $165,262
5. $10,510.20; $16,345
7. $37.62; $639.54
9. $628.40
11. $3,069; $952; $2,117

LU 18–2

1. **a.** $85,700; $143,500; .597; $64,500; $38,507
3. $85,000
5. $342,000; $242,500; 5.85; 6.29
7. $60,000; $100,000; $40,000
9. $70,150
11. $5,970
13. 3.24; 3.05
15. $32,340; $35,280; $49,980; $29,400

LU 19–1

1. **a.** $26.80; $562.80
   b. $718.80; $12,698.80
3. **a.** $20.75; $43.89; $463.64
5. Total is **(a)** $1,023; **(b)** $58.55
7. $5.23; $115.23
9. $2,623.93
11. $26.20
13. $685.50

LU 19–2

1. **a.** $68,250 **b.** $775,450
3. **a.** $7.45; $74.50; 74.50
5. $9.10
7. $8,368.94
9. $42,112
11. $32,547.50

LU 20–1

1. **a.** $9.27; 25; $231.75
3. **a.** $93.00; $387.50; $535.00; $916.50
5. $1,242.90
7. $14,265
9. $47.50 more
11. $68,750

LU 20–2

1. **a.** $488 **b.** $2,912
3. **a.** $68,000; $60,000
   b. $41,600; $45,000
5. $1,463
7. $117,187.50
9. $336,000
11. **a.** $131,250 **b.** $147,000

LU 20–3

1. **a.** $98; $160; $258
3. **a.** $312 **b.** $233
   c. $181 **d.** $59; $20
5. **a.** $647 **b.** $706
7. $601
9. $781
11. $10,000; $8,000
13. $60,000; $20,000
15. $19.50; $110.50

LU 21–1

1. **a.** $43.88 **f.** 49
3. $27.06
5. $1,358.52 gain
7. $18,825.15
9. $7.70

LU 21–2

1. **a.** IBM **b.** $10\frac{1}{4}$ **c.** 2025
   d. $102.50 **e.** 102.375
3. **a.** $1,025 **b.** $1,023.75
5. **a.** $3,075
   b. $307.50

7. **a.** $30 discount
   b. $16.25 premium
   c. $42.50 premium
9. **a.** $625 **b.** $375 discount
   c. $105 **d.** 16.8%
11. 7.8%; 7.2%
13. 8.98%

LU 21–3

1. $11.90
3. $15.20
5. +$.14
7. 7.6%
9. $1.45; $18.45
11. $.56; $14.66
13. $1,573.50
15. $123 loss
17. **a.** 2020; 2022
    b. 9.3% Comp USA; 6.9% GMA
    c. $1,023.75 Comp USA
    $1,016.25 GMA
    d. Both at premium
    e. $1,025 Comp USA;
    $1,028.75 GMA

LU 22–1

1. **a.** 20.4 **b.** 83.75 **c.** 10.07
3. **a.** 59.5
   b. 50
5. **a.** 63.7; 62; 62
7. $1,500,388.50
9. $10.75
11. $9.98

LU 22–2

1. 18: ~~||||~~ || 7
3. 25–30: ~~||||~~ ||| 8
5. 7.2°
7. 145–154: |||| 4
9. 98.4°; 9.9°; 70.5°; 169.2°; 11.9°

Metric System

APPENDIX C

John Sullivan: Angie, I drove into the gas station last night to fill the tank up. Did I get upset! The pumps were not in gallons but in liters. This country (U.S.) going to metric is sure making it confusing.

Angie Smith: Don't get upset. Let me first explain the key units of measure in metric, and then I'll show you a convenient table I keep in my purse to convert metric to U.S. (also called customary system), and U.S. to metric. Let's go on.

The metric system is really a decimal system in which each unit of measure is exactly 10 times as large as the previous unit. In a moment, we will see how this aids in conversions. First, look at the middle column (Units) of this to see the basic units of measure:

U.S.	Thousands	Hundreds	Tens	Units	Tenths	Hundredths	Thousandths
Metric	Kilo- 1,000	Hecto- 100	Deka- 10	Gram Meter Liter 1	Deci- .1	Centi- .01	Milli- .001

- Weight: Gram (think of it as $\frac{1}{30}$ of an ounce).
- Length: Meter (think of it for now as a little more than a yard).
- Volume: Liter (a little more than a quart).

To aid you in looking at this, think of a decimeter, a centimeter, or a millimeter as being "shorter" (smaller) than a meter, whereas a dekameter, hectometer, and kilometer are "larger" than a meter. For example:

1 centimeter = $\frac{1}{100}$ of a meter; or 100 centimeters equals 1 meter.
1 millimeter = $\frac{1}{1{,}000}$ meter; or 1,000 millimeters equals 1 meter.
1 hectometer = 100 meters.
1 kilometer = 1,000 meters.

Remember we could have used the same setup for grams or liters. Note the summary here.

Length	Volume	Mass
1 meter:	1 liter:	1 gram:
= 10 decimeters	= 10 deciliters	= 10 decigrams
= 100 centimeters	= 100 centiliters	= 100 centigrams
= 1,000 millimeters	= 1,000 milliliters	= 1,000 milligrams
= .1 dekameter	= .1 dekaliter	= .1 dekagram
= .01 hectometer	= .01 hectoliter	= .01 hectogram
= .001 kilometer	= .001 kiloliter	= .001 kilogram

Practice these conversions and check solutions.

1 PRACTICE QUIZ

Convert the following:

1. 7.2 meters to centimeters
2. .89 meter to millimeters
3. 64 centimeters to meters
4. 350 grams to kilograms
5. 7.4 liters to centiliters
6. 2,500 milligrams to grams

✓ **Solutions**

1. 7.2 meters = 7.2 × 100 = 720 centimeters (remember, 1 meter = 100 centimeters)
2. .89 meter = .89 × 1,000 = 890 millimeters (remember, 1 meter = 1,000 millimeters)
3. 64 centimeters = 64/100 = .64 meters (remember, 1 meter = 100 centimeters)
4. 350 grams = $\frac{350}{1{,}000}$ = .35 kilogram (remember 1 kilogram = 1,000 grams)
5. 7.4 liters = 7.4 × 100 = 740 centiliters (remember, 1 liter = 100 centiliters)
6. 2,500 milligrams = $\frac{2{,}500}{1{,}000}$ = 2.5 grams (remember, 1 gram = 1,000 milligrams)

Angie: Look at the table of conversions and I'll show you how easy it is. Note how we can convert liters to gallons. Using the conversion from metric to U.S. (liters to gallons), we see that you multiply numbers of liters by .26, so for 37.95 liters we get 37.95 × .26 = 9.84 gallons.

Common conversion factors for U.S./metric

A. To convert from U.S. to	Metric	Multiply by	B. To convert from metric to	U.S.	Multiply by
Length:			*Length:*		
Inches (in)	Meters (m)	.025	Meters (m)	Inches (in)	39.37
Feet (ft)	Meters (m)	.31	Meters (m)	Feet (ft)	3.28
Yards (yd)	Meters (m)	.91	Meters (m)	Yards (yd)	1.1
Miles	Kilometers (km)	1.6	Kilometers (km)	Miles	.62
Weight:			*Weight:*		
Ounces (oz)	Grams (g)	28	Grams (g)	Ounces (oz)	.035
Pounds (lb)	Grams (g)	454	Grams (g)	Pounds (lb)	.0022
Pounds (lb)	Kilograms (kg)	.45	Kilograms (kg)	Pounds (lb)	2.2
Volume or capacity:			*Volume or capacity:*		
Pints	Liters (L)	.47	Liters (L)	Pints	2.1
Quarts	Liters (L)	.95	Liters (L)	Quarts	1.06
Gallons (gal)	Liters (L)	3.8	Liters (L)	Gallons	.26

John: How would I convert 6 miles to kilometers?

Angie: Take the number of miles times 1.6; thus 6 miles × 1.6 = 9.6 kilometers.

John: If I weigh 120 pounds, what is my weight in kilograms?

Angie: 120 times .45 (use the conversion table) equals 54 kilograms.

John: OK. Last night, when I bought 16.6 liters of gas, I really bought 4.3 gallons (16.6 liters times .26).

2 PRACTICE QUIZ

Convert the following:

1. 10 meters to yards
2. 110 quarts to liters
3. 78 kilometers to miles
4. 52 yards to meters
5. 82 meters to inches
6. 292 miles to kilometers

✓ Solutions

1. 10 meters × 1.1 = 11 yards
2. 110 quarts × .95 = 104.5 liters
3. 78 kilometers × .62 = 48.36 miles
4. 52 yards × .91 = 47.32 meters
5. 82 meters = 39.37 = 3,228.34 inches
6. 292 miles × 1.6 = 467.20 kilometers

Appendix C: Problems

DRILL PROBLEMS

Convert:

1. 65 centimeters to meters
2. 7.85 meters to centimeters
3. 44 centiliters to liters
4. 1,500 grams to kilograms
5. 842 millimeters to meters
6. 9.4 kilograms to grams
7. .854 kilogram to grams
8. 5.9 meters to millimeters
9. 8.91 kilograms to grams
10. 2.3 meters to millimeters

Convert, rounding to the nearest tenth:

11. 50.9 kilograms to pounds

12. 8.9 pounds to grams

13. 395 kilometers to miles

14. 33 yards to meters

15. 13.9 pounds to grams

16. 594 miles to kilometers

17. 4.9 feet to meters

18. 9.9 feet to meters

19. 100 yards to meters

20. 40.9 kilograms to pounds

21. 895 miles to kilometers

22. 1,000 grams to pounds

23. 79.1 meters to yards

24. 12 liters to quarts

25. 2.92 meters to feet

26. 5 liters to gallons

27. 8.7 meters to feet

28. 8 gallons to liters

29. 1,600 grams to pounds

30. 310 meters to yards

WORD PROBLEM

31. A metric ton is 39.4 bushels of corn. The Russians bought 450,000 metric tons of U.S. corn, valued at $58 million, for delivery after September 30. Convert the number of bushels purchased from metric tons to bushels of corn.

Glossary/Index

Note: Page numbers followed by n indicate material found in footnotes.

Accelerated Cost Recovery System (ACRS) *Tax law enacted in 1981 for assets put in service from 1981 through 1986,* 454.
Accelerated depreciation *Computes more depreciation expense in the early years of the asset's life than in the later years,* 453–454.
Accounts payable *Amounts owed to creditors for services or items purchased,* 421.
Accounts receivable *Amounts owed by customers to a business from previous sales,* 421.
Accumulated depreciation *Amount of depreciation that has accumulated on plant and equipment assets,* 450.
Acid test *Current assets less inventory less prepaid expenses divided by current liabilities,* 430.
ACRS. *see* **Accelerated Cost Recovery System**
Addends *Numbers that are combined in the addition process,* 9. *Example:* 8 + 9 = 17, in which 8 and 9 are the addends.
Addition
  of decimals, 75
  estimating, 9
  of fractions, 41–43
  of mixed numbers, 43–44
  for solving equations, 124
  of whole numbers, 9–10
Adjustable rate mortgage (ARM) *Rate of mortgage is lower than a fixed rate mortgage. Rates adjusted without refinancing. Caps available to limit how high rate can go for each adjustment period over term of loan,* 400.
Adjusted balance *The balance after partial payment less interest is subtracted from the principal,* 278, 279.
Adjusted bank balance *Current balance of checkbook after reconciliation process,* 102.
Aflac, 543–544
Albertsons, 45–46
Allstate, 519
Alphabet Inc., 540
Amazon, 92, 183, 185, 212, 215, 216, 246, 540, 560
American Express, 212, 381
American Funds, 546
Amortization *Process of paying back a loan (principal plus interest) by equal periodic payments,* 376; *see also* **Amortization schedule**.
  on installment loans, 376, 378–381
  on mortgages, 401–402, 404–405
Amortization schedule *Shows monthly payment to pay back loan at maturity. Payment also includes interest. Note payment is fixed at same amount each month,* 405.
Amortization table *A table that shows each periodic payment on a loan or mortgage,* 401–402.
Amount financed *Cash price less down payment,* 376.
Analysis. *see* **Horizontal analysis**; **Vertical analysis**
  ratio, 430
  trend, 429–430
Annual percentage rate (APR) *True or effective annual interest rate charged by sellers. Required to be stated by Truth in Lending Act,* 377–379.
Annual percentage rate (APR) table *Effective annual rate of interest on a loan or installment purchase as shown by table lookup,* 377–379.
Annual percentage yield (APY) *Truth in Savings law forced banks to report actual interest in form of APY. Interest yield must be calculated on actual number of days bank has the money,* 322–323.
Annuities certain *Annuities that have stated beginning and ending dates,* 345.
Annuity *Stream of equal payments made at periodic times,* 345.
  contingent, 345
  deferred income, 366
  future value, 345, 372
  lump sum vs., 352, 370–371
  ordinary, 345–347, 349–352
  payment periods, 345, 349
  present value, 350–352, 372
  sinking funds, 353
  term, 345
Annuity due *Annuity that is paid (or received) at the beginning of the time period,* 346.
  future value, 348–349
Apple, 77, 150, 163–164, 427, 562
Apps, banking, 99–100, 102, 105
APR. *see* **Annual percentage rate**
APY. *see* **Annual percentage yield**
ARM. *see* **Adjustable rate mortgage**
Artificial intelligence (AI), 100
Assessed value *Value of a property that an assessor sets (usually a percent of property's market value) that is used in calculating property taxes,* 498, 499.
Asset cost *Amount company paid for the asset,* 450.
Assets *Things of value owned by a business,* 421; *see also* **Depreciation**.
  acid test, 430
  on balance sheet, 420, 421
  book value, 450
  current, 421
  debt ratio, 430
  estimated useful life, 450
  land, 421, 450
  plant and equipment, 421
  quick, 430
  residual value, 450, 536
Asset turnover *Net sales divided by total assets,* 430.
AT&T, 540–541
ATM *Automatic teller machine that allows customer of a bank to transfer funds and make deposits or withdrawals,* 94, 95, 117.
Aunt Jemima, 121
Authentication methods, 146
Automatic teller machine. *see* **ATM**
Automobile emissions, 9
Automobile insurance
  claims, 519, 523–524
  compulsory, 519–520, 523
  deductibles, 521, 523
  no-fault, 522
  optional, 521–522
  payouts, 523–524
  premiums, 520, 523, 535
  rates, 510
  shopping for, 535
Average daily balance *Sum of daily balances divided by number of days in billing cycle,* 383–384.
Average day's collection *Number of days to collect your receivables,* 430.
Average inventory *Total of all inventories divided by number of times inventory taken,* 476.
Averages. *see* **Mean**; **Median**
Balance sheet *Financial report that lists assets, liabilities, and equity. Report reflects the financial position of the company as of a particular date,* 419, 420.
  assets on, 420, 421
  elements of, 420–421
  horizontal analysis, 422–424
  liabilities on, 420–422
  owner's equity on, 420, 422
  vertical analysis, 422, 423
Bank discount *The amount of interest charged by a bank on a note (Maturity value × Bank discount rate × Number of days bank holds note ÷ 360),* 298, 301.
Bank discount rate *Percent of interest,* 298.
Banker's Rule *Time is exact days/360 in calculating simple interest,* 275.
Banking apps *Special client application programs like checking account balances, paying bills, and transferring funds for those doing online banking,* 99–100, 102, 105.
Bank of America, 400
Bank reconciliation *Process of comparing the bank balance to the checkbook balance so adjustments can be made regarding checks outstanding, deposits in transit, and the like,* 100–104.
Banks
  debit cards, 95–96
  fees, 95, 96, 98
  industry trends, 99–100
  interest rates, 322–323
  selecting, 117
  technology in, 95, 99–100, 105, 117, 118
Bank statement *Report sent by the bank to the owner of the checking account indicating checks processed, deposits made, and so on, along with beginning and ending balances,* 100–102.
Bar graph *Visual representation using horizontal or vertical bars to make comparison or to show relationship on items of similar makeup,* 568–569.
Base *Number that represents the whole 100%. It is the whole to which something is being compared. Usually follows word of,* 154.
  solving for, 158
BE. *see* **Breakeven point**
Beneficiary *Person(s) designated to receive the face value of the life insurance when insured dies,* 512.
Benefits, employment, 515
Bitcoin, 32
Biweekly *Every 2 weeks (26 times in a year),* 249, 250.
Biweekly mortgage *Mortgage payments made every 2 weeks rather than monthly. This payment method takes years off the life of the mortgage and substantially reduces the cost of interest,* 399–400.
Blank endorsement *Current owner of check signs name on back. Whoever presents checks for payment receives the money,* 98.
Block, Sandra, 508
Blockchain technology, 32
Bodily injury *Auto insurance that pays damages to people injured or killed by your auto,* 519–522.
Boeing, 418
Bond discount *Bond selling for less than the face value,* 543.

Bond premium *Bond selling for more than the face value,* 543.
Bonds *Written promise by a company that borrows money usually with fixed-interest payment until maturity (repayment time),* 543.
 quotations, 543–544
 returns, 543, 544
 savings, 118
Bond yield *Total annual interest divided by total cost,* 544.
Book value *Cost less accumulated depreciation,* 450.
Borrowing. *see* Loans
Boston Celtics, 164
Breakeven analysis, 230–231
Breakeven point (BE) *Point at which seller has covered all expenses and costs and has made no profit or suffered a loss,* 216, 230, 231.
Brokers. *see* **Stockbrokers**
Budgeting, 31, 293, 294, 365, 395, 446
Building *An asset found on a balance sheet,* 421.
Burger King, 129
Business insurance, 515
Buying center *A group of people responsible for making purchases.*

Campbell's Soup, 121, 451
Cancellation of fire insurance, 517
Cancellation *Reducing process that is used to simplify the multiplication and division of fractions,* 48. *Example:*

$$\frac{1}{\cancel{6}_{3}} \times \frac{\cancel{4}^{2}}{7}$$

Capital cost reduction, 536
Capitalized cost, 536
Capital *Owners' investment in the business,* 420.
Car emissions, 9
Car insurance. *see* Automobile insurance
Cash advance *Money borrowed by holder of credit card. It is recorded as another purchase and is used in the calculation of the average daily balance,* 383.
Cash discount *Savings that result from early payment by taking advantage of discounts offered by the seller; discount is not taken on freight or taxes,* 183, 191.
 credit periods, 191–192
 credit terms, 192–197
 word problems, 197
Cash dividend *Cash distribution of company's profit to owners of stock,* 540.
Cash flow statement, 419n
Cash-out refinance *Borrowing more than the current mortgage and taking the difference out in cash,* 403.
Cash *Total cash in checking accounts, savings accounts, and on hand,* 421.
Cash value *Except for term insurance, this indicates the value of the policy when terminated. Options fall under the heading of nonforfeiture values,* 512–514.
Centi- (Appendix C) *Prefix indicating .01 of a basic metric unit,* C-1.
Certificates of deposit, 118
CFA. *see* Consumer Federation of America
Chain discount *Two or more trade discounts that are applied to the balance remaining after the previous discount is taken. Often called a series discount,* 187–188.
Check card. *see* **Debit card**
Checking accounts
 adjusted balances, 102
 fees, 96, 98
 opening, 96
 reconciling, 100–104
 statements, 100–102
 top-yielding accounts, 118
 using, 93–94
Check register *Record-keeping device that records checks paid and deposits made by companies using a checking account,* 97–98, 103.
Check stub *Provides a record of checks written. It is attached to the check,* 97.
Checks *Written documents signed by appropriate person that direct the bank to pay a specific amount of money to a particular person or company,* 96.
 depositing, 96, 97
 endorsements, 98
 outstanding, 102
 overdrafts, 98, 101n
 structure of, 97
 writing, 97
Circle graph *A visual representation of the parts to the whole,* 570–571.
Citibank, 118, 383
Citigroup Inc., 95
Closed-end credit, 375
Closing costs *Costs incurred when property passes from seller to buyer such as for credit reports, recording costs, points, and so on,* 402.
CM. *see* **Contribution margin**
CM *Abbreviation for credit memorandum. The bank is adding to your account. The CM is found on the bank statement,* 103. *Example:* Bank collects a note for you.
Coinsurance *Type of fire insurance in which the insurer and insured share the risk. Usually there is an 80% coinsurance clause,* 518.
COLI. *see* Cost of living index
College education
 earnings based on degree, 248
 saving for, 66
 student loans for, 246, 294, 298
Collision *Optional auto insurance that pays for the repairs to your auto from an accident after deductible is met. Insurance company will only pay for repairs up to the value of the auto (less deductible),* 521, 522.
Commissions *Payments based on established performance criteria,* 252–253.
 broker, 540, 542, 545
 home sales, 398, 415, 416
 overrides, 253
 salary plus, 253
 straight, 252
 variable, 252
Common denominator *To add two or more fractions, denominators must be the same,* 41–43.
Common stock *Units of ownership called shares,* 422, 540, 541.
Comparative statement *Statement showing data from two or more periods side by side,* 422.
 balance sheets, 422–424
 income statements, 428
Complement *100% less the stated percent,* 186. *Example:* 18% → 82% is the complement (100% – 18%).
Compound amount *The future value of loan or investment,* 317.
Compounded annually *Interest on balance calculated once a year,* 318.
Compounded daily *Interest calculated on balance each day,* 318, 323.
Compounded monthly *Interest on balance calculated twelve times a year,* 318.
Compounded quarterly *Interest on balance calculated four times a year,* 318.
Compounded semiannually *Interest on balance calculated two times a year,* 318.
Compounding *Calculating the interest periodically over the life of the loan and adding it to the principal,* 317–318.
 continuous, 323
 formula, 320n
 relationship to present value, 324–327, 370
Compound interest *The interest that is calculated periodically and then added to the principal. The next period the interest is calculated on the adjusted principal (old principal plus interest),* 317–318; *see also* **Annuity**.
 calculating, 319–322
 comparison with simple interest, 318–322
 effective rate, 322, 323
 periods, 318
 rates, 318
 sinking funds, 353
Comprehensive insurance *Optional auto insurance that pays for damages to the auto caused by factors other than from collision (fire, vandalism, theft, and the like),* 521, 522.
Compulsory insurance *Insurance required by law—standard coverage,* 519–520, 523.
Constants *Numbers that have a fixed value such as 3 or –7. Placed on right side of equation; also called* knowns, 121–122.
Consumer Federation of America (CFA), 246
Consumer Financial Protection Bureau, 246
Consumer Price Index (CPI), 571–572
Contingent annuities *Beginning and ending dates of the annuity are uncertain (not fixed),* 345.
Contingent liability *Potential liability that may or may not result from discounting a note,* 301.
Contribution margin (CM) *Difference between selling price and variable cost,* 230.
Conversion periods *How often (a period of time) the interest is calculated in the compounding process. Example: Daily—each day; monthly—12 times a year; quarterly—every 3 months; semiannually—every 6 months.*
Conversions
 to decimal fractions, 73–74
 to decimals, 72–73, 151–152
 foreign currency, 77–78
 fractions, 37–39, 153
 mixed numbers, 37–38, 73
 to percents, 150–153
 whole numbers, 5
Cops & Doughnuts bakery, 492
Corporation *Company with many owners or stockholders. Equity of these owners is called stockholders' equity,* 420.
Costco Wholesale, 560
Cost of living index (COLI), 92
Cost of merchandise (goods available for sale) *Sum of beginning inventory plus cost of net purchases,* 427.
Cost of merchandise (goods) sold *Beginning inventory + Net purchases – Ending inventory,* 426–427, 470.
Cost of net purchases *Cost of purchases less purchase returns and allowances less purchase discounts,* 427.
Cost *Price retailers pay to manufacturer or supplier to bring merchandise into store,* 216; *see also* **Closing costs**; Inventory costs.
 asset, 450
 capitalized, 536
 fixed, 230
 installment, 375–376
 markups based on, 216–219, 224–225
 variable, 230
Council for Community and Economic Research, 92
CPI. *see* Consumer Price Index
Cranor, Lorrie, 146

Credit. *see also* **Bonds**; Loans
closed-end, 375
FICO scores, 375, 396
lines of, 303, 366, 403
open-end, 375, 382
regulations, 377, 382
revolving, 381–384
Credit Card Act *A federal statute passed in 2009 to provide fair and transparent practices to the extension of credit,* 382.
Credit card *A piece of plastic that allows you to buy on credit,* 381–382.
average daily balances, 383–384
balances, 381–384
comparison with debit cards, 96
extended warranties from, 212
interest calculations, 382–384
liability, 96
low-rate, 118
for online purchases, 246
retail rebate, 118
Credit due dates, calculating, 191–192
Credit memo (CM) *Bank transactions that increase customer's account,* 103.
Credit period *Credit days are counted from date of invoice. Has no relationship to the discount period,* 191–192.
due dates, 191–192
end of, 191
Credit reports, 245
Credit scores, 245, 296, 375, 396
Credit terms, 192–197
Credit unions, 117
Cross, Miriam, 396
Cryptocurrency, 32
Cumulative preferred stock *Holders of preferred stock must receive current year's dividends and any dividends in arrears before any dividends are paid out to the holders of common stock,* 540, 541.
Current assets *Assets that are used up or converted into cash within 1 year or operating cycle,* 421.
Current liabilities *Obligations of a company due within 1 year,* 421, 422.
Current ratio *Current assets divided by current liabilities,* 430.
Cybertheft, 94, 95

Daily balance *Calculated to determine customer's finance charge: Previous balance + Any cash advances + Purchases − Payments,* 383–384.
Damcevski, Andrew, 294
Days-in-a-month rules, 191–192
Days-in-a-year calendar, 192, 193
Debit card *Transactions result in money being immediately deducted from customer's checking account,* 95–96.
Debit memo (DM) *A debit transaction bank does for customers,* 103.
Debt, planning for, 293, 294
Debt to assets ratio, 430
Deci- (Appendix C) *Prefix indicating .1 of basic metric unit,* C-1.
Decimal equivalent *Decimal represents the same value as the fraction,* 69. *Example:*

$$.05 = \frac{5}{100}$$

Decimal fraction *Decimal representing a fraction; the denominator has a power of 10,* 72–73.
converting pure and mixed decimals to, 73–74
converting to decimals, 72–73
Decimal point *Center of the decimal system—located between units and tenths. Numbers to left are whole numbers; to the right are decimal numbers,* 4, 69–70.
Decimals *Numbers written to the right of a decimal point,* 69. *Example:* 5.3, 18.22.
adding, 75
applications in foreign currency, 77–78
converting fractions to, 72–73
converting mixed numbers to, 73
converting percents to, 151–152
converting to percents, 150–151
dividing, 76–79
mixed, 73–74
multiplying, 76, 78–79
pure, 73–74
reading, 71
repeating, 71
rounding, 71–72
subtracting, 75, 76
word problems, 79
Decimal system *The U.S. base 10 numbering system that uses the 10 single-digit numbers shown on a calculator,* 4.
Declining-balance method *Accelerated method of depreciation. The depreciation each year is calculated by book value beginning each year times the rate,* 453–454.
Deductibles *Amount insured pays before insurance company pays. Usually the higher the deductible, the lower the premium will be,* 521, 523.
Deductions *Amounts deducted from gross earnings to arrive at net pay,* 251, 253–257, 269.
Deferred income annuities, 366
Deferred payment price *Total of all monthly payments plus down payment,* 376.
Deka- (Appendix C) *Prefix indicating 10 times basic metric unit,* C-1.
Denominator *The number of a common fraction below the division line (bar),* 36. *Example:* $\frac{8}{9}$, in which 9 is the denominator.
least common, 41–43
Deposits in transit *Deposits not received or processed by bank at the time the bank statement is prepared,* 102.
Deposit slip *Document that shows date, name, account number, and items making up a deposit,* 96, 97.
Depreciation *Process of allocating the cost of an asset (less residual value) over the asset's estimated life,* 450.
accelerated, 453–454
accumulated, 450
causes of, 450
declining-balance method, 453–454
for partial years, 451
straight-line method, 451, 453
for tax purposes, 455–456
units-of-production method, 452
Depreciation causes *Normal use, product obsolescence, aging, and so on,* 450.
Depreciation expense *Process involving asset cost, estimated useful life, and residual value (salvage or trade-in value),* 450.
Depreciation schedule *Table showing amount of depreciation expense, accumulated depreciation, and book value for each period of time for a plant asset,* 451.
Deveny, Kathleen, 567
Diamond, Michael, 118
Difference *The resulting answer from a subtraction problem,* 10. *Example:* Minuend less subtrahend equals difference. 215 − 15 = 200.
Differential pay schedule *Pay rate is based on a schedule of units completed,* 252.
Digit *Our decimal number system of 10 characters from 0 to 9,* 4.
Digital currencies, 32
Discount *Amount bond sells below $1,000,* 543.
Discount codes, 212
Discounting a note *Receiving cash from selling a note to a bank before the due date of a note. Steps to discount include (1) calculate maturity value, (2) calculate number of days bank waits for money, (3) calculate bank discount, and (4) calculate proceeds,* 301–302.
Discounting Treasury bills, 300
Discount period *Amount of time to take advantage of a cash discount,* 191, 302.
Discounts. *see* **Cash discount**; **Trade discount**
Discount sheets, 184
Disney, 538, 539, 542
Distribution of overhead *Companies distribute overhead by floor space or sales volume,* 477–478.
Dividend *Number in the division process that is being divided by another,* 14. *Example:* 15 ÷ 5, in which 15 is the dividend.
Dividends *Distribution of company's profit in cash or stock to owners of stock,* 540.
cash, 540
pay outs, 541
Dividends in arrears *Dividends that accumulate when a company fails to pay cumulative dividends to preferred stockholders,* 540.
Division
of decimals, 76–79
estimating, 15
of fractions, 49–50
of mixed numbers, 50
shortcuts, 15, 78–79
for solving equations, 124
of whole numbers, 14–15
Divisor *Number in the division process that is dividing into another,* 14. *Example:* 15 ÷ 5, in which 5 is the divisor.
DM *Abbreviation for debit memorandum. The bank is charging your account. The DM is found on the bank statement,* 103. *Example:* NSF.
Dodson, Mark, 32
Dollar markdown *Original selling price less the reduction to price. Markdown may be stated as a percent of the original selling price,* 228. *Example:*

$$\frac{\text{Dollar markdown}}{\text{Original selling price}}$$

Dollar markup *Selling price less cost. Difference is the amount of the markup. Markup is also expressed in percent,* 216–218, 221–222.
Down payment *Amount of initial cash payment made when item is purchased,* 375.
Drafts *Written orders like checks instructing a bank, credit union, or savings and loan institution to pay your money to a person or organization,* 96.
Drawee *One ordered to pay the check,* 97.
Drawer *One who writes the check,* 97.
Draw *The receiving of advance wages to cover business or personal expenses. Once wages are earned, drawing amount reduces actual amount received,* 252.
Due date *Maturity date, or when the note will be repaid,* 191–192.
Dunkin' Donuts, 15–16, 128, 163, 343–344

Earnings, retained, 422. *see also* Profit; Wages
Earnings per share (EPS) *Annual earnings divided by total number of shares outstanding,* 541.
eBay, 77
Effective rate *True rate of interest. The more frequent the compounding, the higher the effective rate,* 300.
annual percentage yield, 322–323
for compound interest, 322, 323
of simple discount note, 299, 300
of simple interest note, 299

Electronic deposits *Credit card run through terminal which approves (or disapproves) the amount and adds it to company's bank balance,* 102, 118.
Electronic funds transfer (EFT) *A computerized operation that electronically transfers funds among parties without the use of paper checks,* 102.
Emergency funds, 365
Empirical Rule *For a normal distribution, nearly all the data fall within three standard deviations of the mean,* 575.
Employees. *see* **Commissions**; Employment benefits; Payroll; Wages
Employee's Withholding Allowance Certificate (W-4) *Completed by employee to indicate allowance claimed to determine amount of FIT that is deducted,* 254, 255.
Employers
- insurance, 515
- responsibilities of, 257–258
- workers' compensation, 515

Employer sponsored retirement programs, 339
Employment benefits, 515
End of credit period *Last day from date of invoice when customer can take cash discount,* 191.
End of month (EOM) *(also proximo) Cash discount period begins at the end of the month invoice is dated. After the 25th discount period, one additional month results,* 196–197.
Endorse *Signing the back of the check; thus ownership is transferred to another party,* 98.
Endowment life *Form of insurance that pays at maturity a fixed amount of money to insured or to the beneficiary. Insurance coverage would terminate when paid—similar to term life,* 514, 515.
EOM. *see* **End of month**
EPS. *see* **Earnings per share**
Equation equality rule, 124
Equation *Math statement that shows equality for expressions or numbers, or both,* 122.
Equation solving for the unknown, 121–126
- equation equality rule, 124
- like unknowns rule, 126
- multiple processes rule, 125
- opposite process rule, 123
- parentheses rule, 125
- solving for the unknown rule, 123

Equipment. *see* **Plant and equipment**
Equivalent (fractional) *Two or more fractions equivalent in value,* 39.
Escrow account *Lending institution requires that each month* $\frac{1}{12}$ *of the insurance cost and real estate taxes be kept in a special account,* 403.
Esswein, Patricia Mertz, 447, 492
Estate planning, 559
Estimated useful life *How long asset will be in use,* 450.
Estimating
- addition, 9
- division, 15
- multiplication, 13

Exact days-in-a-year calendar, 192, 193
Exact interest *Calculating simple interest using 365 days per year in time,* 275.
Excise tax *Tax that government levies on particular products and services. Tax on specific luxury items or nonessentials,* 497.
Expenses *Cost of doing business; found on the income statement,* 427.
- depreciation, 450
- operating, 216, 425, 427
- overhead, 216, 477
- prepaid, 421

Expression *A meaningful combination of numbers and letters called* terms, 122.
Extended term insurance *Resulting from nonforfeiture, it keeps the policy for the full face value going without further premium payments for a specific period of time,* 514.
Extended warranties, 212

Face amount *Dollar amount stated in policy,* 512.
Face value *Amount of insurance that is stated on the policy. It is usually the maximum amount for which the insurance company is liable,* 513.
- of fire insurance, 518
- of life insurance, 513, 514
- of promissory notes, 298

Fair Credit and Charge Card Disclosure Act of 1988 *Act that tightens controls on credit card companies soliciting new business,* 382.
Fair Labor Standards Act *Federal law has minimum wage standards and the requirement of overtime pay. There are many exemptions for administrative personnel and for others,* 250.
FC. *see* **Fixed costs**
Federal Deposit Insurance Corporation, 118
Federal Housing Administration, 400
Federal income tax (FIT) withholding *Federal tax withheld from paycheck,* 255–257.
Federal Insurance Contribution Act (FICA) *Percent of base amount of each employee's salary. FICA taxes are used to fund retirement, disabled workers, Medicare, and so on. FICA is now broken down into Social Security and Medicare,* 255.
Federal Reserve, 121, 275, 374
Federal Unemployment Tax Act (FUTA) *Tax paid by employer. Current rate is .8% on first $7,000 of earnings,* 257, 258.
Federal withholding tax. *see* **Federal income tax (FIT) withholding**
FedEx, 182, 185
FICA. *see* **Federal Insurance Contribution Act**
FICO scores, 375, 396
FIFO. *see* **First-in, first-out method**
Finance charge *Total payments – Actual loan cost,* 376, 382–384.
Financial fitness, 507
Financial planning, 31, 145, 293, 294, 313, 507
Financial statements. *see also* **Balance sheet; Income statement**
- ratio analysis, 430
- trend analysis, 429–430

Fire insurance *Stipulated percent (normally 80%) of value that is required for insurance company to pay to reimburse one's losses,* 516.
- for businesses, 516–517
- canceling, 517
- coinsurance, 518
- premiums, 516
- short-rate table, 517

First-in, first-out (FIFO) method *This method assumes the first inventory brought into the store will be the first sold. Ending inventory is made up of goods most recently purchased,* 472, 474.
First-time home buying programs, 415
FIT. *see* **Federal income tax withholding**
Fixed costs (FC) *Costs that do not change with increase or decrease in sales,* 230.
Fixed-rate mortgage *Monthly payment fixed over number of years, usually 30 years,* 399, 400.
Flexible spending accounts, 269
FOB destination *Seller pays cost of freight in getting goods to buyer's location,* 185.
FOB shipping point *Buyer pays cost of freight in getting goods to his location,* 185.
Fonda, Daren, 560
Foreclosure *A legal process used by lender to recover balance of the loan from the borrower who has stopped making payments on the loan,* 399, 400.
Foreign currency conversions, 77–78
Formula *Equation that expresses in symbols a general fact, rule, or principle,* 122.
Fraction *Expresses a part of a whole number,* 36. *Example:*

$\frac{5}{6}$ expresses 5 parts out of 6

- adding, 41–43
- converting percents to, 153
- converting to decimals, 72–73
- converting to percents, 152–153
- converting to whole or mixed numbers, 37–38
- denominators, 36
- dividing, 49–50
- higher terms, 39
- improper, 37–38
- like, 41, 44
- lowest terms, 38, 39
- multiplying, 48
- numerators, 36
- proper, 36, 48–50, 73
- subtracting, 44
- types, 36–37
- unlike, 41, 44
- word problems, 45–46, 50

Fraud, 246
Freight terms *Determine how freight will be paid. Most common freight terms are FOB shipping point and FOB destination,* 185.
Frequency distribution *Shows by table the number of times event(s) occurs,* 567.
Fruit of the Loom, Inc., 473
Full endorsement *This endorsement identifies the next person or company to whom the check is to be transferred,* 98.
FUTA. *see* **Federal Unemployment Tax Act**
Future value (FV) *Final amount of the loan or investment at the end of the last period. Also called compound amount,* 317, 318.
- relationship with present value, 325–327, 370

Future value of annuity *Future dollar amount of a series of payments plus interest,* 345.
- annuity due, 348–349
- ordinary annuity, 346–347, 349, 372

Future value table, 320–321
Futurion, 118
FV. *see* **Future value**

Gap Inc., 215–217, 221, 222, 227
Gasoline taxes, 494
General Depreciation System (GDS) *Most common (MACRS) system to calculate depreciation,* 455.
General Mills, 451, 542
General Motors, 16–17
Gerstner, Lisa, 118, 294, 366
Gillette Corporation, 451
Goal setting, 145
Google Inc., 4, 41, 540
Graduated-payment mortgage (GPM) *Borrower pays less at beginning of mortgage. As years go on, the payments increase,* 400.
Gram (Appendix C) *Basic unit of weight in metric system. An ounce equals about 28 grams,* C-1, C-2.
Graphs
- bar, 568–569
- circle, 570–571
- line, 569

Greater fool theory, 32
Greatest common divisor *The largest possible number that will divide evenly into both the numerator and denominator,* 38–39.

Greenway Farms of Georgia, LLC, 447
Gross pay *Wages before deductions,* 251.
Gross profit *Difference between cost of bringing goods into the store and selling price of the goods,* 216.
Gross profit from sales *Net sales – Cost of goods sold,* 427.
Gross profit method *Used to estimate value of inventory,* 475–476.
Gross sales *Total earned sales before sales returns and allowances or sales discounts,* 426.

Hackers, 146
Hamm, Nate, 294
Health insurance, 269, 515
Healthy lifestyle, 591
Hecto- (Appendix C) *Prefix indicating 100 times basic metric unit,* C-1.
Hershey Co., 11, 149
Higher terms *Expressing a fraction with a new numerator and denominator that is equivalent to the original,* 39. *Example:*

$$\frac{2}{9} \rightarrow \frac{6}{27}$$

Home-based businesses, 465
Home Depot, 468, 469
Home equity line of credit *A revolving line of credit secured by your home equity,* 403.
Home equity loan *A loan using the borrower's home equity as collateral,* 400, 403.
Horizontal analysis *Method of analyzing financial reports where each total this period is compared by amount of percent to the same total last period,* 422.
of balance sheet, 422–424
of income statement, 428
Hough, Emily, 246
Housing costs, 402–403, 415. *see also* **Mortgage**
Huang, Nellie S., 32

IBM, 185
Improper fraction *Fraction that has a value equal to or greater than 1; numerator is equal to or greater than the denominator,* 37–38. *Example:*

$$\frac{6}{6}, \frac{14}{9}$$

Income. *see* Payroll; Wages
Income statement *Financial report that lists the revenues and expenses for a specific period of time. It reflects how well the company is performing,* 419, 425.
in assessment of financial standing, 446
depreciation expense, 450
elements of, 425–427
formulas, 427
horizontal analysis, 428
net income, 425, 427
operating expenses, 425, 427
revenue, 425, 426
vertical analysis, 428
Income tax or FIT *Tax that depends on allowances claimed, marital status, and wages earned,* 255–257, 270.
Indemnity *Insurance company's payment to insured for loss,* 518.
Index numbers *Express the relative changes in a variable compared with some base, which is taken as 100,* 571–572.
Individual retirement account (IRA) *An account established for retirement planning,* 547
records of contribution, 416
Roth, 339, 366
traditional, 339, 366
Installment cost *Down payment + (Number of payments × Monthly payment). Also called* deferred payment, 375–376.
Installment loan *Loan paid off with a series of equal periodic payments,* 376.
Installment purchase *Purchase of an item(s) that requires periodic payments for a specific period of time, usually with a high rate of interest,* 375.
amortization, 376, 378–381
amount financed, 376
deferred payment price, 376
down payments, 375
finance charge, 376
monthly payments, 377–379
Instant gratification, 340
Insurance. *see also* Automobile insurance; Life insurance
business, 515
coverage limitations, 491
fire, 516–518
health, 269, 515
Private Mortgage Insurance (PMI), 403
workers' compensation, 515
Insured *Customer or policyholder,* 512.
Insurer *The insurance company that issues the policy,* 512.
Interest-bearing note *Maturity value of note is greater than amount borrowed since interest is added on,* 298, 301–303.
Interest-only mortgage *Type of mortgage where in early years only interest payment is required,* 400.
Interest *Principal × Rate × Time,* 273; *see also* **Compound interest**; **Simple interest**.
annual percentage yield, 322–323
on installment loans, 376
on mortgages, 399–400, 402–403
on revolving charge accounts, 382–384
Interest rates
annual percentage rate, 377–379
banks, 322–323
formula for, 277
growth of, 366
on loans, 273–275
on mortgages, 399, 400, 402–403
nominal, 322, 323
in peer-to-peer lending, 314
on promissory notes, 298
Interviews, 91
Inventory
average, 476
on balance sheet, 421, 470
merchandise, 421, 426, 427
Inventory control
just-in-time system, 477
periodic method of, 469
perpetual method of, 469
physical counts, 469
Inventory costs
comparison of methods, 474
FIFO method, 472, 474
gross profit method, 475–476
LIFO method, 473–474
retail method, 475
specific identification method, 470–471
weighted-average method, 471–472, 474
Inventory turnover *Ratio that indicates how quickly inventory turns:*

$$\frac{\text{Cost of goods sold}}{\text{Average inventory at cost}}, 476\text{–}477.$$

Investments. *see also* **Stocks**
bonds, 543–544
mutual funds, 545–547
principles, 539
returns, 430, 542–544
risk in, 539
strategies, 547
Invoice *Document recording purchase and sales transactions,* 184.
credit periods, 191–192
credit terms, 192–197
partial payments, 198
IRA. *see* **Individual retirement account**

Jajodia, Sid, 314
J.Crew Group Inc., 543
Just-in-time (JIT) inventory system *System that eliminates inventories. Suppliers provide materials daily as manufacturing company needs them,* 477.

Kellogg's, 8, 121
Kilo- (Appendix C) *Prefix indicating 1,000 times basic metric unit,* C-1.
Kmart, 228
Known *A number or fact that is specified,* 122.

L. L. Bean, 130
Land *Asset on balance sheet that does not depreciate,* 421, 450.
Lankford, Kimberly, 366, 416
Last-in, first-out (LIFO) method *This method assumes the last inventory brought into the store will be the first sold. Ending inventory is made up of the oldest goods purchased,* 473–474.
LastPass, 146
LCD. *see* **Least common denominator**
Leary, Elizabeth, 314
Leasing, 536
Least common denominator (LCD) *Smallest nonzero whole number into which all denominators will divide evenly,* 41–43. *Example:*

$$\frac{2}{3} \text{ and } \frac{1}{4} \quad \text{LCD} = 12$$

Lending Club, 314
Lesser, Matthew, 294
Level premium term *Insurance premium that is fixed, say, for 50 years,* 512.
Liabilities *Amounts business owes to creditors,* 421.
on balance sheet, 420–422
contingent, 301
current, 421, 422
long-term, 422
Liability insurance *Insurance for bodily injury to others and damage to someone else's property,* 519–520.
Life insurance
for employees, 515
nonforfeiture values, 514–515
premiums, 512, 513
rates, 513
terminology, 512
types, 512–515
LIFO. *see* **Last-in, first-out method**
Like fractions *Proper fractions with the same denominators,* 41, 44.
Like terms *Terms that are made up with the same variable:* $A + 2A + 3A = 6A$.
Like unknowns rule, 126
Limited payment life (20-payment life) *Premiums are for 20 years (a fixed period) and provide paid-up insurance for the full face value of the policy,* 513–515.
Line graph *Graphical presentation that involves a time element. Shows trends, failures, backlogs, and the like,* 569.
Line of credit *Provides immediate financing up to an approved limit,* 303, 366.
home equity, 403
List price *Suggested retail price paid by customers,* 183, 187.

Liter (Appendix C) *Basic unit of measure in metric, for volume,* C-1, C-2.
Living costs, 92
Living trusts, 559
Loan amortization table *Table used to calculate monthly payments,* 378–381.
Loans. *see also* Credit; **Installment purchase; Interest; Mortgage; Promissory note**
- compounding, 317–318
- home equity, 400, 403
- installment, 376
- interest rates, 273–275, 277
- maturity value, 273–274
- partial payment on, 278–279
- peer-to-peer, 314
- principal, 273, 276
- student, 246, 294, 298
- subprime, 399

Long-term-care insurance premiums, 366
Long-term liabilities *Debts or obligations that company does not have to pay within 1 year,* 422.
Louis Vuitton, 69
Lowest terms *Expressing a fraction when no number divides evenly into the numerator and denominator except the number 1,* 38, 39. *Example:*

$$\frac{5}{10} \rightarrow \frac{1}{2}$$

Low-rate credit cards, 118
Loyalty cards, 212
Lyft, 519

MACRS. *see* **Modified Accelerated Cost Recovery System**
Macy's, 227
Maker *One who writes the note,* 298.
M&M'S®, 35–38, 47, 69–71, 150, 151, 154–158, 160–161
Marcus, David, 560
Margin *Difference between cost of bringing goods into store and selling price of goods,* 216.
Markdowns *Reductions from original selling price caused by seasonal changes, special promotions, and so on,* 228.
Markup *Amount retailers add to cost of goods to cover operating expenses and make a profit,* 216.
- based on cost, 216–219, 224–225
- based on selling price, 221–225
- dollar amount, 216–218, 221–222
- word problems, 217–219, 222–224

Markup percent calculation *Markup percent on cost × Cost = Dollar markup; or Markup percent on selling price × Selling price = Dollar markup,* 216–219, 221–222.
MasterCard, 212
Maturity date *Date the principal and interest are due,* 298.
Maturity value (MV) *Principal plus interest (if interest is charged). Represents amount due on the due date,* 273–274, 298, 302.
Mazda, 448, 449
McDonald's, 3, 430
McGraw-Hill Higher Education, 184, 186, 187
Mean *Statistical term that is found by: Sum of all figures ÷ Number of figures,* 564–565.
Measure of dispersion *Number that describes how the numbers of a set of data are spread out or dispersed,* 573–575.
Median *Statistical term that represents the central point or midpoint of a series of numbers,* 565–566.
Medical power of attorney, 559
Medicare *Part of FICA tax that has no minimum base,* 255.
Merchandise inventory *Cost of goods for resale,* 421, 426, 427; *see also* Inventory.
Meter (Appendix C) *Basic unit of length in metric system. A meter is a little longer than a yard,* C-1, C-2.
Metric system (Appendix C) *A decimal system of weights and measures. The basic units are meters, grams, and liters,* C-1–C-2.
Mill, 470

$\frac{1}{10}$ *of a cent or* $\frac{1}{1,000}$ *of a dollar.*

In decimal, it is .001. In application:

Property tax due = Mills × .001 × Assessed valuation

Milli- (Appendix C) *Prefix indicating .001 of basic metric unit,* C-1.
Millionaires, 592
Minimum wage, 250
Minuend *In a subtraction problem, the larger number from which another is subtracted,* 10. *Example:* 50 – 40 = 10.
Mitek, 118
Mixed decimal *Combination of a whole number and decimal, such as 59.8, 810.85,* 73–74.
Mixed number *Sum of a whole number greater than zero and a proper fraction,* 37. *Example:*

$$2\frac{1}{4}, 3\frac{3}{9}$$

- adding, 43–44
- converting fractions to, 37–38
- converting to decimals, 73
- converting to improper fraction, 38
- dividing, 50
- multiplying, 48–49
- subtracting, 44–45

Mobile banking *Doing banking transactions on a mobile device such as a smartphone or tablet,* 99–100, 105, 117, 118.
Mock interviews, 91
Mode *Value that occurs most often in a series of numbers,* 566.
Modified Accelerated Cost Recovery System (MACRS) *Part of Tax Reform Act of 1986 that revised depreciation schedules of ACRS. Tax Bill of 1989, 2010 updates MACRS,* 454–456.
Money factor, 536
Monthly payment *Amount paid each period to pay off part of the mortgage,* 401–405.
Monthly *Some employers pay employees monthly,* 249, 250.
Morgan Stanley, 342
Mortgage *Cost of home less down payment,* 400.
- adjustable rate, 400
- amortization, 401–402, 404–405
- biweekly, 399–400
- closing costs, 402
- escrow accounts, 403
- fixed-rate, 399, 400
- foreclosures, 399, 400
- graduated-payment, 400
- interest-only, 400
- interest rates, 399, 400, 402–403
- monthly payments, 401–405
- points, 402
- reverse, 366, 400–401
- types of, 399–400

Mortgage note payable *Debt owed on a building that is a long-term liability; often the building is the collateral,* 422.
Motor vehicle insurance. *see* Automobile insurance
Muhlbaum, David, 536
Multiple processes rule, 125
Multiplicand *The first or top number being multiplied in a multiplication problem,* 12. *Example:*

Product	=	Multiplicand	×	Multiplier
12	=	20	×	2

Multiplication
- checking, 13
- of decimals, 76, 78–79
- estimating, 13
- of fractions, 48
- of mixed numbers, 48–49
- by power of 10, 14
- shortcuts, 13, 78
- for solving equations, 124–125
- of whole numbers, 12–14

Multiplier *The second or bottom number doing the multiplication in a problem,* 12. *Example:*

Product	=	Multiplicand	×	Multiplier
12	=	20	×	2

Mutual fund *Investors buy shares in the fund's portfolio (group of stocks and/or bonds),* 545.
- advantages of, 545
- commissions, 545
- net asset value, 545
- quotations, 546

MV. *see* **Maturity value**

National Bureau of Economic Research (NBER), 340
National Collegiate Student Loan Trusts, 246
National Park Service (NPS), 92
National Student Loan System, 246
National Taxpayers Union, 508
Net asset value (NAV) *The dollar value of one mutual fund share; calculated by subtracting current liabilities from current market value of fund's investments and dividing this by number of shares outstanding,* 545.
Net income *Gross profit less operating expenses,* 216, 425, 427.
Net pay *Gross pay less deductions,* 257.
Net price *List price less amount of trade discount. The net price is before any cash discount,* 184, 186–188; *see also* **Trade discount**.
Net price equivalent rate *When multiplied times the list price, this rate or factor produces the actual cost to the buyer. Rate is found by taking the complement of each term in the discount and multiplying them together (do not round off),* 188.
Net proceeds *Maturity value less bank discount,* 298, 301.
Net profit (net income) *Gross profit – Operating expenses,* 216, 425, 427.
Net purchases *Purchases – Purchase discounts – Purchase returns and allowances,* 427.
Net sales *Gross sales – Sales discounts – Sales returns and allowances,* 426, 427, 430.
Net wages. *see* **Net pay**
Net worth *Assets less liabilities,* 422, 446.
No-fault insurance *Involves bodily injury. Damage (before a certain level) that is paid by an insurance company no matter who is to blame,* 522.
Nominal rate *Stated rate,* 322, 323.
Nonforfeiture values *When a life insurance policy is terminated (except term), it represents (1) the available cash value, (2) additional extended term, or (3) additional paid-up insurance,* 514–515.
Non-interest-bearing note *Note where the maturity value will be equal to the amount of money borrowed since no additional interest is charged,* 298, 302.
Nonsufficient funds (NSF) *Drawer's account lacked sufficient funds to pay written amount of check,* 103.

Normal distribution *Data are spread symmetrically about the mean,* 574–575.
Notes. *see also* Loans
discounting, 301–302
interest-bearing, 298, 301–303
non-interest-bearing, 298, 302
simple discount, 298–300
simple interest, 299
NSF. *see* **Nonsufficient funds**
Number of periods *Number of years times number of times interest is compounded per year,* 318.
Numerator *Number of a common fraction above the division line (bar),* 36. *Example:*

$\frac{8}{9}$, in which 8 is the numerator

Odd lot *Less than 100 shares,* 542.
Open-end credit *Set payment period. Also, additional credit amounts can be added up to a set limit. It is a revolving charge account,* 375, 382.
Operating expenses (overhead) *Regular expenses of doing business. These are not costs,* 216, 477.
on income statement, 425, 427.
Opposite process rule, 123
Ordinary annuity *Annuity that is paid (or received) at end of the time period,* 345
future value, 346–347, 349, 372
present value, 350–352, 372
Ordinary dating *Cash discount is available within the discount period. Full amount due by end of credit period if discount is missed,* 192–195.
Ordinary interest *Calculating simple interest using 360 days per year in time,* 275.
Ordinary life insurance. *see* **Straight-life insurance**
Organizational buying process *The decision-making process used by an organization to identify, evaluate, and purchase products and services.*
Organizational buying situation *A variety of purchase scenarios employing the organizational buying process.*
Outstanding balance *Amount left to be paid on a loan,* 382.
Outstanding checks *Checks written but not yet processed by the bank before bank statement preparation,* 102.
Overdraft *Occurs when company or person wrote a check without enough money in the bank to pay for it (NSF check),* 98, 101n.
Overhead expenses *Operating expenses not directly associated with a specific department or product,* 477; *see also* **Operating expenses**.
Override *Commission that managers receive due to sales by people that they supervise,* 253.
Overtime *Time-and-a-half pay for more than 40 hours of work,* 250, 251.
Owens Corning, 468
Owner's equity. *see* **Capital**; **Stockholders' equity**

Paid-up insurance *A certain level of insurance can continue, although the premiums are terminated. This results from the nonforfeiture value (except term). Result is a reduced paid-up policy until death,* 513.
Paperno, Barry, 396
Parentheses rule, 125
Partial payments, 198, 278–279
Partial products *Numbers between multiplier and product,* 12.
Partial quotient *Occurs when divisor doesn't divide evenly into the dividend,* 14.
Partnership *Business with two or more owners,* 420.
Password security, 146
Payee *One who is named to receive the amount of the check,* 97.
of promissory notes, 298
Payment periods *Length of payments used to calculate maturity value,* 345, 349.
Pay periods, 249–250
Payroll. *see also* **Commissions**; Wages
calculations, 251–253
deductions, 251, 253–257, 259
pay periods, 249–250
unemployment taxes, 257–258
Payroll register *Multicolumn form to record payroll data,* 253–254.
Pedersen, Brendan, 246
Peer-to-peer lending, 314
PE ratio. *see* **Price-earnings ratio**
Percent *Stands for hundredths,* 150. *Example:*

4% is 4 parts of one hundred, or $\frac{4}{100}$

converting decimals to, 150–151
converting to decimals, 151–152
converting to fractions, 153
fractional, 152
mixed, 153
portion formula, 155–158
rounding, 151
Percent decrease *Calculated by decrease in price over original amount,* 158–161.
Percent increase *Calculated by increase in price over original amount,* 158–161.
Percent markup on cost *Dollar markup divided by the cost; thus, markup is a percent of the cost,* 216–219, 224–225.
Percent markup on selling price *Dollar markup divided by the selling price; thus, markup is a percent of the selling price,* 221–225
Percentage method *A method to calculate withholdings. Opposite of wage bracket method,* 255–257.
Periodic inventory system *Physical count of inventory taken at end of a time period. Inventory records are not continually updated,* 469.
Periods *Number of years times the number of times compounded per year,* 318, 318n; *see* **Conversion period**.
Perishables *Goods or services with a limited life,* 228–229.
Perpetual inventory system *Inventory records are continually updated; opposite of periodic inventory system,* 469.
Personal financial planning, 507
Personal-injury protection (PIP), 522
Personal property *Items of possession, like cars, home, furnishings, jewelry, and so on. These are taxed by the property tax (don't forget real property is also taxed),* 498.
protection of, 491
Pew Research Center, 146
Piecework *Compensation based on the number of items produced or completed,* 251.
Pie charts. *see* **Circle graph**
PIP. *see* Personal-injury protection
Pitsker, Kaitlin, 92, 146, 246
Place value *The digit value that results from its position in a number,* 4–6, 9, 70.
Plant and equipment *Assets that will last longer than 1 year,* 421.
PMI. *see* Private Mortgage Insurance
Point of sale *Terminal that accepts cards (like those used at ATMs) to purchase items at retail outlets. No cash is physically exchanged.*
Points *Percentage(s) of mortgage that represents an additional cost of borrowing. It is a one-time payment made at closing,* 402.
Policy *Written insurance contract,* 511–515
Policyholder *The insured,* 512.
Portion *Amount or part that results from multiplying the base times the rate. Not expressed as a percent; it is expressed as a number,* 155.
formula, 155–158
solving for, 155–156
Power of attorney, 559
Preferred stock *Type of stock that has a preference regarding a corporation's profits and assets,* 540, 541.
Premium *Periodic payments that one makes for various kinds of insurance protection,* 512.
for automobile insurance, 520, 523, 535
for fire insurance, 516
for life insurance, 512, 513
Premium on bonds *Bond purchase price above $1,000,* 543.
Prepaid expenses *Items a company buys that have not been used are shown as assets,* 421.
Prepaid rent *Rent paid in advance,* 421.
Present value (PV) *How much money will have to be deposited today (or at some date) to reach a specific amount of maturity (in the future),* 317–318, 324–325.
relationship with future value, 325–327, 370
Present value of an ordinary annuity *Amount of money needed today to receive a specified stream (annuity) of money in the future,* 350–352, 372.
Present value table, 325–327
Price-earnings (PE) ratio *Closing price per share of stock divided by earnings per share,* 541.
Price relative *The quotient of the current price divided by some previous year's price—the base year—multiplied by 100,* 571.
Prices. *see also* **Markup**; **Trade discount**
deferred payment, 376
list, 183, 187
net, 184, 186–188
selling, 215, 216, 218, 221–225, 228–230
stock, 540–541
Prime number *Whole number greater than 1 that is only divisible by itself and 1,* 42–43. *Examples:* 2, 3, 5.
Principal *Amount of money that is originally borrowed, loaned, or deposited,* 273, 276.
Private Mortgage Insurance (PMI), 403
Proceeds *Maturity value less the bank charge,* 298, 301.
Procter & Gamble, 158
Product *Answer to a multiplication process,* 12. *Example:*

Product = Multiplicand × Multiplier
50 = 5 × 10

Profit
gross, 216, 427, 475–476
net, 216
on net sales, 430
online vs. physical stores, 214
Profit margin on net sales *Amount of net income as a percent of net sales,* 430.
Promissory note *Written unconditional promise to pay a certain sum (with or without interest) at a fixed time in the future,* 298.
discounting, 301–302
interest-bearing, 298
non-interest-bearing, 298, 302
simple discount notes, 298–300
structure, 297–298
Promotional offers, 211
Proper fractions *Fractions with a value less than 1; numerator is smaller than denominator, such as* $\frac{5}{9}$, 36.
converting to decimals, 73
dividing, 49–50
multiplying, 48
reciprocals, 49

Property damage *Auto insurance covering damages that are caused to the property of others,* 520, 521.
Property tax *Tax that raises revenue for school districts, cities, counties, and the like,* 499.
  assessed value, 498, 499
  calculating, 499
  lowering, 499n, 508
  rates, 498
Property tax due *Tax rate × Assessed valuation,* 499.
Proximo (prox) *Same as end of month,* 196n.
Public Service Loan Forgiveness program, 294
Purchase discounts *Savings received by buyer for paying for merchandise before a certain date,* 427.
Purchase returns and allowances *Cost of merchandise returned to store due to damage, defects, and so on. An allowance is a cost reduction that results when buyer keeps or buys damaged goods,* 426.
Purchases *Merchandise for resale; found on the income statement,* 426.
  online, 246
  researching, 211
Pure decimal *Has no whole number(s) to the left of the decimal point, such as .45,* 73–74.
PV. *see* **Present value**

Quick assets *Current assets – Inventory – Prepaid expenses,* 430.
Quick ratio *(Current assets – Inventory – Prepaid expenses) ÷ Current liabilities,* 430.
Quotient *The answer to a division problem,* 14.

Range *Difference between the highest and lowest values in a group of values or set of data,* 573.
Rate *Percent that is multiplied times the base that indicates what part of the base we are trying to compare to. Rate is not a whole number,* 155.
  solving for, 156–157
Rate for each period *Annual rate divided by number of times interest is compounded in one year,* 318, 318n.
Rate of interest *Percent of interest that is used to compute the interest charge on a loan for a specific time,* 273, 274, 277; *see also* Interest rates.
Ratio analysis *Relationship of one number to another,* 430.
Real estate. *see also* **Mortgage**
  buildings, 422
  land, 421, 450
  property taxes, 498–499
  short sales, 399
Real property *Land, buildings, and so on, which are taxed by the property tax,* 498.
Receipt documentation, 491
Receipt of goods (ROG) *Used in calculating the cash discount period; begins the day that the goods are received,* 195–196.
Reciprocal of a fraction *The interchanging of the numerator and the denominator. Inverted number is the reciprocal,* 49. *Example:*

$$\frac{6}{7} \rightarrow \frac{7}{6}$$

Reconciliation. *see* **Bank reconciliation**
Redfin, 416
Reduced paid-up insurance *Insurance that uses cash value to buy protection, face amount is less than original policy, and policy continues for life,* 514.
Regions Bank, 118
Remainder *Leftover amount in division,* 14.
Repeating decimals *Decimal numbers that repeat themselves continuously and thus do not end,* 71.
Residual value *Estimated value of a plant asset after depreciation is taken (or end of useful life),* 450, 536.
Restrictive endorsement *Check must be deposited to the payee's account. This restricts one from cashing it,* 98.
Retailers. *see also* **Markup**; **Trade discount**
  breakeven analysis, 230–231
  costs, 216
  inventories, 469, 475–476
  markdowns, 228
  online, 214, 215, 246
  perishables, 228–229
  promotional offers, 211
  researching, 211
Retail method *Method to estimate cost of ending inventory. The cost ratio times ending inventory at retail equals the ending cost of inventory,* 475.
Retail rebate credit cards, 118
Retained earnings *Amount of earnings kept in the business,* 422.
Retirement savings, 31, 180, 316, 339–340, 343, 365, 547
Return on equity *Net income divided by stockholders' equity,* 430.
Returns and allowances, 426
Returns on investment, 430, 542–544
Revenues *Total earned sales (cash or credit) less any sales discounts, returns, or allowances,* 425, 426.
Reverse mortgage *Federal Housing Administration makes it possible for older homeowners to live in their homes and get cash or monthly income,* 366, 400–401.
Revolving charge account *Charges for a customer are allowed up to a specified maximum, a minimum monthly payment is required, and interest is charged on balance outstanding,* 382–384.
Richter, Amber, 294
Risk in investments, 539
Robertson, ShirlyAnn, 340
ROG. *see* **Receipt of goods**
Roth IRAs, 339, 366
Rounding
  percents, 151
  whole numbers, 5–6
Rounding all the way *Process to estimate actual answer. When rounding all the way, only one nonzero digit is left. Rounding all the way gives the least degree of accuracy,* 6, 9–10. *Example:* 1,251 to 1,000; 2,995 to 3,000.
Rounding decimals *Reducing the number of decimals to an indicated position, such as 59.59 → 59.6 to the nearest tenth,* 71–72.
Round lot *Multiple of 100 shares,* 542.
Rule of 72, 319
Rule of 115, 543

Salaries. *see* Payroll; Wages
Salaries payable *Obligations that a company must pay within 1 year for salaries earned but unpaid,* 422.
Sales
  actual, 497
  distribution of overhead, 478
  gross, 426
  gross profit, 427
  home, 416
  net, 27, 426, 430
  seasonal, 212
Salesforce.com, 149
Sales (not trade) discounts *Reductions in selling price of goods due to early customer payment,* 426.
Sales returns and allowances *Reductions in price or reductions in revenue due to goods returned because of product defects, errors, and so on. When the buyer keeps the damaged goods, an allowance results,* 426.
Sales tax *Tax levied on consumers for certain sales of merchandise or services by states, counties, or various local governments,* 495.
  calculating, 495–497
  rates, 495, 496
Salvage value *Cost less accumulated depreciation,* 450.
Savings. *see also* Retirement savings
  for college education, 66
  importance of, 179
  yield on accounts, 118
Savings bonds, 118
Schneider, Jeffrey, 416
Sears, 79
Seasonal sales, 212
Selling price *Cost plus markup equals selling price,* 215; *see also* **Markup**.
  breakeven analysis, 230
  formula, 216, 218, 222
  markups based on, 221–225
  of perishables, 228–229
Semiannually *Twice a year,* 318.
Semimonthly *Twice a month,* 249, 250.
Series discount. *see* **Chain discount**
Shareholder. *see* **Stockholder**
Shares. *see* **Stocks**
Shipping fees, 212
Short-rate table *Fire insurance rate table used when insured cancels the policy,* 517.
Short sale *A real estate sale when amount received is less than balance of the debt,* 399.
Short-term policy *Fire insurance policy for less than 1 year.*
Signature card *Information card signed by person opening a checking account,* 96.
Simple discount note *A note in which bank deducts interest in advance,* 298–300.
Simple interest formula, 274

$$\text{Interest} = \text{Principal} \times \text{Rate} \times \text{Time}$$

$$\text{Principal} = \frac{\text{Interest}}{\text{Rate} \times \text{Time}}$$

$$\text{Rate} = \frac{\text{Interest}}{\text{Principal} \times \text{Time}}$$

$$\text{Time} = \frac{\text{Interest}}{\text{Principal} \times \text{Rate}}$$

  finding unknown in, 276–277
Simple interest *Interest is only calculated on the principal. In $I = P \times R \times T$, the interest plus original principal equals the maturity value of an interest-bearing note,* 274–275.
  Banker's Rule, 275
  calculating, 319
  comparison with compound interest, 318–322
  exact interest method, 275
  ordinary interest method, 275
  U.S. Rule, 278–279
Simple interest notes, 299
Single equivalent discount rate *Rate or factor as a single discount that calculates the amount of the trade discount by multiplying the rate times the list price. This single equivalent discount replaces a series of chain discounts. The single equivalent rate is (1 – Net price equivalent rate),* 189.

Single trade discount *Company gives only one trade discount,* 186.
Sinking fund *An annuity in which the stream of deposits with appropriate interest will equal a specified amount in the future,* 353.
SIT. *see* **State income tax**
Slome, Jesse, 366
Slott, Ed, 366
Small businesses, 465
Smith, Anne Kates, 340
Snickers, 127
Social Security *Part of FICA tax that has a minimum base,* 180, 255, 343, 416.
Sole proprietorship *A business owned by one person,* 420.
Solving for the unknown rule, 123
Specific identification method *This method calculates the cost of ending inventory by identifying each item remaining to invoice price,* 470–471.
Standard deviation *Measures the spread of data around the mean,* 573–574.
Star Wars franchise, 12, 14
State Farm Insurance, 519
State income tax (SIT) *Taxation rate imposed by individual states. State rates vary. Some states do not have a state income tax,* 257.
Statement of cash flows, 419n
Statements. *see* **Bank statement**
State Unemployment Tax Act (SUTA) *Tax paid by employer. Rate varies depending on amount of unemployment the company experiences,* 257–258.
Statistician *A person who is skilled at compiling statistics,* 512.
Statistics
  dispersion measures, 573–575
  frequency distributions, 567
  graphs, 568–572
  index numbers, 571–572
  mean, median, and mode, 563–566
  normal distribution, 574–575
Step approach to finding greatest common divisor, 38–39
Stern, Hersh, 366
Stinson, Rivan, 92
Stockbrokers *People who with their representatives do the trading on the floor of the stock exchange,* 540, 542.
Stock certificate *Evidence of ownership in a corporation,* 540.
Stockholder *One who owns stock in a company,* 420, 540.
Stockholders' equity *Assets less liabilities,* 420, 422.
  return on, 430
Stocks *Ownership shares in the company sold to buyers, who receive stock certificates,* 540.
  common, 422, 540, 541
  dividends, 540, 541
  preferred, 540, 541
  quotations, 540–541
  return on investment, 430, 542
  terminology, 540
  trading, 540, 542
Stock yield *Dividend per share divided by the closing price per share,* 540.
Straight commission *Wages calculated as a percent of the value of goods sold,* 252.
Straight-life insurance *Protection (full value of policy) results from continual payment of premiums by insured. Until death or retirement, nonforfeiture values exist for straight life,* 513, 515.
Straight-line method *Method of depreciation that spreads an equal amount of depreciation each year over the life of the assets,* 451, 453.
Straight-line rate (rate of depreciation) *One divided by number of years of expected life,* 453.
Student loans, 246, 294, 298
Subprime loan *A loan with a rate higher than prime due to uncertainty of payment,* 399.
Subtraction
  of decimals, 75, 76
  of fractions, 44
  of mixed numbers, 44–45
  for solving equations, 124, 125
  of whole numbers, 10–11
Subtrahend *In a subtraction problem, smaller number that is being subtracted from another,* 10. *Example:* 30 in

$$150 - 30 = 120$$

Sum *Total in the adding process,* 9.
SUTA. *see* **State Unemployment Tax Act**

Target, 272
Tax Cuts and Jobs Act of 2017, 270
Taxes
  depreciation and, 455–456
  excise, 497
  FICA, 255
  financial implications, 269
  gasoline, 494
  income, 255–257, 270
  property, 498–499, 508
  sales, 495–497
  state, 257–258
  taxable items, 466
  unemployment, 257–258
Tax rate, 470

$$\frac{\text{Budget needed}}{\text{Total assessed value}}$$

Tax Reform Act of 1986, 455
Technology
  apps, 99–100, 102, 105
  in banks, 95, 99–100, 105, 117, 118
  blockchain, 32
  inventory control, 469
  product obsolescence, 450
  security issues, 146
Term
  of annuity, 345
  of promissory notes, 298
Term insurance *Inexpensive life insurance that provides protection for a specific period of time. No nonforfeiture values exist for term,* 512.
Term of the annuity *Time period from the first to last payment of a stream of payments,* 345.
Term policy *Period of time that the policy is in effect,* 512, 512n.
Terms of the sale *Criteria on invoice showing when cash discounts are available, such as rate and time period,* 191.
Terrazas, Aaron, 508
Tesla Inc., 148
Texaco, 542
Time *Expressed as years or fractional years, used to calculate simple interest,* 275, 277.
Tootsie Roll Industries, 7
Total assets *Total of current assets and plant and equipment,* 421.
Total current assets *Total of all assets that the company will consume or convert to cash within 1 year,* 421.
Total current liabilities *Total obligations that the company must pay within 1 year,* 422.
Total debt to total assets ratio *Amount of debt as a percent of total assets,* 430.
Total liabilities and stockholders' equity *Total current liabilities, long-term liabilities, stock, and retained earnings. This total represents all the claims on assets – prior and present claims of creditors, owners' residual claims, and any other claims,* 422.
Total liabilities *Total of current and long-term liabilities,* 422.
Total operating expenses *Total of all the individual expenses,* 427.
Total plant and equipment *Total of building and land, including machinery and equipment,* 421.
Total stockholders' equity *Total of stock plus retained earnings,* 422.
Toyota, 79, 448, 449
Toys "R" Us, 272, 273
Trade discount *Reduction off original selling price (list price) not related to early payment,* 183.
  chain, 187–188
  complement method, 186
  discount sheets, 184
  formula, 184
  net price equivalent rate, 188
  single, 186
  single equivalent discount rate, 189
  word problems, 186–188, 197
Trade discount amount *List price less net price,* 183.
Trade discount rate *Trade discount amount given in percent,* 184.
  complements, 186
Trade-in value *Estimated value of a plant asset after depreciation is taken (or end of useful life),* 450.
Traditional IRAs, 339, 366
Treasury bill *Loan to the federal government for 91 days (13 weeks), 182 days (26 weeks), or 1 year,* 118, 300.
Treasury notes, 118
Trend analysis *Analyzing each number as a percentage of a base year,* 429–430.
Truth in Lending Act *Federal law that requires sellers to inform buyers, in writing, of (1) the finance charge and (2) the annual percentage rate. The law doesn't dictate what can be charged,* 376–377.
Truth in Savings law, 322
Turnover. *see* **Asset turnover**; **Inventory turnover**
20-payment life *Provides permanent protection and cash value, but insured pays premiums for first 20 years,* 513–515.
20-year endowment *Most expensive life insurance policy. It is a combination of term insurance and cash value,* 514, 515.
Twitter, 149

Uber, 519
Unemployment tax *Tax paid by the employer that is used to aid unemployed persons,* 257–258.
United Parcel Service (UPS), 182, 185
Units-of-production method *Depreciation method that estimates amount of depreciation based on usage,* 452.
Universal life *Whole life insurance plan with flexible premium and death benefits. This life plan has limited guarantees,* 514–515.
Unknown *The variable we are solving for,* 122
  solving equations for, 121–126
  solving word problems for, 127–130
Unlike fractions *Proper fractions with different denominators,* 41–42, 44.
UPS. *see* United Parcel Service
U.S. Bank, 118
U.S. Bureau of Labor Statistics, 571

U.S. Rule *Method that allows the borrower to receive proper interest credits when paying off a loan in more than one payment before the maturity date,* 278–279.
U.S. Treasury bill. *see* **Treasury bill**
Useful life *Estimated number of years the plant asset is used,* 450, 451.

Value, assessed. *see* **Assessed value**
Value of an annuity *Sum of series of payments and interest (think of this as the maturity value of compounding),* 345.
VantageScore, 396
Variable commission scale *Company pays different commission rates for different levels of net sales,* 252.
Variable costs (VC) *Costs that do change in response to change in volume of sales,* 230.
Variable rate *Home mortgage rate that is not fixed over its lifetime.*
Variables *Letters or symbols that represent unknowns,* 121–123.
VC. *see* **Variable costs**
Vehicle emissions, 9
Vehicle insurance. *see* Automobile insurance
Vertical analysis *Method of analyzing financial reports where each total is compared to one total,* 422. *Example:* Cash is a percent of total assets.
  of balance sheet, 422, 423
  of income statement, 428
Virtual currencies, 32
Visa Inc., 2, 212
Volkswagen, 5

Wages. *see also* Payroll
  differential schedules, 252
  expectations regarding, 65
  gross pay, 251
  minimum, 250
  net, 257, 425, 427
  overtime, 250, 251
  piecework, 251
Walmart, 183, 212, 249, 272
Walt Disney Co., 538, 539, 542
Warranties, 212
Weekly *Once a week; some employers pay employees weekly,* 249, 250.
Weighted-average method *Calculates the cost of ending inventory by applying an average unit cost to items remaining in inventory for that period of time,* 471–472, 474.
Weighted mean *Used to find an average when values appear more than once,* 564.
Wells Fargo, 375
W-4. *see* **Employee's Withholding Allowance Certificate**
Whole life. *see* **Straight-life insurance**
Whole number *Number that is 0 or larger and doesn't contain a decimal or fraction, such as 10, 55, or 92,* 4.
  adding, 9–10
  converting fractions to, 37–38
  converting parts of, 5
  dividing, 14–15
  multiplying, 12–14
  reading and writing, 4–5
  rounding, 5–6
  subtracting, 10–11
  word problems, 6–8, 11, 15–16
Wholesalers. *see* **Trade discount**
Withholding *Amount of deduction from one's paycheck,* 254–257.
Word problems
  calculating unit and dollar sales, 129–130
  cash discount, 197
  decimals, 79
  difference problems, 129
  finding whole when part is known, 129
  fractions, 45–46, 50
  markup, 217–219, 222–224
  number problems, 128–129
  percent decrease, 161
  percent increase, 160–161
  solving for base, 158
  solving for portion, 155–156
  solving for rate, 156–157
  solving for unknowns, 127–130
  trade discount, 186–188, 197
  whole numbers, 6–8, 11, 15–16
Workers' compensation *Business insurance covering sickness or accidental injuries to employees that result from on-the-job activities,* 515.

XY Planning Network, 294

Yields
  bond, 544
  stock, 540
YouTube, 41